T0257903

IET TELECOMMUNICATIONS SERIES 76

Trusted Communications with Physical Layer Security for 5G and Beyond

Other volumes in this series:

Trusted Communications with Physical Layer Security for 5G and Beyond

Edited by
Trung Q. Duong, Xiangyun Zhou and H. Vincent Poor

The Institution of Engineering and Technology

Published by The Institution of Engineering and Technology, London, United Kingdom

The Institution of Engineering and Technology is registered as a Charity in England & Wales (no. 211014) and Scotland (no. SC038698).

The Institution of Engineering and Technology
Michael Faraday House
Six Hills Way, Stevenage
Herts, SG1 2AY, United Kingdom

www.theiet.org

British Library Cataloguing in Publication Data
A catalogue record for this product is available from the British Library

ISBN 978-1-78561-235-0 (hardback)
ISBN 978-1-78561-236-7 (PDF)

Typeset in India by MPS Limited

Contents

20 Application cases of secrecy coding in communication nodes and terminals

*Christiane Kameni Ngassa, Cong Ling, François Delaveau,
Sandrine Boumard, Nir Shapira, Ling Liu, Renaud Molière,
Adrian Kotelba, and Jani Suomalainen*

About the editors

Trung Q. Duong is currently an Assistant Professor at Queen's University Belfast, UK. He is currently serving on the Editorial Boards of the *IEEE Transactions on Wireless Communications, IEEE Transactions on Communications,* and *IET Communications,* and as a Senior Editor for *IEEE Communications Letters.* He has served as the Lead Guest Editor on special issues of the *IEEE Journal in Selected Areas on Communications* and *IET Communications,* and as a Guest Editor for the *IEEE Journal on Selected Areas in Communications, IEEE Communications Magazine, IEEE Access, IEEE Wireless Communications Magazine, EURASIP Journal on Wireless Communications and Networking,* and *EURASIP Journal on Advances in Signal Processing*; and as an Editor of *Electronics Letters, Emerging Telecommunications Technologies,* and *IEEE Communications Letters.* His research interests are in the field of signal processing for communications and he is the author/co-author of 135 journal articles and 117 conferences papers in these areas. He is a recipient of Best Paper Awards at the 77th IEEE Vehicular Technology Conference (VTC) 2013, IEEE International Conference on Communications (ICC) 2014 and IEEE Global Communications Conference (GLOBECOM) 2016. In 2016, he was awarded a prestigious Royal Academy of Engineering Research Fellowship (2016–2021).

Xiangyun (Sean) Zhou obtained the Bachelor of Engineering (Hon) and the Ph.D. degrees from the Australian National University (ANU) in 2007 and 2010, respectively. He currently works as a Senior Lecturer within the Research School of Engineering at the ANU. His primary research interests are in the fields of communication theory and wireless networks. From 2010 to 2011, he was a postdoctoral fellow at UNIK – University Graduate Centre, University of Oslo in Norway. He has also spent time as a visitor at a number of overseas universities, including University of Texas at Austin, University of California at Irvine, Aalborg University, Hong Kong University of Science and Technology and National Tsing-Hua University. Dr. Zhou currently serves on the Editorial Boards of the *IEEE Transactions on Wireless Communications* and *IEEE Communications Letters.* He was a Guest Editor of the 2015 special issue on Wireless Physical Layer Security of *IEEE Communications Magazine* and the 2014 special issue on Energy Harvesting Wireless Communications of the *EURASIP Journal on Wireless Communications and Networking.* He has served as an organiser and chair of various reputable international workshops including ICC'14, ICC'15 and ICC'16 workshops on Wireless Physical Layer Security. He has also served as a symposium/track chair for major IEEE conferences. He was the Chair of the ACT Chapter of the IEEE Communications Society and Signal Processing Society in 2013 and 2014. He was a recipient of a Best Paper Award at the IEEE ICC'11 and the IEEE Communications Society Asia-Pacific Outstanding Paper Award in 2016.

H. Vincent Poor received the Ph.D. degree in EECS from Princeton University in 1977. From 1977 until 1990, he was on the faculty of the University of Illinois at Urbana-Champaign. Since 1990 he has been on the faculty at Princeton, where he is the Michael Henry Strater University Professor of Electrical Engineering. He has also held visiting appointments at several institutions, including most recently at Berkeley and Cambridge. Dr. Poor's research interests are in the areas of information theory and signal processing, with applications in wireless communications and related fields. A Fellow of the IEEE and the IET, Dr. Poor is also a Member of the U.S. National Academy of Engineering and the U.S. National Academy of Sciences, a Foreign Member of the Royal Society and an International Fellow of the Royal Academy of Engineering, among memberships in other national and international academies. In 1990, he served as President of the IEEE Information Theory Society, and in 2004–2007 he served as the Editor-in-Chief of the *IEEE Transactions on Information Theory*. He received a Guggenheim Fellowship in 2002, and the IEEE Education Medal in 2005. Recent recognition of his work includes the 2016 John Fritz Medal, the 2017 IEEE Alexander Graham Bell Medal, and honorary doctorates from a number of universities in Asia, Europe and North America.

Preface

The unremitting increase in the demand for mobile data capacity is anticipated to further explode in the coming years. The emerging fifth generation (5G) of wireless communication technologies and standards will be tasked to support a diverse range of quality-of-service (QoS) requirements in different application scenarios. 5G is expected to encompass a family of wireless communications approaches that spans low data rate machine-to-machine (M2M)-type communications and very high data rate cellular technologies.

Mobile data traffic is growing daily and this presents a significant challenge as the broadcast nature of wireless channels makes it extremely vulnerable to security breaches. It has been recently reported that 78% of large organisations and 63% of small businesses are attacked annually, and these figures will continue to increase. As a consequence, security and privacy is of utmost concern in future wireless technologies. However, securely transferring confidential information over a wireless network in the presence of eavesdroppers that may intercept the information exchange between legitimate terminals, remains a challenging task. Although security was originally viewed as a high-layer problem to be solved using cryptographic methods, physical layer (PHY) security is now emerging as a promising new (additional) means of defence to realise wireless secrecy in communications. In wireless PHY security, the breakthrough idea is to exploit the characteristics of wireless channels such as fading or noise to transmit a message from the source to the intended receiver while trying to keep this message confidential from both passive and active eavesdroppers. Over the past few years, PHY security has been widely recognised as a key-enabling technique for secure wireless communications in future networks because of its potential for addressing security in new networking paradigms, such as the Internet of Things (IoT), for which more traditional methods of security may be impractical.

Research into new wireless networks to meet the anticipated demands noted above is receiving urgent attention. With the rapid development of new technologies for wireless communications and networks, e.g. millimetre-wave communications, massive MIMO, device-to-device communications, and energy harvesting communications, the first 5G networks are expected to be in place within a few years. However, significant challenges remain, particularly in terms of 5G security. There are serious constraints introduced by the need for security which, unresolved, preclude the technological precursors to 5G and, implicitly, challenge the viability of 5G itself. This book addresses this issue, not only for 5G but also for subsequent generations of wireless networks. It brings together practitioners and researchers from both academia and industry to discuss fundamental and practically relevant questions related to the many challenges arising from secure PHY communications for 5G networks. This book has been inspired by a series of workshops on PHY security, i.e. the IEEE GLOBECOM Workshops on Trusted Communications with Physical Layer Security in 2013–2017, for which the editors have served as the organisers.

The primary goal of this book is to provide comprehensive insights into the theory, models and techniques of PHY security and its applications in 5G and other emerging wireless networks. A brief tour of the book's chapters will indicate how this book addresses the challenges in a number of ways. In **Part I**, the three chapters give a comprehensive overview of the fundamentals of PHY security. *Chapter 1* presents an in-depth discussion of how to measure secrecy performance in wireless networks. Several new performance metrics for PHY security over fading channels are introduced, which can help wireless system engineers to appropriately design secure communications systems. *Chapter 2* considers PHY security for uncertain channels, namely the ergodic secrecy capacity is addressed by taking into account imperfect channel state information (CSI). The energy efficiency for secret communication is addressed in *Chapter 3* in which radio resource allocation techniques can be readily extended to include QoS constraints in the form of minimum secrecy rate guarantees for a legitimate user. As a complementary approach to that in Chapter 1, a new performance metric for secrecy measurement is introduced, namely, secrecy energy efficiency, which is defined as the ratio between the achievable secrecy rate and the consumable power.

Comprising the next three parts of the book, each of Chapters 4 through 16 focuses on PHY security in one of three broad areas: multiple-antenna technologies, emerging 5G technologies and modulation technologies.

In **Part II**, PHY security for co-located and distributed MIMO (relaying) is discussed in Chapters 4 to 8. *Chapter 4* discusses antenna selection as an enhancement scheme for PHY security at the cost of negligible extra feedback overhead. Several important issues relevant to antenna selection for PHY security are reviewed, including Alamouti codes, full duplex transmission, imperfect feedback and channel correlation. *Chapter 5* introduces the concept of PHY security for massive MIMO systems. The secrecy performance of a multi-cell massive MIMO system with pilot contamination is evaluated in terms of the secrecy rate and secrecy outage probability for different pairs of linear data and artificial noise precoders. Another important aspect of PHY security in massive MIMO is jamming, which is addressed in *Chapter 6*. This chapter provides an overview of potential counter-attack strategies, which can enhance the robustness of massive MIMO against malicious nodes. Multi-user scenarios, commonly encountered in cellular networks, are vulnerable to malicious hacks since several active users in the networks can act as eavesdroppers and overhear confidential messages due to the broadcast nature of wireless transmission. To prevent eavesdropping in multi-user networks, *Chapter 7* proposes three multi-user scheduling algorithms to improve the secrecy performance by exploiting the CSI of the main and eavesdropping links. In *Chapter 8*, the context of multi-user PHY security is extended to a more practical scenario in which a realistic multi-user network topology is taken into account by using stochastic geometry. This chapter presents low-complexity transmission schemes for multi-user MIMO systems via spatial multiplexing based on linear precoding that can achieve secrecy at the physical layer.

Several emerging technologies for 5G are addressed in Chapters 9–12 in **Part III**. *Chapter 9* investigates the secrecy performance of power beacon assisted wirelessly powered communication systems. The authors provide an analytical study of the achievable secrecy outage performance assuming a simple maximal ratio transmission beamformer at the information source and further look into the optimal design of the transmit beamformer. *Chapter 10* focuses on device-to-device (D2D) communication, which is regarded as a promising technology for 5G networks. The authors study how D2D communication can affect the secrecy performance of cellular communication in small-scale networks and large-scale networks and examine different transmission schemes through which interference from D2D communication can be exploited to enhance PHY security. For cognitive radio networks, *Chapter 11* presents key design issues and resource allocation problems in PHY-security, introduces a resource allocation framework for such networks and provides examples to show the effectiveness of the proposed methods. *Chapter 12* studies the network-wide physical layer security performance of the downlink transmission in a millimetre wave cellular network using a stochastic geometry framework. An analysis of the secure connectivity probability and average number of perfect communication links per unit area for colluding/non-colluding eavesdroppers is provided.

Part IV brings together the many ways in which modulation technologies can be applied to ensure PHY security. *Chapter 13* provides a review of directional modulation, which is a promising keyless PHY security technique, together with its development and architecture, and future directions. The extension of the directional modulation technology for multi-beam applications and multipath environments is also discussed. *Chapter 14* is devoted to secure waveform design for 5G. PHY security is examined in terms of secure transmission and secret key extraction over a set of parallel narrowband channels, various waveform approaches (i.e. orthogonal frequency division multiplexing (OFDM), filter bank modulation (FBMC), single carrier frequency division multiple access (SC-FDMA), universal filter multicarrier (UFMC) and generalised frequency division multiplexing (GFDM) are explored for 5G systems, and their security performance is assessed. Specific security solutions for the various waveforms are also proposed. Non-orthogonal multiple access (NOMA) is a promising technology to enable high-efficiency wireless transmission in 5G systems. In *Chapter 15*, the authors investigate the PHY security of two types of single-input single-output (SISO) NOMA systems. Depending on the system model, the aim is to maximise the sum of secrecy rates subject to an individual QoS constraint for each legitimate user, or to optimise the power allocation and beamforming design to ensure secure transmissions for each legitimate user. *Chapter 16* considers a general scenario with a hybrid time and spatial artificial noise (AN) injection scheme in a multiple-input multiple-output multi-antenna eavesdropper (MIMOME)-OFDM wiretap channel and provides a comparison between spatial AN-aided (in a non-OFDM system) and temporal AN aided PHY-layer security schemes in terms of average secrecy rates, implementation feasibility and complexity.

Part V is a unique feature of this book, in which the practical aspects of PHY security are presented. The chapters in this part follow two main directions in PHY applications: key generation (Chapters 17–19) and secrecy coding (Chapter 20). Although the reader may find some overlapping parts in Chapters 17–19, each chapter has been structured in a coherent way so as to enable a richer set of illustrations and examples. *Chapter 17* provides a brief discussion of the role of PHY security as a building block for dealing with challenging security issues in practical systems, for example, in the context of the IoT. The authors further focus on channel-based key generation and look at some fundamental aspects of key generation from reciprocal channel properties, suitable architectures, performance indicators, and how channel-based key generation can be implemented under the constraints of practical systems. Two practical implementations based on CSI and a received signal strength indicator (RSSI) are set up and compared in terms of their performance as well as security properties. *Chapter 18* complements *Chapter 17* by reviewing key generation techniques with a focus on their principles (temporal variation, channel reciprocity and spatial de-correlation), evaluation metrics (randomness, key generation rate and key disagreement rate), procedures and applications. An implementation of a received signal strength-based key generation system on a customised field programmable gate array (FPGA)-based hardware platform, namely WARP, is described in order to verify the key generation principles. *Chapter 19* studies explicit key extraction techniques and algorithms. A practical implantation of secret key generation (SKG) schemes is detailed, based on the channel quantisation alternate algorithm aided by channel de-correlation techniques, which is applicable to public networks such as WiFi and radio-cells of fourth generation (LTE, long-term evolution) mobile networks. The security performance of the implemented SKG schemes is further analysed, and significant practical results and perspectives for future embedding into existing and next generation radio standards are discussed. Focusing on secrecy coding, *Chapter 20* proposes practical secrecy coding schemes that are able to provide reliable and confidential wireless communication between legitimate users. Implementation of these practical wiretap codes in WiFi and LTE testbeds is described and their confidentiality performance is evaluated using the bit error rate as a simple and practical metric for secrecy. Further, the potential of these techniques for radio standards relating to privacy of subscribers and confidentiality of data streams is discussed, and ideas for their practical implementation are discussed.

Our intent in this book is to provide the most progressive approach to trusted communications in 5G networks and beyond via PHY security. Whether readers are postgraduate students, researchers or practicing engineers in the field of wireless communications, they will find this book a valuable resource, containing concepts, frameworks and techniques that will be essential to the further development of the potential of PHY security with a view to designing more secure communications in advanced networks of the future.

Looking at the research and development of PHY security from a bird's-eye view, the modern PHY security technologies developed over the past 10–15 years have embraced the unique properties of wireless channels and wireless multi-user networks. This is in contrast to the pioneering work on the foundations of PHY security established in the context of wired networks in the 1970s. The technologies described in the chapters of this book collectively present a convincing story on how PHY security can provide a much-needed additional layer of protection in future wireless networks. Although they are promising, one must recognise that most of the technologies are still in their early stage of theoretical development with limited practical experiments. In addition, security is a very broad concept that consists of many different aspects of a system. Most PHY security technologies, however, aim only at enhancing secrecy (or confidentiality) of communication. To move forward and make PHY security a practical reality, significant effort is needed in the following two areas of research.

First, PHY security is likely to be welcomed by industry leaders for massive practical development only through cross-layer implementation with cryptographic technologies. That is, as almost all security solutions are based on cryptography, industry will adopt PHY security only if it is designed to work hand-in-hand with cryptography. Thus, an important aspect of future work in this area is to demonstrate how PHY security technology can improve the performance of cryptographic methods, e.g. in terms of an increase in the time required for the best known attack to succeed.

Second, PHY security needs to branch out to solve security aspects other than data confidentiality. For example, wireless authentication is an equally important security issue and PHY-based techniques can be applicable here as well. Recent research has yielded techniques that make use of location-based or channel-based identity information in wireless networks to provide robust authentication. In cases where extremely sensitive communication must happen secretly, protecting the very action of communication can be as important as protecting the content of it. PHY–based techniques are naturally relevant to protecting the physical existence of communication. As another application, PHY-based forensic techniques can be designed to identify malicious communications or attacks. Again, a cross-layer approach is often the best way forward.

This book could not have been possible without the help of many people. The editors would like to thank the chapter authors for their contributions. We also acknowledge the many reviewers who contributed to the accuracy of each chapter, and last but not least, the valuable support of the IET staff during the preparation of this book.

Part I

Fundamentals of physical layer security

Chapter 1

Secrecy metrics for physical layer security over fading channels

Biao He[1], Vincent K. N. Lau[2], Xiangyun Zhou[3], and A. Lee Swindlehurst[1]

Physical layer security over fading channels has drawn a considerable amount of attention in the literature. How to measure the secrecy performance is a fundamental but important issue for the study of wireless physical layer security. This chapter gives a comprehensive overview of the classical secrecy metrics as well as several new secrecy metrics for physical layer security over fading channels, which enables one to appropriately design secure communication systems with different views on how secrecy is measured.

1.1 Introduction

The secrecy performance of wireless systems over fading channels is often characterised by ergodic secrecy capacity [1] or secrecy outage probability [2,3]. The ergodic secrecy capacity captures the capacity limit under a classical information-theoretic secrecy constraint for systems, in which the encoded messages span sufficient channel realisations to experience the ergodic features of the fading channels. The secrecy outage probability gives the probability of not achieving classical information-theoretic secrecy for systems over quasi-static fading channels. The secrecy outage probability has two limitations in evaluating secrecy performance. First, the secrecy outage probability does not give any insight into the eavesdropper's ability to decode the confidential messages. Second, the amount of information leaked to the eavesdropper is not characterised. The limitations of secrecy outage probability motivate the introduction of three new secrecy metrics [4], namely, generalised secrecy outage probability, average fractional equivocation, and average information leakage rate. These three new secrecy metrics provide a more comprehensive and in-depth understanding of secrecy performance by giving insights into the eavesdropper's decodability and evaluating how much or how fast the confidential information is leaked.

[1]Center for Pervasive Communications and Computing, University of California, USA
[2]Department of Electronic and Computer Engineering, The Hong Kong University of Science and Technology, Hong Kong
[3]Research School of Engineering, The Australian National University, Australia

In the current chapter, we first provide background information on the information-theoretic secrecy. We then introduce the classical secrecy metrics for fading channels, i.e. the ergodic secrecy capacity and the secrecy outage probability. We further present the new secrecy metrics, i.e. the generalised secrecy outage probability, the average fractional equivocation, and the average information leakage rate. The use of the new secrecy metrics is illustrated by analysing an example wireless system with fixed-rate wiretap codes.

1.2   Information-theoretic secrecy

In this section, the wiretap channel, which is the basic model of physical layer security, is first presented. The notions of classical information-theoretic secrecy and partial secrecy are then introduced.

1.2.1   Wiretap channel model

Wyner presented the wiretap-channel in his seminal work [5]. A transmitter, Alice, sends confidential information, M, to an intended receiver, Bob, in the presence of an eavesdropper, Eve.

The confidential information, M, is encoded into an n-vector X^n. The received vectors at Bob and Eve are denoted by Y^n and Z^n, respectively. The entropy of the source information and the residual uncertainty for the message at the eavesdropper are denoted by $H(M)$ and $H(M \mid Z^n)$, respectively.

1.2.2   Classical information-theoretic secrecy

In this chapter, we use the phrase "classical information-theoretic secrecy" to encompass the concepts of Shannon's perfect secrecy, strong secrecy, and weak secrecy, which are detailed as follows.

The requirement of classical information-theoretic secrecy is that the amount of information leakage to the eavesdropper vanishes and the eavesdropper's optimal attack is to guess the message at random. From Shannon's definition, perfect secrecy requires that the original message and Eve's observation are statistically independent, formulated mathematically as follows:

$$H(M \mid Z^n) = H(M) \quad \text{or, equivalently,} \quad I(M; Z^n) = 0. \tag{1.1}$$

As it is inconvenient to adopt Shannon's definition of perfect secrecy for analysis, current research often investigates strong secrecy or weak secrecy. Strong secrecy requires that the message and Eve's observation are asymptotically statistically independent as the codeword length goes to infinity, represented mathematically as follows:

$$\lim_{n \to \infty} I(M; Z^n) = 0. \tag{1.2}$$

Weak secrecy requires that the rate of information leaked to the eavesdropper vanishes, which is formulated as follows:

$$\lim_{n \to \infty} \frac{1}{n} I(M; Z^n) = 0. \tag{1.3}$$

It is worth pointing out that the weak secrecy condition is not equivalent to the strong secrecy condition. It has been shown that some weakly secure schemes are inappropriate from a cryptographic perspective [6]. For simplicity, we do not explicitly denote the assumption of $n \to \infty$ for the discussions in the rest of this chapter.

The requirement of no information leakage to Eve actually ensures the highest possible decoding error probability at Eve. Considering the following case as pointed out in [6, Remark 3.1], maximum-likelihood decodes to minimise her decoding error probability P_e. No information leakage guarantees that Eve has to guess the original message. The probability of error under maximum-likelihood decoding is $P_e = (K - 1)/K$. Thus, from the decodability point of view, classical information-theoretic secrecy ensures that $P_e \geq (K - 1)/K$. When the entropy of the message is sufficiently large as $K \to \infty$, classical information-theoretic secrecy actually ensures that P_e asymptotically goes to 1:

$$\lim_{K \to \infty} P_e \geq \lim_{K \to \infty} \frac{K - 1}{K} = 1. \tag{1.4}$$

A more rigorous discussion on the relation between the eavesdropper's error probability and the classical information-theoretic secrecy can be found in [7].

1.2.3 Partial secrecy

Partial secrecy can be defined using the concept of equivocation, which quantifies the level at which Eve is confused. The equivocation was defined by Wyner [5] as the conditional entropy $H(M \mid Z^n)$. Leung-Yan-Cheong and Hellman [8] then defined the fractional equivocation as follows:

$$\Delta = \frac{H(M \mid Z^n)}{H(M)}. \tag{1.5}$$

Analysis based on the fractional equivocation has an advantage; in that, it depends on the channel only, whereas Wyner's original definition of the equivocation depends on the source as well as the channel.

Evaluating secrecy based on (fractional) equivocation is related to the conventional requirement on the decodability of messages at Eve [5]. Tight lower and upper bounds on the decoding error probability can be derived from the equivocation [9,10], although there is no one-to-one relation between the equivocation and the error probability.

To ensure the secrecy level of a system, we would particularly like to guarantee that the decoding error probability at the eavesdropper is larger than a certain level. Hence, having the decoding error probability at Eve lower bounded by the equivocation or fractional equivocation is desirable. Consider again the general case in which messages are uniformly taken from a size K set $[1, 2, \ldots, K]$, which achieves the maximal entropy over an alphabet of size K. The entropy of the message is then given by $H(M) = \log_2(K)$. From Fano's inequality [9, Chapter 2.10], we have

$$H(M \mid Z^n) \leq h(P_e) + P_e \log_2(K), \tag{1.6}$$

where $h(x) = -x \log_2(x) - (1-x) \log_2(1-x)$, $0 \le x \le 1$. We can weaken the inequality (1.6) to

$$P_e \ge \frac{H(M \mid Z^n) - 1}{\log_2(K)} = \Delta - \frac{1}{\log_2(K)}. \tag{1.7}$$

When the entropy of the message is sufficiently large as $K \to \infty$, we further derive (1.7) as follows:

$$\lim_{K \to \infty} P_e \ge \Delta - \lim_{K \to \infty} \frac{1}{\log_2(K)} = \Delta. \tag{1.8}$$

From (1.8), we see that P_e is asymptotically lower bounded by Δ.

1.3   Classical secrecy metrics for fading channels

In this section, we present the classical secrecy metrics for wireless systems over fading channels: ergodic secrecy capacity and secrecy outage probability.

1.3.1   Wireless system setup

Consider the wiretap system with fading channels. Bob and Eve's instantaneous channel capacities are given by

$$C_b = \log_2(1 + \gamma_b) \tag{1.9}$$

and

$$C_e = \log_2(1 + \gamma_e), \tag{1.10}$$

respectively, where the subscripts b and e denote Bob and Eve, respectively, $\gamma_b = P|h_b|^2/\sigma_b^2$ and $\gamma_e = P|h_e|^2/\sigma_e^2$ denote the instantaneous received signal-to-noise ratios (SNRs), P denotes the transmit power, h_b and h_e denote the instantaneous channel gains, and σ_b^2 and σ_e^2 denote the receiver noise variances.

If we consider the Rayleigh fading model with fixed transmit power, the instantaneous received SNRs at Bob and Eve would have exponential distributions, given by

$$f_{\gamma_b}(\gamma_b) = \frac{1}{\bar{\gamma}_b} \exp\left(-\frac{\gamma_b}{\bar{\gamma}_b}\right), \quad \gamma_b > 0 \tag{1.11}$$

and

$$f_{\gamma_e}(\gamma_e) = \frac{1}{\bar{\gamma}_e} \exp\left(-\frac{\gamma_e}{\bar{\gamma}_e}\right), \quad \gamma_e > 0, \tag{1.12}$$

where $\bar{\gamma}_b = \mathbb{E}\{\gamma_b\}$ and $\bar{\gamma}_e = \mathbb{E}\{\gamma_e\}$ denote the average received SNRs, and $\mathbb{E}\{\cdot\}$ denotes the expectation operation.

1.3.2 Ergodic secrecy capacity

For a system in which the encoded message spans sufficient channel realisations to experience the ergodic features of the fading channel, the ergodic secrecy capacity captures the capacity limit subject to the constraint of classical information-theoretic secrecy.

For one fading realisation, wireless channels can be regarded as complex AWGN channels. The secrecy performance of an AWGN wiretap channel can be measured by the secrecy capacity [8], which can be regarded as the maximum achievable transmission rate subject to not only the reliability constraint but also the requirement of classical information-theoretic secrecy [5]. The secrecy capacity for a single wiretap fading channel realisation is given by:

$$C_s = \left[\log_2 (1 + \gamma_b) - \log_2 (1 + \gamma_e)\right]^+, \tag{1.13}$$

where $[x]^+ = \max\{x, 0\}$. To achieve the secrecy capacity in (1.13), Alice is required to have perfect knowledge of both γ_b and γ_e.

Taking the average of (1.13) overall fading realisations, the ergodic secrecy capacity with full CSI available at Alice is given by [1]

$$\bar{C}_s^F = \int_0^\infty \int_{\gamma_e}^\infty \left(\log_2 (1 + \gamma_b) - \log_2 (1 + \gamma_e)\right) f_{\gamma_b}(\gamma_b) f_{\gamma_e}(\gamma_e) \mathrm{d}\gamma_b \mathrm{d}\gamma_e. \tag{1.14}$$

It is worth mentioning that the channel model considered in [1] is not the commonly used fast fading channel. Instead, the number of channel uses within each coherence interval is assumed to be sufficiently large to have the ergodic property. Note that with full CSI, Alice can make sure that the transmission occurs only when $\gamma_b > \gamma_e$. In addition, different from traditional ergodic fading scenarios without secrecy considerations, variable-rate transmission is required to achieve the ergodic secrecy rate. A detailed explanation can be found in [1]. When only main channel information (Bob's CSI) is known at Alice, the ergodic secrecy capacity is given by

$$\bar{C}_s^M = \int_0^\infty \int_0^\infty \left[\log_2 (1 + \gamma_b) - \log_2 (1 + \gamma_e)\right]^+ f_{\gamma_b}(\gamma_b) f_{\gamma_e}(\gamma_e) \mathrm{d}\gamma_b \mathrm{d}\gamma_e. \tag{1.15}$$

The system with full CSI can achieve a higher ergodic secrecy capacity than the system with only main channel CSI subject to the same constraint on the average transmit power, as the transmitter with full CSI can suspend the transmission when $\gamma_b \leq \gamma_e$. Calculation of the ergodic secrecy capacity with imperfect CSI at the receiver side remains in general an open problem.

We plot the ergodic secrecy capacity versus the average power constraint in Figure 1.1. As shown in the figure, for the same average power constraint $\bar{P}$, the ergodic secrecy capacity of the system with full CSI is larger than that of the system with only main channel CSI. For a system with full CSI, Alice suspends the transmission when $|h_b|^2 \leq |h_e|^2$, which is a simple on–off power scheme. Thus, Alice with full CSI transmits with a power of $\bar{P}/\mathbb{P}\big(|h_b|^2 > |h_e|^2\big)$, where $\mathbb{P}(\cdot)$ denotes the probability measure. For a system with only main channel CSI, Alice cannot adopt the on–off power scheme due to the lack of information about Eve's channel, and hence, the transmit power is $\bar{P}$.

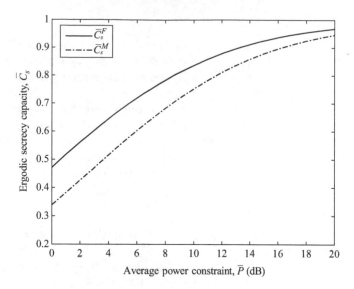

Figure 1.1   Ergodic secrecy capacity versus average power constraint. The parameters are $\mathbb{E}\{|h_b|^2/\sigma_b^2\} = \mathbb{E}\{|h_e|^2/\sigma_e^2\} = 1$

1.3.3   Secrecy outage probability

For a system with quasi-static fading channels in which classical information-theoretic secrecy is not always achievable, the secrecy outage probability measures the probability that classical information theoretic secrecy is not achieved.

Barros and Rodrigues [11] detailed the definition of secrecy outage probability as follows:

$$p_{\text{out}}^{D1} = \mathbb{P}(C_s < R_s) = \mathbb{P}(C_b - C_e < R_s), \tag{1.16}$$

where C_s is the secrecy capacity given in (1.13), and $R_s > 0$ is the target secrecy rate. Such a definition of secrecy outage probability captures the probability of having a reliable and secure transmission. Note that reliability and secrecy are not differentiated, because an outage occurs whenever the transmission is either unreliable or insecure.

From the perspective of system design, an explicit measure of the secrecy level is preferred rather than the joint performance of reliability and secrecy. To this end, the authors in [3] gave an alternative definition of secrecy outage probability, which is formulated by

$$p_{\text{out}}^{D2} = \mathbb{P}(C_e > R_b - R_s \mid \text{message transmission}), \tag{1.17}$$

where R_b and R_s are the rate of the transmitted codeword and the rate of the confidential information of the wiretap code, respectively. The outage probability is conditioned on the message being transmitted. The secrecy outage probability in (1.17) directly measures the probability that a transmitted message does not satisfy the requirement

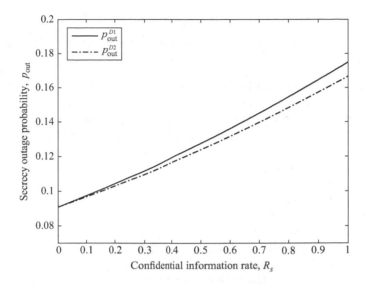

Figure 1.2 Secrecy outage probability versus confidential information rate. The parameters are $\bar{\gamma}_b = 20$ dB and $\bar{\gamma}_e = 10$ dB

of classical information-theoretic secrecy. In addition, we note that p_{out}^{D2} takes into account the system design parameters, i.e. the encoding rates and the condition of the message being transmitted. Therefore, the secrecy outage probability defined by (1.17) is useful for designing transmission schemes that satisfy the secrecy requirement.

We illustrate the secrecy performance of a wireless system measured by p_{out}^{D1} and p_{out}^{D2} in Figure 1.2. The codeword transmission rate is adaptively chosen as $R_b = C_b$. As shown in the figure, p_{out}^{D1} is always larger than p_{out}^{D2}. This is because the definition of p_{out}^{D1} regards not only insecure transmission but also unreliable transmission as outage events. In contrast, the definition of p_{out}^{D2} regards the case of insecure transmission as the only outage event. A detailed comparison between p_{out}^{D1} and p_{out}^{D2} was discussed in [3].

In this section, we have presented two classical secrecy metrics, which have been widely adopted in the literature. It is worth mentioning that there are also some other secrecy metrics which have not been discussed, e.g. secrecy degrees of freedom [12] and effective secrecy throughput [13].

1.4 New secrecy metrics for quasi-static fading channels

As introduced in Section 1.3, the secrecy outage probability is the classical secrecy metric for systems over quasi-static fading channels where classical information-theoretic secrecy is not always achievable. However, the secrecy outage probability has some limitations in evaluating the secrecy performance of wireless systems, which motivates the introduction of new secrecy metrics for quasi-static fading channels in this section.

1.4.1   Limitations of secrecy outage probability

The secrecy outage probability has two major limitations in evaluating secrecy performance:

- The secrecy outage probability does not give any insight into the eavesdropper's ability to decode the confidential messages. The eavesdropper's decodability is an intuitive measure of security in real-world communication systems when classical information-theoretic secrecy is not always achievable, and error-probability-based secrecy metrics are often adopted to quantify secrecy performance in the literature, e.g. [14–19]. A general secrecy requirement for the eavesdropper's decoding error probability can be given as $P_e \geq \varepsilon$, where $0 < \varepsilon \leq 1$ denotes the minimum acceptable value of P_e. In contrast, the current definitions of secrecy outage probability reflect only the extremely stringent requirement where $\varepsilon \to 1$, as classical information-theoretic secrecy guarantees $P_e \to 1$.
- The amount of information leakage to the eavesdropper cannot be characterised. When classical information-theoretic secrecy is not achievable, some information will be leaked to the eavesdropper. Different secure transmission designs that lead to the same secrecy outage probability may actually result in very different amounts of information leakage. Consequently, it is important to know how much or how fast the confidential information is leaked to the eavesdropper to obtain a finer view of the secrecy performance. However, the outage-based approach is not able to evaluate the amount of information leakage when a secrecy outage occurs.

1.4.2   New secrecy metrics: a partial secrecy perspective

Motivated by the secrecy outage probability's limitations, three new secrecy metrics were introduced in [4]:

- Extending the classical definitions of secrecy outage, a generalised formulation of secrecy outage probability was defined. The generalised secrecy outage probability takes the level of secrecy measured by equivocation into account. It establishes a link between the secrecy outage and the eavesdropper's decodability of messages.
- An asymptotic lower bound on the eavesdropper's decoding error probability was introduced. This metric provides a direct link to error-probability-based secrecy metrics that are often used for evaluating the secrecy of practical implementations of wireless systems.
- A metric quantifying the average information leakage rate was defined. This secrecy metric tells how much or how fast the confidential information is leaked to the eavesdropper when classical information-theoretic secrecy is not achieved.

The new secrecy metrics are based on partial secrecy, and more specifically, fractional equivocation. This is different from the secrecy outage probability based on classical information-theoretic secrecy.

For wireless transmissions over fading channels, the fractional equivocation, Δ, is a random quantity due to the fading properties of the channel. A given fading realisation of the wireless channel is equivalent to the Gaussian wiretap channel [20]. The fractional equivocation of a system over a Gaussian wiretap channel depends on the coding and transmission strategies, and hence, there is no general expression for the fractional equivocation which is applicable for all scenarios. Instead, an upper bound on Δ can be derived following closely from [8, Theorem 1] and [20, Corollary 2]. The maximum achievable fractional equivocation for a given fading realisation of the wireless channel is given by

$$\Delta = \begin{cases} 1, & \text{if } C_e \leq C_b - R_s \\ (C_b - C_e)/R_s, & \text{if } C_b - R_s < C_e < C_b \\ 0, & \text{if } C_b \leq C_e, \end{cases} \tag{1.18}$$

where C_b and C_e denote Bob and Eve's channel capacities, respectively, and $R_s = H(M)/n$ denotes the secrecy rate for transmission.

From (1.10), one can obtain the distribution of Δ according to the distribution of γ_e. The new secrecy metrics based on the distribution of Δ are detailed as follows.

(1) Generalised secrecy outage probability
Extending the classical definitions of secrecy outage probability, a generalised definition of secrecy outage probability is given by

$$p_{\text{out}}^G = \mathbb{P}(\Delta < \theta), \tag{1.19}$$

where $0 < \theta \leq 1$ denotes the minimum acceptable value of the fractional equivocation.

As the fractional equivocation is related to the decoding error probability, the generalised secrecy outage probability is applicable to systems with different levels of secrecy requirements measured in terms of Eve's decodability by choosing different values of θ. The classical definitions of secrecy outage probability are equivalent to $\mathbb{P}(\Delta < 1)$ and hence are a special case of the new secrecy outage metric.

There is another way to understand the generalised secrecy outage probability. From (1.5), the information leakage ratio to Eve can be written as $I(M; Z^n)/H(M) = 1 - \Delta$. The information leakage ratio quantifies the percentage of the transmitted confidential information that is leaked to the eavesdropper. Thus, the generalised secrecy outage probability, $p_{\text{out}}^G = \mathbb{P}(\Delta < \theta) = \mathbb{P}(1 - \Delta > 1 - \theta)$, in fact characterises the probability that the information leakage ratio is larger than a certain value, $1 - \theta$.

(2) Average fractional equivocation
The long-term average value of the fractional equivocation is given by

$$\bar{\Delta} = \mathbb{E}\{\Delta\}. \tag{1.20}$$

Note that the expression for the average fractional equivocation, $\bar{\Delta}$, takes the average of the values of fractional equivocation over all fading realisations.

(3) Average information leakage rate

The average information leakage rate is given by

$$R_L = \mathbb{E}\left\{ \frac{I(M;Z^n)}{n} \right\} = \mathbb{E}\left\{ \frac{I(M;Z^n)}{H(M)} \cdot \frac{H(M)}{n} \right\} = \mathbb{E}\{(1 - \Delta)R_s\}, \tag{1.21}$$

where $R_s = H(M)/n$ is the confidential information rate. The average information leakage rate captures how fast the information is leaked to the eavesdropper. Note that the message transmission rate R_s cannot be taken out of the expectation in (1.21), as R_s can be a variable parameter for adaptive-rate transmissions and its distribution may be correlated with the distribution of Δ. For fixed-rate transmission schemes, (1.21) can be further simplified as

$$R_L = \mathbb{E}\{(1 - \Delta)R_s\} = (1 - \bar{\Delta})R_s. \tag{1.22}$$

Remark 1.1. *The aforementioned new secrecy metrics, i.e. (1.19)–(1.21), are generally applicable to evaluating the secrecy performance of different systems. A specific scenario is presented as an example in the next section, wherein the expressions for the new secrecy metrics are derived in terms of transmission rates and channel statistics.*

1.5   Illustrating the use of new secrecy metrics: an example with fixed-rate wiretap codes

In this section, the use of the new secrecy metrics given in Section 1.4 is illustrated by an example wireless system with fixed-rate wiretap codes. It is shown that the new secrecy metrics can provide a more comprehensive and in-depth understanding of the secrecy performance over fading channels. In addition, the impact of the new secrecy metrics on the transmission design is investigated. One can find that the new secrecy metrics lead to very different optimal design parameters that optimise the secrecy performance of the system, compared with the optimal design minimising the classical secrecy outage probability. Simply applying the optimal design that minimises the secrecy outage probability can result in a large secrecy loss, if the actual system requires a low decodability at the eavesdropper and/or a low information leakage rate.

1.5.1   System model

Consider a wiretap channel model over quasi-static Rayleigh fading channels. The instantaneous channel capacities at Bob and Eve, C_b and C_e, are given by (1.9) and (1.10), respectively. The instantaneous received SNRs at Bob and Eve, γ_b and γ_e, follow exponential distributions, given by (1.11) and (1.12), respectively.

Alice adopts the wiretap code [5] for message transmissions. There are two rate parameters, namely, the codeword transmission rate, $R_b = H(X^n)/n$, and the confidential information rate, $R_s = H(M)/n$. A length n wiretap code is constructed by generating 2^{nR_b} codewords $x^n(w, v)$, where $w = 1, 2, \ldots, 2^{nR_s}$ and $v = 1, 2, \ldots, 2^{n(R_b - R_s)}$. For each message index w, we randomly select v from $\{1, 2, \ldots, 2^{n(R_b - R_s)}\}$ with uniform probability and transmit the codeword $x^n(w, v)$. In addition, we consider fixed-rate transmissions where the transmission rates, i.e., R_b and R_s, are fixed over time. Fixed-rate transmissions are often adopted to reduce system complexity. In practice, applications like video streaming in multimedia applications often require fixed-rate transmission.

Bob and Eve are assumed to perfectly know their own channels. Hence, C_b and C_e are known at Bob and Eve, respectively. Alice has statistical knowledge of Bob and Eve's channels but does not know either Bob or Eve's instantaneous CSI. We further assume that Bob provides a one-bit feedback about his channel quality to Alice in order to avoid unnecessary transmissions [3,21]. The one-bit feedback enables an on–off transmission scheme to guarantee that the transmission takes place only when $R_b \leq C_b$. In addition, the on–off transmission scheme incurs a probability of transmission given by

$$p_{\text{tx}} = \mathbb{P}(R_b \leq C_b) = \mathbb{P}\big(R_b \leq \log_2(1 + \gamma_b)\big) = \exp\left(-\frac{2^{R_b} - 1}{\bar{\gamma}_b}\right). \qquad (1.23)$$

1.5.2 Secrecy performance evaluation

Following [20, Corollary 2] and [8, Section III] with $H(X^n)/n = R_b$, one can derive the fractional equivocation for a given fading realisation of the channel as follows. For a given fading realisation of the channel, the maximum achievable fractional equivocation for the wiretap code with $R_b \leq C_b$ and $R_s \leq R_b$ is given by

$$\Delta = \begin{cases} 1, & \text{if } C_e \leq R_b - R_s \\ (R_b - C_e)/R_s, & \text{if } R_b - R_s < C_e < R_b \\ 0, & \text{if } R_b \leq C_e. \end{cases} \qquad (1.24)$$

Then, one can evaluate the secrecy performance of wireless transmissions over fading channels from the distribution of Δ.

(1) Generalised secrecy outage probability
The generalised secrecy outage probability is given by

$$\begin{aligned} p_{\text{out}}^G &= \mathbb{P}(\Delta < \theta) \\ &= \mathbb{P}\big(2^{R_b} - 1 \leq \gamma_e\big) + \mathbb{P}\big(2^{R_b - R_s} - 1 < \gamma_e < 2^{R_b} - 1\big) \\ &\quad \cdot \mathbb{P}\left(\frac{R_b - \log_2(1 + \gamma_e)}{R_s} < \theta \,\middle|\, 2^{R_b - R_s} - 1 < \gamma_e < 2^{R_b} - 1\right) \\ &= \exp\left(-\frac{2^{R_b - \theta R_s} - 1}{\bar{\gamma}_e}\right), \end{aligned} \qquad (1.25)$$

where $0 < \theta \leq 1$.

If $\theta = 1$, we have

$$p_{\text{out}}^G(\theta = 1) = \exp\left(-\frac{2^{R_b - R_s} - 1}{\bar{\gamma}_e}\right). \qquad (1.26)$$

Note that (1.26) is the same as [3, Eq. (8)], which is the expression for the classical secrecy outage probability.

(2) Average fractional equivocation

The average fractional equivocation is given by

$$\bar{\Delta} = \mathbb{E}\{\Delta\}$$

$$= \int_0^{2^{R_b-R_s}-1} f_{\gamma_e}(\gamma_e)d\gamma_e + \int_{2^{R_b-R_s}-1}^{2^{R_b}-1} \left(\frac{R_b - \log_2(1+\gamma_e)}{R_s}\right) f_{\gamma_e}(\gamma_e)d\gamma_e$$

$$= 1 - \frac{1}{R_s \ln 2} \exp\left(\frac{1}{\bar{\gamma}_e}\right)\left(\text{Ei}\left(-\frac{2^{R_b}}{\bar{\gamma}_e}\right) - \text{Ei}\left(-\frac{2^{R_b-R_s}}{\bar{\gamma}_e}\right)\right), \tag{1.27}$$

where $\text{Ei}(x) = \int_{-\infty}^x e^t/t \, dt$ denotes the exponential integral function. The average fractional equivocation is an asymptotic lower bound on the eavesdropper's decoding error probability.

(3) Average information leakage rate

With the fixed-rate transmission scheme, the average information leakage rate is derived from (1.22) as

$$R_L = (1 - \bar{\Delta})R_s = \frac{1}{\ln 2} \exp\left(\frac{1}{\bar{\gamma}_e}\right)\left(\text{Ei}\left(-\frac{2^{R_b}}{\bar{\gamma}_e}\right) - \text{Ei}\left(-\frac{2^{R_b-R_s}}{\bar{\gamma}_e}\right)\right). \tag{1.28}$$

It is worth mentioning that R_L in (1.28) does not depend on the probability of transmission p_{tx}, as R_L characterises how fast on average the information is leaked to the eavesdropper when a message transmission occurs.

In Table 1.1, we show that secrecy performance sometimes cannot be appropriately characterised by the secrecy outage probability, whereas on the other hand it can be quantified by the new secrecy metrics. We consider an extreme case where the confidential information rate is the same as the total codeword rate, $R_b = R_s$. This is equivalent to using an ordinary code instead of the wiretap code for transmission. As shown in Table 1.1, the secrecy performance measured by the classical secrecy outage probability ($\theta = 1$) is not related to Eve's channel condition, as it is always equal to 1. However, we know that the decodability of messages at the receiver is related to the channel condition. Intuitively, with an improvement in Eve's channel quality, the probability of error at Eve should decrease, and the secrecy performance should become worse. Therefore, we see that the secrecy performance cannot be properly characterised by the classical secrecy outage probability. In contrast, we find that

Table 1.1   Secrecy performance measured by different secrecy metrics for systems with different channel qualities for Eve. Results are shown for transmission with $R_b = R_s = 1$

$\bar{\gamma}_e$	−10 dB	−5 dB	0 dB	5 dB	10 dB	15 dB
$p_{out}^G(\theta=1)$	1	1	1	1	1	1
$p_{out}^G(\theta=0.8)$	0.23	0.62	0.86	0.95	0.98	0.99
$\bar{\Delta}$	0.87	0.65	0.33	0.13	0.04	0.01
R_L	0.13	0.35	0.67	0.87	0.96	0.99

the change in secrecy performance with Eve's channel quality can be appropriately quantified by all of the new secrecy metrics. The generalised secrecy outage probability ($\theta = 0.8$) increases as the average SNR at Eve increases. The average fractional equivocation decreases as the average SNR at Eve increases. The average information leakage rate increases as the average SNR at Eve increases. This simple example of transmission with an ordinary code shows that the new secrecy metrics are able to reveal information about the secrecy performance that cannot be captured by the secrecy outage probability.

1.5.3 Impact on system design

In what follows, we examine the significance of the new secrecy metrics given in Section 1.4 from the perspective of a system designer. In particular, we first check whether the new secrecy metrics lead to different system designs that optimise the secrecy performance, compared with the optimal design parameters minimising the classical secrecy outage probability. We then evaluate how large the secrecy loss would be if one applies the optimal transmission design based on the secrecy outage probability, whereas the actual system requires a low decodability at the eavesdropper or a low information leakage rate.

To this end, we study the problem of optimising the secrecy performance subject to a throughput constraint $\eta > \Gamma$, where η denotes the throughput and Γ denotes its minimum required value. We still consider transmission with fixed-rate wiretap codes, and the parameters to design are the encoding rates R_b and R_s. Taking the probability of transmission into account given in (1.23), the throughput is given by

$$\eta = p_{\text{tx}} R_s = \exp\left(-\frac{2^{R_b} - 1}{\bar{\gamma}_b}\right) R_s. \tag{1.29}$$

The three problems for the systems with different secrecy metrics are formulated as follows:

Problem 1: Generalised secrecy outage probability minimisation

$$\min_{R_b, R_s} \ p_{\text{out}}^{G} = \exp\left(-\frac{2^{R_b - \theta R_s} - 1}{\bar{\gamma}_e}\right), \tag{1.30a}$$

$$\text{s.t.} \ \ \eta \geq \Gamma, R_b \geq R_s > 0. \tag{1.30b}$$

Problem 2: Average fractional equivocation maximisation

$$\max_{R_b, R_s} \ \bar{\Delta} = 1 - \frac{1}{R_s \ln 2} \exp\left(\frac{1}{\bar{\gamma}_e}\right) \left(\text{Ei}\left(-\frac{2^{R_b}}{\bar{\gamma}_e}\right) - \text{Ei}\left(-\frac{2^{R_b - R_s}}{\bar{\gamma}_e}\right)\right), \tag{1.31a}$$

$$\text{s.t.} \ \ \eta \geq \Gamma, R_b \geq R_s > 0. \tag{1.31b}$$

Problem 3: Average information leakage rate minimisation

$$\min_{R_b, R_s} \ R_L = \frac{1}{\ln 2} \exp\left(\frac{1}{\bar{\gamma}_e}\right) \left(\text{Ei}\left(-\frac{2^{R_b}}{\bar{\gamma}_e}\right) - \text{Ei}\left(-\frac{2^{R_b - R_s}}{\bar{\gamma}_e}\right)\right), \tag{1.32a}$$

$$\text{s.t.} \ \ \eta \geq \Gamma, R_b \geq R_s > 0. \tag{1.32b}$$

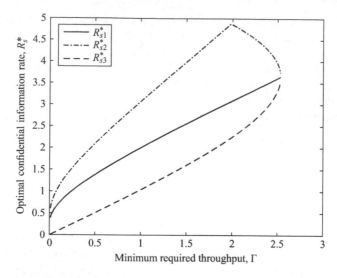

Figure 1.3   For different secrecy metrics: optimal confidential information rate versus minimum required throughput

These three problems can be easily solved. For the sake of brevity, the detailed analytical solutions and derivations are omitted, which however can be found in [4]. We numerically compare the optimal solutions and show the significance of the new secrecy metrics on the system design by Figures 1.3–1.5. For the numerical results in this section, the parameters are set as $\bar{\gamma}_b = \bar{\gamma}_e = 15$ dB and $\theta = 1$. The optimal codeword transmission rates for Problems 1–3 are denoted by R_{b1}^*, R_{b2}^*, and R_{b3}^*, respectively. The optimal confidential information rates for Problems 1–3 are denoted by R_{s1}^*, R_{s2}^*, and R_{s3}^*, respectively.

Figure 1.3 plots the optimal confidential information rate R_s^* versus the throughput constraint Γ. The optimal codeword transmission rate R_b^* is not shown in the figure, as it is equal to $R_b^* = \log_2(1 - \bar{\gamma}_b \ln(\Gamma/R_s^*))$ for all three problems, and the differences between R_{b1}^*, R_{b2}^*, and R_{b3}^* are determined by the differences between R_{s1}^*, R_{s2}^*, and R_{s3}^*. As shown in the figure, the values of R_{s1}^*, R_{s2}^*, and R_{s3}^* are very different from each other. The observation above illustrates that the optimal transmission designs are different when we use different secrecy metrics to evaluate secrecy performance, and hence, the new secrecy metrics lead to different system design choices that optimise the secrecy performance.

Figure 1.4 plots $\bar{\Delta}$ achieved by (R_{b1}^*, R_{s1}^*) and (R_{b2}^*, R_{s2}^*). Figure 1.5 plots R_L achieved by (R_{b1}^*, R_{s1}^*) and (R_{b3}^*, R_{s3}^*). The rate pair (R_{b1}^*, R_{s1}^*) is optimal for minimising the secrecy outage probability, (R_{b2}^*, R_{s2}^*) is optimal for maximising the average fractional equivocation, and (R_{b3}^*, R_{s3}^*) is optimal for minimising the average information leakage rate. As shown in Figure 1.4, adopting an optimal design based on secrecy outage probability, would lead to a considerable loss if the practical secrecy requirement is to ensure a low decodability at the eavesdropper. As shown in

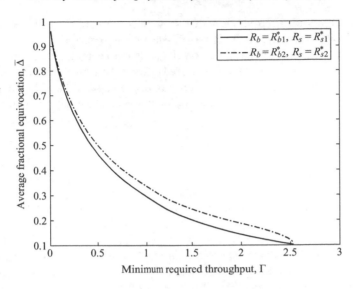

Figure 1.4   Average fractional equivocation versus minimum required throughput

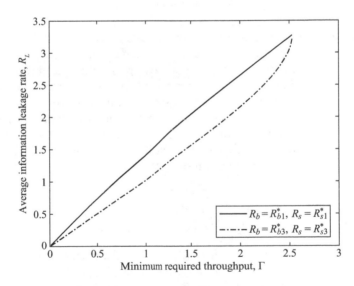

Figure 1.5   Average information leakage rate versus minimum required throughput

Figure 1.5, adopting an optimal design based on secrecy outage probability would also lead to a considerable loss if the practical secrecy requirement is to maintain a low information leakage rate. These observations show that designing the system with the appropriate secrecy metric is important. Applying the transmission design based on secrecy outage probability can result in a large secrecy loss if the actual system requires a low decodability at the eavesdropper or a low information leakage rate.

The results and discussions above show that the new secrecy metrics have a significant value from the perspective of system designers. System designers should not adopt the optimal design based on the classical secrecy outage probability to optimise the secrecy performance measured by the new secrecy metrics, as adopting the optimal design based on the classical secrecy outage probability would lead to a large secrecy loss when the secrecy performance is measured by the new secrecy metrics.

1.6 Conclusion

In this chapter, we have introduced new secrecy metrics for physical layer security over fading channels. Classical secrecy metrics include the ergodic secrecy capacity and the secrecy outage probability. The ergodic secrecy capacity characterises the capacity limit of the system where the encoded messages span sufficient channel realisations to capture the ergodic features of fading channels. For quasi-static fading channels, the secrecy outage probability measures the probability of failing to achieve classical information-theoretic secrecy. We have further introduced three new secrecy metrics for quasi-static fading channels based on partial secrecy, which address the limitations of the secrecy outage probability. The generalised secrecy outage probability establishes a link between the concept of secrecy outage and the decodability of messages at the eavesdropper. The asymptotic lower bound on the eavesdropper's decoding error probability provides a direct error-probability-based secrecy metric. The average information leakage rate characterises how fast the confidential information is leaked to the eavesdropper when classical information-theoretic secrecy is not achieved. We have shown that the new secrecy metrics provide a more comprehensive understanding of physical layer security over fading channels. Furthermore, we have demonstrated that the new secrecy metrics enable appropriate transmission designs for systems with different secrecy requirements.

References

[1] P. K. Gopala, L. Lai, and H. El Gamal, "On the secrecy capacity of fading channels," *IEEE Trans. Inf. Theory*, vol. 54, no. 10, pp. 4687–4698, 2008.

[2] M. Bloch, J. Barros, M. R. D. Rodrigues, and S. W. McLaughlin, "Wireless information-theoretic security," *IEEE Trans. Inf. Theory*, vol. 54, no. 6, pp. 2515–2534, 2008.

[3] X. Zhou, M. R. McKay, B. Maham, and A. Hjørungnes, "Rethinking the secrecy outage formulation: A secure transmission design perspective," *IEEE Commun. Lett.*, vol. 15, no. 3, pp. 302–304, 2011.

[4] B. He, X. Zhou, and A. L. Swindlehurst, "On secrecy metrics for physical layer security over quasi-static fading channels," *IEEE Trans. Wireless Commun.*, vol. 15, no. 10, pp. 6913–6924, 2016.

[5] A. D. Wyner, "The wire-tap channel," *Bell Syst. Tech. J.*, vol. 54, no. Oct. (8), pp. 1355–1387, 1975.

[6] M. Bloch and J. Barros, *Physical-Layer Security: From Information Theory to Security Engineering*. England: Cambridge University Press, 2011.

[7] I. B. H. Boche and J. Sommerfeld, "Capacity results for compound wiretap channels," in *Proc. IEEE ITW*, Paraty, Brazil, Oct. 2011, pp. 60–64.

[8] S. K. Leung-Yan-Cheong and M. E. Hellman, "The Gaussian wire-tap channel," *IEEE Trans. Inf. Theory*, vol. 24, no. 4, pp. 451–456, 1978.

[9] T. M. Cover and J. A. Thomas, *Elements of Information Theory*, 2nd ed. Hoboken, NJ: Wiley, 2006.

[10] M. Feder and N. Merhav, "Relations between entropy and error probability," *IEEE Trans. Inf. Theory*, vol. 40, no. 1, pp. 259–266, 1994.

[11] J. Barros and M. R. D. Rodrigues, "Secrecy capacity of wireless channels," in *Proc. IEEE ISIT*, Seattle, WA, July 2006, pp. 356–360.

[12] X. He, A. Khisti, and A. Yener, "MIMO multiple access channel with an arbitrarily varying eavesdropper: Secrecy degrees of freedom," *IEEE Trans. Inf. Theory*, vol. 59, no. 8, pp. 4733–4745, 2013.

[13] S. Yan, N. Yang, G. Geraci, R. Malaney, and J. Yuan, "Optimization of code rates in SISOME wiretap channels," *IEEE Trans. Wireless Commun.*, vol. 14, no. 11, pp. 6377–6388, 2015.

[14] M. Baldi, G. Ricciutelli, N. Maturo, and F. Chiaraluce, "Performance assessment and design of finite length LDPC codes for the Gaussian wiretap channel," in *Proc. IEEE ICC Workshops*, London, UK, Jun. 2015, pp. 446–451.

[15] D. Klinc, J. Ha, S. W. McLaughlin, J. Barros, and B. J. Kwak, "LDPC codes for the Gaussian wiretap channel," *IEEE Trans. Inf. Forensics Security*, vol. 6, no. 3, pp. 532–540, 2011.

[16] M. Baldi, M. Bianchi, and F. Chiaraluce, "Coding with scrambling, concatenation, and HARQ for the AWGN wire-tap channel: A security gap analysis," *IEEE Trans. Inf. Forensics Security*, vol. 7, no. 3, pp. 883–894, 2012.

[17] R. Soosahabi and M. Naraghi-Pour, "Scalable PHY-layer security for distributed detection in wireless sensor networks," *IEEE Trans. Inf. Forensics Security*, vol. 7, no. 4, pp. 1118–1126, 2012.

[18] A. S. Khan, A. Tassi, and I. Chatzigeorgiou, "Rethinking the intercept probability of random linear network coding," *IEEE Commun. Lett.*, vol. 19, no. 10, pp. 1762–1765, 2015.

[19] J. E. Barcelo-Llado, A. Morell, and G. Seco-Granados, "Amplify-and-forward compressed sensing as a physical-layer secrecy solution in wireless sensor networks," *IEEE Trans. Inf. Forensics Security*, vol. 9, no. 5, pp. 839–850, 2014.

[20] Y. Liang, H. V. Poor, and S. Shamai, "Secure communication over fading channels," *IEEE Trans. Inf. Theory*, vol. 54, no. 6, pp. 2470–2492, 2008.

[21] B. He and X. Zhou, "Secure on-off transmission design with channel estimation errors," *IEEE Trans. Inf. Forensics Security*, vol. 8, no. 12, pp. 1923–1936, 2013.

Chapter 2

Secure data networks with channel uncertainty

Amal Hyadi[1], Zouheir Rezki[2], and Mohamed-Slim Alouini[1]

Recent years have been marked by an enormous growth of wireless communication networks and an extensive use of wireless applications. In return, this phenomenal expansion is inducing more concerns about the privacy and the security of the users. For many years, the security challenge has been mainly addressed at the application layer using cryptographic techniques. However, with the emergence of ad-hoc and decentralised networks and the deployment of 5G and beyond wireless communication systems, the need for less complex securing techniques had become a necessity. It is mainly for this reason that wireless physical layer security has gained much attention from the research community. What distinguishes information-theoretic security compared to other high-layer cryptographic techniques is that it exploits the randomness and the fluctuations of the wireless channel to achieve security at a remarkably reduced computational complexity. However, these technical virtues rely heavily on perhaps idealistic channel state information assumptions. In this chapter, we look at the physical layer security paradigm from the channel uncertainty perspective. In particular, we discuss the ergodic secrecy capacity of wiretap channels when the transmitter is hampered by the imperfect knowledge of the channel state information (CSI).

2.1 Introduction

Information-theoretic security dates back to 1949 when Shannon introduced his pioneer work on cipher systems [1]. Shannon's work considers the secure transmission of confidential information when a random secret key is shared between the legitimate parties, and a passive eavesdropper is intercepting the communication. To guarantee perfect secrecy, Shannon showed that the entropy of the shared secret key should exceed the entropy of the message, or in other words, this requires the key to be at least as long as the confidential message itself. Many years later, Wyner came up with a new relaxed information-theoretic security constraint that requires the information leakage rate to asymptotically vanish for a sufficiently large codeword length [2]. This constraint is known as Wyner's weak secrecy constraint. Also, Wyner's model,

[1]Computer, Electrical, and Mathematical Sciences & Engineering (CEMSE) Division, King Abdullah University of Science and Technology (KAUST), Saudi Arabia
[2]Electrical and Computer Engineering Department, University of Idaho, USA

known as the wiretap channel, considers the case when the signal received at the eavesdropper is a degraded version of the signal received at the legitimate receiver and exploits this particular structure of the channel to transmit a message reliably to the intended receiver while asymptotically leaking no information to the eavesdropper. Ulterior works generalised Wyner's work to the case of non-degraded channels [3], Gaussian channels [4], and fading channels [5–7], to cite only a few.

In particular, securing fading channels from potential wiretapping attacks is of crucial interest, especially in regard to the unprecedented growth of wireless communication applications and devices. The fading wiretap channel has opened new research directions for information-theoretic security. What is unique about the fading model is that it takes advantage of the randomness of the channel gain fluctuations to secure the transmission against potential eavesdroppers, at the physical layer itself. As a result, even if the eavesdropper has a better average signal-to-noise ratio (SNR) than the legitimate receiver, physical layer security can still be achieved over fading channels without requiring the sharing of a secret key. To make the most of what fading has to offer, the knowledge of the CSI at the transmitter (CSIT) is of primordial importance.

The vast majority of research works on physical layer security assume that the transmitter has perfect knowledge of the legitimate receiver's CSI, usually referred to as the main CSI, or even of both the main and the eavesdropper CSI. Although this assumption reduces the complexity of the analysis and allows the characterisation of the full potential of the fading wiretap channel, it does not capture the practical aspect of the transmission model. In a wireless communication system, acquiring the CSIT requires the receiver to feedback its CSI constantly to the transmitter. This feedback process is typically accompanied by the introduction of uncertainty into the CSIT. Different phenomena can cause the CSIT to be imperfect. Most commonly, the uncertainty comes from an error of estimation at the transmitter who ends up with a noisy version of the CSI, or from a feedback link with a limited capacity which requires the transmission of quantised CSI, or also from a delayed feedback causing outdated CSIT. A synopsis of how different levels of CSIT impact the system's security is provided in [8] and a detailed state-of-the-art review of physical layer security with CSIT uncertainty is presented in [9].

In this chapter, we provide an overview of the impact of partial CSIT on the ergodic secrecy capacity of wiretap channels. We consider two prevalent causes of CSIT uncertainty; either an error of estimation occurs at the transmitter, and he can only base the coding and the transmission strategy on a noisy version of the CSI, or the CSI feedback link has a limited capacity and the legitimate receivers can only inform the transmitter about the quantised CSI. We investigate both the single-user and the multi-user transmissions. In the latter scenario, we distinguish between common message transmission, where the source broadcasts the same information to all the legitimate receivers, and multiple messages transmission, where the transmitter broadcasts multiple independent messages to the legitimate receivers. We also discuss the broadcast channel with confidential messages (BCC) where the transmitter has one common message to be transmitted to two users and one secret message intended to only one of them.

It should also be noted that this chapter investigates the information-theoretic security of wiretap fading channels from the weak secrecy constraint perspective. At the difference of Shannon's perfect secrecy, which requires the exact information leakage to be zero, i.e. $I(W; Y_E^n) = 0$, where W is the confidential information, and Y_E^n is the n-length received signal at the eavesdropper; the weak secrecy constraint only requires the rate of the information leaked to the eavesdropper to asymptotically vanish, i.e. $\lim_{n \to \infty} \frac{1}{n} I(W; Y_E^n) = 0$, where n is the length of the transmitted codeword. The weak secrecy condition can be further straightened to the strong secrecy constraint which requires the absolute amount of secrecy leaked to the eavesdropper to go to zero as the length of the transmitted codeword becomes very large, i.e. $\lim_{n \to \infty} I(W; Y_E^n) = 0$. Generally, a specific code achieving a secure communication under the weak secrecy constraint does not necessarily achieve strong secrecy [10–11]. Yet, both weak and strong secrecy constraints result in the same secrecy capacity [12–13]. Another interesting secrecy condition is the semantic security constraint that was firstly introduced in cryptography and was lately extended to the wiretap channel context [14]. This secrecy condition could be more appropriate in real-life scenarios as it alleviates the assumptions on the transmitted confidential message, i.e. it does not assume that the message is random and uniformly distributed. This is a new challenging and promising direction for research on information-theoretic security.

The rest of this chapter is organised as follows. Section 2.2 considers single-user wiretap channels with CSIT uncertainty. First, the adopted system model is introduced. Then, the ergodic secrecy capacity is characterised in two scenarios; when the transmitter has a noisy version of the main CSI and when the CSI feedback link has a limited capacity. In the same vein, Section 2.3 investigates the ergodic secrecy capacity of multi-user wiretap channels with CSIT imperfections. Finally, Section 2.4 offers some concluding remarks and briefly outlines some possible future directions.

Notations: Throughout this chapter, we use the following notational conventions. The expectation operation is denoted by $\mathbb{E}[.]$, the modulus of a scalar x is expressed as $|x|$, and we define $\{v\}^+ = \max(0, v)$. The function $P(.)$ is used to describe the power profile adopted at the transmitter. The argument of this function can be a scalar or a vector. The entropy of a discrete random variable X is denoted by $H(X)$, the mutual information between random variables X and Y is denoted by $I(X; Y)$, and a sequence of length n is denoted by X^n, i.e. $X^n = \{X(1), X(2), \ldots, X(n)\}$. In addition, we use $f_X(.)$ and $F_X(.)$ to denote the probability density function (PDF) and the cumulative distribution function (CDF) of the random variable X. We also use the notation $X \sim \mathcal{CN}(0, \sigma^2)$ to indicate that X is a circularly symmetric complex-valued Gaussian random variable with zero mean and variance σ^2.

2.2 Secure single-user transmission with channel uncertainty

We start by considering the case of single-user transmission when the confidential information is intended to only one legitimate receiver in the presence of an eavesdropper. We are particularly interested in the ergodic secrecy capacity of the system when the transmitter has partial knowledge of the main CSI.

2.2.1   System model

We consider a discrete-time memoryless single-user wiretap channel consisting of a transmitter, a legitimate receiver, and an eavesdropper. Each terminal is equipped with a single antenna. The outputs at both the legitimate destination and the eavesdropper at time instant t, $t \in \{1, \ldots, n\}$, are, respectively, expressed by

$$\begin{cases} Y_R(t) = h_R(t)X(t) + z_R(t), \\ Y_E(t) = h_E(t)X(t) + z_E(t), \end{cases} \tag{2.1}$$

where $X(t)$ is the transmitted signal, $h_R(t)$ and $h_E(t)$ are zero-mean, unit-variance complex channel gain coefficients, respectively, corresponding to the main channel and the eavesdropper's channel, and $z_R(t)$ and $z_E(t)$ represent zero-mean, unit-variance circularly symmetric additive white Gaussian noises at the legitimate receiver and the eavesdropper, respectively. An average transmit power constraint is imposed at the transmitter such that $\frac{1}{n} \sum_{t=1}^{n} \mathbb{E}[|X(t)|^2] \leq P_{avg}$, where the expectation is over the input distribution.

The channel gains h_R and h_E are assumed to be independent and identically distributed (i.i.d.), stationary and ergodic random variables. We consider that the legitimate receiver can perfectly estimate its channel gain and that it sends a feedback information about its CSI to the transmitter before data transmission. The received feedback information, at the transmitter, is subject to channel imperfections; because of either an estimation error or a feedback link with limited capacity. Details on what is known at the transmitter and CSIT uncertainties are provided in the following sub-sections. As for the eavesdropper, we assume that it is perfectly aware of its channel gain and that it tracks the feedback link between the legitimate receiver and the transmitter. Consequently, the CSI cannot be used as a source of randomness for key generation. The statistics of both h_R and h_E are available to all terminals. We are interested in the secrecy capacity of such a channel when $n \to \infty$.

The transmitter wishes to send a secret message W to the legitimate receiver. A $(2^{n\mathcal{R}_s}, n)$ code consists of the following elements: a message set $\mathcal{W} = \{1, 2, \ldots, 2^{n\mathcal{R}_s}\}$ with the messages $W \in \mathcal{W}$ independent and uniformly distributed over $\mathcal{W}$, a stochastic encoder $f: \mathcal{W} \to \mathcal{X}^n$ that maps each message w to a codeword $x^n \in \mathcal{X}^n$, and a decoder at the legitimate receiver $g: \mathcal{Y}^n \to \mathcal{W}$ that maps a received sequence $y_R^n \in \mathcal{Y}^n$ to a message $\hat{w} \in \mathcal{W}$.

A rate $\mathcal{R}_s$ is an *achievable secrecy rate* if there exists a sequence of $(2^{n\mathcal{R}_s}, n)$ code such that both the average error probability, $P_e = \frac{1}{2^{n\mathcal{R}_s}} \sum_{w=1}^{2^{n\mathcal{R}_s}} \Pr[W \neq \hat{W} | W = w]$, and the leakage rate at the eavesdropper, $\frac{1}{n} I(W; Y_E^n, h_E^n, h_R^n)$, go to zero as n goes to infinity. The *secrecy capacity* $\mathcal{C}_s$ is defined as the maximum achievable secrecy rate, i.e. $\mathcal{C}_s \triangleq \sup \mathcal{R}_s$.

2.2.2   Wiretap channel with noisy CSIT

In this sub-section, we consider that the legitimate receiver feeds back its CSI over a feedback link with infinite capacity and that an error of estimation occurs at the transmitter. The main channel gain estimation model can be formulated as

$$h_R(t) = \sqrt{1-\alpha} \, \hat{h}_R(t) + \sqrt{\alpha} \, \tilde{h}_R(t), \tag{2.2}$$

where $\hat{h}_R(t)$ is the noisy version of the CSI available at the transmitter at time instant t, $\tilde{h}_R(t)$ is the channel estimation error, and α is the estimation error variance ($\alpha \in [0, 1]$). The case $\alpha = 0$ corresponds to the perfect main CSIT scenario, whereas $\alpha = 1$ corresponds to the no main CSIT case.

2.2.2.1 Secrecy capacity analysis

The ergodic secrecy capacity of fast fading channels under imperfect main channel estimation at the transmitter was studied in [15,16], where the following results are presented.

Theorem 2.1. *The ergodic secrecy capacity of the single-user single-antenna fading wiretap channel with noisy main CSIT is characterised as*

$$\mathscr{C}_s^- \leq \mathscr{C}_s \leq \mathscr{C}_s^+, \tag{2.3}$$

where $\mathscr{C}_s^-$ and $\mathscr{C}_s^+$ are given by

$$\mathscr{C}_s^- = \max_{\substack{P(\tau) \\ |h_E|^2, |h_R|^2, |\hat{h}_R|^2 \geq \tau}} \mathbb{E} \left[\log\left(\frac{1 + |h_R|^2 P(\tau)}{1 + |h_E|^2 P(\tau)} \right) \right], \tag{2.4}$$

$$\mathscr{C}_s^+ = \max_{P(\hat{h}_R)} \mathbb{E}_{\hat{h}_R, \tilde{h}_R} \left[\left\{ \log\left(\frac{1 + |\sqrt{1-\alpha}\hat{h}_R + \sqrt{\alpha}\tilde{h}_R|^2 P(\hat{h}_R)}{1 + |\tilde{h}_R|^2 P(\hat{h}_R)} \right) \right\}^+ \right], \tag{2.5}$$

with $P(\tau) = \dfrac{P_{\text{avg}}}{1 - F_{|\hat{h}_R|^2}(\tau)}$ and $\mathbb{E}[P(\hat{h}_R)] \leq P_{\text{avg}}$.

Proof. The achievability of the lower bound $\mathscr{C}_s^-$ follows by using wiretap coding along with a Gaussian input and an on–off power scheme that adapts the transmission power according to the estimated channel gain $\hat{h}_R$. The optimal transmission threshold τ is chosen to satisfy:

$$\mathbb{E}_{|h_R|^2, |\hat{h}_R|^2 \geq \tau} \left[\frac{|h_R|^2 P'(\tau)}{1 + |h_R|^2 P(\tau)} \right] - \mathbb{E}_{|h_E|^2} \left[\frac{|h_E|^2 P'(\tau)}{1 + |h_E|^2 P(\tau)} \right] (1 - F_{|\hat{h}_R|^2}(\tau))$$

$$= f_{|\hat{h}_R|^2}(\tau) \left(\mathbb{E}_{|h_R|^2 | |\hat{h}_R|^2} \left[\log(1 + |h_R|^2 P(\tau)) | |\hat{h}_R|^2 = \tau \right] - \mathbb{E}_{|h_E|^2} \left[\log(1 + |h_E|^2 P(\tau)) \right] \right). \tag{2.6}$$

where $P'(\tau)$ is the derivative of $P(\tau)$ with respect to τ. Obviously, the achievable secrecy rate can be directly improved by optimising over all power policies satisfying the average power constraint. However, the optimal procedure is rather complex and time-consuming and does not seem to provide a substantial gain.

The upper bound $\mathscr{C}_s^+$ follows by properly correlating the main and the eavesdropper's channel gains. Indeed, as the estimation error $\tilde{h}_R$ is identically distributed as h_E, and as the transmitter is only aware of $\hat{h}_R$, which means that $\tilde{h}_R$ is independent of the transmitted signal X, then substituting h_E by $\tilde{h}_R$ is a valid choice that provides a tight upper bound. The presented upper bound has the following interpretation. In order to increase the information leakage, the eavesdropper sticks to the component of the main channel that is unknown to the transmitter. We note also that the optimisation problem is concave in this case. That is, the optimum power profile $P(\hat{h}_R)$ is the solution of the following optimality condition:

$$\mathbb{E}_{\tilde{h}_R \in \mathscr{D}_{\hat{h}_R}} \left[\frac{|\sqrt{1-\alpha}\hat{h}_R + \sqrt{\alpha}\tilde{h}_R|^2}{1 + |\sqrt{1-\alpha}\hat{h}_R + \sqrt{\alpha}\tilde{h}_R|^2 P(\hat{h}_R)} - \frac{|\tilde{h}_R|^2}{1 + |\tilde{h}_R|^2 P(\hat{h}_R)} \right] - \lambda = 0, \quad (2.7)$$

where λ is Lagrange multiplier corresponding to the average power constraint, and $\mathscr{D}_{\hat{h}_R} = \left\{ \tilde{h}_R : |\sqrt{1-\alpha}\hat{h}_R + \sqrt{\alpha}\tilde{h}_R|^2 \geq |\tilde{h}_R|^2 \right\}$. □

Remarks:

- Although the lower and the upper bounds in Theorem 2.1 do not generally coincide, they provide the best available characterisation of the ergodic secrecy capacity over i.i.d. fading channels with imperfect main channel estimation at the transmitter.
- In the case of perfect main CSIT, the lower and the upper bounds coincide with the ones derived in [17]. Also, in this case, the upper bound coincides with the secrecy capacity of a wiretap fading channel under the assumption of asymptotically long coherence intervals derived in [7, Theorem 2]. When specialised to the no CSIT case, the bounds coincide providing a trivial secrecy capacity.
- Even a poor main channel estimator at the transmitter can help establish a secure communication. This fact has also been demonstrated in e.g. [18], although in a slightly different setting. Furthermore, a simple constant rate on–off power scheme is enough to achieve a positive secrecy rate.
- The presented upper bound relies on the eavesdropper's channel having the same statistics as the main channel estimation error. This leaves open the problem of determining a generic upper bound.

The effect of changing the estimation error variance α on the lower and the upper bounds on the secrecy capacity is illustrated in Figure 2.1 for i.i.d. Rayleigh fading channels. We consider two different values of the average power constraint $P_{avg} = 10$ dB, and $P_{avg} = 30$ dB. We can see that the secrecy throughput decreases gradually as the error variance increases. However, a positive secrecy rate can still be achieved as long as the transmitter has some knowledge about the main CSI. Clearly, when no main CSI is available at the transmitter, i.e. $\alpha = 1$, the secrecy capacity is equal to zero.

2.2.2.2 Asymptotic results

The following result considers the high-SNR regime and follows from Theorem 2.1.

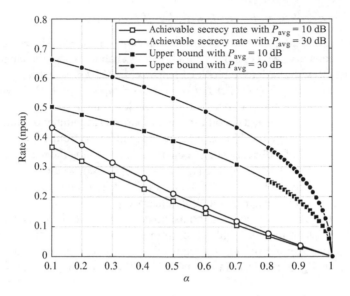

*Figure 2.1   Lower and upper bounds on the secrecy capacity for i.i.d. Rayleigh
fading channels with noisy CSIT in function of the error variance α*

Corollary 2.1. *At high SNR, the ergodic secrecy capacity of the single-user fading
wiretap channel with noisy main CSIT is bounded by*

$$\mathscr{C}_{s\text{-}HSNR}^{-} \leq \mathscr{C}_{s\text{-}HSNR} \leq \mathscr{C}_{s\text{-}HSNR}^{+}, \tag{2.8}$$

where $\mathscr{C}_{s\text{-}HSNR}^{-}$ and $\mathscr{C}_{s\text{-}HSNR}^{+}$ are given by

$$\mathscr{C}_{s\text{-}HSNR}^{-} = \mathop{\mathbb{E}}_{|h_E|^2,|h_R|^2,|\hat{h}_R|^2 \geq \tau} \left[\log\left(\left| \frac{h_R}{h_E} \right|^2 \right) \right], \tag{2.9}$$

$$\mathscr{C}_{s\text{-}HSNR}^{+} = \mathop{\mathbb{E}}_{\hat{h}_R,\tilde{h}_R} \left[\left\{ \log\left(\frac{|\sqrt{1-\alpha}\hat{h}_R + \sqrt{\alpha}\tilde{h}_R|^2}{|\tilde{h}_R|^2} \right) \right\}^{+} \right], \tag{2.10}$$

and where τ is optimised to maximise $\mathscr{C}_{s\text{-}HSNR}^{-}$.

Clearly, Corollary 2.1 states that the ergodic secrecy capacity is bounded at high
SNR confirming that the secret multiplexing gain is equal to zero, regardless of the
main channel estimation quality.

Low-SNR Regime: Although the high-SNR analysis provides somehow a negative result in the sense that the capacity is bounded no matter how P_{avg} increases,
at low SNR, the ergodic secrecy capacity is asymptotically equal to the capacity of
the main channel as if there is no secrecy constraint. Hence, the low-SNR analysis
reveals the potential capacity gain provided by partial CSIT for any non-null channel
estimation quality, i.e. $\alpha \neq 1$.

2.2.3 Wiretap channel with limited CSI feedback

In this case, we consider that the CSI feedback is sent over an error-free feedback link with limited capacity. We consider a block-fading channel where the channel gains remain constant within a fading block of length κ, i.e. $h_R(\kappa l) = h_R(\kappa l - 1) = \cdots = h_R(\kappa l - \kappa + 1)$ and $h_E(\kappa l) = h_E(\kappa l - 1) = \cdots = h_E(\kappa l - \kappa + 1)$, where $l = 1, \ldots, L$, and L is the total number of fading blocks, implying $n = \kappa L$. We assume that the channel encoding and decoding frames span a large number of fading blocks, i.e. L is large, and that the blocks change independently from a fading block to another. The adopted feedback strategy consists of partitioning the main channel gain support into Q intervals $[\tau_1, \tau_2), \ldots, [\tau_q, \tau_{q+1}), \ldots, [\tau_Q, \infty)$, where $Q = 2^b$, and b is the number of available bits for CSI feedback. That is, during each fading block, the legitimate receiver determines in which interval $[\tau_q, \tau_{q+1})$, with $q = 1, \ldots, Q$, its channel gain lies in and feeds back the associated index q to the transmitter. At the transmitter side, each feedback index q corresponds to a power transmission strategy P_q satisfying the average power constraint. We assume that all nodes are aware of the main channel gain partition intervals and the corresponding power transmission strategies.

2.2.3.1 Secrecy capacity analysis

The secrecy capacity of block-fading wiretap channels with limited CSI feedback was studied in [19,20]. In what follows, we summarise the obtained results.

Theorem 2.2. *The ergodic secrecy capacity of the block-fading single-user wiretap channel with an error free b-bits CSI feedback sent by the legitimate receiver, at the beginning of each fading block, is characterised as*

$$\mathscr{C}_s^- \leq \mathscr{C}_s \leq \mathscr{C}_s^+, \tag{2.11}$$

where $\mathscr{C}_s^-$ and $\mathscr{C}_s^+$ are given by

$$\mathscr{C}_s^- = \max_{\{\tau_q; P_q\}_{q=1}^Q} \sum_{q=1}^Q \Pr\left[|h_R|^2 \in \mathscr{A}_q\right] \underset{|h_E|^2}{\mathbb{E}} \left[\left\{\log\left(\frac{1 + \tau_q P_q}{1 + |h_E|^2 P_q}\right)\right\}^+\right], \tag{2.12}$$

$$\mathscr{C}_s^+ = \max_{\{\tau_q; P_q\}_{q=0}^Q} \sum_{q=0}^Q \Pr\left[|h_R|^2 \in \mathscr{A}_q\right] \underset{|h_R|^2, |h_E|^2}{\mathbb{E}} \left[\left\{\log\left(\frac{1 + |h_R|^2 P_q}{1 + |h_E|^2 P_q}\right)\right\}^+ \Big| |h_R|^2 \in \mathscr{A}_q\right], \tag{2.13}$$

with $Q = 2^b$, $\{P_q\}_{q=1}^Q$ are the power transmission strategies satisfying the average power constraint, $\{\tau_q \,|\, 0 = \tau_0 < \tau_1 < \cdots < \tau_Q\}_{q=1}^Q$ represent the reconstruction points describing the support of the squared main channel norm $|h_R|^2$, $\tau_{Q+1} = \infty$ for convenience, and $\mathscr{A}_q = \{|h_R|^2 : \tau_q \leq |h_R|^2 < \tau_{q+1}\}$ for all $q \in \{1, \ldots, Q\}$.

Proof. The key point in the proof of achievability of $\mathscr{C}_s^-$ is to fix the transmission rate during each coherence block. As the transmission is controlled by the feedback information, we consider that during each fading block, if the main channel gain falls

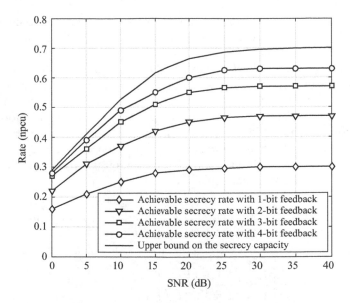

Figure 2.2 Lower and upper bounds on the secrecy capacity for i.i.d. Rayleigh fading channels with various b-bit CSI feedback, b = 1, 2, 3, 4

within the interval $[\tau_q, \tau_{q+1})$, $q \in \{1, \ldots, Q\}$, the transmitter conveys the codewords at rate $\mathscr{R}_q = \log(1 + \tau_q P_q)$. Rate $\mathscr{R}_q$ changes only periodically and is held constant over the duration interval of a fading block. This setup guarantees that when $|h_E|^2 > \tau_q$, the mutual information between the transmitter and the eavesdropper is upper bounded by $\mathscr{R}_q$. Otherwise, this mutual information is equal to $\log(1 + |h_E|^2 P_q)$. Details on the codebook generation, the encoding, decoding, and the secrecy analysis, along with the proof of the upper bound $\mathscr{C}_s^+$ can be found in [20]. $\qquad\square$

Figure 2.2 illustrates the lower and the upper bounds on the secrecy capacity for i.i.d. Rayleigh fading channels when the CSI feedback link has limited capacity. We can clearly see that, as the capacity of the feedback link grows, i.e. the number of bits b increases, the achievable secrecy rate increases too. Also, we note that with a 4-bit CSI feedback, almost 90% of the available capacity may be achieved.

Remarks:
- The achievable secrecy rate in Theorem 2.2 is an increasing function of Q. Also, when $Q \to \infty$, the bounds coincide, hence fully characterising the secrecy capacity in this case. Note that letting Q goes to ∞ may be interpreted as if there is a noiseless public link with infinite capacity.
- When transmitting over Rayleigh fading channels, it is shown that a 4-bit CSI feedback achieves 90% of the secrecy capacity with perfect CSIT.

Guaranteeing a positive secrecy rate is also possible when the feedback is sent at the end of each coherence block. In fact, let us assume that the legitimate receiver only feeds back a 1-bit automatic repeat request (ARQ) to the transmitter at the end of each coherence block informing him whether the actual frame has been correctly decoded (ACK), or not (NACK). The legitimate receiver keeps retransmitting the same block until he gets an ACK, then moves on to the next frame. Obviously, as some of the frames are transmitted more than once, this scheme leaks some information to the eavesdropper. Ultimately, one can assume that the blocks repeated because of the NACK feedback are completely revealed to the eavesdropper as a worst case scenario. Fortunately, even such a conservative scheme guarantees a positive secrecy rate as formalised in the following theorem.

Theorem 2.3. *A lower bound on the ergodic secrecy capacity of the block-fading single-user wiretap channel, with an error free 1-bit ARQ feedback sent at the end of each coherence block, is given by*

$$
\mathcal{C}_s^- = \max_{\{\tau;P\}} \theta^2 \, \mathbb{E}_{|h_E|^2}\left[\left\{\log\left(\frac{1+\tau P}{1+|h_E|^2 P}\right)\right\}^+\right],
\tag{2.14}
$$

where θ is the probability of success defined by $\theta = \Pr[|h_R|^2 \geq \tau]$.

The secrecy rate in Theorem 2.3 can be immediately improved by accounting for the contribution of the fading blocks that have been repeated, say once, cf. [20, Corollary 1]. However, one should be very careful in this case to the fact that, for each repeated transmission, the eavesdropper will try to access the same secret information over two independent coherence blocks.

2.2.3.2 Asymptotic results
The following result considers the high-SNR regime and follows from Theorem 2.2.

Corollary 2.2. *At high SNR, the ergodic secrecy capacity of the block-fading single-user wiretap channel with an error free b-bits CSI feedback sent by the legitimate receiver, at the beginning of each fading block, is bounded by*

$$
\mathcal{C}_{s\text{-HSNR}}^- \leq \mathcal{C}_{s\text{-HSNR}} \leq \mathcal{C}_{s\text{-HSNR}}^+,
\tag{2.15}
$$

where $\mathcal{C}_{s\text{-HSNR}}^-$ and $\mathcal{C}_{s\text{-HSNR}}^+$ are given by

$$
\mathcal{C}_{s\text{-HSNR}}^- = \max_{0 \leq \tau_1 < \cdots < \tau_Q} \sum_{q=1}^{Q} \Pr[|h_R|^2 \in \mathscr{A}_q] \, \mathbb{E}_{|h_E|^2}\left[\left\{\log\left(\frac{\tau_q}{|h_E|^2}\right)\right\}^+\right],
\tag{2.16}
$$

$$
\mathcal{C}_{s\text{-HSNR}}^+ = \mathbb{E}_{|h_R|^2,|h_E|^2}\left[\left\{\log\left(\left|\frac{h_R}{h_E}\right|^2\right)\right\}^+\right],
\tag{2.17}
$$

with $\mathscr{A}_q = \{|h_R|^2 : \tau_q \leq |h_R|^2 < \tau_{q+1}\}$ for all $q \in \{1,\ldots,Q\}$.

Clearly, the secrecy capacity is bounded at high-SNR regardless of the number of bits available for CSI feedback.

2.3 Secure multi-user transmission with channel uncertainty

In this section, we examine the impact of CSIT uncertainty on the secrecy throughput of broadcast wiretap channels. In particular, we consider both cases when a common secret message and when independent confidential messages are broadcasted to multiple legitimate receivers in the presence of an eavesdropper under the assumption of imperfect main CSIT.

2.3.1 System model

We consider a broadcast wiretap channel where a transmitter communicates with K legitimate receivers in the presence of an eavesdropper. During every coherence interval, $t \in \{1, \ldots, n\}$, the received signals by each legitimate receiver is given by

$$Y_k(t) = h_k(t)X(t) + z_k(t), \quad k \in \{1, \ldots, K\}, \tag{2.18}$$

where $X(t)$ is the transmitted signal, $h_k(t) \in \mathbb{C}$ are zero-mean, unit-variance complex Gaussian channel gains corresponding to each legitimate channel, and $z_k(t) \in \mathbb{C}$ represent zero-mean, unit-variance circularly symmetric additive white Gaussian noises. The received signal at the eavesdropper can be expressed similarly as in (2.1). Here also, we impose an average transmit power constraint at the transmitter such that

$$\frac{1}{n} \sum_{t=1}^{n} \mathbb{E}[|X(t)|^2] \leq P_{\text{avg}}.$$

The same assumptions on the channel gains h_R and h_E, considered in the previous section for single-user transmission, apply here for h_k and h_E, with $k \in \{1, \ldots, K\}$.

2.3.2 Secure broadcasting with noisy CSIT

We assume that the transmitter is only provided with a noisy version of each $h_k(t)$, say $\hat{h}_k(t)$, such that the main channel estimation model can be written as

$$h_k(t) = \sqrt{1 - \alpha}\hat{h}_k(t) + \sqrt{\alpha}\tilde{h}_k(t), \quad k \in \{1, \ldots, K\},$$

where α is the estimation error variance ($\alpha \in [0, 1]$) and $\tilde{h}_k(i)$ is the zero-mean unit-variance channel estimation error.

The ergodic secrecy capacity of the broadcast wiretap channels under imperfect main channel estimation at the transmitter was studied in [21,22], where the following results are presented.

2.3.2.1 Transmission of a common message

In this case, we consider that the transmitter broadcasts a common message to all the legitimate receivers while keeping it secret from the eavesdropper.

Theorem 2.4. *The ergodic common message secrecy capacity of the multi-user fading broadcast wiretap channel with noisy main CSIT is characterised as*

$$\mathscr{C}_s^- \leq \mathscr{C}_s \leq \mathscr{C}_s^+, \tag{2.19}$$

where $\mathscr{C}_s^-$ and $\mathscr{C}_s^+$ are given by

$$\mathscr{C}_s^- = \max_{P(\tau)} \min_{1 \leq k \leq K} \mathbb{E}_{|h_E|^2, |h_k|^2, |\hat{h}_k|^2 \geq \tau} \left[\log\left(\frac{1 + |h_k|^2 P(\tau)}{1 + |h_E|^2 P(\tau)} \right) \right], \tag{2.20}$$

$$\mathscr{C}_s^+ = \min_{1 \leq k \leq K} \max_{P(\hat{h}_k)} \mathbb{E}_{\hat{h}_k, \tilde{h}_k} \left[\left\{ \log\left(\frac{1 + |\sqrt{1-\alpha}\hat{h}_k + \sqrt{\alpha}\tilde{h}_k|^2 P(\hat{h}_k)}{1 + |\tilde{h}_k|^2 P(\hat{h}_k)} \right) \right\}^+ \right], \tag{2.21}$$

with $P(\tau) = \dfrac{P_{\mathrm{avg}}}{1 - F_{|\hat{h}_k|^2}(\tau)}$ and $\mathbb{E}[P(\hat{h}_k)] \leq P_{\mathrm{avg}}$.

Proof. The two essential points in the proof of achievability of the lower bound $\mathscr{C}_s^-$ are the construction of K independent Gaussian wiretap codebooks and the adoption of a probabilistic transmission model where the communication is constrained by the quality of the legitimate channels. The transmission scheme guarantees that all the legitimate receivers can decode the secret message, whereas no extra information is leaked to the eavesdropper. The proof in details can be found in [22]. As for the upper bound, it can be deduced from the single-user case with noisy CSIT. □

Remarks:

- When a common message is transmitted to all the legitimate receivers, the secrecy rate is limited by the legitimate receiver having, on average, the worst main channel link. Also, from the obtained results, we can see that a nonzero secrecy rate can still be achieved even when the CSI at the transmitter is noisy.
- When the transmitter has perfect knowledge of the legitimate receivers CSI, i.e. $\alpha = 0$, the presented bounds in Theorem 2.4 coincides with the ones presented in [23], whereas when no main CSI is available at the transmitter, the secrecy capacity is equal to zero.

2.3.2.2 Transmission of independent messages

Here, we consider the independent messages case when multiple confidential messages are transmitted to the legitimate receivers while being kept secret from the eavesdropper.

Theorem 2.5. *The ergodic secrecy sum-capacity of the multi-user fading broadcast wiretap channel with noisy main CSIT is characterised as*

$$\mathscr{C}_s^- \leq \mathscr{C}_s \leq \mathscr{C}_s^+, \tag{2.22}$$

where $\mathscr{C}_s^-$ and $\mathscr{C}_s^+$ are given by

$$\mathscr{C}_s^- = \max_{P(\tau)\ \ |h_E|^2,|h_{\max}^{est}|^2,|\hat{h}_{\max}|^2 \geq \tau} \mathbb{E}\left[\log\left(\frac{1 + |h_{\max}^{est}|^2 P(\tau)}{1 + |h_E|^2 P(\tau)}\right)\right], \tag{2.23}$$

$$\mathscr{C}_s^+ = \min\left\{ \max_{P(\hat{h}_1,\ldots,\hat{h}_K)\ |h_{\max}|^2,|\hat{h}_1|^2,\ldots,|\hat{h}_K|^2,|\tilde{h}_R|^2} \mathbb{E}\left[\left\{\log\left(\frac{1 + |h_{\max}|^2 P(\hat{h}_1,\ldots,\hat{h}_K)}{1 + |\tilde{h}_R|^2 P(\hat{h}_1,\ldots,\hat{h}_K)}\right)\right\}^+\right],\right.$$

$$\left. K \max_{P(\hat{h}_R)\ |h_R|^2,|\hat{h}_R|^2,|\tilde{h}_R|^2} \mathbb{E}\left[\left\{\log\left(\frac{1 + |h_R|^2 P(\hat{h}_R)}{1 + |\tilde{h}_R|^2 P(\hat{h}_R)}\right)\right\}^+\right]\right\}, \tag{2.24}$$

with $h_{\max}^{est} = \sqrt{1 - \alpha}\hat{h}_{\max} + \sqrt{\alpha}\tilde{h}$, $|\hat{h}_{\max}|^2 = \max_{1 \leq k \leq K}|\hat{h}_k|^2$, $P(\tau) = \dfrac{P_{\mathrm{avg}}}{1 - F_{|\hat{h}_{\max}|^2}(\tau)}$,
$\mathbb{E}[P(\hat{h}_{\max})] \leq P_{\mathrm{avg}}$, *and* $|h_{\max}|^2 = \max_{1 \leq k \leq K}|h_k|^2$.

Proof. The lower bound is achieved using a time division multiplexing scheme that selects instantaneously one receiver to transmit to. That is, at each time, the source only transmits to the user with the best-estimated channel gain $\hat{h}_{\max}$. As we are transmitting to only one legitimate receiver at a time, the achieving coding scheme consists of using independent standard single-user wiretap codebooks.

The upper bound on the secrecy sum-capacity $\mathscr{C}_s^+$ is given as the minimum between two upper bounds. The second upper bound can be proved easily, whereas the first one requires the conception of a new genie-aided channel whose capacity upper bounds the capacity of the K-receivers channel with noisy CSIT. The receiver in the new channel needs to instantaneously get the signal transmitted over the strongest channel, and the transmitter has to know the estimated gains of all K channels of the original K-receivers channel. As a matter of fact, if we only consider that the transmission is intended for the strongest receiver at each time, the capacity of this channel cannot be proven to upper bound the capacity of our K-receivers channel as the transmitter will have the estimated gain of only the strong channel. That is, the new channel needs to observe all the K channels and to account for the strongest one at each time. This is why, we consider a genie-aided channel with a selection combining receiver equipped with a number of antennas equivalent to the number of legitimate receivers in the K-receivers channel. The selection combiner chooses the signal with the highest instantaneous gain and uses it for decoding. A detailed proof is provided in [22]. □

Remark:

The upper bound on the secrecy sum-capacity, in Theorem 2.5, is given as the minimum between two upper bounds. The reason behind choosing this particular representation was to ensure having the tightest possible upper bound for all the values of the error variance α. We would note that the second bound is a loose upper bound for the secrecy sum-rate for most values of α, especially when the number of users K is large. However, when the CSIT gets very noisy, i.e. $\alpha \rightarrow 1$, this bound becomes tighter.

Corollary 2.3. *At high SNR, the ergodic secrecy sum-capacity of the fading broadcast wiretap channel with noisy main CSIT is bounded by*

$$\mathscr{C}_{s\text{-}HSNR}^{-} \leq \mathscr{C}_{s\text{-}HSNR} \leq \mathscr{C}_{s\text{-}HSNR}^{+}, \tag{2.25}$$

where $\mathscr{C}_{s\text{-}HSNR}^{-}$ and $\mathscr{C}_{s\text{-}HSNR}^{+}$ are given by

$$\mathscr{C}_{s\text{-}HSNR}^{-} = \mathop{\mathbb{E}}_{|h_{\mathrm{E}}|^2, |h_{\max}^{est}|^2, |\hat{h}_{\max}|^2 \geq \tau} \left[\log\left(\left| \frac{h_{\max}^{est}}{h_{\mathrm{E}}} \right|^2 \right) \right], \tag{2.26}$$

$$\mathscr{C}_{s\text{-}HSNR}^{+} = \min\left\{ \mathop{\mathbb{E}}_{|h_{\max}|^2, |\tilde{h}_R|^2} \left[\left\{ \log\left(\frac{|h_{\max}|^2}{|\tilde{h}_R|^2} \right) \right\}^+ \right], K \mathop{\mathbb{E}}_{|h_R|^2, |\tilde{h}_R|^2} \left[\left\{ \log\left(\frac{|h_R|^2}{|\tilde{h}_R|^2} \right) \right\}^+ \right] \right\}, \tag{2.27}$$

and where τ is optimised to maximise $\mathscr{C}_{s\text{-}HSNR}^{-}$.

The asymptotic high-SNR expressions can be deduced directly from Theorem 2.5. From Corollary 2.3, we can see that the asymptotic bounds depend on the number of legitimate receivers K. Now, letting K go to ∞, we characterise the scaling law of the system.

Corollary 2.4. *The asymptotic high-SNR secrecy sum-capacity when broadcasting independent messages to a large number of legitimate receivers, i.e. $K \to \infty$, with noisy main CSIT is bounded by*

$$\log\left((1-\alpha)\log(K)\right) \leq \mathscr{C}_s \leq \log\log K, \quad \text{for all } \alpha \neq 1. \tag{2.28}$$

Figure 2.3 considers the case when broadcasting independent messages to K legitimate receivers over i.i.d. Rayleigh fading channels with an estimation error variance $\alpha = 0.5$ and two different values of the average power constraint, $P_{\mathrm{avg}} = 10$ dB and $P_{\mathrm{avg}} = 30$ dB. We can see that both the achievable secrecy sum-rate and the upper bound on the secrecy sum-capacity scale with the number of users K. That is, and in accordance with the multiuser diversity aim, the proposed achievable scheme is asymptotically optimal as the number of legitimate receivers grows. The figure also shows that the difference between the lower and the upper bounds approaches $\log(1-\alpha)$ as the number of users increases.

2.3.3 Secure broadcasting with limited CSI feedback

In this case, we consider that each legitimate receiver provides the transmitter with b-bits CSI feedback sent through an error-free orthogonal channel with limited capacity. This feedback is transmitted at the beginning of each fading block and is also tracked by the other legitimate receivers, i.e. all communicating nodes are aware of each and every feedback information. The eavesdropper knows all channels and also track the feedback links so that they are not sources of secrecy. The adopted feedback strategy is similar to the one described in Section 2.2.3.

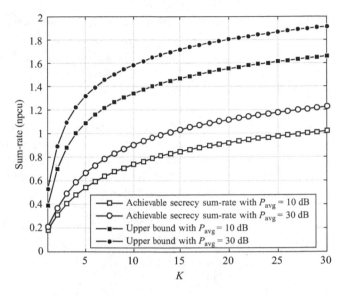

Figure 2.3 Lower and upper bounds on the secrecy sum-capacity for i.i.d. Rayleigh fading channels versus the number of users K with α = 0.5 and two values of P_{avg}

The secrecy analysis of the block-fading wiretap channel with finite feedback was studied in [24,25]. In what follows, we summarise the obtained results.

2.3.3.1 Transmission of a common message

Theorem 2.6. *The ergodic common message secrecy capacity of the block-fading multi-user wiretap channel with an error free b-bits CSI feedback sent by each legitimate receiver, at the beginning of each fading block, is characterised as*

$$\mathscr{C}_s^- \le \mathscr{C}_s \le \mathscr{C}_s^+, \tag{2.29}$$

where $\mathscr{C}_s^-$ and $\mathscr{C}_s^+$ are given by

$$\mathscr{C}_s^- = \min_{1 \le k \le K} \max_{\{\tau_q; P_q\}_{q=1}^Q} \sum_{q=1}^Q \Pr[|h_k|^2 \in \mathscr{A}_q] \mathop{\mathbb{E}}_{|h_E|^2}\left[\left\{\log\left(\frac{1 + \tau_q P_q}{1 + |h_E|^2 P_q}\right)\right\}^+\right], \tag{2.30}$$

$$\mathscr{C}_s^+ = \min_{1 \le k \le K} \max_{\{\tau_q; P_q\}_{q=0}^Q} \sum_{q=0}^Q \Pr[|h_k|^2 \in \mathscr{A}_q] \mathop{\mathbb{E}}_{|h_k|^2, |h_E|^2}\left[\left\{\log\left(\frac{1 + |h_k|^2 P_q}{1 + |h_E|^2 P_q}\right)\right\}^+ \Big| |h_k|^2 \in \mathscr{A}_q\right], \tag{2.31}$$

with $Q = 2^b$, $\{P_q\}_{q=1}^{Q}$ *are the power transmission strategies satisfying the aver-age power constraint,* $\{\tau_q \,|\, 0 = \tau_0 < \tau_1 < \cdots < \tau_Q\}_{q=1}^{Q}$ *represent the reconstruction points describing the support of the squared main channel norm* $|h_k|^2$, $\tau_{Q+1} = \infty$ *for convenience, and* $\mathscr{A}_q = \{|h_k|^2 : \tau_q \leq |h_k|^2 < \tau_{q+1}\}$ *for all* $q \in \{1, \ldots, Q\}$ *and* $k \in \{1, \ldots, K\}$.

It is worth mentioning that the main difference between the lower and the upper bounds in Theorem 2.6 is that the transmission scheme, for the achievable secrecy rate, uses the feedback information to adapt both the rate and the power in such a way that the transmission rate is fixed during each fading block. When $Q \to \infty$, the presented bounds coincide, yielding the following result.

Corollary 2.5. *The ergodic common message secrecy capacity of the block-fading multi-user wiretap channel, with perfect main CSIT, is given by*

$$\mathscr{C}_s = \min_{1 \leq k \leq K} \max_{P(h_k)} \mathbb{E}_{|h_k|^2, |h_E|^2} \left[\left\{ \log \left(\frac{1 + |h_k|^2 P(h_k)}{1 + |h_E|^2 P(h_k)} \right) \right\}^+ \right], \tag{2.32}$$

with $\mathbb{E}[P(h_k)] \leq P_{\mathrm{avg}}$.

Note that when $Q \to \infty$, the set of reconstruction points, $\{\tau_1, \ldots, \tau_Q\}$, becomes infinite and each legitimate receiver is basically forwarding its channel gain to the transmitter.

2.3.3.2 Transmission of independent messages

Theorem 2.7. *The ergodic secrecy sum-capacity of the block-fading multi-user wire-tap channel with an error free b-bits CSI feedback sent by each legitimate receiver, at the beginning of each fading block, is characterised as*

$$\mathscr{C}_s^- \leq \mathscr{C}_s \leq \mathscr{C}_s^+, \tag{2.33}$$

where $\mathscr{C}_s^-$ *and* $\mathscr{C}_s^+$ *are given by*

$$\mathscr{C}_s^- = \max_{\{\tau_q; P_q\}_{q=1}^{Q}} \sum_{q=1}^{Q} \Pr\left[|h_{\max}|^2 \in \mathscr{A}_q\right] \mathbb{E}_{|h_E|^2} \left[\left\{ \log \left(\frac{1 + \tau_q P_q}{1 + |h_E|^2 P_q} \right) \right\}^+ \right], \tag{2.34}$$

$$\mathscr{C}_s^+ = \max_{\{\tau_q; P_q\}_{q=0}^{Q}} \sum_{q=0}^{Q} \Pr\left[|h_{\max}|^2 \in \mathscr{A}_q\right] \mathbb{E}_{|h_{\max}|^2, |h_E|^2} \left[\left\{ \log \left(\frac{1 + |h_{\max}|^2 P_q}{1 + |h_E|^2 P_q} \right) \right\}^+ \Big| |h_{\max}|^2 \in \mathscr{A}_q \right], \tag{2.35}$$

with $Q = 2^b$, $\{P_q\}_{q=1}^{Q}$ *are the power transmission strategies satisfying the average power constraint,* $\{\tau_q | 0 = \tau_0 < \tau_1 < \cdots < \tau_Q\}_{q=1}^{Q}$ *represent the reconstruction points describing the support of the squared main channel norm* $|h_{\max}|^2$, $\tau_{Q+1} = \infty$ *for convenience,* $|h_{\max}|^2 = \max_{1 \leq k \leq K} |h_k|^2$, *and* $\mathscr{A}_q = \{|h_{\max}|^2 : \tau_q \leq |h_{\max}|^2 < \tau_{q+1}\}$ *for all* $q \in \{1, \ldots, Q\}$.

The proposed achievability scheme has a time sharing interpretation to it and even if the result is given in terms of the secrecy sum-rate, the secrecy rate $\mathscr{R}_k$ of each legitimate receiver $k \in \{1, \ldots, K\}$ can also be characterised. In fact, we can write $\mathscr{R}_k \leq \mathscr{C}_s^- \times$ Prob[user k is the best]. When $Q \to \infty$, the presented bounds coincide, yielding the following result.

Corollary 2.6. *The ergodic secrecy sum-capacity of the block-fading multi-user wiretap channel, with perfect main CSIT, is given by*

$$\mathscr{C}_s = \max_{P(h_{\max})} \mathbb{E}_{|h_{\max}|^2, |h_E|^2} \left[\left\{ \log \left(\frac{1 + |h_{\max}|^2 P(h_{\max})}{1 + |h_E|^2 P(h_{\max})} \right) \right\}^+ \right], \tag{2.36}$$

with $|h_{\max}|^2 = \max_{1 \leq k \leq K} |h_k|^2$, *and* $\mathbb{E}[P(h_{\max})] \leq P_{\text{avg}}$.

Remarks:
- The presented results, for both common message and independent messages transmissions, are also valid in the case when multiple non-colluding eavesdroppers conduct the attack. In the case when the eavesdroppers collude, the results can be extended by replacing the channel gain h_E with the vector of channel gains of the colluding eavesdroppers.
- In the analysed system, we assumed unit-variance Gaussian noises at all receiving nodes. The results can be easily extended to a general setup where the noise variances are different.

2.3.3.3   Broadcast channel with confidential messages

Let us now look at the case when the transmitter has both common and independent messages to transmit. In particular, let us consider the case of a broadcast channel where a transmitter communicates with two receivers R_1 and R_2 with respective channel gains h_1 and h_2. The transmitter wants to send a common message W_0 to both receivers and a confidential message W_1 to R_1 only. Message W_1 has to be kept secret from R_2. We assume perfect CSI at the receiving nodes. That is, each receiver is instantaneously aware of its channel gain. Further, we assume that the transmitter is not aware of the instantaneous channel realisations of neither channel. However, the receivers provide the transmitter with a 1-bit CSI feedback sent at the beginning of each fading block. The 1-bit feedback is sent either by one of the receivers, if they share their CSI, or by a central controller who is aware of the CSI of both receivers. The last setting is possible as both receivers are interested by a common message. Hence, they both belong to the same network and their channels are more likely to be known by a controller centre.

Theorem 2.8. *The ergodic secrecy capacity region of the block-fading BCC with a 1-bit CSI feedback, sent at the beginning of each coherence block over an error-free link, is given by*

$$
\mathscr{C}_s = \bigcup_{(p_{01},p_{02},p_1)\in\mathscr{P}}
\begin{cases}
(\mathscr{R}_0,\mathscr{R}_1): \\[4pt]
\mathscr{R}_0 \le \min\Big\{ \mathbb{E}_{\underline{\gamma}}\Big[\log\Big(1+\dfrac{p_{01}|h_1|^2}{1+p_1|h_1|^2}\Big)\,\Big|\underline{\gamma}\in\mathscr{A}\Big]\,Pr[\underline{\gamma}\in\mathscr{A}] \\[6pt]
\qquad +\mathbb{E}_{\underline{\gamma}}\Big[\log(1+p_{02}|h_1|^2)\,\Big|\underline{\gamma}\in\mathscr{A}^c\Big]\,Pr[\underline{\gamma}\in\mathscr{A}^c]; \\[8pt]
\qquad\quad \mathbb{E}_{\underline{\gamma}}\Big[\log\Big(1+\dfrac{p_{01}|h_2|^2}{1+p_1|h_2|^2}\Big)\,\Big|\underline{\gamma}\in\mathscr{A}\Big]\,Pr[\underline{\gamma}\in\mathscr{A}] \\[6pt]
\qquad\quad +\mathbb{E}_{\underline{\gamma}}\Big[\log(1+p_{02}|h_2|^2)\,\Big|\underline{\gamma}\in\mathscr{A}^c\Big]\,Pr[\underline{\gamma}\in\mathscr{A}^c]\Big\} \\[8pt]
\mathscr{R}_1 \le \mathbb{E}_{\underline{\gamma}}\Big[\log(1+p_1|h_1|^2)-\log(1+p_1|h_2|^2)\,\Big|\underline{\gamma}\in\mathscr{A}\Big]\,Pr[\underline{\gamma}\in\mathscr{A}],
\end{cases}
$$

$$(2.37)$$

where $\underline{\gamma} = \big[|h_1|^2; |h_2|^2\big]$, $\mathscr{A} = \Big\{\underline{\gamma} : |h_1|^2 > |h_2|^2\Big\}$ *and*

$$
\mathscr{P} = \Big\{(p_{01},p_{02},p_1): (p_{01}+p_1)Pr[\underline{\gamma}\in\mathscr{A}]+p_{02}\,Pr[\underline{\gamma}\in\mathscr{A}^c]\le P_{\mathrm{avg}}\Big\}.
$$

We can see, from Theorem 2.8, that the common message W_0 is sent over all coherence blocks, whereas the confidential message W_1 is transmitted only over the fading blocks where the channel to receiver R_1 is better than the one to receiver R_2, i.e. $\gamma \in \mathscr{A}$. That is, when $\gamma \in \mathscr{A}$, we decode the common message considering the secure message as noise, whereas when $\gamma \in \mathscr{A}^c$, as the confidential message is not sent, the common message is decoded at a single-user rate. The minimisation is due to a bottleneck argument. Also, Theorem 2.8 states that even with a 1-bit CSI feedback sent at the beginning of each fading block, and as long as event $\mathscr{A}$ is not a zero probability event, a positive secrecy rate can still be achieved.

At the difference of the perfect CSIT case [26], the power cannot be instantaneously adapted to the channel realisations and will only depend on the received 1-bit CSI feedback according to a deterministic mapping. It is worth mentioning that p_{01} and p_{02} in Theorem 2.8 correspond to the power allocated to common message transmissions in $\mathscr{A}$ and $\mathscr{A}^c$, respectively, whereas p_1 is the power allocated to the confidential message. When the feedback link has a larger capacity, i.e. more feedback bits can be sent, one bit should be used as an indication bit to point out which channel is better, whereas the remaining bits should be used to adapt the transmission power.

Figure 2.4 illustrates the secrecy capacity region for Rayleigh BCC when 1-bit CSI feedback is sent over an error-free link with $h_1 \sim \mathscr{CN}(0,1)$, $h_2 \sim \mathscr{CN}(0,\sigma_2^2)$, and $P_{\mathrm{avg}} = 5\,\mathrm{dB}$. The boundary of the secrecy capacity region when perfect CSI is available at the transmitter is also presented as a benchmark. We can see that when

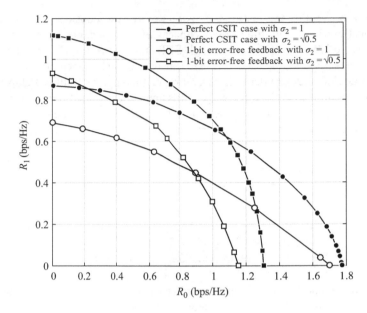

Figure 2.4   Secrecy capacity regions for Rayleigh BCC with an error-free 1-bit CSI feedback

the channel to receiver R_1 is better, in average, than the channel to receiver R_2, i.e. when $\sigma_2^2 = 0.5$, the confidential rate $\mathscr{R}_1$ improves, whereas the common rate $\mathscr{R}_0$ decreases.

2.4   Conclusion

In this chapter, we discussed the impact of CSIT uncertainty on the ergodic secrecy capacity of fading wiretap channels. We showed that even though the secrecy performance of the system deteriorates compared to the case when the transmitter has perfect CSI, a positive secrecy rate can still be achieved as long as the transmitter has some knowledge of the main channel gain. We considered two common causes of CSIT imperfections, namely, the occurrence of an estimation error of the CSI at the transmitter and the limited capacity of the CSI feedback link. In both cases, we notice that the more the transmitter knows about the main CSI, the better the secrecy performances are.

It is of particular importance to notice that the upper bounds on the secrecy capacity for the noisy CSIT case are not general and still assume that the main channel estimation error $\tilde{h}_R$ and the eavesdropper gain h_E have the same statistics. Also, the lower and the upper bounds on the ergodic secrecy capacity do not generally coincide, leaving the characterisation of the secrecy capacity under imperfect main CSI an open problem. Another interesting extension would be to consider multiple antenna communications as neither the secrecy capacity of multiple antenna systems nor the secret degree of freedom (which is a high-SNR characterisation) is known when the transmitter has imperfect main CSI. Furthermore, among the open problems that still have to be tackled, the consideration of a more practical model for the wiretap channel, where no assumption on the adversary CSI is imposed, is one of the most important. An appealing treatment, in this case, could be found in the framework of arbitrary varying eavesdropper channel. However, only a few related results are reported in the literature.

References

[1]   Shannon CE. Communication theory of secrecy systems. Bell Systems Technical Journal. 1949 Oct;28:656–719.
[2]   Wyner AD. The wiretap channel. Bell System Technical Journal. 1975;54(8):1355–1387.
[3]   Csiszár I, and Körner J. Broadcast channels with confidential messages. IEEE Transactions on Information Theory. 1978;24(3):339–348.
[4]   Leung-Yan-Cheong SK, and Hellman ME. The Gaussian wiretap channel. IEEE Transactions on Information Theory. 1978 Jul;24(4):451–456.
[5]   Barros J, and Rodrigues MRD. Secrecy capacity of wireless channels. In: Proc. International Symposium on Information Theory (ISIT'2006). Seattle, WA, US; 2006. p. 356–360.
[6]   Bloch M, Barros J, Rodrigues M, and McLaughlin S. Wireless information-theoretic security. IEEE Transactions on Information Theory. 2008 Jun;54(6):2515–2534.
[7]   Gopala PK, Lai L, and Gamal HE. On the secrecy capacity of fading channels. IEEE Transactions on Information Theory. 2008 Oct;54(10):4687–4698.
[8]   Liu TY, Lin PH, Hong YWP, and Jorswieck E. To avoid or not to avoid CSI leakage in physical layer secret communication systems. IEEE Communications Magazine. 2015 Dec;53(12):19–25.
[9]   Hyadi A, Rezki Z, and Alouini MS. An overview of physical layer security in wireless communication systems with CSIT uncertainty. IEEE Access. 2016 Sep;4:6121–6132.
[10]   Hayashi M. General nonasymptotic and asymptotic formulas in channel resolvability and identification capacity and their application to the wiretap channels. IEEE Transactions on Information Theory. 2006 Apr;52(4):1562–1575.
[11]   Bloch M, and Laneman J. Strong secrecy from channel resolvability. IEEE Transactions on Information Theory. 2013 Dec;59(12):8077–8098.
[12]   Maurer UM, and Wolf S. Information-Theoretic Key Agreement: From Weak to Strong Secrecy for Free. In: Advances in Cryptology – Eurocrypt 2000, Lecture Notes in Computer Science. B. Preneel; 2000. p. 351.
[13]   Csiszar I. Almost independence and secrecy capacity. Problems of Information Transmission. 1996 Jan;32(1):40–47.

[14] Bellare M, Tessaro S, and Vardy A. A cryptographic treatment of the wiretap channel. In: Advances in Cryptology – (CRYPTO 2012). Santa Barbara, CA, US; 2012. p. 351.

[15] Rezki Z, Khisti A, and Alouini MS. On the ergodic secrecy capacity of the wiretap channel under imperfect main channel estimation. In: Proc. Forty Fifth Asilomar Conference on Signals, Systems and Computers (ASILO-MAR'2011). Pacific Grove, CA, US; 2011. p. 952–957.

[16] Rezki Z, Khisti A, and Alouini MS. On the secrecy capacity of the wiretap channel under imperfect main channel estimation. IEEE Transactions on Communications. 2014 Sep;62(10):3652–3664.

[17] Khisti A, and Womell G. Secure transmission with multiple antennas. Part I: The MISOME wiretap channel. IEEE Transactions on Information Theory. 2010 Jul;56(7):3088–3104.

[18] Bloch M, and Laneman J. Exploiting partial channel state information for secrecy over wireless channels. IEEE Journal on Selected Areas of Communication. 2013 Sep;31(9):1840–1849.

[19] Rezki Z, Khisti A, and Alouini MS. On the ergodic secret message capacity of the wiretap channel with finite-rate feedback. In: Proc. IEEE International Symposium on Information Theory (ISIT'2012). Cambridge, MA, US; 2012. p. 239–243.

[20] Rezki Z, Khisti A, and Alouini MS. Ergodic secret message capacity of the wiretap channel with finite-rate feedback. IEEE Transactions on Wireless Communications. 2014 Jun;13(6):3364–3379.

[21] Hyadi A, Rezki Z, Khisti A, and Alouini MS. On the secrecy capacity of the broadcast wiretap channel with imperfect channel state information. In: IEEE Global Communications Conference (GLOBECOM'2014). Austin, TX, US; 2014. p. 1608–1613.

[22] Hyadi A, Rezki Z, Khisti A, and Alouini MS. Secure broadcasting with imperfect channel state information at the transmitter. IEEE Transactions on Wireless Communications. 2016 Mar;15(3):2215–2230.

[23] Khisti A, Tchamkerten A, and Wornell GW. Secure broadcasting over fading channels. IEEE Transactions on Information Theory. 2008 Jun;54(6): 2453–2469.

[24] Hyadi A, Rezki Z, and Alouini MS. On the secrecy capacity of the broadcast wiretap channel with limited CSI feedback. In: Proc. IEEE Information Theory Workshop (ITW'2016). Cambridge, UK; 2016. p. 36–40.

[25] Hyadi A, Rezki Z, and Alouini MS. On the secrecy capacity region of the block-fading BCC with limited CSI feedback. In: Proc. IEEE Global Communications Conference (Globecom'2016). Washington, DC, US; 2016.

[26] Liang Y, Poor H, and Shamai S. Secure communication over fading channels. IEEE Transactions on Information Theory. 2008 Jun;54(6):2470–2492.

Chapter 3
Confidential and energy-efficient communications by physical layer security

Alessio Zappone[1], Pin-Hsun Lin[1], and Eduard A. Jorswieck[1]

3.1 Introduction

Traditionally, the security of data transmission has been ensured by key-based enciphering. However, this application-layer approach to cryptography requires additional complexity and resources to perform the encryption/decryption tasks, which increases the communication delay. Moreover, a centralized network infrastructure is required to either distribute the cryptographic keys (for private-key approaches) or to verify public keys and digital signatures (for public-key approaches). Finally, the need for key distribution/verification leads to undesirable communication and feedback overheads. These drawbacks make traditional cryptography not suited to future cellular networks [1,2], which will have to serve more than 50 billion of nodes by 2020, and where self-organization, high-data rates, low latency, and low energy consumptions will be critical design goals.

   A promising approach to overcome these issues and make future networks secure is to place the security module directly at the physical layer. This approach, known as physical layer security [3,4], has the potential to replace and complement traditional application layer cryptography, without requiring the use of any cryptographic key, and dispensing with central authorities. Instead of protecting the communication by encrypting the communication by cryptographic keys, physical layer security exploits the unpredictability and randomness of the wireless communication channel to prevent eavesdropping.

   The fundamental building block of the theory of physical layer security is represented by the wiretap channel, defined as a three-node system in which a legitimate transmitter, conventionally labeled Alice, wishes to communicate to a legitimate receiver, labeled Bob, whereas a malicious node, labeled Eve, eavesdrops the communication. In [3], the discrete memoryless degraded wiretap channel is characterized, and the largest rate at which Alice and Bob can communicate with Eve eavesdropping *no information* is determined. This fundamental performance limit is defined as the channel *secrecy capacity*, and defines the maximum amount of information which

[1]Institute of Communication Technology, Dresden University of Technology, Germany

can be reliably and securely transmitted between two nodes in the presence of an eavesdropper. In [4], the results of [3] are extended to nondegraded wiretap channels.

More recently, with the advent of multiple antenna techniques, physical layer security has started being investigated also with reference to multiple-antenna communication systems. In [5], the secrecy capacity of a Gaussian channel with a 2-by-2 legitimate channel and one single-antenna-eavesdropper is characterized. Following works extended this result to the case in which the legitimate nodes, as well as the eavesdropper, have an arbitrary number of antennas [6–10]. Such a scenario is commonly referred to as a Gaussian multiple-input multiple-output multiple-antenna-eavesdropper channel (MIMOME). A common assumption made by above works, which is critical to obtain the derived capacity results, is the knowledge of perfect channel state information (CSI) at the legitimate transmitter (CSIT) of both Bob's and Eve's channels. This assumption might be not fulfilled in real-world systems, especially as far as the channel to the eavesdropper is concerned. Unavoidable channel estimation errors and the limited capacity of feedback channels can cause the imperfect knowledge of the legitimate channel, and even more, the eavesdropper can be a hidden node unwilling to send any feedback signal to the legitimate transmitter. Moreover, in fast-fading scenarios, tracking the instantaneous channel realizations is not feasible. Therefore, the results from previously cited papers, which require perfect CSIT, represent a performance upper bound for practical system designs.

On the other hand, when perfect CSI is lacking, the derivation of the secrecy capacity is an open problem in general. Only a few results partially solve this problem, which will be discussed later. In the context of imperfect CSI, several kinds of CSI models have been proposed [11]: statistical CSIT [12–20] in which only the statistics of the channel are assumed known, but not the actual realization; quantized CSIT [21] where each channel is assumed known up to a quantization error; CSIT with uncertainty region [22,23], where an estimate of the channel is available, subject to an error which lies in a given region. The focus of this chapter will be on the statistical CSIT model, which naturally arises for example in fast-fading scenarios. Furthermore, statistical CSIT can be feasibly collected offline at some indoor and outdoor places or from former transmissions. It is also possible for Alice to ensure the security of the transmission in a given area, where the eavesdropper is assumed to lie.

In the context of CSIT, a promising approach which has gained momentum in recent years is the so-called artificial noise (AN) technique. This approach consists of superimposing an interference signal to the useful message, to disrupt the eavesdropper's reception [12,13]. Earlier works on AN exploit beamforming techniques to confine the AN into the orthogonal complement of the space used for communication with the legitimate receiver. This approach is simple and allows for AN suppression at the legitimate receiver. However, it was shown to be suboptimal in general [24], as restricting the message bearing signal and the AN signal in two orthogonal subspaces reduces the degrees of freedom of the legitimate transmission.

As briefly mentioned at the beginning of this section, one critical requirement of future cellular communications will be to ensure high data-rates, while at the same time reducing the energy consumptions. More specifically, sustainable growth and ecological concerns are putting forth the issue of energy efficiency as one key performance indicator for communication networks. It is foreseen that the number of connected nodes will reach 50 billion by 2020 and the corresponding energy demand will soon become unmanageable [25]. Moreover, the resulting greenhouse gas emissions and electromagnetic pollution will exceed safety thresholds. There is general consensus that future 5G networks will be required to provide a $1,000\times$ data-rate increase, but at half of today's the energy consumption. This requires a $2,000\times$ increase of the bit-per-Joule energy efficiency, defined as the number of bits which can be reliably transmitted per Joule of consumed energy [26,27]. This performance metric has received much attention for communication networks without confidentiality constraints [28], but its study in networks with secure communications is still open and very little prior works have appeared which jointly consider energy-efficient and secrecy issues. In [29], the secrecy-energy trade-off in single-antenna, single-carrier Gaussian wiretap channels is studied in terms of actual communication rate and consumed energy, whereas [30] provides an information-theoretic analysis of secret communications per transmission cost. In [31], the ratio between the system outage capacity and the consumed energy is optimized for a downlink orthogonal frequency division multiple access (OFDMA) network, whereas [32,33] consider the ratio between the secrecy outage capacity and the consumed power is analyzed for multiple-antenna systems.

When confidentiality constraints are added to the communication, a natural extension of the energy-efficient concept is to consider the number of bits which can be reliably *and confidentially* transmitted per Joule of consumed energy, which leads to defining the *secure bit-per-Joule* energy efficiency secrecy energy efficiency (SEE), as the ratio between the system secrecy capacity (or achievable rate) and the consumed power. Given their fractional nature, energy-efficient metrics are naturally optimized by means of fractional programming theory, the branch of optimization theory concerned with the properties and optimization of ratios [28]. The SEE is no exception, and indeed this chapter will show how fractional programming can be used to perform radio resource allocation for SEE maximization in a MIMOME channel with statistical CSI and fully accounting for the most general form of AN.

The rest of this chapter is organized as follows. The problem formulation and optimization approach are described in Section 3.3. However, before delving into the technical details, some preliminaries about secrecy capacity and fractional programming are provided in Section 3.2. Finally, a numerical performance assessment is carried out in Section 3.4. Concluding remarks are provided in Section 3.5.

3.2 Preliminaries

3.2.1 Physical layer security and secrecy measures

The concept of secrecy capacity has been already mentioned. Here, this notion is formally introduced and the main results concerning this fundamental performance metric are reviewed.

The notion of secrecy capacity can be formally defined by considering a $(2^{nR}, n)$-code with an encoder that maps the message $W \in \mathcal{W}_n = \{1, 2, \ldots, 2^{nR}\}$ into a length-n

codeword, and a decoder at the Bob that maps the received sequence y_r^n (the collections of y_r over the code length n) from Bob's channel to an estimated message $\hat{W} \in \mathscr{W}_n$. The received signal at Eve is denoted by y_e^n.

Traditionally, the following four secrecy measures have been considered:

1. Sematic secrecy

$$
S_1 : \max_{f, P_W} \left(\sum_{y_e^n} P_{Y_e^n}(y_e^n) \max_{f_i \in \text{supp}(f)} P(f(W) = f_i | y_e^n) - \max_{f_i \in \text{supp}(f)} P(f(W) = f_i) \right) = 0.
$$

2. Strong secrecy

$$
S_2 : \lim_{n \to \infty} I(W; Y_e^n) = 0.
$$

3. Secrecy in variational distance

$$
S_3 : \sup_{A \subseteq \mathscr{X}} |P_{WY_e^n}[A] - P_W[A]P_{Y_e^n}[A]| = 0.
$$

4. Weak secrecy

$$
S_4 : \lim_{n \to \infty} \frac{1}{n} I(W; Y_e^n) = 0.
$$

In [34,35], it is shown that S_1 implies S_2, S_2 implies S_3, and S_3 implies S_4. Among these four alternatives, the most widely used metric is S_4, due to its mathematical tractability. Then, we have the following definition.

Definition 3.1. (Secrecy capacity). *A communication rate R is achievable with the weak secrecy constraint if, for any $\varepsilon > 0$, there exists a sequence of $(2^{nR}, n)-$ codes and an integer n_0 such that for any $n > n_0$, S_4 and also $Pr(\hat{w} \neq w) \leq \varepsilon$ are valid. The secrecy capacity C_S is the supremum of all achievable secrecy rates.*

In other words, the secrecy capacity is the maximum communication rate at which reliable communication is possible, while at the same time fulfilling the weak secrecy property. In this chapter, the focus will be on the secrecy capacity and secrecy achievable rate as defined in Definition 3.1.

It is also useful to review the following definitions, in which X denotes the channel input, Y_r and Y_e denote the channel output at Bob and Eve, respectively, and U denotes an auxiliary random variable.

Definition 3.2. Degradedness: *A wiretap channel is physically degraded if the transition distribution satisfies $P_{Y_r Y_e | X}(\cdot | \cdot) = P_{Y_r | X}(\cdot | \cdot) P_{Y_e | Y_r}(\cdot | \cdot)$, i.e., X, Y_r, and Y_e form a Markov chain $X \to Y_r \to Y_e$. Such channel is stochastically degraded if its conditional marginal distribution is the same as that of a physically degraded channel, i.e., there exists a distribution $\tilde{P}_{Y_e | Y_r}(\cdot | \cdot)$ such that $P_{Y_e | X}(y_e | x) = \sum_{y_r} P_{Y_r | X}(y_r | x) \tilde{P}_{Y_e | Y_r}(y_e | y_r)$.*

Definition 3.3. [Less noisy] *A wiretap channel is less noisy if $I(U; Y_r) \geq I(U; Y_e)$.*

Note that the distribution of U is a degree of freedom to be optimized, as will be clarified in the coming Theorem 3.1. Moreover, it is worth pointing out that finding the optimal U is a hard problem unless the channel is not degraded.

Definition 3.4. **[More capable]** *A wiretap channel is more capable if* $I(X; Y_r) \geq I(X; Y_e)$.

Given the above definitions, the following results about the secrecy capacity of wiretap channels are known:

Theorem 3.1. *[4]. The secrecy capacity of a nondegraded discrete memoryless wiretap channel under weak secrecy constraint* $\$_4$ *can be represented as*

$$C_S = \max_{p(x|u), p(u)} I(U; Y_r) - I(U; Y_e), \tag{3.1}$$

where $p(x|u)$ *is the channel prefixing. When the wiretap channel is degraded,* (3.1) *simplifies as*

$$C_S = \max_{p(x)} I(X; Y_r) - I(X; Y_e), \tag{3.2}$$

which means $U = X$ *is optimal. Moreover,* $U = X$ *is optimal also if the channel is less noisy or more capable.*

Moreover, if Alice has perfect knowledge of Bob's channel but statistical knowledge of Eve's channel, and Bob perfectly knows his channel and Eve perfectly knows both Bob's and Eve's channel, then we can restate (3.1) as

$$C_S = \max_{p(x|u,h_r), p(u|h_r)} I(U; Y_r|H_r) - I(U; Y_e|H_r, H_e). \tag{3.3}$$

From the previous discussions, we know that the identification of the degradedness is important as it can lead to a simpler optimization problem. However, for wiretap channels with statistical CSIT, which is the main scenario considered in this chapter, it is difficult to see the degradedness property by definition directly. In particular, the order relationship between the squares of Bob's and Eve's channel gains may change at each symbol time, thus implying that the channel is not degraded. In such a case, a more complex optimization problem should be solved. However, if the distributions of the two random channels satisfy some conditions, it is possible to construct equivalent Bob's and Eve's channels which fulfill the degradedness definition, while at the same time having the same secrecy capacity as the original channels. One of the conditions which make this possible is the stochastic order defined as follows:

Definition 3.5. *[36, (1.A.3)] For random variables* X *and* Y, $X \leq_{st} Y$ *if and only if* $\bar{F}_X(x) \leq \bar{F}_Y(x)$ *for all* x, *where* $\bar{F}$ *is the complementary cumulative distribution function.*

A more detailed discussion of this approach can be found in [37], where the coupling theorem from [38] is exploited to show the almost sure existence of an equivalent degraded channel.

3.2.2 Fractional programming theory

The basics of fractional programming theory are reviewed here. For a more detailed review with applications to wireless communications, we refer to [28].

Definition 3.6 ((Strict) Pseudoconcavity). *Let $\mathscr{C} \subseteq \mathbb{R}^n$ be a convex set. Then $r :$ $\mathscr{C} \to \mathbb{R}$ is pseudoconcave if and only if, for all $\mathbf{x}_1, \mathbf{x}_2 \in \mathscr{C}$, it is differentiable and*

$$r(\mathbf{x}_2) < r(\mathbf{x}_1) \Rightarrow \nabla(r(\mathbf{x}_2))^T(\mathbf{x}_1 - \mathbf{x}_2) > 0. \tag{3.4}$$

Strict pseudoconcavity holds if, for all $\mathbf{x}_1 \neq \mathbf{x}_2 \in \mathscr{C}$,

$$r(\mathbf{x}_2) \leq r(\mathbf{x}_1) \Rightarrow \nabla(r(\mathbf{x}_2))^T(\mathbf{x}_1 - \mathbf{x}_2) > 0. \tag{3.5}$$

The interest for pseudoconcave functions stems from the following result.

Proposition 3.1. *Let $r : \mathscr{C} \to \mathbb{R}$ be a pseudoconcave function.*

(a) *If $\mathbf{x}^*$ is a stationary point for r, then it is a global maximizer for r;*
(b) *The Karush Kuhn Tucker (KKT) conditions for the problem of maximizing r subject to convex constraints are necessary and sufficient conditions for optimality;*
(c) *If r is strictly pseudoconcave, then a unique maximizer exists.*

Pseudoconcavity plays a key-role in the optimization of fractional functions, due to the following result.

Proposition 3.2. *Let $r(\mathbf{x}) = \dfrac{f(\mathbf{x})}{g(\mathbf{x})}$, with $f : \mathscr{C} \subseteq \mathbb{R}^n \to \mathbb{R}$ and $g : \mathscr{C} \subseteq \mathbb{R}^n \to \mathbb{R}_+$. If f is nonnegative, differentiable, and concave, whereas g is differentiable and convex, then r is pseudoconcave. If g is affine, the nonnegativity of f can be relaxed. Strict pseudoconcavity holds if either f is strictly concave, or g is strictly convex.*

Finally, let us introduce the definition of fractional program.

Definition 3.7 (Fractional program). *Let $\mathscr{X} \subseteq \mathbb{R}^n$, and consider the functions $f : \mathscr{X} \to \mathbb{R}_0^+$ and $g : \mathscr{X} \to \mathbb{R}^+$. A fractional program is the optimization problem:*

$$\max_{x \in \mathscr{X}} \frac{f(\mathbf{x})}{g(\mathbf{x})}. \tag{3.6}$$

The following result relates the solution of (3.6) to the auxiliary function $F(\beta) = \max_{x \in \mathscr{X}} \{f(\mathbf{x}) - \beta g(\mathbf{x})\}$.

Theorem 3.2 ([39]). *An $\mathbf{x}^* \in \mathscr{X}$ solves (3.6) if and only if $\mathbf{x}^* = \arg\max_{x \in \mathscr{X}} \{f(\mathbf{x}) - \beta^* g(\mathbf{x})\}$, with β^* being the unique zero of $F(\beta)$. Moreover, β^* coincides with the global maximum of (3.6), while $F(\beta) > 0$ for any $\beta < \beta^*$ and $F(\beta) < 0$ for all $\beta > \beta^*$.*

This result allows us to solve (3.6) by finding the zero of $F(\beta)$. An efficient algorithm to do so is the Dinkelbach's algorithm, which is reported here in Algorithm 1.

Algorithm 1 Dinkelbach's algorithm

1: Set $\varepsilon > 0$; $\beta = 0$; $F > \varepsilon$;
2: **repeat**
3:      $x^* = \arg\max_{x \in \mathcal{X}} \{f(x) - \beta g(x)\}$
4:      $F = f(x^*) - \beta g(x^*)$;
5:      $\beta = f(x^*)/g(x^*)$;
6: **until** $F \geq \varepsilon$

If $f(x)$ and $g(x)$ are concave and convex, respectively, and if $\mathcal{X}$ is a convex set, then the Dinkelbach's algorithm requires to solve one convex problem in each iteration. If one of these assumptions is not fulfilled, the auxiliary problem to be solved in each iteration is nonconvex, which in general requires an exponential complexity to be solved. It should be also stressed that in this case Dinkelbach's algorithm is still guaranteed to converge to the global solution, assuming however that the auxiliary problem is *globally* solved in each iteration. More in general, in this scenario, no computationally efficient algorithm is known to find the global solution of (3.6).

As for the number I of iterations required for Algorithm 1 to converge, Dinkelbach's algorithm enjoys a super-linear convergence rate [40]. However, a closed-form expression for the number of iterations required to obtain a given tolerance ε is not available. Instead, this can be obtained by a simple modification of Algorithm 1, which makes use of the bisection method to determine the zero of the auxiliary function. Denoting by U and L the initial positive and negative values for $F(\beta)$, a very similar algorithm as Algorithm 1 can be considered, with the only difference that β is updated in each iteration according to the bisection rule. Then, the number of required iterations to obtain the tolerance ε can be expressed as

$$I = \left\lceil \log_2 \frac{U - L}{\varepsilon} \right\rceil. \tag{3.7}$$

3.3 Radio resource allocation for SEE maximization

3.3.1 MIMOME system model

Let us consider a MIMOME wiretap channel with N_A, N_B, and N_E denoting the number of antennas at Alice, Bob, and Eve, respectively. The received signals at Bob and Eve can be described as

$$y_r = H_r x + z_r,$$

$$y_e = H_e x + z_e,$$

where $H_r \in \mathbb{C}^{N_B \times N_A}$, $H_e \in \mathbb{C}^{N_E \times N_A}$, $z_r \in CN(0, I_{N_B})$, and $z_e \in CN(0, I_{N_E})$ are Gaussian vectors modeling the thermal noise at Bob and Eve, respectively, which are assumed independent. With this notation, the system secrecy achievable rate and ergodic secrecy achievable rate are expressed as

$$R_s = \log_2 \left| \mathbf{I}_{N_B} + \mathbf{H}_r \mathbf{Q} \mathbf{H}_r^H \right| - \log_2 \left| \mathbf{I}_{N_E} + \mathbf{H}_e \mathbf{Q} \mathbf{H}_e^H \right|, \tag{3.8}$$

$$R_{s,erg} = \mathbb{E}_{\mathbf{H}_r} \left[\log_2 \left| \mathbf{I}_{N_B} + \mathbf{H}_r \mathbf{Q} \mathbf{H}_r^H \right| \right] - \mathbb{E}_{\mathbf{H}_e} \left[\log_2 \left| \mathbf{I}_{N_E} + \mathbf{H}_e \mathbf{Q} \mathbf{H}_e^H \right| \right], \tag{3.9}$$

with $\mathbf{Q}$ being Alice's covariance matrix. In a scenario in which perfect CSI is available at Alice, (3.8) can be used for resource allocation purposes, whereas in fast-fading channels, or in general when only statistical CSI is available at the transmitter, only (3.9) can be optimized at Alice. As mentioned, this latter scenario will be the focus of this work. Nevertheless, we briefly summarize available results for the perfect CSI case in which (3.8) is optimized. In [13], assuming multiple antennas at Alice and Eve and a single antenna at Bob, Gaussian input is proved to be optimal, and the optimal transmit direction is the generalized eigenvector of the two terms inside the logarithms of the secrecy capacity expression. When each node has multiple-antenna, Gaussian input is proved to be optimal [6–8]. In particular, [6,7] adopted matrix algebra with optimization theory to solve the problem, whereas [8] used a seminal information theoretic tool, namely *channel enhancement*, which is used to prove the capacity region of Gaussian MIMO broadcast channel, to obtain the result.

When there is only statistical CSI, determining whether the channel is degraded is not easy [37], and the approaches used for the perfect CSI case, e.g., channel enhancement, do not work in general.[1] In the general case, AN appears as the most effective technique in statistical CSI scenarios.

Injecting AN in the transmit signal means splitting the channel input x into two parts, namely:

$$x = u + v, \tag{3.10}$$

where $u \in \mathbb{C}^{N_A}$ is the message bearing signal vector, which follows a Gaussian distribution with zero mean and covariance matrix $\mathbf{Q}_U$, whereas $v \in \mathbb{C}^{N_A}$ is the AN vector, which is independent of u and also follows a Gaussian distribution with zero mean and covariance matrix $\mathbf{Q}_V$, such that $\mathbf{Q} = \mathbf{Q}_U + \mathbf{Q}_V$.[2] The first approaches to AN constrained v to lie in the orthogonal subspace of the legitimate channel $\mathbf{H}_r$, with the aim of suppressing the interference at Bob and thus proposing the use of AN only in multiple-antenna systems where beamforming could be performed. However, [15] showed that AN can prove beneficial also in single-antenna scenarios when Eve's fading variance is large enough relative to Bob's channel gain. In addition, [15] proposes to use discrete signaling instead of Gaussian signal, providing numerical evidence that M-QAM can outperform Gaussian signaling when Bob's channel gain is small enough relative to Eve's channel variance. Finally, in [24], it is shown that transmitting AN in all directions, instead of only the orthogonal complement of the legitimate

---

[1]Channel enhancement can be used in some rare cases to obtain equivalent degraded channels as shown in [41].

[2]The use of Gaussian signaling for both the message bearing and the AN signal might not be optimal in general, but it is the most common choice in the literature due to its mathematical tractability [12,13,15, 24,42,43].

channel, can lead to higher secrecy rates. The intuition behind this result is that the secrecy rate is a difference of two logarithmic functions, and even if injecting AN in all directions reduces the rate at Bob, it also reduces the rate at Eve. This form of AN is labeled *generalized AN* (GAN) in [24]. Considering this more general form of AN, (3.9) becomes:

$$
R_{s,erg} = \mathbb{E}_{\mathbf{H}_r}\left[\log_2\frac{\left|\mathbf{I}_{N_B}+\mathbf{H}_r(\mathbf{Q}_U+\mathbf{Q}_V)\mathbf{H}_r^H\right|}{\left|\mathbf{I}_{N_B}+\mathbf{H}_r\mathbf{Q}_V\mathbf{H}_r^H\right|}\right] - \mathbb{E}_{\mathbf{H}_e}\left[\log_2\frac{\left|\mathbf{I}_{N_E}+\mathbf{H}_e(\mathbf{Q}_U+\mathbf{Q}_V)\mathbf{H}_e^H\right|}{\left|\mathbf{I}_{N_E}+\mathbf{H}_e\mathbf{Q}_V\mathbf{H}_e^H\right|}\right]
$$

$$(3.11)$$

On the other hand, the power that needs to be consumed to make (3.9) achievable is expressed as

$$
P_T = \mu\mathrm{tr}(\mathbf{Q}_U+\mathbf{Q}_V)+P_c,
\tag{3.12}
$$

wherein $\mathrm{tr}(\mathbf{Q}_U+\mathbf{Q}_V)$ is the radiated power, which is scaled by the coefficient $\mu > 1$ denoting the inverse of the transmit amplifier efficiency, whereas P_c is the circuit power dissipated in all other hardware blocks to operate the legitimate system. It should be stressed that (3.12) naturally includes the presence of multiple transmit–receive chains, as the term P_c is a global static power consumption term, which includes the part of the hardware power which scales with the number of deployed antennas, plus any other fixed power term which does not depend on $\mathbf{Q}_U$ and $\mathbf{Q}_V$.

Finally, given (3.11) and (3.12), the system SEE is written as

$$
\mathrm{SEE} = W\frac{R_{s.erg}(\mathbf{Q}_U,\mathbf{Q}_V)}{P_T(\mathbf{Q}_U,\mathbf{Q}_V)} \text{ [bit/J]},
\tag{3.13}
$$

with W denoting the communication bandwidth. It is seen that (3.13) is measured in bit per Joule, thus representing a measure of the amount of information reliably and confidentially transmitted to the receiver, per Joule of consumed energy. It is important to emphasize that the maximization of the SEE is quite different from the more traditional maximization of the secrecy rate at the numerator. The main difference is that unlike the secrecy rate, the SEE is not monotonically increasing with the transmit power but actually tends to zero for increasing transmit powers. Indeed, although the numerator grows at most logarithmically with the transmit power, the denominator grows linearly. This implies that while secrecy rate maximization typically leads to using all the available transmit power, the SEE function is maximized by a finite power level. Once the maximum feasible power is larger than this finite power level, then using full power is a suboptimal choice as far as SEE maximization is concerned. On the other hand, for lower maximum feasible powers, it is possible that using full power is optimal also for SEE maximization, and in this case, optimizing the SEE is equivalent to optimizing the secrecy rate at the numerator.

The radio resource allocation problem is thus formulated as the maximization problem:

$$\max_{\boldsymbol{Q}_U, \boldsymbol{Q}_V} \frac{R_{s,erg}(\boldsymbol{Q}_U, \boldsymbol{Q}_V)}{\mu \text{tr}(\boldsymbol{Q}_U + \boldsymbol{Q}_V) + P_c} \tag{3.14a}$$

$$\text{s.t. } \text{tr}(\boldsymbol{Q}_U + \boldsymbol{Q}_V) \le P_{\max} \tag{3.14b}$$

$$\boldsymbol{Q}_U \succeq \boldsymbol{0}, \boldsymbol{Q}_V \succeq \boldsymbol{0}. \tag{3.14c}$$

The rest of this section will describe how Problem (3.14) can be tackled, considering first the general case described here, and then the special case in which Bob and Eve are equipped with a single-antenna.

3.3.2   Radio resource allocation for MIMOME systems

The main challenge posed by Problem (3.14) is that it is a fractional program with nonconcave numerator. Thus, the fractional programming framework described in Section 3.2 cannot be directly employed with affordable complexity. To circumvent this problem, fractional programming will be complemented with the framework of sequential optimization [44–46], to develop an optimization approach able to trade-off optimality and complexity, being able to yield a resource allocation enjoying strong optimality properties, while at the same time requiring affordable complexity. Sequential optimization has been used for resource allocation in communication networks in several works, for both rate and energy efficiency optimization [47–50].

The idea behind the sequential optimization tool is to tackle a difficult maximization problem by solving a sequence of easier maximization problems. More formally, consider a maximization problem $\mathscr{P}$ with differentiable objective f_0 and constraint functions $\{f_i\}_{i=1}^I$. Then, let us consider a sequence of approximate problems $\{\mathscr{P}_\ell\}_\ell$ with differentiable objectives $\{f_{0,\ell}\}_\ell$ and constraint functions $\{f_{\ell,i}\}_\ell, i = 1, \ldots, I$, such that the following three properties are fulfilled, for all ℓ:

(P1) $f_{i,\ell}(\mathbf{x}) \le f_i(\mathbf{x})$, for all $\mathbf{x}$, $i = 0, 1, \ldots, I$;
(P2) $f_{i,\ell}(\mathbf{x}^{(\ell-1)}) = f_i(\mathbf{x}^{(\ell-1)})$, with $\mathbf{x}^{(\ell-1)}$ the maximizer of $f_{\ell-1}$, $i = 0, 1, \ldots, I$;
(P3) $\nabla f_{i,\ell}(\mathbf{x}^{(\ell-1)}) = \nabla f_i(\mathbf{x}^{(\ell-1)})$, $i = 0, 1, \ldots, I$.

Now, if Properties **(P1)**, **(P2)**, and **(P3)** are fulfilled, the sequence $\{f(\mathbf{x}^{(\ell)})\}_\ell$ is monotonically increasing and convergent. Moreover, upon convergence of the sequence $\{f(\mathbf{x}^{(\ell)})\}_\ell$, the resulting point $\mathbf{x}^*$ fulfills the KKT conditions of the original Problem $\mathscr{P}$ [44].

Some recent refinements of the above results are also available in the literature. If the feasible set of the original problem is not modified, then every limit point of the sequence $\{\mathbf{x}^{(\ell)}\}_\ell$ is a KKT point of the original Problem $\mathscr{P}$ [46]. Another result from [45] states that if the original objective is strictly concave,[3] then again every

---

[3] Reference [45] assumes strict convexity, as minimization problems are discussed instead of maximization problems.

limit point of the sequence $\{x^{(\ell)}\}_\ell$ is a KKT point of the original Problem $\mathscr{P}$, also when the original feasible set is approximated in each iteration.

This approach appears to be very powerful, but the critical point is to find suitable functions $\{f_\ell\}_\ell$ fulfilling Properties **(P1)**, **(P2)**, and **(P3)**, and that can also be maximized with limited complexity. This latter constraint is not directly related to the theoretical properties of the method, and indeed was not mentioned in the description above, but clearly it is a practical requirement which must be fulfilled in order for the method to be of practical relevance. For the case of Problem (3.14), all these requirements can be fulfilled as described next.

To begin with, let us observe that the numerator of (3.14a) can be expressed as

$$R_{s,erg} = \underbrace{\mathbb{E}_{\mathbf{H}_r}\left[\log_2\left|\mathbf{I}_{N_B} + \mathbf{H}_r(\mathbf{Q}_U + \mathbf{Q}_V)\mathbf{H}_r^H\right|\right] + \mathbb{E}_{\mathbf{H}_e}\left[\left|\mathbf{I}_{N_E} + \mathbf{H}_e\mathbf{Q}_V\mathbf{H}_e^H\right|\right]}_{R^+(\mathbf{Q}_U, \mathbf{Q}_V)}$$
$$- \underbrace{\left\{\mathbb{E}_{\mathbf{H}_r}\left[\log_2\left|\mathbf{I}_{N_E} + \mathbf{H}_e(\mathbf{Q}_U + \mathbf{Q}_V)\mathbf{H}_e^H\right|\right] + \mathbb{E}_{\mathbf{H}_r}\left[\log_2\left|\mathbf{I}_{N_B} + \mathbf{H}_r\mathbf{Q}_V\mathbf{H}_r^H\right|\right]\right\}}_{R^-(\mathbf{Q}_U, \mathbf{Q}_V)},$$

$$(3.15)$$

which is the difference of the two concave functions $R^+(\mathbf{Q}_U, \mathbf{Q}_V)$ and $R^-(\mathbf{Q}_U, \mathbf{Q}_V)$. Thus, for any given point $(\mathbf{Q}_{U,0}, \mathbf{Q}_{V,0})$, the numerator of (3.14a) can be lower bounded by replacing $R^-(\mathbf{Q}_U, \mathbf{Q}_V)$ by its Taylor expansion around $(\mathbf{Q}_{U,0}, \mathbf{Q}_{V,0})$, which yields:

$$\begin{aligned} R_{s,erg} &= R^+(\mathbf{Q}_U, \mathbf{Q}_V) - R^-(\mathbf{Q}_U, \mathbf{Q}_V) \\ &\geq R^+(\mathbf{Q}_U, \mathbf{Q}_V) - \Big\{R^-(\mathbf{Q}_{U,0}, \mathbf{Q}_{V,0}) \\ &\quad + 2\Re\Big\{\mathrm{tr}\Big(\nabla_{\mathbf{Q}_U}^H R^-|_{\substack{\mathbf{Q}_{U,0}\\\mathbf{Q}_{V,0}}}(\mathbf{Q}_U - \mathbf{Q}_{U,0}) + \nabla_{\mathbf{Q}_V}^H R^-|_{\substack{\mathbf{Q}_{U,0}\\\mathbf{Q}_{V,0}}}(\mathbf{Q}_V - \mathbf{Q}_{V,0})\Big)\Big\}\Big\} \\ &= \tilde{R}_{s,erg} \end{aligned}$$

$$(3.16)$$

It can be seen that $\tilde{R}_{s,erg}$ is a concave lower bound of $R_{s,erg}$. Moreover, the two functions and their gradients coincide at $(\mathbf{Q}_{U,0}, \mathbf{Q}_{V,0})$. Thus, we can fulfill all the requirements of the sequential framework, by replacing the numerator of (3.14a) by $\tilde{R}_{s,erg}$, which yields the generic approximate problem:

$$\max \quad \frac{\tilde{R}_{s,erg}(\mathbf{Q}_U, \mathbf{Q}_V)}{\mu\mathrm{tr}(\mathbf{Q}_U + \mathbf{Q}_V) + P_c} \tag{3.17a}$$

$$\text{s.t.} \quad \mathrm{tr}(\mathbf{Q}_U + \mathbf{Q}_V) \leq P_{\max} \tag{3.17b}$$

$$\mathbf{Q}_U \succeq 0, \mathbf{Q}_V \succeq 0. \tag{3.17c}$$

Problem (3.17) is still a fractional program, but unlike (3.14), it has a concave numerator of the objective. This enables us to globally solve (3.17) with polynomial complexity by means of standard fractional programming approaches, as described in Section 3.2. More specifically, the overall resource allocation procedure can be stated as

---

Algorithm 2 SEE maximization for MIMOME wiretap channels

$\ell = 0$; Select a feasible point $(\boldsymbol{Q}_{U,0}^{(\ell)}, \boldsymbol{Q}_{V,0}^{(\ell)})$;
repeat
    Solve Problem (3.17) by fractional programming and set $(\boldsymbol{Q}_U^{(\ell)}, \boldsymbol{Q}_V^{(\ell)})$ as the solution.
    $(\boldsymbol{Q}_{U,0}, \boldsymbol{Q}_{V,0}) = (\boldsymbol{Q}_U^{(\ell)}, \boldsymbol{Q}_V^{(\ell)})$;
    $\ell = \ell + 1$;
until Convergence

The following result follows from the described results on sequential optimization and fractional programming.

Proposition 3.3. *Algorithm 2 monotonically increases the SEE value in (3.14a) until convergence. Moreover, every limit point of the sequence $\{(\boldsymbol{Q}_U^{(\ell)}, \boldsymbol{Q}_V^{(\ell)})\}_\ell$ is a KKT point of (3.14). Finally, upon convergence in the objective, the output of the algorithm fulfills the KKT optimality conditions of Problem (3.14).*

3.3.3   SEE maximization with QoS constraints

The approach described above can be extended to include quality of service (QoS) constraints in the form of a minimum secrecy rate to be guaranteed to the legitimate user. In this case, Problem (3.14) becomes:

$$\max_{\boldsymbol{Q}_U, \boldsymbol{Q}_V} \frac{R_{s,erg}(\boldsymbol{Q}_U, \boldsymbol{Q}_V)}{\mu \mathrm{tr}(\boldsymbol{Q}_U + \boldsymbol{Q}_V) + P_c} \tag{3.18a}$$

$$\text{s.t. } \mathrm{tr}(\boldsymbol{Q}_U + \boldsymbol{Q}_V) \le P_{max} \tag{3.18b}$$

$$\boldsymbol{Q}_U \succeq 0, \boldsymbol{Q}_V \succeq 0 \tag{3.18c}$$

$$R_{s,erg}(\boldsymbol{Q}_U, \boldsymbol{Q}_V) \ge R_{min}. \tag{3.18d}$$

The additional, nonconvex, constraint (3.18d) can be handled by using the same approximation as used for the secrecy rate at the numerator of the SEE. Indeed, following the same approach which led to Algorithm 2, it is possible to consider the approximate problem:

$$\max \frac{\tilde{R}_{s,erg}(\boldsymbol{Q}_U, \boldsymbol{Q}_V)}{\mu \mathrm{tr}(\boldsymbol{Q}_U + \boldsymbol{Q}_V) + P_c} \tag{3.19a}$$

$$\text{s.t. } \mathrm{tr}(\boldsymbol{Q}_U + \boldsymbol{Q}_V) \le P_{max} \tag{3.19b}$$

$$\boldsymbol{Q}_U \succeq 0, \boldsymbol{Q}_V \succeq 0 \tag{3.19c}$$

$$\tilde{R}_{s,erg}(\boldsymbol{Q}_U, \boldsymbol{Q}_V) \ge R_{min}, \tag{3.19d}$$

which is again a fractional problem with concave numerator, convex denominator, and convex constraints, and thus can be globally solved with polynomial complexity by fractional programming theory. Moreover, Problem (3.19) fulfills all theoretical properties required by the sequential method with respect to Problem (3.18), thereby implying that a similar algorithm as Algorithm 2 can be developed.

Before closing this section, the following remarks are in order.

Remark 3.1. *The results in this section can be specialized or extended in several directions. First, it is possible to consider different CSI scenarios in which only one of the two channels is statistically known, whereas perfect CSI is available for the other channel. Second, it is possible to consider arbitrary distributions for the channel matrices* $\mathbf{H}_r$ *and* $\mathbf{H}_e$, *as no particular assumption has been made on the two channel matrices.*

Remark 3.2. *Although the focus of this chapter has been on the maximization of the SEE, the presented framework can be readily specialized to perform secrecy rate maximization. This can be simply accomplished by fixing* $\mu = 0$ *and* $P_c = 1$ *in (3.14), thus obtaining a denominator identically equal to 1, and the running Algorithm 2. Clearly, in this special case, each approximate problem (3.17) will no longer be a fractional problem, but rather a simpler convex problem that can be solved by standard methods.*

Another obvious extension of the scenario considered in this section is the so-called MISOSE system, in which a single antenna is deployed at Bob and Eve (multiple antennas are still present at Alice). Although simpler, this scenario deserves a dedicated treatment, which is done in the coming section, because, unlike the general MIMOME case, it allows for the closed-form derivation of the optimal transmit directions of the matrices $\mathbf{Q}_U$ and $\mathbf{Q}_V$, thus reducing the radio resource allocation problem to a simpler power control program. This point is detailed in the next section.

3.3.4   Radio resource allocation for MISOSE systems

Assume that perfect CSI is available at Bob as to the legitimate channel $\mathbf{h}$. However, statistical CSI is available as to the channel to the eavesdropper $\mathbf{g}$. If Bob and Eve are also both equipped with a single antenna, Problem (3.14) reduces to

$$\max \quad \frac{\log\left(1 + \frac{h^H Q_U h}{1 + h^H Q_V h}\right) - \mathbb{E}_g\left[\log\left(1 + \frac{g^H Q_U g}{1 + g^H Q_V g}\right)\right]}{\mu \mathrm{tr}(Q_U + Q_V) + P_c} \tag{3.20a}$$

$$\text{s.t. } \mathrm{tr}(Q_U + Q_V) \le P_{\max} \tag{3.20b}$$

$$Q_U \succeq 0, Q_V \succeq 0, \tag{3.20c}$$

with $\mathbf{h}$ and $\mathbf{g}$ being the $N_A \times 1$ channel vectors to Bob and Eve, respectively. Clearly, Problem (3.20) could be tackled by means of the same approach described in the more general MIMOME case. However, by making the additional assumption that the channel vectors **h** and **g** follow a Gaussian distribution with uncorrelated entries, it possible to determine the eigenvectors of **Q** in closed-form, as shown in the following result.

Theorem 3.3. *Assume* **h** *and* **g** *are complex Gaussian random vectors with (possibly scaled) identity covariance matrix. Then, the optimal* $(\mathbf{Q}_U^*, \mathbf{Q}_V^*)$ *are such that*

$$\mathbf{Q}_U^* = P_U \mathbf{u}_1 \mathbf{u}_1^H, \tag{3.21}$$

with $\mathbf{u}_1 = \mathbf{h}/\|\mathbf{h}\|$, *and*

$$\mathbf{Q}_V^* = V_V \mathbf{\Lambda}_V V_V^H, \tag{3.22}$$

with $V_V = [\mathbf{u}_1, \mathbf{u}_1^\perp]$ *and* $\mathbf{\Lambda}_V = \mathrm{diag}(P_{V_s}, P_{V_r}, \ldots, P_{V_r})$ *a diagonal* $N_A \times N_A$ *matrix with* P_{V_s} *and* P_{V_r} *variables to be determined. Moreover, Problem* (3.20) *can be recast as in* (3.23), *with* $\tilde{G}_i = |\mathbf{g}^H \mathbf{u}_i|^2$, *for all* $i =, 1, \ldots, N_A$.

$$\max_{(P_U, P_{V_s}, P_{V_r})} \frac{\log\left(1 + \frac{\|\mathbf{h}\|^2 P_U}{1 + \|\mathbf{h}\|^2 P_{V_s}}\right) - \mathbb{E}\left[\log\left(1 + \frac{\tilde{G}_1 P_U}{1 + \tilde{G}_1 P_{V_s} + \left(\sum_{i=2}^{N_A} \tilde{G}_i\right) P_{V_r}}\right)\right]}{\mu(P_U + P_{V_s} + (N_A - 1)P_{V_r}) + P_c} \tag{3.23a}$$

s.t. $P_U + P_{V_s} + (N_A - 1)P_{V_r} \leq P_{\max}$ \hfill (3.23b)

$P_{V_s} \geq 0, P_{V_r}, \geq 0, P_U \geq 0.$ \hfill (3.23c)

Proof. The result follows from [24,51]. \hfill $\square$

Thanks to Theorem 3.3, the original matrix-variate optimization problem (3.20) is equivalently reformulated into a vector-variate problem. Moreover, the number of optimization variables is constantly equal to 3 and does not scale with the number of antennas N_A at Alice. Thus, the complexity of Problem (3.23) is significantly lower than that of Problem (3.20).

Despite this, Problem (3.23) poses a formally similar challenge as the general Problem (3.14) in the MIMOME case, in the sense that the numerator of (3.23a) is again a nonconcave function of the optimization variables. However, it is possible to resort to a similar approach as in the MIMOME case, merging sequential optimization with fractional programming. In this case, the resulting complexity is even lower, since each approximate problem has only three scalar optimization variables, unlike the $N_A(N_A - 1)$ optimization variables of Problem[4] (3.17). Following the same approach as in Section 3.3.2, a similar algorithm as Algorithm 2 can be developed, enjoying similar optimality claims with respect to Problem (3.23a).

---

[4]A generic $N_A \times N_A$ Hermitian matrix has $N_A(N_A - 1)$ free variables.

3.4   Numerical experiments

Numerical simulations have been performed for a MISOSE wiretap channel with $N_A = 3$. The main and eavesdropper channels $\boldsymbol{h}$ and $\boldsymbol{g}$ have been generated as realizations of zero-mean Gaussian random vectors with covariance matrices $\sigma_h^2 \mathbf{I}_{N_A}$ and $\sigma_g^2 \mathbf{I}_{N_A}$, wherein σ_h^2 and σ_g^2 represent the channel power path-loss coefficients normalized to the receive noise power. The path-loss follows the model from [52], with power decay factor $\eta = 3.5$, whereas the noise power is computed as $F \mathcal{N}_0 W$, with $F = 3$ dB the receive noise figure, $\mathcal{N}_0 = -174$ dBm/Hz the noise power spectral density, and $W = 180$ kHz. The power consumption parameters are $P_c = 10$ dBm and $\mu = 1$. All results have been obtained by averaging over 1,000 independent channel realizations.

Figure 3.1 illustrates the average SEE vs. $P_{\max}$, achieved by

- SEE maximization by the described Algorithm with AN;
- SEE maximization without the use of AN. In this case, we set $P_{V_s} = P_{V_r} = 0$ and only P_U is optimized;
- SEE maximization assuming perfect CSI of $\boldsymbol{G}$ at Alice, which serves as a benchmark scenario. In this case, $\boldsymbol{Q}_V = \boldsymbol{0}$ as AN is not required for secrecy capacity maximization when perfect CSI of $\boldsymbol{G}$ is available at Alice. For each of the 1,000 realization of $\boldsymbol{G}$, the optimal $\boldsymbol{Q}_U$ is determined, and then, the average of the 1,000 optimal SEE values is computed;
- Secrecy rate maximization with statistical CSI and AN, by modifying the described Algorithm as explained in Remark 3.2;

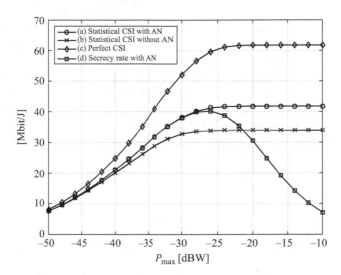

Figure 3.1 $N_A = 3$, $P_c = 10$ dBm. Achieved SEE vs. P_{max}; (a) SEE maximization by fractional programming plus sequential optimization, with statistical CSI and AN; (b) statistical CSI without AN; (c) perfect CSI; (d) secrecy rate maximization by sequential optimization, with statistical CSI and AN

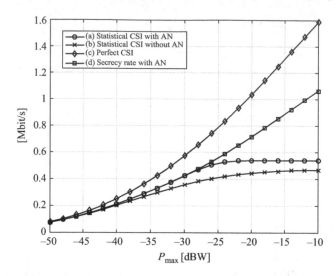

Figure 3.2 $N_A = 3$, $P_c = 10$ dBm. Achieved SEE vs. P_{max}; (a) SEE maximization by
fractional programming plus sequential optimization, with statistical
CSI and AN; (b) statistical CSI without AN; (c) perfect CSI; (d) secrecy
rate maximization by sequential optimization, with statistical CSI
and AN

It is seen that with only statistical CSI of G, using AN grants a significant performance gain compared to the case in which no AN is used. It is interesting to remark that all gaps among the curves corresponding to SEE maximizations remain constant for increasing P_{max}, due to the saturation of the SEE. The reason for this behavior is that for high P_{max} the SEE-maximizing resource allocation strategy is not to transmit at maximum power. As a consequence, for large P_{max}, the actual transmit power does not grow large, but remains constantly equal to the finite maximizer of the SEE. For the simulated scenario, this happens at $P_{max} = -28$ dBW. If P_{max} is further increased, the excess power is not used as it would lead to a decrease of the SEE. This is true both with perfect and statistical CSI, with and without AN. Instead, it can be seen that the SEE achieved by secrecy rate maximization is increasing and equal to that obtained by SEE maximization up to $P_{max} = -28$ dBW, and then starts decreasing because, unlike the SEE, secrecy rate maximization uses the excess power, but this degrades the SEE value. Instead, for low P_{max}, full power transmission is the optimal strategy for both SEE and secrecy rate maximization.

A similar situation is illustrated in Figure 3.2, with the difference that the reported metric is the secrecy rate instead of the SEE. In accordance with Figure 3.1, the secrecy rate obtained by SEE maximization eventually saturates with P_{max} both with perfect and statistical CSI. Instead, the secrecy rate obtained by secrecy rate maximization increases with P_{max}, showing that the secrecy rate is monotonically increasing with P_{max}.

Table 3.1 $N_A = 3$, $P_c = 10$ dBm; number of iterations required by
Algorithm 2 to reach convergence

P_{max} [dBW]	Average number of iterations
−50	2.186
−45	2.346
−40	2.694
−35	3.121
−30	3.509
−25	4.721
−20	7.339

Finally, Table 3.1 shows the average number of outer iterations required for the optimization algorithm to converge vs. P_{max}, i.e., the number of approximate fractional problems to be solved before convergence is reached. Convergence is checked with respect to the SEE value, and the tolerance is set to $\varepsilon = 10^{-3}$. The results show that only a few iterations are required, which supports the use of the described framework in practical systems. It is also seen that the number of iterations slightly increases with P_{max}, as a higher P_{max} implies a larger feasible set to search. The values of P_{max} are reported up to $P_{max} = -20$ dBW, as larger P_{max} falls in the saturation region of the SEE according to Figure 3.1, and therefore should not be considered as system operating points.

3.5   Conclusions

After a review of the main performance metrics and methodologies in the field of physical layer security, this chapter has been focused on the issue of radio resource allocation in systems employing physical layer security techniques. Unlike most previous contributions, which focus on traditional and nonenergy-efficient performance metrics like secrecy capacity and secrecy achievable rates, this book chapter has described a radio resource allocation framework to optimize both the confidentiality and the energy efficiency of a communication system. To this end, a key step has been the introduction of the new performance metric of SEE, defined as the ratio between the system secrecy capacity (or achievable rate) and the consumed power, including both radiated and hardware power.

The SEE metric measures the amount of information that can be reliably and confidentially transmitted per Joule of consumed energy, thus being a natural measure of both the confidentiality and energy efficiency of the communication. Both aspects are anticipated to be particularly relevant in future communication networks, which will be composed of an unprecedented amount of devices, which poses both confidentiality and energy issues.

The maximization of the SEE has been carried out assuming statistical CSI is available at the legitimate transmitter, and considering the general scenario of a MIMOME systems, where all nodes are equipped with multiple antennas. Moreover, the use of the most general form of AN transmission has been accounted for. The resulting fractional and NP-hard optimization problems have been tackled by a combination of fractional programming and sequential optimization theories. The described iterative algorithm, although not being theoretically guaranteed to be globally optimal, is able to monotonically increase the SEE value after each iteration, also enjoying first-order KKT optimality properties, while at the same time enjoying limited complexity.

The proposed optimization framework is very general and encompasses many relevant special cases. In particular, although it is conceived for SEE maximization, it can be readily specialized to perform secrecy rate optimization by simply fixing the power consumption parameters. Moreover, no assumption on the channel distributions has been made, thus allowing the optimization framework to work regardless of the distributions followed by the channel matrices. Finally, the framework can be readily extended to include QoS constraints in the form of a minimum secrecy rate guarantee for the legitimate user.

The chapter also features the application of the proposed framework to the special case of a MISOSE system, in which the legitimate receiver and the eavesdropper are equipped with only one antenna. Although less general than the MIMOME case, this scenario is interesting as it allows for the closed-form optimization of the transmit directions, thus significantly further lowering the complexity of the radio resource allocation procedure.

References

[1]   Y. Liang and H. V. Poor, "Multiple access channels with confidential messages," *IEEE Trans. Inf. Theory*, vol. 54, pp. 976–1002, Mar. 2008.

[2]   X. Zhou, R. K. Ganti, and J. G. Andews, "Secure wireless network connectivity with multi-antenna transmission," *IEEE Transactions on Wireless Communications*, vol. 10, no. 2, pp. 425–430, Feb. 2011.

[3]   A. D. Wyner, "The wiretap channel," *Bell Syst. Tech. J.*, vol. 54, pp. 1355–1387, 1975.

[4]   I. Csiszár and J. Korner, "Broadcast channels with confidential messages," *IEEE Trans. Inf. Theory*, vol. 24, no. 3, pp. 339–348, 1978.

[5]   S. Shafiee and S. Ulukus, "Towards the secrecy capacity of the Gaussian MIMO wire-tap channel: the 2-2-1 channel," *IEEE Trans. Inf. Theory*, vol. 55, no. 9, pp. 4033–4039, Sept. 2009.

[6]   A. Khisti and G. W. Wornell, "Secure transmission with multiple antennas-II: The MIMOME wiretap channel," *IEEE Trans. Inf. Theory*, vol. 56, no. 11, pp. 5515–5532, Nov. 2010.

[7]   F. Oggier and B. Hassibi, "The secrecy capacity of the MIMO wiretap channel," *IEEE Trans. Inf. Theory*, vol. 57, no. 8, pp. 4961–4972, Aug. 2011.

[8]   T. Liu and S. Shamai (Shitz), "A note on the secrecy capacity of the multi-antenna wiretap channel," *IEEE Trans. Inf. Theory*, vol. 55, no. 6, pp. 2547–2553, Jun. 2009.

[9]   Y. Liang, V. Poor, and S. Shamai (Shitz), "Secure communication over fading channels," *IEEE Trans. Inf. Theory*, vol. 54, no. 6, pp. 2470–2492, Jun. 2008.

[10] S. Bashar, Z. Ding, and C. Xiao, "On secrecy rate analysis of MIMO wiretap channels driven by finite-alphabet input," *IEEE Trans. Commun.*, vol. 60, no. 12, pp. 3816–3825, Dec. 2012.

[11] E. A. Jorswieck, S. Tomasin, and A. Sezgin, "Broadcasting into the uncertainty: Authentication and confidentiality by physical-layer processing," in *Proceedings of the IEEE*, vol. 103, no. 10, pp. 1702–1724, Oct. 2015.

[12] S. Goel and R. Negi, "Guaranteeing secrecy using artificial noise," *IEEE Trans. Wireless Commun.*, vol. 7, no. 6, pp. 2180–2189, Jun. 2008.

[13] A. Khisti and G. W. Wornell, "Secure transmission with multiple antennas-I: The MISOME wiretap channel," *IEEE Trans. Inf. Theory*, vol. 56, no. 7, pp. 3088–3104, Jul. 2010.

[14] J. Li and A. Petropulu, "On ergodic secrecy rate for Gaussian MISO wiretap channels," *IEEE Trans. Wireless Commun.*, vol. 10, no. 4, pp. 1176–1187, Apr. 2011.

[15] Z. Li, R. Yates, and W. Trappe, "Achieving secret communication for fast Rayleigh fading channels," *IEEE Trans. Wireless Commun.*, vol. 9, no. 9, pp. 2792 – 2799, Sep. 2010.

[16] P. Gopala, L. Lai, and H. El Gamal, "On the secrecy capacity of fading channels," *IEEE Trans. Inf. Theory*, vol. 54, no. 10, pp. 4687–4698, Oct. 2008.

[17] S.-C. Lin and P.-H. Lin, "On ergodic secrecy capacity of multiple input wiretap channel with statistical CSIT," *IEEE Trans. Inf. Forensics Security*, vol. 8, no. 2, pp. 414–419, Feb. 2013.

[18] X. Zhou and M. R. McKay, "Secure transmission with artificial noise over fading channels: achievable rate and optimal power allocation," *IEEE Trans. Veh. Technol.*, vol. 59, no. 8, pp. 3831–3842, Oct. 2010.

[19] T. Nguyen and H. Shin, "Power allocation and achievable secrecy rates in MISOME wiretap channels," *IEEE Commun. Lett.*, vol. 15, no. 11, pp. 1196–1198, Nov. 2011.

[20] S. Gerbracht, A. Wolf, and E. Jorswieck, "Beamforming for Fading Wiretap Channels with Partial Channel Information," in *Proc. of International ITG Workshop on Smart Antennas (WSA)*, Bremen, Germany, Feb. 2010.

[21] S.-C. Lin, T. H. Chang, Y. L. Liang, Y. W. P. Hong, and C. Y. Chi, "On the impact of quantized channel feedback in guaranteeing secrecy with artificial noise: The noise leakage problem," *IEEE Trans. Wireless Commun.*, vol. 10, no. 3, pp. 901–915, Mar. 2011.

[22] A. Wolf, E. A. Jorswieck, and C. R. Janda, "Worst-case secrecy rates in MIMOME systems under input and state constraints," in *Proc. of IEEE International Workshop on Information Forensics and Security (WIFS) Rome, Italy*, Nov. 2015.

[23]   A. Wolf and E. A. Jorswieck, "Maximization of worst-case secrecy rates in MIMO wiretap channels," in *Proc. of the Asilomar Conference on Signals, Systems, and Computers Pacific Grove, USA*, Nov. 2010.

[24]   P.-H. Lin, S.-H. Lai, S.-C. Lin, and H.-J. Su, "On optimal artificial-noise assisted secure beamforming for the fading eavesdropper channel," *IEEE J. Sel. Areas Commun.*, vol. 31, no. 9, pp. 1728–1740, Sept. 2013.

[25]   G. Auer, V. Giannini, C. Desset, *et al.*, "How much energy is needed to run a wireless network?" *IEEE Wireless Commun.*, vol. 18, no. 5, pp. 40–49, Oct. 2011.

[26]   *NGMN 5G White Paper*, NGMN Alliance, March 2015.

[27]   S. Buzzi, C.-L. I, T. E. Klein, H. V. Poor, C. Yang, and A. Zappone, "A survey of energy-efficient techniques for 5G networks and challenges ahead," *IEEE J. Sel. Areas Commun.*, vol. 34, no. 4, pp. 697–709, 2016.

[28]   A. Zappone and E. Jorswieck, "Energy efficiency in wireless networks via fractional programming theory," *Found. Trends Commun. Inf. Theory*, vol. 11, no. 3–4, pp. 185–396, 2015.

[29]   C. Comaniciu and H. V. Poor, "On energy-secrecy trade-offs for Gaussian wiretap channels," *IEEE Trans. Inf. Forensics Security*, vol. 8, no. 2, pp. 314–323, Feb. 2013.

[30]   M. El-Halabi, T. Liu, and C. N. Georghiades, "Secrecy capacity per unit cost," *IEEE J. Sel. Areas Commun.*, vol. 31, no. 9, pp. 1909–1920, Sept. 2013.

[31]   D. W. K. Ng, E. S. Lo, and R. Schober, "Energy-efficient resource allocation for secure OFDMA systems," *IEEE Trans. Veh. Technol.*, vol. 61, no. 6, pp. 2572–2585, Jul. 2012.

[32]   X. Chen and L. Lei, "Energy-efficient optimization for physical layer security in multi-antenna downlink networks with QoS guarantee," *IEEE Commun. Lett.*, vol. 17, no. 4, pp. 637–640, Apr. 2013.

[33]   X. Chen, C. Zhonga, C. Yuen, and H.-H. Chen, "Multi-antenna relay aided wireless physical layer security," *IEEE Commun. Mag.*, vol. 53, no. 12, pp. 40–46, Dec. 2015.

[34]   M. R. Bloch and J. N. Laneman, "Strong secrecy from channel resolvability," *IEEE Trans. Inf. Theory*, vol. 59, no. 12, pp. 8077–8098, Dec. 2013.

[35]   M. R. Bloch, M. Hayashi, and A. Thangaraj, "Error-control coding for physical-layer secrecy," in *Proc. IEEE*, vol. 103, no. 10, pp. 1725–1746, Oct. 2015.

[36]   M. Shaked and J. G. Shanthikumar, *Stochastic Orders*. Springer, 2007.

[37]   P.-H. Lin and E. Jorswieck, "On the fading Gaussian wiretap channel with statistical channel state information at transmitter," *IEEE Trans. Inf. Forensics Security*, vol. 11, no. 1, pp. 46–58, Jan. 2016.

[38]   H. Thorisson, *Coupling, Stationarity, and Regeneration*. Springer-Verlag New York, 2000.

[39]   W. Dinkelbach, "On nonlinear fractional programming," *Manage. Sci.*, vol. 13, no. 7, pp. 492–498, Mar. 1967.

[40]   J. P. Crouzeix, J. A. Ferland, and S. Schaible, "An algorithm for generalized fractional programs," *J. Optim. Theory Appl.*, vol. 47, no. 1, pp. 35–49, 1985.

[41] P.-H. Lin, E. A. Jorswieck, R. F. Schaefer, and M. Mittelbach, "On the degradedness of fast fading Gaussian multiple-antenna wiretap channels with statistical channel state information at the transmitter," in *Proceedings of IEEE Globecom Workshop on Trusted Communications with Physical Layer Security*, 2015.

[42] N. Yang, S. Yan, J. Yuan, R. Malaney, R. Subramanian, and I. Land, "Artificial noise: Transmission optimization in multi-input single-output wiretap channels," *IEEE Trans. Commun.*, vol. 63, no. 5, pp. 1771–1783, May 2015.

[43] N. Yang, M. Elkashlan, T. Q. Duong, J. Yuan, and R. Malaney, "Optimal transmission with artificial noise in MISOME wiretap channels," *IEEE Trans. Veh. Technol.*, vol. 65, no. 4, pp. 2170–2181, April 2016.

[44] B. R. Marks and G. P. Wright, "A general inner approximation algorithm for non-convex mathematical programs," *Oper. Res.*, vol. 26, no. 4, pp. 681–683, 1978.

[45] A. Beck, A. Ben-Tal, and L. Tetruashvili, "A sequential parametric convex approximation method with applications to non-convex truss topology design problems," *J. Global Optim.*, vol. 47, no. 1, 2010.

[46] M. Razaviyayn, M. Hong, and Z.-Q. Luo, "A unified convergence analysis of block successive minimization methods for nonsmooth optimization," *SIAM J. Optim.*, vol. 23, no. 2, 2013.

[47] M. Chiang, C. Wei, D. P. Palomar, D. O'Neill, and D. Julian, "Power control by geometric programming," *IEEE Trans. Wireless Commun.*, vol. 6, no. 7, pp. 2640–2651, Jul. 2007.

[48] L. Venturino, N. Prasad, and X. Wang, "Coordinated scheduling and power allocation in downlink multicell OFDMA networks," *IEEE Trans. Veh. Technol.*, vol. 58, no. 6, pp. 2835–2848, Jul. 2009.

[49] L. Venturino, A. Zappone, C. Risi, and S. Buzzi, "Energy-efficient scheduling and power allocation in downlink OFDMA networks with base station coordination," *IEEE Trans. Wireless Commun.*, vol. 14, no. 1, pp. 1–14, Jan. 2015.

[50] A. Zappone, E. A. Jorswieck, and S. Buzzi, "Energy efficiency and interference neutralization in two-hop MIMO interference channels," *IEEE Trans. Signal Proc.*, vol. 62, no. 24, pp. 6481–6495, Dec. 2014.

[51] A. Zappone, P.-H. Lin, and E. A. Jorswieck, "Energy efficiency of confidential multi-antenna systems with artificial noise and statistical CSI," *IEEE J. Sel. Topics Signal Proc.*, vol. 10, no. 8, pp. 1462–1477, Dec 2016.

[52] G. Calcev, D. Chizhik, B. Goransson, *et al.*, "A wideband spatial channel model for system-wide simulations," *IEEE Trans. Veh. Technol.*, vol. 56, no. 2, March 2007.

Part II

Physical layer security for multiple-antenna technologies

Chapter 4
Antenna selection strategies for wiretap channels

Shihao Yan[1], Nan Yang[1], Robert Malaney[2], and Jinhong Yuan[2]

Antenna selection, which has been recognised as an important technique to enhance physical layer security in multiple-input multiple-output wiretap channels, requires low feedback overhead and low hardware complexity. In this chapter, antenna selection strategies in different application scenarios for wiretap channels are reviewed in order to demonstrate their benefits in terms of enhancing secure transmissions. The basic idea of transmit antenna selection (TAS) is first detailed and then TAS with Alamouti coding is provided. Furthermore, antenna selection strategies in full-duplex wiretap channels are presented, followed by a discussion on the impact of imperfect feedback and correlation on the secrecy performance of TAS.

4.1 Introduction

Motivated by emerging wireless applications with multiple antenna terminals, physical layer security in multiple-input multiple-output (MIMO) wiretap channels is of growing interest [1–11]. In previous works, transmit beamforming in the direction of the receiver was investigated as a practical method of performing secure transmission [12–14]. In [12], beamforming was proposed to minimise the transmit power constrained by a pre-specified signal-to-interference-plus-noise ratio (SINR) at the receiver. In [13], artificial noise was incorporated in the beamforming weights to constrain the maximum SINRs of the eavesdroppers. Applying linear pre-coding at the transmitter, the authors in [14] adopted a game-theoretic formulation to balance performance and fairness. However, these beamforming methods mandated precise channel state information (CSI) at the transmitter. In general, such a mandate incurs high feedback overhead and computational cost of signal processing, especially when the nodes are equipped with a large number of transmit antennas [15]. Against this background, antenna selection was applied at a multi-antenna transmitter in order to enhance security with reduced feedback overhead and hardware complexity [16,17].

[1] Research School of Engineering, The Australian National University, Australia
[2] School of Electrical Engineering and Telecommunications, The University of New South Wales, Australia

4.2   Single transmit antenna selection

In this section, we focus on the traditional/fundamental TAS strategy in MIMO wiretap channels, in which only a single antenna is selected as active at the transmitter. We
refer to this antenna selection strategy as single TAS. We consider a general MIMO
wiretap channel as shown in Figure 4.1, in which the transmitter (Alice) is equipped
with N_A antennas, the intended receiver (Bob) is equipped with N_B antennas, and the
eavesdropper (Eve) is equipped with N_E antennas. In this MIMO wiretap channel,
we consider a passive eavesdropping scenario, where there is no CSI feedback from
Eve to Alice. As such, the CSI of the eavesdropper's channel (i.e. the channel from
Alice to Eve) is not known at Alice. Alice encodes her messages and transmits the
resulting codewords to Bob. Eve overhears the information conveyed from Alice to
Bob without inducing any interference in the channel between Alice and Bob (i.e.
the main channel) [18,19]. We assume that both the main channel and the eavesdropper's channel experience independent and slow block fading where the fading
coefficients are invariant during one fading block (or equivalently, the coherence
time of the channel). We also assume that the block length is sufficiently long to
allow for capacity-achieving codes within each block. Furthermore, the main channel
and the eavesdropper's channel are assumed to have the same fading block length.

4.2.1   Index of the selected antenna

In the passive eavesdropping scenario, there is no CSI feedback from Eve to Alice,
since Eve does not cooperate with Alice. As such, the selected antenna in single TAS
is the strongest one that maximises the instantaneous signal-to-noise-ratio (SNR)
between Alice and Bob. As Bob is equipped with multiple antennas, different diversity combining techniques lead to different SNRs at Bob. Therefore, the index of the
selected antenna depends on the adopted diversity combining technique at Bob. We

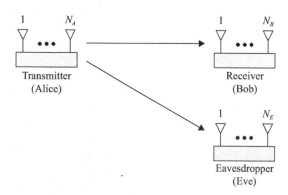

*Figure 4.1   Illustration of a MIMO wiretap channel with N_A, N_B, and N_E antennas
at Alice, Bob, and Eve, respectively*

next detail how to determine this index for three practical diversity combining techniques: maximal-ratio combining (MRC), selection combining (SC), and generalised selection combining (GSC).

MRC is used to coherently combine the received signals when the number of implemented radio frequency (RF) chains is the same as the number of antennas. When MRC is adopted at Bob, the index of the selected antenna is given by [20,21]:

$$n^*_{\text{MRC}} = \underset{1 \leq n \leq N_A}{\text{argmax}} \left\| \mathbf{h}_{n,B} \right\|, \tag{4.1}$$

where $\mathbf{h}_{n,B}$ denotes the channel vector between the nth transmit antenna at Alice and the N_B antennas at Bob and is given by $\mathbf{h}_{n,B} = [h_{n,1}, h_{n,2}, \ldots, h_{n,N_B}]^T$. Here, $[\cdot]^T$ denotes the transpose operation and $\| \cdot \|$ denotes the Euclidean norm.

SC is used to select the received signal with the highest instantaneous SNR when there is only one RF chain (because of size and complexity constraints). As such, when SC is adopted at Bob, the index of the selected antenna is given by [20,21]:

$$n^*_{\text{SC}} = \underset{1 \leq n \leq N_A, 1 \leq m \leq N_B}{\text{argmax}} \left| h_{n,m} \right|, \tag{4.2}$$

where m denotes the antenna index at Bob.

GSC is used to select and combine the signals of L_B strongest antennas out of the available antennas when the number of implemented RF chains is L_B for $1 \leq L_B \leq N_B$. When GSC is adopted at Bob, the received signals at the L_B strongest antennas (out of the N_B available antennas) are combined. On the basis of the rules of GSC, let $\left| h_{n,1} \right|^2 \geq \left| h_{n,2} \right|^2 \geq \cdots \geq \left| h_{n,N_B} \right|^2$ be the order statistics from arranging $\{ \left| h_{n,m} \right|^2 \}_{m=1}^{N_B}$ in descending order of magnitude. Combining the first L_B variable(s) in the order statistics, Bob obtains $\theta_n = \sum_{m=1}^{L_B} \left| h_{n,m} \right|^2$. As such, when GSC is adopted at Bob, the index of the selected antenna is given by [21,22]:

$$n^*_{\text{GSC}} = \underset{1 \leq n \leq N_A}{\text{argmax}} \{ \theta_n \}. \tag{4.3}$$

Among the aforementioned three diversity combining techniques, MRC is the optimal in terms of maximising the SNR at Bob while its complexity is the highest (i.e. it requires N_B RF chains at Bob). In contrast, SC is the one with worst performance but lowest complexity. GSC is a generalised diversity combining technique, which covers MRC ($L_B = N_B$) and SC ($L_B = 1$) as special cases. We note that the diversity combining technique used at Eve depends on the number of RF chains at Eve. MRC is optimal at Eve because the benefits of multiple antennas are fully exploited, which leads to the fact that the probability of successful eavesdropping is maximised.

4.2.2 Secrecy performance metrics

The achievable secrecy rate C_S in a wiretap channel is expressed as [4]:

$$C_s = \begin{cases} C_B - C_E, & \gamma_B > \gamma_E \\ 0, & \gamma_B \leq \gamma_E, \end{cases} \tag{4.4}$$

where $C_B = \log_2(1 + \gamma_B)$ is the capacity of the main channel, and $C_E = \log_2(1 + \gamma_E)$ is the capacity of the eavesdropper's channel. In (4.4), γ_B and γ_E denote the instantaneous SNRs of the main channel and the eavesdropper's channel, respectively.

In the wiretap channel with TAS, Alice has no CSI about the main channel since Bob only feeds back the strongest antenna index to Alice. For passive eavesdropping, Alice and Bob have no CSI about the eavesdropper's channel. As such, Alice sets a constant secrecy rate R_s to transmit confidential information to Bob. Therefore, the secrecy outage probability is adopted as the main performance metric in the context of physical layer security with TAS [17,23–27], which is given by:

$$P_{out}(R_s) = \Pr(C_s < R_s).\tag{4.5}$$

Specifically, the secrecy outage probability is the probability that either there is an outage between Alice and Bob (i.e. the conventional outage probability where the message is not decodable at Bob) or Eve can eavesdrop on data such that perfect secrecy is compromised. We note that Bob may also have to feedback the capacity of the main channel corresponding to the strongest antenna in order to enable Alice to adopt different coding schemes (e.g. [26]).

In order to seek simplicity, the asymptotic secrecy outage probability in the high SNR regime with $\overline{\gamma}_B \to \infty$ (where $\overline{\gamma}_B$ is the average SNR of the main channel) is widely used to illustrate the benefits of TAS in MIMO wiretap channels. The asymptotic secrecy outage probability $P_{out}^{\infty}(R_s)$ can be written as [20]:

$$P_{out}^{\infty}(R_s) = \left(\Psi\overline{\gamma}_B\right)^{-\Phi} + o\left(\overline{\gamma}_B^{-\Phi}\right),\tag{4.6}$$

where $o(\cdot)$ denotes higher order terms. The asymptotic outage probability facilitates valuable insights via the secrecy diversity order Φ, which determines the slope of the asymptotic outage probability curve versus $\overline{\gamma}_B$, and the secrecy array gain Ψ, which characterises the SNR advantage of the asymptotic outage probability relative to the reference curve $\overline{\gamma}_B^{-\Phi}$.

Besides the secrecy outage probability, the probability of non-zero secrecy capacity and the ε-outage secrecy capacity are also adopted to evaluate the secrecy performance of TAS. The probability of non-zero secrecy capacity is defined as the probability that the secrecy capacity is larger than zero, which is given by

$$P_{non} = \Pr(C_s > 0) = \Pr(\gamma_B > \gamma_E).\tag{4.7}$$

We note that the probability of non-zero secrecy capacity is a special case of the secrecy outage probability, i.e. $P_{non} = P_{out}(0)$. The ε-outage secrecy capacity is defined as the maximum secrecy rate at which the secrecy outage probability is no larger than ε, which is given by

$$C_{out}(\varepsilon) = \underset{P_{out}(R_s)\leq\varepsilon}{\mathrm{argmax}}\ R_s.\tag{4.8}$$

4.2.3   Secrecy performance of single TAS

In this sub-section, we first outline the main steps to derive the secrecy outage probability and then refer the reader to related works for detailed derivations. We also summarise the main results on the secrecy performance of single TAS with different diversity combining techniques and in different fading channels.

Following from (4.5), the secrecy outage probability can be written as

$$P_{\text{out}}(R_s) = \underbrace{\Pr(C_s < R_s | \gamma_B > \gamma_E)\Pr(\gamma_B > \gamma_E)}_{V_1}$$

$$+ \underbrace{\Pr(C_s < R_s | \gamma_B < \gamma_E)}_{V_2}\underbrace{\Pr(\gamma_B < \gamma_E)}_{V_3}, \tag{4.9}$$

where V_1 is

$$V_1 = \int_0^\infty \int_{\gamma_E}^{2^{R_s}(1+\gamma_E)-1} f_{\gamma_E}(\gamma_E)f_{\gamma_B}(\gamma_B)d\gamma_B d\gamma_E$$

$$= \underbrace{\int_0^\infty f_{\gamma_E}(\gamma_E)\left[\int_0^{2^{R_s}(1+\gamma_E)-1} f_{\gamma_B}(\gamma_B)d\gamma_B\right]d\gamma_E}_{U_1}$$

$$- \underbrace{\int_0^\infty f_{\gamma_E}(\gamma_E)\left[\int_0^{\gamma_E} f_{\gamma_B}(\gamma_B)d\gamma_B\right]d\gamma_E}_{U_2}. \tag{4.10}$$

We note that $V_2 = 1$ since $C_s = 0$ when $\gamma_B < \gamma_E$, and we also have

$$V_3 = \int_0^\infty \int_0^{\gamma_E} f_{\gamma_E}(\gamma_E)f_{\gamma_B}(\gamma_B)d\gamma_B d\gamma_E = U_2. \tag{4.11}$$

As such, $P_{\text{out}}(R_s) = U_1$, which leads to

$$P_{\text{out}}(R_s) = \int_0^\infty f_{\gamma_E}(\gamma_E)\left[\int_0^{2^{R_s}(1+\gamma_E)-1} f_{\gamma_B}(\gamma_B)d\gamma_B\right]d\gamma_E. \tag{4.12}$$

In (4.10)–(4.12), $f_{\gamma_B}(\cdot)$ and $f_{\gamma_E}(\cdot)$ denote the probability density functions (PDFs) of γ_B and γ_E, respectively. Therefore, we can derive the secrecy outage probability through first determining the PDFs of both γ_B and γ_E, and then substituting these PDFs into (4.12).

Instead of repeating the derivations of the secrecy outage probability, we refer the reader to the following related works for the detailed secrecy outage probabilities of single TAS in different application scenarios. The secrecy outage probability of single TAS was initially derived in a multiple-input–single-output (MISO) Rayleigh fading wiretap channel, where Alice and Eve were equipped with multiple antennas while Bob was equipped with a single antenna, and all channels were subject to quasi-static Rayleigh fading [16,17]. Then, the secrecy outage probability was generalised into a MIMO Nakagami fading wiretap channel, where all terminals were equipped with multiple antennas and all channels were subject to Nakagami fading [20]. In [20], the secrecy diversity order and the secrecy array gain were initially introduced to evaluate the secrecy performance of single TAS, and two different diversity combining

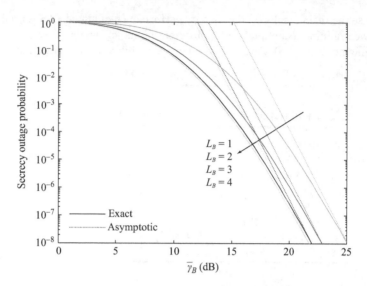

Figure 4.2   The secrecy outage probability of single TAS versus $\overline{\gamma}_B$ for $N_A = 2$,
$N_B = 4$, $\overline{\gamma}_E = 5$ dB, $N_E = 3$, and $L_E = 2$

techniques (i.e. SC and MRC) were considered at both Bob and Eve. The generalised diversity combining technique (i.e. GSC) was considered in the context of single TAS and the associated secrecy outage probability was derived for Rayleigh fading channels in [22]. Furthermore, the secrecy outage probability of single TAS with GSC at Bob was generalised into Nakagami fading wiretap channels by [28]. Finally, the secrecy performance of single TAS in a MIMO wiretap channel with large-scale fading was examined [29]. We note that the secrecy performance of single TAS in cooperative (e.g. relaying) networks was also examined in the literature (e.g. [30,31]).

We next provide some main results on the secrecy performance of single TAS. In Figure 4.2, we plot the secrecy outage probability of single TAS with GSC at both Bob and Eve (i.e. the received signals at the L_E strongest antennas are combined at Eve) for a Rayleigh fading wiretap channel as a function of $\overline{\gamma}_B$. In this figure, we first observe that the secrecy outage diversity gain is $N_A N_B = 8$ regardless of L_B. This means that single TAS with different diversity combining techniques (i.e. MRC, SC, and GSC) can achieve the full secrecy diversity order (i.e. $N_A N_B$ in the Rayleigh fading wiretap channel). We also observe that the secrecy outage probability improves with increasing L_B, which also means that MRC ($L_B = 4$) outperforms SC ($L_B = 1$). This can be explained by the analysis in [22], which is detailed as follows. Assuming that GSC is adopted at the eavesdropper, the SNR gap between TAS/GSC (i.e. single TAS with GSC at Bob) and TAS/SC (i.e. single TAS with SC at Bob) in the main channel is $\Delta_1 = (10/N_B) \log \left(L_B^{N_B - L_B} L_B! \right)$ dB. We confirm that $\Delta_1 > 0$ and the SNR gap increases as L_B increases. We also present the SNR

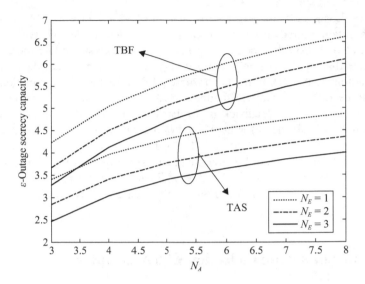

Figure 4.3 Comparison of ε-outage secrecy rate between TAS and TBF versus N_A for $\overline{\gamma}_B = 20$ dB, $N_B = 1$, $\overline{\gamma}_E = 0$ dB, and $\epsilon = 0.01$

gap between TAS/GSC and TAS/MRC (i.e. single TAS with MRC at Bob) as $\Delta_2 = (10/N_B) \log \left(L_B^{N_B - L_B} L_B!/N_B! \right)$ dB. We confirm that the SNR gap decreases as L_B increases. We highlight that TAS/GSC brings a significant SNR advantage relative to TAS/SC. We also highlight that TAS/GSC provides comparable secrecy outage to TAS/MRC. Since GSC has a lower complexity than MRC and a higher complexity than SC, this figure confirms that TAS/GSC provides a cost-performance trade-off in physical layer security enhancement.

In summary, we repeat the main results on the secrecy performance of single TAS as follows.

- Single TAS always achieves the maximum secrecy diversity order regardless of the diversity combining techniques adopted at Bob.
- The maximum secrecy diversity order of a wiretap channel is fully determined by the main channel and is independent of the eavesdropper's channel, which is $N_A N_B$ for the Rayleigh fading wiretap channel and $N_A N_B m_B$ for the Nakagami-fading wiretap channel, where m_B is the Nakagami-fading parameter of the main channel.
- The secrecy outage probability of single TAS decreases as L_B increases when GSC is adopted at Bob. This means TAS/MRC is the best in terms of minimising the secrecy outage probability.

We next provide a comparison between single TAS and the transmit beamforming (TBF) in Figure 4.3. As expected, in this figure, we observe that the ε-outage secrecy capacity increases as N_A increases for both single TAS and TBF. Notably, we observe that the ε-outage secrecy capacity of single TAS is close to that of TBF when N_A is small. We also observe that the rate advantage of TBF over single TAS increases with N_A. We note that single TAS only requires $\lceil \log_2 N_A \rceil$ bits to feedback the index of the selected antennas from Bob to Alice (where $\lceil \cdot \rceil$ denotes the ceiling function), while TBF requires Bob to accurately feedback $N_A N_B$ complex numbers to Alice, which costs significantly more feedback bits relative to single TAS. The extra feedback bits required by TBF increases as N_A or N_B increases. As such, we can conclude that the secrecy performance gain of TBF relative to single TAS comes at the cost of significantly higher feedback overhead and signal processing complexity (e.g. only one antenna is active in single TAS, while all N_A antennas are active in TBF). Importantly, the feedback overhead for TBF increase with N_B, while that for TAS remains unchanged.

4.3   Transmit antenna selection with Alamouti coding

Considering the same MIMO wiretap channel as Section 4.2, we focus on a new TAS scheme which examines the trade-off between feedback overhead and secrecy performance. The new TAS scheme is carried out in two steps. First, the transmitter selects the first two strongest antennas to maximise the instantaneous SNR of the main channel. Second, Alamouti coding is employed at the two selected antennas in order to perform secure data transmission. When equal power is applied to the two selected antennas, we refer to our new scheme as TAS-Alamouti. We also show how the optimal power allocation (OPA) across the two selected antennas at Alice leads to a new scheme, which we refer to as TAS-Alamouti-OPA, that outperforms single TAS and TAS-Alamouti.

4.3.1   Indices of the two selected antennas

As detailed in Section 4.2, among different diversity combining techniques, MRC is the best in terms of minimising the secrecy outage probability. As such, in this section, we consider MRC at Bob for both TAS-Alamouti and TAS-Alamouti-OPA. Given that Bob employs MRC to combine the received signals, the index of the first strongest antenna is given by [32,33]:

$$n_1^* = \operatorname*{argmax}_{0 \le n \le N_A} \left\| \mathbf{h}_{n,B} \right\|, \tag{4.13}$$

and the index of the second strongest antenna is determined by [32,33]:

$$n_2^* = \operatorname*{argmax}_{0 \le n \le N_A, n \ne n_1^*} \left\| \mathbf{h}_{n,B} \right\|. \tag{4.14}$$

To conduct transmit antenna selection, Alice sends Bob pilot symbols prior to data transmission. Using these symbols, Bob determines the CSI of the main channel and determines n_1^* and n_2^* according to (4.13) and (4.14), respectively. After this, Bob feeds back n_1^* and n_2^* to Alice via a low-rate feedback channel. As such, TAS-Alamouti reduces the feedback overhead compared with TBF, since only $\left\lceil \log_2 \frac{N_A(N_A-1)}{2} \right\rceil$ bits are required to feedback the antenna indices. Compared with single TAS, TAS-Alamouti requires $\left(\left\lceil \log_2 \frac{N_A(N_A-1)}{2} \right\rceil - \lceil \log_2 N_A \rceil \right)$ extra feedback bits. For example, when $N_A = 3$, TAS-Alamouti requires no extra feedback bit. For $4 \leq N_A \leq 6$, TAS-Alamouti requires only one extra feedback bit. We note that the antenna indices n_1^* and n_2^* are entirely dependent on the main channel. Due to the independence of the main channel and the eavesdropper's channel, it follows that TAS-Alamouti scheme improves the quality of main channel relative to the eavesdropper's channel, which in turn promotes the secrecy of the MIMO wiretap channel.

4.3.2 Transmission with Alamouti coding

After selecting the two strongest antennas, Alice adopts Alamouti coding to perform secure transmission. During the transmission, Alice allocates a percentage α of its total transmit power to the first strongest antenna and allocates a percentage β of its total transmit power to the second strongest antenna. Due to the total power constraint, we have $\beta = 1 - \alpha$.

As per the rules of Alamouti coding, the $N_B \times 1$ received signal vectors at Bob in the first and second time slots are given by [32,33]:

$$\mathbf{y}_B(1) = \left[\sqrt{\alpha}\mathbf{h}_{n_1^*,B}, \sqrt{\beta}\mathbf{h}_{n_2^*,B} \right] \begin{bmatrix} x_1 \\ x_2 \end{bmatrix} + \mathbf{w}(1), \tag{4.15}$$

and

$$\mathbf{y}_B(2) = \left[\sqrt{\alpha}\mathbf{h}_{n_1^*,B}, \sqrt{\beta}\mathbf{h}_{n_2^*,B} \right] \begin{bmatrix} -x_2^\dagger \\ x_1^\dagger \end{bmatrix} + \mathbf{w}(2), \tag{4.16}$$

respectively, where $\left[\mathbf{h}_{n_1^*,B}, \mathbf{h}_{n_2^*,B} \right]$ is the $N_B \times 2$ main channel matrix after TAS, $[x_1, x_2]^T$ is the transmit signal vector in the first-time slot, $[-x_2^\dagger, x_1^\dagger]^T$ is the transmit signal vector in the second time slot, $\mathbf{w}$ is the zero-mean circularly symmetric complex Gaussian noise vector satisfying $\mathbb{E}[\mathbf{w}\mathbf{w}^\dagger] = \mathbf{I}_{N_B}\sigma_{AB}^2$, σ_{AB}^2 is the noise variance for each receive antenna at Bob, and $\mathbb{E}[\cdot]$ denotes expectation. Under the power constraint, we have $\mathbb{E}[|x_1|^2] = \mathbb{E}[|x_2|^2] = P_A$, where P_A is the total transmit power at Alice.

By performing MRC and space-time signal processing, the signals containing x_1 and x_2 at Bob can be expressed as [32,33]:

$$y_B(x_1) = \left(\alpha \mathbf{h}_{n_1^*,B}^\dagger \mathbf{h}_{n_1^*,B} + \beta \mathbf{h}_{n_2^*,B}^\dagger \mathbf{h}_{n_2^*,B} \right) x_1$$
$$+ \sqrt{\alpha}\mathbf{h}_{n_1^*,B}^\dagger \mathbf{w}(1) + \sqrt{\beta}\mathbf{w}(2)^\dagger \mathbf{h}_{n_2^*,B}, \tag{4.17}$$

and

$$y_B(x_2) = \left(\alpha \mathbf{h}_{n_1^*,B}^\dagger \mathbf{h}_{n_1^*,B} + \beta \mathbf{h}_{n_2^*,B}^\dagger \mathbf{h}_{n_2^*,B} \right) x_2$$
$$+ \sqrt{\alpha}\mathbf{h}_{n_2^*,B}^\dagger \mathbf{w}(1) - \sqrt{\beta}\mathbf{w}(2)^\dagger \mathbf{h}_{n_1^*,B}, \tag{4.18}$$

respectively. The instantaneous SNR at Bob is written as

$$\gamma_B = \frac{(\alpha \|\mathbf{h}_{n_1^*,B}\|^2 + \beta \|\mathbf{h}_{n_2^*,B}\|^2) P_A}{\sigma_{AB}^2}. \tag{4.19}$$

Likewise, the instantaneous SNR at Eve is written as

$$\gamma_E = \frac{(\alpha \|\mathbf{g}_{n_1^*,E}\|^2 + \beta \|\mathbf{g}_{n_2^*,E}\|^2) P_A}{\sigma_{AE}^2}, \tag{4.20}$$

where $[\mathbf{g}_{n_1^*,E}, \mathbf{g}_{n_2^*,E}]$ is the $N_E \times 2$ eavesdropper's channel matrix after TAS and σ_{AE}^2 is the noise variance for each receive antenna at Eve.

4.3.3 Secrecy performance of TAS-Alamouti and TAS-Alamouti-OPA

In TAS-Alamouti, equal power is allocated to the selected two antennas, i.e. $\alpha = \beta = 0.5$. The secrecy outage probability of TAS-Alamouti was derived in [32,33]. To determine the optimal power allocation for TAS-Alamouti, a closed-form expression for the secrecy outage probability with general power allocation, i.e. $0.5 \leq \alpha \leq 1$, was derived in [33] as well. On the basis of this expression, the optimal power allocation for TAS-Alamouti was determined and the secrecy performance of TAS-Alamouti-OPA was achieved. In TAS-Alamouti-OPA, the optimal α that minimises the secrecy outage probability is determined at Alice based on the knowledge of $\overline{\gamma}_B$ and $\overline{\gamma}_E$. In order to determine the optimal α, Alice requires to know her strongest and second strongest antennas. This is different from TAS-Alamouti in which Alice only requires to know her two strongest antennas, but does not need to know which of these two is the strongest. As such, TAS-Alamouti-OPA requires one extra feedback bit relative to TAS-Alamouti.

We next provide some numerical results on the secrecy performance of TAS-Alamouti and TAS-Alamouti-OPA. Figure 4.4 plots the secrecy outage probability versus $\overline{\gamma}_B$ for different values of N_E, where $P_{\text{out}}(R_s)$ and $P_{\text{out}}^{\infty}(R_s)$ denote the exact and asymptotic secrecy outage probabilities, respectively. In this figure, we first observe that the asymptotic secrecy outage curves of single TAS and TAS-Alamouti are parallel, which demonstrates that TAS-Alamouti achieves the same secrecy diversity order as single TAS (i.e. full secrecy diversity order $N_A N_B$). We also observe that TAS-Alamouti has an SNR gain relative to single TAS at the same secrecy outage probability. This SNR gain is due to the fact that TAS-Alamouti has a higher secrecy array gain than single TAS. Notably, this SNR gain increases with N_A. We further observe that the *cross-over point* at which TAS-Alamouti and single TAS achieve the same secrecy outage probability moves to higher $\overline{\gamma}_B$ when N_E increases. Finally, we observe that although $P_{\text{out}}(R_s)$ increases as N_E increases, the asymptotic curves of TAS-Alamouti for different values of N_E are parallel, which confirms that the secrecy diversity order is not affected by N_E. The aforementioned observations demonstrate the advantage of TAS-Alamouti over single TAS.

Figure 4.5 plots the secrecy outage probabilities of single TAS, TAS-Alamouti, and TAS-Alamouti-OPA. In this figure, we first observe that TAS-Alamouti-OPA always achieves a lower secrecy outage probability than TAS-Alamouti. We also

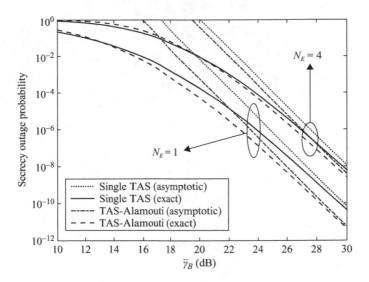

Figure 4.4   The secrecy outage probability versus $\overline{\gamma}_B$ for $R_s = 1$, $\overline{\gamma}_E = 10\,dB$, $N_A = 4$, and $N_B = 2$

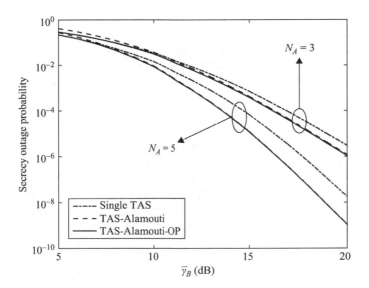

Figure 4.5   Secrecy outage probability versus $\overline{\gamma}_B$ for $R_s = 1$, $\overline{\gamma}_E = 5\,dB$, and $N_B = 2$

observe that TAS-Alamouti-OPA always achieves a better secrecy performance than single TAS regardless of the values of $\overline{\gamma}_B$. We note that as $\overline{\gamma}_B \to 0$, the secrecy performance of TAS-Alamouti-OPA becomes the same as that of single TAS, which means that as $\overline{\gamma}_B \to 0$, all the transmit power at Alice is allocated to the first

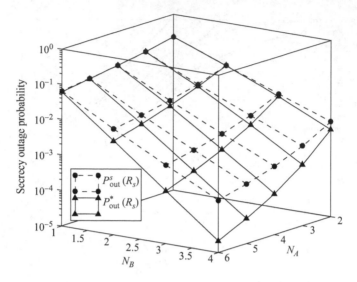

Figure 4.6   *Secrecy outage probability versus N_A and N_B for $R_s = 1$, $\overline{\gamma}_B = 15$ dB, $\overline{\gamma}_E = 10$ dB, and $N_E = 1$*

strongest antenna. Furthermore, we observe that as $\overline{\gamma}_B$ increases, the secrecy performance gap between TAS-Alamouti and TAS-Alamouti-OPA disappears, which demonstrates that the transmit power at Alice is equally allocated to the selected two strongest antennas in the high regime of $\overline{\gamma}_B$. Finally, the secrecy outage probabilities of single TAS, TAS-Alamouti, and TAS-Alamouti-OPA are parallel in the high regime of $\overline{\gamma}_B$, which confirms that they all achieve the full secrecy diversity order.

Figure 4.6 presents 3D plots to compare the secrecy outage probability of single TAS with that of TAS-Alamouti-OPA in order to examine the joint impact of N_A and N_B on the secrecy performance. In Figure 4.6, $P_{out}^s(R_s)$ is the secrecy outage probability of single TAS, and $P_{out}^*(R_s)$ is the secrecy outage probability of TAS-Alamouti-OPA. In Figure 4.6, we first observe that when $N_B = 1, N_A < 6$ or $N_A = 2, N_B < 3$, TAS-Alamouti-OPA does not achieve a significantly lower secrecy outage probability relative to single TAS. However, when $N_A = 3$, as long as $N_B > 2$ TAS-Alamouti-OPA achieves a notable advantage over single TAS. We note that, when $N_A = 3$, TAS-Alamouti-OPA requires only 1 extra feedback bit relative to single TAS. We also observe that the gap between single TAS and TAS-Alamouti-OPA increases as N_A or N_B increases. We also note that the extra feedback overhead required by TAS-Alamouti or TAS-Alamouti-OPA is not a function of N_B.

In summary, we repeat the main results on the secrecy performance of TAS-Alamouti and TAS-Alamouti-OPA with single TAS as the benchmark as follows.

- TAS-Alamouti outperforms the traditional single TAS conditioned on the SNR of the main channel being larger than a specific value. The extra feedback bits required by TAS-Alamouti relative to single TAS is given by $\left(\left\lceil \log_2 \frac{N_A(N_A-1)}{2} \right\rceil - \lceil \log_2 N_A \rceil \right)$, which is 0 when $N_A = 3$.

- TAS-Alamouti-OPA outperforms single TAS unconditionally, which requires 1 additional feedback bit relative to TAS-Alamouti.
- Both TAS-Alamouti and TAS-Alamouti-OPA achieve the full secrecy diversity order as single TAS. The secrecy performance gap between TAS-Alamouti-OPA and single TAS increases as N_A or N_B increases, while this gap decreases as N_E increases. The extra feedback overhead required by TAS-Alamouti or TAS-Alamouti-OPA is independent of N_B.

We note that the Alamouti code is the only space time block code (STBC) that achieves full rate and full diversity with linear receiver algorithms. The selection of more than two antennas and an appropriate STBC can achieve a lower secrecy outage probability at the cost of reducing the rate (or increasing the decoding complexity). This is the main reason that we have focused on TAS-Alamouti scheme with two transmit antennas. We also note that the secrecy performance of a space-time transmission (without antenna selection) was shown to be worse than that of TAS in [34]. Besides STBC, the secrecy performance of space-time network coding with TAS was also examined in the literature (e.g. [35,36]).

In future, it is possible to explore the trade-off between the feedback overhead and secrecy performance further. For example, one could probe the use of additional antennas combined with other coding schemes, at an increased cost in feedback overhead. Potential coding schemes that could be considered in this context include those discussed in [37,38]. Another potential avenue of research along these lines could be a system in which Bob quantises his CSI feedback using a predetermined number of bits. This estimate of the CSI could be modelled as quantisation error on the true CSI of the main channel. Future work in this area may wish to consider such possibilities.

4.4 Antenna selection in full-duplex wiretap channels

In this section, we consider antenna selection in a full-duplex wiretap channel, in which Bob operates in a full-duplex mode such that he simultaneously receives signals from Alice and transmits jamming signals to confuse Eve. Bob's antennas have to be divided into two groups: transmit and receive. We will discuss how to determine the size of each group and how to perform the division between transmit and receive antennas.

4.4.1 Transmit and receive antenna switching

In this sub-section, we focus on a specific full-duplex wiretap channel as shown in Figure 4.7, where Alice and Eve are equipped with a single antenna whereas Bob is equipped with two antennas. We assume that Bob operates in the full-duplex mode such that Bob uses one antenna to receive signals from Alice and uses the other

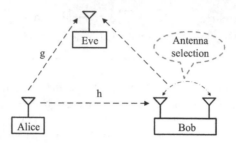

Figure 4.7   A full-duplex wiretap channel where Bob operates in the full-duplex mode and applies antenna switching

antenna to transmit jamming signals in order to deliberately confuse Eve. We further assume that the functions of Bob's two antennas are not predetermined. In particular, Bob selects the receive antenna out of the two antennas and utilises the remaining antenna as the transmit antenna.

In the passive eavesdropping scenario, the instantaneous CSI of the jamming channel (i.e. the channel from Bob to Eve) is not available at Bob, and thus Bob selects the receive antenna and the transmit antenna based on the knowledge of the instantaneous CSI of the main channel. Specifically, Bob selects the antenna that maximises the main channel gain as the receive antenna, and selects the remaining antenna as the transmit antenna. As such, the index of the receive antenna at Bob is determined through:

$$n^* = \underset{n \in \{1,2\}}{\operatorname{argmax}} \left| h_{1,n} \right|. \tag{4.21}$$

Since we consider a block fading channel, Bob only needs to switch his antenna at a frequency that is dependent on the covariance time of the main channel. The received signal at Bob is given by

$$y_B = h_{1,n^*} x + \sqrt{\varphi} f_s u + w, \tag{4.22}$$

where x denotes the transmit signal satisfying the transmit power constraint at Alice given by $\mathbb{E}[|x|^2] = P_A$ (P_A is the transmit power of Alice), f_s denotes the complex channel gain of the self-interference channel (i.e. the channel from Bob's transmit antenna to his receive antenna), φ denotes the residual self-interference cancellation parameter [39,40], u is the zero-mean complex Gaussian jamming signal satisfying the transmit power constraint at Bob given by $\mathbb{E}[|u|^2] = P_B$ (P_B is the transmit power of Bob), and w is the additive white complex Gaussian noise at Bob with zero mean and variance σ_{AB}^2. On the basis of the definition of φ, we note that $\sqrt{\varphi} f_s$ is the effective complex channel gain of the self-interference channel after performing self-interference cancellation. On the basis of (4.22), the SINR at Bob is given by

$$\gamma_B = \frac{\left| h_{1,n^*} \right|^2 P_A}{\varphi |f_s|^2 P_B + \sigma_{AB}^2}. \tag{4.23}$$

Likewise, the SINR at Eve is given by

$$\gamma_E = \frac{|g_{1,1}|^2 P_A}{|f_j|^2 P_B + \sigma_{AE}^2},\tag{4.24}$$

where f_j denotes the jamming channel.

The secrecy outage probability of the full-duplex wiretap channel with antenna switching for a given P_B was derived in [41]. In the full-duplex wiretap channel, the self-interference between Bob's transmit and receive antennas cannot be totally cancelled, i.e. $\varphi \neq 0$. This leads to the fact that the SINR of the main channel can be decreased if Bob transmits jamming signals with a higher power. Such a higher power, nevertheless, depresses the SINR of the main channel. Therefore, there exists a trade-off between reducing γ_E and depressing γ_B when Bob adjusts the value of P_B. We define the optimal value of P_B as the one that minimises the secrecy outage probability for a given $\overline{\gamma}_B$ and $\overline{\gamma}_E$. Mathematically, this is determined through:

$$P_B^* = \underset{P_B \geq 0}{\arg\min}\, P_{\text{out}}\,(R_s)\,.\tag{4.25}$$

4.4.2 Secrecy performance of the full-duplex wiretap channel with antenna switching

In this sub-section, we provide some numerical comparisons between the half-duplex wiretap channel and full-duplex wiretap channel with and without antenna switching.

Figure 4.8 plots the secrecy outage probability versus φ. We note that the secrecy outage probability of full-duplex wiretap channel without antenna switching was also derived in [41] and the secrecy outage probability of the half-duplex wiretap channel was obtained from [24, Eq. (6)]. In the half-duplex wiretap channel considered by [24], Bob's two antennas are both used for receiving signals and MRC is adopted to combine the received signals. In this figure, we first observe that both the secrecy outage probabilities of the full-duplex wiretap channel with and without antenna switching increase as φ increases (in this figure, the transmit power at Bob has been optimised). This can be explained by the fact that a smaller φ indicates less self-interference at Bob. We also observe that the full-duplex wiretap channel with antenna switching outperforms the full-duplex wiretap channel without antenna switching, which demonstrates the benefits of introducing antenna selection into full-duplex wiretap channels. Furthermore, we observe that the secrecy outage probability of the full-duplex wiretap channel with antenna switching is lower than that of the half-duplex wiretap channel when φ is less than some specific value. This observation indicates that a lower secrecy outage probability can be achieved in the full-duplex wiretap channel relative to the half-duplex channel. Also, this observation demonstrates that the lower secrecy outage probability is achieved by the full-duplex wiretap channel with and without antenna switching only when the self-interference cancellation satisfies specific requirements.

In Figure 4.9, we plot the secrecy outage probability versus $\overline{\gamma}_B$, where $\overline{\gamma}_S$ and $\overline{\gamma}_J$ are the average SNRs of the self-interference channel and jamming channel, respectively. As expected, we observe that the secrecy outage probabilities of the half-duplex

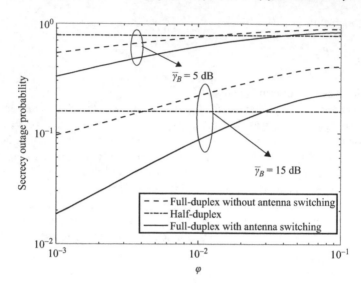

Figure 4.8   Secrecy outage probability versus φ for $\overline{\gamma}_E = 15$ dB, $\overline{\gamma}_S/\overline{\gamma}_J = 10$ dB, and $R_s = 1$ bits/channel

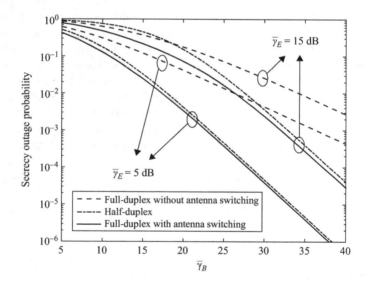

Figure 4.9   Secrecy outage probability versus $\overline{\gamma}_B$ for $\overline{\gamma}_S/\overline{\gamma}_J = 10$ dB, $\varphi = 0.01$, $R_s = 1$ (bits/channel), and different values of $\overline{\gamma}_E$

wiretap channel and the full-duplex wiretap channel with and without antenna switching decrease as $\overline{\gamma}_B$ increases. The slope of the secrecy outage probability curve in Figure 4.9 is known as the secrecy diversity order as detailed in Section 4.2. In this figure, we observe that the secrecy diversity order of the full-duplex wiretap channel with antenna switching is 2, which is the same as that of the half-duplex wiretap channel. Notably, this diversity order is higher than that of the full-duplex wiretap channel without antenna switching. We also observe that the secrecy performance gap between the full-duplex wiretap channel with antenna switching and the half-duplex wiretap channel increases as $\overline{\gamma}_B$ decreases. This can be explained by the fact that the SNR reduction, due to the loss of one receive antenna in the main channel, decreases as $\overline{\gamma}_B$ decreases. We further observe that the secrecy performance gain of the full-duplex wiretap channel with antenna switching relative to that without antenna switching increases as $\overline{\gamma}_E$ increases.

In summary, we repeat the main results on the secrecy performance of the full-duplex wiretap channel with antenna switching as follows:

- The full-duplex wiretap channel with antenna switching outperforms that without antenna switching in terms of achieving a lower secrecy outage probability.
- The full-duplex wiretap channel with antenna switching can still outperform the half-duplex wiretap channel, even when the full-duplex wiretap channel without antenna switching cannot.
- The full-duplex wiretap channel with antenna switching achieves the full secrecy diversity order of 2, which is the same as that of the half-duplex wiretap channel. Notably, this diversity order is higher than that of the full-duplex wiretap channel without antenna switching.

4.4.3 Other antenna selection problems in full-duplex wiretap channels

In this sub-section, we review and discuss some other antenna selection problems in full-duplex wiretap channels.

Considering multiple antennas at Alice, Bob, and Eve, the antenna allocation problem at Bob when the total number of antennas for transmitting and receiving at Bob was fixed was studied in [42]. The authors focused on the high SNR regimes and adopted the maximum achievable secure degrees of freedom (SDF) as the performance metric. As such, the antenna allocation considered in [42] did not depend on any CSI, and thus, only the number of receive or transmit antennas had to be determined. The optimal number of antennas allocated for receiving at Bob the maximises the SDF was derived in [42] as

$$N_r = \min\left\{ \left\lfloor \frac{N_A + N_B - N_E}{2} \right\rfloor^+, N_B, N_A \right\}. \tag{4.26}$$

The analysis in [42] showed that the full-duplex wiretap channel achieves a higher SDF than the corresponding half-duplex wiretap channel for the following two cases: (i) $N_A \leq N_E$; and (ii) $N_A > N_E$ with $N_B > N_A - N_E$. In these two cases, $N_r < N_B$, i.e. Bob operates in the full-duplex mode. For case (iii), $N_A > N_E$ and $N_B \leq N_A - N_E$, the full-duplex and half-duplex wiretap channels achieve the same SDF. In case (iii), $N_r = N_B$, i.e. Bob operates in the half-duplex mode and all his antennas are used for receiving.

In [43], the antenna selection at a full-duplex secondary destination node was considered in the context of secure cognitive radio networks. In this work, the authors assumed that the number of receive antennas and the number of transmit antennas were predetermined and fixed. Then, the optimal single antenna used for receiving signals that maximises the capacity from the secondary source node to the secondary destination node was determined. In addition, two strategies to determine the optimal single antenna used for transmitting AN were proposed. The analysis provided in [43] showed that antenna selection could lead to diversity gain for secure data transmission in the secondary network.

In full-duplex wiretap channels, the self-interference between transmit and receive antennas at Bob has significant impact on the secrecy performance. Many techniques have been developed in the literature to suppress the self-interference [44–46]. Some self-interference cancellation techniques require the CSI of the self-interference channel, which are known as channel-aware self-interference cancellation techniques. Considering channel estimation, the antenna selection problem in full-duplex wiretap channels may have to be revisited. For example, when Bob has limited resources (e.g. transmit power, symbol periods) used for channel estimation and AN transmission, it may not be optimal to utilise all the antennas available at Bob for receiving signals and transmitting AN. Therefore, revisiting antenna selection with channel estimation and self-interference cancellation would be a worthwhile endeavour in full-duplex wiretap channels.

4.5 Single TAS with imperfect feedback and correlation

As discussed in previous sections, the procedure of TAS in wiretap channels required feedback from Bob to Alice, which was assumed to be free of any delay or error. However, in practice this feedback is imperfect with a high probability due to channel uncertainty or channel estimation error [47,48]. As such, in this section, we first review and discuss single TAS with imperfect feedback in the context of wiretap channels. Besides the imperfect feedback, correlation between antennas or channels also have impact on the secrecy performance of single TAS, and thus, we also discuss single TAS with correlation in antennas or channels in this section.

4.5.1 Single TAS with imperfect feedback

In wiretap channels with passive eavesdropping, only Bob feeds back information to Alice. As such, we only consider imperfect feedback of the main channel, which may affect the antenna selection of single TAS (and thus its secrecy performance). With imperfect feedback, the CSI of the main channel for antenna selection is different from that for data transmission. Specifically, we adopt the widely used model of [49] for the imperfect feedback. In this model, we have:

$$\widetilde{\mathbf{h}}_{n,B} = \sqrt{\rho}\mathbf{h}_{n,B} + \sqrt{1-\rho}\mathbf{e},$$

(4.27)

where $\tilde{\mathbf{h}}_{n,B}$ is the CSI when data transmission is performed, $\mathbf{h}_{n,B}$ is the CSI when antenna selection is conducted, $\mathbf{e}$ denotes a random vector that is independent and identically distributed (i.i.d.) as $\mathbf{h}_{n,B}$. If the imperfect feedback is caused by the delay in transmission due to antenna selection, ρ in (4.27) is given by

$$\rho = [J_0(2\pi q_d \tau_d)]^2,\qquad\qquad(4.28)$$

where $J_0(\cdot)$ denotes the zero-order Bessel function of the first kind, q_d is the maximum Doppler frequency, and τ_d is the delayed time. If the imperfect feedback is due to channel estimation errors, this ρ can be determined based on specific channel estimation algorithms. We note that $\rho \in [0, 1]$, $\rho = 0$ represents the scenario where $\tilde{\mathbf{h}}_{n,B}$ is independent of $\mathbf{h}_{n,B}$ (i.e. $\tilde{\mathbf{h}}_{n,B}$ is fully outdated or no channel estimation is conducted for antenna selection), and $\rho = 1$ represents the case where $\tilde{\mathbf{h}}_{n,B}$ and $\mathbf{h}_{n,B}$ are identical (i.e. no time delay or no channel estimation error exists).

In order to examine the impact of imperfect feedback on single TAS, the secrecy performance of single TAS with imperfect feedback has been analysed for different channel models in the literature (e.g. [49–52]). Considering the MISO Rayleigh fading wiretap channel, [49] initially examined the effects of outdated CSI on the secrecy performance of single TAS, in which the exact and asymptotic secrecy outage probabilities were derived. The authors of [50] extended the analysis presented in [49] into the MIMO Rayleigh fading wiretap channel, where MRC was adopted at both Bob and Eve. Then, the authors in [51] further generalised this analysis into the Nakagami MIMO wiretap channel while considering multiple eavesdroppers, in which the average secrecy capacity was also derived when Alice obtained the CSI of the eavesdropper's channel. With imperfect feedback, the data transmission of single TAS may incur connection outage (i.e. Alice transmits but Bob cannot decode the message correctly) in addition to secrecy outage. To avoid the connection outage, the authors in [52] proposed a new secure transmission scheme in the presence of imperfect feedback and antenna correlation.

We next provide some numerical results to illustrate the impact of imperfect feedback on the secrecy performance of single TAS. Figure 4.10 plots the secrecy outage probability of single TAS with imperfect feedback for different values of ρ. As expected, we first observe that the secrecy outage probability increases as ρ decreases, which demonstrates that imperfect feedback leads to a dramatic performance reduction of single TAS. On the basis of the asymptotic curves, we observe that the full secrecy diversity order cannot be achieved when $\rho < 1$ (i.e. imperfect feedback exists). It is interesting to observe that the secrecy diversity order for $\rho = 0$ is the same as that for $\rho = 0.8$ (which is 2 instead of the full secrecy diversity order 6). This is confirmed by the analysis provided in [52], which shows that the secrecy diversity order of single TAS with any imperfect feedback is N_B, not $N_A N_B$.

We further demonstrate the impact of imperfect feedback in Figure 4.11. In this figure, we first observe that the improvement in the secrecy performance brought by the increase in N_A increases as ρ increases and that this improvement completely disappears when $\rho = 0$. This observation indicates that a lower ρ (i.e. more severe imperfect feedback) reduces the benefits of TAS in security enhancement. We note

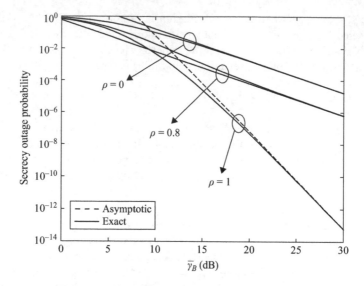

Figure 4.10   *Secrecy outage probability versus $\overline{\gamma}_B$ for $R_s = 1$ (bits/channel), $N_A = 3$, $N_B = N_E = 2$, $\overline{\gamma}_E = 0$ dB*

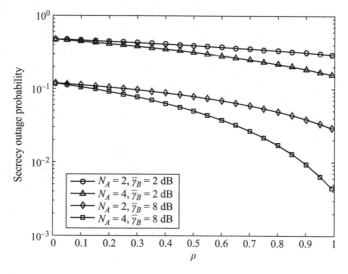

Figure 4.11   *Secrecy outage probability versus ρ for $R_s = 1$ (bits/channel), $N_B = N_E = 3$, $\overline{\gamma}_E = 0$ dB*

that $\rho = 0$ corresponds to the specific case where a random antenna is selected for secure transmission and TAS does not offer secrecy performance gain. As per this observation, we conclude that the improvement brought by TAS to secrecy performance with a higher N_A is sensitive to the accuracy of the feedback.

In summary, we repeat the main results on the impact of imperfect feedback on single TAS as follows:

- Imperfect feedback (e.g. time-delayed feedback, feedback suffering from channel estimation errors) can lead to a dramatic secrecy performance reduction of single TAS. The performance reduction increases as the imperfect feedback parameter ρ increases.
- The secrecy diversity order of single TAS with any imperfect feedback is N_B for Rayleigh fading and $m_B N_B$ for Nakagami fading, which means that no transmitter diversity is achieved.
- The secrecy performance improvement brought by increasing N_A decreases as the feedback becomes more imperfect (i.e. as ρ decreases) and it completely disappears when $\rho = 0$ (i.e. when $\rho = 0$ TAS cannot lead to any benefits).

As discussed in [52], the secrecy performance of wiretap channel with imperfect feedback can be further improved through selecting a subset of antennas at Alice [53]. The specific design required by seeking this improvement is a potential future work. Along this line, whether TAS-Alamouti or TAS-Alamouti-OPA can lead to a higher secrecy diversity order relative to single TAS in wiretap channel with imperfect feedback is a question worth pursuing.

4.5.2 Single TAS with antenna correlation or channel correlation

In this sub-section, we review and discuss the impact of receive antenna correlation and channel correlation on the secrecy performance of single TAS.

The impact of antenna correlation on the secrecy performance of single TAS was initially studied in [54], in which MRC was adopted at both Bob and Eve. With antenna correlation, $\mathbf{h}_{n,B}$ is updated to $\Phi_B^{1/2}\mathbf{h}_{n,B}$, where $\Phi_B^{1/2}$ is the $N_B \times N_B$ antenna correlation matrix at Bob, and $\mathbf{g}_{n,E}$ is updated to $\Phi_E^{1/2}\mathbf{g}_{n,E}$, where $\Phi_E^{1/2}$ is the $N_E \times N_E$ antenna correlation matrix at Eve. The exact and asymptotic secrecy outage probabilities of single TAS with such arbitrary correlations were derived in closed-form expressions [54]. We next present a figure based on the analysis provided in [54] to show the impact of antenna correlation on single TAS.

Figure 4.12 plots the secrecy outage probability versus $\overline{\gamma}_B$, in which the exponential correlation is adopted at both Bob and Eve, i.e. the (i,j)th element in $\Phi_B^{1/2}$ and $\Phi_E^{1/2}$ is given by $\xi_B^{|i-j|}$ and $\xi_E^{|i-j|}$, respectively. In this exponential correlation, $\xi_B \in [0, 1]$ and $\xi_E \in [0, 1]$ are the correlation parameters specified by system settings (e.g. signal frequency) at Bob and Eve, respectively, where $|i - j|$ is the distance between the ith and jth antennas. In this figure, we first observe that asymptotic curves are parallel, which demonstrates that the secrecy outage diversity order is not affected by ξ_B or ξ_E. This figure also indicates that correlation is detrimental to the secrecy performance when $\overline{\gamma}_B$ is in the medium and high regimes. In these regimes (i.e. $\overline{\gamma}_B > -1$ dB with $\overline{\gamma}_B = -5$ dB; $\overline{\gamma}_B > 9$ dB with $\overline{\gamma}_B = 5$ dB), the secrecy outage probability increases as ξ_B or ξ_E increases. On the one hand, higher ξ_B indicates the degraded quality of the

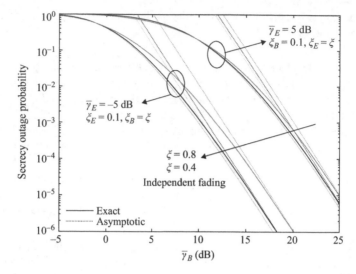

Figure 4.12   *The secrecy outage probability for $N_A = N_B = N_E = 2$ and $R_s = 1$*
with independent fading, exponential correlation with $\xi_B = \xi$ and
$\xi_E = 0.1$ for $\overline{\gamma}_E = -5$ dB, and exponential correlation with $\xi_B = 0.1$
and $\xi_E = \xi$ for $\overline{\gamma}_E = 5$ dB

main channel, which results in poorer secrecy performance. On the other, $\overline{\gamma}_E$ is relatively low compared with $\overline{\gamma}_B$. Moreover, we observe that the performance degradation
caused by increasing ξ_B is larger than that brought by increasing ξ_E.

We note that the impact of antenna correlation together with imperfect feedback
was also examined in [52] and similar conclusions were drawn. In summary, we repeat
the main results on the impact of antenna correlation on single TAS as follows:

- Antenna correlation does not affect the secrecy diversity order of single TAS.
- For low average SNR of the main channel, higher correlation at Eve brings greater
  performance improvement than higher correlation at Bob.
- For medium and high average SNR of the main channel, higher correlation at Eve
  imposes less performance degradation than higher correlation at Bob.

The impact of the correlation between the main channel and the eavesdropper's
channel on the secrecy performance of single TAS was examined in a MISO wiretap
channel [55], in which the eavesdropper's channel was modelled as

$$g_{n,1} = \eta h_{n,1} + \sqrt{1 - \eta^2}e, \tag{4.29}$$

where e represented a random variable that was i.i.d. as $h_{n,1}$, and $0 \leq \eta \leq 1$ was the
correlation coefficient between the main and eavesdropper's channels, with $\eta = 0$
representing independence and $\eta = 1$ representing full correlation. The analysis

provided in [55] provided the following conclusions on the impact of the correlation between the main and eavesdropper's channels on single TAS.

- The correlation between the main and eavesdropper's channels does not affect the secrecy diversity order of single TAS.
- In the low average SNR regime of the main channel, this correlation degrades the secrecy performance of single TAS.
- In the medium and high average SNR regimes of the main channel, this correlation leads to secrecy performance improvement for single TAS.

On the basis of the above discussions, we can see that antenna correlation and the correlation between the main and eavesdropper's channels have contradictory impacts on the secrecy performance of single TAS. We note that these two correlations have no effect on the secrecy diversity order of single TAS.

4.6 Conclusion

In this chapter, antenna selection in MIMO wiretap channels was reviewed and discussed. Specifically, the secrecy performance of single TAS was fully reviewed and investigated within different fading channels. Our examination showed that single TAS achieved full secrecy diversity order as long as the feedback required by antenna selection did not suffer from any error. TAS-Alamouti and TAS-Alamouti-OPA could outperform single TAS at the cost of negligible extra feedback overhead. Furthermore, we discussed how antenna selection can be exploited in order to enhance the secrecy performance of full-duplex wiretap channels. Finally, the effects of imperfect feedback and channel correlation on the secrecy performance of TAS were reviewed and discussed.

References

[1] T. Liu and S. Shamai (Shitz), "A note on the secrecy capacity of the multiple-antenna wiretap channel," *IEEE Trans. Inf. Theory*, vol. 55, no. 6, pp. 2547–2553, Jun. 2009.

[2] A. Khisti and G. W. Wornell, "Secure transmission with multiple antennas Part I: The MISOME wiretap channel," *IEEE Trans. Inf. Theory*, vol. 56, no. 7, pp. 3088–3104, Jul. 2010.

[3] A. Khisti and G. W. Wornell, "Secure transmission with multiple antennas—Part II: The MIMOME wiretap channel," *IEEE Trans. Inf. Theory*, vol. 56, no. 11, pp. 5515–5532, Nov. 2010.

[4] F. Oggier and B. Hassibi, "The secrecy capacity of the MIMO wiretap channel," *IEEE Trans. Inf. Theory*, vol. 57, no. 8, pp. 4961–4972, Aug. 2011.

[5] T. F. Wong, M. Bloch, and J. M. Shea, "Secret sharing over fast fading MIMO wiretap channels," *EURASIP J. Wireless Commun. Netw.*, vol. 2009, pp. 506973/1–17, Dec. 2009.

[6]   M. Kobayashi, P. Piantanida, S. Yang, and S. Shamai (Shitz), "On the secrecy degrees of freedom of the multi-antenna block fading wiretap channels," *IEEE Trans. Inf. Forensics Security*, vol. 6, no. 3, pp. 703–711, Sep. 2011.

[7]   Y.-W. Hong, P.-C. Lan, and C.-C. Kuo, "Enhancing physical-layer secrecy in multiantenna wireless systems: an overview of signal processing approaches," *IEEE Signal Proc. Mag.*, vol. 30, no. 5, pp. 29–40, Sep. 2013.

[8]   X. Zhou, L. Song, and Y. Zhang, *Physical Layer Security in Wireless Communications*, CRC Press, 2013.

[9]   N. Yang, L. Wang, G. Geraci, M. Elkashlan, J. Yuan, and M. Di Renzo, "Safeguarding 5G wireless communication networks using physical layer security," *IEEE Commun. Mag.*, vol. 53, no. 4, pp. 20–27, Apr. 2015.

[10]   S. Yan, X. Zhou, N. Yang, B. He, and T. D. Abhayapala, "Artificial-noise-aided secure transmission in wiretap channels with transmitter-side correlation," *IEEE Trans. Wireless Commun.*, vol. 15, no. 12, pp. 8286–8297, Dec. 2016.

[11]   B. He, N. Yang, S. Yan, and X. Zhou, "Linear precoding for simultaneous confidential broadcasting and power transfer," *IEEE J. Sel. Topics Signal Process.*, vol. 10, no. 8, pp. 1404–1416, Dec. 2016.

[12]   A. Mukherjee and A. L. Swindlehurst, "Robust beamforming for security in MIMO wiretap channels with imperfect CSI," *IEEE Trans. Signal Process.*, vol. 59, no. 1, pp. 351–361, Jan. 2011.

[13]   W.-C. Liao, T.-H. Chang, W.-K. Ma, and C.-Y. Chi, "QoS-based transmit beamforming in the presence of eavesdroppers: An optimized artificial-noise-aided approach," *IEEE Trans. Signal Process.*, vol. 59, no. 3, pp. 1202–1216, Mar. 2011.

[14]   S. A. A. Fakoorian and A. L. Swindlehurst, "MIMO interference channel with confidential messages: Achievable secrecy rates and precoder design," *IEEE Trans. Inf. Forensics Security*, vol. 6, no. 3, pp. 640–649, Sep. 2011.

[15]   T. Gucluoglu and T. M. Duman, "Performance analysis of transmit and receive antenna selection over flat fading channels," *IEEE Trans Wireless Commun.*, vol. 7, no. 8, pp. 3056–3065, Aug. 2008.

[16]   H. Alves, R. D. Souza, and M. Debbah, "Enhanced physical layer security through transmit antenna selection", in *IEEE GlobeCOM 2011 Workshops*, pp. 879–883, Dec. 2011.

[17]   H. Alves, R. D. Souza, M. Debbah, and M. Bennis, "Performance of transmit antenna selection physical layer security schemes," *IEEE Signal Process. Lett.*, vol. 19, no. 6, pp. 372–375, Jun. 2012.

[18]   D. W. K. Ng, E. S. Lo, and R. Schober, "Secure resource allocation and scheduling for OFDMA decode-and-forward relay networks," *IEEE Trans. Wireless Commun.*, vol. 10, no. 10, pp. 3528–3540, Oct. 2011.

[19]   N. Romero-Zurita, M. Ghogho, and D. McLernon, "Outage probability based power distribution between data and artificial noise for physical layer security," *IEEE Signal Process. Lett.*, vol. 19, no. 2, pp. 71–74, Feb. 2012.

[20]   N. Yang, P. L. Yeoh, M. Elkashlan, R. Schober, and I. B. Collings, "Transmit antenna selection for security enhancement in MIMO wiretap channels," *IEEE Trans. Commun.*, vol. 61, no. 1, pp. 144–154, Jan. 2013.

[21] N. Yang, M. Elkashlan, P. L. Yeoh, and J. Yuan, "An introduction to transmit antenna selection in MIMO wiretap channels," *ZTE Communications*, vol. 11, no. 3, pp. 26–32, Sep. 2013 (Invited Paper).

[22] N. Yang, P. L. Yeoh, M. Elkashlan, R. Schober, and J. Yuan, "MIMO wiretap channels: Secure transmission using transmit antenna selection and receive generalized selection combining," *IEEE Commun. Lett.*, vol. 17, no. 9, pp. 1754–1757, Sep. 2013.

[23] M. Bloch, J. Barros, M. Rodrigues, and S. McLaughlin, "Wireless information-theoretic security," *IEEE Trans. Inf. Theory*, vol. 54, no. 6, pp. 2515–2534, Jun. 2008.

[24] F. He, H. Man, and W. Wang, "Maximal ratio diversity combining enhanced security," *IEEE Comm. Lett.*, vol. 15, no. 5, pp. 509–511, May 2011.

[25] V. U. Prabhu and M. R. D. Rodrigues, "On wireless channels with M-antenna eavesdroppers: Characterization of the outage probability and ε-outage secrecy capacity," *IEEE Trans. Inf. Forensics Security*, vol. 6, no. 3, pp. 853–860, Sep. 2011.

[26] S. Yan, N. Yang, G. Geraci, R. Malaney, and J. Yuan, "Optimization of code rates in SISOME wiretap channels," *IEEE Trans. Wireless Commun.*, vol. 14, no. 11, pp. 6377–6388, Nov. 2015.

[27] S. Yan and R. Malaney, "Location-based beamforming for enhancing secrecy in Rician wiretap channels," *IEEE Trans. Wireless Commun.*, vol. 15, no. 4, pp. 2780–2791, Apr. 2016.

[28] L. Wang, M. Elkashlan, J. Huang, R. Schober, and R. K. Mallik, "Secure transmission with antenna selection in MIMO Nakagami-m channels," *IEEE Trans. Wireless Commun.*, vol. 13, no. 11, pp. 6054–6067, Nov. 2014.

[29] S. Hessien, F. S. Al-Qahtani, R. M. Radaydeh, C. Zhong, and H. Alnuweiri, "On the secrecy enhancement with low-complexity largescale transmit selection in MIMO generalized composite fading," *IEEE Wireless Commun. Lett.*, vol. 4, no. 4, pp. 429–432, Aug. 2015.

[30] Z. Ding, Z. Ma, and P. Fan, "Asymptotic studies for the impact of antenna selection on secure two-way relaying communications with artificial noise," *IEEE Trans. Wireless Commun.*, vol. 13, no. 4, pp. 2189–2203, Apr. 2014.

[31] G. Brante, H. Alves, R. Souza, and M. Latva-aho, "Secrecy analysis of transmit antenna selection cooperative schemes with no channel state information at the transmitter," *IEEE Trans. on Commun.*, vol. 63, no. 4, pp. 1330–1342, Apr. 2015.

[32] S. Yan, N. Yang, R. Malaney, and J. Yuan, "Transmit antenna selection with Alamouti scheme in MIMO wiretap channels," in *Proc. IEEE GlobeCOM*, pp. 687–692, Dec. 2013.

[33] S. Yan, N. Yang, R. Malaney, and J. Yuan, "Transmit antenna selection with Alamouti coding and power allocation in MIMO wiretap channels," *IEEE Trans. Wireless Commun.*, vol. 13, no. 3, pp. 1656–1667, Mar. 2014.

[34] J. Zhu, Y. Zou, G. Wang, Y.-D. Yao, and G. K. Karagiannidis, "On secrecy performance of antenna selection aided MIMO systems against eavesdropping," *IEEE Trans. Veh. Technol.*, vol. 65, no. 1, pp. 214–225, Jan. 2016.

[35]   K. Yang, N. Yang, C. Xing, J. Wu, and Z. Zhang, "Space-time network coding with transmit antenna selection and maximal-ratio combining," *IEEE Trans. Wireless Commun.*, vol. 14, no. 4, pp. 2106–2117, Apr. 2015.

[36]   K. Yang, N. Yang, C. Xing, J. Wu, and J. An, "Space-time network coding with antenna selection," *IEEE Trans. Wireless Commun.*, vol. 65, no. 7, pp. 5264–5274, Jul. 2016.

[37]   V. Tarokh, H. Jafarkhani, and A. R. Calderbank, "Space-time block coding for wireless communications: performance results," *IEEE J. Sel. Areas Commun.*, vol. 17, no. 3, pp. 451–460, Mar. 1999.

[38]   V. Tarokh, H. Jafarkhani, and A. R. Calderbank, "Space-time block codes from orthogonal designs," *IEEE Trans. Inf. Theory*, vol. 45, no. 5, pp. 1456–1467, Jul. 1999.

[39]   A. Mukherjee and A. L. Swindlehurst, "A full-duplex active eavesdropper in MIMO wiretap channels: Construction and countermeasures," in *Proc. Asilomar Conf. Sign. Syst. Comput.*, pp. 265–269, Nov. 2011.

[40]   G. Zheng, I. Krikidis, J. Li, A. P. Petropulu, and B. Ottersten, "Improving physical layer secrecy using full-duplex jamming receivers," *IEEE Trans. Signal Process.*, vol. 61, no. 20, pp. 4962–4974, Oct. 2013.

[41]   S. Yan, N. Yang, R. Malaney, and J. Yuan, "Full-duplex wiretap channels: security enhancement via antenna switching," in *Proc. IEEE GlobeCOM TCPLS Workshop*, pp. 1412–1417, Dec. 2014.

[42]   L. Li, Z. Chen, and J. Fang, "A full-duplex Bob in the MIMO Gaussian wiretap channel: Scheme and performance," *IEEE Signal Process. Lett.*, vol. 21, no. 1, pp. 107–111, Jan. 2016.

[43]   G. Chen, Y. Gong, P. Xiao, and J. Chambers, "Dual antenna selection in secure cognitive radio networks," *IEEE Trans. Veh. Technol.*, vol. 65, no. 10, pp. 7993–8002, Oct. 2016.

[44]   M. Duarte, C. Dick, and A. Sabharwal, "Experiment-driven characterization of full-duplex wireless systems," *IEEE Trans. Wireless Commun.*, vol. 11, no. 12, pp. 4296–4307, Dec. 2012.

[45]   D. Bharadia, E. McMilin, and S. Katti, "Full duplex radios," in *Proc. SIGCOMM*, pp. 375–386, Aug. 2013.

[46]   A. Sabharwal, P. Schniter, D. Guo, D. Bliss, S. Rangarajan, and R. Wichman, "In-band full-duplex wireless: Challenges and opportunities," *IEEE J. Sel. Areas Commun.*, vol. 32, no. 9, pp. 1637–1652, Sep. 2014.

[47]   T. R. Ramya and S. Bhashyam, "Using delayed feedback for antenna selection in MIMO systems," *IEEE Trans. Wireless Commun.*, vol. 8, no. 12, pp. 6059–6067, Dec. 2009.

[48]   R. M. Radaydeh, "Impact of delayed arbitrary transmit antenna selection on the performance of rectangular QAM with receive MRC in fading channels," *IEEE Commun. Lett.*, vol. 13, no. 6, pp. 390–392, Jun. 2009.

[49]   N. S. Ferdinand, D. Benevides da Costa, and M. Latva-Aho, "Effects of outdated CSI on the secrecy performance of MISO wiretap channels with transmit antenna selection," *IEEE Commun. Lett.*, vol. 17, no. 5, pp. 864–867, May 2013.

[50] J. Xiong, Y. Tang, D. Ma, P. Xiao, and K. K. Wong, "Secrecy performance analysis for TAS-MRC system with imperfect feedback," *IEEE Trans. Inf. Forensics Security*, vol. 10, no. 8, pp. 1617–1629, Aug. 2015.

[51] Y. Huang, F. S. Al-Qahtani, T. Q. Duong, and J. Wang, "Secure transmission in MIMO wiretap channels using general-order transmit antenna selection with outdated CSI," *IEEE Trans. Commun.*, vol. 63, no. 8, pp. 2959–2971, Aug. 2015.

[52] J. Hu, Y. Cai, N. Yang, and W. Yang, "A new secure transmission scheme with outdated antenna selection," *IEEE Trans. Inf. Forensics Security*, vol. 10, no. 11, pp. 2435–2446, Nov. 2015.

[53] H. Cui, R. Zhang, L. Song, and B. Jiao, "Relay selection for bidirectional AF relay network with outdated CSI," *IEEE Trans. Veh. Technol.*, vol. 62, no. 9, pp. 4357–4365, Nov. 2013.

[54] N. Yang, H. A. Suraweera, I. B. Collings, and C. Yuen, "Physical layer security of TAS/MRC with antenna correlation," *IEEE Trans. Inf. Forensics Security*, vol. 8, no. 1, pp. 254–259, Jan. 2013.

[55] N. S. Ferdinand, D. B. da Costa, A. L. F. de Almeida, and M. Latvaaho, "Physical layer secrecy performance of TAS wiretap channels with correlated main and eavesdropper channels," *IEEE Wireless Commun. Lett.*, vol. 3, no. 1, pp. 86–89, Feb. 2014.

Chapter 5
Physical layer security for massive MIMO systems
Jun Zhu[1], Robert Schober[2], and Vijay K. Bhargava[3]

The fifth-generation (5G) wireless systems are expected to create a paradigm shift compared to the current long-term evolution (LTE)/LTE-Advanced systems in order to meet the unprecedented demands for future wireless applications, including the tremendous throughput and massive connectivity. Massive multiple-input multiple-output (MIMO) [1–8], an architecture employing large-scale multi-user MIMO processing using the array of hundreds or even thousands of antennas, simultaneously serving tens or hundreds of mobile users, has been identified as a promising air-interface technology to address a significant portion of the above challenges. Besides, security is a vital issue in wireless networks due to the broadcast nature of the medium [9]. Despite the great efforts on massive MIMO from both the academia and industry, the security paradigms guaranteeing the confidentiality of wireless communications in 5G networks have scarcely been stated. The objective of this chapter is to exploit physical layer security for massive MIMO systems.

The chapter is organised as follows. In Sections 5.1 and 5.2, we briefly review the fundamentals of massive MIMO and physical layer security, respectively. In Section 5.3, we motivate the chapter by illustrating why we consider physical layer security for massive MIMO systems. The considered model and the performance evaluation metrics for secure massive MIMO systems are introduced in Sections 5.4 and 5.5, respectively. Section 5.6 studies linear precoding for downlink secure massive MIMO transmission. In Section 5.7, we conclude with a brief summary of this chapter.

5.1 Fundamentals of massive MIMO

Massive multiple-input multiple-output (MIMO) systems, also known as large-scale antenna or very large MIMO systems, refer to systems with base stations (BSs) deploying an order of magnitude more elements than what is used in current systems, i.e. a hundred antennas or more and simultaneously serve low-complexity single-antenna mobile terminals (MTs) [1–8]. Massive MIMO enjoys all the benefits of

[1]Qualcomm Inc, San Diego, USA
[2]Institute for Digital Communications, Friedrich-Alexander-Universität Erlangen-Nürnberg, Germany
[3]3Department of Electrical and Computer Engineering, University of British Columbia, Canada

conventional multi-user MIMO, such as improved data rate, reliability and reduced interference, but at a much larger scale and with simple linear precoding/detection schemes [1–3]. Remarkable improvements in spectral and power efficiency can be achieved by focusing the radiating power onto the MTs with very large antenna arrays [6]. Massive MIMO is therefore capable of achieving robust performance at low signal-to-interference-plus-noise ratio (SINR), as the effects of noise and interference vanish completely in the limit of an infinite number of antennas [5]. Other benefits of massive MIMO include, but are not limited to, the extensive use of inexpensive low-power components, reduced latency, simplification of the media access control (MAC) layer, and robustness to intentional jamming [1–3].

5.1.1 Time-division duplex and uplink pilot training

It is well understood that the acquisition of channel state information (CSI) is essential for signal processing at the BS. Most current cellular systems work on frequency-division duplex (FDD) mode, where the CSI is typically acquired via feedback (full or limited) [10]. However, when the BS is equipped with a large excess of antennas compared with the number of terminals, which is customary for massive MIMO systems, the time-division duplex (TDD) mode provides the main solution to acquire CSI. This is because the training burden for uplink pilots in a TDD system is proportional to the number of MTs, but independent of the number of BS antennas, while conversely the training burden for downlink pilots in an FDD system is proportional to the number of BS antennas [5]. The adoption of an FDD system imposes a severe limitation on the number of antennas deployed at the BS. By exploiting the reciprocity between uplink and downlink channels for TDD systems, the BS is able to eliminate the need for feedback, and uplink pilot training is sufficient for providing the desired uplink and downlink CSI.

5.1.2 Downlink linear precoding

With the desired downlink CSI available via uplink training by exploiting the channel reciprocity for TDD operation, the BS performs precoding in order to simultaneously serve multiple single-antenna MTs. Most precoding techniques are identical to those used for conventional multi-user MIMO schemes, but at a much larger scale. The theoretical sum-capacity optimal dirty paper coding (DPC) technique [11] is too complex to be implemented in practice even in a conventional MIMO system and is thus not considered here. In contrast, linear precoding is typically adopted in massive MIMO systems. The most popular scheme is matched-filter (MF) precoding, due to its simple processing [2,5,8]. However, MF precoding results in a performance degradation with an increasing number of serving MTs. This is because when more MT channels exist, the near orthogonality between the MT channels becomes weak, which increases the level of multi-user interference. In this case, zero-forcing (ZF)/regularised channel inversion (RCI) precoding are preferable [12–14]. Similar as in the conventional MIMO systems, the former suppresses the multi-user interference, while the latter strikes a balance between MF and ZF precoding. Unfortunately, they require high-dimensional matrix inversions, which lead to a high computational complexity, especially when the number of BS antennas and MTs are both large.

5.1.3 Multi-cell deployment and pilot contamination

In massive MIMO systems, each terminal is ideally assigned to an orthogonal uplink pilot sequence. However, the maximum number of orthogonal pilot sequences is limited by the coherence block length. In a multi-cell network, the available orthogonal pilot sequences are quickly exhausted. As such, pilot sequences have to be re-used from one cell to another. The negative effect incurred by the pilot re-use is generally called pilot contamination [1,8]. More precisely, when the BS estimates the channel of a specific MT, it correlates the received pilot signal with the pilot sequence of that MT. In the case of pilot re-use between cells, it actually acquires a channel estimate that is contaminated by a linear combination of the channels associated with other MTs that share the same pilot sequence. The downlink precoding based on the contaminated channel estimates introduces interference which is directed to the MTs in other cells that share the same pilot sequence. The directed interference grows with the number of BS antennas at the same speed as the desired signal. Similar interference also exists for uplink data transmission.

5.2 Physical layer security basics

Security is a vital issue in wireless networks due to the broadcast nature of the medium. Traditionally, security has been achieved through cryptographic encryption implemented at the application layer, which requires a certain form of information (e.g. key) shared between the legitimate entities [9,15]. This approach ignores the behaviour of the communication channels and relies on the theoretical assumption that communication between the legitimate entities is error free. More importantly, all cryptographic measures assume that it is computationally infeasible for them to be deciphered without knowledge of the secret key, which remains mathematically unproven. Ciphers that were considered potentially unbreakable in the past are continually defeated due to the increasingly growth of computational power. Moreover, error-free communication cannot be always guaranteed in non-deterministic wireless channels [9]. A novel approach for wireless security taking advantage of the characteristics of physical channels was proposed by Wyner in [16] and is referred to as physical layer security. The concept was originally developed for the classical wire-tap channel [16], cf. Figure 5.1 (left). Wyner showed that a source (Alice)–destination (Bob) pair can exchange perfectly secure messages with a positive rate if the desired receiver enjoys better channel conditions than the eavesdropper (Eve). However, this condition cannot always hold in practice, especially in wireless fading channels. To make things worse, Eve enjoys a better average channel gain than Bob as long as he/she is located closer to Alice than Bob. Therefore, perfectly secure communication seems impossible, and techniques to enhance Bob's channel condition while degrading Eve's are needed. One option is to utilise artificial noise (AN) to perturb Eve's reception [17], as shown in Figure 5.1 (right). Eves are typically passive so as to hide their existence, and thus, their CSI cannot be obtained by Alice. In this case, multiple-transmit antennas can be exploited to enhance secrecy by simultaneously transmitting both the information-bearing signal and AN.

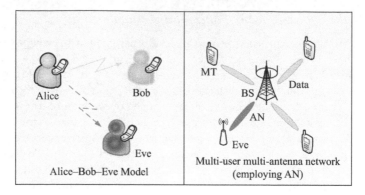

Figure 5.1   Physical layer security model

Specifically, precoding is used to make the AN invisible to Bob while degrading the decoding performance of possibly present Eves [17,18]. In [19], the authors investigated the secrecy outage probability for the AN-aided secrecy system, where only Alice has multiple antennas. When Eve is also equipped with multiple antennas, the work in [20] proposes an AN precoder to achieve a near-optimal performance in high signal-to-noise (SNR) regime. The contribution extends to a secrecy system where all nodes have multiple antennas in [21].

More recent studies have considered physical layer security provisioning in multi-user networks [22–29]. Although the secrecy capacity region for multi-user networks remains an open problem, it is interesting to investigate the achievable secrecy rates of such networks for certain practical transmission strategies. All aforementioned work generally assumed that Alice can acquire perfect CSI of Bob, which seems too ideal. Robust beamforming designs with estimated CSI were reported in [30–34].

5.3   Motivations

The emerging massive MIMO technology offers tremendous performance gains in terms of network throughput and energy efficiency by employing simple coherent processing on a large-scale antenna array. However, very little attention has been given to the security issue in massive MIMO systems. In order to address this concern, we need first to consider two fundamental questions: (1) Is massive MIMO secure? and (2) if not, how can we improve security in massive MIMO systems? In this section, we illustrate the main motivation of this chapter by providing brief and general responses to the two questions.

5.3.1   Is massive MIMO secure?

Compared with conventional MIMO, massive MIMO is inherently more secure, as the large-scale antenna array deployed at the transmitter (Alice) can accurately focus a narrow and directional information beam on the intended terminal (Bob), such that the received signal power at Bob is several orders of magnitude higher than that at any incoherent passive eavesdropper (Eve) [35]. Unfortunately, this benefit may vanish if Eve also employs a massive antenna array for eavesdropping. The following scenarios further deteriorate the security of the massive MIMO system:

- As Eve is passive, it is able to move arbitrarily close to Alice without being detected by either Alice or Bob. In this case, the signal received by Eve can be strong.
- In a ultra-dense multi-cell network, Bob suffers from severe multi-user interference (both pilot contaminated and uncontaminated), while Eve may have access to the information of all other MTs, e.g. by collaborating with them, and remove their interference when decoding Bob's information. This becomes practical when those MTs are also malicious.

In the aforementioned scenarios, unless additional measures to secure the communication are taken by Alice, even a single passive Eve is able to intercept the signal intended for Bob [36]. Furthermore, we note that Eve could emit its own pilot symbols to impair the channel estimates obtained at Alice to improve his ability to decode Bob's signals during downlink transmission [37]. However, this would also increase the chance that the presence of the eavesdropper is detected by Alice [38]. Therefore, in this chapter, we limit ourselves to passive eavesdropping.

5.3.2 How to improve security for massive MIMO?

Massive MIMO systems offer an abundance of BS antennas, while multiple-transmit antennas can be exploited for secrecy enhancement, e.g. by emitting AN. Therefore, the combination of both concepts seems natural and promising. There arise several challenges and open problems for physical layer security pertaining to massive MIMO systems that are not present for conventional MIMO systems. We summarise them as follows:

- In a conventional massive MIMO system (without security), pilot contamination constitutes a limit on performance in terms of data throughput [5]. However, its effects on the AN design, as well as wireless security have not been considered.
- In conventional MIMO systems, AN is transmitted in the null space (NS) of the channel matrix [17]. The complexity associated with computing the NS may not be affordable in the case of massive MIMO, and thus, simpler AN precoding methods are essential.

This chapter will provide detailed and insightful solutions to the aforementioned challenges and problems. As massive MIMO will serve as an essential enabling technology for the emerging 5G wireless networks, it is expected that its design from physical layer security perspective opens a new and promising research path. Related contributions will be summarised in Section 5.3.3.

5.3.3 State-of-the-art

In this section, we summarise the related work on the topic of physical layer security for massive MIMO systems. In [39], the authors overviewed the possible research options for the design of physical layer security in the emerging massive MIMO systems. Large system secrecy analysis of MIMO systems achieved by RCI precoding was provided in [40–42]. In [43], the authors adopted the channel between Alice and Bob as a secret key and showed that the complexity required by Eve to decode Alice's message is at least of the same order as a worst case lattice problem. AN-aided secure transmission in multi-cell massive MIMO systems with pilot contamination was studied in [36] for the first time. The authors in [36] considered simple MF precoder, while secure transmission for different pairs of data and AN precoders was investigated in [44]. The effects of hardware impairments on secure massive MIMO transmission were studied in [45]. AN-aided jamming for Rician fading massive MIMO channels was investigated in [46], where the power allocation is optimised between messages and AN for both uniform and directional jamming; the authors in [47] investigated power scaling law for secure massive MIMO systems without the help of AN, while the authors in [48,49] developed new data and AN precoding frameworks for massive MIMO systems with limited number of radio-frequency (RF) chains and with constant envelope constraint, respectively, in order to enhance the system security. In the context of massive MIMO relaying, the work presented in [50,51] compared two classic relaying schemes, i.e. amplify-and-forward (AF) and decode-and-forward (DF), for physical layer security provisioning with imperfect CSI at the massive MIMO relay. The authors in [52] provided a large system secrecy rate analysis for simultaneous wireless information and power transfer (SWIPT) MIMO wiretap channels. Although [46–52] and the contributions in this chapter all assumed that Eve is passive, the so-called pilot contamination attack [37], a form of active eavesdropping, has also been considered in the literature. In particular, several techniques for detection of the pilot contamination attack were proposed in [35], including a detection scheme based on random pilots and a cooperative detection scheme. Moreover, the authors in [53] developed a secret key agreement protocol under the pilot contamination attack, and the authors in [54] proposed to encrypt the pilot sequence in order to hide it from the attacker. The encryption enables the MTs to achieve the performance as if they were under no attack. Methods for combating the pilot contamination attack in a multi-cell network were reported in [55], which exploited the low-rank property of the transmit correlation matrices of massive MIMO channels.

5.4 System models for secure massive MIMO

In this section, we outline the adopted system model for the considered secure massive MIMO systems. We consider a flat-fading multi-cell system consisting of M cells, as depicted in Figure 5.2. Each cell comprises an N_T-antenna BS and K

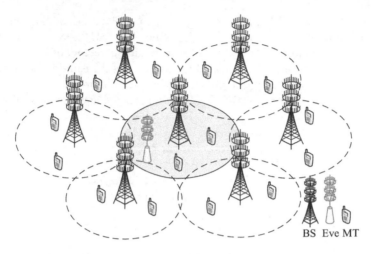

Figure 5.2   Multi-cell massive MIMO system in the presence of a multi-antenna eavesdropper

single-antenna MTs[1]. An eavesdropper equipped with N_E antennas (equivalent to N_E cooperative single-antenna eavesdroppers) is located in cell n of the considered multi-cell region. The eavesdropper is passive and seeks to acquire the information transmitted to MT k in cell n. We note that neither the BSs nor the MTs are assumed to know which MT is targeted by the eavesdropper. Let $\mathbf{G}_{mn} = \mathbf{D}_{mn}^{1/2}\mathbf{H}_{mn} = [(\mathbf{g}_{mn}^1)^T, \ldots, (\mathbf{g}_{mn}^K)^T]^T \in \mathbb{C}^{K \times N_T}$ and $\mathbf{G}_{mE} = \sqrt{\beta_{mE}}\mathbf{H}_{mE} \in \mathbb{C}^{N_E \times N_T}$ denote the matrices modelling the channels from BS m to the K MTs and the eavesdropper in cell n, respectively. Thereby, $\mathbf{D}_{mn} = \mathrm{diag}\{\beta_{mn}^1, \ldots, \beta_{mn}^K\}$ and β_{mE} represent the path-losses from BS m to the K MTs and the eavesdropper in cell n, respectively. Matrix $\mathbf{H}_{mn} \in \mathbb{C}^{K \times N_T}$, with row vector $\mathbf{h}_{mn}^k \in \mathbb{C}^{1 \times N_T}$ in the kth row, and matrix $\mathbf{H}_{mE} \in \mathbb{C}^{N_E \times N_T}$ represent the corresponding small-scale fading components. Their elements are modelled as mutually independent and identically distributed (i.i.d.) complex Gaussian random variables (r.v.s) with zero mean and unit variance[2].

5.4.1   Channel estimation and pilot contamination

We assume that the BSs are perfectly synchronised and operate in the TDD mode with universal frequency re-use. Furthermore, we assume that the path-losses between all

---

[1]We note that the results derived in this chapter can be easily extended to multi-antenna MTs if the BS transmits one independent data stream per MT receive antenna and receive combining is not performed at the MTs. In this case, each MT receive antenna can be treated as one (virtual) MT, and the results derived in this chapter are applicable. For example, the secrecy rate of a multi-antenna MT can be obtained by summing up the secrecy rates of its receive antennas.

[2]Log-normal shadowing is not considered here, as it does not add any technical insights to the analytical results, but only complicates the results.

MTs in the system and BS n, β_{nm}^k, $m = 1, \ldots, M$, $k = 1, \ldots, K$, are known at BS n, whereas the small-scale fading vectors $\mathbf{h}_{nm}^k$, $m = 1, \ldots, M$, $k = 1, \ldots, K$, are not known and BS n estimates only the small-scale fading vectors of the MTs within cell n. These assumptions are motivated by the fact that the path-losses change on a much slower time scale than the small-scale fading vectors, and thus, their estimation creates a comparatively low overhead. BS n estimates the downlink CSI of all MTs, $\mathbf{h}_{nn}^k$, $k = 1, \ldots, K$, by exploiting reverse training and channel reciprocity for TDD operations [1–8]. The K pilot sequences used in a cell are still orthogonal but all cells re-use the identical pilot sequences. Let $\sqrt{\tau}\omega_k \in \mathbb{C}^{\tau \times 1}$ denote the pilot sequence of length τ transmitted by MT k in each cell in the training phase, where $\omega_k^H \omega_k = 1$ and $\omega_k^H \omega_j = 0$, $\forall, j, k = 1, \ldots, K$, $k \neq j$. Assuming perfect synchronisation, the training signal received at BS n, $\mathbf{Y}_n^{\text{pilot}} \in \mathbb{C}^{\tau \times N_T}$, can be expressed as:

$$
\mathbf{Y}_n^{\text{pilot}} = \sum_{m=1}^M \sum_{k=1}^K \sqrt{p_\tau \tau \beta_{nm}^k}\, \omega_k \mathbf{h}_{nm}^k + \mathbf{N}_n, \tag{5.1}
$$

where p_τ denotes the pilot power and $\mathbf{N}_n \in \mathbb{C}^{\tau \times N_T}$ is a Gaussian noise matrix having zero mean, unit variance elements. Assuming minimum mean-square error (MMSE) channel estimation [7,8], the estimate of $\mathbf{h}_{nn}^k$ given $\mathbf{Y}_n^{\text{pilot}}$ is obtained as:

$$
\hat{\mathbf{h}}_{nn}^k = \sqrt{p_\tau \tau \beta_{nn}^k}\, \omega_k^H \left(\mathbf{I}_\tau + \omega_k \left(p_\tau \tau \sum_{m=1}^M \beta_{nm}^k \right) \omega_k^H \right)^{-1} \mathbf{Y}_n^{\text{pilot}}
$$

$$
= \frac{\sqrt{p_\tau \tau \beta_{nn}^k}}{1 + p_\tau \tau \sum_{m=1}^M \beta_{nm}^k}\, \omega_k^H \mathbf{Y}_n^{\text{pilot}}. \tag{5.2}
$$

For MMSE estimation, we can express the channel as $\mathbf{h}_{nn}^k = \hat{\mathbf{h}}_{nn}^k + \tilde{\mathbf{h}}_{nn}^k$, where the estimate $\hat{\mathbf{h}}_{nn}^k$ and the estimation error $\tilde{\mathbf{h}}_{nn}^k \in \mathbb{C}^{1 \times N_T}$ are mutually independent. Hence, considering (5.2) we can statistically characterise $\hat{\mathbf{h}}_{nn}^k$ and $\tilde{\mathbf{h}}_{nn}^k$ as $\hat{\mathbf{h}}_{nn}^k \sim \mathbb{CN}\left(\mathbf{0}_{N_T}^T, \frac{p_\tau \tau \beta_{nn}^k}{1 + p_\tau \tau \sum_{m=1}^M \beta_{nm}^k} \mathbf{I}_{N_T} \right)$ and $\tilde{\mathbf{h}}_{nn}^k \sim \mathbb{CN}\left(\mathbf{0}_{N_T}^T, \frac{1 + p_\tau \tau \sum_{m \neq n} \beta_{nm}^k}{1 + p_\tau \tau \sum_{m=1}^M \beta_{nm}^k} \mathbf{I}_{N_T} \right)$, respectively. Still from (5.2), we also observe that $\omega_k^H \mathbf{Y}_n^{\text{pilot}}$ is proportional to the MMSE estimate of $\mathbf{h}_{nm}^k$ for any m, i.e.:

$$
\frac{\hat{\mathbf{h}}_{nm}^k}{\|\hat{\mathbf{h}}_{nm}^k\|} = \frac{\omega_k^H \mathbf{Y}_n^{\text{pilot}}}{\|\omega_k^H \mathbf{Y}_n^{\text{pilot}}\|}, \forall m. \tag{5.3}
$$

Equation (5.3) implies that the estimate of MT k in each cell is simply a scaled version of the same vector $\omega_k^H \mathbf{Y}_n^{\text{pilot}}$. Hence, the BS is not able to distinguish between the channel to its own MT k and to MT k in other cells [8]. In the same manner, we also expand the channel $\mathbf{h}_{mn}^k = \hat{\mathbf{h}}_{mn}^k + \tilde{\mathbf{h}}_{mn}^k$,[3] where $\hat{\mathbf{h}}_{mn}^k$ and $\tilde{\mathbf{h}}_{mn}^k$ are mutually independent. We

---

[3]In this chapter, BS n only needs to estimate $\mathbf{h}_{nn}^k$. The role of this expansion is to facilitate a mathematical simplification in deriving the achievable rate in Section 5.5, by decomposing the inter-cell interference/AN leakage from cell m into correlated terms $\hat{\mathbf{h}}_{mn}^k$ and uncorrelated terms $\tilde{\mathbf{h}}_{mn}^k$ with respect to (w.r.t.) the desired MT's channel estimate.

also have $\hat{\mathbf{h}}_{mn}^k \sim \mathbb{CN}\left(\mathbf{0}_{N_T}^T, \frac{p_\tau \tau \beta_{mn}^k}{1+p_\tau \tau \sum_{l=1}^M \beta_{ml}^k}\mathbf{I}_{N_T}\right)$ and $\tilde{\mathbf{h}}_{mn}^k \sim \mathbb{CN}\left(\mathbf{0}_{N_T}^T, \frac{1+p_\tau \tau \sum_{l\neq n}\beta_{ml}^k}{1+p_\tau \tau \sum_{l=1}^M \beta_{ml}^k}\mathbf{I}_{N_T}\right)$,
respectively. For future reference, we collect the estimates and the estimation errors at BS n corresponding to all K MTs in cell m in matrices $\hat{\mathbf{H}}_{nm} = [(\hat{\mathbf{h}}_{nm}^1)^T, \ldots, (\hat{\mathbf{h}}_{nm}^K)^T]^T \in \mathbb{C}^{K\times N_T}$ and $\tilde{\mathbf{H}}_{nm} = [(\tilde{\mathbf{h}}_{nm}^1)^T, \ldots, (\tilde{\mathbf{h}}_{nm}^K)^T]^T \in \mathbb{C}^{K\times N_T}$, respectively.

5.4.2 Downlink data and AN transmission

In cell n, the BS intends to transmit a confidential signal s_{nk} to MT k. The signal vector for the K MTs is denoted by $\mathbf{s}_n = [s_{n1}, \ldots, s_{nK}]^T \in \mathbb{C}^{K\times 1}$ with $\mathbb{E}[\mathbf{s}_n \mathbf{s}_n^H] = \mathbf{I}_K$. Each signal vector $\mathbf{s}_n$ is multiplied by a transmit beamforming matrix, $\mathbf{F}_n = [\mathbf{f}_{n1}, \ldots, \mathbf{f}_{nk}, \ldots, \mathbf{f}_{nK}] \in \mathbb{C}^{N_T\times K}$, before transmission. Furthermore, we assume that the eavesdropper's CSI is not available at BS n. Hence, assuming that there are $K < N_T$ MTs, the BS may use the remaining $N_T - K$ degrees of freedom offered by the N_T transmit antennas for emission of AN to degrade the eavesdropper's ability to decode the data intended for the MTs [17,30,31]. The AN vector, $\mathbf{z}_n = [z_{n1}, \ldots, z_{n(N_T-K)}]^T \sim \mathbb{CN}(\mathbf{0}_{N_T-K}, \mathbf{I}_{N_T-K})$, is multiplied by an AN precoding matrix $\mathbf{A}_n = [\mathbf{a}_{n1}, \ldots, \mathbf{a}_{ni}, \ldots, \mathbf{a}_{n(N_T-K)}] \in \mathbb{C}^{N_T\times(N_T-K)}$ with $\|\mathbf{a}_{ni}\| = 1$, $i = 1, \ldots, N_T - K$. The function of AN is to introduce extra noise in order to confuse the eavesdropper. The considered choices for the AN precoding matrix will be discussed in Section 5.6.2. The signal vector transmitted by BS n is given by:

$$\mathbf{x}_n = \sqrt{p}\mathbf{F}_n \mathbf{s}_n + \sqrt{q}\mathbf{A}_n \mathbf{z}_n = \sum_{k=1}^K \sqrt{p}\mathbf{f}_{nk} s_{nk} + \sum_{i=1}^{N_T-K} \sqrt{q}\mathbf{a}_{ni} z_{ni}, \tag{5.4}$$

where p and q denote the transmit power allocated to each MT and each AN signal, respectively, i.e. for simplicity, we assume uniform power allocation across users and AN signals, respectively. Let the total transmit power be denoted by P_T. Then, p and q can be represented as $p = \frac{\phi P_T}{K}$ and $q = \frac{(1-\phi)P_T}{N_T-K}$, respectively, where the power allocation factor ϕ, $0 < \phi \le 1$, strikes a power balance between the information-bearing signal and the AN.

The $M - 1$ cells adjacent to cell n transmit their own signals and AN. In this work, in order to be able to gain some fundamental insights, we assume that all cells employ identical values for p and q as well as ϕ. Accordingly, the received signals at MT k in cell n, y_{nk}, and at the eavesdropper, $\mathbf{y}_E$, are given by:

$$y_{nk} = \sqrt{p}\mathbf{g}_{nn}^k \mathbf{f}_{nk} s_{nk} + \sum_{\{m,l\}\neq\{n,k\}} \sqrt{p}\mathbf{g}_{mn}^k \mathbf{f}_{ml} s_{ml} + \sum_{m=1}^M \sqrt{q}\mathbf{g}_{mn}^k \mathbf{A}_m \mathbf{z}_m + n_{nk} \tag{5.5}$$

and

$$\mathbf{y}_E = \sqrt{p}\sum_{m=1}^M \mathbf{G}_{mE} \mathbf{F}_m \mathbf{s}_m + \sqrt{q}\sum_{m=1}^M \mathbf{G}_{mE} \mathbf{A}_m \mathbf{z}_m + \mathbf{n}_E, \tag{5.6}$$

respectively, where $n_{nk} \sim \mathbb{CN}(0, \sigma_{nk}^2)$ and $\mathbf{n}_E \sim \mathbb{CN}(\mathbf{0}_{N_E}, \sigma_E^2 \mathbf{I}_{N_E})$ are the Gaussian noises at MT k and at the eavesdropper, respectively. The first term on the right hand side of (5.5) is the signal intended for MT k in cell n with effective channel gain $\sqrt{p}\mathbf{g}_{nn}^k\mathbf{f}_{nk}$. The second and the third terms on the right hand side of (5.5) represent intra-cell/inter-cell interference and AN leakage, respectively. On the other hand, the eavesdropper observes an $MN_T \times N_E$ MIMO channel comprising K in-cell user signals, $(M-1)K$ out-of-cell user signals, $N_T - K$ in-cell AN signals, and $(N_T - K)(M-1)$ out-of-cell AN signals. In order to obtain a lower bound on the achievable secrecy rate, we assume that the eavesdropper can acquire perfect knowledge of the effective channels of all MTs, i.e. $\mathbf{H}_{mE}\mathbf{f}_{mk}, \forall m, k$. We note however that this is a quite pessimistic assumption because the uplink training performed in massive MIMO [8] makes it difficult for the eavesdropper to perform accurate channel estimation.

5.5 Achievable ergodic secrecy rate and secrecy outage probability for secure massive MIMO systems

In this section, we first show that the achievable ergodic secrecy rate of MT k in cell n can be expressed as the difference between the achievable ergodic rate of the MT and the ergodic capacity of the eavesdropper. Subsequently, we provide a simple lower bound on the achievable ergodic rate of the MT, a closed-form expression for the ergodic capacity of the eavesdropper, and a simple and tight upper bound for the ergodic capacity of the eavesdropper. We then define the secrecy outage probability of MT k in cell n. For convenience, we define the ratio of the number of eavesdropper antennas and the number of BS antennas as $\alpha = N_E/N_T$, and the ratio of the number of users and the number of BS antennas as $\beta = K/N_T$. In the following, we are interested in the asymptotic regime where $N_T \to \infty$ but α and β are constant.

5.5.1 Achievable ergodic secrecy rate

The ergodic secrecy rate is an appropriate performance measure if delays can be afforded and coding over many independent channel realisations (i.e. over many coherence intervals) is possible [18]. Considering MT k in cell n, the considered channel is an instance of a multiple-input single-output multiple-eavesdropper (MISOME) wiretap channel [20]. In the following lemma, we provide an expression for an achievable ergodic secrecy rate of MT k in cell n.

Lemma 5.1. *An achievable ergodic secrecy rate of MT k in cell n is given by:*

$$R_{nk}^{sec} = [R_{nk} - C_{nk}^{eve}]^+, \tag{5.7}$$

where $[x]^+ = \max\{0, x\}$, R_{nk} *is an achievable ergodic rate of MT k in cell n, and* C_{nk}^{eve} *is the ergodic capacity between BS n and the eavesdropper seeking to decode the information of MT k in cell n. Thereby, it is assumed that the eavesdropper is able to cancel the received signals of all in-cell and out-of-cell MTs except the signal intended for the MT of interest, i.e.:*

$$C_{nk}^{eve} = \mathbb{E}\left[\log_2\left(1 + p\mathbf{f}_{nk}^H\mathbf{G}_{nE}^H\mathbf{X}^{-1}\mathbf{G}_{nE}\mathbf{f}_{nk}\right)\right], \tag{5.8}$$

where $\mathbf{X} = q \sum_{m=1}^{M} \mathbf{A}_m^H \mathbf{G}_{mE}^H \mathbf{G}_{mE} \mathbf{A}_m$ denotes the noise correlation matrix at the eavesdropper under the worst-case assumption that the receiver noise is negligible, i.e. $\sigma_E^2 \to 0$.

Proof. Please refer to Appendix A.1.1.   □

Equation (5.7) reveals that the achievable ergodic secrecy rate of MT k in cell n has the subtractive form typical for many wiretap channels [9,16,20–31], i.e. it is the difference of an achievable ergodic rate of the user of interest and the capacity of the eavesdropper.

5.5.1.1 Lower bound on the achievable user rate

On the basis of (5.5) an achievable ergodic rate of MT k in cell n is given by $R_{nk} =$

$$\mathbb{E}\left[\log_2\left(1 + \frac{|\sqrt{p}\mathbf{g}_{nn}^k \mathbf{f}_{nk}|^2}{\sum_{m=1}^{M}\sum_{i=1}^{N_T-K} |\sqrt{q}\mathbf{g}_{mn}^k \mathbf{a}_{mi}|^2 + \sum_{\{m,l\}\neq\{n,k\}} |\sqrt{p}\mathbf{g}_{mn}^k \mathbf{f}_{ml}|^2 + \sigma_{nk}^2}\right)\right]. \quad (5.9)$$

Unfortunately, evaluating the expected value in (5.9) analytically is cumbersome. Therefore, we derive a lower bound on the achievable ergodic rate of MT k in cell n by following the same approach as in [8]. In particular, we rewrite the received signal at MT k in cell n as:

$$y_{nk} = \mathbb{E}[\sqrt{p}\mathbf{g}_{nn}^k \mathbf{f}_{nk}]s_{nk} + n'_{nk}, \quad (5.10)$$

where n'_{nk} represents an effective noise, which is given by:

$$n'_{nk} = \left(\sqrt{p}\mathbf{g}_{nn}^k \mathbf{f}_{nk} - \mathbb{E}[\sqrt{p}\mathbf{g}_{nn}^k \mathbf{f}_{nk}]\right) s_{nk} + \sum_{m=1}^{M} \mathbf{g}_{mn}^k \sqrt{q}\mathbf{A}_m \mathbf{z}_m$$

$$+ \sum_{\{m,l\}\neq\{n,k\}} \sqrt{p}\mathbf{g}_{mn}^k \mathbf{f}_{ml}s_{ml} + n_{nk}. \quad (5.11)$$

Equation (5.10) can be interpreted as an equivalent single-input single-output channel with constant gain $\mathbb{E}[\sqrt{p}\mathbf{g}_{nn}^k \mathbf{f}_{nk}]$ and AWGN n'_{nk}. Hence, we can apply Theorem 1 in [8] to obtain a computable lower bound for the achievable rate of MT k in cell n as $\underline{R}_{nk} = \log_2(1 + \gamma_{nk}) \leq R_{nk}$, where γ_{nk} denotes the received signal-to-interference-plus-noise ratio (SINR), given by

$$\gamma_{nk} = \frac{\overbrace{|\mathbb{E}[\sqrt{p}\mathbf{g}_{nn}^k \mathbf{f}_{nk}]|^2}^{\text{desired signal}}}{\underbrace{\text{var}[\sqrt{p}\mathbf{g}_{nn}^k \mathbf{f}_{nk}]}_{\text{signal leakage}} + \underbrace{\sum_{m=1}^{M}\sum_{i=1}^{N_T-K} \mathbb{E}[|\sqrt{q}\mathbf{g}_{mn}^k \mathbf{a}_{mi}|^2]}_{\text{AN leakage}} + \underbrace{\sum_{\{m,l\}\neq\{n,k\}} \mathbb{E}[|\sqrt{p}\mathbf{g}_{mn}^k \mathbf{f}_{ml}|^2]}_{\text{intra- and inter-cell interference}} + \sigma_{nk}^2}$$

$$(5.12)$$

with $\text{var}[\sqrt{p}g_{nn}^k f_{nk}] = \mathbb{E}[|\sqrt{p}g_{nn}^k f_{nk} - \mathbb{E}[\sqrt{p}g_{nn}^k f_{nk}]|^2]$. The SINR in (5.12) is underestimated by assuming that only the BS has channel estimates, while the MTs only know the mean of the effective channel gain $|\mathbb{E}[\sqrt{p}g_{nn}^k f_{nk}]|$ and employ it for signal detection. The deviation from the average effective channel gain is treated as Gaussian noise having variance $\text{var}[\sqrt{p}g_{nn}^k f_{nk}]$ [8]. The closed-form expressions for γ_{nk} under different pairs of data and AN precoders will be provided in Table 5.1. The tightness of the lower bound will be confirmed by our results in Section 5.6.5.

5.5.1.2 Ergodic capacity of the eavesdropper

In this section, we provide a closed-form expression for the ergodic capacity of the eavesdropper. The results are provided in the following theorem.

Theorem 5.1. *For $N_T \to \infty$, the ergodic capacity of the eavesdropper in (5.8) can be written as:*

$$C_{nk}^{\text{eve}} = \frac{1}{\ln 2} \sum_{i=0}^{N_E-1} \lambda_i \times \frac{1}{\mu_0} \sum_{j=1}^{2} \sum_{l=2}^{b_j} \omega_{jl} I(1/\mu_j, l), \tag{5.13}$$

where $\lambda_i = \binom{M(N_T-K)}{i}$, $\mu_0 = \prod_{j=1}^{2} \mu_j^{b_j}$,

$$(\mu_j, b_j) = \begin{cases} (\eta, N_T - K), & j = 1 \\ (\beta_{mE}/\beta_{nE}\eta, (M-1)(N_T - K)), & j = 2, \end{cases} \tag{5.14}$$

$\eta = q/p$,

$$\omega_{jl} = \frac{1}{(b_j - l)!} \frac{d^{b_j-l}}{dx^{b_j-l}} \left(\frac{x^i}{\prod_{s \neq j}(x + \frac{1}{\mu_s})^{b_s}} \right) \Big|_{x=-\frac{1}{\mu_j}}, \tag{5.15}$$

and $I(a, n) = \int_0^{\infty} \frac{1}{(x+1)(x+a)^n} dx$, $a, n > 0$. A closed-form expression for $I(\cdot, \cdot)$ is given in [56, Lemma 3].

Proof. Please refer to Appendix A.1.2.   □

A lower bound on the achievable ergodic secrecy rate of MT k in cell n is obtained by combining (5.7), (5.12), and (5.13). However, the expression for the ergodic capacity of the eavesdropper in (5.13) is somewhat cumbersome and offers little insight into the impact of the various system parameters. Hence, in the next subsection, we derive a simple and tight upper bound for C_{nk}^{eve}.

5.5.1.3 Tight upper bound on the ergodic capacity
of the eavesdropper

In the following theorem, we provide a tight upper bound for the ergodic capacity of the eavesdropper.

Theorem 5.2. *For $N_T \to \infty$, the ergodic capacity of the eavesdropper in (5.8) is upper bounded by*[4]

$$C_{nk}^{\text{eve}} < \overline{C}_{nk}^{\text{eve}} \approx \log_2\left(1 + \frac{\alpha}{\eta a(1-\beta) - c\eta\alpha/a}\right) = \log_2\left(\frac{(1-\zeta)\phi + \zeta}{-\zeta\phi + \zeta}\right), \quad (5.16)$$

if $\beta < 1 - c\alpha/a^2$, where we introduce the definitions $a = 1 + \sum_{m\neq n}^{M} \beta_{mE}/\beta_{nE}$, $c = 1 + \sum_{m\neq n}^{M}(\beta_{mE}/\beta_{nE})^2$ and $\zeta = \frac{a\beta}{\alpha} - \frac{\beta c}{a(1-\beta)}$.

Proof. Please refer to Appendix A.1.3. □

Remark 5.1. *We note that a finite eavesdropper capacity results only if matrix $\mathbf{X}$ in (5.8) is invertible. Since $\mathbf{G}_{mE}$, $m = 1, \ldots, M$, are independent matrices with i.i.d. entries, $\mathbf{X}$ is invertible if $M(N_T - K) \leq N_E$ or equivalently $\beta \leq 1 - \alpha/M$. Regardless of the values of M and ρ, we have:*

$$1 - \alpha/c \leq 1 - c\alpha/a^2 \leq 1 - \alpha/M. \quad (5.17)$$

For $M = 1$ or $\rho = 1$, equality holds in (5.17). For $M > 1$ and $\rho < 1$, the condition for β in Theorem 5.2 is in general stricter than the invertibility condition for $\mathbf{X}$. Nevertheless, the typical operating region for a massive MIMO system is $\beta \ll 1$ [2,5], where the upper bound in Theorem 5.2 is applicable.

Equation (5.16) reveals that $\overline{C}_{nk}^{\text{eve}}$ is monotonically increasing in α, i.e. as expected, the eavesdropper can enhance his eavesdropping capability by deploying more antennas. Furthermore, in the relevant parameter range, $0 < \beta < 1 - c\alpha/a^2$, $\overline{C}_{nk}^{\text{eve}}$ is not monotonic in β but a decreasing function for $\beta \in (0, 1 - \sqrt{c\alpha}/a)$ and an increasing function for $\beta \in (1 - \sqrt{c\alpha}/a, 1 - c\alpha/a^2)$. Hence, $\overline{C}_{nk}^{\text{eve}}$ has a minimum at $\beta = 1 - \sqrt{c\alpha}/a$. Assuming N_T and N_E are fixed, this behaviour can be explained as follows. For small K (corresponding to small β), the capacity of the eavesdropper is large because the amount of power allocated to the intercepted MT, $\phi P_T/K$, is large. As K increases, the power allocated to the MT decreases, which leads to a decrease in the capacity. However, if K is increased beyond a certain point, $\mathbf{X}$ becomes increasingly ill-conditioned, leading to an increase in the eavesdropper capacity.

Combining now (5.7), (5.12), and (5.16) gives a tight lower bound on the ergodic secrecy rate of MT k in cell n. This will allow us to further simplify the SINR expression of MT k in cell n and the resulting ergodic secrecy rate expression.

Lower bound on the achievable ergodic secrecy rate: Combining now (5.7), $\underline{R}_{nk} = \log_2(1 + \gamma_{nk})$, with γ_{nk} given in (5.12), (5.13) (for C_{nk}^{eve}), and (5.16) (for $\overline{C}_{nk}^{\text{eve}}$) gives two tight lower bounds on the achievable ergodic secrecy rate of MT k in cell n, as:

$$\underline{R}_{nk}^{\text{sec}} = [\underline{R}_{nk} - C_{nk}^{\text{eve}}]^+ \quad \text{and} \quad \underline{\underline{R}}_{nk}^{\text{sec}} = [\underline{R}_{nk} - \overline{C}_{nk}^{\text{eve}}]^+. \quad (5.18)$$

---

[4]We note that, strictly speaking, we have not proved that (5.16) is a bound since we used an approximation for its derivation, see Appendix A.1.3. However, this approximation is known to be very accurate [57] and comparisons of (5.16) with simulation results for various system parameters suggest that (5.16) is indeed an upper bound.

5.5.2 Secrecy outage probability analysis

In delay limited scenarios, where one codeword spans only one channel realisation, outages are unavoidable since Alice does not have the CSI of the eavesdropper channel and the secrecy outage probability has to be used to characterise the performance of the system instead of the ergodic rate. For the considered multi-cell massive MIMO system, the rate of the desired user, R_{nk}, becomes deterministic as $N_T \to \infty$, but the instantaneous capacity of the eavesdropper channel remains a random variable. A secrecy outage occurs whenever the target secrecy rate R_0 exceeds the actual instantaneous secrecy rate. Thus, the secrecy outage probability of MT k in cell n is given by:

$$\varepsilon_{\text{out}} = \Pr\{R_{nk} - \log_2(1 + \gamma_E) \leq R_0\} = \Pr\{\gamma_E \geq 2^{R_{nk}-R_0} - 1\}$$

$$= 1 - F_{\gamma_E}(2^{R_{nk}-R_0} - 1), \tag{5.19}$$

where $\gamma_E = p\mathbf{f}_{nk}^H \mathbf{G}_{nE}^H \mathbf{X}^{-1} \mathbf{G}_{nE} \mathbf{f}_{nk}$ and $F_{\gamma_E}(x)$ is given in Appendix A.1.2. A closed-form upper bound on the secrecy outage probability is obtained finally.

5.6 Linear data and AN precoding in massive MIMO systems

5.6.1 Linear data precoders for secure massive MIMO

In this section, we analyse the achievable rate of MF/ZF/RCI data precoding. In contrast to existing analysis and designs of data precoders for massive MIMO, e.g. [13, 14], the results presented in this section account for the effect of AN leakage, which is only present if AN is introduced by the BS for secrecy enhancement. We are interested in the asymptotic regime where $K, N_T \to \infty$ but $\beta = K/N_T$ and $\alpha = N_E/N_T$ are finite.

For $N_T \to \infty$, analysing the achievable rate is equivalent to analysing the SINR in (5.12). Thereby, the effect of the AN precoder can be captured by the term:

$$Q = \sum_{m=1}^{M} \sum_{i=1}^{N_T-K} \mathbb{E}[|\sqrt{\beta_{mn}^k}\mathbf{h}_{mn}^k \mathbf{a}_{mi}|^2] = \sum_{m=1}^{M} \beta_{mn}^k \mathbb{E}[\mathbf{h}_{mn}^k \mathbf{A}_m \mathbf{A}_m^H (\mathbf{h}_{mn}^k)^H] \tag{5.20}$$

in the denominator of (5.12), which represents the inter-cell and intra-cell AN leakage. This term is assumed to be given in this section and will be analysed in detail for different AN precoders in Section 5.6.

5.6.1.1 RCI data precoding

The RCI data precoder for cell n is given by:

$$\mathbf{F}_n = \gamma \mathbf{L}_{nn} \hat{\mathbf{H}}_{nn}^H, \tag{5.21}$$

where $\mathbf{L}_{nn} = (\hat{\mathbf{H}}_{nn}^H \hat{\mathbf{H}}_{nn} + \kappa \mathbf{I}_{N_T})^{-1}$, γ is a scalar normalisation constant, and κ is a regularisation constant. In the following proposition, we provide the resulting SINR of MT k in cell n.

Proposition 5.1. *For RCI data precoding, the received SINR at MT k in cell n is given by:*

$$\gamma_{nk}^{\mathrm{RCI}} = \frac{1}{\frac{\sum_{m=1}^{M} \hat{\Gamma}_{\mathrm{RCI}}^{m} + (1+\mathscr{G}(\beta,\kappa))^2}{\mathscr{G}(\beta,\kappa)\left(\hat{\Gamma}_{\mathrm{RCI}}^{n} + \frac{\hat{\Gamma}_{\mathrm{RCI}}^{n}\kappa}{\beta}(1+\mathscr{G}(\beta,\kappa_1))^2\right)} + \sum_{m\neq n} \lambda_{mk}/\lambda_{nk}}, \tag{5.22}$$

$$\mathscr{G}(\beta,\kappa) = \frac{1}{2}\left[\sqrt{\frac{(1-\beta)^2}{\kappa^2} + \frac{2(1+\beta)}{\kappa} + 1} + \frac{1-\beta}{\kappa} - 1\right], \tag{5.23}$$

and

$$\hat{\Gamma}_{\mathrm{RCI}}^{m} = \frac{\Gamma_{\mathrm{RCI}}\lambda_{mk}}{\Gamma_{\mathrm{RCI}}\sum_{m=1}^{M}\mu_{mk}+1} \tag{5.24}$$

with $\Gamma_{\mathrm{RCI}} = \frac{K}{\eta Q + \frac{K}{\phi P_T}}$, $\lambda_{mk} = \beta_{mn}^k \frac{p_\tau \tau \beta_{mn}^k}{\theta_{mn}^k + p_\tau \tau \beta_{mn}^k}$, *and* $\mu_{mk} = \beta_{mn}^k \frac{\theta_{mn}^k}{\theta_{mn}^k + p_\tau \tau \beta_{mn}^k}$.

Proof. Please refer to Appendix A.1.4. □

The regularisation constant κ can be optimised for maximisation of the lower bound on the secrecy rate in (5.18), which is equivalent to maximising the SINR in (5.22). Setting the derivative of $\gamma_{nk}^{\mathrm{RCI}}$ with respect to κ to zero, the optimal regularisation parameter is found as $\kappa_{\mathrm{opt}} = \beta/\sum_{m=1}^{M}\hat{\Gamma}_{\mathrm{RCI}}^{m}$ in Appendix A.1.5, and the corresponding maximum SINR is given by:

$$\gamma_{nk}^{\mathrm{RCI}} = \frac{1}{\hat{\Gamma}_{\mathrm{RCI}}^{n}/\sum_{m=1}^{M}\hat{\Gamma}_{\mathrm{RCI}}^{m}\mathscr{G}(\beta,\kappa_{1,\mathrm{opt}}) + \sum_{m\neq n}\lambda_{mk}/\lambda_{nk}}. \tag{5.25}$$

On the other hand, for $\kappa \to 0$ and $\kappa \to \infty$, the RCI data precoder in (5.21) reduces to the ZF and MF data precoder, respectively. The corresponding received SINR is provided in the following two corollaries.

Corollary 5.1. *Assuming $\beta < 1$, for ZF data precoding, the received SINR at MT k in cell n is given by:*

$$\gamma_{nk}^{\mathrm{ZF}} = \lim_{\kappa=0} \gamma_{nk}^{\mathrm{RCI}} = \frac{1}{\frac{\beta}{(1-\beta)\hat{\Gamma}_{\mathrm{RCI}}^{n}} + \sum_{m\neq n}\lambda_{mk}/\lambda_{nk}}. \tag{5.26}$$

Corollary 5.2. *Assuming $\beta < 1$, for MF data precoding, the received SINR at MT k in cell n is given by:*

$$\gamma_{nk}^{\mathrm{MF}} = \lim_{\kappa\to\infty} \gamma_{nk}^{\mathrm{RCI}} = \frac{1}{\beta\left(1+\frac{1}{\hat{\Gamma}_{\mathrm{RCI}}^{n}}\right) + \sum_{m\neq n}\lambda_{mk}/\lambda_{nk}}. \tag{5.27}$$

5.6.1.2 Computational complexity of data precoding

We compare the computational complexity of the considered data precoders in terms of the number of floating point operations (FLOPs) [58]. Each FLOP represents one scalar complex addition or multiplication. We assume that the coherence time of the channel is T symbol intervals of which τ are used for training and $T - \tau$ are used for data transmission. Hence, the complexity required for precoding in one coherence interval, consist of the complexity required for generating one precoding matrix and $T - \tau$ pre-coded vectors. A similar complexity analysis was conducted in [59, Section IV] for various selfish data precoders without AN injection at the BS. Since the AN injection does not affect the structure of the data precoders, we can directly adopt the results from [59, Section IV] to the case at hand. In particular, the MF and the ZF/RCI require $(2K - 1)N_T(T - \tau)$, and $0.5(K^2 + K)(2N_T - 1) + K^3 + K^2 + K + N_T K(2K - 1) + (2K - 1)N_T(T - \tau)$ FLOPs per coherence interval, respectively, see [59, Section IV].

5.6.2   Linear AN precoders for secure massive MIMO

In this section, we investigate the performance of NS and random AN precoders.

5.6.2.1   Analysis of AN precoders

For a given dimensionality of the AN precoder, $N_T - K$, the secrecy rate depends on the AN precoder only via the AN leakage, Q, given in (5.20), which affects the SINR of the MT. In this sub-section, for $N_T \to \infty$, we will provide closed-form expressions for Q for the NS and random AN precoders.

The NS AN precoder of BS n is given by [17]:

$$\mathbf{A}_n = \mathbf{I}_{N_T} - \hat{\mathbf{H}}_{nn}^H \left(\hat{\mathbf{H}}_{nn} \hat{\mathbf{H}}_{nn}^H \right)^{-1} \hat{\mathbf{H}}_{nn}, \tag{5.28}$$

which has rank $N_T - K$ and exists only if $\beta < 1$. For the NS AN precoder, Q_{NS} is obtained as:

$$Q_{\mathrm{NS}} = \sum_{m=1}^{M} \beta_{mn}^k \mathbb{E}\left[\mathbf{h}_{mn}^k \mathbf{A}_m \mathbf{A}_m^H (\mathbf{h}_{mn}^k)^H \right] = (N_T - K) \sum_{m=1}^{M} \mu_{mk}, \tag{5.29}$$

where we exploited [60, Lemma 11] and the independence of $\mathbf{A}_m$ and $\tilde{\mathbf{h}}_{mn}^k$. For the random precoder, all elements of $\mathbf{A}_n$ are i.i.d. r.v.s independent of the channel [36]. Hence, $\mathbf{h}_{mn}^k$ and $\mathbf{A}_m$, $\forall m$, are mutually independent, and we obtain:

$$Q_{\mathrm{random}} = \sum_{m=1}^{M} \beta_{mn}^k \mathbb{E}\left[\mathbf{h}_{mn}^k \mathbf{A}_m \mathbf{A}_m^H (\mathbf{h}_{mn}^k)^H \right] = (N_T - K) \sum_{m=1}^{M} \beta_{mn}^k. \tag{5.30}$$

5.6.2.2   Computational complexity of AN precoding

Similarly to the data precoders, the complexity of the AN precoders is evaluated in terms of the number of FLOPs required per coherence interval T. For the NS AN precoder, the computation of $\mathbf{A}_n$ in (5.28) requires the computation and inversion of a $K \times K$ positive definite matrix, which entails $0.5(K^2 + K)(2N_T - 1) + K^3 + K^2 + K$ FLOPs [58], and the multiplication of an $N_T \times K$, an $K \times K$, and an $K \times N_T$ matrix, which entails $N_T(N_T + K)(2K - 1)$ FLOPs [58]. Furthermore, the $T - \tau$ vector-matrix multiplications required for AN precoding entail a complexity of $(2N_T - 1)N_T$ FLOPs [58], respectively. Hence, the overall complexity is $0.5(K^2 + K)(2N_T - 1) + K^3 + K^2 + K + N_T(N_T + K)(2K - 1) + (2N_T - 1)N_T(T - \tau)$ FLOPs, whereas the random AN precoder entails a complexity of $(2N_T - 1)N_T(T - \tau)$ FLOPs as only the AN vector-matrix multiplications are required.

5.6.3   Comparison of linear data and AN precoders

In this sub-section, we compare the secrecy performances of the considered data and AN precoders. Thereby, in order to get tractable results, we focus on the relative performances of ZF and MF, cf. data precoders and NS and random AN precoders. The performances of RCI precoder will be investigated via numerical and simulation results in Section 5.6.5.

In order to gain some insights for system design and analysis, we adopt a simplified path-loss model. In particular, we assume the path losses are given by:

$$\beta_{mn}^k = \begin{cases} 1, & m = n \\ \rho, & \text{otherwise} \end{cases} \tag{5.31}$$

where $\rho \in [0, 1]$ denotes the inter-cell interference factor. For this simplified model, a and c in Theorem 5.2 simplify to $a = 1 + (M - 1)\rho$ and $c = 1 + (M - 1)\rho^2$. Furthermore, the SINR expressions of the linear data precoders considered in Section 5.3.1 can be simplified considerably and are provided in Table 5.1. For this model, $\hat{\Gamma}_{RCI}^n$ simplifies to $\hat{\Gamma}_{RCI}^n = \frac{\Gamma_{RCI}\lambda}{\Gamma_{RCI}a\mu+1}$, where $\Gamma_{RCI} = \frac{\beta\phi}{(1-\phi)\beta Q/(N_T-K)+\beta/P_T}$, $\lambda = \frac{p_\tau \tau}{1+ap_\tau \tau}$, and $\mu = \frac{1+(a-1)p_\tau \tau}{1+ap_\tau \tau}$. Moreover, in Table 5.1, Q simplifies to $a(1 - \lambda)(N_T - K)$ for NS AN precoder and $a(N_T - K)$ for random AN precoder.

5.6.3.1   Comparison of ZF and MF data precoders

We compare the performances achieved with ZF and MF data precoders for a given AN precoder, i.e. Q are fixed. Since the upper bound on the capacity of the eavesdropper channel is independent of the adopted data precoder, cf. Section 5.3.1, we compare the considered data precoders based on their SINRs. Exploiting the results in Table 5.1, we obtain the following relations between γ_{nk}^{ZF} and γ_{nk}^{MF}:

$$\frac{\gamma_{nk}^{ZF}}{\gamma_{nk}^{MF}} = 1 + \beta(c\gamma_{nk}^{ZF} - 1). \tag{5.32}$$

Table 5.1   SINR of MT k in cell n for linear data precoding and the simplified path-loss model in (5.31)

Data precoder	γ_{nk}
RCI	$\dfrac{1}{1/c\mathcal{G}(\beta, \beta/c\hat{\Gamma}_{SRCI}) + c - 1}$
ZF	$\dfrac{\lambda\phi(1 - \beta)}{(1 - \phi)\beta Q/(N_T - K) + \beta\phi(a - c\lambda) + (c - 1)\lambda\phi(1 - \beta) + \beta/P_T}$
MF	$\dfrac{\lambda\phi}{(1 - \phi)\beta Q/(N_T - K) + \beta\phi a + (c - 1)\lambda\phi + \beta/P_T}$

Hence, for $\gamma_{nk}^{\text{ZF}} > \gamma_{nk}^{\text{MF}}$, we require $\gamma_{nk}^{\text{ZF}} > 1/c = 1/(1 + \rho^2(M - 1))$. As expected, (5.32) suggests that for a lightly loaded system, i.e. $\beta \to 0$, both precoders have a similar performance, i.e. $\gamma_{nk}^{\text{ZF}} \approx \gamma_{nk}^{\text{MF}}$. On the other hand, by recalling Section 5.5, ZF precoding requires much higher computational complexity compared with its MF counterpart, especially when N_T and K are large.

5.6.3.2   Comparison of NS and random AN precoding

We analyse the impact of the AN precoders on the secrecy rate. AN precoders affect the achievable rate of the MT via the leakage, Q. We observe $Q_{\text{random}} \geq Q_{\text{NS}}$. Since according to Table 5.1 the SINRs for all data precoders are decreasing functions of Q, for a given data precoder, we obtain for the lower bound on the ergodic rate of MT k in cell n: $R_{nk}|_{\text{random}} \leq R_{nk}|_{\text{NS}}$. Considering the independence between the ergodic capacity of the eavesdropper and the choice of AN precoder, and the expression for the ergodic secrecy rate, $R_{nk}^{\text{sec}} = [R_{nk} - C_{nk}^{\text{eve}}]^+$, we simply have $R_{nk}^{\text{sec}}|_{\text{random}} \leq R_{nk}^{\text{sec}}|_{\text{NS}}$, whereas random AN precoder requires much less computational complexity than NS AN precoder, especially when N_T and K are large, cf. Section 5.5.

5.6.4   Optimal power splitting

Another interesting aspect is to find the optimal power splitting between data transmission and AN emission, for the maximisation of the achievable secrecy rate. By solving the equation $\frac{\partial R_{nk}^{\text{sec}}}{\partial \phi} = 0$ with respect to $\phi \in (0, 1)$, it can be shown that $\underline{R}_{nk}^{\text{sec}}$ given in (5.18) as a function of ϕ has a total number of two stationary points. One of them is less than zero, and the other is greater than zero. As $\lim_{\phi \to 0} \underline{R}_{nk}^{\text{sec}} = 0$ and $\lim_{\phi \to 1} \underline{R}_{nk}^{\text{sec}} = 0$, as long as there exists some $\varepsilon \in (0, 1)$ such that $\underline{R}_{nk}^{\text{sec}}(\varepsilon) > 0$, it suffices to prove that the second stationary point of $\underline{R}_{nk}^{\text{sec}}(\phi)$ falls in $(0, 1)$. The function $\underline{R}_{nk}^{\text{sec}}$ is therefore unimodal on $(0, 1)$ and the stationary point in this interval is the optimal power splitting factor ϕ^* that maximises $\underline{R}_{nk}^{\text{sec}}$.

5.6.5   Numerical examples

In this section, we evaluate the performance of the considered secure multi-cell massive MIMO system. We consider cellular systems with $M = 7$ hexagonal cells, and to gain insight for system design, we adopt the simplified path-loss model introduced in Section 5.6.3, i.e. the severeness of the inter-cell interference is only characterised by the parameter $\rho \in (0, 1]$. The simulation results for the ergodic secrecy rate of MT k in cell n are based on (5.7)–(5.9), and are averaged over 5,000 random channel realisations. Note that, in this chapter, we consider the ergodic secrecy rate of a certain MT, i.e. MT k in cell n. The cell sum secrecy rate can be obtained by multiplying the secrecy rate of MT k by the number of MTs, K, as for the considered channel model, all MTs in cell n achieve the same secrecy rate. The values of all relevant system parameters are provided in the captions of the figures. To enable a fair comparison, throughout this section, we adopted the NS AN precoder when we compare

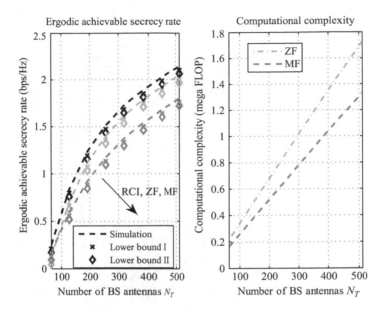

Figure 5.3   *Ergodic secrecy rate (left-hand side) and computational complexity (right-hand side) of various linear data precoders for a network employing $P_T = 10$ dB, $N_T = 128$ (right hand side), $p_\tau = P_T/\tau$, $\rho = 0.1$, $T = 500$, $\tau = K$, optimal ϕ^*, and an NS AN precoder. Lower bound I is obtained from $\underline{R}^{sec}_{nk}$, and lower bound II is obtained from $\underline{\underline{R}}^{sec}_{nk}$ in (5.18)*

different data precoders and the ZF data precoder when we compare different AN precoders.

In Figure 5.3, we plot the achievable secrecy rate and required computational complexity by RCI, ZF, and MF data precoders, as functions of the number of BS antennas, N_T. Figure 5.3 reveals that the derived bounds for the ergodic secrecy rate is accurate. As expected, for the ergodic secrecy rate, lower bound I is somewhat tighter than lower bound II. Furthermore, increasing the number of BS antennas N_T improves the ergodic secrecy rate. Moreover, as expected, the RCI data precoder outperforms ZF data precoder, which outperforms MF data precoder when the number of BS antennas goes large enough. On the other hand, observed from the right subfigure, MF precoder requires the least computational complexity among three data precoders.

Similarly, Figure 5.4 depicts the achievable secrecy rate and required computational complexity by NS and random AN precoders, as functions of the number of BS antennas, N_T. As expected, NS AN precoder outperforms its random counterpart, at the expense of higher computational complexity.

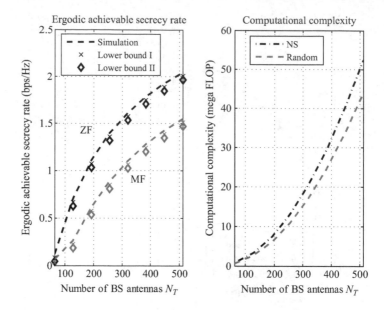

Figure 5.4   Ergodic secrecy rate (left hand side) and computational complexity (right hand side) of various linear AN precoders for a network employing $P_T = 10 \, dB$, $N_T = 128$ (right hand side), $p_\tau = P_T/\tau$, $\rho = 0.1$, $T = 500$, $\tau = K$, optimal ϕ^*, and a ZF data precoder. Lower bound I is obtained from $\underline{R}_{nk}^{sec}$, and lower bound II is obtained from $\underline{\underline{R}}_{nk}^{sec}$ in (5.18)

5.7   Conclusions and future prospects

In this chapter, we introduced the concept of physical layer security for the emerging massive MIMO systems. Specifically, the secrecy performance of a multi-cell massive MIMO with pilot contamination was evaluated in terms of achievable ergodic secrecy rate and secrecy outage probability, for different pair of linear data (RCI, ZF, and MF) and AN (NS and random) precoders. The choice of precoders strikes a trade-off between system performance and required computational complexity. Thanks to the tools from random matrix theory, the expressions of achievable ergodic secrecy rates are in closed-form, which provided sufficient design insights to practical secure massive MIMO system design.

For future work, as the hardware cost increases linearly with the number of antennas, more practical concerns are needed to be considered for real-world secure massive MIMO system design. This includes non-ideal transceivers, reduced number of radio-frequency (RF) chains, low-resolution analog–digital converters (ADCs), and constant-envelope transmission. As massive MIMO has been identified as the key technology for 5G wireless networks, and proposals on physical layer security have been considered in the emerging 5G standards, it is foreseeable that research on secure massive MIMO systems will attract more attention from academia and industry when launching 5G to practice by 2020.

A.1 Appendix

A.1.1 Proof of Lemma 5.1

The proof closely follows [29]. We first derive an expression for the secrecy rate for given realisations of $\mathbf{h}_{mk}$ and $\mathbf{H}_m^{\text{eve}}$, $k = 1, \ldots, K$, $m = 1, \ldots, M$. Since the multiple-input single-output multiple-eavesdropper (MISOME) channel in (5.5) and (5.6) is a non-degraded broadcast channel [20], the secrecy capacity is given by [29]:

$$C_{nk}^{\text{sec}}(\mathbf{h}) = \max_{s_{nk} \to \mathbf{w}_{nk}s_{nk} \to y_{nk}, \mathbf{y}_{\text{eve}}} I(s_{nk}; y_{nk}|\mathbf{h}) - I(s_{nk}; \mathbf{y}_{\text{eve}}|\mathbf{h}), \tag{A.1}$$

where vector $\mathbf{h}$ contains the CSI of all user and eavesdropper channels and $I(x; y|\mathbf{h})$ is the mutual information between two r.v.s x and y conditioned on the CSI vector. $C_{nk}^{\text{sec}}(\mathbf{h})$ is achieved by maximising over all joint distributions such that a Markov chain $s_{nk} \to \mathbf{w}_{nk}s_{nk} \to y_{nk}, \mathbf{y}_{\text{eve}}$ results, where s_{nk} is an arbitrary input variable [29]. Specifically, for $s_{nk} \sim \mathbb{CN}(0, 1)$ an achievable secrecy rate for MT k in cell n, $R_{nk}^{\text{sec}}(\mathbf{h})$, is given by:

$$R_{nk}^{\text{sec}}(\mathbf{h}) = \left[I(s_{nk}; y_{nk}|\mathbf{h}) - I(s_{nk}; \mathbf{y}_{\text{eve}}|\mathbf{h})\right]^+ \stackrel{(a)}{=} \left[I(\mathbf{w}_{nk}s_{nk}; y_{nk}|\mathbf{h}) - I(\mathbf{w}_{nk}s_{nk}; \mathbf{y}_{\text{eve}}|\mathbf{h})\right]^+$$

$$\stackrel{(b)}{\geq} \left[R_{nk}(\mathbf{h}) - C_{nk}^{\text{eve}}(\mathbf{h})\right]^+ \tag{A.2}$$

where (a) follows since $\mathbf{w}_{nk}s_{nk}$ is a deterministic function of s_{nk}. Furthermore, $R_{nk}(\mathbf{h}) \leq \max I(\mathbf{w}_{nk}s_{nk}; y_{nk}|\mathbf{h})$ is an achievable rate of MT k in cell n and $C_{nk}^{\text{eve}}(\mathbf{h}) = \log_2(1 + p\mathbf{w}_{nk}^H \mathbf{H}_n^{\text{eveH}} \mathbf{X}^{-1} \mathbf{H}_n^{\text{eve}} \mathbf{w}_{nk}) \geq I(\mathbf{w}_{nk}s_{nk}; \mathbf{y}_{\text{eve}}|\mathbf{h})$ is an upper bound on the mutual information $I(\mathbf{w}_{nk}s_{nk}; \mathbf{y}_{\text{eve}}|\mathbf{h})$. Thus, follows (b). We note that for computation of $C_{nk}^{\text{eve}}(\mathbf{h})$ we made the worst-case assumption that the eavesdropper can decode and cancel the signals of all MTs except the signal intended for the MT of interest [61, Chapter 10.2].

Finally, to arrive at the ergodic secrecy rate, we average $R_{nk}^{\text{sec}}(\mathbf{h})$ over all channel realisations, which results in [18]:

$$\mathbb{E}\left[R_{nk}^{\text{sec}}(\mathbf{h})\right] = \mathbb{E}\left[\left[R_{nk}(\mathbf{h}) - C_{nk}^{\text{eve}}(\mathbf{h})\right]^+\right] \geq \left[\mathbb{E}\left[R_{nk}(\mathbf{h})\right] - \mathbb{E}\left[C_{nk}^{\text{eve}}(\mathbf{h})\right]\right]^+ = R_{nk}^{\text{sec}}.$$

$$\tag{A.3}$$

Introducing the definitions of the achievable ergodic secrecy rate, $R_{nk} = \mathbb{E}[R_{nk}(\mathbf{h})]$, and the ergodic eavesdropper capacity, $C_{nk}^{\text{eve}} = \mathbb{E}\left[C_{nk}^{\text{eve}}(\mathbf{h})\right]$, completes the proof.

A.1.2 Proof of Theorem 5.1

We first recall that the entries of $\mathbf{H}_m^{\text{eve}}$, $m = 1, \ldots, M$, are mutually independent complex Gaussian r.v.s. On the other hand, for $N_T \to \infty$ and both AN shaping matrix designs, the vectors $\mathbf{v}_{ml}, l = 1, \ldots, N_T - K$, form an orthonormal basis. Hence, $\mathbf{H}_m^{\text{eve}} \mathbf{V}_m$, $m = 1, \ldots, M$, also has independent complex Gaussian entries, which are independent from the complex Gaussian entries of $\mathbf{H}_n^{\text{eve}} \mathbf{w}_{nk}$. Thus, the term $\gamma_{\text{eve}} = p\mathbf{w}_{nk}^H \mathbf{H}_n^{\text{eveH}} \mathbf{X}^{-1} \mathbf{H}_n^{\text{eve}} \mathbf{w}_{nk}$ in (5.8) is equivalent to the SINR of an N_E-branch MMSE diversity combiner with $M(N_T - K)$ interferers [18,62]. As a result, for the considered simplified path-loss model, the cumulative density function (CDF) of the received SINR, γ_{eve}, at the eavesdropper is given by [62]:

$$F_{\gamma_{\text{eve}}}(x) = \frac{\sum_{i=0}^{N_E-1} \lambda_i x^i}{\prod_{j=1}^{2} (1 + \mu_j x)^{b_j}},$$ (A.4)

where λ_i, μ_j, and b_j are defined in Theorem 5.1. Exploiting (A.4), we can rewrite (5.8) as:

$$C_{\text{eve}} \overset{(a)}{=} \frac{1}{\ln 2} \int_0^\infty (1+x)^{-1} F_{\gamma_{\text{eve}}}(x) dx$$

$$= \frac{1}{\ln 2} \sum_{i=0}^{N_E-1} \lambda_i \times \int_0^\infty \frac{x^i}{(1+x) \prod_{j=1}^{2} (1 + \mu_j x)^{b_j}} dx$$

$$\overset{(b)}{=} \frac{1}{\ln 2} \sum_{i=0}^{N_E-1} \lambda_i \times \frac{1}{\mu_0} \sum_{j=1}^{2} \sum_{l=1}^{b_j} \int_0^\infty \frac{\omega_{jl}}{(x+1)(x+\frac{1}{\mu_j})^l} dx$$

$$\overset{(c)}{=} \frac{1}{\ln 2} \sum_{i=0}^{N_E-1} \lambda_i \times \frac{1}{\mu_0} \sum_{j=1}^{2} \sum_{l=2}^{b_j} \omega_{jl} I(1/\mu_j, l),$$ (A.5)

where μ_0, ω_{jl}, and $I(\cdot, \cdot)$ are defined in Theorem 5.1. Here, (a) is obtained using integration by parts, (b) holds if the order of x in the denominator of (A.4) is not smaller than that in the numerator, i.e. $N_T - K \geq N_E/M$ or equivalently $1 - \beta \geq \alpha/M$, which is also the condition to ensure invertibility of $\mathbf{X}$ in (5.8), and (c) is obtained using the definition of $I(\cdot, \cdot)$ given in Theorem 5.1. This completes the proof.

A.1.3 Proof of Theorem 5.2

Using Jensen's inequality and the mutual independence of $\tilde{\mathbf{f}}_{nk} = \mathbf{H}_n^{\text{eve}} \mathbf{f}_{nk}$ and $\mathbf{H}_m^{\text{eve}} \mathbf{A}_m$, $m = 1, \ldots, M$, C_{nk}^{eve} in (5.8) is upper bounded by:

$$C_{nk}^{\text{eve}} \leq \log_2 \left(1 + \mathbb{E}_{\tilde{\mathbf{f}}_{nk}} \left[p \tilde{\mathbf{f}}_{nk}^H \mathbb{E} \left[\mathbf{X}^{-1}\right] \tilde{\mathbf{f}}_{nk}\right]\right).$$ (A.6)

Let us first focus on the term $\mathbb{E}\left[\mathbf{X}^{-1}\right]$ in (A.6) and note that $\mathbf{X}$ is statistically equivalent to a weighted sum of two scaled Wishart matrices [63]. Specifically, we have $\mathbf{X} = q\mathbf{X}_1 + \rho q \mathbf{X}_2$ with $\mathbf{X}_1 \sim \mathscr{W}_{N_E}(N_T - K, \mathbf{I}_{N_E})$ and $\mathbf{X}_2 \sim \mathscr{W}_{N_E}((M-1)(N_T - K), \mathbf{I}_{N_E})$, where $\mathscr{W}_A(B, \mathbf{I}_A)$ denotes an $A \times A$ Wishart matrix with B degrees of freedom. Strictly speaking, $\mathbf{X}$ is not a Wishart matrix, and the exact distribution of $\mathbf{X}$ seems intractable. However, $\mathbf{X}$ may be accurately approximated as a single scaled Wishart matrix, $\mathbf{X} \sim \mathscr{W}_{N_E}(\varphi, \xi \mathbf{I}_{N_E})$, where parameters ξ and φ are chosen such that the first two moments of $\mathbf{X}$ and $q\mathbf{X}_1 + \rho q \mathbf{X}_2$ are identical [57,64]. Equating the first two moments of the traces of these matrices yields [64]:

$$\xi \varphi = q(N_T - K) + \rho q(M-1)(N_T - K),$$ (A.7)

and

$$\xi^2 \varphi = q^2(N_T - K) + \rho^2 q^2 (M-1)(N_T - K).$$ (A.8)

By exploiting the expectation of an inverse Wishart matrix given in [64, Eq. (12)], we obtain $\mathbb{E}[\mathbf{X}^{-1}] = \frac{1}{\xi(\varphi - N_E - 1)}\mathbf{I}_{N_E}$ with $\xi = cq/a$ if $\varphi - N_E > 1$ or equivalently if $\beta < 1 - c\alpha/a^2$ for $N_T \to \infty$. Plugging this result and $\mathbb{E}[\tilde{\mathbf{w}}_{nk}^H \tilde{\mathbf{w}}_{nk}] = N_E$ into (A.6), we finally obtain the result in (5.16). This completes the proof.

A.1.4 Proof of Proposition 5.1

Considering $\mathbf{h}_{mn}^k = \hat{\mathbf{h}}_{mn}^k + \tilde{\mathbf{h}}_{mn}^k$ and (5.21), the effective signal power, i.e. the numerator in (5.12), can be expressed as [14]:

$$\mathbb{E}^2[\mathbf{h}_{nn}^k \mathbf{f}_{nk}] = \gamma^2 \mathbb{E}^2[\mathbf{h}_{nn}^k \mathbf{L}_{nn}(\hat{\mathbf{h}}_{nn}^k)^H] = \gamma^2 \mathbb{E}^2\left[\frac{\mathbf{h}_{nn}^k \mathbf{L}_{n,k}(\hat{\mathbf{h}}_{nn}^k)^H}{1 + \hat{\mathbf{h}}_{nn}^k \mathbf{L}_{n,k}(\hat{\mathbf{h}}_{nn}^k)^H}\right] = \frac{\gamma^2 \lambda_{nk}(X_{nk} + A_{nk})^2}{\beta_{nn}^k(1 + X_{nk})^2},$$

(A.9)

where $\mathbf{L}_{n,k} = (\hat{\mathbf{H}}_{nn}\hat{\mathbf{H}}_{nn}^H - (\hat{\mathbf{h}}_{nn}^k)^H \hat{\mathbf{h}}_{nn}^k + \kappa\mathbf{I}_{N_T})^{-1}$, $X_{nk} = \mathbb{E}[\hat{\mathbf{h}}_{nn}^k \mathbf{L}_{n,k}(\hat{\mathbf{h}}_{nn}^k)^H]$, and $A_{nk} = \mathbb{E}[\tilde{\mathbf{h}}_{nn}^k \mathbf{L}_{n,k}(\hat{\mathbf{h}}_{nn}^k)^H]$. On the other hand, the intra-cell interference term in the denominator of (5.12) can be expressed as:

$$\mathbb{E}\left[\sum_{l \neq k}|\mathbf{h}_{nn}^k \mathbf{f}_{nl}|^2\right] = \gamma^2 \mathbb{E}\left[\frac{\mathbf{h}_{nn}^k \mathbf{L}_{n,k}\hat{\mathbf{H}}_{n,k}^H \hat{\mathbf{H}}_{n,k}\mathbf{L}_{n,k}(\mathbf{h}_{nn}^k)^H}{\left(1 + \hat{\mathbf{h}}_{nn}^k \mathbf{L}_{n,k}(\hat{\mathbf{h}}_{nn}^k)^H\right)^2}\right] = \frac{\gamma^2 \lambda_{nk}(Y_{nk} + B_{nk})}{\beta_{nn}^k(1 + X)^2}, \quad (A.10)$$

where $\hat{\mathbf{H}}_{n,k}$ is equal to $\hat{\mathbf{H}}_{nn}$ with the kth row removed, and:

$$Y_{nk} = \mathbb{E}[\hat{\mathbf{h}}_{nn}^k \mathbf{L}_{n,k}\hat{\mathbf{H}}_{n,k}^H \hat{\mathbf{H}}_{n,k}\mathbf{L}_{n,k}(\hat{\mathbf{h}}_{nn}^k)^H], \quad B_{nk} = \mathbb{E}[\tilde{\mathbf{h}}_{nn}^k \mathbf{L}_{n,k}\hat{\mathbf{H}}_{n,k}^H \hat{\mathbf{H}}_{n,k}\mathbf{L}_{n,k}(\tilde{\mathbf{h}}_{nn}^k)^H]. \quad (A.11)$$

Due to pilot contamination, the data precoding matrix of BS m is a function of the channel vectors between BS m and the MTs in all cells with re-used pilots. Hence, the inter-cell interference from the BSs in adjacent cells is obtained as:

$$\mathbb{E}[|\mathbf{h}_{mn}^k \mathbf{f}_{mk}|^2] = \frac{\gamma^2 \lambda_{mk}(X_{mk} + A_{mk})^2}{\beta_{mn}^k(1 + X_{mk})^2} + \frac{1 + p_\tau \tau \sum_{l \neq n}\beta_{ml}^k}{1 + p_\tau \tau \sum_{l=1}^M \beta_{ml}^k}$$

(A.12)

and

$$\mathbb{E}\left[\sum_{l \neq k}|\mathbf{h}_{mn}^k \mathbf{f}_{ml}|^2\right] = \gamma^2 \mathbb{E}\left[\frac{\mathbf{h}_{mn}^k \mathbf{L}_{m,k}\hat{\mathbf{H}}_{m,k}^H \hat{\mathbf{H}}_{m,k}\mathbf{L}_{m,k}(\mathbf{h}_{mn}^k)^H}{\left(1 + \hat{\mathbf{h}}_{mm}^k \mathbf{L}_{m,k}(\hat{\mathbf{h}}_{mm}^k)^H\right)^2}\right] = \frac{\gamma^2 \lambda_{mk}(Y_{mk} + B_{mk})}{\beta_{mn}^k(1 + X_{mk})^2},$$

(A.13)

respectively. Meanwhile, by exploiting (A.9), (A.12), and the definition of the variance, i.e. $\text{var}[x] = \mathbb{E}[x^2] - \mathbb{E}^2[x]$, we obtain for the first term of the denominator of (5.12):

$$\text{var}[\mathbf{h}_{nn}^k \mathbf{f}_{nk}] = \frac{1 + p_\tau \tau \sum_{l \neq n}\beta_{nl}^k}{1 + p_\tau \tau \sum_{l=1}^M \beta_{nl}^k}.$$

(A.14)

According to [14, Eq. (16)], for $N_T \to \infty$ and constant β, X_{mk} converges to $\mathscr{G}(\beta, \kappa)$ defined in (5.23) and $A_{mk} \to 0$. Similarly, Y_{mk} and B_{mk} approach:

$$Y_{mk} \overset{N_T \to \infty}{=} \mathscr{G}(\beta, \kappa) + \kappa \frac{\partial}{\partial \kappa} \mathscr{G}(\beta, \kappa) \tag{A.15}$$

and

$$B_{mk} \overset{N_T \to \infty}{=} \frac{\mu_{mk}}{\lambda_{mk}} (1 + \mathscr{G}(\beta, \kappa))^2 \left(\mathscr{G}(\beta, \kappa) + \kappa \frac{\partial}{\partial \kappa} \mathscr{G}(\beta, \kappa) \right), \tag{A.16}$$

respectively, where $\frac{\partial}{\partial \kappa} \mathscr{G}(\beta, \kappa) = -\frac{\mathscr{G}(\beta,\kappa)(1+\mathscr{G}(\beta,\kappa))^2}{\beta+\kappa(1+\mathscr{G}(\beta,\kappa))^2}$.

On the other hand, the constant scaling factor γ for RCI precoding is given by [14, Eq. (22)]:

$$\gamma^2 = \frac{\phi P}{\mathscr{G}(\beta, \kappa) + \kappa \frac{\partial}{\partial \kappa} \mathscr{G}(\beta, \kappa)}. \tag{A.17}$$

Hence, employing (A.9)–(A.17) in (5.12), the received SINR in (5.22) is obtained as:

$$
\begin{aligned}
\gamma_{nk}^{\text{RCI}} &= \frac{\dfrac{\gamma^2 \lambda_{nk} X_{nk}^2}{(1 + X_{nk})^2}}{\dfrac{\gamma^2 \sum_{m=1}^{M} \lambda_{mk}(Y_{mk} + B_{mk})}{(1 + X_{mk})^2} + \sum_{m \neq n} \dfrac{\gamma^2 \lambda_{mk} X_{mk}^2}{(1 + X_{mk})^2} + qQ + 1} \\[2ex]
&= \frac{\dfrac{K \lambda_{nk}}{g + \kappa \frac{\partial g}{\partial \kappa}} \dfrac{g^2}{(1 + g)^2}}{K \sum_{m=1}^{M} \lambda_{mk} \left(\dfrac{1 + \frac{\mu_{mk}}{\lambda_{mk}}(1 + g)^2}{(1 + g)^2} \right) + \sum_{m \neq n} \dfrac{\gamma^2 \lambda_{mk} g^2}{(1 + g)^2 p} + \sigma^2} \\[2ex]
&= \frac{1}{\dfrac{\sum_{m=1}^{M} \hat{\Gamma}_{\text{RCI}}^m + (1 + g)^2}{g \left(\hat{\Gamma}_{\text{RCI}}^n + \frac{\hat{\Gamma}_{\text{RCI}}^n \kappa}{\beta}(1 + g)^2 \right)} + \sum_{m \neq n} \lambda_{mk}/\lambda_{nk}}, \tag{A.18}
\end{aligned}
$$

where we denote $g = \mathscr{G}(\beta, \kappa)$ for notational simplicity, $\sigma^2 = \eta Q + \frac{K}{\phi P_T}$ and $\hat{\Gamma}_{\text{RCI}}^m$ is defined in Proposition 5.1. This completes the proof.

A.1.5 Derivation of κ_{opt}

We first denote $\gamma_{nk}^{\text{RCI}} = \frac{1}{1/\Gamma + \sum_{m \neq n} \lambda_{mk}/\lambda_{nk}}$ in (5.22), where:

$$\Gamma = \frac{\hat{\Gamma}_{\text{RCI}}^n}{\beta} \cdot \mathscr{G}(\beta, \kappa) \cdot \frac{\beta + \kappa(1 + \mathscr{G}(\beta, \kappa))^2}{\Upsilon + (1 + \mathscr{G}(\beta, \kappa))^2}. \tag{A.19}$$

with $\Upsilon = \sum_{m=1}^{M} \hat{\Gamma}_{\text{RCI}}^m$. From (A.19), it is obvious that the optimal κ to maximise γ_{nk}^{RCI} is equivalent to the one that maximises Γ.

In order to obtain the optimal κ_{opt}, we need the following steps:

$$
\begin{aligned}
\frac{\partial \Gamma}{\partial \kappa} &= \frac{\hat{\Gamma}^n_{\mathrm{RCI}}}{\beta}\left(\frac{\partial g}{\partial \kappa}\cdot\frac{\beta+(1+g)^2}{\Upsilon+(1+g)^2}+g\cdot\frac{\partial}{\partial \kappa}\left(\frac{\beta+(1+g)^2}{\Upsilon+(1+g)^2}\right)\right) \\
&= \frac{\hat{\Gamma}^n_{\mathrm{RCI}}g}{\beta}\cdot\frac{\beta+(1+g)^2}{\Upsilon+(1+g)^2}\left(\frac{2\kappa(1+g)\partial g/\partial \kappa}{\beta+\kappa(1+g)^2}+\frac{2(1+g)\partial g/\partial \kappa}{\Upsilon+(1+g)^2}\right) \\
&= \frac{2\Upsilon^2 g(1+g)^2}{\beta\left(\Upsilon+(1+g)^2\right)^2}\frac{\partial g}{\partial \kappa}\left(\kappa-\frac{\beta}{\Upsilon}\right)=0, \qquad\qquad\qquad\text{(A.20)}
\end{aligned}
$$

where we denote $g=\mathscr{G}(\beta,\kappa)$. This finally gives $\kappa_{\mathrm{opt}}=\beta/\Upsilon$, which completes the derivation.

References

[1]   E. G. Larsson, F. Tufvesson, O. Edfors, and T. L. Marzetta, "Massive MIMO for next generation wireless systems," *IEEE Commun. Mag.*, vol. 52, no. 2, pp. 186–195, Feb. 2014.

[2]   F. Rusek, D. Persson, B. K. Lau, *et al.*, "Scaling up MIMO: Opportunities and challenges with very large arrays," *IEEE Sig. Proc. Mag.*, vol. 30, no. 1, pp. 40–46, Jan. 2013.

[3]   E. G. Larsson, and F. Tufvesson, *ICC 2013 tutorial on Massive MIMO*, part I and part II, Jun. 2013.

[4]   E. Björnson, E. G. Larsson, and T. L. Marzetta, "Massive MIMO: Ten myths and one critical question," *IEEE Commun. Mag.*, vol. 54, no. 2, pp. 114–123, Feb. 2016.

[5]   T. L. Marzetta, "Noncooperative cellular wireless with unlimited numbers of BS antennas," *IEEE Trans. Wireless Commun.*, vol. 9, no. 11, pp. 3590–3600, Nov. 2010.

[6]   H. Q. Ngo, E. G. Larsson, and T. L. Marzetta, "Energy and spectral efficiency of very large multiuser MIMO systems," *IEEE Trans. Commun.*, vol. 61, no. 4, pp. 1436–1449, Apr. 2013.

[7]   J. Hoydis, S. ten Brink, and M. Debbah, "Massive MIMO in UL/DL cellular systems: How many antennas do we need," *IEEE J. Sel. Areas Commun.*, vol. 31, no. 2, pp. 160–171, Feb. 2013.

[8]   J. Jose, A. Ashikhmin, T. L. Marzetta, and S. Vishwanath, "Pilot contamination and precoding in multi-cell TDD systems," *IEEE Trans. Wireless Commun.*, vol. 10, no. 8, pp. 2640–2651, Aug. 2011.

[9]   A. Mukherjee, S. A. A. Fakoorian, J. Huang, and A. L. Swindlehurst, "Principles of physical-layer security in multiuser wireless networks: A survey," *IEEE Commun. Surv. Tutorials*, vol. 16, no. 3, pp. 1550–1573, Aug. 2014.

[10] D. J. Love, R. W. Heath Jr., V. K. N. Lau, D. Gesbert, B. D. Rao, and M. Andrews, "An overview of limited feedback in wireless communication systems," *IEEE J. Sel. Areas Commun.*, vol. 26, no. 8, pp. 1341–1365, Oct. 2008.

[11]   M. Costa, "Writing on dirty paper," *IEEE Trans. Inform. Theory*, vol. 39, no. 3, pp. 439–441, May 1983.

[12]   X. Gao, F. Tufvesson, O. Edfors, and F. Rusek, "Measured propagation characteristics for very-large MIMO at 2.6 GHz," in *Proc. 46th Annual Asilomar Conference on Signals, Systems, and Computers*, Pacific Grove, CA, US, pp. 295–299, Nov. 2012.

[13]   H. Yang and T. L. Marzetta, "Performance of conjugate and zero-forcing beamforming in large-scale antenna systems," *IEEE J. Sel. Areas Commun.*, vol. 31, no. 2, pp. 172–179, Feb. 2013.

[14]   V. K. Nguyen and J. S. Evans, "Multiuser transmit beamforming via regularized channel inversion: A large system analysis," in *Proc. IEEE Global Communications Conference*, New Orleans, LO, US, pp. 1–4, Dec. 2008.

[15]   J. L. Massey, "An introduction to contemporary cryptology," *Proc. IEEE*, vol. 76, no. 5, pp. 533–549, May 1988.

[16]   A. D. Wyner, "The wire-tap channel," *Bell Syst. Tech. J.*, vol. 54, no. 8, pp. 1355–1387, Oct. 1975.

[17]   S. Goel and R. Negi, "Guaranteeing secrecy using artificial noise," *IEEE Trans. Wireless Commun.*, vol. 7, no. 6, pp. 2180–2189, Jun. 2008.

[18]   X. Zhou and M. R. McKay, "Secure transmission with artificial noise over fading channels: achievable rate and optimal power allocation," *IEEE Trans. Veh. Tech.*, vol. 59, pp. 3831–3842, Jul. 2010.

[19]   S. Gerbracht, C. Scheunert, and E. A. Jorswieck, "Secrecy outage in MISO systems with partial channel information," *IEEE Trans. Inform. Forensics Sec.*, vol. 7, no. 2, pp. 704–716, Apr. 2012.

[20]   A. Khisti and G. Wornell, "Secure transmission with multiple antennas I: The MISOME wiretap channel," *IEEE Trans. Inform. Theory*, vol. 56, no. 7, pp. 3088–3104, Jul. 2010.

[21]   A. Khisti and G. Wornell, "Secure transmission with multiple antennas II: The MIMOME wiretap channel," *IEEE Trans. Inform. Theory*, vol. 56, no. 11, pp. 5515–5532, Nov. 2010.

[22]   E. Ekrem and S. Ulukus, "The secrecy capacity region of the Gaussian MIMO multi-receiver wiretap channel", *IEEE Trans. Inform. Theory*, vol. 57, no. 4, pp. 2083–2114, Apr. 2011.

[23]   F. Oggier and B. Hassibi, "The secrecy capacity of the MIMO wiretap channel," *IEEE Trans. Inform. Theory*, vol. 57, no. 8, pp. 4961–4972, Aug. 2011.

[24]   D. W. K. Ng, E. S. Lo, and R. Schober, "Multi-objective resource allocation for secure communication in cognitive radio networks with wireless information and power transfer," *IEEE Trans. Veh. Tech.*, vol. 65, no. 5, pp. 3166–3184, May 2016.

[25]   D. W. K. Ng and R. Schober, "Secure and green SWIPT in distributed antenna networks with limited backhaul capacity," *IEEE Trans. Wireless Commun.*, vol. 14, no. 9, pp. 5082–5097, Sept. 2015.

[26]   H. Wang, C. Wang, D. W. K. Ng, M. H. Lee, and J. Xiao, "Artificial noise assisted secure transmission under training and feedback," *IEEE Trans. Sig. Proc.*, vol. 63, no. 23, pp. 6285–6298, Dec. 2015.

[27] C. Wang, H. Wang, D. W. K. Ng, X.-G. Xia, and C. Liu, "Joint beamforming and power allocation for security in peer-to-peer relay networks," *IEEE Trans. Wireless Commun.*, vol. 14, no. 6, pp. 3280–3293, Jun. 2015.

[28] N. Zhao, F. R. Yu, M. Li, Q. Yan, and V. C. M. Leung, "Physical layer security issues in interference alignment (IA)-based wireless networks," *IEEE Commun. Mag.*, to appear, Jun. 2016.

[29] G. Geraci, M. Egan, J. Yuan, A. Razi, and I. Collings, "Secrecy sum-rates for multi-user MIMO regularized channel inversion precoding," *IEEE Trans. Commun.*, vol. 60, no. 11, pp. 3472–3482, Nov. 2012.

[30] M. Pei, J. Wei, K.-K. Wong, and X. Wang, "Masked beamforming for multiuser MIMO wiretap channels with imperfect CSI," *IEEE Trans. Wireless Commun.*, vol. 11, no. 2, pp. 544–549, Feb. 2012.

[31] A. Mukherjee, and A. L. Swindlehurst, "Robust beamforming for security in MIMO wiretap channels with imperfect CSI," *IEEE Trans. Sig. Proc.*, vol. 59, no. 1, pp. 351–361, Jan. 2011.

[32] Z. Peng, W. Xu, J. Zhu, H. Zhang, and C. Zhao, "On performance and feedback strategy of secure multiuser communications with MMSE channel estimate," *IEEE Trans. Wireless Commun.*, vol. 15, no. 2, pp. 1602–1616, Feb. 2016.

[33] Y. Sun, D. W. K. Ng, J. Zhu, and R. Schober, "Multi-objective optimization for robust power efficiency and secure full-duplex wireless communications systems," *IEEE Trans. Wireless Commun.*, vol. 15, no. 8, pp. 5511–5526, Aug. 2016.

[34] J. Zhu, W. Xu, and V. K. Bhargava, "Relay precoding in multi-user MIMO channels for physical layer security," in *Proc. IEEE/CIC International Communications Conference in China (ICCC 2014)*, Shanghai, P.R. China, Oct. 2014.

[35] D. Kapetanovic, G. Zheng, and F. Rusek, "Physical layer security for massive MIMO: An overview on passive eavesdropping and active attacks," *IEEE Commun. Mag.*, vol. 53, no. 6, pp. 21–27, Jun. 2015.

[36] J. Zhu, R. Schober, and V. K. Bhargava, "Secure transmission in multicell massive MIMO systems, " *IEEE Trans. Wireless Commun.*, vol. 13, no. 9, pp. 4766–4781, Sept. 2014.

[37] X. Zhou, B. Maham, and A. Hjorungnes, "Pilot contamination for active evesdropping," *IEEE Trans. Wireless Commun.*, vol. 11, no. 3, pp. 903–907, Mar. 2012.

[38] D. Kapetanovic, G. Zheng, K.-K. Wong, and B. Ottersten, "Detection of pilot contamination attack using random training in massive MIMO," in *Proc. IEEE Intern. Symp. Personal, Indoor and Mobile Radio Commun. (PIMRC)*, pp. 13–18, London, UK, Sept. 2013.

[39] N. Yang, L. Wang, G. Geraci, M. Elkashlan, J. Yuan, and M. D. Renzo, "Safeguarding 5G wireless communication networks using physical layer security," *IEEE Commun. Mag.*, vol. 53, no. 4, pp. 20–27, Apr. 2015.

[40] G. Geraci, M. Egan, J. Yuan, A. Razi, and I. B. Collings, "Secrecy sum-rates for multi-user MIMO regularized channel inversion precoding," *IEEE Trans. Commun.*, vol. 60, no. 11, pp. 3472–3482, Nov. 2012.

[41] G. Geraci, H. S. Dhillon, J. G. Andrews, J. Yuan, and I. B. Collings, "Physical layer security in downlink multi-antenna cellular networks," *IEEE Trans. Commun.*, vol. 62, no. 6, pp. 2006–2021, Jun. 2014.

[42] G. Geraci, J. Yuan, and I. B. Collings, "Large system analysis of linear precoding in MISO broadcast channels with confidential messages," *IEEE J. Sel. Areas Commun.*, vol. 31, no. 9, pp. 1660–1671, Sept. 2013.

[43] T. Dean and A. Goldsmith, "Physical layer cryptography through massive MIMO," *Proc. IEEE Inform. Theory Workshop*, Sevilla, pp. 1–5, Sept. 2013.

[44] J. Zhu, R. Schober, and V. K. Bhargava, "Linear precoding of data and artificial noise in secure massive MIMO systems," *IEEE Trans. Wireless Commun.*, vol. 15, no. 3, pp. 2245–2261, Mar. 2016.

[45] J. Zhu, D. W. K. Ng, N. Wang, R. Schober, and V. K. Bhargava, "Analysis and design of secure massive MIMO systems in the presence of hardware impairments," *IEEE Trans. Wireless Commun.*, vol. 16, no. 3, pp. 2001–2016, Mar. 2017.

[46] J. Wang, J. Lee, F. Wang, and T. Q. S. Quek, "Jamming-aided secure communication in massive MIMO Rician channels," *IEEE Trans. Wireless Commun.*, vol. 14, no. 12, pp. 6854–6868, Dec. 2015.

[47] J. Zhu and W. Xu, "Securing massive MIMO via power scaling," *IEEE Commun. Lett.*, vol. 20, no. 5, pp. 1014–1017, May 2016.

[48] J. Zhu, W. Xu, and N. Wang, "Secure massive MIMO systems with limited RF chains," to appear in *IEEE Trans. Veh. Tech.*, Oct. 2016. DOI: 10.1109/TVT.2016.2615885.

[49] J. Zhu, N. Wang, and V. K. Bhargava, "Per-antenna constant envelope precoding for secure transmission in large-scale MISO systems," in *Proc. IEEE/CIC International Communications Conference in China (ICCC 2015)*, Shenzhen, P.R. China, Nov. 2015.

[50] X. Chen, L. Lei, H. Zhang, and C. Yuen, "Large-scale MIMO relaying techniques for physical layer security: AF or DF?" *IEEE Trans. Wireless Commun.*, vol. 14, no. 9, pp. 5135–5146, Sept. 2015.

[51] J. Chen, X. Chen, W. Gerstacker, and D. W. K. Ng, "Resource allocation for a massive MIMO relay aided secure communication," *IEEE Trans. Inform. Forensics Sec.*, vol. 11, no. 8, pp. 1700–1711, Aug. 2016.

[52] J. Zhang, C. Yuen, C.-K. Wen, S. Jin, K.-K. Wong, and H. Zhu, "Large system secrecy rate analysis for SWIPT MIMO wiretap channels," *IEEE Trans. Inform. Forensics Sec.*, vol. 11, no. 1, pp. 74–85, Jan. 2016.

[53] S. Im, H. Jeon, J. Choi, and J. Ha, "Secret key agreement with large antenna arrays under the pilot contamination attack," *IEEE Trans. Wireless Commun.*, vol. 14, no. 12, pp. 6579–6594, Dec. 2015.

[54] Y. Basciftci, C. Koksal, and A. Ashikhmin, "Securing massive MIMO at the physical layer," *Proc. IEEE CNS*, Florence, Italy, Sep. 2015.

[55] Y. Wu, R. Schober, D. W. K. Ng, C. Xiao, and G. Caire, "Secure massive MIMO transmission with an active eavesdropper," *IEEE Trans. Inform. Theory*, vol. 62, no. 7, pp. 3880–3900, Jul. 2016.

[56] J. Zhang, R. W. Heath Jr., M. Koutouris, and J. G. Andrews, "Mode switching for MIMO broadcast channel based on delay and channel quantization," *EURASIP J. Adv. Sig. Proc.*, vol. 2009, Feb. 2009 doi:10.1155/2009/802548.

[57] Q. T. Zhang and D. P. Liu, "A simple capacity formula for correlated diversity Rician channels," *IEEE Commun. Lett.*, vol. 6, no. 11, pp. 481–483, Nov. 2002.

[58] R. Hunger, "Floating point operations in matrix-vector calculus," Technische Universität München, Associate Institute for Signal Processing, Tech. Rep., 2007.

[59] S. Zarei, W. Gerstacker, R. R. Muller, and R. Schober, "Low-complexity linear precoding for downlink large-scale MIMO systems," in *Proc. IEEE Int. Symp. Personal, Indoor and Mobile Radio Commun. (PIMRC)*, London, UK, Sept. 2013.

[60] A. Müller, A. Kammoun, E. Björnson, and M. Debbah, "Linear precoding based on polynomial expansion: Reducing complexity in massive MIMO," *EURASIP J. Wireless Commun. Networking*, 2016:63, Feb. 2016. DOI: 10.1186/s13638-016-0546-z.

[61] D. Tse and P. Viswanath, "Fundamentals of wireless communications," *Cambridge University Press*, 2005.

[62] H. Gao, P. J. Smith, and M. V. Clark, "Theoretical reliability of MMSE linear diversity combining in Rayleigh-fading additive interference channels," *IEEE Trans. Commun.*, vol. 46, no. 5, pp. 666–672, May 1998.

[63] A. M. Tulino and S. Verdu, "Random matrix theory and wireless communications," *Found. Trends Commun. Inf. Theory*, vol. 1, no. 1, pp. 1–182, Jun. 2004.

[64] S. W. Nydick, "The Wishart and Inverse Wishart Distributions," May 2012, [online] http://www.tc.umn.edu/nydic001/docs/unpubs/Wishart_Distribution.pdf.

Chapter 6

Physical layer security for massive MIMO with anti-jamming

Tan Tai Do[1], Hien Quoc Ngo[2], and Trung Q. Duong[2]

Massive multiple-input multiple-output (MIMO) is one of the most promising technologies for the next generation of wireless systems. While many aspects of massive MIMO have been intensively investigated, physical layer security in massive MIMO has not attracted much attention. Among the very few research studies on massive MIMO physical layer security, only some of them have focused on jamming aspects although jamming exists as a critical problem for reliable communications. Understanding the effect of jamming attacks is especially important for massive MIMO systems, which are sensitive to channel estimation errors and pilot contamination. This chapter provides some basic concepts of the physical layer security for massive MIMO and jamming attacks. The effect of the jamming attacks on pilot contamination is analysed. Potential counter-attack strategies, which can enhance the robustness of massive MIMO against jamming, are also presented.

6.1 Introduction

This section provides a brief introduction on massive multiple-input multiple-output (MIMO) systems, physical layer security, and jamming attacks.

6.1.1 Massive MIMO

One of the well-known truths of wireless communications is that the demand for data throughput will always grow. According to Cisco, global mobile data throughput has grown 4,000-fold over the past 10 years, and is expected to increase to 32 exabytes per month by 2020, which is approximately an eight-fold increase over 2015 [1]. The demand for data throughput will be even much more in the future since the number of wireless devices with many real-time applications is continuously increasing. Let R (bits/s) be the data throughput, W (Hz) be the available bandwidth, and S (bits/s/Hz)

[1]Department of Electrical Engineering (ISY), Linköping University, Sweden
[2]School of Electronics, Electrical Engineering and Computer Science (EEECS), Queen's University Belfast, UK

be the spectral efficiency. The data throughput is then defined as:

$$R = W \cdot S \text{ (bits/s)}. \tag{6.1}$$

Accordingly, to improve the throughput, new technologies and solutions that increase the bandwidth or/and the spectral efficiency, should be deployed. One emerging technology which can improve the spectral efficiency significantly is massive MIMO [2–5]. In massive MIMO systems, the base station with many service antennas (hundreds or thousands of antennas) can be collocated in a compact area [2] or distributed in a large area [6] to simultaneously serve as many as tens or hundreds of users in the same time-frequency resource. Some advantages of these systems are as follows:

- Favourable propagation: According to the law of large numbers and the central limit theorem, for most propagation environments, when the number of antennas at the base station is large, the channel vectors from the base station to the users are (nearly) pairwisely orthogonal. This property, called favourable propagation, significantly reduces the uncorrelated interference and noise, and more importantly, results in a simplified processing system. With linear processing such as maximum-ratio or zero-forcing processing, we can obtain nearly optimal performance.

- Channel hardening: Channel hardening is a phenomenon where the norms of the channel vectors fluctuate little (i.e. very close to their mean values). This phenomenon appears in massive MIMO thanks to the law of large numbers and the randomness of the channels. With channel hardening property, the system performance depends only on the large-scale fading. As a result, the system design including the scheduling, power control, and interference management can be done over the large-scale fading time scale. Thus, the corresponding overhead is reduced significantly.

- Array gain and multiplexing gain: The large number of users and antennas being used allow massive MIMO systems to achieve large array gain and multiplexing gain, which result in huge energy efficiency and huge spectral efficiency. As quantitatively shown in [4], massive MIMO systems with a 100-antenna base station and simple maximum-ratio processing can offer, at the same time, a 100-fold improvement in spectral efficiency and 100-fold improvement in energy efficiency, compared with single-input–single-output systems.

- Low-latency: Besides data throughput, low-latency is another primary target of the future wireless communication systems. Fading is the main cause of latency in wireless communication. If a receiver suffers a bad (deep fading) channel, it needs to wait until the channel quality is improved, resulting in latency. Massive MIMO can overcome this fading dip by relying on the large number of antennas (which reduces the fluctuation of the fading), and hence, it can offer low-latency communications.

- All complexity being at the base station: The design of massive MIMO involves a large number of antennas and radio chains. However, since all the complexity can be allocated at the base station, massive MIMO is easy to deploy and compatible with the current systems and users' devices.

Massive MIMO is maturing. Many practical aspects behind the design of massive MIMO have been investigated in depth. While channel estimation and pilot

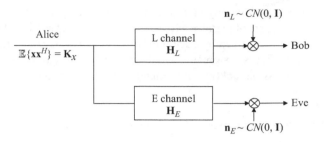

Figure 6.1   A MIMO Gaussian wiretap channel

contamination are well-known as the main sources of limitations in massive MIMO, securing transmission is also an important issue that needs to be addressed in the design of massive MIMO.

6.1.2 Physical layer security on massive MIMO

Although massive MIMO has received a lot of attention recently, there has been very little work on physical layer security in massive MIMO [7–13], mainly due to the common belief that physical layer security techniques in regular MIMO systems, which have been intensively studied, can be straight-forwardly extended to massive MIMO networks. However, as it has been shown in [10,11], massive MIMO brings challenges as well as opportunities to the physical layer security, which is fundamentally different from conventional MIMO systems.

For instance, let us consider a MIMO Gaussian wiretap channel as in Figure 6.1. A legitimate user (Bob) communicates with a base station (Alice) in the presence of an adversary (Eve), who wants to eavesdrop the information of the legitimate link.

Let us define the secrecy capacity of a wiretap channel as the maximum transmission rate such that the information can be reliably decoded at the legitimate receiver but cannot be inferred at any positive rate by the eavesdropper. The secrecy capacity C_s of the MIMO Gaussian wiretap channel is then given by [14]:

$$C_s = \max\left\{0, \max_{\mathbf{K}_x \geq 0}\left[\log\det\left(\mathbf{I} + \mathbf{H}_L\mathbf{K}_x\mathbf{H}_L^H\right) - \log\det\left(\mathbf{I} + \mathbf{H}_E\mathbf{K}_x\mathbf{H}_E^H\right)\right]\right\}, \quad (6.2)$$

where $\mathbf{K}_x$ is the covariance matrix of the transmit signal, $\mathbf{H}_L$ and $\mathbf{H}_E$ are the channel matrices from the base station to the legitimate user and the eavesdropper, respectively. This secrecy capacity can be interpreted as the difference between the channel capacity of the legitimate channel and the capacity of the eavesdropping channel. In the conventional MIMO systems, these capacities are of a similar order, resulting in a relative small secrecy capacity. However, in massive MIMO systems, thanks to the array gain and the channel hardening effect, the transmitter can design $\mathbf{K}_x$ such that the capacity corresponding to the legitimate link is much higher than the one corresponding to the overhearing link. Thus, one can achieve a very good secrecy capacity in massive MIMO without extra effort. In other words, massive MIMO is naturally good for physical layer security against passive attackers (eavesdroppers).

While passive eavesdropping can be easily overcome, active attacks (jamming) are a big issue in the design of massive MIMO systems. When a massive MIMO system is attacked by a jammer, especially during the training phase, the resulting jamming-pilot contamination causes a significant reduction in the system performance [11]. This chapter will focus on tackling the jamming at the physical security in massive MIMO.

6.1.3 Jamming on massive MIMO systems

Depending on how much knowledge the jammer can learn about a massive MIMO system, it can attack the system in various ways. Some common attacks can be listed as follows:

- Attack on the data transmission phase only: In this case, the pilot signal of the system is well protected (e.g. using a special transmission band or pilot encryption), the jammer can decide to attack only the data transmission phase.
- Attack on both the training and data transmission phases: In this case, the jammer attacks the system as long as there is communication within the legitimate system.
- Attack with random signals: If the jammer does not have knowledge on the pilot signals used by the legitimate system, it will send random signals to attack the legitimate system. Random signal with Gaussian distribution is a natural choice since it is the most harmful noise in general.
- Attack with deterministic signals: This occurs, for example, when the jammer has partial knowledge of the pilot signals used by the legitimate link, such as length and pilot sequence codebook. It can try to attack using a signal which is a deterministic function (e.g. a linear combination) of those pilot sequences.

Pilot contamination appears when the pilot signal, transmitted for estimation of the desired channel, is interfered by other transmissions. The typical effect is that the base station cannot use the estimated channel to amplify the desired signal, without also coherently combining the interference. Pilot contamination between users within the system itself is already a big challenge in massive MIMO. The problem with jamming-pilot contamination is even trickier since it is caused from the attack of the jammer, which aims to maximise the pilot contamination rather than minimise it. Moreover, the information related to the jamming attack such as the jamming channel and signal is basically unknown by the system. Therefore, the conventional interference rejection techniques cannot be applied to eliminate the effect of jamming signal.

When the massive MIMO systems are attacked by the jammer, from a system design perspective, some natural questions would be:

(i)   How bad the pilot contamination is under the jamming attacks?
(ii) Which signal processing techniques should be used to mitigate the effect of pilot contamination due to the jamming attacks?
(iii) How can we exploit the advantages of massive MIMO to deal with jamming attacks?

In this chapter, we aim to address these questions.

6.2 Uplink massive MIMO with jamming

Consider a single user massive MIMO uplink with the presence of a jammer as depicted in Figure 6.2. The jammer aims at disrupting the uplink transmission from the legitimate user to the base station. We assume that the base station is equipped with M antennas, while the legitimate user and the jammer have a single antenna. Although this chapter focuses on a single-user setup, the anti-jamming framework is still somehow similar to the one in a multi-user system. The reason is that, if multiple users are considered, an additional term contributed to the system is inter-user interference, which can be negligible when M is large [2].

Let us denote $\mathbf{g}_u \in \mathbb{C}^{M \times 1}$ and $\mathbf{g}_j \in \mathbb{C}^{M \times 1}$ as the channel vectors from the user and the jammer to the base station, respectively. The channel model consists of the effects of small-scale fading and large-scale fading (path loss and shadowing). We consider a block-fading model, in which the small-scale fading remains constant during a coherence block of T symbols, and varies independently from one coherence block to the next. The large-scale fading changes much more slowly and stays constant for several tens of coherence intervals. More precisely, $\mathbf{g}_u$ and $\mathbf{g}_j$ are modelled as:

$$\mathbf{g}_u = \sqrt{\beta_u}\mathbf{h}_u, \tag{6.3}$$

$$\mathbf{g}_j = \sqrt{\beta_j}\mathbf{h}_j, \tag{6.4}$$

where $\mathbf{h}_u$ and $\mathbf{h}_j$ represent small-scale fading, while β_u and β_j represent large-scale fading. We assume that the elements of $\mathbf{h}_u$ and $\mathbf{h}_j$ are independent and identically distributed (i.i.d.) zero mean circularly symmetric complex Gaussian (ZMCSCG) random variables.

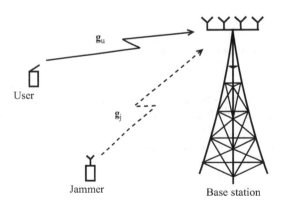

Figure 6.2 Massive MIMO uplink with jamming attack

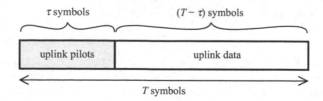

Figure 6.3   Allocation of the symbols for the uplink pilot and uplink payload data transmission in a coherence interval

The communication between the legitimate user and the base station is done under time-division duplex (TDD) operation. During each coherence interval, there are three phases:

1. Uplink training: The legitimate user sends pilot sequences to the base station for channel estimation.
2. Uplink payload data transmission: The user sends data to the base station. The base station uses the channel acquired during the uplink training phase to detect the desired data.
3. Downlink payload data transmission: The base station uses the channel estimate to process the data intended to the user. Then it sends the processed signal to the user. We assume that the calibration of the hardware chains is perfect so that the channel is reciprocal, i.e. the channel gains of the uplink and of the downlink are the same.

Since our focus is the uplink, the downlink data transmission phase can be neglected. The two-phase TDD transmission protocol of the uplink is shown in Figure 6.3.

We assume that the jammer attacks the uplink transmission in both the training and data payload transmission phases.

6.2.1 Training phase

The channel is estimated at the base station once for each coherence interval. During the first τ symbols ($\tau < T$), the user sends to the base station a pilot sequence $\sqrt{\tau p_t}\mathbf{s}_u$, where p_t is the transmit pilot power and $\mathbf{s}_u \in \mathbb{C}^{\tau \times 1}$ originates from a pilot codebook $\mathscr{S}$ containing τ orthogonal unit-power vectors. At the same time, the jammer sends $\sqrt{\tau q_t}\mathbf{s}_j$ to interfere the channel estimation, where $\mathbf{s}_j \in \mathbb{C}^{\tau \times 1}$ satisfies $\mathbb{E}\{\|\mathbf{s}_j\|^2\} = 1$ and q_t is the transmit power of the jammer during the training phase. Accordingly, the received pilot signal at the base station is given by:

$$\mathbf{Y}_t = \sqrt{\tau p_t}\mathbf{g}_u\mathbf{s}_u^T + \sqrt{\tau q_t}\mathbf{g}_j\mathbf{s}_j^T + \mathbf{N}_t, \tag{6.5}$$

where $\mathbf{N}_t \in \mathbb{C}^{M \times \tau}$ is the additive noise matrix with unit power i.i.d. ZMCSCG elements, i.e. $\text{Vec}(\mathbf{N}_t) \sim CN(0, \mathbf{I}_{M\tau})$.

The base station knows the pilot sequence $\mathbf{s}_u$, it uses this information to estimate the channel $\mathbf{g}_u$ from its received pilot signal $\mathbf{Y}_t$. The channel estimation is done via two steps:

1. The base station first performs a de-spreading operation:

$$
\begin{aligned}
\mathbf{y}_t &= \mathbf{Y}_t \mathbf{s}_u^* \\
&= \sqrt{\tau p_t} \mathbf{g}_u + \sqrt{\tau q_t} \mathbf{g}_j \mathbf{s}_j^T \mathbf{s}_u^* + \tilde{\mathbf{n}}_t,
\end{aligned}
\tag{6.6}
$$

   where $\tilde{\mathbf{n}}_t \triangleq \mathbf{N}_t \mathbf{s}_u^*$. Since $\|\mathbf{s}_u\|^2 = 1$, we have $\tilde{\mathbf{n}}_t \sim CN(0, \mathbf{I}_M)$.

2. Then the base station uses the minimum mean-square error (MMSE) channel estimation technique to estimate $\mathbf{g}_u$. With MMSE criterion, the base station wants to estimate the channel which minimises the mean-square error:

$$
\begin{aligned}
\hat{\mathbf{g}}_u &= \arg \min_{\theta \in \mathbb{C}^{M \times 1}} \mathbb{E}\left\{ \|\theta - \mathbf{g}_u\|^2 \right\} \\
&= \mathbb{E}\left\{ \mathbf{g}_u | \mathbf{y}_t \right\}.
\end{aligned}
\tag{6.7}
$$

Since $\mathbf{g}_u$ and $\mathbf{y}_t$ are jointly Gaussian, MMSE estimator is linear MMSE estimator. Therefore, the MMSE estimate of $\mathbf{g}_u$ given $\mathbf{y}_t$ is [15]:

$$
\hat{\mathbf{g}}_u = \left(\mathbb{E}\left\{ \mathbf{g}_u \mathbf{y}_t^H \right\} \mathbb{E}\left\{ \mathbf{y}_t \mathbf{y}_t^H \right\} \right)^{-1} \mathbf{y}_t,
$$

where $\mathbb{E}\left\{ \mathbf{g}_u \mathbf{y}_t^H \right\} = \sqrt{\tau p_t} \beta_u$ and $\mathbb{E}\left\{ \mathbf{y}_t \mathbf{y}_t^H \right\} = \tau p_t \beta_u + \tau q_t \beta_j \left| \mathbf{s}_j^T \mathbf{s}_u^* \right|^2 + 1$. Thus,

$$
\begin{aligned}
\hat{\mathbf{g}}_u &= \frac{\sqrt{\tau p_t} \beta_u}{\tau p_t \beta_u + \tau q_t \beta_j \left| \mathbf{s}_j^T \mathbf{s}_u^* \right|^2 + 1} \mathbf{y}_t \\
&= c_u \sqrt{\tau p_t} \mathbf{g}_u + c_u \sqrt{\tau q_t} \mathbf{g}_j \mathbf{s}_j^T \mathbf{s}_u^* + c_u \tilde{\mathbf{n}}_t,
\end{aligned}
\tag{6.8}
$$

with

$$
c_u \triangleq \frac{\sqrt{\tau p_t} \beta_u}{\tau p_t \beta_u + \tau q_t \beta_j |\mathbf{s}_j^T \mathbf{s}_u^*|^2 + 1}.
\tag{6.9}
$$

In order to perform the MMSE estimation in (6.8), the base station has to know β_u, β_j, and $|\mathbf{s}_j^T \mathbf{s}_u^*|^2$. Since β_u and β_j are large-scale fading coefficients which change very slowly with time (some 40 times slower than the small-scale fading coefficients), they can be estimated at the base station easily [16]. The quantity $|\mathbf{s}_j^T \mathbf{s}_u^*|^2$ includes the jamming sequence $\mathbf{s}_j$ which is unknown at the base station. However, by exploiting asymptotic properties of the massive MIMO, the base station can estimate $|\mathbf{s}_j^T \mathbf{s}_u^*|^2$ from the received pilot signal $\mathbf{Y}_t$. The details for the estimation of $|\mathbf{s}_j^T \mathbf{s}_u^*|^2$ are discussed in Section 6.5.

Let $\mathbf{e}_u$ be the channel estimation error, i.e.:

$$
\mathbf{e}_u = \mathbf{g}_u - \hat{\mathbf{g}}_u.
\tag{6.10}
$$

From the properties of MMSE estimation, $\hat{\mathbf{g}}_u$ and $\mathbf{e}_u$ are uncorrelated. Since $\hat{\mathbf{g}}_u$ and $\mathbf{e}_u$ are jointly Gaussian, they are independent. From (6.8), we have:

$$
\hat{\mathbf{g}}_u \sim CN(0, \gamma_u \mathbf{I}_M),
\tag{6.11}
$$

and

$$\mathbf{e_u} \sim CN\left(0, (\beta_u - \gamma_u)\, \mathbf{I}_M\right), \tag{6.12}$$

where

$$\gamma_u \triangleq \frac{\tau p_t \beta_u^2}{\tau p_t \beta_u + \tau q_t \beta_j \left|\mathbf{s}_j^T \mathbf{s}_u^*\right|^2 + 1}. \tag{6.13}$$

6.2.2 Data transmission phase

During the remaining part of the coherence interval, the user transmits the payload data to the base station. The jammer continues to interfere the legitimate link with its jamming signal. Let x_u, where $\mathbb{E}\{|x_u|^2\} = 1$, and x_j, where $\mathbb{E}\{|x_j|^2\} = 1$, be the signals transmitted from the user and the jammer, respectively. The received signal at the base station is then given by:

$$\mathbf{y}_d = \sqrt{p_d}\mathbf{g}_u x_u + \sqrt{q_d}\mathbf{g}_j x_j + \mathbf{n}_d, \tag{6.14}$$

where the noise vector $\mathbf{n}_d$ is assumed to have i.i.d. $CN(0, 1)$ elements, p_d and q_d are the transmit powers from the user and jammer in the data transmission phase, respectively. We assume that the transmit powers of the user and jammer satisfy:

$$\tau p_t + (T - \tau)p_d \leq TP,$$

$$\tau q_t + (T - \tau)q_d \leq TQ,$$

where P and Q are the average power constraints of the legitimate user and jammer, respectively.

To detect x_u, the base station performs the maximal-ratio combining (MRC) using the channel estimate $\hat{\mathbf{g}}_u$ as follows:

$$\begin{aligned} y &= \hat{\mathbf{g}}_u^H \mathbf{y}_d \\ &= \sqrt{p_d}\hat{\mathbf{g}}_u^H \mathbf{g}_u x_u + \sqrt{q_d}\hat{\mathbf{g}}_u^H \mathbf{g}_j x_j + \hat{\mathbf{g}}_u^H \mathbf{n}_d. \end{aligned} \tag{6.15}$$

Note that, with perfect channel state information (CSI) and without interference from the jammer, the MRC is optimal in the sense that it maximises the signal-to-noise ratio (SNR).

6.3 Jamming-pilot contamination

We can see from (6.8) that the channel estimate includes the true channel $\mathbf{g}_u$ contaminated by the jamming channel $\mathbf{g}_j$, which causes a degradation on the system performance of the legitimate link. This effect is called jamming-pilot contamination. In this section, we will show that the jamming-pilot contamination causes the ultimate limit on the system performance.

Using (6.8), (6.15) can be rewritten as:

$$y = \left(c_u\sqrt{\tau p_t}\mathbf{g}_u + c_u\sqrt{\tau q_t}\mathbf{g}_j \mathbf{s}_j^T \mathbf{s}_u^* + c_u\tilde{\mathbf{n}}_t\right)^H \left(\sqrt{p_d}\mathbf{g}_u x_u + \sqrt{q_d}\mathbf{g}_j x_j + \mathbf{n}_d\right). \tag{6.16}$$

By dividing y by M,

$$\frac{y}{M} = c_u\sqrt{\tau p_t p_d}\frac{\|\mathbf{g}_u\|^2}{M}x_u + c_u\sqrt{\tau q_t q_d}\mathbf{s}_u^T\mathbf{s}_j^*\frac{\|\mathbf{g}_j\|^2}{M}x_j + c_u\sqrt{\tau p_t p_d}\frac{\mathbf{g}_u^H\left(\sqrt{q_d}\mathbf{g}_j x_j + \mathbf{n}_d\right)}{M}$$

$$+ c_u\sqrt{\tau q_t q_d}\mathbf{s}_u^T\mathbf{s}_j^*\frac{\mathbf{g}_j^H\left(\sqrt{p_d}\mathbf{g}_u x_u + \mathbf{n}_d\right)}{M} + c_u\frac{\tilde{\mathbf{n}}_t^H\left(\sqrt{p_d}\mathbf{g}_u x_u + \sqrt{q_d}\mathbf{g}_j x_j + \mathbf{n}_d\right)}{M},$$

$$\text{(6.17)}$$

and using the law of large numbers, as $M \to \infty$:

$$\frac{\|\mathbf{g}_u\|^2}{M} \overset{\text{a.s.}}{\to} \beta_u,$$

$$\frac{\|\mathbf{g}_j\|^2}{M} \overset{\text{a.s.}}{\to} \beta_j,$$

$$\frac{\mathbf{g}_u^H\left(\sqrt{q_d}\mathbf{g}_j x_j + \mathbf{n}_d\right)}{M} \overset{\text{a.s.}}{\to} 0,$$

$$\frac{\mathbf{g}_j^H\left(\sqrt{p_d}\mathbf{g}_u x_u + \mathbf{n}_d\right)}{M} \overset{\text{a.s.}}{\to} 0,$$

$$\frac{\tilde{\mathbf{n}}_t^H\left(\sqrt{p_d}\mathbf{g}_u x_u + \sqrt{q_d}\mathbf{g}_j x_j + \mathbf{n}_d\right)}{M} \overset{\text{a.s.}}{\to} 0, \qquad \text{(6.18)}$$

we obtain

$$\frac{y}{M} \overset{\text{a.s.}}{\to} c_u\sqrt{\tau p_t p_d}\beta_u x_u + c_u\sqrt{\tau q_t q_d}\mathbf{s}_u^T\mathbf{s}_j^*\beta_j x_j, \text{ as } M \to \infty, \qquad \text{(6.19)}$$

where $\overset{\text{a.s.}}{\to}$ denotes the almost sure convergence. Two important remarks can be made:

(i)   If the jammer does not attack during the training phase ($q_t = 0$) or the attacked signal $\mathbf{s}_j$ is orthogonal with the pilot sequence $\mathbf{s}_u$, i.e. $\mathbf{s}_j^T\mathbf{s}_u^* = 0$, then (6.19) becomes:

$$\frac{y}{M} \overset{\text{a.s.}}{\to} c_u\sqrt{\tau p_t p_d}\beta_u x_u, \text{ as } M \to \infty. \qquad \text{(6.20)}$$

Even if the system is attacked during the data transmission phase, the received signal (normalised by M) at the base station includes only the desired signal without the interference and noise effects. This is thank to the favourable propagation of the massive MIMO channels. Indeed, when the number of antennas at the base station is large, the legitimate channel vector is nearly orthogonal with the jamming channel vector. Therefore, if the training phase is not attacked and the estimation of the legitimate channel is good enough, the received filter can amplify the desired signal without coherently combining the jamming signal.

(ii)   If the jammer attacks during the training phase and $\mathbf{s}_j^T\mathbf{s}_u^* \neq 0$, we have:

$$\frac{y}{M} \overset{\text{a.s.}}{\to} c_u\sqrt{\tau p_t p_d}\beta_u x_u + c_u\sqrt{\tau q_t q_d}\mathbf{s}_u^T\mathbf{s}_j^*\beta_j x_j, \text{ as } M \to \infty. \qquad \text{(6.21)}$$

The corresponding signal-to-interference-plus-noise ratio (SINR) converges to:

$$\text{SINR} \to \frac{p_t p_d \beta_u^2}{q_t q_d \beta_j^2 \left| \mathbf{s}_u^T \mathbf{s}_j^* \right|^2}. \tag{6.22}$$

When training phase is attacked, as the number of antennas at the base station grows without bound, the effects of small-scale fading and uncorrelated noise disappear. However, the jamming-pilot contamination effect persists, which dictates the ultimate limit on the system performance of the legitimate link.

6.4 Achievable rate

We next study the impact of jamming attacks on the system with a finite number of base station antennas through the achievable rate.

The substitution of $\mathbf{g}_u = \hat{\mathbf{g}}_u + \mathbf{e}_u$ into (6.15) yields:

$$y = \sqrt{p_d} \|\hat{\mathbf{g}}_u\|^2 x_u + \sqrt{p_d} \hat{\mathbf{g}}_u^H \mathbf{e}_u x_u + \sqrt{q_d} \hat{\mathbf{g}}_u^H \mathbf{g}_j x_j + \hat{\mathbf{g}}_u^H \mathbf{n}_d. \tag{6.23}$$

Since we are considering massive MIMO systems where M is large, the effective channel gain becomes hardening, i.e. $\|\hat{\mathbf{g}}_u\|^2$ is very close to its mean $\mathbb{E}\{\|\hat{\mathbf{g}}_u\|^2\}$. Therefore, we apply the bounding technique in [17] to derive a capacity lower bound (achievable rate). The nice things of this bounding technique are: (i) it yields a simple closed-form expression which can be used for further analysis and system designs, and (ii) the resulting bound is very tight due to the hardening property of the channel in massive MIMO. To this end, we decompose the received signal in (6.23) as:

$$y = \underbrace{\sqrt{p_d} \mathbb{E}\{\|\hat{\mathbf{g}}_u\|^2\} x_u}_{\text{desired signal}} + \underbrace{\sqrt{p_d} \left(\|\hat{\mathbf{g}}_u\|^2 - \mathbb{E}\{\|\hat{\mathbf{g}}_u\|^2\} \right) x_u + \hat{\mathbf{g}}_u^H \mathbf{e}_u x_u + \sqrt{q_d} \hat{\mathbf{g}}_u^H \mathbf{g}_j x_j + \hat{\mathbf{g}}_u^H \mathbf{n}_d}_{\triangleq n_{\text{eff}} - \text{effective noise}}.$$

$$\tag{6.24}$$

Since x_u is independent of $\hat{\mathbf{g}}_u, \mathbf{e}_u, \mathbf{g}_j, x_j$, and $\hat{\mathbf{g}}_u$ is independent of $\mathbf{e}_u$, we have:

$$\mathbb{E}\left\{ x_u^* \left(\|\hat{\mathbf{g}}_u\|^2 - \mathbb{E}\left\{ \|\hat{\mathbf{g}}_u\|^2 \right\} \right) x_u \right\} = 0, \tag{6.25}$$

$$\mathbb{E}\left\{ x_u^* \hat{\mathbf{g}}_u^H \mathbf{e}_u x_u \right\} = 0, \tag{6.26}$$

$$\mathbb{E}\left\{ x_u^* \hat{\mathbf{g}}_u^H \mathbf{g}_j x_j \right\} = 0, \tag{6.27}$$

$$\mathbb{E}\left\{ x_u^* \hat{\mathbf{g}}_u^H \mathbf{n}_d \right\} = 0. \tag{6.28}$$

Thus, the desired signal and the effective noise are uncorrelated. By using the fact that, with Gaussian signalling, if the noise is uncorrelated with the desired signal, Gaussian noise is the worst case noise, we obtain the following achievable rate:

$$R = \left(1 - \frac{\tau}{T} \right) \log_2 \left(1 + \frac{p_d \left| \mathbb{E}\left\{ \|\hat{\mathbf{g}}_u\|^2 \right\} \right|^2}{\mathbb{E}\left\{ |n_{\text{eff}}|^2 \right\}} \right). \tag{6.29}$$

The pre-log factor $\left(1 - \frac{\tau}{T}\right)$ accounts for the channel estimation overhead. More precisely, for each coherence interval of length T symbols, we spend τ symbols for the uplink training and remaining $T - \tau$ symbols for the payload data transmission.

Using the identity $\mathbb{E}\left\{\|\hat{\mathbf{g}}_u\|^2\right\} = M\gamma_u$, (6.29) can be rewritten as:

$$R = \left(1 - \frac{\tau}{T}\right)\log_2\left(1 + \frac{p_d M^2 \gamma_u^2}{E_1 + E_2 + E_3 + E_4}\right), \tag{6.30}$$

where E_1, E_2, E_3, and E_4 represents the effects of channel uncertainty, channel estimation error, interference from the jammer, and noise, respectively, given by:

$$E_1 \triangleq p_d \mathbb{E}\left\{\left|\|\hat{\mathbf{g}}_u\|^2 - \mathbb{E}\left\{\|\hat{\mathbf{g}}_u\|^2\right\}\right|^2\right\},$$

$$E_2 \triangleq p_d \mathbb{E}\left\{\left|\hat{\mathbf{g}}_u^H \mathbf{e}_u\right|^2\right\},$$

$$E_3 \triangleq q_d \mathbb{E}\left\{\left|\hat{\mathbf{g}}_u^H \mathbf{g}_j\right|^2\right\},$$

and

$$E_4 \triangleq \mathbb{E}\left\{\left|\hat{\mathbf{g}}_u^H \mathbf{n}_d\right|^2\right\}.$$

By using the identity $\mathbb{E}\{\|\hat{\mathbf{g}}_u\|^4\} = M(M+1)\gamma_u^2$, we obtain:

$$\begin{aligned}
E_1 &= p_d \mathbb{E}\{\|\hat{\mathbf{g}}_u\|^4\} - p_d(\mathbb{E}\{\|\hat{\mathbf{g}}_u\|^2\})^2 \\
&= M(M+1)p_d\gamma_u^2 - M^2 p_d \gamma_u^2 \\
&= M p_d \gamma_u^2. \tag{6.31}
\end{aligned}$$

To compute E_2, we use the property that $\hat{\mathbf{g}}_u$ and $\mathbf{e}_u$ are independent,

$$\begin{aligned}
E_2 &= p_d \mathbb{E}\{\hat{\mathbf{g}}_u^H \mathbf{e}_u \mathbf{e}_u^H \hat{\mathbf{g}}_u\} \\
&= p_d(\beta_u - \gamma_u)\mathbb{E}\left\{\|\hat{\mathbf{g}}_u\|^2\right\} \\
&= M p_d \gamma_u(\beta_u - \gamma_u). \tag{6.32}
\end{aligned}$$

From (6.8), and using the fact that $\mathbf{g}_u, \mathbf{g}_j, \tilde{\mathbf{n}}_t$ are independent and zero mean random vectors, we have:

$$\begin{aligned}
E_3 &= q_d c_u^2 \mathbb{E}\left\{\left|\sqrt{\tau p_t}\mathbf{g}_u^H \mathbf{g}_j + \sqrt{\tau q_t}\|\mathbf{g}_j\|^2 \mathbf{s}_u^T \mathbf{s}_j^* + \tilde{\mathbf{n}}_t^H \mathbf{g}_j\right|^2\right\} \\
&= q_d c_u^2 \left(\tau p_t \mathbb{E}\left\{|\mathbf{g}_u^H \mathbf{g}_j|^2\right\} + \tau q_t |\mathbf{s}_u^T \mathbf{s}_j^*|^2 \mathbb{E}\{\|\mathbf{g}_j\|^4\} + \mathbb{E}\left\{|\tilde{\mathbf{n}}_t^H \mathbf{g}_j|^2\right\}\right) \\
&= q_d c_u^2 (\tau p_t M \beta_u \beta_j + \tau q_t M(M+1)\beta_j^2 |\mathbf{s}_u^T \mathbf{s}_j^*|^2 + M\beta_j).
\end{aligned}$$

Then by using (6.13),

$$E_3 = M q_d \gamma_u \left(\beta_j + M\gamma_u \frac{q_t}{p_t}\frac{\beta_j^2}{\beta_u^2}|\mathbf{s}_j^T \mathbf{s}_u^*|^2\right). \tag{6.33}$$

Similarly,

$$E_4 = \mathbb{E}\{|\hat{\mathbf{g}}_u^H \mathbf{n}_d|^2\} = M\gamma_u. \tag{6.34}$$

The substitution of (6.31)–(6.34) into (6.30) yields:

$$R = \left(1 - \frac{\tau}{T}\right) \log_2 \left(1 + \frac{M p_d \gamma_u}{p_d \beta_u + q_d \beta_j + M \frac{q_d q_t}{p_t} \left(\frac{\beta_j}{\beta_u}\right)^2 |s_j^T s_u^*|^2 \gamma_u + 1}\right). \quad (6.35)$$

As expected, if the jammer does not attack during the training phase,

$$M \frac{q_d q_t}{p_t} \left(\frac{\beta_j}{\beta_u}\right)^2 |s_j^T s_u^*|^2 \gamma_u = 0,$$

then

$$R = \left(1 - \frac{\tau}{T}\right) \log_2 \left(1 + \frac{M p_d \gamma_u}{p_d \beta_u + q_d \beta_j + 1}\right) \to \infty, \text{ as } M \to \infty. \quad (6.36)$$

The achievable rate increases without bound as $M \to \infty$, even when the jammer attacks during the data transmission phase. By contrast, if the jammer attacks during the training phase and $s_j^T s_u^* \neq 0$, the system suffers from jamming-pilot contamination, and hence, the achievable rate is rapidly saturated when M goes to infinity:

$$R \to \left(1 - \frac{\tau}{T}\right) \log_2 \left(1 + \frac{p_d p_t \beta_u^2}{q_d q_t \beta_j^2 |s_j^T s_u^*|^2}\right), \text{ as } M \to \infty. \quad (6.37)$$

Accordingly, the effective SINR is $\frac{p_d p_t \beta_u^2}{q_d q_t \beta_j^2 |s_j^T s_u^*|^2}$ which is identical to (6.22).

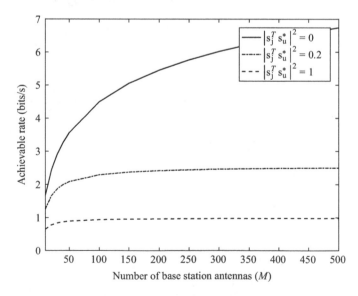

Figure 6.4   Achievable rate versus the number of antennas at the base station. Here $p_t = q_t = p_d = q_d = 0$ dB, $T = 200$, $\tau = 2$, $\beta_u = \beta_j = 1$

To see the effect of jamming-pilot contamination on the performance of the legitimate link quantitatively, we numerically evaluate the achievable rate as a function of M for different values of $|\mathbf{s}_j^T \mathbf{s}_u^*|^2$. The results are given in Figure 6.4. We can see that $|\mathbf{s}_j^T \mathbf{s}_u^*|^2$ (which represents the jamming-pilot contamination effect) significantly affects the system performance. When $|\mathbf{s}_j^T \mathbf{s}_u^*|^2 = 0$, the achievable rate increases as M increases. However, when $|\mathbf{s}_j^T \mathbf{s}_u^*|^2 = 0.2$ or 1, the achievable rate converges to a finite limit as M increases.

6.5 Anti-jamming pilot re-transmission

As discussed in the previous section, jamming attacks during the training phase highly affect the system performance. In order to mitigate this effect, a natural approach is to construct counter strategies on the training phase to resist the jamming-pilot contamination. Let us consider the achievable rate given in (6.35), we can see that R is a decreasing function with respect to $|\mathbf{s}_j^T \mathbf{s}_u^*|^2$. Since the jammer aims to attack the system, it will select the jamming-pilot sequence $\mathbf{s}_j$ so that $|\mathbf{s}_j^T \mathbf{s}_u^*|^2$ is large. On the other hand, the massive MIMO system wants to minimise the effect of jamming, it will choose the pilot sequence $\mathbf{s}_u$ so that $|\mathbf{s}_j^T \mathbf{s}_u^*|^2$ is as small as possible.

Typically, the jamming-pilot sequence $\mathbf{s}_j$ is unknown by the legitimate system. However, by using the favourable propagation and channel hardening properties of massive MIMO systems, the base station can estimate the information related to the jamming-pilot signal of the previous transmissions. On the basis of this information, the user can select a proper pilot sequence $\mathbf{s}_u$ so that $|\mathbf{s}_j^T \mathbf{s}_u^*|^2$ is small. In the following, we first develop the schemes to measure $|\mathbf{s}_j^T \mathbf{s}_u^*|^2$ and $\mathbf{s}_j^* \mathbf{s}_j^T$. Then, we present two pilot re-transmission algorithms, in which the pilots will be retransmitted when the estimated jamming-pilot contamination is high, i.e. $|\mathbf{s}_j^T \mathbf{s}_u^*|^2$ is larger than a certain threshold.

6.5.1 Estimation of $|\mathbf{s}_j^T \mathbf{s}_u^*|^2$ and $\mathbf{s}_j^* \mathbf{s}_j^T$

We show that by exploiting asymptotic properties of the massive MIMO systems, the base station can estimate $|\mathbf{s}_j^T \mathbf{s}_u^*|^2$ and $\mathbf{s}_j^* \mathbf{s}_j^T$. These estimates are used for the pilot re-transmission schemes which are discussed in detail in Sections 6.5.2 and 6.5.3. The estimations perform based on the received pilot signals $\mathbf{y}_t$ and $\mathbf{Y}_t$.

6.5.1.1 Estimation of $|\mathbf{s}_j^T \mathbf{s}_u^*|^2$

Consider the power of the received pilot signal (normalised by the number of antennas), which can be expressed as:

$$\frac{1}{M}\|\mathbf{y}_t\|^2 = \tau p_t \frac{\|\mathbf{g}_u\|^2}{M} + \tau q_t |\mathbf{s}_j^T \mathbf{s}_u^*|^2 \frac{\|\mathbf{g}_j\|^2}{M} + \frac{\|\tilde{\mathbf{n}}_t\|^2}{M} + \sqrt{\tau p_t} \frac{\mathbf{g}_u^H \left(\sqrt{\tau q_t}\mathbf{g}_j \mathbf{s}_j^T \mathbf{s}_u^* + \tilde{\mathbf{n}}_t\right)}{M}$$

$$+ \sqrt{\tau q_t} \mathbf{s}_j^T \mathbf{s}_u^* \frac{\mathbf{g}_j^H \left(\sqrt{\tau p_t}\mathbf{g}_u + \tilde{\mathbf{n}}_t\right)}{M} + \frac{\tilde{\mathbf{n}}_t^H \left(\sqrt{\tau p_t}\mathbf{g}_u + \sqrt{\tau q_t}\mathbf{g}_j \mathbf{s}_j^T \mathbf{s}_u^*\right)}{M}.$$

$$(6.38)$$

By the law of large numbers, as $M \to \infty$,

$$\frac{\|\tilde{\mathbf{n}}_t\|^2}{M} \overset{a.s.}{\to} 1,$$

$$\sqrt{\tau p_t} \frac{\mathbf{g}_u^H \left(\sqrt{\tau q_t} \mathbf{g}_j \mathbf{s}_j^T \mathbf{s}_u^* + \tilde{\mathbf{n}}_t\right)}{M} \overset{a.s.}{\to} 0,$$

$$\sqrt{\tau q_t} \mathbf{s}_j^T \mathbf{s}_u^* \frac{\mathbf{g}_j^H \left(\sqrt{\tau p_t} \mathbf{g}_u + \tilde{\mathbf{n}}_t\right)}{M} \overset{a.s.}{\to} 0,$$

$$\frac{\tilde{\mathbf{n}}_t^H \left(\sqrt{\tau p_t} \mathbf{g}_u + \sqrt{\tau q_t} \mathbf{g}_j \mathbf{s}_j^T \mathbf{s}_u^*\right)}{M} \overset{a.s.}{\to} 0. \tag{6.39}$$

The substitution of (6.18) and (6.39) into (6.38) yields:

$$\frac{1}{M} \|\mathbf{y}_t\|^2 \overset{a.s.}{\to} \tau p_t \beta_u + \tau q_t |\mathbf{s}_j^T \mathbf{s}_u^*|^2 \beta_j + 1, \text{ as } M \to \infty. \tag{6.40}$$

Result (6.40) implies that, when M is large,

$$|\mathbf{s}_j^T \mathbf{s}_u^*|^2 \approx \frac{1}{\tau q_t M \beta_j} \|\mathbf{y}_t\|^2 - \frac{p_t \beta_u}{q_t \beta_j} - \frac{1}{\tau q_t \beta_j}. \tag{6.41}$$

Since the base station knows M, τ, p_t, β_u, $q_t \beta_j$, and $\mathbf{y}_t$, it can compute the right hand-side of (6.41). We also note that $\mathbf{y}_t$ is a random vector, it can happen that the right-hand side of (6.41) is negative even the probability of this event is almost zero. Therefore, an estimate of $|\mathbf{s}_j^T \mathbf{s}_u^*|^2$ can be obtained from (6.41) as:

$$\widehat{|\mathbf{s}_j^T \mathbf{s}_u^*|^2} = \max\left\{0, \frac{1}{\tau q_t M \beta_j} \|\mathbf{y}_t\|^2 - \frac{p_t \beta_u}{q_t \beta_j} - \frac{1}{\tau q_t \beta_j}\right\}. \tag{6.42}$$

6.5.1.2   Estimation of $\mathbf{s}_j^* \mathbf{s}_j^T$

From the received signal $\mathbf{Y}_t$ in (6.5),

$$\frac{1}{M} \mathbf{Y}_t^H \mathbf{Y}_t = \tau p_t \frac{\|\mathbf{g}_u\|^2}{M} \mathbf{s}_u^* \mathbf{s}_u^T + \tau q_t \frac{\|\mathbf{g}_j\|^2}{M} \mathbf{s}_j^* \mathbf{s}_j^T + \mathbf{N}_t^H \mathbf{N}_t + \frac{\sqrt{\tau p_t} \mathbf{s}_u^* \mathbf{g}_u^H (\sqrt{\tau q_t} \mathbf{g}_j \mathbf{s}_j^T + \mathbf{N}_t)}{M}$$

$$+ \frac{\sqrt{\tau q_t} \mathbf{s}_j^* \mathbf{g}_j^H (\sqrt{\tau p_t} \mathbf{g}_u \mathbf{s}_u^T + \mathbf{N}_t)}{M} + \frac{\mathbf{N}_t (\sqrt{\tau p_t} \mathbf{g}_u \mathbf{s}_u^T + \sqrt{\tau q_t} \mathbf{g}_j \mathbf{s}_j^T)}{M}. \tag{6.43}$$

Again from the law of large numbers, as $M \to \infty$, we have:

$$\frac{1}{M} \mathbf{Y}_t^H \mathbf{Y}_t \overset{a.s.}{\to} \tau p_t \beta_u \mathbf{s}_u^* \mathbf{s}_u^T + \tau q_t \beta_j \mathbf{s}_j^* \mathbf{s}_j^T + \mathbf{I}_\tau. \tag{6.44}$$

Thus, the base station can estimate $\mathbf{s}_j^* \mathbf{s}_j^T$ as:

$$\widehat{\mathbf{s}_j^* \mathbf{s}_j^T} = \frac{1}{\tau q_t \beta_j M} \mathbf{Y}_t^H \mathbf{Y}_t - \frac{p_t \beta_u}{q_t \beta_j} \mathbf{s}_u^* \mathbf{s}_u^T - \frac{1}{\tau q_t \beta_j} \mathbf{I}_\tau. \tag{6.45}$$

When the number of antennas M is large, which is the case in the massive MIMO systems, the estimates in (6.42) and (6.45) are very close to the true values of $|\mathbf{s}_j^T \mathbf{s}_u^*|^2$ and $\mathbf{s}_j^* \mathbf{s}_j^T$. On the basis of the estimates of $|\mathbf{s}_j^T \mathbf{s}_u^*|^2$ and $\mathbf{s}_j^* \mathbf{s}_j^T$, we design two pilot re-transmission protocols to deal with two common jamming cases: random jamming and deterministic jamming.

6.5.2 Pilot re-transmission under random jamming

When the jammer does not have the prior knowledge of the pilot sequences used by the legitimate user, it will attack the system using random jamming sequences. During the training phase, the user sends a pilot sequence $\mathbf{s}_u$, while the jammer sends a random jamming sequence $\mathbf{s}_j$. For this case, the base station can mitigate the jamming-pilot contamination by using the estimate of $|\mathbf{s}_j^T \mathbf{s}_u^*|^2$. More precisely, the base station first estimates $|\mathbf{s}_j^T \mathbf{s}_u^*|^2$ using the received pilot signal $\mathbf{y}_t$ together with (6.42). Then it compares the estimate of $|\mathbf{s}_j^T \mathbf{s}_u^*|^2$ with a threshold ε. If $\widehat{|\mathbf{s}_j^T \mathbf{s}_u^*|^2} > \varepsilon$, the base station requests the user to retransmit a new pilot sequence. The re-transmission is processed until $\widehat{|\mathbf{s}_j^T \mathbf{s}_u^*|^2} \leq \varepsilon$ or the number of pilot transmissions exceeds the maximum number $N_{\max}$.

The pilot re-transmission algorithm can be summarised as follows.

Algorithm 6.1 (Under random jamming).

---

1. *Initialisation: set $N = 1$, choose the values of pilot length τ, threshold ε, and $N_{\max}$ ($N_{\max}\tau < T$).*
2. *User transmits a random $\tau \times 1$ pilot sequence $\mathbf{s}_u \in \mathscr{S}$.*
3. *The base station estimates $|\mathbf{s}_j^T \mathbf{s}_u^*|^2$ using (6.42). If $\widehat{|\mathbf{s}_j^T \mathbf{s}_u^*|^2} \leq \varepsilon$ or $N = N_{\max} \rightarrow$ Stop. Otherwise, set $N = N + 1$ and go to step 2.*

In order to effectively exploit the benefit of the pilot re-transmissions scheme, the base station buffers the received pilot signals then processes with the best one, i.e. the pilot transmission with minimal $|\mathbf{s}_j(n)^T \mathbf{s}_u(n)^*|$, instead of using the last transmission. In the case that N is large, the buffer memory can be used efficiently by updating the best transmission candidate after each re-transmission and discarding the others instead of saving the received pilot signals of all N transmissions.

Let $\mathbf{s}_u(n)$ and $\mathbf{s}_j(n)$ be the pilot and jamming sequences respectively, corresponding to the nth re-transmission, $n = 1, \ldots, N$. By following similar steps as in Section 6.4, an achievable rate of the massive MIMO uplink using the pilot re-transmission protocol for random jamming is given by:

$$R_{rj} = \left(1 - \frac{N\tau}{T}\right) \log_2 \left(1 + \frac{Mp_d\gamma_u}{p_d\beta_u + q_d\beta_j + \alpha_{rj} + 1}\right), \tag{6.46}$$

where

$$\alpha_{rj} = M \frac{q_d q_t}{p_t} \frac{\beta_j^2}{\beta_u^2} \min_n |\mathbf{s}_j(n)^T \mathbf{s}_u(n)^*|^2 \gamma_u. \tag{6.47}$$

The achievable rate R_{rj} in (6.46) is similar to the one without the pilot re-transmission protocol in (6.35) with differences on the pre-log factor and the effective SINR. The pre-log factor of R_{rj} is smaller due to the re-transmissions, which decreases the achievable rate. However, this decrease is compensated by the increase of the effective SINR resulted from the pilot re-transmission protocol. It is expected that the improvement from the increase of SINR overcomes the effect of smaller pre-log factor, and thus the system performance is improved overall.

6.5.3 Pilot re-transmission under deterministic jamming

Next, we assume that the jamming sequences are deterministic during the training phase, i.e. $\mathbf{s}_j(1) = \cdots = \mathbf{s}_j(N)$. Such scenario can happen, for instance, when the jammer has the prior knowledge of the pilot length and pilot sequence codebook. In order to maximise the jamming-pilot contamination, the jammer tries to transmit a jamming sequence which is as similar to the user's pilot sequence as possible. However, since the jammer only knows the pilot codebook and does not know which pilot sequence is used, it will attack using a jamming sequence as a deterministic function of all possible pilot sequences [11]. In this case, the massive MIMO system can outsmart the jammer by adapting the training sequences using the knowledge obtained from the last pilot transmission instead of just randomly retransmitting them as in the previous case.

We observe that $|\mathbf{s}_j^T \mathbf{s}_u^*|^2$ can be decomposed as:

$$|\mathbf{s}_j^T \mathbf{s}_u^*|^2 = \mathbf{s}_u^T \mathbf{s}_j^* \mathbf{s}_j^T \mathbf{s}_u^*. \tag{6.48}$$

Thus, if the base station knows $\mathbf{s}_j^* \mathbf{s}_j^T$, it can select $\mathbf{s}_u$ for the next transmission, such that $|\mathbf{s}_j^T \mathbf{s}_u^*|^2$ is minimal. In Section 6.5.1.2, we show that the base station can estimate $\mathbf{s}_j^* \mathbf{s}_j^T$ from $\mathbf{Y}_t$. From this observation, we consider the following pilot re-transmission scheme:

Algorithm 6.2 (Under deterministic jamming).

---

1. *Initialisation: choose the values of pilot length τ and threshold ε.*
2. *User sends a $\tau \times 1$ pilot sequence $\mathbf{s}_u \in \mathcal{S}$.*
3. *The base station estimates $|\mathbf{s}_j^T \mathbf{s}_u^*|^2$ using (6.42). If $\widehat{|\mathbf{s}_j^T \mathbf{s}_u^*|^2} \leq \varepsilon \rightarrow$ Stop. Otherwise, go to step 4.*
4. *The base station estimates $\mathbf{s}_j^* \mathbf{s}_j^T$ using (6.45). Then, the base station finds $\mathbf{s}_u^{opt}$ so that $\mathbf{s}_u^{optT} \widehat{\mathbf{s}_j^* \mathbf{s}_j^T} \mathbf{s}_u^{opt*}$ is minimal. The user will retransmit this new pilot.*

In this pilot re-transmission scheme, the base station requests the user to retransmit its pilot only if $|\mathbf{s}_j^T \mathbf{s}_u^*|^2$ of the first transmission exceeds the threshold ε. Unlike the re-transmission scheme in Algorithm 6.1, the re-transmission process is completed after only one re-transmission even if $|\mathbf{s}_j^T \mathbf{s}_u^*|^2$ is still larger than the threshold ε. This is because we can select the best pilot sequence $\mathbf{s}_u^{opt}$ after the first transmission by exploiting the estimation in (6.45). The maximal number of retransmissions in this case is one. Any additional retransmission cannot improve the performance for the given pilot codebook $\mathscr{S}$.

Once again, by following similar steps as in Section 6.4, an achievable rate of the massive MIMO uplink using the pilot retransmission protocol for deterministic jamming can be obtained as:

$$R_{dj} = \left(1 - \frac{N\tau}{T}\right) \log_2 \left(1 + \frac{Mp_d \gamma_u}{p_d \beta_u + q_d \beta_j + \alpha_{dj} + 1}\right), \tag{6.49}$$

where
$$\begin{cases} \alpha_{dj} = M \frac{q_d q_t}{p_t} \left(\frac{\beta_j}{\beta_u}\right)^2 |\mathbf{s}_j^T \mathbf{s}_u^*|^2 \gamma_u, N = 1, \text{ if } |\mathbf{s}_j^T \mathbf{s}_u^*| \le \varepsilon \\ \alpha_{dj} = M \frac{q_d q_t}{p_t} \left(\frac{\beta_j}{\beta_u}\right)^2 |\mathbf{s}_j^T \mathbf{s}_u^{opt*}|^2 \gamma_u, N = 2, \text{ otherwise.} \end{cases}$$

When the retransmission occurs, the received SINR depends on $|\mathbf{s}_j^T \mathbf{s}_u^{opt*}|$. It is expected that the jamming-pilot contamination effect reduces with the higher cardinality of the pilot codebook $\mathscr{S}$ since we have more possibilities to select $\mathbf{s}_u$ from $\mathscr{S}$. However, the pilot length τ increases with the cardinality of $\mathscr{S}$ assuming that orthogonal pilot sequences are used. When τ increases, the pre-log factor decreases. Therefore, in order to improve the system performance, one has to find a proper τ to balance the pre-log factor and the received SINR in (6.49).

6.5.4 Numerical examples

In order to exemplify the performance of the above pilot retransmission protocols, let us consider some numerical examples of the achievable rates in (6.46) and (6.49). We consider the set up with $T = 200$ symbols and maximum number of transmissions $N_{max} = 2$.

Figure 6.5 illustrates the achievable rates for different pilot retransmission protocols according to the training payload (τ/T). For comparison, the achievable rate for the conventional protocol without pilot retransmission is also included. It is shown that in order to achieve the best performances, the training payloads should be selected properly to balance the channel estimation quality (τ is large enough) and the resource allocated for data transmission (τ is not too large). As expected, the pilot retransmission protocols outperform the conventional protocol. When the pilot sequences are very long, i.e. τ/T is large, the performance of the pilot retransmission protocols are close to the conventional one. This can be explained from the fact that when the pilot sequences are long, the probability of requiring the pilot retransmission is very small because the channel estimation quality is often good enough after the first pilot transmission.

Figure 6.6 shows the achievable rates versus the number of antennas at the base station. Without pilot retransmission, the jamming-pilot contamination severely

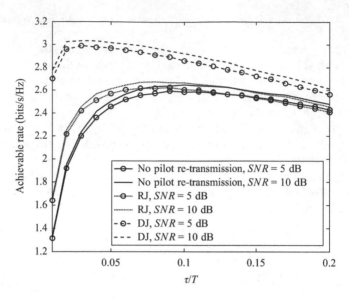

Figure 6.5 *Achievable rates of different anti-jamming schemes for $\varepsilon = 0.1$, $q_t = p_t = q_d = p_d = SNR$, and $M = 50$. The solid curves, dotted curves (with label "RJ"), and dashed curves (with label "DJ") denote the achievable rates without pilot re-transmission (cf. (6.35)), with counter strategy for random jamming (cf. (6.46)), and with counter strategy for deterministic jamming (cf. (6.49)), respectively*

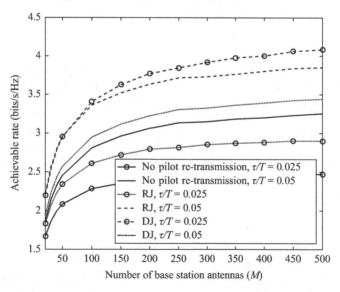

Figure 6.6 *Achievable rates according to the number of antennas M for $\varepsilon = 0.1$, and $p_t = q_t = p_d = q_d = SNR = 5$ dB. The solid curves, dotted curves (with label "RJ"), and dashed curves (with label "DJ") denote the achievable rates without pilot re-transmission (cf. (6.35)), with counter strategy for random jamming (cf. (6.46)), and with counter strategy for deterministic jamming (cf. (6.49)), respectively*

harms the system performance and obstructs the scaling of achievable rate with M. This is consistent with the analysis in Section 6.4. As expected, the studied pilot retransmission protocols, especially with the deterministic jamming, remarkably improve the achievable rate. Particularly, for the case of the deterministic jamming, the pilot retransmission scheme can overcome the jamming-pilot contamination bottleneck and allows the achievable rates scale with $\log_2(M)$ if there exists a pilot sequence s_u^{opt} in the pilot codebook, which is orthogonal with s_j.

6.6 Chapter conclusion

In this chapter, we discussed the physical layer security in massive MIMO. Particularly, we focused on the jamming aspect and show that jamming attacks during the training phase cause jamming-pilot contamination. This jamming-pilot contamination effect could severely degrade the system performance and persists even when the number of base station antennas increases without bound.

By exploiting the asymptotic properties of large antenna arrays, the pilot retransmission protocols can be designed to reduce the effect of jamming-pilot contamination. In the considered pilot retransmission protocols, the pilot sequences and training payload could flexibly be adjusted to enhance the robustness of the system against jamming.

Although this chapter focuses on a single-user setup, the anti-jamming framework can be applied to multi-user networks. This can be done by addressing two main concerns: (i) which users should be chosen for retransmitting pilots and (ii) how the inter-user interference affects the performance of the retransmission schemes. The first concern can be addressed by the max–min fairness criterion, for example, we design the pilot re-transmission protocol based on the worst user which has the smallest achievable rate. The second concern can be easily addressed in the massive MIMO systems thanks to the fact that when multiple users are considered, an additional term contributed to the system is inter-user interference, which can be negligible when the number of antennas is large.

References

[1]   Cisco. Cisco Visual Networking Index: Global Mobile Data Traffic Forecast Update, 2015–2020. Cisco Systems, Inc.; Feb. 2016.

[2]   Marzetta TL. Noncooperative cellular wireless with unlimited numbers of base station antennas. IEEE Trans Wireless Commun. 2010 Nov;9(11):3590–3600.

[3]   Rusek F, Persson D, Lau BK, *et al.* Scaling up MIMO: Opportunities and challenges with very large arrays. IEEE Signal Process Mag. 2013 Jan;30(1):40–60.

[4]   Ngo HQ, Larsson EG, and Marzetta TL. Energy and spectral efficiency of very large multiuser MIMO systems. IEEE Trans Commun. 2013 Apr;61(4): 1436–1449.

[5]   Larsson EG, Edfors O, Tufvesson F, and Marzetta TL. Massive MIMO for next generation wireless systems. IEEE Commun Mag. 2014 Feb;52(2):186–195.

[6]   Ngo HQ, Ashikhmin A, Yang H, Larsson EG, and Marzetta TL. Cell-Free Massive MIMO versus Small Cells. IEEE Trans Wireless Commun. 2017 Mar;16(3):1834–1850.

[7]   Zhu J, Schober R, and Bhargava VK. Secure Transmission in Multicell Massive MIMO Systems. IEEE Trans Wireless Commun. 2014 Sep;13(9):4766–4781.

[8]   Zhu J, and Xu W. Securing massive MIMO via power scaling. IEEE Commun Lett. 2016 May;20(5):1014–1017.

[9]   Zhu J, Schober R, and Bhargava VK. Linear precoding of data and artificial noise in secure massive MIMO systems. IEEE Trans Wireless Commun. 2016 Mar;15(3):2245–2261.

[10]   Kapetanovic D, Zheng G, and Rusek F. Physical layer security for massive MIMO: An overview on passive eavesdropping and active attacks. IEEE Commun Mag. 2015 Jun;53(6):21–27.

[11]   Basciftci YO, Koksal CE, and Ashikhmin A. Securing massive MIMO at the physical layer. In: IEEE Conf. on Commun. and Net. Sec. (CNS) 2015. Florence, Italy; 2015. p. 272–280.

[12]   Pirzadeh H, Razavizadeh SM, and Björnson E. Subverting massive MIMO by smart jamming. IEEE Wireless Commun Lett. 2016 Feb;5(1):20–23.

[13]   Wang J, Lee J, Wang F, and Quek TQS. Jamming-aided secure communication in massive MIMO Rician channels. IEEE Trans Wireless Commun. 2015 Dec;14(12):6854–6868.

[14]   Oggier F, and Hassibi B. The secrecy capacity of the MIMO wiretap channel. IEEE Trans Inf Theory. 2011 Aug;57(8):4961–4972.

[15]   Kay SM. Fundamentals of statistical signal processing: Estimation theory. NJ, USA: Prentice-Hall; 1993.

[16]   Ashikhmin A, Marzetta TL, and Li L. Interference reduction in multi-cell massive MIMO systems I: Large-scale fading precoding and decoding; 2014. Available from https://arxiv.org/abs/1411.4182.

[17]   Jose J, Ashikhmin A, Marzetta TL, and Vishwanath S. Pilot contamination and precoding in multi-cell TDD systems. IEEE Trans Wireless Commun. 2011 Aug;10(8):2640–2651.

Chapter 7
Physical layer security for multiuser relay networks

Lisheng Fan[1] and Trung Q. Duong[2]

The scenario of multiuser communication networks is often encountered in wireless communication systems. An example of this is the communication of cellular networks, where multiple users intend to communicate with the base station. However, the eavesdroppers in the network can overhear the message from the users due to the broadcast nature of wireless transmission, which brings out the severe issue of information wiretap. To prevent the wiretap, many works have been done for the physical layer security of multiuser networks. For example, the authors in [1] exploited multiuser diversity to enhance the physical layer security in transmit antenna selection systems and found that the secrecy outage in the throughput-based policy gets improved with the increased number of users. The secure communication of a multiuser downlink system was studied in [2] by using the regularised zero-forcing precoding based on imperfect channel estimation and an improvement in the ergodic secrecy sum-rate was achieved. The authors investigated the secure communications in multiuser uplink transmissions, and two low-complexity user selection schemes were proposed under different assumptions on the eavesdropper's channel state information, in order to achieve a multiuser gain in [3]. In [4], the authors employed jamming technique with user selection to enhance the security of multiuser networks and studied the secure degrees of freedom and the associated jammer-scaling law.

Relaying technique can increase the transmission capacity and network coverage without requiring additional transmit power, and hence it has become one of the most promising techniques for the next-generation communication systems. For multiuser communication systems, relays can be used to assist the data transmission between users and base station, forming into multiuser multi-relay networks. In this chapter, we will present criteria for user selection and relay selection to enhance the physical layer security in multiuser multi-relay networks. Specifically, for multiuser networks with multiple amplify-and-forward (AF) relays [5], we present three criteria to select the best relay and user pair. Criteria I and II study the received signal-to-noise ratio (SNR) at the receivers and perform the selection by maximising the SNR ratio of

[1]School of Computer Science and Educational Software, Guangzhou University
[2]School of Electronics, Electrical Engineering and Computer Science, Queen's University Belfast, UK

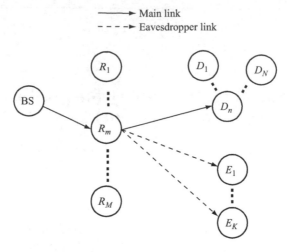

Figure 7.1　A two-phase multiuser multi-relay cooperative network with multiple eavesdroppers. © IEEE 2016. Reprinted with permission from [5]

the user to the eavesdroppers. To this end, criterion I relies on both the main and eavesdropper links, while criterion II relies on the main links only. Criterion III is the standard max–min selection criterion, which maximises the minimum of the dual-hop channel gains of main links. For the three selection criteria, we examine the system secrecy performance by deriving the analytical expressions for the secrecy outage probability. We also derive the asymptotic analysis for the secrecy outage probability with high main-to-eavesdropper ratio (MER). From the asymptotic analysis, an interesting observation is reached: for each criterion, the system diversity order is equivalent to the number of relays regardless of the number of users and eavesdroppers. For multiuser networks with multiple decode-and-forward (DF) relays, we investigate two criteria for user and relay selection and examine the achievable secrecy performance in [6]. Criterion I performs a joint user and relay selection, while Criterion II performs separate user and relay selections, with a lower implementation complexity. We derive a tight lower bound on the secrecy outage probability for Criterion I and an accurate analytical expression for the secrecy outage probability for Criterion II. We further derive the asymptotic secrecy outage probabilities at high transmit signal-to-noise ratios and high main-to-eavesdropper ratios for both criteria.

7.1　AF relaying

For AF relaying, a two-phase multiuser multi-relay cooperative network with multiple eavesdroppers is demonstrated in Figure 7.1. This network is composed of a base station BS, M trusted AF relays, N desired users and K eavesdroppers. The data transmission from the BS to the users can only be via the relays, in the presence of the K eavesdroppers. We assume that the eavesdroppers can cooperate with each other by using maximal ratio combining (MRC) technique to increase the wiretap possibility. To prevent the wiretap, the best relay and user pair (R_{m*}, D_{n*}) is selected to enhance the transmission security, while the other relays and users keep silent. All nodes in the network have only one antenna due to size limitation, and they operate in a time-division half-duplex mode.

Suppose that the mth relay and nth user have been selected for data transmission. In the first phase, BS sends normalised signal s to R_m, while R_m receives:

$$y_m^R = \sqrt{P_S} h_{BS,R_m} s + n_R, \tag{7.1}$$

where P denotes the transmit power at the base station, $h_{BS,R_m} \sim \mathscr{CN}(0, \alpha)$ is the channel of the BS→R_m link, and $n_R \sim \mathscr{CN}(0, 1)$ represents the additive white noise at the relay. Then relay R_m amplifies the received signal y_m^R with power P_R by a factor κ:

$$\kappa = \sqrt{\frac{P_R}{P_S |h_{BS,R_m}|^2 + 1}}, \tag{7.2}$$

and forwards the resultant signal in the second phase. User D_n and eavesdropper E_k respectively receive:

$$y_{m,n}^D = h_{R_m,D_n} \kappa y_m^R + n_D, \tag{7.3}$$

$$y_{m,k}^E = h_{R_m,E_k} \kappa y_m^R + n_E, \tag{7.4}$$

where $h_{R_m,D_n} \sim \mathscr{CN}(0, \beta)$ and $h_{R_m,E_k} \sim \mathscr{CN}(0, \varepsilon)$ are the channels of R_m→D_n and R_m→E_k links, respectively. Notations $n_D \sim \mathscr{CN}(0, 1)$ and $n_E \sim \mathscr{CN}(0, 1)$ are the additive white noise at the user and eavesdropper, respectively. From (7.1) to (7.3), the received SNR at D_n is obtained as:

$$\mathsf{SNR}_{m,n}^D = \frac{P_S P_R u_m v_{m,n}}{P_S u_m + P_R v_{m,n} + 1}, \tag{7.5}$$

where $u_m = |h_{BS,R_m}|^2$ and $v_{m,n} = |h_{R_m,D_n}|^2$ denote the channel gains of the BS→R_m and R_m→D_n links, respectively. To increase the wiretap probability, the eavesdroppers cooperate to combine the received signals $y_{m,k}^E$ in the MRC manner to obtain a scalar symbol as [7]:

$$y_n^E = \sum_{k=1}^K h_{R_m,E_k}^\dagger y_{m,k}^E = \sum_{k=1}^K |h_{R_m,E_k}|^2 \kappa y_m^R + h_{R_m,E_k}^\dagger n_E, \tag{7.6}$$

where $\dagger$ denotes the conjugate transpose operation. From (7.6), the received SNR of the K eavesdroppers with MRC is obtained as [7]:

$$\mathsf{SNR}_m^E = \frac{P_S P_R u_m w_m}{P_S u_m + P_R w_m + 1}, \tag{7.7}$$

where $w_m = \sum_{k=1}^K |h_{R_m,E_k}|^2$ denotes the sum channel gain of the K eavesdropper links.

For the considered system with target secrecy data rate R_s, the secrecy outage probability with the mth relay and nth user is given by:

$$P_{\text{out},m,n} = \Pr\left[\frac{1}{2}\log_2\left(1 + \text{SNR}^{\text{D}}_{m,n}\right) - \frac{1}{2}\log_2\left(1 + \text{SNR}^{\text{E}}_m\right) < R_s\right] \tag{7.8}$$

$$= \Pr\left(\frac{1 + \text{SNR}^{\text{D}}_{m,n}}{1 + \text{SNR}^{\text{E}}_m} < \gamma_{\text{th}}\right), \tag{7.9}$$

where $\gamma_{\text{th}} = 2^{2R_s}$ denotes the secrecy SNR threshold.

7.1.1 Relay and user selection

For the considered system, the best relay and user pair (R_{m^*}, D_{n^*}) is selected to minimise the secrecy outage probability:

$$(m^*, n^*) = \arg\min_{m=1,\dots,M}\ \min_{n=1,\dots,N} P_{\text{out},m,n} \tag{7.10}$$

$$= \arg\min_{m=1,\dots,M}\ \min_{n=1,\dots,N} \Pr\left(\frac{1 + \text{SNR}^{\text{D}}_{m,n}}{1 + \text{SNR}^{\text{E}}_m} < \gamma_{\text{th}}\right). \tag{7.11}$$

It holds that

$$\frac{1 + \text{SNR}^{\text{D}}_{m,n}}{1 + \text{SNR}^{\text{E}}_m} \simeq \frac{\text{SNR}^{\text{D}}_{m,n}}{\text{SNR}^{\text{E}}_m} \tag{7.12}$$

$$= \frac{P_S P_R u_m v_{m,n} / (P_S u_m + P_R v_{m,n} + 1)}{P_S P_R u_m w_m / (P_S u_m + P_R w_m + 1)} \tag{7.13}$$

$$\simeq \frac{P_S P_R u_m v_{m,n} / (P_S u_m + P_R v_{m,n})}{P_S P_R u_m w_m / (P_S u_m + P_R w_m)} \tag{7.14}$$

where in (7.12) the approximation of $(1 + x)/(1 + y) \simeq x/y$ is applied. Let $\eta = \frac{P_R}{P_S}$ be the transmit power ratio of the relay to the BS. Then we can summarise:

$$\frac{1 + \text{SNR}^{\text{D}}_{m,n}}{1 + \text{SNR}^{\text{E}}_m} \simeq \frac{(u_m + \eta w_m) v_{m,n}}{(u_m + \eta v_{m,n}) w_m}, \tag{7.15}$$

Accordingly, we can approximate $P_{\text{out},m,n}$ using (7.15) as:

$$P_{\text{out},m,n} \simeq \Pr\left[\frac{(u_m + \eta w_m) v_{m,n}}{(u_m + \eta v_{m,n}) w_m} < \gamma_{\text{th}}\right], \tag{7.16}$$

$$= \Pr\left[\frac{u_m v_{m,n}}{\gamma_{\text{th}} u_m + (\gamma_{\text{th}} - 1)\eta v_{m,n}} < w_m\right]. \tag{7.17}$$

Then a relay and user pair selection criterion is proposed as:

$$(m^*, n^*) = \arg\max_{m=1,\dots,M}\ \max_{n=1,\dots,N}\left(\frac{u_m v_{m,n} / (\gamma_{\text{th}} u_m + (\gamma_{\text{th}} - 1)\eta v_{m,n})}{w_m}\right). \tag{7.18}$$

It can be found that the criterion in (7.18) is equivalent to maximising the received SNR ratio of the user to the eavesdroppers based on both the main and eavesdropper links, and a near-optimal secrecy outage performance can be achieved.

Note that the near-optimal selection in (7.18) requires the knowledge of the channel parameters of both the main and eavesdropper's links. In some communication scenarios, it may be difficult or even impossible to acquire the channel parameters of the eavesdropper's links. In this case, the relay and user selection can rely on the channel parameters of main links only. By applying:

$$u_m v_{m,n}/(\gamma_{th} u_m + (\gamma_{th} - 1)\eta v_{m,n}) \le \min\left(\frac{u_m}{(\gamma_{th} - 1)\eta}, \frac{v_{m,n}}{\gamma_{th}}\right) \tag{7.19}$$

into (7.18), a sub-optimal relay and user selection criterion is proposed as:

$$(m^*, n^*) = \arg \max_{m=1,...,M} \max_{n=1,...,N} \min\left(\frac{u_m}{(\gamma_{th} - 1)\eta}, \frac{v_{m,n}}{\gamma_{th}}\right), \tag{7.20}$$

which maximises the received SNR ratio of the user to the eavesdroppers based on the main links only.

In addition, according to the standard max–min criterion, the relay and user pair is selected by maximising the minimum of the dual-hop channel gains of main links, i.e.:

$$(m^*, n^*) = \arg \max_{m=1,...,M} \max_{n=1,...,N} \min(u_m, v_{m,n}). \tag{7.21}$$

For convenience of notation, the selection criteria in (7.18), (7.20) and (7.21) are referred to as criteria I, II and III, respectively.

7.1.2 Lower bound

In this section, the analytical expression of secrecy outage probability for criteria I, II and III will be derived. From (7.17), the network secrecy outage probability with selected R_{m^*} and D_{n^*} with a high transmit power is:

$$P_{\text{out},m^*,n^*} \simeq \Pr(Z_{m^*,n^*} < w_{m^*}), \tag{7.22}$$

where $Z_{m,n} = \frac{u_m v_{m,n}}{\gamma_{th} u_m + (\gamma_{th} - 1)\eta}$.

7.1.2.1 Criterion I

Since $Z_{m,n}$ increases with respect to $v_{m,n}$, the best user $D_{n_m^*}$ with a given relay R_m should be chosen to maximise $v_{m,n}$:

$$n_m^* = \arg \max_{n=1,...,N} v_{m,n}. \tag{7.23}$$

The probability density function (PDF) of v_{m,n_m^*} is [8, (9E.2)]:

$$f_{v_{m,n_m^*}}(v) = \sum_{n=1}^{N} (-1)^{n-1} \binom{N}{n} \frac{n}{\beta} e^{-\frac{nv}{\beta}}. \tag{7.24}$$

The cumulative density function (CDF) of Z_{m,n_m^*} can be written as:

$$F_{Z_{m,n_m^*}}(z) = \Pr\left(\frac{u_m v_{m,n_m^*}}{\gamma_{\text{th}} u_m + (\gamma_{\text{th}} - 1)\eta v_{m,n_m^*}} < z\right) \tag{7.25}$$

$$= \Pr\left[u_m(v_{m,n_m^*} - \gamma_{\text{th}} z) < (\gamma_{\text{th}} - 1)\eta v_{m,n_m^*} z\right]. \tag{7.26}$$

By applying the PDF of v_{m,n_m^*} in (7.24) and $f_{u_m}(u) = \frac{1}{\alpha}e^{-\frac{u}{\alpha}}$ into the above equation, and then solving the required integral, we obtain the CDF of Z_{m,n_m^*} as:

$$F_{Z_{m,n_m^*}}(z) = 1 - \sum_{n=1}^{N}(-1)^{n-1}\binom{N}{n}b_n e^{-\left(\frac{n\gamma_{\text{th}}}{\beta} + \frac{n(\gamma_{\text{th}}-1)}{\alpha}\right)z} z\mathcal{K}_1(b_n z), \tag{7.27}$$

where we apply [9, (3.324)] and $b_n = \sqrt{\frac{4n\eta\gamma_{\text{th}}(\gamma_{\text{th}}-1)}{\alpha\beta}}$. From (7.22) and (7.27), we derive the closed-form expression of the secrecy outage probability with the mth relay for large transmit power as:

$$P_{\text{out},m,n_m^*} \simeq \Pr(Z_{m,n_m^*} < w_m) \tag{7.28}$$

$$= \int_0^\infty f_{w_m}(w)F_{Z_{m,n_m^*}}(w)dw, \tag{7.29}$$

where the approximation sign in (7.28) comes from the assumption of large transmit power which was previously used in equation (7.15). Note that $f_{w_m}(w) = \frac{w^{K-1}}{\Gamma(K)\varepsilon^K}e^{-\frac{w}{\varepsilon}}$ is the PDF of w_m [8, (9.5)], we can obtain the secrecy outage probability with the mth relay by applying [9, (6.621.3)] as:

$$P_{\text{out},m,n_m^*} \simeq 1 - \sum_{n=1}^{N}(-1)^{n-1}\binom{N}{n}\frac{2\sqrt{\pi}b_n^2\Gamma(K+2)}{\varepsilon^K(b_n + c_n)^{K+2}\Gamma\left(K + \frac{3}{2}\right)}$$

$$\times {}_2F_1\left(K+2, \frac{3}{2}, K+\frac{3}{2}; \frac{c_n - b_n}{c_n + b_n}\right), \tag{7.30}$$

where $c_n = \frac{1}{\varepsilon} + \frac{n\gamma_{\text{th}}}{\beta} + \frac{n(\gamma_{\text{th}}-1)}{\alpha}$ and ${}_2F_1(\cdot)$ denotes the Gauss hypergeometric function [9, (9.100)].

Since the statistic $Z_{m^*,n^*}/w_{m^*}$ is the maximum of M independent variables of $\{Z_{m,n_m^*}/w_m\}$, and hence we can obtain the secrecy outage probability for Criterion I with high transmit power as:

$$P_{\text{out},m^*,n^*} \simeq \left[1 - \sum_{n=1}^{N}\binom{N}{n}\frac{2(-1)^{n-1}\sqrt{\pi}b_n^2\Gamma(K+2)}{\varepsilon^K(b_n + c_n)^{K+2}\Gamma\left(K + \frac{3}{2}\right)}\right.$$

$$\left.\times {}_2F_1\left(K+2, \frac{3}{2}, K+\frac{3}{2}; \frac{c_n - b_n}{c_n + b_n}\right)\right]^M. \tag{7.31}$$

7.1.2.2   Criteria II and III

In this subsection, we derive the secrecy outage probability for Criteria II and III in a unified manner. Note that Criteria II and III in (7.20) and (7.21) can be unified as:

$$(m^*, n^*) = \arg \max_{m=1,\ldots,M} \max_{n=1,\ldots,N} \min(u_m, \rho v_{m,n}), \tag{7.32}$$

where $\rho = \rho_{\mathrm{II}}$ and $\rho = \rho_{\mathrm{III}}$ correspond to Criteria II and III, respectively, with $\rho_{\mathrm{II}} = \frac{(\gamma_{\mathrm{th}}-1)\eta}{\gamma_{\mathrm{th}}}$ and $\rho_{\mathrm{III}} = 1$. According to (7.32), we obtain the CDFs of u_{m^*} and v_{m^*,n^*} in the following theorem.

Theorem 7.1. *The CDFs of u_{m^*} and v_{m^*,n^*} are given by*

$$\begin{cases} F_{u_{m^*}}(x) = 1 - \sum_{n=1}^{N} \widetilde{\sum_i} \left(q_{1i}e^{-q_{2i}x} + q_{3i}e^{-\frac{x}{\alpha}} \right) \\ \\ F_{v_{m^*,n^*}}(x) = 1 - \sum_{n=1}^{N} \widetilde{\sum_i} \left(\frac{q_{4i}}{q_{2i}\rho}e^{-q_{2i}\rho x} + \frac{q_{5i}\beta}{n}e^{-\frac{n}{\beta}x} \right) \end{cases}, \tag{7.33}$$

where

$$\begin{cases} q_{1i} = M(-1)^{n-1}\binom{N}{n}\dfrac{d_i c_i}{\alpha\left(e_i + \frac{n}{\rho\beta}\right)\left(e_i + \frac{1}{\alpha} + \frac{n}{\rho\beta}\right)} \\ \\ q_{2i} = e_i + \dfrac{1}{\alpha} + \dfrac{n}{\rho\beta}, \quad q_{3i} = M(-1)^{n-1}\binom{N}{n}\dfrac{nd_i}{n + e_i\rho\beta} \\ \\ q_{4i} = M(-1)^{n-1}\binom{N}{n}\dfrac{nd_i e_i}{\beta(e_i + \frac{1}{\alpha})}, \quad q_{5i} = M(-1)^{n-1}\binom{N}{n}\dfrac{nd_i}{\beta(1 + \alpha e_i)} \end{cases}, \tag{7.34}$$

with

$$\begin{cases} \widetilde{\sum_i} = \sum_{i_1=0}^{M-1}\sum_{i_2=0}^{i_1}\sum_{i_3=0}^{i_2}\cdots\sum_{i_N=0}^{i_{N-1}} \\ d_i = (-1)^{i_1+i_2+\cdots+i_N}\binom{M-1}{i_1}\binom{i_1}{i_2}\cdots\binom{i_{N-1}}{i_N}\left(\binom{N}{1}\right)^{i_1-i_2}\left(\binom{N}{2}\right)^{i_2-i_3} \\ \quad \cdots\left(\binom{N}{N-1}\right)^{i_{N-1}-i_N} \\ e_i = \dfrac{i_1}{\alpha} + \dfrac{i_1 + \cdots + i_N}{\rho\beta} \end{cases}. \tag{7.35}$$

Proof. See Appendix A.   □

From Theorem 7.1, we now analyse the CDF of $Z_{m^*,n^*} = \dfrac{u_{m^*}v_{m^*,n^*}}{\gamma_{\mathrm{th}}u_{m^*}+(\gamma_{\mathrm{th}}-1)\eta v_{m^*,n^*}}$ as:

$$F_{Z_{m^*,n^*}}(z) = \Pr\left(\frac{u_{m^*}v_{m^*,n^*}}{\gamma_{\mathrm{th}}u_{m^*} + (\gamma_{\mathrm{th}} - 1)\eta v_{m^*,n^*}} < z\right) \tag{7.36}$$

$$= \Pr\left[u_{m^*}(v_{m^*,n^*} - \gamma_{\mathrm{th}}z) < (\gamma_{\mathrm{th}} - 1)\eta v_{m^*,n^*}z\right]. \tag{7.37}$$

Applying the results of Theorem 7.1 into the above equation yields the CDF of Z_{m^*,n^*} as:

$$
F_{Z_{m^*,n^*}}(z) = 1 - \sum_{n_1}^{N} \sum_{n_2=1}^{N} \widetilde{\sum_i} \widetilde{\sum_j} \left[\frac{q_{5i}q_{1j}\beta\psi_1}{n_1} ze^{-\left[\frac{n_1\gamma_{th}}{\beta} + q_{2j}\eta(\gamma_{th}-1)\right]z} \mathcal{K}_1(\psi_1 z) \right.
$$
$$
+ \frac{q_{5i}q_{3j}\beta\psi_2}{n_1} ze^{-\left[\frac{n_1\gamma_{th}}{\beta} + \frac{(\gamma_{th}-1)\eta}{\alpha}\right]z} \mathcal{K}_1(\psi_2 z)
$$
$$
+ \frac{q_{4i}q_{1j}\psi_3}{\rho q_{2i}} ze^{-\left[q_{2i}\rho\gamma_{th}+q_{2j}\eta(\gamma_{th}-1)\right]z} \mathcal{K}_1(\psi_3 z)
$$
$$
\left. + \frac{q_{4i}q_{3j}\psi_4}{\rho q_{2i}} ze^{-\left[\rho q_{2i}\gamma_{th} + \frac{(\gamma_{th}-1)\eta}{\alpha}\right]z} \mathcal{K}_1(\psi_4 z) \right],
\tag{7.38}
$$

with

$$
\begin{cases}
\psi_1 = \sqrt{\dfrac{4n_1 q_{2j}\eta\gamma_{th}(\gamma_{th}-1)}{\beta}}, & \psi_2 = \sqrt{\dfrac{4n_1\eta\gamma_{th}(\gamma_{th}-1)}{\alpha\beta}} \\
\psi_3 = \sqrt{4q_{2i}q_{2j}\rho\eta\gamma_{th}(\gamma_{th}-1)}, & \psi_4 = \sqrt{\dfrac{4\rho q_{2i}\eta\gamma_{th}(\gamma_{th}-1)}{\alpha}}
\end{cases}.
\tag{7.39}
$$

The system secrecy outage probability is then derived as:

$$
P_{out,m^*,n^*} \simeq \Pr(Z_{m^*,n^*} < w_{m^*})
\tag{7.40}
$$
$$
= \int_0^\infty f_{w_{m^*}}(w) F_{Z_{m^*,n^*}}(w) dw.
\tag{7.41}
$$

Note that for Criteria II and III, the eavesdropper links are not involved in the relay and user selection. Hence we obtain that $f_{w_{m^*}}(w) = \frac{w^{K-1}}{\Gamma(K)\varepsilon^K} e^{-\frac{w}{\varepsilon}}$ [8, (9.5)]. Applying $f_{w_{m^*}}(w)$ into (7.41) yields:

$$
P_{out,m^*,n^*} \simeq 1 - \sum_{n_1=1}^{N} \sum_{n_2=1}^{N} \widetilde{\sum_i} \widetilde{\sum_j} \frac{2\sqrt{\pi}\Gamma(K+2)}{\Gamma(K+\frac{3}{2})\varepsilon^K} \left[\frac{q_{5i}q_{1j}\beta\psi_1^2}{n_1(\psi_1+\tau_1)^{K+2}} {}_2F_1 \right.
$$
$$
\times \left(K+2, \frac{3}{2}, K+\frac{3}{2}; \frac{\tau_1-\psi_1}{\tau_1+\psi_1} \right) + \frac{q_{5i}q_{3j}\beta\psi_2^2}{n_1(\psi_2+\tau_2)^{K+2}} {}_2F_1
$$
$$
\times \left(K+2, \frac{3}{2}, K+\frac{3}{2}; \frac{\tau_2-\psi_2}{\tau_2+\psi_2} \right) + \frac{q_{4i}q_{1j}\psi_3^2}{\rho q_{2i}(\psi_3+\tau_3)^{K+2}} {}_2F_1
$$
$$
\times \left(K+2, \frac{3}{2}, K+\frac{3}{2}; \frac{\tau_3-\psi_3}{\tau_3+\psi_3} \right) + \frac{q_{4i}q_{3j}\psi_4^2}{\rho q_{2i}(\psi_4+\tau_4)^{K+2}} {}_2F_1
$$
$$
\left. \times \left(K+2, \frac{3}{2}, K+\frac{3}{2}; \frac{\tau_4-\psi_4}{\tau_4+\psi_4} \right) \right],
\tag{7.42}
$$

where:

$$\begin{cases} \tau_1 = \dfrac{1}{\varepsilon} + \dfrac{n_1 \gamma_{\text{th}}}{\beta} + q_{2j}\eta(\gamma_{\text{th}} - 1), \quad \tau_2 = \dfrac{1}{\varepsilon} + \dfrac{n_1 \gamma_{\text{th}}}{\beta} + \dfrac{(\gamma_{\text{th}} - 1)\eta}{\alpha} \\[2mm] \tau_3 = \dfrac{1}{\varepsilon} + q_{2i}\rho\gamma_{\text{th}} + q_{2j}\eta(\gamma_{\text{th}} - 1), \quad \tau_4 = \dfrac{1}{\varepsilon} + q_{2i}\rho\gamma_{\text{th}} + \dfrac{(\gamma_{\text{th}} - 1)\eta}{\alpha} \end{cases}. \tag{7.43}$$

By setting $\rho = \rho_{\text{II}}$ and $\rho = \rho_{\text{III}}$ into (7.42), we can obtain the analytical expression of secrecy outage probability for Criteria II and III, respectively.

7.1.3 Asymptotic analysis

In this section, we analyse the asymptotic secrecy outage probability for the three selection criteria with high MER. From the asymptotic expressions, we further reveal the system diversity order for the three criteria.

7.1.3.1 Criterion I

To analyse the diversity gain of Criterion I, we first consider the upper bound of Z_{m,n_m^*} as:

$$Z_{m,n_m^*}^b \leq \min\left(\frac{u_m}{(\gamma_{\text{th}} - 1)\eta}, \frac{v_{m,n_m^*}}{\gamma_{\text{th}}} \right). \tag{7.44}$$

By applying the approximation of $(1 + x)^{-1} \simeq 1 - x$ [9], we can obtain the asymptotic expression of P_{out,m,n_m^*} in the high MER region:

$$P_{\text{out},m,n_m^*}^{\text{asy}} = \begin{cases} \dfrac{K}{\lambda}\left[\dfrac{(\gamma_{\text{th}} - 1)\eta\beta}{\alpha} + \dfrac{\gamma_{\text{th}}}{} \right], & \text{if } N = 1 \\[3mm] \dfrac{K}{\lambda} \dfrac{(\gamma_{\text{th}} - 1)\eta\beta}{\alpha}, & \text{if } N \geq 2 \end{cases}, \tag{7.45}$$

where $\lambda = \frac{\beta}{\varepsilon}$ denotes the MER [10], defined as the ratio of average channel gain from the relay to the users to that from the relay to the eavesdroppers. It follows from (7.45) that we can obtain the asymptotic secrecy outage probability with high MER for criterion I as:

$$P_{\text{out},m^*,n^*}^{\text{asy}} = \begin{cases} \dfrac{K^M}{\lambda^M}\left[\dfrac{(\gamma_{\text{th}} - 1)\eta\beta}{\alpha} + \dfrac{\gamma_{\text{th}}}{} \right]^M, & \text{if } N = 1 \\[3mm] \dfrac{K^M}{\lambda^M}\left(\dfrac{(\gamma_{\text{th}} - 1)\eta\beta}{\alpha} \right)^M, & \text{if } N \geq 2 \end{cases}, \tag{7.46}$$

Inspired by the asymptotic expression, we find that the diversity order for Criterion I is equal to M, regardless of the number of users and eavesdroppers. Moreover, the asymptotic secrecy outage probability is irrespective of the number of users when $N \geq 2$, indicating that no gain is achieved from increasing the number of users with high MER. This is due to the fact that when $N \geq 2$, the first hop from the BS to the relays becomes the bottleneck for the dual-hop data transmission.

7.1.3.2 Criteria II and III

To derive the asymptotic secrecy outage probability for Criteria II and III, we first give the asymptotic CDFs of u_{m^*} and v_{m^*,n^*} as:

$$F_{u_{m^*}}(x) \simeq \begin{cases} \left(1 + \dfrac{\rho\beta}{\alpha}\right)^{M-1} \dfrac{\rho\beta}{\alpha} \dfrac{x^M}{(\rho\beta)^M}, & \text{if } N = 1 \\[3mm] \dfrac{x^M}{\alpha^M}, & \text{if } N \geq 2 \end{cases} , \qquad (7.47)$$

$$F_{v_{m^*,n^*}}(x) \simeq \begin{cases} \left(1 + \dfrac{\rho\beta}{\alpha}\right)^{M-1} \dfrac{x^M}{\beta^M}, & \text{if } N = 1 \\[3mm] \dfrac{MN}{M+N-1} \dfrac{\rho^{M-1}x^{M+N-1}}{\alpha^{M-1}\beta^N}, & \text{if } N \geq 2 \end{cases} , \qquad (7.48)$$

where we apply the approximation of $e^{-x} \simeq \sum_{m=0}^{M}(-1)^m x^m$ for small value of $|x|$ [9]. We then derive the CDFs of Z_{m^*,n^*}^b as:

$$F_{Z_{m^*,n^*}^b}(z) = \Pr\left[\min\left(\frac{u_{m^*}}{(\gamma_{\text{th}}-1)\eta}, \frac{v_{m^*,n^*}}{\gamma_{\text{th}}}\right) < z\right] \qquad (7.49)$$

$$\simeq \begin{cases} \mu_1\left(\dfrac{z}{\beta}\right)^M, & \text{if } N = 1 \\[3mm] \mu_{21}\left(\dfrac{z}{\beta}\right)^M + \mu_{22}\left(\dfrac{z}{\beta}\right)^{M+N-1}, & \text{if } N \geq 2 \end{cases} , \qquad (7.50)$$

with

$$\begin{cases} \mu_1 = \left(1 + \dfrac{\rho\beta}{\alpha}\right)^{M-1}\left[\dfrac{\rho\beta}{\alpha}\left(\dfrac{(\gamma_{\text{th}}-1)\eta}{\rho}\right)^M + \gamma_{\text{th}}^M\right] \\[3mm] \mu_{21} = \dfrac{\beta^M}{\alpha^M}((\gamma_{\text{th}}-1)\eta)^M, \quad \mu_{22} = \dfrac{MN}{M+N-1}\left(\dfrac{\rho\beta}{\alpha}\right)^{M-1}\gamma_{\text{th}}^{M+N-1} \end{cases} . \qquad (7.51)$$

By applying the asymptotic CDF of Z_{m^*,n^*}^b into (7.41) and then solving the resultant equation, we can obtain the asymptotic secrecy outage probability with high MER for Criteria II and III as:

$$P_{\text{out},m^*,n^*}^{asy} \simeq \begin{cases} \dfrac{\mu_1\Gamma(M+K)}{\Gamma(K)}\dfrac{1}{(\lambda)^M}, & \text{if } N = 1 \\[3mm] \dfrac{\mu_{21}\Gamma(M+K)}{\Gamma(K)}\dfrac{1}{(\lambda)^M} + \dfrac{\mu_{22}\Gamma(M+N+K-1)}{\Gamma(K)} \\[3mm] \times \dfrac{1}{(\lambda)^{M+N-1}}, & \text{if } N \geq 2 \end{cases} , \qquad (7.52)$$

where $\rho = \rho_{II}$ and $\rho = \rho_{III}$ correspond to the asymptotic secrecy outage probabilities of Criteria II and III, respectively. Note that when $N \geq 2$, the first term on the right-hand side (RHS) of (7.52) will dominate with large MER, while the second term will become marginal. Hence, one can conclude that for Criteria II and III, the system diversity order is also equal to M, regardless of the number of users and eavesdroppers. Moreover, the first term in RHS of (7.52) is irrespective of the number of users when $N \geq 2$, indicating that no gain can be achieved from increasing the number of users with high MER. Once again this is due to the bottleneck effect of the first hop from the BS to the relays when $N \geq 2$.

7.1.4 Numerical and simulation results

This section presents numerical and simulation results to verify the proposed studies. All the links in the system experience Rayleigh flat fading. The pathloss model is adopted with loss factor of four to determine the average channel gains. The distance between the base station and the desired users is set to unity. The relays are located between the base station and desired users, and the distance between the base station and relays is denoted by D, so that $\alpha = D^{-4}$ and $\beta = (1 - D)^{-4}$. In addition, we set a high transmit power at the base station with $P_S = 30$ dB, since we focus on the effect of MER on the system secrecy outage probability.

Figure 7.2 demonstrates the effect of the number of relays on the secrecy outage probability of Criteria I, II and III versus MER, where $D = 0.5$, $R_s = 0.2$ bps/Hz, $N = 2$, $K = 2$, and M varies from 1 to 3. For comparison, we plot the simulation

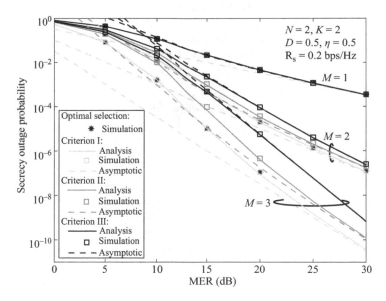

Figure 7.2   Effect of number of relays on the secrecy outage probability versus MER. © IEEE 2016. Reprinted with permission from [5]

results of the three selection criteria as well as the optimal selection performed in (7.11). As observed from the figure, one can find that for different values of MER and M, the analytical results for Criteria I–III match well the simulation, which validates the derived analytical expressions of the secrecy outage probability in (7.31) and (7.42). In addition, the asymptotic results converge with the exact at high MER, which verifies the derived asymptotic expressions. Moreover, the slopes of the curve of the secrecy outage probability are in parallel with M, which verifies the system diversity order of M for all three criteria. Further, Criterion I achieves a comparable performance to the optimal selection and outperforms Criteria II and III. This is because Criterion I performs the selection by incorporating both the main and eavesdropper links. Criterion II exhibits better performance than Criterion III, as the former incorporates different impact from the two relay hops on the system security. One can also find that the performance gap between the three criteria increases with the number of relays.

7.2   DF relaying

For the multiuser multi-relay networks with DF relaying, a secure system model is depicted in Figure 7.3, where the communication between M users S_m, $m \in \{1, \ldots, M\}$, and the BS D is assisted by N DF relays R_n, $n \in \{1, \ldots, N\}$. We assume that an eavesdropper E exists in this network and overhears the transmitted messages, bringing out the important issue of information wiretap. As assumed in AF relaying, all nodes in the network are equipped with a single antenna, due to the size limitation, and operate in a half-duplex time-division mode.

To enhance the transmission security, both user selection and relay selection will be performed in this work to select one best user S_{m^*} among M users to communicate with D, with the help of a selected R_{n^*} out of the N relays. We assume that the direct

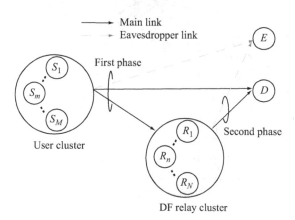

Figure 7.3   *The illustration of a two-phase secure multiuser multi-relay network with direct links. © IEEE 2016. Reprinted with permission from [6]*

link between the source and the destination is available. In practice, the direct link is available if the source and the destination are not placed remotely, or the destination does not fall within heavily shadowed areas. To present the relay and user selection criteria, we first detail the two-phase transmission process, as follow:

Suppose that the user S_m and the relay R_n have been selected for data transmission. In the first phase, S_m sends the encoded signal of unit-variance, x_s, with transmit power P. The received signal at R_n, D and E in the first phase are given by:

$$y_{R_n} = h_{S_m,R_n}\sqrt{P}x_s + n_R, \tag{7.53}$$

$$y_D^{(1)} = h_{S_m,D}\sqrt{P}x_s + n_D^{(1)}, \tag{7.54}$$

$$y_E^{(1)} = h_{S_m,E}\sqrt{P}x_s + n_E^{(1)}, \tag{7.55}$$

respectively, where $h_{S_m,R_n} \sim \mathscr{CN}(0,\alpha)$, $h_{S_m,D} \sim \mathscr{CN}(0,\varepsilon_1)$ and $h_{S_m,E} \sim \mathscr{CN}(0,\varepsilon_2)$ denote the channel coefficients of the S_m–R_n link, S_m–D link and S_m–E link, respectively. We also denote $n_R \sim \mathscr{CN}(0,\sigma^2)$, $n_D^{(1)} \sim \mathscr{CN}(0,\sigma^2)$ and $n_E^{(1)} \sim \mathscr{CN}(0,\sigma^2)$ as the additive white Gaussian noise (AWGN) at R_n, D and E in the first phase. If R_n correctly decodes the message received in the first phase, it then re-encodes the signal with the same code book at S_m and forwards it to D in the second phase. Accordingly, the received signal at D and E in the second phase are given by:

$$y_D^{(2)} = h_{R_n,D}\sqrt{P}x_s + n_D^{(2)} \tag{7.56}$$

$$y_E^{(2)} = h_{R_n,E}\sqrt{P}x_s + n_E^{(2)}, \tag{7.57}$$

respectively, where $h_{R_n,D} \sim \mathscr{CN}(0,\beta_1)$ and $h_{R_n,E} \sim \mathscr{CN}(0,\beta_2)$ denote the channel coefficients of the R_n–D link and R_n–E link, respectively. We also denote $n_D^{(2)} \sim \mathscr{CN}(0,\sigma^2)$ and $n_E^{(2)} \sim \mathscr{CN}(0,\sigma^2)$ as the AWGN at D and E, respectively, in the second phase. We denote $u_{mn} = |h_{S_m,R_n}|^2$, $v_{1n} = |h_{R_n,D}|^2$, $v_{2n} = |h_{R_n,E}|^2$, $w_{1m} = |h_{S_m,D}|^2$ and $w_{2m} = |h_{S_m,E}|^2$ as the channel gains of the S_m–R_n link, R_n–D link, R_n–E link, S_m–D link and S_m–E link, respectively. The end-to-end SNR at D for the repetition-coded fixed DF relaying can be written as [11, Eq. (15)]:

$$\mathsf{SNR}_D = \bar{\gamma}\min(u_{mn}, v_{1n} + w_{1m}), \tag{7.58}$$

where $\bar{\gamma} = P/\sigma^2$ is the transmit SNR. According to [12,13], we note that the secrecy outage occurs when the system achievable secrecy data rate falls below a predetermined secrecy rate R_s, i.e.:

$$\frac{1}{2}\log_2(1 + \bar{\gamma}\min(u_{mn}, v_{1n} + w_{1m})) - \frac{1}{2}\log_2(1 + \bar{\gamma}(v_{2n} + w_{2m})) < R_s. \tag{7.59}$$

After some mathematical manipulations, we re-express (7.59) as

$$\frac{1 + \bar{\gamma}\min(u_{mn}, v_{1n} + w_{1m})}{1 + \bar{\gamma}(v_{2n} + w_{2m})} < \gamma_s, \tag{7.60}$$

where $\gamma_s = 2^{2R_s}$ is the secrecy SNR threshold.

7.2.1   User and relay selection criteria

We consider a practical passive eavesdropping scenario, where only the statistical information of the eavesdropper's channel is known, while the instantaneous information of the eavesdropper's channel is unknown. This indicates that the eavesdropper's channel coefficients, i.e., $h_{S_m,E}$ and $h_{R_n,E}$, are not available at the users, relays and BS. Under this consideration, we propose two-user and relay-selection criteria that select the best user and relay pair to carry out the secure transmission in the network. These criteria are described as follows.

7.2.1.1   Criterion I

In this criterion, the joint user and relay selection is performed by maximising the achievable rate of the main links. Mathematically, the indices of the selected user and the relay are expressed as:

$$(m^*, n^*) = \arg \max_{1 \leq m \leq M} \max_{1 \leq n \leq N} \min(u_{mn}, v_{1n} + w_{1m}). \tag{7.61}$$

This criterion achieves the optimal secrecy performance in the passive eavesdropping scenario.

In Criterion I, the term w_{1m} from the direct link is incorporated into the term $\min(u_{mn}, v_{1n})$ from the relay links. As such, the two terms cannot be separated from each other. It follows that Criterion I is a joint selection scheme where the user selection interacts with the relay selection.

7.2.1.2   Criterion II

Different from Criterion I, Criterion II involves separate user and relay selections. Specifically, the best user is firstly selected based on the direct links [14], and then the best relay is selected based on the two-hop relay links [15]. The indices of the selected user and the relay can be expressed as:

$$m^* = \arg \max_{1 \leq m \leq M} w_{1m}, \tag{7.62}$$

and

$$n^* = \arg \max_{1 \leq n \leq N} \min(u_{m^*n}, v_{1n}). \tag{7.63}$$

It is evident from (7.62) and (7.63) that user selection and relay selection are performed separately in Criterion II. Compared with Criterion I which needs an adder and a comparator to perform joint selection, Criterion II only requires a simple comparator to perform separate selections, bringing about a lower implementation complexity.

Based on (7.60), the secrecy outage probability with the selected user S_{m^*} and relay R_{n^*} is given by:

$$P_{\text{out}} = \Pr\left[\frac{1 + \bar{\gamma}\min(u_{m^*n^*}, v_{1n^*} + w_{1m^*})}{1 + \bar{\gamma}(v_{2n^*} + w_{2m^*})} < \gamma_s\right]$$

$$= \Pr[Z < \gamma_s(v_{2n^*} + w_{2m^*}) + \gamma_s'], \tag{7.64}$$

where $\gamma_s' = (\gamma_s - 1)/\bar{\gamma}$ and $Z \triangleq \min(u_{m^*n^*}, v_{1n^*} + w_{1m^*})$. Evidently, the statistical characterization of Z is the key for the evaluation of P_{out}.

7.2.2 Closed-form analysis

In this section, we derive new exact and asymptotic expressions for the secrecy outage probability, where both criteria are considered.

7.2.2.1 A tight lower bound for Criterion I

For Criterion I, we first rewrite Z as:

$$Z = \max_{1 \leq m \leq M} \max_{1 \leq n \leq N} \min(u_{mn}, v_{1n} + w_{1m}). \tag{7.65}$$

Since multiple users have the common variable v_{1n} and multiple relays have the common variable w_{1m}, it is difficult to provide an exact analytical expression for the secrecy outage probability. Hence, we turn to derive two upper bounds on Z by exchanging the sequence of max and min operations. The first upper bound is given by:

$$
\begin{aligned}
Z_1 &= \max_{1 \leq m \leq M} \min\left(\max_{1 \leq n \leq N} (u_{mn}, v_{1n} + w_{1m}) \right) \\
&= \max_{1 \leq m \leq M} \min\left(\max_{1 \leq n \leq N} u_{mn}, \left(\max_{1 \leq n \leq N} v_{1n} \right) + w_{1m} \right),
\end{aligned} \tag{7.66}
$$

and the second upper bound by:

$$
\begin{aligned}
Z_2 &= \max_{1 \leq n \leq N} \min\left(\max_{1 \leq m \leq M} (u_{mn}, v_{1n} + w_{1m}) \right) \\
&= \max_{1 \leq n \leq N} \min\left(\max_{1 \leq m \leq M} u_{mn}, v_{1n} + \max_{1 \leq m \leq M} (w_{1m}) \right).
\end{aligned} \tag{7.67}
$$

Built upon Z_1 in (7.66) and Z_2 in (7.67), we next derive the lower bounds on the secrecy outage probability, i.e., $P_{1,\text{out}}^{LB}$ and $P_{2,\text{out}}^{LB}$. We now begin to derive $P_{1,\text{out}}^{LB}$ associated with Z_1. Let $v_1 = \max_{1 \leq n \leq N} v_{1n}$ and $u_m = \max_{1 \leq n \leq N} u_{mn}$, and then we rewrite Z_1 as:

$$Z_1 = \max_{1 \leq m \leq M} \underbrace{\min(u_m, v_1 + w_{1m})}_{Z_{1m}}. \tag{7.68}$$

From (7.68), Z_{1m} is correlated with each other, because of the common variable v_1. To deal with this, we first derive the conditional CDF of Z_{1m} with respect to v_1, i.e., $F_{Z_{1m}}(z|v_1)$. By statistically averaging $F_{Z_{1m}}^M(z|v_1)$ with respect to v_1, we then obtain the analytical CDF of Z_1, i.e., $F_{Z_1}(z)$. We further statistically average $F_{Z_1}(\gamma_s(v_{2n^*} + w_{2m^*}) + \gamma_s')$ with respect to both v_{2n^*} and w_{2m^*} to obtain $P_{1,\text{out}}^{LB}$. The CDF of Z_1 is derived and presented in the following theorem.

Theorem 7.2. *The CDF of Z_1 can be expressed as*

$$F_{Z_1}(z) = \sum_{n_1=1}^{N} \sum_{n_2=0}^{MN} b_{1,n_1,n_2} e^{-c_{1,n_1,n_2} z} + \sum_{m=0}^{M} \sum_{n=1}^{N} \widetilde{\sum_{\{i\}}} q_{1i} b_{2,m,n} \left(e^{-q_{2i} z} - e^{-(q_{2i}+c_{2,m,n})z} \right),$$

(7.69)

where

$$b_{1,n_1,n_2} = (-1)^{n_1+n_2+1} \binom{N}{n_1} \binom{MN}{n_2}, \quad c_{1,n_1,n_2} = \frac{n_1}{\beta_1} + \frac{n_2}{\alpha},$$

$$\widetilde{\sum_{\{i\}}} = \sum_{i_1=0}^{m} \sum_{i_2=0}^{i_1} \cdots \sum_{i_{N-1}=0}^{i_{N-2}},$$

$$q_{1i} = \binom{m}{i_1} \binom{i_1}{i_2} \cdots \binom{i_{N-2}}{i_{N-1}} b_{3,1}^{m-i_1} b_{3,2}^{i_1-i_2} \cdots b_{3,N-1}^{i_{N-2}-i_{N-1}} b_{3,N}^{i_{N-1}},$$

$$b_{2,m,n} = (-1)^{m+n-1} \binom{N}{n} \binom{M}{m} \frac{n\varepsilon_1}{n\varepsilon_1 - m\beta_1}, \quad b_{3,n} = \binom{N}{n}(-1)^{n-1},$$

$$q_{2i} = c_{3,1}(m - i_1) + c_{3,2}(i_1 - i_2) + \cdots + c_{3,N-1}(i_{N-2} - i_{N-1}) + c_{3,N} i_{N-1},$$

and

$$c_{2,m,n} = \frac{n}{\beta_1} - \frac{m}{\varepsilon_1}, \quad c_{3,n} = \frac{n}{\alpha} + \frac{1}{\varepsilon_1},$$

Proof. See Appendix B.   □

From Theorem 7.2 and (7.64), we can derive the first lower bound of P_{out} as

$$P_{1,\text{out}}^{LB} = \int_0^\infty \int_0^\infty F_{Z_1}\left(\gamma_s(v_{2n*} + w_{2m*}) + \gamma_s'\right) f_{v_{2n*}}(v_{2n*}) f_{w_{2m*}}(w_{2m*}) dv_{2n*} dw_{2m*},$$

(7.70)

$$= \sum_{n_1=1}^{N} \sum_{n_2=0}^{MN} b_{1,n_1,n_2} \mathcal{L}(c_{1,n_1,n_2}) + \sum_{m=0}^{M} \sum_{n=1}^{N} \widetilde{\sum_{\{i\}}} q_{1i} b_{2,m,n} \left(\mathcal{L}(q_{2i}) \right.$$

$$\left. - \mathcal{L}(q_{2i} + c_{2,m,n}) \right),$$

(7.71)

where $\mathcal{L}(x) = \frac{e^{-\gamma_s' x}}{(1+\beta_2\gamma_s x)(1+\varepsilon_2\gamma_s x)}$, and we have applied the PDF of v_{2n*}, given by $f_{v_{2n*}}(x) = \frac{1}{\beta_2} e^{-\frac{x}{\beta_2}}$, and the PDF of w_{2m*}, given by $f_{w_{2m*}}(x) = \frac{1}{\varepsilon_2} e^{-\frac{x}{\varepsilon_2}}$,

The second lower bound of P_{out}, $P_{2,\text{out}}^{LB}$, has the same form as $P_{1,\text{out}}^{LB}$ in (7.71) after replacing M with N and β_1 with ε_1. This is because of the symmetry existed in (7.66) and (7.67). Finally, the tight lower bound on the secrecy outage probability for Criterion I can be obtained as

$$P_{\text{out}}^{LB} = \max\left(P_{1,\text{out}}^{LB}, P_{2,\text{out}}^{LB}\right).$$

(7.72)

We highlight that (7.72) consists of elementary functions only, and as such, it can be easily evaluated.

7.2.2.2   Exact formula for Criterion II

To derive the secrecy outage probability for Criterion II, we first derive the CDFs of w_{1m*}, u_{m*n*} and v_{1n*}, as per the selection criterion characterised by (7.62) and (7.63), and then we obtain the CDF of Z, $F_Z(z)$. Furthermore, by averaging $F_Z(z)$ with respect to w_{2m*} and v_{2n*}, we derive an exact expression for P_{out}.

Theorem 7.3. *The CDF of Z can be expressed by*

$$
F_Z(z) = 1 - \sum_{m=1}^{M}\sum_{n_1=0}^{N-1}\sum_{n_2=0}^{N-1} \left(t_{1,n_1} t_{3,m,n_2} e^{-\frac{z}{\zeta}} + t_{2,n_1} t_{3,m,n_2}\, e^{-\left(\frac{n_1+1}{\zeta}+\frac{1}{\beta_1}\right)z} \right.
$$
$$
+ t_{1,n_1} t_{4,m,n_2} e^{-\left(\frac{1}{\alpha}+\frac{m}{\varepsilon_1}\right)z} + t_{1,n_1} t_{5,m,n_2} e^{-\left(\frac{1}{\alpha}+\frac{n_2+1}{\zeta}\right)z} + t_{2,n_1} t_{4,m,n_2}
$$
$$
\left. \times e^{-\left(\frac{n_1+1}{\zeta}+\frac{m}{\varepsilon_1}\right)z} + t_{2,n_1} t_{5,m,n_2} e^{-\left(\frac{n_1+n_2+2}{\zeta}\right)z} \right], \tag{7.73}
$$

where

$$
\zeta = \frac{\alpha\beta_1}{\alpha+\beta_1},
$$

$$
t_{1,n} = \frac{b_{4,n}\zeta}{\zeta+n\beta_1}, \quad t_{2,n} = b_{4,n}\left(\frac{1}{n+1}-\frac{\zeta}{\zeta+n\beta_1}\right),
$$

$$
t_{3,m,n} = (-1)^{m-1}\binom{M}{m}\frac{b_{4,n}\zeta m\beta_1}{(\zeta+n\alpha)(m\beta_1-\varepsilon_1)},
$$

$$
t_{4,m,n} = (-1)^{m-1}\binom{M}{m}\left[\frac{1}{N}-\frac{b_{4,n}\zeta m\beta_1}{(\zeta+n\alpha)(m\beta_1-\varepsilon_1)}\right.
$$
$$
\left. -b_{4,n}\left(\frac{1}{n+1}-\frac{\zeta}{\zeta+n\alpha}\right)\frac{m\zeta}{m\zeta-(n+1)\varepsilon_1}\right],
$$

$$
t_{5,m,n} = (-1)^{m-1}\binom{M}{m}\left(\frac{1}{n+1}-\frac{\zeta}{\zeta+n\alpha}\right)\frac{b_{4,n}m\zeta}{m\zeta-(n+1)\varepsilon_1},
$$

and

$$
b_{4,n} = N(-1)^n\binom{N-1}{n}.
$$

Proof. See Appendix C.   □

From Theorem 7.3 and (7.64), we can now derive an exact expression for P_{out} as

$$P_{\text{out}} = \int_0^\infty \int_0^\infty F_Z\big(\gamma_s(v_{2n*} + w_{2m*}) + \gamma_s'\big) f_{v_{2n*}}(v_{2n*}) f_{w_{2m*}}(w_{2m*}) dv_{2n*} dw_{2m*},$$

(7.74)

$$= 1 - \sum_{m=1}^{M} \sum_{n_1=0}^{N-1} \sum_{n_2=0}^{N-1} \left[t_{1,n_1} t_{3,m,n_2} \mathscr{L}\left(\frac{1}{\zeta}\right) + t_{2,n_1} t_{3,m,n_2} \mathscr{L}\left(\frac{n_1+1}{\zeta} + \frac{1}{\beta_1}\right) \right.$$

$$+ t_{1,n_1} t_{4,m,n_2} \mathscr{L}\left(\frac{1}{\alpha} + \frac{m}{\varepsilon_1}\right) + t_{1,n_1} t_{5,m,n_2} \mathscr{L}\left(\frac{1}{\alpha} + \frac{n_2+1}{\zeta}\right)$$

$$\left. + t_{2,n_1} t_{4,m,n_2} \mathscr{L}\left(\frac{n_1+1}{\zeta} + \frac{m}{\varepsilon_1}\right) + t_{2,n_1} t_{5,m,n_2} \mathscr{L}\left(\frac{n_1+n_2+2}{\zeta}\right) \right].$$

(7.75)

Notably, (7.75) only consists of elementary functions and, thus, can be easily evaluated.

7.2.3 Asymptotic analysis

7.2.3.1 Asymptotic P_{out} for Criterion I

We now derive the asymptotic secrecy outage probability for Criterion I, in the presence of high transmit SNRs and MERs. By using the approximation of $e^{-x} \simeq 1 - x$ for small value of $|x|$, the asymptotic CDF of Z_1 can be written as:

$$F_{Z_1}(z) \simeq \frac{\rho_{M,N} z^{M+N}}{\varepsilon_1^M \beta_1^N} + \left(\frac{z}{\alpha}\right)^{MN},$$

(7.76)

where

$$\rho_{M,N} = \begin{cases} \displaystyle\sum_{m=0}^{M} \binom{M}{m} \frac{1}{m+1} \left(\frac{z}{\alpha}\right)^{M-m}, & \text{if } N = 1, \\[2em] \displaystyle\sum_{m=0}^{M} \frac{(-1)^m N}{m+N} \binom{M}{m}, & \text{if } N \geq 2. \end{cases}$$

(7.77)

From the asymptotic expression for $F_{Z_1}(z)$, we can compute the asymptotic secrecy outage probability associated with Z_1 as:

$$P_{1,\text{out}} \simeq \int_0^\infty \int_0^\infty F_{Z_1}\big(\gamma_s(v_{2n*} + w_{2m*}) + \gamma_s'\big) f_{v_{2n*}}(v_{2n*}) f_{w_{2m*}}(w_{2m*}) dv_{2n*} dw_{2m*}$$

$$\simeq \int_0^\infty \int_0^\infty F_{Z_1}\big(\gamma_s(v_{2n*} + w_{2m*})\big) f_{v_{2n*}}(v_{2n*}) f_{w_{2m*}}(w_{2m*}) dv_{2n*} dw_{2m*}$$

$$\simeq \frac{\gamma_s^{M+N}(M+N)! \rho_{M,N}}{\lambda_1^M \lambda_2^N} \sum_{k=0}^{M+N} \left(\frac{\beta_2}{\varepsilon_2}\right)^{M-k}$$

$$+ \frac{\gamma_s^{MN}(MN)!}{\lambda_2^{MN}} \left(\frac{\beta_1}{\alpha}\right)^{MN} \sum_{k=0}^{MN} \left(\frac{\varepsilon_2}{\beta_2}\right)^{MN-k},$$

(7.78)

where $\lambda_1 = \frac{\varepsilon_1}{\varepsilon_2}$ and $\lambda_2 = \frac{\beta_1}{\beta_2}$ are the MERs for the direct and relay links, respectively. Due to the symmetry between Z_1 and Z_2, we can readily obtain the asymptotic secrecy outage probability associated with Z_2, which has the same form as (7.78) after replacing M with N and β_1 with ε_1.

From these two asymptotic expressions, we conclude that the secrecy diversity order for Criterion I is equal to $\min(MN, M + N)$. This reveals that the security of the network can be significantly improved by increasing the number of users or the number of relays. In addition, we provide some valuable insights in the following two remarks:

Remark 7.1. *For a single user or single relay communication system, the secrecy diversity order is MN. In particular, the secrecy diversity order is M for a single relay system with $N = 1$. This is due to the fact that only multiuser diversity can be exploited. On the other hand, the secrecy diversity order is N for a single-user system with $M = 1$, since only the multi-relay diversity can be exploited.*

Remark 7.2. *For a multiuser and multi-relay communication system with $M \geq 2$ and $N \geq 2$, the secrecy diversity order is equivalent to $M + N$. This indicates that both multiuser diversity and multi-relay diversity can be fully exploited for secure communication.*

7.2.3.2 Asymptotic P_{out} for Criterion II

By using the Taylor's series approximation of $e^{-x} \simeq 1 - x + \frac{x^2}{2} + \cdots + \frac{(-x)^N}{N!}$ [9], we can write the asymptotic CDF of Z as:

$$F_Z(z) \simeq \left(\frac{\beta_1}{\alpha + \beta_1}\right) \frac{z^N}{\zeta^N}. \tag{7.79}$$

From this asymptotic CDF, we compute the asymptotic secrecy outage probability for Criterion II as:

$$
\begin{aligned}
P_{out} &\simeq \int_0^\infty \int_0^\infty F_Z\big(\gamma_s(v_{2n^*} + w_{2m^*}) + \gamma_s'\big) f_{v_{2n^*}}(v_{2n^*}) f_{w_{2m^*}}(w_{2m^*}) dv_{2n^*} dw_{2m^*} \\
&\simeq \int_0^\infty \int_0^\infty F_Z(\gamma_s(v_{2n^*} + w_{2m^*})) f_{v_{2n^*}}(v_{2n^*}) f_{w_{2m^*}}(w_{2m^*}) dv_{2n^*} dw_{2m^*} \\
&\simeq \frac{\gamma_s^N N!}{\lambda_2^N} \left(1 + \frac{\alpha}{\beta_1}\right)^{N-1} \sum_{n=0}^N \left(\frac{\varepsilon_2}{\beta_2}\right)^n.
\end{aligned} \tag{7.80}
$$

From the asymptotic secrecy outage probability in (7.80), we provide some valuable insights in the following two remarks:

Remark 7.3. *Criterion II achieves the secrecy diversity order of N, which comes from the multi-relay diversity. This indicates that the multiuser diversity is not efficiently exploited in Criterion II. In particular, increasing the number of users does not affect the secrecy diversity order nor the secrecy coding gain of Criterion II. This can be explained by the fact that the asymptotic secrecy outage probability is irrespective of M.*

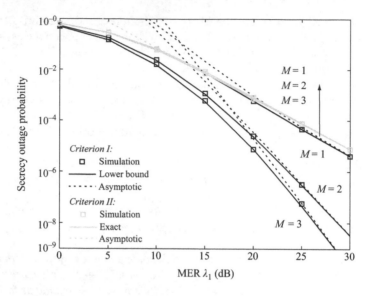

Figure 7.4   *Secrecy outage probability versus MER for $N = 2$, $D = 0.5$, $\bar{\gamma} = 30$ dB and $\lambda_1 = \lambda_2$. © IEEE 2016. Reprinted with permission from [6]*

Remark 7.4. *The secrecy performance of the network can be significantly enhanced by increasing the number of relays. This can be explained by the fact that the relay links are the bottleneck of the secure transmission in Criterion II.*

7.2.4   Numerical and simulation results

In this section, we provide numerical and simulation results to verify the presented analysis of the proposed selection criteria and to examine the impact of the network parameters on the secrecy outage probability. Throughout this section, we assume that all links in the network experience Rayleigh flat fading. Without loss of generality, the distance between the users and the BS is set to unity, and the relays are placed between in them. Let D denote the distance from the users to relay. Accordingly, the average channel gains of the main links are set to $\alpha = D^{-4}$, $\beta_1 = (1 - D)^{-4}$ and $\varepsilon_1 = 1$, where the pathloss model with a loss factor of 4 is used. The secrecy data rate R_s is set to 0.2 bps/Hz and, thus, the secrecy SNR threshold γ_s is equal to 1.32.

Figures 7.4 and 7.5 plot the secrecy outage probabilities for Criteria I and II versus the MER λ_1 for $D = 0.5$, $\bar{\gamma} = 30$ dB and $\lambda_1 = \lambda_2$. From both figures, the asymptotic result for each criterion accurately approximates the corresponding simulation result in the high MER regime for various values of M and N. This validates our asymptotic results derived for Criteria I and II. Moreover, the secrecy diversity order increases

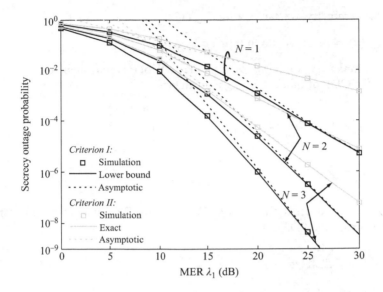

Figure 7.5   Secrecy outage probability versus MER for M = 2, D = 0.5, $\bar{\gamma}$ = 30 dB, and $\lambda_1 = \lambda_2$. © IEEE 2016. Reprinted with permission from [6]

with N for both criteria. This implies that increasing the number of relays significantly improves the secrecy outage probability, especially in the medium and high MER regime. In contrast, the secrecy diversity order increases with M for Criterion I, but remains unchanged for Criterion II, as indicated by our asymptotic results. This implies that increasing the number of users leads to a prominent reduction in the secrecy outage probability for Criterion I, but a minor reduction in the secrecy outage probability for Criterion II.

Appendix A

A.1   Proof of Theorem 1

The CDF of u_{m*} is defined as:

$$F_{u_{m*}}(x) = \Pr(u_{m*} < x) \tag{A.1}$$

$$= \sum_{m=1}^{M} \Pr[u_m < x, \min(u_m, \rho v_{m,n_m^*}) > \max_{m_1=1,\ldots,M, m_1 \neq m} \min(u_{m_1}, \rho v_{m_1,n_{m_1}^*})]. \tag{A.2}$$

Due to the symmetry, we can rewrite $F_{u_{m^*}}(x)$ as:

$$F_{u_{m^*}}(x) = M \Pr[u_1 < x, \min(u_1, \rho v_{1,n_1^*}) > \theta], \tag{A.3}$$

where $\theta = \max_{m=2,\dots,M} \min(u_m, \rho v_{m,n_m^*})$. The CDF of θ is equivalent to the $(M-1)$-th power of the CDF of $\min(u_m, \rho v_{m,n_m^*})$, given by:

$$F_\theta(\theta) = \left[1 - \sum_{n=1}^{N} (-1)^{n-1} \binom{N}{n} e^{-(\frac{1}{\alpha} + \frac{n}{\rho\beta})\theta} \right]^{M-1} \tag{A.4}$$

$$= \widetilde{\sum_i} d_i e^{-e_i\theta}, \tag{A.5}$$

where $\widetilde{\sum}_i$, d_i and e_i are defined in (7.35). By setting $F_\theta(\theta)$ derivative with respect to θ, we can obtain the PDF of θ as:

$$f_\theta(\theta) = -\widetilde{\sum_i} d_i e_i e^{-e_i\theta}. \tag{A.6}$$

Then, we further derive $F_{u_{m^*}}(x)$ from (A.3) as:

$$F_{u_{m^*}}(x) = M \Pr\left(\theta < u_1 < x, v_{1,n_1^*} > \frac{\theta}{\rho}, 0 < \theta < x \right) \tag{A.7}$$

$$= M \int_0^x f_\theta(\theta) \left[\int_\theta^x f_{u_1}(u_1) du_1 \int_{\frac{\theta}{\rho}}^\infty f_{v_{1,n_1^*}(v_1)} dv_1 \right] d\theta. \tag{A.8}$$

Applying the PDFs of θ, u_1 and v_{1,n_1^*} into the above equation leads to the CDF of u_{m^*}, as shown in (7.33) of Theorem 7.1.

Similarly, we derive the CDF of v_{m^*,n^*} as:

$$F_{v_{m^*,n^*}}(x) = \Pr(v_{m^*,n^*} < x) \tag{A.9}$$

$$= \sum_{m=1}^{M} \Pr[v_{m,n_m^*} < x, \min(u_m, \rho v_{m,n_m^*}) > \max_{m_1=1,\dots,M, m_1 \neq m} \min(u_{m_1}, \rho v_{m_1,n_{m_1}^*})] \tag{A.10}$$

$$= M \Pr[v_{1,n_1^*} < x, \min(u_1, \rho v_{1,n_1^*}) > \theta]. \tag{A.11}$$

Note that the condition of $v_{1,n_1^*} < x$ and $\min(u_1, \rho v_{1,n_1^*}) > \theta$ can be written as $u_1 > \theta$, $v_{1,n_1^*} > \frac{\theta}{\rho}$ and $v_{1,n_1^*} < x$, which is equivalent to $u_1 > \theta$, $\frac{\theta}{\rho} < v_{1,n_1^*} < x$ and $0 < \theta < \rho x$. Hence, we can further write $F_{v_{m^*,n^*}}(x)$ as:

$$F_{v_{m^*,n^*}}(x) = M \Pr\left(u_1 > \theta, \frac{\theta}{\rho} < v_{1,n_1^*} < x, 0 < \theta < \rho x \right) \tag{A.12}$$

$$= M \int_0^{\rho x} f_\theta(\theta) \left[\int_\theta^\infty f_{u_1}(u_1) du_1 \int_{\frac{\theta}{\rho}}^x f_{v_{1,n_1^*}}(v_1) dv_1 \right] d\theta. \tag{A.13}$$

By applying the PDFs of θ, u_1 and v_{1,n_1^*} into the above equation, we can obtain the CDF of v_{m^*,n^*}, as shown in (7.33) of Theorem 7.1. Hence, we complete the proof of Theorem 7.1.

Appendix B

B.1　Proof of Theorem 2

To derive the CDF of Z_1, we first derive the conditional CDF of Z_{1m} with respect to v_1 as:

$$
\begin{aligned}
F_{Z_{1m}}(z|v_1) &= \Pr[\min(u_m, v_1 + w_{1m}) < z] \\
&= 1 - \Pr[u_m \geq z] \cdot \Pr[v_1 + w_{1m} \geq z].
\end{aligned}
\tag{B.1}
$$

Due to the fact that $u_m = \max_{1 \leq n \leq N} u_{mn}$, the CDF of u_m is given by [8]:

$$
F_{u_m}(x) = \left(1 - e^{-\frac{x}{\alpha}}\right)^N.
\tag{B.2}
$$

Accordingly, we obtain $\Pr[u_m \geq z]$ as:

$$
\Pr[u_m \geq z] = 1 - \left(1 - e^{-\frac{z}{\alpha}}\right)^N.
\tag{B.3}
$$

To derive $F_{Z_{1m}}(z|v_1)$, we consider $\Pr[v_1 + w_{1m} \geq z]$ for two cases, namely, $z < v_1$ and $z \geq v_1$. When $z < v_1$, $v_1 + w_{1m}$ is always larger than z. As such, we obtain:

$$
F_{Z_{1m}}(z|v_1) = (1 - e^{-\frac{z}{\alpha}})^N.
\tag{B.4}
$$

On the other hand, when $z \geq v_1$, we have:

$$
\Pr[v_1 + w_{1m} \geq z] = \Pr[w_{1m} \geq z - v_1]
\tag{B.5}
$$

$$
= e^{-\frac{z - v_1}{\varepsilon_1}}.
\tag{B.6}
$$

Accordingly, we obtain:

$$
\begin{aligned}
F_{Z_{1m}}(z|v_1) &= 1 - \left(1 - e^{-\frac{z}{\alpha}}\right)^N e^{-\frac{z - v_1}{\varepsilon_1}} \\
&= 1 - e^{\frac{v_1}{\varepsilon_1}} \sum_{n=0}^{N} \binom{N}{n} (-1)^{n-1} e^{-\left(\frac{n}{\alpha} + \frac{1}{\varepsilon_1}\right)z}.
\end{aligned}
\tag{B.7}
$$

From (B.4) and (B.7), the CDF of Z_1 can be written as:

$$
\begin{aligned}
F_{Z_1}(z) &= \int_0^{\infty} F_{Z_{1m}}^M(z|v_1) f_{v_1}(v_1) dv_1 \\
&= \int_0^z F_{Z_{1m}}^M(z|v_1) f_{v_1}(v_1) dv_1 + \int_z^{\infty} F_{Z_{1m}}^M(z|v_1) f_{v_1}(v_1) dv_1.
\end{aligned}
\tag{B.8}
$$

By using the PDF of v_1, given by $f_{v_1}(x) = \sum_{n=1}^{N}(-1)^{n-1}\binom{N}{n}\frac{n}{\beta_1}e^{-\frac{nx}{\beta_1}}$, and the binomial expansion into (B.8), we obtain the desired CDF of Z_1 as shown in (7.69) of Theorem 7.2, which completes the proof.

Appendix C

C.1   Proof of Theorem 3

Since w_{1m^*} is the maximum of M variables $\{w_{1m}|1 \le m \le M\}$, as per the selection criterion characterised by (7.62), it holds that its distribution is given by [8]:

$$f_{w_{1m^*}}(x) = \sum_{m=1}^{M}(-1)^{m-1}\binom{M}{m}\frac{m}{\varepsilon_1}e^{-\frac{mx}{\varepsilon_1}}. \tag{C.1}$$

We next derive the CDF of $u_{m^*n^*}$ and v_{1n^*}, as per the selection criterion characterised by (7.63). We first write the CDF of $u_{m^*n^*}$ as:

$$F_{u_{m^*n^*}}(x) = \Pr[u_{m^*n^*} < x]$$

$$= \sum_{n=1}^{N}\Pr[u_{m^*n} < x, \min(u_{m^*n}, v_{1n}) > \theta_n], \tag{C.2}$$

where θ_n is defined as:

$$\theta_n = \max_{n_1=1,\dots N, n_1 \ne n}\min(u_{m^*n_1}, v_{1n_1}). \tag{C.3}$$

Due to the symmetry among N end-to-end paths, $F_{u_{m^*n^*}}(x)$ in (C.2) is written as:

$$F_{u_{m^*n^*}}(x) = N\Pr[u_{m^*1} < x, u_{m^*1} > \theta_1, v_{11} > \theta_1]$$

$$= N\int_0^x\int_{\theta_1}^x\int_{\theta_1}^{\infty}f_{\theta_1}(\theta_1)f_{u_{m^*1}}(u_{m^*1})f_{v_{11}}(v_{11})dv_{11}du_{m^*1}d\theta_1. \tag{C.4}$$

We first note that u_{m^*1} and v_{11} follow exponential distribution with mean α and β_1, respectively. We also note that the CDF of θ_1 is given by:

$$F_{\theta_1}(x) = \Pr[\theta_1 < x]$$

$$= (\Pr[\min(u_{m^*2}, v_{12}) < x])^{N-1}$$

$$= \sum_{n=0}^{N-1}(-1)^n\binom{N-1}{n}e^{-\frac{nx}{\zeta}}, \tag{C.5}$$

where $\zeta = \alpha\beta_1/(\alpha + \beta_1)$. By applying these results and solving the integral in (C.4), we obtain $F_{u_{m^*n^*}}(x)$ as:

$$F_{u_{m^*n^*}}(x) = 1 - \sum_{n=0}^{N-1}b_{4,n}\left[\frac{\zeta}{\zeta + n\beta_1}e^{-\frac{x}{\alpha}} + \left(\frac{1}{n+1} - \frac{\zeta}{\zeta + n\beta_1}\right)e^{-\frac{(n+1)x}{\zeta}}\right], \tag{C.6}$$

where $b_{4,n} = N(-1)^n \binom{N-1}{n}$.

Similarly, we derive the CDF of v_{1n*} as:

$$F_{v_{1n*}}(x) = 1 - \sum_{n=0}^{N-1} b_{4,n} \left[\frac{\zeta}{\zeta + n\alpha} e^{-\frac{x}{\beta_1}} + \left(\frac{1}{n+1} - \frac{\zeta}{\zeta + n\alpha} \right) e^{-\frac{(n+1)x}{\zeta}} \right]. \quad \text{(C.7)}$$

Using (C.1), (C.6) and (C.7), the CDF of Z is given by:

$$\begin{aligned} F_Z(z) &= \Pr[\min(u_{m*n*}, v_{1n*} + w_{1m*}) < z] \\ &= 1 - \Pr[u_{m*n*} \geq z] \cdot \Pr[v_{1n*} + w_{1m*} \geq z]. \end{aligned} \quad \text{(C.8)}$$

Note that $\Pr[u_{m*n*} \geq z] = 1 - F_{u_{m*n*}}(z)$ can be easily obtained using (C.6). Therefore, $\Pr[v_{1n*} + w_{1m*} \geq z]$ is calculated as:

$$\begin{aligned} \Pr[v_{1n*} + w_{1m*} \geq z] &= 1 - \Pr[v_{1n*} + w_{1m*} < z] \\ &= 1 - \int_0^z F_{v_{1n*}}(z - w_{1m*}) f_{w_{1m*}}(w_{1m*}) dw_{1m*}. \end{aligned} \quad \text{(C.9)}$$

Applying (C.1) and (C.7) into (C.9), we are able to obtain $\Pr[v_{1n*} + w_{1m*} \geq z]$. This leads to the analytical CDF of Z, as shown in (7.73) in Theorem 7.3. This completes the proof.

References

[1] Y. Hu and X. Tao, "Secrecy outage analysis of multiuser diversity with unequal average SNR in transmit antenna selection systems," *IEEE Commun. Lett.*, vol. 19, no. 3, pp. 411–414, Mar. 2015.

[2] J. Zhang, C. Yuen, C.-K. Wen, S. Jin, and X. Gao, "Ergodic secrecy sum-rate for multiuser downlink transmission via regularized channel inversion: Large system analysis," *IEEE Commun. Lett.*, vol. 18, no. 9, pp. 1627–1630, Sep. 2014.

[3] H. Deng, H.-M. Wang, J. Yuan, W. Wang, and Q. Yin, "Secure communication in uplink transmissions: User selection and multiuser secrecy gain," *IEEE Trans. Commun.*, vol. 64, no. 8, pp. 3492–3506, Aug. 2016.

[4] J. H. Lee and W. Choi, "Multiuser diversity for secrecy communications using opportunistic jammer selection: Secure DoF and jammer scaling law," *IEEE Trans. Sig. Proc.*, vol. 62, no. 4, pp. 828–839, Feb. 2014.

[5] L. Fan, X. Lei, T. Q. Duong, M. Elkashlan, and G. K. Karagiannidis, "Secure multiuser communications in multiple amplify-and-forward relay networks," *IEEE Trans. Commun.*, vol. 62, no. 9, pp. 3299–3310, Sep. 2014.

[6] L. Fan, N. Yang, T. Q. Duong, M. Elkashlan, and G. K. Karagiannidis, "Exploiting direct links for physical layer security in multi-user multi-relay networks," *IEEE Trans. Commun.*, vol. 15, no. 6, pp. 3856–3867, Jun. 2016.

[7] P. L. Yeoh, M. Elkashlan, and I. B. Collings, "Exact and asymptotic SER of distributed TAS/MRC in MIMO relay networks," *IEEE Trans. Wireless Commun.*, vol. 10, no. 3, pp. 751–756, Mar. 2011.

[8]   M. K. Simon and M. S. Alouini, *Digital Communication over Fading Channels*, 2nd ed. Hoboken, NJ, USA: John Wiley, 2005.

[9]   I. S. Gradshteyn and I. M. Ryzhik, *Table of Integrals, Series, and Products*, 7th ed. San Diego, CA: Academic, 2007.

[10]   Y. Zou, X. Wang, and W. Shen, "Optimal relay selection for physical-layer security in cooperative wireless networks," *IEEE J. Select. Areas Commun.*, vol. 31, no. 10, pp. 2099–2111, Oct. 2013.

[11]   J. N. Laneman, D. N. C. Tse, and G. W. Wornell, "Cooperative diversity in wireless networks: Efficient protocols and outage behavior," *IEEE Trans. Inf. Theory*, vol. 50, no. 12, pp. 3062–3080, Dec. 2004.

[12]   C. Jeong and I.-M. Kim, "Optimal power allocation for secure multicarrier relay systems," *IEEE Trans. Sig. Proc.*, vol. 59, no. 11, pp. 5428–5442, Nov. 2011.

[13]   C. Wang, H.-M. Wang, and X.-G. Xia, "Hybrid opportunistic relaying and jamming with power allocation for secure cooperative networks," *IEEE Trans. Wireless Commun.*, vol. 14, no. 12, pp. 589–605, Feb. 2015.

[14]   H. Ding, J. Ge, D. B. da Costa, and Z. Jiang, "Two birds with one stone: Exploiting direct links for multiuser two-way relaying systems," *IEEE Trans. Wireless Commun.*, vol. 11, no. 1, pp. 54–59, Jan. 2012.

[15]   M. Ju, H.-K. Song, and I.-M. Kim, "Joint relay-and-antenna selection in multi-antenna relay networks," *IEEE Trans. Commun.*, vol. 58, no. 12, pp. 3417–3422, Dec. 2010.

Chapter 8

Trusted wireless communications with spatial multiplexing

Giovanni Geraci[1] and Jinhong Yuan[2]

This chapter will introduce trusted communications in multiuser multiple-input multiple-output (MIMO) wireless systems via spatial multiplexing. We will start by considering the multiple-input single-output broadcast channel with confidential messages (BCC) under Rayleigh fading, where a multiantenna base station (BS) simultaneously transmits independent confidential messages to several spatially dispersed malicious users that can eavesdrop on each other. We will then present the broadcast channel with confidential messages and external eavesdroppers (BCCE), where a multiantenna BS simultaneously communicates to multiple malicious users, in the presence of a Poisson point process (PPP) of external eavesdroppers. Unlike the BCC, in the BCCE not just malicious users, but also randomly located external nodes can act as eavesdroppers. We will finally turn our attention to cellular networks where, unlike the case of isolated cells, multiple BSs generate intercell interference, and malicious users of neighboring cells can cooperate to eavesdrop. For these involved scenarios, we will present low-complexity transmission schemes based on linear precoding that can achieve secrecy at the physical layer, discuss their performance, and quantify the penalties imposed by the secrecy requirements.

8.1 Introduction to multiuser MIMO systems

Supporting the ever increasing wireless throughput demand is the primary factor that drives the industry and academia alike toward the fifth-generation (5G) wireless systems, which will have to provide a much larger aggregate capacity than current 4G networks [1,3,20]. Multiuser multiple-input multiple-output (MIMO) wireless techniques have received much attention as a way to meet such demand by achieving high spectral efficiency through spatial multiplexing [15,19]. In a multiuser MIMO wireless system, a central multiantenna base station (BS) simultaneously communicates to several users over the same time/frequency resource. While it is known that the sum-capacity of multiuser MIMO systems is achieved by using dirty paper coding [2], the latter requires high-complexity coding schemes that make it difficult to be implemented [14].

[1] Nokia Bell Labs, Ireland
[2] University of New South Wales, Australia

On the other hand, suboptimal precoding schemes have proven to be practical and effective in controlling interuser interference for the downlink of multiuser MIMO networks. Among those, linear precoding has been widely recognized as a low-complexity alternative to DPC for multiuser MIMO downlink implementations [12,23]. A popular and practical linear precoding scheme to control interuser interference is channel inversion (CI) precoding, sometimes known as zero forcing precoding [28,31]. To increase the sum-rate performance of the CI precoder, the regularized channel inversion (RCI) precoder was proposed to tradeoff the interuser interference and the desired signal through a regularization parameter [21,26].

As well as spectral efficiency, security is regarded as a critical concern in wireless multi-user networks, since users rely on these networks to transmit sensitive data [30]. Due to the broadcast nature of the physical medium, wireless multi-user communications are very susceptible to eavesdropping, and even some intended users can themselves act maliciously as eavesdroppers [17]. As a result, the study of physical layer security, which exploits the characteristics of wireless channels to guarantee trusted communications without requiring encryption keys, has been recently extended to multi-user systems.

In the rest of this chapter, we will survey the research in the field of physical layer security for multi-user MIMO communications, especially focusing on low-complexity downlink transmission schemes based on linear precoding. We will start by studying a simplified model with an isolated cell, and will then account for the presence of an additional random field of eavesdropping nodes, before analyzing a more involved cellular network. Most mathematical derivations in this chapter make use of tools from random matrix theory and stochastic geometry and have been omitted for the sake of brevity. Detailed proofs of the theorems and corollaries in Sections 8.2–8.4 can be found in [5–8].

8.2 Physical layer security in an isolated cell

In this section, we introduce the multiple-input single-output (MISO) broadcast channel with confidential messages (BCC), where a multi-antenna base station simultaneously transmits independent confidential messages to several single-antenna users. While early work, e.g., [16,18], studied the theoretical performance limits of a BCC with two users, in this section we consider a generic MISO BCC with any number of users. We derive expressions for the secrecy rates achievable by RCI precoding and quantify the rate reduction caused by introducing secrecy requirements.

8.2.1 The broadcast channel with confidential messages

We consider the downlink of a narrowband multi-user MISO system, consisting of a BS with N antennas which simultaneously transmits K independent confidential messages to K spatially dispersed single-antenna users, as depicted in Figure 8.1. In this model, denoted as the MISO BCC, transmission takes place over a block fading channel, and the transmitted signal is $\mathbf{x} = [x_1, \ldots, x_N]^T \in \mathbb{C}^{N \times 1}$. We assume

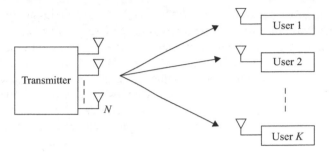

Figure 8.1   The MISO broadcast channel with confidential messages (BCC)

homogeneous users, i.e., each user experiences the same received signal power on average, thus the model assumes that their distances from the transmitter are the same and unitary. This approximates a scenario where users are located at equal distances from the BS or where the BS employs power control to guarantee the same average received signal power. Without loss of generality, such power can be assumed unitary. The conclusions drawn in this chapter can be extended to the scenario where the mobile users are equipped with multiple antennas and have different distances from the serving BS [29].

The received signal at user k is given by:

$$y_k = \sum_{j=1}^{N} h_{k,j} x_j + n_k \tag{8.1}$$

where $h_{k,j} \sim \mathscr{CN}(0, 1)$ is the independent and identically distributed (i.i.d.) Rayleigh fading channel between the jth transmit antenna element and the kth user, and $n_k \sim \mathscr{CN}(0, \sigma^2)$ is the noise seen at the kth receiver. The corresponding vector equation is:

$$\mathbf{y} = \mathbf{H}\mathbf{x} + \mathbf{n} \tag{8.2}$$

where $\mathbf{H} = [\mathbf{h}_1, \ldots, \mathbf{h}_K]^\dagger$ is the $K \times N$ channel matrix. We assume $\mathbb{E}[\mathbf{n}\mathbf{n}^\dagger] = \sigma^2 \mathbf{I}_K$, where $\mathbf{I}_K$ is the $K \times K$ identity matrix, define the transmit signal-to-noise ratio (SNR) as $\rho \triangleq 1/\sigma^2$, and impose the long-term power constraint $\mathbb{E}[\|\mathbf{x}\|^2] = 1$.

In the BCC, it is required that the BS securely transmits each confidential message, ensuring that the unintended users receive no information. This is performed at the secrecy rate R, defined as follows. Let $\mathbb{P}(\mathscr{E}_n)$ be the probability of error at the intended user, m be a confidential message, $\mathbf{y}_e^n$ be the vector of all signals received by the unintended users, and $H(m|\mathbf{y}_e^n)$ be the corresponding equivocation. Then a (weak) secrecy rate R for the intended user is achievable if there exists a sequence of $(2^{nR}, n)$ codes such that $\mathbb{P}(\mathscr{E}_n) \to 0$ and $\frac{1}{n}H(m|\mathbf{y}_e^n) \le \frac{1}{n}H(m) - \varepsilon_n$ with ε_n approaching zero as $n \to \infty$ [13].

For each user k, we denote by $\mathscr{M}_k = \{1, \ldots, k-1, k+1, \ldots, K\}$ the set of remaining users. In general, the behavior of the users cannot be determined by the BS. As a worst-case scenario, we assume that for each user k, all users in $\mathscr{M}_k$ can cooperate to jointly eavesdrop on the kth message. This assumption reflects the fact that the confidentiality of the messages must be ensured in all cases, including the worst case. Since the set of malicious users $\mathcal{M}_k$ can perform joint processing, they can be seen as a single equivalent malicious user M_k with $K - 1$ receive antennas. Due to the assumption of cooperating malicious users, interference cancelation can be performed at M_k, which does not see any undesired signal term apart from the received noise. It will be shown that despite this conservative assumption, a linear precoder can achieve a per-user secrecy rate which is close to the rate achievable in the absence of secrecy requirements.

8.2.2 Achievable secrecy rates in the BCC

We now derive achievable secrecy rates for the MISO BCC by using a linear precoder. Although suboptimal, linear precoding schemes are of particular interest because of their low-complexity implementations and because they can control the amount of crosstalk between the users [11,21]. We then specialize and obtain the secrecy rates achievable by the RCI precoder. RCI is a linear precoding scheme that was proposed to serve multiple users in the MISO broadcast channel (BC). RCI precoding has better performance than plain channel inversion, especially at low SNR [21].

8.2.2.1 Linear precoding

In linear precoding, the transmitted vector $\mathbf{x}$ is derived from the vector containing the confidential messages $\mathbf{u} = [u_1, \ldots, u_k]^T$ through a deterministic linear transformation (precoding) [24,31]. We assume that the entries of $\mathbf{u}$ are chosen independently, satisfying $\mathrm{E}[|u_k|^2] = 1$, $\forall k$.

Let $\mathbf{W} = [\mathbf{w}_1, \ldots, \mathbf{w}_k]$ be the $N \times K$ precoding matrix, where $\mathbf{w}_k$ is the kth column of $\mathbf{W}$. Then the transmitted signal and the power constraint are, respectively:

$$\mathbf{x} = \mathbf{W}\mathbf{u} = \sum_{k=1}^{K} \mathbf{w}_k u_k, \tag{8.3}$$

$$\mathrm{E}\left[\|\mathbf{x}\|^2\right] = \mathrm{E}\left[\|\mathbf{W}\mathbf{u}\|^2\right] = \sum_{k=1}^{K} \|\mathbf{w}_k\|^2 = 1. \tag{8.4}$$

By employing linear precoding as in (8.3), the signals observed at the legitimate user k and at the equivalent malicious user M_k are, respectively:

$$y_k = \mathbf{h}_k^\dagger \mathbf{w}_k u_k + \sum_{j \neq k} \mathbf{h}_k^\dagger \mathbf{w}_j u_j + n_k$$

$$\mathbf{y}_{M,k} = \sum_{j} \mathbf{H}_k \mathbf{w}_j u_j + \mathbf{n}_k \tag{8.5}$$

where $\mathbf{n}_k = [n_1, \ldots, n_{k-1}, n_{k+1}, \ldots, n_K]^T$, $\mathbf{h}_k^\dagger$ is the kth row of $\mathbf{H}$, and $\mathbf{H}_k$ is a matrix obtained from $\mathbf{H}$ by eliminating the kth row. The channel in (8.5) is a multi-input, single-output, multi-eavesdropper (MISOME) wiretap channel [13]. The transmitter, the intended receiver, and the eavesdropper of this MISOME wiretap channel are equipped with N, 1 and $K - 1$ virtual antennas, respectively. Due to the simultaneous transmission of the K messages, user k experiences noise and interference from all the u_j, $j \neq k$.

We now consider RCI precoding for the MISO BCC. For each message u_k, RCI precoding achieves a tradeoff between the signal power at the kth legitimate user and the crosstalk at the other $(K - 1)$ unintended users for each signal. The crosstalk causes interference to the unintended users. In the case when the unintended users are acting maliciously, the crosstalk also causes information leakage. Therefore, RCI achieves a tradeoff between signal power, interference, and information leakage.

With RCI precoding, linear processing exploiting regularization is applied to the vector of messages **u** [21]. The RCI precoding matrix is given by:

$$\mathbf{W} = \frac{1}{\sqrt{\zeta}} \mathbf{H}^\dagger \left(\mathbf{H}\mathbf{H}^\dagger + N\xi \mathbf{I}_K \right)^{-1} \tag{8.6}$$

where

$$\zeta = \mathrm{tr}\left\{ \mathbf{H}^\dagger \mathbf{H} \left(\mathbf{H}^\dagger \mathbf{H} + N\xi \mathbf{I}_K \right)^{-2} \right\} \tag{8.7}$$

is the power normalization constant. The transmitted signal **x** after RCI precoding can be written as:

$$\mathbf{x} = \mathbf{W}\mathbf{u} = \frac{1}{\sqrt{\zeta}} \mathbf{H}^\dagger \left(\mathbf{H}\mathbf{H}^\dagger + N\xi \mathbf{I}_K \right)^{-1} \mathbf{u}. \tag{8.8}$$

The latter passes through the channel, producing the vector of received signals:

$$\mathbf{y} = \frac{1}{\sqrt{\zeta}} \mathbf{H} \left(\mathbf{H}^\dagger \mathbf{H} + N\xi \mathbf{I}_K \right)^{-1} \mathbf{H}^\dagger \mathbf{u} + \mathbf{n}. \tag{8.9}$$

The function of the nonnegative regularization parameter ξ is to improve the behavior of the channel inverse, although it also produces nonzero crosstalk terms in (8.9). Throughout this chapter, perfect knowledge of the channel matrix **H** is assumed at the BS. The results provided can be generalized accounting for imperfect channel state information, as discussed in [5].

8.2.2.2 Achievable secrecy rates with linear precoding

By using a code construction based on independent codebooks and linear precoding, it is possible to show that an achievable secrecy sum-rate S_{BCC} for the MISO BCC is given by:

$$S_{\mathrm{BCC}} \triangleq \sum_{k=1}^{K} R_{\mathrm{BCC},k}, \tag{8.10}$$

where $R_{\mathrm{BCC},k}$ is an achievable secrecy rate for the kth MISOME wiretap channel in (8.5), $k = 1, \ldots, K$, and it is given by:

$$R_{\mathrm{BCC},k} = \left[\log_2 \left(1 + \gamma_k \right) - \log_2 \left(1 + \gamma_{M,k} \right) \right]^+, \tag{8.11}$$

where γ_k and $\gamma_{M,k}$ are the signal-to-interference-plus-noise ratios (SINR) for the message u_k at the legitimate receiver k and at the equivalent malicious user M_k, respectively, and we have used the notation $[x]^+ \triangleq \max(x, 0)$.

From Equation (8.11), it is clearly observed that an efficient tradeoff between maximizing γ_k and minimizing $\gamma_{M,k}$ is required for high-performance linear precoder design.

Using RCI precoding, the SINRs in (8.11) at the legitimate user k and at the equivalent malicious user M_k become, respectively:

$$\gamma_k = \frac{\left| \mathbf{h}_k^\dagger \left(\mathbf{H}^\dagger \mathbf{H} + N\xi \mathbf{I}_K \right)^{-1} \mathbf{h}_k \right|^2}{\zeta \sigma^2 + \sum_{j \neq k} \left| \mathbf{h}_k^\dagger \left(\mathbf{H}^\dagger \mathbf{H} + N\xi \mathbf{I}_K \right)^{-1} \mathbf{h}_j \right|^2}, \tag{8.12}$$

$$\gamma_{M,k} = \frac{\left\| \mathbf{H}_k \left(\mathbf{H}^\dagger \mathbf{H} + N\xi \mathbf{I}_K \right)^{-1} \mathbf{h}_k \right\|^2}{\zeta \sigma^2}. \tag{8.13}$$

Therefore, a secrecy sum-rate achievable by linear precoding in the MISO BCC is given by:

$$S_{\text{BCC}} = \sum_{k=1}^{K} \left[\log_2 \frac{1 + \frac{\left| \mathbf{h}_k^\dagger \mathbf{w}_k \right|^2}{\sigma^2 + \sum_{j \neq k} \left| \mathbf{h}_k^\dagger \mathbf{w}_j \right|^2}}{1 + \frac{\| \mathbf{H}_k \mathbf{w}_k \|^2}{\sigma^2}} \right]^+. \tag{8.14}$$

In the remainder of this chapter, we refer to Equation (8.14) as the secrecy sum-rate in the BCC. We note that the secrecy sum-rate depends on the choice of the precoding matrix $\mathbf{W}$, as well as on the channel $\mathbf{H}$ and the noise variance σ^2.

In the following we provide a deterministic approximation of the secrecy rates $R_{\text{BCC},k}$, which is almost surely exact as $N \to \infty$.

Theorem 8.1. *Let $\rho > 0$ and $\beta \triangleq K/N > 0$. Let $R_{\text{BCC},k}$ be the secrecy rate achievable in the BCC by user k with RCI precoding defined in (8.11). Then*

$$\left| R_{\text{BCC},k} - R_{\text{BCC}}^\circ \right| \xrightarrow{a.s.} 0, \quad as \quad N \to \infty, \ \forall k, \tag{8.15}$$

where R_{BCC}° denotes the secrecy rate in the large-system regime, given by

$$R_{\text{BCC}}^\circ = \left[\log_2 \frac{1 + \gamma^\circ}{1 + \gamma_M^\circ} \right]^+, \tag{8.16}$$

with

$$\gamma^\circ = g(\beta, \xi) \frac{\rho + \frac{\rho \xi}{\beta} [1 + g(\beta, \xi)]^2}{\rho + [1 + g(\beta, \xi)]^2}, \tag{8.17}$$

$$\gamma_M^\circ = \frac{\rho}{(1 + g(\beta, \xi))^2}, \tag{8.18}$$

and

$$g\left(\beta,\xi\right)=\frac{1}{2}\left[\sqrt{\frac{(1-\beta)^2}{\xi^2}+\frac{2\left(1+\beta\right)}{\xi}+1}+\frac{1-\beta}{\xi}-1\right].\qquad(8.19)$$

The secrecy sum-rate S_{BCC} can be therefore approximated by the large-system secrecy sum-rate S_{BCC}°, given by

$$S_{\mathrm{BCC}}^{\circ}=KR_{\mathrm{BCC}}^{\circ}.\qquad(8.20)$$

It should be noted that both the approximated secrecy rate and secrecy sum-rate are given in closed form, and that they are functions of only few variables, namely, the SNR ρ, the network load β, and the regularization parameter ξ. While the value of ξ may have a significant impact on the secrecy sum-rate S_{BCC}°, the derivation of the secrecy-rate maximizing value ξ is here omitted and can be found in [5].

For $\beta=1$, it can be shown that S_{BCC}° is monotonically increasing with the SNR ρ. It can be shown that the same is true for $\beta<1$. However when $\beta>1$, the secrecy sum-rate does not monotonically increase with ρ. There is an optimal value of the SNR, beyond which the achievable secrecy sum-rate S_{BCC}° starts decreasing, until it becomes zero for large SNR. When $\beta\geq2$ no positive secrecy sum-rate is achievable at all.

These results can be explained as follows. In the worst-case scenario, the alliance of cooperating malicious users can cancel the interference, and its received SINR is the ratio between the signal leakage and the thermal noise. In the limit of large SNR, the thermal noise vanishes, and the only means for the transmitter to limit the SINR at the malicious users is by reducing the signal leakage to zero by inverting the channel matrix. This can only be accomplished when the number of transmit antennas is larger than or equal to the number of users, hence only if $\beta\leq1$. When $\beta>1$ this is not possible, and no positive secrecy sum-rate can be achieved. When $\beta\geq2$, the eavesdroppers are able to drive the secrecy sum-rate to zero irrespective of ρ. This result is expected and consistent with the ones obtained for a single-user system in [13].

Figure 8.2 compares the large-system secrecy sum-rate S_{BCC}° of the RCI precoder to the simulated ergodic secrecy sum-rate S_{BCC} with a finite number of users, for different values of β. The value of S_{BCC}° was obtained by (8.20). In both cases, the regularization parameter that maximizes the average secrecy sum-rate was used. We observe that as N increases, the large-system result becomes more accurate for all values of SNR.

The following theorem provides a high-SNR approximation of S_{BCC}°. Note that for the case $\beta>1$, the transmit power is capped to a value that yields the rate-maximizing SNR.

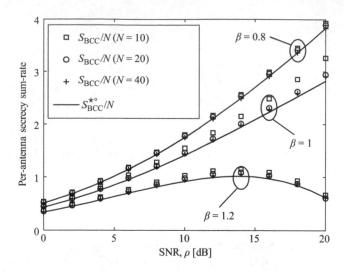

Figure 8.2   Comparison between the secrecy sum-rate with RCI precoding in the large-system regime (8.20) and the simulated ergodic secrecy sum-rate for finite N. Three sets of curves are shown, each one corresponds to a different value of β

Theorem 8.2. *In the high-SNR regime, we have* $\lim_{\rho \to \infty} \frac{S^{\circ}_{BCC} - S^{\circ\circ\circ}_{BCC}}{S^{\circ}_{BCC}} = 0$, *where* $S^{\circ\circ\circ}_{BCC}$ *approximates the large-system secrecy sum-rate* S°_{BCC} *achieved by the RCI precoder, and it is given by*

$$
S^{\circ\circ\circ}_{BCC} = \begin{cases} K \log_2 \frac{1-\beta}{\beta} + K \log_2 \rho & \text{for } \beta < 1 \\ \frac{K}{2} \log_2 \frac{27}{64} + \frac{K}{2} \log_2 \rho & \text{for } \beta = 1 \\ K \log_2 \frac{\beta^2}{4(\beta-1)} & \text{for } 1 < \beta < 2 \\ 0 & \text{for } \beta \geq 2 \end{cases} \tag{8.21}
$$

From (8.21), we can conclude that the high-SNR behavior of the RCI precoder in a BCC can be classified into four regions. When $\beta < 1$, any secrecy sum-rate can be achieved, as long as the transmitter has enough power available, and the secrecy sum-rate scales linearly with the factor K. When $\beta = 1$, the linear scaling factor reduces to $K/2$. When $1 < \beta < 2$, the cooperating eavesdroppers have more antennas than the transmitter, and thus they can limit the achievable secrecy sum-rate regardless of how much power is available at the transmitter. When $\beta \geq 2$, the eavesdroppers are able to prevent secret communications, and the secrecy sum-rate is zero even if unlimited power is available.

8.2.3 The price of secrecy

We now compare the large-system secrecy sum-rate S°_{BCC} achieved by the RCI precoder in the BCC to the large-system sum-rate S°_{BC} achieved in the BC without secrecy requirements. The gap between S°_{BCC} and S°_{BC} represents the *secrecy loss*, i.e., how much the secrecy requirements cost in terms of the achievable sum-rate.

The optimal sum-rate S°_{BC} in the MISO BC without secrecy requirements is given by [10,26]

$$S^\circ_{\mathrm{BC}} = K \log_2 \left[1 + g\left(\beta, \xi^\circ_{\mathrm{BC}}\right) \right], \tag{8.22}$$

with $\xi^\circ_{\mathrm{BC}} = \beta/\rho$. It is easy to show that $S^\circ_{\mathrm{BC}} \geq 0$ for all values of β and ρ, with equality only for $\rho = 0$. Hence, there is no limit to the number of users per transmit antenna β that the system can accommodate with a nonzero sum-rate. However, if we impose the secrecy requirements, the secrecy sum-rate S°_{BCC} is zero for $\beta \geq 2$. Therefore, introducing the secrecy requirements will limit the number of users that can be served with a nonzero rate to two times the number of transmit antennas.

We now compare the secrecy sum-rate S°_{BCC} to the sum-rate S°_{BC} in the limit of large SNR. We obtain $\lim_{\rho \to \infty} \frac{S^\circ_{\mathrm{BC}} - S^{\circ\circ\circ}_{\mathrm{BC}}}{S^\circ_{\mathrm{BC}}} = 0$, with:

$$S^{\circ\circ\circ}_{\mathrm{BC}} = \begin{cases} K \log_2 \frac{1-\beta}{\beta} + K \log_2 \rho & \text{for } \beta < 1 \\ \frac{K}{2} \log_2 \rho & \text{for } \beta = 1 \\ K \log_2 \frac{\beta}{\beta-1} & \text{for } \beta > 1 \end{cases} \tag{8.23}$$

By comparing (8.23) to (8.21), we can draw the following conclusions regarding the large-SNR regime. If the number of transmit antennas N is larger than the number of users K, then $S^{\circ\circ\circ}_{\mathrm{BCC}} = S^{\circ\circ\circ}_{\mathrm{BC}}$ and the secrecy requirements do not decrease the sum-rate of the network. Therefore by using RCI precoding, one can achieve secrecy while maintaining the same sum-rate, i.e., there is no secrecy loss. If $N = K$, then the secrecy loss is $\frac{1}{2} \log_2 \left(\frac{64}{27}\right) \approx 0.62$ bits per user, but the linear scaling factor $K/2$ remains unchanged. Alternatively, one can achieve secrecy while maintaining the same sum-rate, by increasing the transmit power by a factor $64/27 \approx 3.75$ dB. If $N < K < 2N$, then the secrecy loss is $(2 - \log_2 \beta)$ bits per user. Finally, if $K \geq 2N$, then the secrecy requirements force the sum-rate to zero, whereas the sum-rate S°_{BC} remains positive, though it also tends to zero for large β.

In Figure 8.3, we compare the simulated per-user ergodic secrecy rate S_{BCC}/K of the RCI-PR precoder in the MISO BCC to the rate S_{BC}/K in the MISO BC without secrecy requirements. For $\beta < 1$, the difference between S_{BCC}/K and S_{BC}/K becomes negligible at large SNR, and secrecy can be achieved without additional costs. For $\beta = 1$, the two curves tend to have the same slope at large SNR, but there is a residual gap between them. Therefore, secrecy can be achieved at a lower rate. We note that in order to achieve secrecy without decreasing the rate, the required additional power is less than 4 dB at all SNRs. For $1 < \beta < 2$, the sum-rate S_{BC} tends to saturate for large SNR, and so does the secrecy sum-rate S_{BCC}. In the simulations, for $\beta = 1.2$ and $\rho = 25$dB, the gap is about 1.79 bits, close to $2 - \log_2 \beta \approx 1.74$ bits. All these numerical results confirm the ones obtained in the large-system regime.

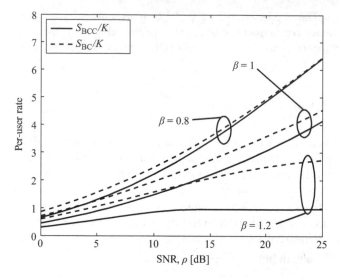

Figure 8.3   *Comparison between the simulated ergodic per-user secrecy rate with RCI (solid) and the per-user rate without secrecy requirements (dashed), for $K = 12$ users. Three values of β are considered: 0.8, 1, and 1.2, corresponding to $N = 15$, 12, and 10 antennas, respectively*

8.3   Physical layer security in a random field of eavesdroppers

In this section, we introduce the MISO broadcast channel with confidential messages and external eavesdroppers (BCCE), where a multi-antenna base station simultaneously communicates to multiple malicious users, in the presence of randomly located external eavesdroppers. We study the performance of RCI precoding in the BCCE and provide explicit expressions for the large-system probability of secrecy outage and mean secrecy rate with respect to the spatial distribution of the nodes and to the fluctuations of their channels. We note that RCI precoding, and linear precoding in general, can be combined with the transmission of artificial noise (AN) to reduce the signal-to-noise ratio perceived by external eavesdroppers. While this chapter will not consider such transmission technique, we refer the interested reader to [9,27,32] and references therein.

8.3.1   *The broadcast channel with confidential messages and external eavesdroppers*

In a MISO BCCE, not only malicious users but also nodes external to the network can act as eavesdroppers. This can be the case in a practical system, where external nodes are randomly scattered in space. These nodes must be regarded as potential eavesdroppers, otherwise the system would be vulnerable to secrecy outage. Similarly

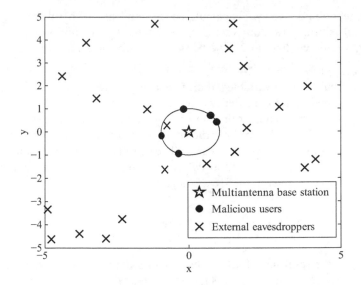

Figure 8.4   *Example of a BCCE with K = 5 malicious users and a density of external eavesdroppers $\lambda_e = 0.2$. © IEEE 2016. Reprinted with permission from Reference [8]*

to the BCC, we here assume homogeneous users, i.e., each user experiences the same average received signal power.

The BCCE can be obtained from the BCC by including external single-antenna eavesdroppers in the system. For the sake of tractability, here the external eavesdroppers are assumed to be distributed on the two-dimensional plane according to a Poisson point process (PPP) Φ_e of density λ_e [25]. Figure 8.4 shows an example of BCCE, where the BS is at the origin, and the users lie on a disc of radius 1. As a worst-case scenario, we assume that each eavesdropper can cancel the interference caused by the remaining $K - 1$ messages. Assuming that the BS lies at the origin, the SINR $\gamma_{e,k}$ for the kth message at a generic eavesdropper located in e is then given by:

$$\gamma_{e,k} = \frac{\left|\mathbf{h}_e^\dagger \mathbf{w}_k\right|^2}{\|e\|^\eta \sigma^2} \tag{8.24}$$

where $\mathbf{w}_k$ is the precoding vector for user k, $\mathbf{h}_e^\dagger$ is the channel vector between the base station and the eavesdropper in e, and it takes into account the Rayleigh fading, and η is the path loss exponent. Some of the results provided in the following assume a path loss exponent $\eta = 4$. In this special case, which is a reasonable value for η in a shadowed urban area [22], it is possible to obtain more compact expressions for quantities of interest, such as the probability of secrecy outage and the mean secrecy rate. Nevertheless, an extension to a generic value of η is possible at the expense of compactness.

The precoding vector $\mathbf{w}_k$ is calculated independently of $\mathbf{h}_e^\dagger$; therefore, they are independent isotropic random vectors. The channel $\mathbf{h}_e^\dagger$ has unit norm, whereas the precoding vector $\mathbf{w}_k$ has norm $\frac{1}{\sqrt{K}}$ because it is obtained after the normalization $\|\mathbf{W}\|^2 = \sum_{k=1}^K \|\mathbf{w}_k\|^2 = 1$. The inner product $\mathbf{h}_e^\dagger \mathbf{w}_k$ is a linear combination of N complex normal random variables, therefore $|\mathbf{h}_e^\dagger \mathbf{w}_k|^2 \sim \exp(\frac{1}{K})$.

In the following, we consider two types of external eavesdroppers, namely noncolluding eavesdroppers and colluding eavesdroppers. In the noncolluding case, the eavesdroppers individually overhear the communication without centralized processing. In the colluding eavesdroppers case, all eavesdroppers are able to jointly process their received messages at a central data processing unit. The secrecy rate $R_{\text{BCCE},k}$ achievable by the kth user in the BCCE is given by:

$$R_{\text{BCCE},k} = \left[\log_2(1 + \gamma_k) - \log_2\left(1 + \max\left(\gamma_{M,k}, \gamma_{E,k}\right)\right)\right]^+, \tag{8.25}$$

where $\gamma_{E,k}$ is the resulting SINR of the PPP of external eavesdroppers for the kth message. The secrecy rate $R_{\text{BCCE},k}$ is therefore affected by the maximum of the SINR $\gamma_{M,k}$ at the alliance of malicious users and the SINR $\gamma_{E,k}$ at the external eavesdroppers. In the case of noncolluding eavesdroppers, $\gamma_{E,k}$ is the SINR at the strongest eavesdropper. In the case of colluding eavesdroppers, all eavesdroppers can perform joint processing, and they can, therefore, be seen as a single multi-antenna eavesdropper. After interference cancelation, each eavesdropper receives the useful signal embedded in noise, and the optimal receive strategy at the colluding eavesdroppers is maximal ratio combining (MRC) which yields to an SINR $\gamma_{E,k} = \sum_{e \in \Phi_e} \gamma_{e,k}$ given by the sum of the SINRs $\gamma_{e,k}$ at all eavesdroppers generated by the PPP Φ_e.

The achievable secrecy sum-rate is denoted by S_{BCCE} and defined as:

$$S_{\text{BCCE}} = \sum_{k=1}^K R_{\text{BCCE},k}. \tag{8.26}$$

8.3.2   Probability of secrecy outage

The secrecy outage probability for user k is defined as:

$$\mathcal{O}_{\text{BCCE},k} \triangleq \mathbb{P}(R_{\text{BCCE},k} = 0) = \begin{cases} 1 & \text{if } \gamma_k \leq \gamma_{M,k} \\ \mathbb{P}(\gamma_{E,k} \geq \gamma_k \mid \gamma_k) & \text{otherwise} \end{cases} \tag{8.27}$$

As discussed in Section 8.2, RCI precoding ensures $\gamma_k > \gamma_{M,k}$ in most cases, as long as $\beta < 1$. Therefore, the secrecy outage probability is often given by the probability that $R_{\text{BCCE},k}$ is driven to zero by the presence of external eavesdroppers.

The results provided in this chapter can be extended to obtain the probability that the achievable secrecy rate $R_{\text{BCCE},k}$ is smaller than a target rate R_T. In this case, the probability of outage becomes a function of R_T.

In the case of noncolluding eavesdroppers, $\gamma_{E,k}$ is the SINR at the strongest eavesdropper E, given by:

$$\gamma_{E,k} = \max_{e \in \Phi_e} \gamma_{e,k} = \max_{e \in \Phi_e} \frac{|\mathbf{h}_e^\dagger \mathbf{w}_k|^2}{\|e\|^\eta \sigma^2}, \tag{8.28}$$

and $\mathscr{O}_{\text{BCCE},k}$ is the probability that any eavesdropper has an SINR greater than or equal to the SINR of the legitimate user k. We obtain the following results.

Theorem 8.3. *The secrecy outage probability in the BCCE in the presence of noncolluding external eavesdroppers satisfies*

$$|\mathscr{O}_{\text{BCCE},k} - \mathscr{O}_{\text{BCCE}}^\circ| \xrightarrow{a.s.} 0, \quad \text{as } N \to \infty, \quad \forall k \tag{8.29}$$

where

$$\mathscr{O}_{\text{BCCE}}^\circ = \begin{cases} 1 & \text{if } \gamma^\circ \leq \gamma_M^\circ \\ 1 - \exp\left[-\frac{2\pi\lambda_e \Gamma\left(\frac{2}{\eta}\right)}{\eta(N\beta\sigma^2\gamma^\circ)^{\frac{2}{\eta}}}\right] & \text{otherwise} \end{cases} \tag{8.30}$$

and with γ° and γ_M° given by (8.17) and (8.18), respectively.

Corollary 8.1. *If $\gamma^\circ > \gamma_M^\circ$ and $\eta = 4$, then (i) the number of transmit antennas required in order to guarantee a large-system secrecy outage probability $\mathscr{O}_{\text{BCCE}}^\circ < \varepsilon$ in the presence of noncolluding eavesdroppers is $N > \left(\frac{\mu\lambda_e}{\varepsilon\sqrt{\gamma^\circ}}\right)^2$ and (ii) the large-system secrecy outage probability $\mathscr{O}_{\text{BCCE}}^\circ$ decays as $\frac{1}{\sqrt{N}}$.*

In the case of colluding eavesdroppers, all eavesdroppers can perform joint processing, and they can therefore be seen as a single multi-antenna eavesdropper. This yields to an SINR $\gamma_{E,k}$ at the colluding eavesdroppers given by

$$\gamma_{E,k} = \frac{1}{\sigma^2} \sum_{e \in \Phi_e} \|e\|^{-\eta} |\mathbf{h}_e^\dagger \mathbf{w}_k|^2. \tag{8.31}$$

We obtain the following results for the secrecy outage probability $\mathscr{O}_{\text{BCCE}}^\circ$ in the presence of colluding eavesdroppers.

Theorem 8.4. *The secrecy outage probability in the BCCE in the presence of colluding external eavesdroppers, under a path loss exponent $\eta = 4$, satisfies*

$$|\mathscr{O}_{\text{BCCE},k} - \mathscr{O}_{\text{BCCE}}^\circ| \xrightarrow{a.s.} 0, \quad \text{as } N \to \infty, \quad \forall k \tag{8.32}$$

where

$$\mathscr{O}_{\text{BCCE}}^\circ = \begin{cases} 1 & \text{if } \gamma^\circ \leq \gamma_M^\circ \\ 1 - 2Q\left(\mu\lambda_e\sqrt{\frac{\pi}{2N\gamma^\circ}}\right) & \text{otherwise} \end{cases} \tag{8.33}$$

and with γ° and γ_M° given by (8.17) and (8.18), respectively.

Corollary 8.2. *Let $\gamma° > \gamma_M°$ and $\eta = 4$, then (i) the number of transmit antennas required in order to guarantee a large-system secrecy outage probability $\mathcal{O}_{\text{BCCE}}° < \varepsilon$ in the presence of colluding eavesdroppers is $N > \left(\frac{\mu\lambda_e}{\varepsilon\sqrt{\gamma°}}\right)^2$ and (ii) the large-system outage probability $\mathcal{O}_{\text{BCCE}}°$ decays as $\frac{1}{\sqrt{N}}$.*

By comparing the results in Corollaries 8.1 and 8.2, we can conclude that (i) the collusion among eavesdroppers does not significantly affect the number of transmit antennas N required to meet a given probability of secrecy outage in the large-system regime and (ii) increasing the density of eavesdroppers λ_e by a factor n requires increasing N by a factor n^2 in order to meet a given probability of secrecy outage.

In Figure 8.5 we compare the simulated probability of outage $\mathcal{O}_{\text{BCCE},k}$ under non-colluding and colluding eavesdroppers, respectively, to the large-system results $\mathcal{O}_{\text{BCCE}}°$ provided in Theorem 8.3 and Theorem 8.4, respectively. We observe that for $\lambda_e = 0.1$ and small probabilities of secrecy outage, (i) $N > \left(\frac{\mu\lambda_e}{0.1\sqrt{\gamma°}}\right)^2 = 34$ yields to a secrecy outage probability smaller than 0.1, (ii) the secrecy outage probability decays as $\frac{1}{\sqrt{N}}$, and (iii) the collusion of eavesdroppers does not significantly affect the probability of secrecy outage. All these observations are consistent with Corollaries 8.1 and 8.2.

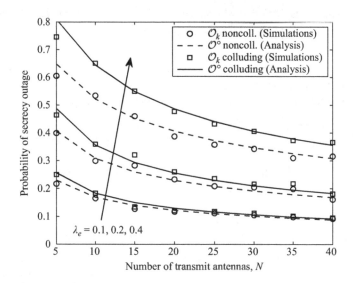

Figure 8.5   *Comparison between the simulated probability of outage $\mathcal{O}_{\text{BCCE},k}$ and the large-system results $\mathcal{O}_{\text{BCCE}}°$ provided in Theorems 8.3 and 8.4, for a network load $\beta = 1$, an SNR $\rho = 10\,dB$, and various values of λ_e. © IEEE 2016. Reprinted with permission from Reference [8]*

8.3.3 Mean secrecy rates

We now analyze the mean secrecy rates, averaged over the location of the external eavesdroppers, achievable by RCI precoding in the BCCE, for both cases of noncolluding and colluding eavesdroppers.

Theorem 8.5. *The mean secrecy rate achievable for user k by RCI precoding in the BCCE satisfies*

$$\left| \mathbb{E}_{\Phi_e}[R_{\text{BCCE},k}] - R_{\text{BCCE}}^\circ \right| \xrightarrow{a.s.} 0, \quad \text{as} \quad N \to \infty, \quad \forall k, \tag{8.34}$$

where R_{BCCE}° denotes the mean secrecy rate in the large-system regime, given by

$$R_{\text{BCCE}}^\circ = \begin{cases} 0 & \text{if } \gamma^\circ \leq \gamma_M^\circ \\ \log_2 \frac{(1+\gamma^\circ)^{1-\mathcal{O}_{\text{BCCE}}^\circ}}{(1+\gamma_M^\circ)^{1-\mathcal{P}_{\text{BCCE}}^\circ}} - \int_{\gamma_M^\circ}^{\gamma^\circ} \log_2 (1+y) f_{\gamma_{E,k}}(y)\, dy & \text{otherwise} \end{cases} \tag{8.35}$$

In (8.35), $\mathcal{P}_{\text{BCCE}}^\circ$ is the probability that the SINR $\gamma_{E,k}$ at the external eavesdroppers is greater than or equal to the large-system SINR γ_M° at the malicious users, and for $\eta = 4$, it is given by

$$\mathcal{P}_{\text{BCCE}}^\circ \triangleq \mathbb{P}(\gamma_{E,k} \geq \gamma_M^\circ) = \begin{cases} 1 - \exp\left(-\frac{\mu \lambda_e}{\sqrt{N \gamma_M^\circ}}\right) & \text{n.c.e.} \\ 1 - 2Q\left(\mu \lambda_e \sqrt{\frac{\pi}{2N\gamma_M^\circ}}\right) & \text{c.e.} \end{cases} \tag{8.36}$$

where "n.c.e" and "c.e." stand for "noncolluding eavesdroppers" and "colluding eavesdroppers," respectively,

By comparing the large-system mean secrecy rate of the MISO BCCE in (8.35) to the large-system secrecy rate of the MISO BCC without external eavesdroppers in (8.16), we can evaluate the secrecy rate loss Δ_e due to the presence of external eavesdroppers, defined as

$$\Delta_e \triangleq R_{\text{BCC}}^\circ - R_{\text{BCCE}}^\circ. \tag{8.37}$$

In particular, we can obtain an upper bound on the secrecy rate loss Δ_e as follows.

Corollary 8.3. *The secrecy rate loss Δ_e due to the presence of external eavesdroppers satisfies*

$$\Delta_e \leq \Delta_e^{UB} \triangleq \frac{C_\mu \lambda_e}{\sqrt{N}}, \tag{8.38}$$

where C_μ is a constant independent of N, λ_e, and of the cooperation strategy at the eavesdroppers (i.e., whether they cooperate or not), given by

$$C_\mu = \mu \left[\frac{R_{\text{BCC}}^\circ}{\sqrt{\gamma^\circ}} + \left(\sqrt{\gamma^\circ} - \sqrt{\gamma_M^\circ}\right)^+ \right]. \tag{8.39}$$

It follows from Corollary 8.3 that, irrespective of the collusion strategy at the external eavesdroppers, (i) as the number N of transmit antennas grows, the secrecy rate loss Δ_e tends to zero as $\frac{1}{\sqrt{N}}$ and (ii) increasing the density of eavesdroppers λ_e by a factor n requires increasing N by a factor n^2 in order to meet a given value of Δ_e^{UB}.

In Figure 8.6, we compare the simulated per-user secrecy rate of (i) the BCCE with noncolluding eavesdroppers, (ii) the BCCE with colluding eavesdroppers, and (iii) the BCC without external eavesdroppers, for $\beta = 1$, $\rho = 10$ dB, and various values of λ_e. We note that in the BCC, the per-user secrecy rate is almost constant with N, for a fixed network load β. On the other hand, the per-user secrecy rate of the BCCE increases with N. Again, this happens because the mean received power at each external eavesdropper scales as $\frac{1}{\beta N}$, hence having more transmit antennas makes the system more robust against external eavesdroppers. We also note that for higher densities of eavesdroppers λ_e, larger values of N are required to achieve a given per-user secrecy rate of the BCCE. More precisely, increasing λ_e by a factor 2 requires increasing N by a factor 4. Moreover, the collusion of external eavesdroppers does not affect the scaling law of the mean rate. These observations are consistent with Corollary 8.3.

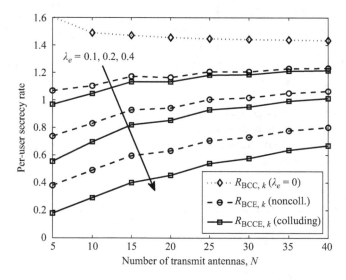

Figure 8.6   *Comparison between the simulated ergodic per-user secrecy rates of:*
            (i) the BCCE with noncolluding eavesdroppers, (ii) the BCCE with
            colluding eavesdroppers, and (iii) the BCC without external
            eavesdroppers, for a network load $\beta = 1$, an SNR $\rho = 10$ dB, and
            various values of λ_e

8.4 Physical layer security in a multi-cell system

In this section, we consider physical layer security in the downlink of cellular networks, where each BS simultaneously transmits confidential messages to several users, and where the confidential messages transmitted to each user can be eavesdropped by both (i) other users in the same cell and (ii) users in other cells.

8.4.1 Cellular networks with malicious users

We consider the downlink of a cellular network, as depicted in Figure 8.7. Each BS transmits at power P and is equipped with N antennas. For tractability, we assume that the locations of the BSs are drawn from a homogeneous PPP Φ_b of density λ_b. We consider single-antenna users, and assume that each user is connected to the closest BS. The locations of the users are drawn from an independent PPP Φ_u of density λ_u. We denote by $\mathcal{K}_b$ and by $K_b = |\mathcal{K}_b|$ the set of users and the number of users connected to the BS b, respectively. We denote by:

$$\mathbf{H}_b = \left[\|b - b_1\|^{-\eta} \mathbf{h}_{b,1}, \ldots, \|b - b_{K_b}\|^{-\eta} \mathbf{h}_{b,K_b} \right]^\dagger \tag{8.40}$$

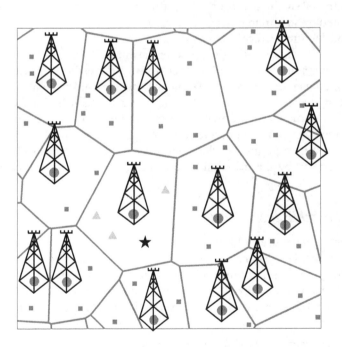

Figure 8.7 *Illustration of a cellular network. The star denotes a typical user. The circles, squares, and triangles denote BSs, out-of-cell users, and in-cell users, respectively*

the $K_b \times N$ channel matrix for the BS b, where $\mathbf{h}_{b,j} \sim \mathscr{CN}(\mathbf{0}, \mathbf{I})$ is the normalized channel vector that accounts only for the fading between the BS b and the user $j \in \mathscr{K}_b$.

Similarly to the case of a BCC, transmission takes place over a block fading channel. The signal transmitted by the generic BS b is $\mathbf{x}_b = [x_{b,1}, \dots, x_{b,N}]^T \in \mathbb{C}^{N \times 1}$. The vector $\mathbf{x}_b$ is obtained via RCI precoding from the vector of confidential messages $\mathbf{m}_b = [m_{b,1}, \dots, m_{b,K_b}]^T$, whose entries are chosen independently, satisfying $\mathbb{E}[|m_{b,j}|^2] = 1, \forall j$. The transmitted signal $\mathbf{x}_b$ after RCI precoding can be written as $\mathbf{x}_b = \sqrt{P}\mathbf{W}_b\mathbf{m}_b$, where $\mathbf{W}_b = [\mathbf{w}_{b,1}, \dots, \mathbf{w}_{b,K_b}]$ is the $N \times K_b$ RCI precoding matrix, given by [21]:

$$\mathbf{W}_b = \frac{1}{\sqrt{\zeta_b}} \mathbf{H}_b^\dagger \left(\mathbf{H}_b \mathbf{H}_b^\dagger + N\xi \mathbf{I}_{K_b} \right)^{-1}, \tag{8.41}$$

and where $\zeta_b = \mathrm{tr} \left\{ \mathbf{H}_b^\dagger \mathbf{H}_b \left(\mathbf{H}_b^\dagger \mathbf{H}_b + N\xi \mathbf{I}_N \right)^{-2} \right\}$ is a long-term power normalization constant.

For a user o connected to the BS b, the set of $K_b - 1$ malicious users within the same cell is denoted by $\mathscr{M}_o^I = \mathscr{K}_b \backslash o$, and the set formed by the rest of the malicious users in the network is denoted by $\mathscr{M}_o^E = \Phi_u \backslash \mathscr{K}_b$. In Figure 8.7, the legitimate user o, the set of (intracell) malicious users $\mathscr{M}_o^I$, and the set of (external) malicious users $\mathscr{M}_o^E$ are represented by star, triangles, and squares, respectively. The total set of malicious users for the legitimate receiver o is denoted by $\mathscr{M}_o = \mathscr{M}_o^I \cup \mathscr{M}_o^E = \Phi_u \backslash o$. It is important to make such a distinction between the intracell malicious users in $\mathscr{M}_o^I$ and the external malicious users in $\mathscr{M}_o^E$. In fact, the BS b can estimate the channels of the intracell malicious users in $\mathscr{M}_o^I \subset \mathscr{K}_b$ and exploits this information by choosing an RCI precoding matrix $\mathbf{W}_b$ which is a function of these channels. The RCI precoding thus controls the amount of information leakage at the malicious users in $\mathscr{M}_o^I$. On the other hand, the BS b cannot in general estimate the channels of all the other external malicious users in $\mathscr{M}_o^E$, and $\mathbf{W}_b$ does not depend upon these channels. Therefore, the signal received by the malicious users in $\mathscr{M}_o^E$ is not directly affected by RCI precoding.

8.4.2 Achievable secrecy rates

We consider a typical user o located at the origin, and connected to the closest BS, located in $c \in \Phi_b$. The distance between the typical user and the closest BS is given by $\|c\|$. The typical user receives self-interference caused by the other messages $m_{c,u}, u \neq o$ transmitted by the BS c, and intercell interference caused by the signal transmitted by all the other BSs $b \in \Phi_b \backslash c$. The signal received by the typical user is given by:

$$y_o = \sqrt{P \|c\|^{-\eta}} \, \mathbf{h}_{c,o}^\dagger \mathbf{w}_{c,o} m_{c,o} + \sqrt{P \|c\|^{-\eta}} \sum_{u \in \mathscr{K}_c \backslash o} \mathbf{h}_{c,o}^\dagger \mathbf{w}_{c,u} m_{c,u}$$

$$+ \sum_{b \in \Phi_b \backslash c} \sqrt{P \|b\|^{-\eta}} \sum_{j=1}^{K_b} \mathbf{h}_{b,o}^\dagger \mathbf{w}_{b,j} m_{b,j} + n_o \tag{8.42}$$

where $\|b\|$ is the distance between the typical user and the generic BS b, and η is the path loss exponent. The four terms in (8.42) represent the useful signal, the crosstalk (or self-interference), the intercell interference, and the thermal noise seen at the typical user, respectively. The latter is given by $n_o \sim \mathcal{CN}(0, \sigma^2)$, and we define the transmit SNR as $\rho \triangleq P/\sigma^2$.

We assume that the legitimate receiver at o treats the interference power as noise. The SINR γ_o at the legitimate receiver o is given by:

$$\gamma_o = \frac{\rho \|c\|^{-\eta} \left| \mathbf{h}_{c,o}^\dagger \mathbf{w}_{c,o} \right|^2}{\rho \|c\|^{-\eta} \sum_{u \in \mathcal{K}_c \setminus o} \left| \mathbf{h}_{c,o}^\dagger \mathbf{w}_{c,u} \right|^2 + \rho \sum_{b \in \Phi_b \setminus c} \frac{g_{b,o}}{K_b} \|b\|^{-\eta} + 1}, \tag{8.43}$$

where we define $\tilde{\mathbf{w}}_{b,j} \triangleq \sqrt{K_b} \mathbf{w}_{b,j}$ and

$$g_{b,o} \triangleq \sum_{j=1}^{K_b} \left| \mathbf{h}_{b,o}^\dagger \tilde{\mathbf{w}}_{b,j} \right|^2 . \tag{8.44}$$

The cell where the typical user o is located is referred to as the *tagged cell*. For the typical user o, the set of malicious users is denoted by $\mathcal{M}_o = \mathcal{M}_o^I \cup \mathcal{M}_o^E$, where $\mathcal{M}_o^I = \mathcal{K}_c \setminus o$ is the set of remaining users in the tagged cell, and $\mathcal{M}_o^E = \Phi_u \setminus \mathcal{K}_c$ is the set of all users in other cells.

In the following, we consider the worst-case scenario where all the malicious users in $\mathcal{M}_o$ can cooperate to eavesdrop on the message intended for the typical user in o. Since each malicious user is likely to decode its own message, it can indirectly pass this information to all the other malicious users. In the worst case scenario, all the malicious users in $\mathcal{M}_o$ can therefore subtract the interference generated by all the messages $m_j, j \neq o$.

After interference cancelation, the signal received at a malicious user $i \in \mathcal{M}_o^I$ in the tagged cell is given by:

$$y_i = \sqrt{P \|i - c\|^{-\eta}} \, \mathbf{h}_{c,i}^\dagger \mathbf{w}_{c,o} m_{c,o} + n_i \tag{8.45}$$

where $\|i - c\|$ is the distance between the BS c and the malicious user $i \in \mathcal{M}_o^I$. The signal received at a malicious user $e \in \mathcal{M}_o^E$ outside the tagged cell is given by:

$$y_e = \sqrt{P \|e - c\|^{-\eta}} \, \mathbf{h}_{c,e}^\dagger \mathbf{w}_{c,o} m_{c,o} + n_e. \tag{8.46}$$

We denote by γ_i and γ_e the SINRs at the malicious users $i \in \mathcal{M}_o^I$ and $e \in \mathcal{M}_o^E$, respectively.

Due to the cooperation among all malicious users in $\mathcal{M}_o = \mathcal{M}_o^I \cup \mathcal{M}_o^E$, the set $\mathcal{M}_o$ can be seen as a single equivalent multi-antenna malicious user, denoted by M_o.

After interference cancelation, M_o sees the useful signal embedded in noise, therefore applying maximal ratio combining (which is optimal) yields to an SINR given by:

$$\gamma_{M,o} = \sum_{i \in \mathcal{M}_o^I} \gamma_i + \sum_{e \in \mathcal{M}_o^E} \gamma_e \tag{8.47}$$

$$= \rho \sum_{i \in \mathcal{M}_o^I} \|i - c\|^{-\eta} \left| \mathbf{h}_{c,i}^\dagger \mathbf{w}_{c,o} \right|^2 + \frac{\rho}{K_c} \sum_{e \in \mathcal{M}_o^E} g_{c,e} \|e - c\|^{-\eta}, \tag{8.48}$$

where

$$g_{c,e} \triangleq \left| \mathbf{h}_{c,e}^\dagger \tilde{\mathbf{w}}_{c,o} \right|^2, \tag{8.49}$$

with $\tilde{\mathbf{w}}_{c,o} \triangleq \sqrt{K_c} \mathbf{w}_{c,o}$ and $n_i, n_e \sim \mathcal{CN}(0, \sigma^2)$.

The secrecy rate R_{CELL} achievable by RCI precoding for the typical user o of a downlink cellular network is therefore given by:

$$R_{\text{CELL}} \triangleq \left\{ \log_2 \left(1 + \frac{\rho \|c\|^{-\eta} \left| \mathbf{h}_{c,o}^\dagger \mathbf{w}_{c,o} \right|^2}{\rho \|c\|^{-\eta} \sum_{i \in \mathcal{M}_o^I} \left| \mathbf{h}_{c,o}^\dagger \mathbf{w}_{c,i} \right|^2 + \rho I + 1} \right) \right.$$

$$\left. - \log_2 \left(1 + \rho \sum_{i \in \mathcal{M}_o^I} \|i - c\|^{-\eta} \left| \mathbf{h}_{c,i}^\dagger \mathbf{w}_{c,o} \right|^2 + \rho L \right) \right\}^+, \tag{8.50}$$

where I and L represent the interference and leakage term, respectively, given by:

$$I = \sum_{b \in \Phi_b \setminus c} \frac{g_{b,o}}{K_b} \|b\|^{-\eta}$$

$$L = \frac{1}{K_c} \sum_{e \in \mathcal{M}_o^E} g_{c,e} \|e - c\|^{-\eta}. \tag{8.51}$$

8.4.3　Probability of secrecy outage and mean secrecy rate

The probability of secrecy outage for the typical user o of a cellular network is defined as:

$$\mathcal{O}_{\text{CELL}} \triangleq \mathbb{P}(R_{\text{CELL}} \le 0), \tag{8.52}$$

and it also denotes the fraction of time for which a BS cannot transmit to a typical user at a nonzero secrecy rate. We now obtain an approximation for the probability of secrecy outage with RCI precoding.

Theorem 8.6. *The probability of secrecy outage with RCI precoding can be approximated as*

$$\mathcal{O}_{\text{CELL}} \approx \hat{\mathcal{O}}_{\text{CELL}} = \int_0^\infty \int_{-\infty}^\infty \int_{-\infty}^\infty \mathbb{I}_{(z \ge \tau(x,y))} f_{\hat{L}}(z) \, dz \, f_{\hat{I}}(x,y) \, dx \, 2\lambda_b \pi y e^{-\lambda_b \pi y^2} \, dy,$$

$$\tag{8.53}$$

where $\mathbb{I}$ is the indicator function, $f_{\hat{I}}(x,y)$ is the probability density function of the interference $\hat{I}$ for $\|c\| = y$, $f_{\hat{L}}(z)$ is the probability density function of the leakage $\hat{L}$, and we have defined

$$\tau(x,y) \triangleq \frac{\alpha y^{-\eta}}{\rho \chi y^{-\eta} + \rho x + 1} - \chi y^{-\eta}. \tag{8.54}$$

While the expressions of $f_{\hat{I}}(x,y)$ and $f_{\hat{L}}(z)$ have been here omitted for brevity, their characterization can be found in [6].

In the previous section, we showed that for an isolated cell in a random field of eavesdroppers, i.e., in a BCCE, a sufficient number of transmit antennas allows the BS to cancel the intracell interference and leakage, and to drive the probability of secrecy outage to zero. However, in a cellular network, the secrecy outage is also caused by the intercell interference and leakage, which cannot be controlled by the BS. It is easy to show that $\lim_{\rho \to \infty} \tau(x,y) \le 0$, which from Theorem 8.6 implies $\lim_{\rho \to \infty} \hat{\mathscr{O}}_{\text{CELL}} = 1$. We can therefore conclude the following.

In cellular networks, RCI precoding can achieve confidential communication with probability of secrecy outage $\hat{\mathscr{O}}_{\text{CELL}} < 1$. However unlike an isolated cell, under the assumptions made in this section, cellular networks tend to be in secrecy outage with probability 1 if the transmit power grows unbounded, irrespective of the number of transmit antennas.

In Figure 8.8, we plot the simulated probability of secrecy outage versus the transmit SNR, for $K = 10$ users per BS and three values of the number of transmit

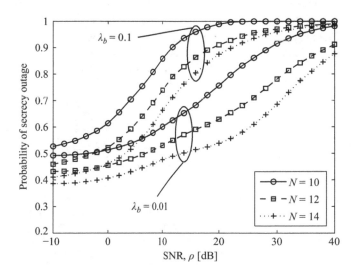

Figure 8.8 *Simulated probability of secrecy outage versus transmit SNR, for $K = 10$ users per BS and various values of the number of transmit antennas N and density of BSs λ_b*

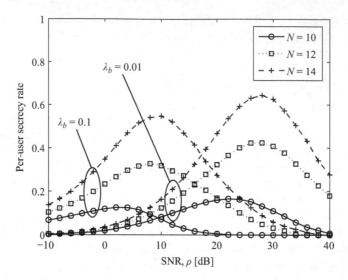

*Figure 8.9   Simulated ergodic secrecy rate versus transmit SNR, for K =10 users
per BS and various values of the number of transmit antennas N and
density of BSs λ_b*

antennas N. In this figure, two cases are considered for the density of BSs λ_b, namely
0.01 and 0.1, while the density of users is given by $\lambda_u = K\lambda_b$. Figure 8.8 shows
that RCI precoding achieves confidential communications in cellular networks with
$\hat{\mathcal{O}}_{\mathrm{CELL}} < 1$, and that having more transmit antennas is beneficial as it reduces the
probability of secrecy outage. However, Figure 8.8 also shows that $\lim_{\rho\to\infty} \hat{\mathcal{O}}_{\mathrm{CELL}} = 1$
irrespective of N. These observations are consistent with the analytical findings.

The mean secrecy rate for the typical user o is defined as:

$$R^{\circ}_{\mathrm{CELL}} \triangleq \mathbb{E}\left[R_{\mathrm{CELL}}\right] \tag{8.55}$$

and it can be approximated as follows.

Theorem 8.7. *The mean secrecy rate achievable by RCI precoding in a cellular
network can be approximated as* $R^{\circ}_{\mathrm{CELL}} \approx \hat{R}^{\circ}_{\mathrm{CELL}}$, *with*

$$
\hat{R}^{\circ}_{\mathrm{CELL}} = \int_0^\infty \int_{-\infty}^{\infty} e^{\frac{\alpha}{\rho\chi} - \frac{1}{\rho} - \chi y^{-\eta}} \left\{ \log_2\left(1 + \frac{\rho\alpha y^{-\eta}}{\rho\chi y^{-\eta} + \rho x + 1}\right) \int_{-\infty}^{\tau(x,y)} f_{\hat{L}}(z)\right.
$$
$$
\left. - \int_{-\infty}^{\tau(x,y)} \log_2\left(1 + \rho\chi y^{-\eta} + \rho z\right) f_{\hat{L}}(z)\,\mathrm{d}z \right\} f_{\hat{l}}(x,y)\,\mathrm{d}x\, 2\lambda_b \pi y e^{-\lambda_b \pi y^2}\,\mathrm{d}y. \tag{8.56}
$$

Details on the approximations involved in (8.56) can be found in [6].

In the earlier sections of this chapter, we showed that in an isolated cell, even in the presence of a random field of eavesdroppers, a sufficient number of transmit antennas allows the BS to cancel the intracell interference and leakage, and the secrecy rate increases monotonically with the SNR. In a cellular network, the secrecy rate is also affected by the intercell interference and leakage, which cannot be controlled by the BS. It is easy to show that $\lim_{\rho \to \infty} \frac{\alpha}{\rho \chi} - \frac{1}{\rho} - \chi y^{-\eta} \leq 0$, which from Theorem 8.7 implies $\lim_{\rho \to \infty} \hat{R}_{\text{CELL}}^{\circ} = 0$. We can therefore conclude the following.

In cellular networks, RCI precoding can achieve a nonzero secrecy rate $\hat{R}_{\text{CELL}}^{\circ}$. However unlike an isolated cell, the secrecy rate in a cellular network is interference-and-leakage-limited, and it cannot grow unbounded with the transmit SNR, irrespective of the number of transmit antennas. The above is confirmed by Figure 8.9, where we plot the simulated per-user ergodic secrecy rate versus the transmit SNR, for $K = 10$ users per BS and three values of the number of transmit antennas N. In this figure, again, two cases are considered for the density of BSs λ_b, namely 0.01 and 0.1, while the density of users is given by $\lambda_u = K\lambda_b$. Similar to Figure 8.8, Figure 8.9 validates the conclusions drawn from the analytical results presented in this section.

8.5 Conclusions

In this chapter, we studied trusted communications via spatial multiplexing in multi-user MIMO wireless systems.

We started by considering the MISO BCC under Rayleigh fading, where a multi-antenna BS simultaneously transmits independent confidential messages to several spatially dispersed malicious users that can eavesdrop on each other. For this system set-up, we showed that although the secrecy requirements result in a rate loss, the RCI precoder achieves a secrecy rate with the same high-SNR scaling factor as the one of a MISO BC without secrecy requirements.

We then introduced the BCCE, where a multi-antenna BS simultaneously communicates to multiple malicious users, in the presence of a PPP of external eavesdroppers. Unlike the BCC, in the BCCE not just malicious users, but also randomly located external nodes can act as eavesdroppers. We showed that, irrespective of the collusion strategy at the external eavesdroppers, a large number of transmit antennas N drives both the probability of secrecy outage and the rate loss due to the presence of external eavesdroppers to zero. Increasing the density of eavesdroppers λ_e by a factor n, requires n^2 as many antennas to meet a given probability of secrecy outage and a given mean secrecy rate.

We finally turned our attention to cellular networks where, unlike the case of isolated cells, multiple BSs generate inter-cell interference, and malicious users of neighboring cells can cooperate to eavesdrop. Under the assumptions made, we found that RCI precoding can achieve a nonzero secrecy rate with probability of outage smaller than one. However we also found that unlike isolated cells, the secrecy rate in a cellular network does not grow monotonically with the SNR, and the network tends to be in secrecy outage if the transmit power grows unbounded.

In order to maintain tractability, this chapter focused on scenarios where mobile users are equipped with a single receive antenna and have similar path-loss. More general results that apply to multi-antenna receivers under unequal path loss are provided in [29]. Moreover, while equal power allocation among the users was here assumed, a power allocation algorithm that can increase the secrecy rate performance of RCI precoding is designed in [7]. Finally, a generalization to practical channels that accounts for imperfect channel state information at the BS and for transmit antenna correlation can be found in [4,5], respectively.

References

[1] J. G. Andrews, S. Buzzi, W. Choi, *et al.*, "What will 5G be?" *IEEE J. Sel. Areas Commun.*, vol. 32, no. 6, pp. 1065–1082, Jun. 2014.

[2] G. Caire and S. Shamai, "On the achievable throughput of a multiantenna Gaussian broadcast channel," *IEEE Trans. Inf. Theory*, vol. 49, no. 7, pp. 1691–1706, Jul. 2003.

[3] Ericsson, "5G radio access – Capabilities and technologies," *White paper*, Apr. 2016.

[4] G. Geraci, A. Y. Al-nahari, J. Yuan, and I. B. Collings, "Linear precoding for broadcast channels with confidential messages under transmit-side channel correlation," *IEEE Commun. Lett.*, vol. 17, no. 6, pp. 1164–1167, Jun. 2013.

[5] G. Geraci, R. Couillet, J. Yuan, M. Debbah, and I. B. Collings, "Large system analysis of linear precoding in MISO broadcast channels with confidential messages," *IEEE J. Sel. Areas Commun.*, vol. 31, no. 9, pp. 1660–1671, Sept. 2013.

[6] G. Geraci, H. S. Dhillon, J. G. Andrews, J. Yuan, and I. B. Collings, "Physical layer security in downlink multi-antenna cellular networks," *IEEE Trans. Commun.*, vol. 62, no. 6, pp. 2006–2021, Jun. 2014.

[7] G. Geraci, M. Egan, J. Yuan, A. Razi, and I. B. Collings, "Secrecy sum-rates for multi-user MIMO regularized channel inversion precoding," *IEEE Trans. Commun.*, vol. 60, no. 11, pp. 3472–3482, Nov. 2012.

[8] G. Geraci, S. Singh, J. G. Andrews, J. Yuan, and I. B. Collings, "Secrecy rates in the broadcast channel with confidential messages and external eavesdroppers," *IEEE Trans. Wireless Commun.*, vol. 13, no. 5, pp. 2931–2943, May 2014.

[9] S. Goel and R. Negi, "Guaranteeing secrecy using artificial noise," *IEEE Trans. Wireless Commun.*, vol. 7, no. 6, pp. 2180–2189, Jun. 2008.

[10] B. Hochwald and S. Vishwanath, "Space-time multiple access: Linear growth in the sum rate," in *Proc. Allerton Conf. on Commun., Control, and Computing*, Monticello, IL, Oct. 2002.

[11] M. Joham, W. Utschick, and J. Nossek, "Linear transmit processing in MIMO communications systems," *IEEE Trans. Signal Process.*, vol. 53, no. 8, pp. 2700–2712, Aug. 2005.

[12] A. Kammoun, A. Müller, E. Björnson, and M. Debbah, "Linear precoding based on polynomial expansion: Large-scale multi-cell MIMO systems," *IEEE J. Sel. Topics Signal Process.*, vol. 8, no. 5, pp. 861–875, Oct. 2014.

[13] A. Khisti and G. Wornell, "Secure transmission with multiple antennas I: The MISOME wiretap channel," *IEEE Trans. Inf. Theory*, vol. 56, no. 7, pp. 3088–3104, Jul. 2010.

[14] Q. Li, G. Li, W. Lee, *et al.*, "MIMO techniques in WiMAX and LTE: a feature overview," *IEEE Comms. Mag.*, vol. 48, no. 5, pp. 86–92, May 2010.

[15] C. Lim, T. Yoo, B. Clerckx, B. Lee, and B. Shim, "Recent trend of multiuser MIMO in LTE-advanced," *IEEE Comms. Mag.*, vol. 51, no. 3, pp. 127–135, Mar. 2013.

[16] R. Liu, I. Maric, P. Spasojevic, and R. D. Yates, "Discrete memoryless interference and broadcast channels with confidential messages: Secrecy rate regions," *IEEE Trans. Inf. Theory*, vol. 54, no. 6, pp. 2493–2507, Jun. 2008.

[17] R. Liu, T. Liu, H. V. Poor, and S. Shamai (Shitz), "Multiple-input multiple-output Gaussian broadcast channels with confidential messages," *IEEE Trans. Inf. Theory*, vol. 56, no. 9, pp. 4215–4227, Sept. 2010.

[18] R. Liu and H. Poor, "Secrecy capacity region of a multiple-antenna Gaussian broadcast channel with confidential messages," *IEEE Trans. Inf. Theory*, vol. 55, no. 3, pp. 1235–1249, Mar. 2009.

[19] T. L. Marzetta, "Noncooperative cellular wireless with unlimited numbers of base station antennas," *IEEE Trans. Wireless Commun.*, vol. 9, no. 11, pp. 3590–3600, Nov. 2010.

[20] Nokia Networks, "Ten key rules of 5G deployment – Enabling 1 Tbit/s/km^2 in 2030," *White paper*, Apr. 2015.

[21] C. B. Peel, B. M. Hochwald, and A. L. Swindlehurst, "A vector-perturbation technique for near-capacity multiantenna multiuser communication – Part I: Channel inversion and regularization," *IEEE Trans. Commun.*, vol. 53, no. 1, pp. 195–202, Jan. 2005.

[22] T. S. Rappaport, *Wireless Communications: Principles and Practice*, 1st ed. IEEE Press, 1996.

[23] Q. H. Spencer, C. B. Peel, A. L. Swindlehurst, and M. Haardt, "An introduction to the multi-user MIMO downlink," *IEEE Comms. Mag.*, vol. 42, no. 10, pp. 60–67, Oct. 2004.

[24] Q. H. Spencer, A. L. Swindlehurst, and M. Haardt, "Zero-forcing methods for downlink spatial multiplexing in multiuser MIMO channels," *IEEE Trans. Signal Process.*, vol. 52, no. 2, pp. 461–471, Feb. 2004.

[25] D. Stoyan, W. Kendall, and J. Mecke, *Stochastic geometry and its applications*, 2nd ed. New York, NY: John Wiley & Sons Ltd., 1996.

[26] S. Wagner, R. Couillet, M. Debbah, and D. T. M. Slock, "Large system analysis of linear precoding in correlated MISO broadcast channels under limited feedback," *IEEE Trans. Inf. Theory*, vol. 58, no. 7, pp. 4509–4537, Jul. 2012.

[27]   N. Yang, M. Elkashlan, T. Q. Duong, J. Yuan, and R. Malaney, "Optimal transmission with artificial noise in MISOME wiretap channels," *IEEE Trans. Veh. Technol.*, vol. 65, no. 4, pp. 2170–2181, Apr. 2016.

[28]   H. H. Yang, G. Geraci, T. Q. S. Quek, and J. G. Andrews, "Cell-edge-aware precoding for downlink massive MIMO cellular networks," *IEEE Trans. Signal Process.*, vol. 65, no. 13, pp. 3344–3358, Jul. 2017.

[29]   N. Yang, G. Geraci, J. Yuan, and R. Malaney, "Confidential broadcasting via linear precoding in non-homogeneous MIMO multiuser networks," *IEEE Trans. Commun.*, vol. 62, no. 7, pp. 2515–2530, Jul. 2014.

[30]   N. Yang, L. Wang, G. Geraci, M. Elkashlan, J. Yuan, and M. D. Renzo, "Safeguarding 5G wireless communication networks using physical layer security," *IEEE Comms. Mag.*, vol. 53, no. 4, pp. 20–27, Apr. 2015.

[31]   T. Yoo and A. Goldsmith, "On the optimality of multiantenna broadcast scheduling using zero-forcing beamforming," *IEEE J. Sel. Areas Commun.*, vol. 24, no. 3, pp. 528–541, Mar. 2006.

[32]   X. Zhou and M. McKay, "Secure transmission with artificial noise over fading channels: Achievable rate and optimal power allocation," *IEEE Trans. Veh. Technol.*, vol. 59, no. 8, pp. 3831–3842, Oct. 2010.

Part III

Physical layer security with emerging 5G technologies

Chapter 9

Physical layer security for wirelessly powered communication systems

Caijun Zhong[1] and Xiaoming Chen[1]

9.1 Introduction

9.1.1 Background

The proliferation of portable mobile devices such as mobile phones and tablets, along with the commercialization of the fourth-generation mobile communications systems, has promoted the fast development of mobile internet, enabling a plethora of new types of mobile services such as mobile social networking, mobile payment, online gaming and mobile video. As the penetration of the mobile internet into the daily life of people at an unprecedented pace, how to prolong the operation time of mobile terminals has become one of the most critical problems to be urgently addressed in order to improve user experiences. The traditional way of tackling this issue is to increase the battery capacity. Unfortunately, the technology improvement for battery capacity has been rather slow over the last decades. As such, seeking novel and effective power solutions has emerged as one of the major tasks in the area of information technology.

One of the promising candidate technologies to address the energy bottleneck issue is energy harvesting, where the mobile terminals can scavenge energy from ambient environment such as solar, wind, hydroelectric or piezoelectric. However, the major limitation of harvesting energy from these resources is the unpredictability due to weather, location and many other factors, and thus, it is challenging to guarantee stable energy output, making it undesirable for wireless communication applications. To resolve this challenge, radio frequency (RF)-based wireless energy transfer has been recently proposed as a promising means to power wireless devices. The advantages of RF signal based wirelessly powered communications are two-fold. First, since RF signals can be fully controlled, it is possible to provide a reliable energy supply by properly adjusting some key parameters of the energy signal such as power, direction and duration. Second, since RF signals can carry both energy and information, wireless power and information transfer can be jointly designed to optimize the performance of wirelessly powered communications.

[1]Institute of Information and Communication Engineering, Zhejiang University, China

Due to the broadcast nature of the wireless medium, wireless communications are extremely vulnerable to eavesdropping. The conventional approach to safeguard the wireless transmission is to adopt the cryptographic techniques at the upper layer, which nevertheless incurs high complexity and is energy inefficient. On the other hand, energy is a critical resource in wirelessly powered communication systems. In particular, the amount of harvested energy by the wireless devices is in general limited due to path-loss; hence, complicated cryptographic techniques may be undesirable for wirelessly powered communications. Therefore, physical layer security, which is more energy efficient, becomes an ideal candidate to provide security for wirelessly powered communication systems.

9.1.2   Literature review

Physical layer security aims to prevent the eavesdropper from correctly decoding the message sent to the legitimate receiver, by exploiting the physical characteristics of wireless channels, i.e. fading, noise and interference. Compared to the wirelessly powered communication systems without secrecy requirement, the design of secrecy wirelessly powered communication systems is much more complicated, since it involves two unaligned and even conflicting objectives, namely maximizing the energy transfer efficiency and enhancing the information transmission security [1–3]. On the one hand, energy transfer and information transmission compete for the same wireless resources. On the other hand, the two processes may benefit from each other through proper coordination. For instance, the energy signal can be tuned to interfere with the eavesdropper, while the information signal might be exploited as an extra energy source.

In order to achieve the two goals with limited resource, an effective way is to use the multi-antenna techniques [4,5]. Specifically, by exploiting the spatial degree of freedom, it is possible to align the energy signal to the direction of the energy receiver while reducing the information leakage to the eavesdropper [6]. Intuitively, the secrecy performance of multi-antenna techniques in wirelessly powered communications is heavily dependent on the accuracy of channel state information (CSI) at the transmitter. With full CSI, the authors in [7] addressed the problem of optimal beamformer design in a multiuser multiple-input–single-output system where the transmitter is equipped with multiple antennas, while the legitimate information and energy receivers and the eavesdropper are equipped with a single antenna each. Later on, the work [8] considered a more general case with multi-antenna legitimate receivers and eavesdroppers. However, obtaining full CSI in secrecy wirelessly powered communication systems is a non-trivial task due to the following reasons. First, the CSI of the eavesdropper channel and energy transfer channel may be imperfect or even unavailable, since the eavesdropper and energy receiver may remain silent [9,10]. Second, the CSI corresponding to the information receiver may be imperfect due to channel estimation error or feedback delay [11]. Therefore, it is of great interest to understand the secrecy performance of multi-antenna wirelessly powered communication systems with imperfect CSI. Responding to this, assuming that the uncertainty of the channel is bounded, robust beamforming for multi-antenna-based wirelessly powered communication has been studied with the aim of maximizing the worst-case secrecy rate. Robust beamforming schemes for the case with an external eavesdropper who acted as energy receiver were designed in [12–14]. An advantage of secrecy wirelessly powered communications is that the energy signal can be used as artificial noise to confuse the eavesdropper [15]. As such, it is important to maintain a fine balance between enhancing energy harvesting efficiency at the energy receiver, minimizing the interference at the information receiver, and maximizing the interference at the eavesdropper. The joint design of transmit beamforming vector and artificial noise with perfect and imperfect CSI were studied in [16,17], respectively. In the scenario where the artificial noise originates from an external jammer, more spatial degrees of freedom were exploited to improve the secrecy performance [18,19].

9.1.3   Organization of the chapter

The main purpose of this chapter is to present the achievable secrecy performance of power beacon (PB)-assisted wirelessly powered communication systems. We first introduce the PB-assisted wiretap channel model and then present an analytical study on the achievable secrecy outage performance assuming a simple maximum ratio transmission (MRT) beamformer at the information source. Later on, the optimal design of the transmit beamformer is investigated. Next, the PB is exploited to act as a friendly jammer to further enhance the secrecy performance of the system. We conclude the chapter with a summary and present potential interesting directions in the area.

9.2   Secrecy performance of wirelessly powered wiretap channels

9.2.1   System model

Consider a four-node wirelessly powered communication system consisting of one PB, one information source Alice, one legitimate user Bob and one eavesdropper Eve, as illustrated in Figure 9.1. It is assumed that the information source is equipped with N_S antennas, while the other three nodes are equipped with a single antenna each. Quasi-static fading is assumed, such that the channel coefficients remain unchanged during each transmission block, but vary independently between different blocks.

The system adopts the time-sharing protocol proposed in [20]. Hence, a complete transmission slot with time duration of T is divided into two orthogonal sub-slots with unequal lengths, i.e. the first one for energy transfer with time duration of θT where $\theta(0 < \theta < 1)$ is the time switching ratio, and the second one for information transmission with time duration of $(1 - \theta)T$.

At the beginning of each time block, the PB transmits an energy signal to Alice; hence, the received energy signal at Alice can be expressed as:

$$\mathbf{y}_s = \sqrt{P}\mathbf{g}x_s + \mathbf{n}_s, \tag{9.1}$$

where P denotes the transmit power of the PB, x_s is the energy signal with unit power, $\mathbf{n}_s$ is an N-dimensional additive white Gaussian noise (AWGN) vector with $\mathrm{E}\{\mathbf{n}_s\mathbf{n}_s^\dagger\} = \sigma_s^2\mathbf{I}$. The $N_S \times 1$ vector $\mathbf{g}$ denotes the power transfer channel from PB to Alice.

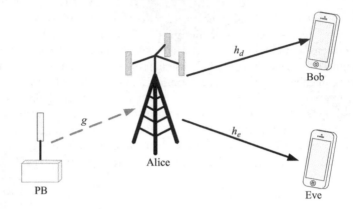

Figure 9.1   A schematic diagram of the system model consisting of one PB, one information source Alice, one legitimate user Bob and one eavesdropper Eve

Hence, at the end of the first phase, the total harvested energy is given by:

$$E = \eta P \|\mathbf{g}\|^2 \theta T, \tag{9.2}$$

where $0 < \eta \le 1$ denotes the energy conversion efficiency.

Then, in the second phase with duration $(1-\theta)T$, Alice performs secrecy communications using the harvested energy. Hence, the transmit power can be computed as:

$$P_a = \frac{E}{(1-\theta)T} = \eta P \|\mathbf{g}\|^2 \frac{\theta}{1-\theta}. \tag{9.3}$$

To exploit the benefits of multiple antennas at Alice, beamforming schemes can be used. Therefore, the received signals at Bob and Eve can be expressed as:

$$y_b = \sqrt{P_a} \mathbf{h}_d \mathbf{w} x + n_d, \tag{9.4}$$

and

$$y_e = \sqrt{P_a} \mathbf{h}_e \mathbf{w} x + n_e, \tag{9.5}$$

respectively, where x is the Gaussian distributed transmit signal with unit power, vectors $\mathbf{h}_d$ and $\mathbf{h}_e$ denote the legitimate channel from Alice to Bob and the eavesdropper channel from Alice to Eve respectively, $\mathbf{w}$ is the normalized transmit beamforming vector, n_d and n_e are the AWGN with variance σ_d^2 and σ_e^2 at Bob and Eve, respectively. For simplicity, we assume that $\sigma_d^2 = \sigma_e^2 = N_0$. Hence, the instantaneous signal-to-noise ratio (SNR) at Bob and Eve can be respectively expressed as:

$$\gamma_b = \frac{\eta P \|\mathbf{g}\|^2 |\mathbf{h}_d \mathbf{w}|^2}{N_0} \frac{\theta}{1-\theta}, \tag{9.6}$$

and

$$\gamma_e = \frac{\eta P ||\mathbf{g}||^2 |\mathbf{h}_e \mathbf{w}|^2}{N_0} \frac{\theta}{1 - \theta}. \tag{9.7}$$

Now, according to [21], the secrecy rate C_S is given by the difference of the legitimate channel capacity and the eavesdropper channel capacity as:

$$C_S = \begin{cases} \log_2(1 + \gamma_b) - \log_2(1 + \gamma_e) & \gamma_b > \gamma_e, \\ 0 & \gamma_b \leq \gamma_e. \end{cases} \tag{9.8}$$

9.2.2 Secrecy performance analysis

We now analyse the achievable secrecy performance of the considered wirelessly powered communication system. Before delving into the detailed analysis, we would like to make the following few clarifications on the system model:

- Due to a relatively short distance between the PB and Alice, it is likely that the line-of-sight propagation exists. Hence, the Nakagami-m distribution is used to model the power transfer channel, i.e. the amplitude of each element of $\mathbf{g}$ follows Nakagami-m distribution with shape parameter m and average power λ_P.
- Rayleigh fading is assumed for the information transmission channels, i.e. the elements of $\mathbf{h}_d$ and $\mathbf{h}_e$ are zero mean circularly symmetric complex Gaussian random variables (RVs) with variance λ_d and λ_e, respectively.
- A delay constrained communication scenario is considered, where Alice transmits at a constant rate R_S. According to [22], perfect secrecy is achievable when $R_S < C_S$, otherwise, secrecy is compromised. As such, secrecy outage probability becomes an appropriate performance metric.

From (9.6) and (9.7), it is obvious that the secrecy performance hinges on the beamforming vector adopted at Alice. In this section, we consider a simple and low-complexity maximum-ratio-transmission (MRT) scheme, i.e. the beamforming vector $\mathbf{w}$ is given by:

$$\mathbf{w} = \frac{\mathbf{h}_d^\dagger}{||\mathbf{h}_d||}. \tag{9.9}$$

As such, the instantaneous SNR at Bob γ_b and at Eve γ_b are given by:

$$\gamma_b = \frac{\eta P ||\mathbf{g}||^2 ||\mathbf{h}_d||^2}{N_0} \frac{\theta}{1 - \theta}, \tag{9.10}$$

and

$$\gamma_e = \frac{\eta P ||\mathbf{g}||^2 \frac{|\mathbf{h}_e \mathbf{h}_d^\dagger|^2}{||\mathbf{h}_d||^2}}{N_0} \frac{\theta}{1 - \theta}, \tag{9.11}$$

respectively, where $(\cdot)^\dagger$ denotes the conjugate and transpose operator.

According to the definition, the secrecy outage probability can be expressed mathematically as [6]:

$$P_{out}(R_S) = P(C_S < R_S). \tag{9.12}$$

We now proceed to the analysis of the exact secrecy outage probability of the MRT scheme.

Theorem 9.1. *The exact secrecy outage probability of the MRT scheme can be expressed in closed-form as*

$$P_{out}(R_S) = 1 - \frac{2}{\Gamma(mN_S)} \sum_{k=0}^{N_S-1} \sum_{p=0}^{k} \frac{\lambda_d(k_2\lambda_e)^{k-p}}{p!(\lambda_d + k_2\lambda_e)^{k-p+1}}$$

$$\times \left(\frac{(k_2-1)m}{k_1\lambda_d\lambda_P}\right)^{\frac{mN_S+p}{2}} K_{mN_S-p}\left(2\sqrt{\frac{(k_2-1)m}{k_1\lambda_d\lambda_P}}\right), \tag{9.13}$$

where $k_1 = \frac{\eta P}{N_0} \frac{\theta}{1-\theta}$ and $k_2 = 2^{R_S}$. Also, $\Gamma(x)$ is the gamma function [23, Eq. (8.31)], and $K_v(x)$ is the vth order modified Bessel function of the second kind [23, Eq. (8.407.1)].

Proof. We start by expressing the SNR given in (9.10) and (9.11) as

$$\gamma_b = k_1 y_g y_{h_d}, \quad \text{and} \quad \gamma_e = k_1 y_g y_{h_e}, \tag{9.14}$$

where $y_g = ||\mathbf{g}||^2$, $y_{h_d} = ||\mathbf{h}_d||^2$ and $y_{h_e} = \frac{|\mathbf{h}_e\mathbf{h}_d^\dagger|^2}{||\mathbf{h}_d||^2}$. It is straightforward to show that the probability density function (pdf) of y_g follows a gamma distribution with shape parameter mN_S and scale parameter λ_P/m given by [24]

$$f_{y_g}(x) = \frac{1}{\Gamma(mN_S)}\left(\frac{m}{\lambda_P}\right)^{mN_S} x^{mN_S-1} e^{-\frac{m}{\lambda_P}x}, \tag{9.15}$$

and the pdf of y_{h_d} follows a chi-square distribution with $2N_S$ degrees of freedom given by [25]

$$f_{y_{h_d}}(x) = \frac{x^{N_S-1}}{\lambda_d^{N_S}\Gamma(N_S)} e^{-\frac{x}{\lambda_d}}. \tag{9.16}$$

In addition, according to [26], y_{h_e} follows an exponential distribution with pdf

$$f_{y_{h_e}}(x) = \frac{1}{\lambda_e} e^{-\frac{x}{\lambda_e}}, \tag{9.17}$$

and is independent of y_{h_d}. As such, the secrecy outage probability can be written as

$$P_{out}(R_S) = 1 - P\left(\frac{1 + k_1 y_g y_{h_d}}{1 + k_1 y_g y_{h_e}} \geq k_2\right). \tag{9.18}$$

Conditioned on y_g and y_{h_e}, with the help of [23, Eq. (3.351.2)], we obtain

$$
P_{\text{out}}(R_S|y_g, y_{h_e}) = 1 - \int_{\frac{k_2-1}{k_1 y_g} + k_2 y_{h_e}}^{\infty} \frac{x^{N_S-1}}{\lambda_d^{N_S} \Gamma(N_S)} e^{-\frac{x}{\lambda_d}} \, dx
$$

$$
= 1 - e^{-\frac{k_2-1}{k_1 \lambda_d y_g} - \frac{k_2 y_{h_e}}{\lambda_d}} \sum_{k=0}^{N_S-1} \frac{1}{k!} \left(\frac{k_2-1}{k_1 \lambda_d y_g} + \frac{k_2 y_{h_e}}{\lambda_d} \right)^k . \tag{9.19}
$$

By applying the binomial expansion $(x_1 + x_2)^n = \sum_{k=0}^{n} \binom{n}{k} x_1^k x_2^{n-k}$, (9.19) can be further expressed as

$$
P_{\text{out}}(R_S|y_g, y_{h_e}) = 1 - \sum_{k=0}^{N_S-1} \sum_{p=0}^{k} \frac{1}{p!(k-p)!} e^{-\frac{k_2-1}{k_1 \lambda_d y_g}} \left(\frac{k_2-1}{k_1 \lambda_d y_g} \right)^p e^{-\frac{k_2 y_{h_e}}{\lambda_d}} \left(\frac{k_2 y_{h_e}}{\lambda_d} \right)^{k-p} .
$$

$$
\tag{9.20}
$$

Noticing that the RV y_g is decoupled with y_{h_e}, the expectation over y_g and y_{h_e} can be taken separately. The desired result can be obtained after some algebraic manipulations. □

Theorem 1 presents an exact closed-form expression for the secrecy outage probability, which can be efficiently evaluated. However, the expression is too complicated to yield any insight. Motivated by this, we now look into the asymptotic regime, where simple expressions can be obtained.

For the asymptotic high SNR regime, we assume that $\lambda_d \to \infty$ with an arbitrary λ_e. Such a scenario has been widely adopted in the literature, see for instance [27–30]. In practice, this occurs when the quality of the legitimate channel is much better than the eavesdropper channel, i.e. Bob is relatively close to Alice, while Eve is far away from Alice or the eavesdropper channel undergoes severe small-scale and large-scale fading effects. In the following, we characterize the two key-performance parameters governing the secrecy outage probability in the high SNR regime, i.e. secrecy diversity order G_d and secrecy array gain G_a defined by [31]:

$$
P_{\text{out}}^{\infty}(R_S) = (G_a \lambda_d)^{-G_d} . \tag{9.21}
$$

Proposition 9.1. *In the high SNR regime, i.e. $\lambda_d \to \infty$, the secrecy outage probability of the MRT scheme can be approximated by*

$$
P^{\infty}(R_S) = \sum_{k=0}^{N_S} \frac{1}{k!} \frac{\Gamma(mN_S - k)}{\Gamma(mN_S)} \left(\frac{m(k_2-1)}{k_1 k_2 \lambda_e \lambda_P} \right)^k \left(\frac{k_2 \lambda_e}{\lambda_d} \right)^{N_S} . \tag{9.22}
$$

Proof. Starting from (9.18), conditioned on y_g and y_{h_d}, we have

$$
P_{\text{out}}(R_S | y_{h_g}, y_{h_d}) = 1 - \text{Prob}\left(y_{h_d} > \frac{k_2 - 1}{k_1 y_g}\right) \times \int_0^{\frac{y_{h_d}}{k_2} - \frac{k_2 - 1}{k_1 k_2 y_g}} \frac{1}{\lambda_e} e^{-\frac{x}{\lambda_e}} dx
$$

$$
= 1 - \text{Prob}\left(y_{h_d} > \frac{k_2 - 1}{k_1 y_g}\right) \times \left(1 - e^{-\frac{y_{h_d}}{k_2 \lambda_e} + \frac{k_2 - 1}{k_1 k_2 \lambda_e y_g}}\right). \tag{9.23}
$$

With the help of [23, Eq. (3.351.2)] and $e^{\frac{k_2 - 1}{k_1 \lambda_d y_g}} = \sum_{k=0}^{\infty} \frac{1}{k!}\left(\frac{k_2 - 1}{k_1 \lambda_d y_g}\right)^k$, conditioned on y_g, the outage probability can be expressed as

$$
P_{\text{out}}(R_S | y_g) = e^{-\frac{k_2 - 1}{k_1 \lambda_d y_g}} \sum_{k=N_S}^{\infty} \frac{1}{k!}\left(\frac{k_2 - 1}{k_1 \lambda_d y_g}\right)^k + e^{-\frac{k_2 - 1}{k_1 \lambda_d y_g}} \sum_{k=0}^{N_S - 1} \frac{1}{k!} \frac{\left(\frac{k_2 - 1}{k_1 \lambda_d y_g}\right)^k}{\left(1 + \frac{\lambda_d}{k_2 \lambda_e}\right)^{N_S - k}}. \tag{9.24}
$$

Then, averaging over y_g, with the help of [23, Eq. (3.471.9)], the secrecy outage probability can be computed as

$$
P_{\text{out}}(R_S) = \sum_{k=N_S}^{\infty} \frac{1}{k!}\left(\frac{m(k_2 - 1)}{k_1 \lambda_d \lambda_P}\right)^k \frac{2}{\Gamma(mN_S)}\left(\frac{m(k_2 - 1)}{k_1 \lambda_d \lambda_P}\right)^{\frac{mN_S - k}{2}} K_{mN_S - k}\left(2\sqrt{\frac{m(k_2 - 1)}{k_1 \lambda_d \lambda_P}}\right)
$$

$$
+ \sum_{k=0}^{N_S - 1} \frac{1}{k!} \frac{\left(\frac{m(k_2 - 1)}{k_1 \lambda_d \lambda_P}\right)^k}{\left(1 + \frac{\lambda_d}{k_2 \lambda_e}\right)^{N_S - k}} \frac{2}{\Gamma(mN_S)}\left(\frac{m(k_2 - 1)}{k_1 \lambda_d \lambda_P}\right)^{\frac{mN_S - k}{2}} K_{mN_S - k}\left(2\sqrt{\frac{m(k_2 - 1)}{k_1 \lambda_d \lambda_P}}\right).
$$

$$\tag{9.25}$$

Expanding the Bessel function by [23, Eq. (8.446)] and omitting the high order items yield

$$
P^{\infty}(R_S) = \sum_{k=N_S}^{\infty} \frac{1}{k!}\left(\frac{m(k_2 - 1)}{k_1 \lambda_d \lambda_P}\right)^k \frac{\Gamma(mN_S - k)}{\Gamma(mN_S)} + \sum_{k=0}^{N_S - 1} \frac{1}{k!} \frac{\left(\frac{m(k_2 - 1)}{k_1 \lambda_d \lambda_P}\right)^k}{\left(1 + \frac{\lambda_d}{k_2 \lambda_e}\right)^{N - k}} \frac{\Gamma(mN_S - k)}{\Gamma(mN_S)}.
$$

$$\tag{9.26}$$

By omitting the high order items, the desired result can be obtained. □

It is evident from (9.22) that the system achieves a secrecy diversity order of N. In addition, we observe the intuitive effect of the position of nodes on the secrecy outage probability. For instance, the secrecy outage probability decreases when the PB is close to the source, i.e. large λ_P. It is also easy to see that the high SNR secrecy outage probability $P^{\infty}(R_S)$ is a decreasing function with respect to $\frac{P}{N_0}$, indicating that increasing the transmit power of the PB is always beneficial.

9.2.3 Resource allocation

In the previous section, we have analysed the secrecy outage probability of the MRT scheme with a fixed time switching ratio θ. However, the MRT scheme is in general suboptimal and a proper choice of θ also has a significant impact on the achievable secrecy performance. Therefore, it is of paramount importance to optimize the available resources such as $\mathbf{w}$ and θ for secrecy performance enhancement.

Mathematically, the joint resource allocation can be formulated as the following optimization problem:

$$\begin{aligned} \text{OP1}: \max_{\mathbf{v},\theta} \ (1-\theta)\log_2\left(1 + \frac{\theta\eta\|\mathbf{g}\|^2|\mathbf{h}_d\mathbf{v}|^2}{(1-\theta)\sigma_d^2}\right) \\ -(1-\theta)\log_2\left(1 + \frac{\theta\eta\|\mathbf{g}\|^2|\mathbf{h}_e\mathbf{v}|^2}{(1-\theta)\sigma_e^2}\right) \end{aligned} \tag{9.27}$$

$$\text{s.t. C1}: \|\mathbf{v}\|^2 = P \leq P_{\max}$$
$$\text{C2}: 0 \leq \theta \leq 1,$$

where $\mathbf{v} = \sqrt{P}\mathbf{w}$ with $\mathbf{w}$ being a unit-norm vector, C1 denotes the transmit power constraint and $P_{\max}$ is the maximum transmit power at the PB, and C2 is the constraint on the time switching ratio.

After some simple calculations, the optimization problem OP1 can be transformed into the following optimization problem:

$$\text{OP2}: \max_{\mathbf{v},\theta} \ (1-\theta)\log_2 \frac{(1-\theta)\sigma_d^2 + \theta\eta\|\mathbf{g}\|^2|\mathbf{h}_d\mathbf{v}|^2 \, \sigma_e^2}{(1-\theta)\sigma_e^2 + \theta\eta\|\mathbf{g}\|^2|\mathbf{h}_e\mathbf{v}|^2 \, \sigma_d^2} \tag{9.28}$$

$$\text{s.t. C1, \ C2.}$$

Unfortunately, the optimization problem OP2 is still non-convex with respect to $\mathbf{v}$ and θ, thus it is challenging to obtain the optimal solution. To circumvent this problem, we propose an alternating optimization method capitalizing the following property $\inf_{x,y} f(x,y) = \inf_x \tilde{f}(x)$, where $\tilde{f}(x) = \inf_y f(x,y)$ [32]. Specifically, we tackle the original problem by alternatingly solving two sub-problems, namely, finding the optimal $\mathbf{v}$ for a fixed θ and finding the optimal θ for a fixed $\mathbf{v}$. We start by the optimization of beamforming vector $\mathbf{v}$ for a given θ.

A. Optimal design of v

With a fixed θ, the optimization problem OP2 is equivalent to the following optimization problem:

$$\text{OP3}: \max_{\mathbf{v}} \ (1-\theta)\log_2 \frac{(1-\theta)\sigma_d^2 + \theta\eta\|\mathbf{g}\|^2|\mathbf{h}_d\mathbf{v}|^2 \, \sigma_e^2}{(1-\theta)\sigma_e^2 + \theta\eta\|\mathbf{g}\|^2|\mathbf{h}_e\mathbf{v}|^2 \, \sigma_d^2} \tag{9.29}$$

$$\text{s.t. C1.}$$

Define $a = \frac{\sigma_d^2}{\sigma_e^2}$, $b = \theta\eta\|\mathbf{g}\|^2$, $c = (1 - \theta)\sigma_e^2$, the optimization problem OP3 can be rewritten as:

$$\text{OP4}: \quad \max_{\mathbf{V}} \frac{ac + b\text{tr}(\mathbf{H}_d\mathbf{V})}{c + b\text{tr}(\mathbf{H}_e\mathbf{V})} \tag{9.30}$$

$$\text{C3}: \quad \text{tr}(\mathbf{V}) \le P_{\max}$$

$$\text{C4}: \quad \mathbf{V} \succeq 0$$

$$\text{C5}: \quad \text{Rank}(\mathbf{V}) = 1,$$

where $\mathbf{V} = \mathbf{v}\mathbf{v}^\dagger$, $\mathbf{H}_d = \mathbf{h}_d^\dagger\mathbf{h}_d$, $\mathbf{H}_e = \mathbf{h}_e^\dagger\mathbf{h}_e$, and $\succeq$ denotes positive semi-definite. The optimization problem OP4 is a fractional programming problem, which is in general non-convex. Fortunately, it can be reformulated to a semi-definite programming (SDP) problem through the Charnes–Cooper transformation [33]. Let:

$$\xi = \frac{1}{c + b\text{tr}(\mathbf{H}_e\mathbf{V})} \quad \text{and} \quad \bar{\mathbf{V}} = \xi\mathbf{V}, \tag{9.31}$$

the optimization problem OP4 can be rewritten as:

$$\text{OP5}: \quad \max_{\bar{\mathbf{V}}} ac\xi + b\text{tr}(\mathbf{H}_d\bar{\mathbf{V}}) \tag{9.32}$$

$$\text{C6}: \quad c\xi + b\text{tr}(\mathbf{H}_e\bar{\mathbf{V}}) = 1$$

$$\text{C7}: \quad \text{tr}(\bar{\mathbf{V}}) \le \xi P_{\max}$$

$$\text{C8}: \quad \bar{\mathbf{V}} \succeq 0$$

$$\text{C9}: \quad \xi > 0$$

$$\text{C10}: \quad \text{Rank}(\bar{\mathbf{V}}) = 1.$$

Due to the rank constraint condition C10, the optimization problem OP5 is still non-convex. However, removing the constraint C10, OP5 becomes a convex SDP problem, which can then be efficiently solved by standard optimization software, such as CVX [34]. If the optimal solution of the relaxed SDP problem is rank-one, then it is also the optimal solution of the original optimization problem OP5. In the following, we prove that the optimal solution $\bar{\mathbf{V}}$ of the relaxed optimization problem is always rank-one.

Without the rank-one constraint, the Lagrangian dual function of the optimization problem can be written as:

$$\begin{aligned}\mathscr{L}(\bar{\mathbf{V}}, \lambda_1, \lambda_2, \mathbf{Q}) &= ac\xi + b\text{tr}(\mathbf{H}_d\bar{\mathbf{V}}) - \lambda_1[c\xi + b\text{tr}(\mathbf{H}_e\bar{\mathbf{V}}) - 1] \\ &\quad + \lambda_2[\xi P_{\max} - \text{tr}(\bar{\mathbf{V}})] + \text{tr}(\mathbf{Q}\bar{\mathbf{V}}),\end{aligned} \tag{9.33}$$

where λ_1 and λ_2 are the Lagrangian dual variables for C6 and C7, respectively, and $\mathbf{Q}$ is the Lagrangian dual variable for the constraint $\bar{\mathbf{V}} \succeq 0$. Then, the corresponding KKT conditions are given by:

$$b\mathbf{H}_d - \lambda_1 b\mathbf{H}_e - \lambda_2\mathbf{I} + \mathbf{Q} = 0, \tag{9.34}$$

$$\mathbf{Q}\bar{\mathbf{V}} = 0, \tag{9.35}$$

$$\bar{\mathbf{V}} \succeq 0. \tag{9.36}$$

Right-multiplying (9.34) by $\bar{\mathbf{V}}$ and making use of (9.35) yields:

$$b\mathbf{H}_d\bar{\mathbf{V}} = (\lambda_1 b\mathbf{H}_e + \lambda_2\mathbf{I})\bar{\mathbf{V}}, \tag{9.37}$$

which implies that:

$$\begin{aligned}
\text{Rank}((\lambda_1 b\mathbf{H}_e + \lambda_2\mathbf{I})\bar{\mathbf{V}}) &= \text{Rank}(\mathbf{H}_d\bar{\mathbf{V}}) \\
&= \text{Rank}(\mathbf{h}_d\mathbf{h}_d\bar{\mathbf{V}}) \\
&= 1. \tag{9.38}
\end{aligned}$$

Since $\lambda_1 b\mathbf{H}_e + \lambda_2\mathbf{I} \succ 0$, the following relation holds:

$$\text{Rank}(\bar{\mathbf{V}}) = \text{Rank}((\lambda_1 b\mathbf{H}_e + \lambda_2\mathbf{I})\bar{\mathbf{V}}), \tag{9.39}$$

which establishes the claim $\text{Rank}(\bar{\mathbf{V}}) = 1$.

Now, assume that $(\bar{\mathbf{V}}^*, \xi^*)$ is the optimal solution of the relaxed optimization problem OP5, the optimal solution of the original optimization problem OP1 can be computed by:

$$P^* = \text{tr}(\mathbf{V}^*), \tag{9.40}$$

and

$$\mathbf{w}^*(\mathbf{w}^*)^\dagger = \mathbf{V}^*/P^*. \tag{9.41}$$

B. Optimal design of θ

With a fixed $\mathbf{V}^*$, the optimization problem OP2 reduces to:

$$\text{OP6} : \max_{\theta} (1 - \theta)\log_2 \frac{(1 - \theta)\sigma_d^2 + \theta\eta\|\mathbf{g}\|^2\text{tr}(\mathbf{H}_d\mathbf{V}^*)\,\sigma_e^2}{(1 - \theta)\sigma_e^2 + \theta\eta\|\mathbf{g}\|^2\text{tr}(\mathbf{H}_e\mathbf{V}^*)\,\sigma_d^2} \tag{9.42}$$

$$\text{s.t. C2.}$$

Let us define:

$$t_d = \frac{\eta\|\mathbf{g}\|^2\text{tr}(\mathbf{H}_d\mathbf{V}^*)}{\sigma_d^2}, \tag{9.43}$$

and

$$t_e = \frac{\eta\|\mathbf{g}\|^2\text{tr}(\mathbf{H}_e\mathbf{V}^*)}{\sigma_e^2}. \tag{9.44}$$

Then, the optimization problem OP6 can be rewritten as:

$$\text{OP7}: \max_{\theta} (1 - \theta) \log_2 \frac{1 - \theta + \theta t_d}{1 - \theta + \theta t_e} \tag{9.45}$$

s.t. C2.

The remaining task is to establish the convexity of the objective function of optimization problem OP7. To this end, let $f(\theta) = (1 - \theta) \log_2 \frac{1-\theta+\theta t_d}{1-\theta+\theta t_e}$, then we have:

$$f(\theta) = (1 - \theta) \log_2 \frac{\dfrac{1 - \theta + \theta t_d}{1 - \theta}}{\dfrac{1 - \theta + \theta t_e}{1 - \theta}}$$

$$= (1 - \theta) \log_2 \frac{\dfrac{\theta(t_d - 1) + (1 - t_d) + t_d}{1 - \theta}}{\dfrac{\theta(t_e - 1) + (1 - t_e) + t_e}{1 - \theta}}$$

$$= (1 - \theta) \log_2 \frac{\dfrac{(1 - t_d)(1 - \theta) + t_d}{1 - \theta}}{\dfrac{(1 - t_e)(1 - \theta) + t_e}{1 - \theta}}. \tag{9.46}$$

Making a change of variable, $x = 1 - \theta$, $(0 \le x \le 1)$, $f(x)$ can be rewritten as:

$$f(x) = x \log_2 \frac{(1 - t_d)x + t_d}{x} - x \log_2 \frac{(1 - t_e)x + t_e}{x}. \tag{9.47}$$

Taking the first derivative of $f(x)$ with respect to x yields:

$$f'(x) = \frac{1}{\ln 2} \left[\left(\ln \frac{(1 - t_d)x + t_d}{x} + \frac{-t_d}{(1 - t_d)x + t_d} \right) \right.$$

$$\left. - \left(\ln \frac{(1 - t_e)x + t_e}{x} + \frac{-t_e}{(1 - t_e)x + t_e} \right) \right]. \tag{9.48}$$

Similarly, the second derivative of $f(x)$ with respect to x can be obtained as:

$$f''(x) = \frac{1}{\ln 2} \left[\frac{-t_d^2}{[(1 - t_d)x + t_d]^2 x} - \frac{-t_e^2}{[(1 - t_e)x + t_e]^2 x} \right]. \tag{9.49}$$

Now let $g(t) = \frac{-t^2}{[(1-t)x+t]^2 x}$, the first derivative of $g(t)$ with respect to t is given by:

$$g'(t) = \frac{-2t[(1 - t)x + t] + 2(1 - x)t^2}{[(1 - t)x + t]^3 x}$$

$$= \frac{-2xt}{[(1 - t)x + t]^3 x} < 0, \tag{9.50}$$

which implies that $g(t)$ is a monotonic decreasing function of t. Since the secrecy rate is positive, which is equivalent to the fact that $t_d > t_e$, we have $f''(x) < 0$. Hence, $f(\theta)$ is a concave function with respect to θ. Therefore, the optimal solution of the optimization problem OP7 can be efficiently obtained.

Having obtained the optimal solutions of the two sub-problems, we now summarize the alternating algorithm for joint resource allocation as follows:

Algorithm: Joint Resource Allocation Scheme

---

1. Initiation: Given N, $\mathbf{g}$, $\mathbf{h}_d$, $\mathbf{h}_e$, η, T and $P_{\max}$. Let $\mathbf{H}_d = \mathbf{h}_d^\dagger \mathbf{h}_d$, $\mathbf{H}_e = \mathbf{h}_e^\dagger \mathbf{h}_e$, $n = 1$, $\theta(1) = 1/2$, $\xi(1) = 1$, $\bar{\mathbf{V}}(1)$ is an arbitrary $N \times N$ matrix with $\mathrm{tr}(\bar{\mathbf{V}}(1)) = P_{\max}$, $\delta_{\mathbf{V}}$ and δ_θ are small positive real numbers.
2. Fix $\theta(n)$, and compute $\bar{\mathbf{V}}(n+1)$ and $\xi(n+1)$ by solving the relaxed optimization problem OP5. Then, set $\mathbf{V}(n+1) = \bar{\mathbf{V}}(n+1)/\xi(n+1)$.
3. Fix $\mathbf{V}(n+1)$, and compute the optimal solution $\theta(n+1)$ of the optimization problem OP7.
4. If $\|\mathbf{V}(n+1) - \mathbf{V}(n)\| > \delta_{\mathbf{V}}$ or $|\theta(n+1) - \theta(n)| > \delta_\theta$, let $n = n+1$, and then go to step 2.
5. Compute the optimal P^* and $\mathbf{w}^*$ from $\mathbf{V}(n+1)$ according to (9.40) and (9.41), and let $\theta^* = \theta(n+1)$.

Please note, since the secrecy rate increases after each iteration, the proposed iterative algorithm for joint resource allocation is guaranteed to converge. Yet, it is worth noting that the proposed alternating algorithm is not optimal and may settle at a local optimum point.

9.2.4 Numerical results

We now present several numerical results to examine the effectiveness of the proposed scheme. Unless otherwise specified, the following set of parameters are used: $N_S = 4$, $\sigma_d^2 = \sigma_e^2 = 1$, $P_{\max} = 10$ W, and $\eta = 1$. We use λ_P, λ_d and λ_e to denote the path loss of $\mathbf{g}$, $\mathbf{h}_d$ and $\mathbf{h}_e$. For convenience, we let $\lambda_P = \lambda_d = 1$, and use λ_e to represent the relative path loss. Specifically, $\lambda_e > 1$ implies short-distance interception. All the simulation curves are obtained by averaging over 1,000 independent channel realizations.

We first compare the average secrecy rate of the proposed joint resource allocation scheme with the fixed resource allocation scheme. For the fixed resource allocation scheme, we set $P = P_{\max}$, $\theta = 1/2$, and $\mathbf{w} = \frac{\mathbf{h}_d^\dagger}{\|\mathbf{h}_d\|}$. As illustrated in Figure 9.2, the proposed scheme performs better than the fixed scheme in the whole λ_e region. More importantly, the performance gain becomes larger as λ_e increases. Thus, the proposed joint resource allocation scheme can effectively solve the challenge of short-distance interception.

Then, we investigate the impact of maximum available transmit power at the PB on the average secrecy rates with different numbers of antennas N_S at the information

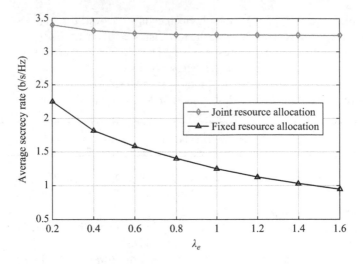

Figure 9.2   Average secrecy rate performance comparison of various resource allocation schemes with different λ_e

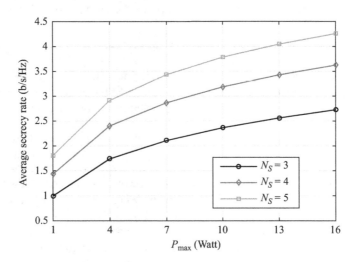

Figure 9.3   Performance comparison of the proposed scheme with different numbers of antennas at the information source

source when $\lambda_e = 1$. As depicted in Figure 9.3, for a given N_S, the secrecy rate increases as $P_{\max}$ becomes larger. Moreover, as N_S increases, the performance of the proposed scheme is improved. This is intuitive since more antennas at the information source can enhance both the energy transfer efficiency and information transmission security.

9.3 Secrecy performance of wirelessly powered wiretap channels with a friendly jammer

In the previous section, we have studied the secrecy outage performance and optimal resource allocation scheme of a wirelessly powered communication system, where the PB is purely used to power the energy constrained information source, i.e. the PB is active during the energy transfer phase and remains silent during the information transfer phase. To fully unleash the potential of the PB, in this section, we propose the idea of using the PB as a friendly jammer during the information transfer phase for secrecy performance enhancement.

9.3.1 System model

We consider a four-node wirelessly powered communication system consisting one PB, one information source Alice, one legitimate user Bob, and one passive eavesdropper Eve, as depicted in Figure 9.4. To further improve the energy transfer efficiency, the PB is equipped with N_J antennas and Alice has N_S antennas, while both Bob and Eve are equipped with a single antenna. Quasi-static Rayleigh fading is assumed, such that the channel conditions do not change during each transmission block but vary independently between different blocks.

Similarly, a two-stage time division communication protocol is adopted. During the first stage, the PB performs wireless power transfer to power Alice. During the second stage, Alice uses the harvested energy to conduct secure communications with Bob, while PB sends artificial noise to confuse the eavesdropper. Assuming an entire

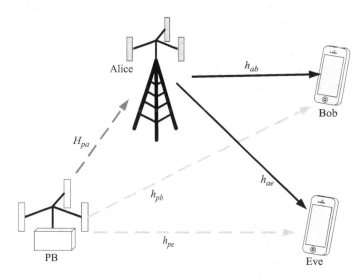

Figure 9.4   A schematic diagram of the system model consisting of one PB, one information source Alice, one legitimate user Bob and one eavesdropper Eve

transmission block of length T, the first portion of length θT, $0 < \theta < 1$, is used for wireless power transfer, while the remaining time period $(1 - \theta)T$ is used for information transmission.

Hence, the received energy signal at Alice can be expressed as:

$$\mathbf{y}_s = \sqrt{\frac{P}{N_J}} \mathbf{H}_{pa} \mathbf{x}_s + \mathbf{n}_s, \tag{9.51}$$

where P is the transmit power at PB, $\mathbf{H}_{pa}$ denotes the power-transfer channel, which is an $N_S \times N_J$ matrix with elements being independently and identically distributed (i.i.d.) zero-mean complex Gaussian random variables with variance λ_1. $\mathbf{x}_s$ is the $N_J \times 1$ energy signal vector satisfying $\mathrm{E}\{\mathbf{x}_s \mathbf{x}_s^\dagger\} = \mathbf{I}$, $\mathbf{n}_s$ is an $N_S \times 1$ vector and denotes the AWGN with $\mathrm{E}\{\mathbf{n}_s \mathbf{n}_s^\dagger\} = N_0 \mathbf{I}$.

Hence, the total harvested energy at the end of the first phase can be computed as:

$$E = \frac{\eta P \|\mathbf{H}_{pa}\|^2 \theta T}{N_J}, \tag{9.52}$$

where $\| \cdot \|$ denotes the Frobenius norm.

Assuming that all the harvested energy is used during the information transmission phase, the transmit power of Alice is given by:

$$P_a = \frac{E}{(1 - \theta)T} = \frac{\eta P \|\mathbf{H}_{pa}\|^2}{N_J} \frac{\theta}{1 - \theta}. \tag{9.53}$$

We assume that Alice is aware of the legitimate channel CSI but does not have the knowledge of the eavesdropper channel CSI. As such, the MRT scheme is used to exploit the available multiple antennas at Alice. As mentioned earlier, the PB acts as a friendly jammer during the information transmission stage. Since PB is also equipped with multiple antennas, different types of jamming schemes can be designed, according to the requirement on the system performance, implementation cost and complexity, and the available CSI at the PB. One particularly interesting scheme is transmit beamforming. As such, the received signal at Bob, y_d, can be written as:

$$y_d = \sqrt{P_a} \mathbf{h}_{ab}^T \mathbf{w}_1 x + \sqrt{P} \mathbf{h}_{pb}^T \mathbf{w}_2 v + n_d, \tag{9.54}$$

where the $N_S \times 1$ vector $\mathbf{h}_{ab}$ denotes the legitimate channel, while the $N_J \times 1$ vector $\mathbf{h}_{pb}$ denotes the jamming channel from the PB to Bob. The elements of $\mathbf{h}_{ab}$ and $\mathbf{h}_{pb}$ are i.i.d. zero-mean complex Gaussian random variables with variance λ_2 and λ_4, respectively. $\mathbf{w}_1 = \frac{\mathbf{h}_{ab}^\dagger}{\|\mathbf{h}_{ab}\|}$ is the $N_S \times 1$ MRT beamforming vector used by Alice, while x denotes the source symbol with unit power. $\mathbf{w}_2$ is the $N_J \times 1$ beamforming vector for the jamming signal with $\|\mathbf{w}_2\|^2 = 1$, while v is the jamming signal with unit power. And n_d is the AWGN at Bob with variance N_0.

Similarly, the received signal at Eve, y_e, can be expressed as:

$$y_e = \sqrt{P_a} \mathbf{h}_{ae}^T \mathbf{w}_1 x + \sqrt{P} \mathbf{h}_{pe}^T \mathbf{w}_2 v + n_e, \tag{9.55}$$

where $\mathbf{h}_{ae}$ denotes the eavesdropper channel, and the $N_J \times 1$ vector $\mathbf{h}_{pe}$ denotes the jamming channel from the PB to Eve. Note that $\mathbf{h}_{ae}$ and $\mathbf{h}_{pe}$ are i.i.d. zero-mean complex Gaussian random variables with variance λ_3 and λ_5, respectively, while n_e is the AWGN at the eavesdropper with variance N_0.

Hence, the end-to-end signal-to-interference-and-noise ratio (SINR) at Bob, γ_b, can be computed as:

$$\gamma_b = \frac{P_a |\mathbf{h}_{ab}^T \mathbf{w}_1|^2}{P |\mathbf{h}_{pb}^T \mathbf{w}_2|^2 + N_0} = \frac{a ||\mathbf{h}_{ab}||^2 ||\mathbf{H}_{pa}||^2}{|\mathbf{h}_{pb}^T \mathbf{w}_2|^2 + b}, \tag{9.56}$$

where $a = \frac{\eta}{N_J} \frac{\theta}{1-\theta}$ and $b = \frac{N_0}{P}$. For analytical tractability, we neglect the noise at Eve. Such an assumption is reasonable since the jamming signal dominates at the eavesdropper in a properly designed system. In addition, this can be regarded as a worst-case pessimistic analysis, an approach that has been commonly adopted in prior works [35]. Therefore, the end-to-end signal-to-interference ratio (SIR) at Eve, γ_e, is given by:

$$\gamma_e = \frac{P_a |\mathbf{h}_{ae}^T \mathbf{w}_1|^2}{P |\mathbf{h}_{pe}^T \mathbf{w}_2|^2} = a \frac{\frac{|\mathbf{h}_{ae}^T \mathbf{h}_{ab}^\dagger|^2}{||\mathbf{h}_{ab}||^2} ||\mathbf{H}_{pa}||^2}{|\mathbf{h}_{pe}^T \mathbf{w}_2|^2}. \tag{9.57}$$

Depending on the available CSI of $\mathbf{h}_{pb}$ and $\mathbf{h}_{pe}$ at the PB, the corresponding optimal beamforming vector $\mathbf{w}_2$ can be designed, which will be discussed in the next section.

9.3.2 Transmit beamforming design

Case 1: Perfect CSI of both $\mathbf{h}_{pb}$ and $\mathbf{h}_{pe}$

With perfect CSI of $\mathbf{h}_{pb}$, it is possible to completely avoid interference at Bob by adopting the zero-forcing (ZF) protocol. Meanwhile, we want to maximize the interference at Eve by exploiting the knowledge of $\mathbf{h}_{pe}$. Hence, the optimal beamforming vector $\mathbf{w}_2$ should be the solution of the following maximization problem:

$$\mathbf{w}_2 = \arg \max_{\mathbf{w}_2} |\mathbf{h}_{pe}^T \mathbf{w}_2|^2$$

$$\text{s.t.} \quad \mathbf{h}_{pb}^T \mathbf{w}_2 = 0 \quad \& \quad ||\mathbf{w}_2|| = 1. \tag{9.58}$$

According to [36], the solution of the above optimization problem can be written as:

$$\mathbf{w}_2 = \frac{\Pi_{\mathbf{h}_{pb}} \mathbf{h}_{pe}^\dagger}{\sqrt{\mathbf{h}_{pe}^T \Pi_{\mathbf{h}_{pb}} \mathbf{h}_{pe}^\dagger}}, \tag{9.59}$$

where the $N_J \times N_J$ matrix $\Pi_{\mathbf{h}_{pb}}$ is given by:

$$\Pi_{\mathbf{h}_{pb}} = \mathbf{I}_N - \mathbf{h}_{pb}^\dagger (\mathbf{h}_{pb}^T \mathbf{h}_{pb}^\dagger)^{-1} \mathbf{h}_{pb}^T, \tag{9.60}$$

which is the orthogonal complement of the column space of $\mathbf{h}_{pb}$.

Case 2: Perfect CSI of $\mathbf{h}_{pb}$, no CSI of $\mathbf{h}_{pe}$

With perfect CSI of $\mathbf{h}_{pb}$, ZF protocol can still be applied such that the jamming signal results in no interference at Bob. However, since there is no CSI of $\mathbf{h}_{pe}$, no further effective actions can be taken to increase the interference level at Eve. Consider the singular value decomposition (SVD) of the matrix $\Pi_{\mathbf{h}_{pb}}$ as follows:

$$\Pi_{\mathbf{h}_{pb}} = \mathbf{U}_{pb}\Delta_{pb}\mathbf{V}_{pb}^*. \tag{9.61}$$

Then, the $N_J - 1$ left singular vectors $\mathbf{u}_i$ associated with $N_J - 1$ non-zero singular values constitute the span of the column space of $\Pi_{\mathbf{h}_{pb}}$. As such, $\mathbf{u}_i$ can be expressed as the linear combination of the column vectors of $\Pi_{\mathbf{h}_{pb}}$, hence,

$$\mathbf{h}_{pb}^T\mathbf{u}_i = 0. \tag{9.62}$$

Therefore, the beamforming vector $\mathbf{w}_2$ can be arbitrarily selected from the $N_J - 1$ left singular vectors $\mathbf{u}_i$.

9.3.3 Performance analysis

In this section, we present a detailed analysis on the achievable secrecy performance of the above-mentioned jamming protocols. We focus on the scenario where Alice transmits at a constant rate R_S to communicate with Bob. In such a scenario, secrecy outage probability is an appropriate metric to characterize the secrecy performance of the system.

Case 1: Perfect CSI of both $\mathbf{h}_{pb}$ and $\mathbf{h}_{pe}$

We start with the perfect CSI scenario, and we have the following key result:

Theorem 9.2. *The secrecy outage probability of the system can be expressed in closed-form as*

$$P_{\text{out}}(R_S) = 1 - \sum_{k=0}^{N_S-1}\sum_{p=0}^{k}\sum_{q=0}^{N_J-1}\frac{2(N_J-1)\lambda_5^{q-k+p-N_J+1}(f\lambda_3)^{k-p}}{\Gamma(N_JN_S)\Gamma(p+1)\Gamma(q+1)\Gamma(N_J-q)}\left(\frac{f-1}{a\lambda_1}\right)^p\left(\frac{b}{\lambda_2}\right)^k$$
$$\times\left(-\frac{bf\lambda_3}{\lambda_2}\right)^{N_J-q-1}e^{\frac{bf\lambda_3}{\lambda_2\lambda_5}}\Gamma\left(q-k+p,\frac{bf\lambda_3}{\lambda_2\lambda_5}\right)\left(\frac{b(f-1)}{a\lambda_1\lambda_2}\right)^{\frac{N_JN_S-p}{2}}K_{N_JN_S-p}\left(2\sqrt{\frac{b(f-1)}{a\lambda_1\lambda_2}}\right), \tag{9.63}$$

where $f = 2^{R_S}$, and $\Gamma(\alpha,x)$ is the upper incomplete gamma function [23, Eq. (8.350.2)].

Proof. We start by expressing the end-to-end SINR (SIR) given in (9.56) and (9.57) as

$$\gamma_b = a\frac{y_{ab}y_{pa}}{y_{pb}+b}, \quad \text{and} \quad \gamma_e = a\frac{y_{ae}y_{pa}}{y_{pe}}, \tag{9.64}$$

where $y_{ab} = ||\mathbf{h}_{ab}||^2$, $y_{pa} = ||\mathbf{H}_{pa}||^2$, $y_{pb} = |\mathbf{h}_{pb}^T \mathbf{w}_2|^2$, $y_{ae} = \frac{|\mathbf{h}_{ae}^T \mathbf{h}_{ab}^\dagger|^2}{||\mathbf{h}_{ab}||^2}$ and $y_{pe} = |\mathbf{h}_{pe}^T \mathbf{w}_2|^2$. It is straightforward to show that y_{pa} and y_{ab} follow the chi-square distribution with $2N_J N_S$ and $2N_S$ degrees of freedom, with pdf given by [25]

$$f_{y_{pa}}(x) = \frac{x^{N_J N_S - 1}}{\lambda_1^{N_J N_S} \Gamma(N_J N_S)} e^{-\frac{x}{\lambda_1}}, \quad \text{and} \quad f_{y_{ab}}(x) = \frac{x^{N_S - 1}}{\lambda_2^{N_S} \Gamma(N_S)} e^{-\frac{x}{\lambda_2}}, \tag{9.65}$$

respectively. In addition, according to [26], y_{ae} follows an exponential distribution with pdf

$$f_{y_{ae}}(x) = \frac{1}{\lambda_3} e^{-\frac{x}{\lambda_3}}. \tag{9.66}$$

For Case 1, $y_{pb} = 0$ and $y_{pe} = |\mathbf{h}_{pe}^T \Pi_{\mathbf{h}_{pb}} \mathbf{h}_{pe}^\dagger|$, and the pdf of y_{pe} can be expressed as [36]

$$f_{y_{pe}}(x) = \frac{x^{N_J - 2}}{\lambda_5^{N_J - 1} \Gamma(N_J - 1)} e^{-\frac{x}{\lambda_5}}. \tag{9.67}$$

As such, the secrecy outage probability can be written as

$$P_{\text{out}}(R_S) = 1 - \text{Prob}\left(\frac{1 + \frac{a}{b} y_{ab} y_{pa}}{1 + a \frac{y_{ae} y_{pa}}{y_{pe}}} \geq f \right). \tag{9.68}$$

Conditioned on y_{pa}, y_{ae} and y_{pe}, utilizing [23, Eq. (3.351.2)] yields

$$P_{\text{out}}(R_S) = 1 - e^{-\frac{b(f-1)}{a\lambda_2 y_{pa}} - \frac{bfy_{ae}}{\lambda_2 y_{pe}}} \sum_{k=0}^{N_S - 1} \frac{1}{k!} \left(\frac{b(f-1)}{a\lambda_2 y_{pa}} + \frac{bfy_{ae}}{\lambda_2 y_{pe}} \right)^k. \tag{9.69}$$

Applying the binomial expansion, (9.69) can be further expressed as

$$P_{\text{out}}(R_S) = 1 - \sum_{k=0}^{N_S - 1} \sum_{p=0}^{k} \frac{f^{k-p}}{p!(k-p)!} \left(\frac{f-1}{a} \right)^p \left(\frac{b}{\lambda_2} \right)^k \frac{e^{-\frac{b(f-1)}{a\lambda_2} \frac{1}{y_{pa}}}}{y_{pa}^p} e^{-\frac{bf}{\lambda_2} \frac{y_{ae}}{y_{pe}}} \left(\frac{y_{ae}}{y_{pe}} \right)^{k-p}. \tag{9.70}$$

Noticing that the random variable y_{pa} is decoupled with y_{pe} and y_{ae}, the expectation can be taken separately. Hence, with the help of [23, Eq. (3.471.9)], we obtain

$$\int_0^\infty \frac{e^{-\frac{b(f-1)}{a\lambda_2} \frac{1}{x}}}{x^p} \frac{x^{N_J N_S - 1}}{\lambda_1^{N_J N_S} \Gamma(N_J N_S)} e^{-\frac{x}{\lambda_1}} dx$$

$$= \frac{2}{\Gamma(N_J N_S)\lambda_1^p} \left(\frac{b(f-1)}{a\lambda_1\lambda_2} \right)^{\frac{N_J N_S - p}{2}} K_{N_J N_S - p}\left(2\sqrt{\frac{b(f-1)}{a\lambda_1\lambda_2}} \right). \tag{9.71}$$

Similarly, invoking [23, Eq. (3.326.2)], we have

$$\int_0^\infty e^{-\frac{bf}{\lambda_2}\frac{x}{y_{pe}}}\left(\frac{x}{y_{pe}}\right)^{k-p}\frac{1}{\lambda_3}e^{-\frac{x}{\lambda_3}}\,dx = \lambda_3^{k-p}\Gamma(k-p+1)\frac{y_{pe}}{\left(y_{pe}+\frac{bf\lambda_3}{\lambda_2}\right)^{k-p+1}}.$$

(9.72)

To this end, making a change of variable $t = x + \frac{bf\lambda_3}{\lambda_2}$ and applying the binomial expansion, we obtain

$$\int_0^\infty \frac{x\lambda_3^{k-p}\Gamma(k-p+1)}{\left(x+\frac{bf\lambda_3}{\lambda_2}\right)^{k-p+1}}\frac{x^{N_J-2}}{\lambda_5^{N_J-1}\Gamma(N_J-1)}e^{-\frac{x}{\lambda_5}}\,dx$$

$$= \frac{\lambda_3^{k-p}\Gamma(k-p+1)e^{\frac{bf\lambda_3}{\lambda_2\lambda_5}}}{\lambda_5^{N_J-1}\Gamma(N_J-1)}\times\sum_{q=0}^{N_J-1}\binom{N_J-1}{q}\left(-\frac{bf\lambda_3}{\lambda_2}\right)^{N_J-q-1}$$

$$\times \int_{\frac{bf\lambda_3}{\lambda_2}}^\infty t^{q-k+p-1}e^{-\frac{t}{\lambda_5}}\,dt.$$

(9.73)

With the help of [23, Eq. (3.381.3)], (9.73) can be further expressed as

$$\frac{\lambda_3^{k-p}\Gamma(k-p+1)e^{\frac{bf\lambda_3}{\lambda_2\lambda_5}}}{\lambda_5^{N_J-1}\Gamma(N_J-1)}\sum_{q=0}^{N_J-1}\binom{N_J-1}{q}\left(-\frac{bf\lambda_3}{\lambda_2}\right)^{N_J-q-1}\int_{\frac{bf\lambda_3}{\lambda_2}}^\infty t^{q-k+p-1}e^{-\frac{t}{\lambda_5}}\,dt$$

$$= \frac{\Gamma(k-p+1)\lambda_3^{k-p}}{\Gamma(N_J-1)\lambda_5^{N_J-1}}\sum_{q=0}^{N_J-1}\binom{N_J-1}{q}\left(-\frac{bf\lambda_3}{\lambda_2}\right)^{N_J-q-1}$$

$$\times \lambda_5^{q-k+p}e^{\frac{bf\lambda_3}{\lambda_2\lambda_5}}\Gamma\left(q-k+p,\frac{bf\lambda_3}{\lambda_2\lambda_5}\right).$$

(9.74)

To this end, pulling everything together yields the desired result.

□

Theorem 2 presents a closed-form expression for the secrecy outage probability of the system, which is valid for arbitrary system configuration and provides an efficient means to evaluate the system's secrecy outage probability. Nevertheless, the expression is too complicated to gain more insightful information. Motivated by this, we now look into the high SNR regime and derive an asymptotic approximation for the secrecy outage probability, which enables the characterization of the achievable diversity order.

For the asymptotic high SNR regime, we assume that $\lambda_2 \to \infty$ while λ_3 remains finite.

Lemma 9.1. *In the high SNR regime, i.e. $\lambda_2 \to \infty$, the secrecy outage probability of the system can be approximated as*
$N_J - 1 > N_S$:

$$P_{\text{out}}^{\infty}(R_S) =$$

$$\sum_{k=0}^{N_S} \frac{b^{N_S}}{k!} \frac{\Gamma(N_J N_S - k)}{\Gamma(N_J N_S)} \frac{\Gamma(N_J - N_S - 1 + k)}{\Gamma(N_J - 1)} \left(\frac{f \lambda_3}{\lambda_5}\right)^{N_S - k} \left(\frac{f - 1}{a \lambda_1}\right)^{k} \times \left(\frac{1}{\lambda_2}\right)^{N_S}.$$

$$(9.75)$$

$N_J - 1 = N_S$:

$$P_{\text{out}}^{\infty}(R_S) =$$

$$\left(\sum_{k=0}^{N_J-3} \binom{N_J - 2}{k} \frac{(-1)^{N_J-k}}{N_J - k - 2} + \sum_{j=1}^{N_J-1} \frac{\Gamma(N_J N_S - j)}{j \Gamma(N_J N_S)} \left(\frac{\lambda_5(f-1)}{af \lambda_1 \lambda_3} \right)^{j} - \mathbf{C} + \ln \frac{\lambda_2 \lambda_5}{bf \lambda_3} \right)$$

$$\times \frac{1}{\Gamma(N_J - 1)} \left(\frac{bf \lambda_3}{\lambda_5} \right)^{N_J - 1} \times \left(\frac{1}{\lambda_2} \right)^{N_J - 1}.$$

$$(9.76)$$

$N_J - 1 < N_S$:

$$P_{\text{out}}^{\infty}(R_S) = \frac{1}{\Gamma(N_J - 1)} \left(\sum_{k=0}^{N_J-2} \binom{N_J - 2}{k} \frac{(-1)^{N_J-k}}{N_S - k - 1} \right) \left(\frac{bf \lambda_3}{\lambda_5} \right)^{N_J - 1} \times \left(\frac{1}{\lambda_2} \right)^{N_J - 1}.$$

$$(9.77)$$

Lemma 9.1 indicates that the system achieves a secrecy diversity order of $\min (N_J - 1, N_S)$, which implies that the maximum diversity order cannot exceed N_S regardless of the number of antennas at PB.

Case 2: Perfect CSI of $\mathbf{h}_{pb}$ and no CSI of $\mathbf{h}_{pe}$

We now move to Case 2, where the PB has perfect CSI of $\mathbf{h}_{pb}$, but has no CSI of $\mathbf{h}_{pe}$.

Theorem 9.3. *The secrecy outage probability of the system can be expressed in closed-form as*

$$P_{\text{out}}(R_S) = 1-$$

$$\sum_{k=0}^{N_S-1} \sum_{p=0}^{k} \frac{2e^{\frac{bf\lambda_3}{\lambda_2\lambda_5}}}{\Gamma(N_J N_S)\Gamma(p+1)} \left(\frac{\lambda_3 f}{\lambda_5}\right)^{k-p} \left(\frac{f-1}{a\lambda_1}\right)^{p} \left(\frac{b}{\lambda_2}\right)^{k} \left(\frac{b(f-1)}{a\lambda_1\lambda_2}\right)^{\frac{N_J N_S - p}{2}}$$

$$K_{N_J N_S - p}\left(2\sqrt{\frac{b(f-1)}{a\lambda_1\lambda_2}}\right)\left(\Gamma\left(1-k+p, \frac{bf\lambda_3}{\lambda_2\lambda_5}\right) - \frac{bf\lambda_3}{\lambda_2\lambda_5}\Gamma\left(-k+p, \frac{bf\lambda_3}{\lambda_2\lambda_5}\right)\right).$$

(9.78)

Proof. The desired result can be obtained by following the similar lines as in Case 1, hence omitted here. □

Since the above expression is also quite complicated, and does not allow for easy extraction of useful insights, we now look into the high SNR regime.

Lemma 9.2. *In the high SNR regime, i.e. $\lambda_2 \to \infty$, the secrecy outage probability of the system can be approximated as*

$$P_{\text{out}}^{\infty}(R_S) = \frac{bf\lambda_3}{\lambda_5(N_S-1)}\frac{1}{\lambda_2}.$$

(9.79)

Lemma 9.2 indicates that only unit diversity order is achieved. Interestingly, we observe that the number of antennas at PB N_J does not affect the asymptotic secrecy outage probability. This can be explained as follows: Recall the transmit beamforming vector adopted at PB is $\mathbf{w}_2 = \mathbf{u}_i$; hence, the effective jamming power at Eve can be expressed as $|\mathbf{h}_{pe}^T \mathbf{u}_i|^2$, which follows the exponential distribution with mean λ_5 regardless of N_J. Also, we observe the intuitive effect of the position of nodes on the secrecy outage probability. For instance, the secrecy outage probability increases when Eve is close to the legitimate user, i.e. large λ_3, and the secrecy outage probability decreases when the PB is close to Eve, i.e. large λ_5.

9.3.4 Numerical results

We now present simulation results to verify the theoretical analysis. Unless otherwise specify, we set the source transmission rate as $R_S = 1$ bit/s/Hz, the energy conversion efficiency as $\eta = 0.8$, the transmit power of the PB to the noise ratio as $\frac{P_S}{N_0} = 10$ dB, while the channel variances are $\lambda_1 = \lambda_4 = \lambda_5 = 1$, and $\lambda_3 = 10$. Also, we set, $\rho = \frac{P_S}{N_0}\lambda_2$, to denote the average SNR of the legitimate channel.

Figure 9.5 plots the secrecy outage probability with different N_J and N_S when $\theta = 0.5$. It can be readily observed that for Case 1, an increase of N_S decreases the secrecy outage probability by improving the secrecy diversity order, and the maximum achievable diversity order is $\min(N_J - 1, N_S)$, as proposed in Lemma 1. However,

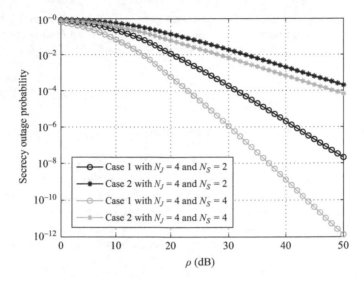

Figure 9.5   Secrecy outage probability of the system with $\theta = 0.5$

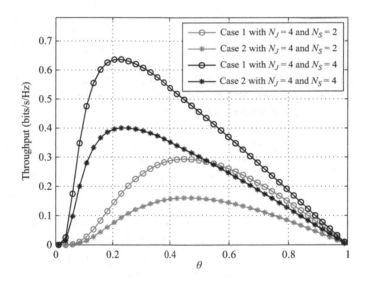

Figure 9.6   Effective throughput versus θ with $\lambda_2 = 0.1$ and $\lambda_3 = 1$

for Case 2, unit diversity order can be achieved, and increasing N_S provides only array gain.

Figure 9.6 depicts the impact of time split parameter θ on the secrecy performance for Cases 1 and 2. Specifically, we adopt the effective secrecy throughput as the performance measure which is given by $R = (1 - P_{\text{out}})R_S(1 - \theta)$. As expected, we observe that Case 1 attains higher effective throughput than Case 2. Also, for both cases, the effective throughput first increases along with θ, and then start to decrease after reaching the maximum point, indicating that there exists a unique optimal time split parameter θ. Therefore, a proper choice of θ is of significant importance to enhance the secrecy performance.

9.4   Conclusion and future directions

We investigated the secrecy performance of the PB-assisted wirelessly powered communication systems, and demonstrated that joint design of time split ratio and transmit beamforming vector is of critical importance to enhance the secrecy performance of the system. In addition, the PB can be further exploited to act as a friendly jammer during the information transmission phase to improve the secrecy performance. However, the potential performance gain depends heavily on the available CSI at the PB.

9.4.1   Future directions

- Most of the results presented in this chapter assume perfect CSI. However, in practice, perfect CSI is difficult to obtain in wirelessly powered communication systems. Therefore, understanding the effect of imperfect CSI on the secrecy performance is an important question to be tackled. Moreover, how to design efficient resource allocation schemes in the presence of imperfect CSI is an interesting research topic of great practical value.
- The massive MIMO technology can generate very high-resolution spatial beamformers, hence, can substantially reduce the information leakage to unintended nodes and significantly improve the energy transfer efficiency. Therefore, it is a promising solution for secrecy wirelessly powered communications to achieve the both goals of energy transfer efficiency and information transmission security with limited energy. However, there are many challenges remain to be solved. For instance, how to obtain accurate CSI at the transmitter with a massive antenna array is an important issue. Also, on the one hand, increasing the antenna number improves the energy harvesting efficiency. On the other, it costs more energy consumption due to additional hardware. Therefore, characterizing the impact of antenna number on the energy efficiency by considering realistic power consumption model is an important issue.
- The current chapter mainly deals with PB-assisted three-node wiretap channels. However, there are many different variants of wiretap channels. In particular, the relaying technique has been introduced to improve the secrecy performance of conventional wiretap channels. With additional relaying nodes, an immediate question is how to properly design the energy beamforming vector such that the harvested energy at different nodes is in fine balance. In addition, how to design efficient jamming strategy should be addressed.

References

[1] X. Chen, D. W. K. Ng, and H-H. Chen, "Secrecy wireless information and power transfer: challenges and opportunities," *IEEE Wireless Commun.*, vol. 23, no. 2, pp. 54–61, Apr. 2016.

[2] H. Xing, L. Liu, and R. Zhang, "Secrecy wireless information and power transfer in fading wiretap channel," *IEEE Trans. Veh. Technol.*, vol. 65, no. 1, pp. 180–190, Jan. 2016.

[3] D. W. K. Ng, E. S. Lo, and R. Schober, "Multiobjective resource allocation for secure communication in cognitive radio networks with wireless information and power transfer," *IEEE Trans. Veh. Technol.*, vol. 65, no. 5, pp. 3166–3184, May 2016.

[4] Z. Ding, C. Zhong, D. W. K. Ng, *et al.*, "Application of smart antenna technologies in simultaneous wireless information and power transfer," *IEEE Commun. Mag.*, vol. 53, no. 4, pp. 86–93, Apr. 2015.

[5] X. Chen, Z. Zhang, H-H. Chen, and H. Zhang, "Enhancing wireless information and power transfer by exploiting multi-antenna techniques," *IEEE Commun. Mag.*, vol. 53, no. 4, pp. 133–141, Apr. 2015.

[6] X. Jiang, C. Zhong, X. Chen, and Z. Zhang, "Secrecy outage probability of wirelessly powered wiretap channels," in *Proc. EUSIPCO 2016*, pp. 1–5, Aug. 2016.

[7] L. Liu, R. Zhang, and K-C. Chua, "Secrecy wireless information and power transfer with MISO beamforming," *IEEE Trans. Signal Process.*, vol. 62, no. 7, pp. 1850–1863, Apr. 2014.

[8] Q. Shi, W. Xu, J. Wu, E. Song, and Y. Wang, "Secure beamforming for MIMO broadcasting with wireless information and power transfer," *IEEE Trans. Wireless Commun.*, vol. 14, no. 5, pp. 2841–2853, May 2015.

[9] X. Chen and H-H. Chen, "Physical layer security in multi-cell MISO downlink with incomplete CSI-a unified secrecy performance loss," *IEEE Trans. Signal Process.*, vol. 62, no. 23, pp. 6286–6297, Dec. 2014.

[10] Y. Zeng and R. Zhang, "Optimized training design for wireless energy transfer," *IEEE Trans. Commun.*, vol. 63, no. 2, pp. 536–550, Feb. 2015.

[11] X. Chen, C. Yuen, and Z. Zhang, "Wireless energy and information transfer tradeoff for limited feedback multi-antenna systems with energy beamforming," *IEEE Trans. Veh. Technol.*, vol. 63, no. 1, pp. 407–412, Jan. 2014.

[12] R. Feng, Q. Li, Q. Zhang, and J. Qin, "Robust secure transmission in MISO simultaneous wireless information and power transfer system," *IEEE Trans. Veh. Technol.*, vol. 64, no. 1, pp. 400–405, Jan. 2015.

[13] S. Wang and B. Wang, "Robust secure transmit design in MIMO channels with simultaneous wireless information and power transfer," *IEEE Signal Process. Lett.*, vol. 22. no. 11, pp. 2147–2151, Nov. 2015.

[14] D. W. K. Ng, E. S. Lo, and R. Schober, "Robust beamforming for secure communication in systems with wireless information and power transfer," *IEEE Trans. Wireless Commun.*, vol. 13, no. 8, pp. 4599–4615, Aug. 2014.

[15] A. El Shafie, D. Niyato and N. Al-Dhahir, "Security of rechargeable energy-harvesting transmitters in wireless networks," *IEEE Wireless Commun. Lett.*, vol. 5, no. 4, pp. 384–387, Aug. 2016.

[16] X. Zhao, J. Xiao, Q. Li, Q. Zhang, and J. Qin, "Joint optimization of AN-Aided transmission and power splitting for MISO secure communications with SWIPT," *IEEE Commun. Lett.*, vol. 19, no. 11, pp. 1969–1972, Nov. 2015.

[17] M. Tian, X. Huang, Q. Zhang, and J. Qin, "Robust AN-aided secure transmission scheme in MISO channels with simultaneous wireless information and power transfer," *IEEE Signal Process. Lett.*, vol. 22, no. 6, pp. 723–726, Jun. 2015.

[18] Q. Zhang, X. Huang, Q. Li, and J. Qin, "Cooperative jamming aided robust secure transmission for wireless information and power transfer in MISO channels," *IEEE Trans. Commun.*, vol. 63, no. 3, pp. 906–915, Mar. 2015.

[19] A. El Shafie, D. Niyato and N. Al-Dhahir, "Artificial-noise-aided secure MIMO full-duplex relay channels with fixed-power transmissions," *IEEE Commun. Lett.*, vol. 20, no. 8, pp. 1591–1594, Aug. 2016.

[20] A. A. Nasir, X. Zhou, S. Durrani, and R. Kennedy, "Relaying protocols for wireless energy harvesting and information processing," *IEEE Trans. Wireless Commun.*, vol. 12, no. 7, pp. 3622–3636, Jul. 2013.

[21] M. Bloch, J. Barros, M. R. D. Rodrigues, and S. W. McLaughlin, "Wireless information-theoretic security," *IEEE Trans. Inf. Theory*, vol. 54, no. 6, pp. 2515–2534, June 2008.

[22] A. Wyner, "The wire-tap channel," *Bell Syst. Tech. J.*, vol. 54, no. 8, pp. 1355–1387, Oct. 1975.

[23] I. S. Gradshteyn and I. M. Ryzhik, *Tables of Integrals, Series and Products*, 6th ed. San Diego: Academic Press, 2000.

[24] A. M. Magableh and M. M. Matalgah, "Capacity of SIMO systems over non-identically independent Nakagami-*m* channels," in *Proc. IEEE Sarroff Symposium*, Nassau Inn, Princeton, NJ, pp. 1–5, Apr. 2007.

[25] M. K. Simon and M. S. Alouini, "Digital Communication over Fading Channels: A Unified Approach to Performance Analysis," Hoboken, NJ: Wiley, 2000.

[26] A. Shah and A. M. Haimovich, "Performance analysis of maximal ratio combining and comparison with optimum combining for mobile radio communications with cochannel interference," *IEEE Trans. Veh. Technol.*, vol. 49, no. 4, pp. 1454–1463, Jul. 2000.

[27] Y. Zou, X. Wang, and W. Shen, "Optimal relay selection for physical-layer security in cooperative wireless networks," *IEEE J. Sel. Areas Commun.*, vol. 31, no. 10, pp. 2099–2111, Oct. 2013.

[28] Y. Huang, F. S. Al-Qahtani, T. Q. Duong, and J. Wang, "Secure transmission in MIMO wiretap channels using general-order transmit antenna selection with outdated CSI," *IEEE Trans. Commun.*, vol. 63, no. 8, pp. 2959–2971, Aug. 2015.

[29] S. Hessien, F. S. Al-Qahtani, R. M. Radaydeh, C. Zhong, and H. Alnuweiri, "On the secrecy enhancement with low-complexity large-scale transmit selection in

MIMO generalized composite fading," *IEEE Wireless Commun. Lett.*, vol. 4, no. 4, pp. 429–432, Aug. 2015.

[30] F. S. Al-Qahtani, C. Zhong, and H. M. Alnuweiri, "Opportunistic relay selection for secrecy enhancement in cooperative networks," *IEEE Trans. Commun.*, vol. 63, no. 5, pp. 1756–1770, May 2015.

[31] L. Wang, N. Yang, M. Elkashlan, P. L. Yeoh, and J. Yuan, "Physical layer security of maximal ratio combining in two-wave with diffuse power fading channels," *IEEE Trans. Inform. Foren. Sec.*, vol. 9, no. 2, pp. 247–258, Feb. 2014.

[32] S. Boyd and L. Vandenberghe, *Convex Optimization*, Cambridge, UK: Cambridge University Press, 2008.

[33] A. Charnes and W. W. Copper, "Programming with linear fractional functionals," *Naval Res. Logist. Quarter.*, vol. 9, pp. 181–186, Dec. 1962.

[34] M. Grant and S. Boyd, CVX: Matlab Software for Disciplined Convex Programming. [Online]: http://cvxr.com/cvx.

[35] W. Liu, X. Zhou, S. Durrani, and P. Popovski, "Secure communication with a wireless powered friendly jammer," *IEEE Trans. Wireless Commun.*, vol. 15, no. 1, pp. 401–415, Jan. 2016.

[36] Z. Ding, K. K. Leung, D. L. Goeckel, and D. Towsley, "On the application of cooperative transmission to secrecy communications," *IEEE J. Sel. Areas Commun.*, vol. 30, no. 2, pp. 359–368, Feb. 2012.

Chapter 10

Physical layer security for D2D-enabled cellular networks

Chuan Ma[1], Jianting Yue[2], Hui Yu[2], and Xiaoying Gan[2]

Device-to-device (D2D) communication, which enables direct communication between two mobile devices that are in proximity, is regarded as a promising technology for the next generation cellular networks. In this chapter, we focus on the physical layer security issues for D2D-enabled cellular networks. In Section 10.1, we introduce the background of D2D communication, and in Section 10.2, we review the state-of-the-art research on physical layer security for D2D-enabled cellular networks. In Sections 10.3 and 10.4, we study how D2D communication can affect the secrecy performance of cellular communication in small-scale networks and large-scale networks, respectively.

10.1 D2D communication in cellular networks

Recently, there has been a rapid increase in the demand for local area services and proximity services (ProSe) among the highly capable mobile devices in cellular networks. Consequently, device-to-device (D2D) communication, which enables direct communication between two mobile devices that are in proximity, has been proposed as a competitive technology component for the next generation cellular networks. The typical scenarios of D2D communication in cellular networks include one-to-one direct communication, one-to-many direct communication, and relaying communication, as shown in Figure 10.1.

The integration of D2D communication into cellular networks holds the promise of many types of advantages [1]:

- *Proximity gain*: The proximity of D2D devices can increase the data rate and decrease the delay and power consumption.
- *Reuse gain*: D2D devices can reuse the radio resources of cellular links, which tightens the reuse factor of the network.

[1] Nokia Bell Labs, China
[2] Department of Electronic Engineering, Shanghai Jiao Tong University, China

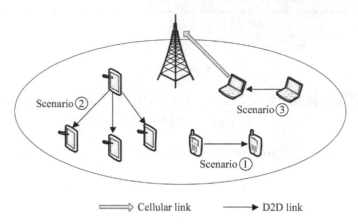

Figure 10.1   D2D communication in cellular networks

- *Hop gain*: The mobile devices use a single link in the D2D mode rather than using both an uplink and a downlink in the cellular mode.
- *Coverage gain*: D2D communication can extend the cellular coverage by relaying the mobile devices in out-of-coverage area to the network.

For these reasons, D2D communication has been strongly appealing to both the academia and industry.

In the academia, the idea of enabling multi-hop direct communication in cellular networks was first proposed in [2], and the concept of D2D communication was formally established in [3,4], in which it was shown that the adoption of D2D communication in cellular networks can improve their overall spectrum efficiency. Extensive research has been undertaken on the design of D2D communication for various applications, for example, multi-cast communication, machine-type communication, vehicle-to-vehicle communication, video storage and delivery, and cellular offloading. In the industry, Qualcomm proposed the implementation of D2D communication in LTE networks and designed a PHY/MAC network architecture known as FlashLinQ [5] that enables proximity-aware communication among mobile devices.

Standards organisations such as 3GPP and IEEE have developed a series of standards to address the need for D2D operation in cellular networks. 3GPP has specified D2D communication and D2D discovery as a support for ProSe in scenarios related to public safety and critical communications [6]. The core features supported by D2D technologies in 3GPP include direct discovery, direct 1:1 communication, and direct 1:many communication. IEEE has also introduced D2D technologies, known as high reliability mobile station direct communication (HR-MS DC), in 802.16n [7].

The introduction of D2D communication to cellular networks has brought a number of technical challenges, such as device discovery, mode selection, intra-cell interference management, physical layer security, multi-access scheme, channel estimation, energy efficiency, and link adaptation. In this chapter, we focus on the aspect of physical layer security for D2D-enabled cellular networks.

10.2 Physical layer security for D2D-enabled cellular networks

Privacy and security are key issues for D2D communication in cellular networks. Different from traditional cellular networks or ad hoc networks where there is only one type of mobile users, i.e. either cellular users or ad hoc users, in D2D-enabled cellular networks there exist both of the aforementioned types of mobile users. Therefore, the physical layer security problems in D2D-enabled cellular networks are quite different from those for traditional networks. In the literature, the physical layer security issues are taken into consideration from different perspectives:

- *Securing the cellular communication in the hybrid network, where third-party nodes eavesdrop the cellular communication [8,9].* In D2D-enabled cellular networks, the D2D users can act as friendly jammers of the cellular users to provide efficient jamming service to the cellular network.
- *Securing the cellular communication in the hybrid network, where the D2D nodes eavesdrop the cellular communication [10–12].* In D2D-enabled cellular networks, the cellular users share their spectrum with the D2D users among which some may be potential eavesdroppers, resulting in unsecured cellular communication. Therefore, effective scheduling schemes are needed to prevent the cellular users from being eavesdropped by the D2D users.
- *Securing the D2D communication in the hybrid network, where third-party nodes eavesdrop the D2D communication [13–17].*
- *Securing both the cellular and D2D communications in the hybrid network, where the third-party nodes eavesdrop both the cellular and D2D communications [18].*
- *Securing the communication between two users through communication mode selection [19].*

10.2.1 Securing cellular communication against third-party eavesdroppers

In [8], the co-operation issue was studied for physical layer security enhancement of D2D-enabled cellular networks. The main idea of this paper is that the D2D links and cellular links can co-operate to improve the security of cellular links and the throughput of D2D links. The authors formulated a co-operation problem among cellular links and D2D links as a coalitional game and proposed a merge-and-split-based coalition formation algorithm that can achieve improved system secrecy rate and social welfare.

In [9], the authors investigated the robust secrecy rate optimisation problems for a secrecy channel with multiple D2D links, where the D2D links help to improve the secrecy rate of cellular links by confusing the eavesdroppers, and the cellular links guarantee required data rates of D2D links by sharing its spectrum. Two robust secrecy rate optimisation problems were studied: the robust power minimisation problem and the robust secrecy rate maximisation problem, both under the cellular secrecy rate and the D2D data rate constraints.

10.2.2   Securing cellular communication against D2D-type eavesdroppers

In [10], the interference link from the cellular user to the D2D receiver was considered the eavesdropping link in the sense that the D2D receiver may try to decode the cellular user's message. The authors considered the secure communication of cellular users acting as the reuse partner for a D2D pair and introduced the concept of secure region for the reuse partner candidates for the D2D pair to meet the secrecy capacity requirement. The D2D link is not allowed to share the spectrum with the cellular user residing outside the secure region. An optimal power allocation algorithm was designed to optimise the secrecy capacity of the cellular users.

In [11], the authors considered the scenario that the D2D receivers are untrusted to the cellular users and required that the D2D transmitters maintain a target secrecy rate of the cellular users. The authors formulated a joint spectrum and power allocation problem to maximise the energy efficiency of the D2D users while guaranteeing the physical layer security of the cellular users and accordingly proposed an optimal resource allocation strategy for solving this problem.

In [12], the authors focused on the co-operative communication where the D2D users serve as relays to assist the two-way transmissions between cellular users while the cellular users want to keep their messages secret from the D2D users. To address the security issue, a security-embedded interference avoidance scheme was proposed. The proposed scheme can create interference-free links for both D2D and cellular communications and also provide an inherent secrecy protection at the physical layer.

10.2.3   Securing D2D communication

In [13], the authors investigated how to select jamming partners for D2D users and allocate transmit power for both source and jammer nodes, in order to thwart eavesdroppers by exploiting the social relationship among D2D users. In [14], the authors proposed a resource allocation scheme that can achieve the maximum secrecy capacity of D2D users while guaranteeing the basic capacity of cellular users. In [15], the authors developed a Stackelberg game framework to model D2D communication in cellular networks and proposed a power control and channel access scheme to maximise the data rate of cellular users and the secrecy rate of D2D users.

In [16], the authors considered the secure D2D communication in energy harvesting large-scale cognitive cellular networks. On the basis of the time switching receiver and the concept of power beacons, the authors proposed three wireless power transfer policies in the power transfer model – co-operative power beacons power transfer, best power beacon power transfer, and nearest power beacon power transfer – and analysed the secrecy outage probability and the secrecy throughput under these policies.

In [17], the authors studied the problem of multi-hop D2D communication in the presence of eavesdroppers. The authors proposed a game-theoretic formulation that enables each D2D user to choose its preferred path to reach the base station, while optimising physical layer security-related utilities. To solve the game, a distributed algorithm that enables the D2D users to engage in pairwise negotiation so as to decide on the graph structure that will inter-connect them was proposed in the paper.

10.2.4 Securing both cellular and D2D communications

In [18], the authors studied the overall secrecy capacity of a D2D-enabled cellular network. The authors formulated the problem of maximising the system secrecy capacity as a matching problem in the weighted bipartite graph and introduced the Kuhn–Munkres algorithm to obtain the optimal solution. The results showed that the system secrecy capacity can be greatly improved by introducing D2D communication underlaying cellular networks.

10.2.5 Physical layer security in different communication modes

In [19], the authors compared the security performance when the mobile users operate in cellular mode and D2D mode. This paper showed that the D2D paradigm can provide significantly improved security at the physical layer, by reducing exposure of the information to eavesdroppers from two relatively high-power transmissions to a single low-power hop, while the cellular paradigm is only seen to have an advantage in certain cases, e.g. when the AP has a large number of antennas and perfect channel state information.

In this chapter, we focus on the first category of physical layer security problem for D2D-enabled cellular networks, i.e. securing cellular communication against the third-party eavesdroppers, and investigate the impact of D2D communication on the secrecy performance of cellular communication. Next, we first study a small-scale network (i.e. point-to-point model) in Section 10.3, and then extend the analysis to a large-scale network via stochastic geometry in Section 10.4.

10.3 Secure transmission schemes for small-scale D2D-enabled cellular networks

10.3.1 System model

Consider a hybrid network consisting of one cellular link, multiple D2D links and one eavesdropper that overhears the transmission of the cellular link, as shown in Figure 10.2. We focus on the downlink scenario for cellular communication, in which the base station (BS) transmits with power P_0. Denote the channel gains of the cellular link (i.e. the link between BS and the cellular user) and the cellular eavesdropping link (i.e. the link between BS and the eavesdropper) by g_1 and g_2, respectively. Then, the secrecy capacity of the cellular link [20] without D2D communication in the network can be expressed as:

$$R_s = \left[\log_2 \left(1 + \frac{P_0 g_1}{\sigma^2} \right) - \log_2 \left(1 + \frac{P_0 g_2}{\sigma^2} \right) \right]^+,$$

where σ^2 is the noise power at the receiver and $[x]^+ = \max\{0, x\}$.

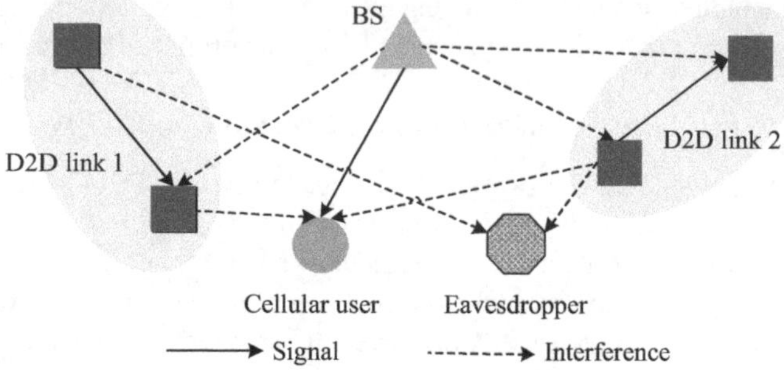

Figure 10.2   A small-scale D2D-enabled cellular network with an eavesdropper

Assume there are N D2D links in the network, and at each time instant, at most one D2D link can reuse the spectrum of the cellular link. The transmission power of D2D link i ($i = 1, 2, \ldots, N$) is denoted by P_i, and the channel gains of D2D link i and the link between BS and the receiver of D2D link i are denoted by $h_{i,0}$ and $g_{i,0}$, respectively. Then, if D2D link i is allowed to underlay the cellular communication, the data rate of the D2D link can be expressed as:

$$R_i = \log_2\left(1 + \frac{P_i h_{i,0}}{P_0 g_{i,0} + \sigma^2}\right), \quad i = 1, 2, \ldots, N,$$

and the secrecy rate of the cellular link can be expressed as:

$$R'_s = \left[\log_2\left(1 + \frac{P_0 g_1}{P_i h_{i,1} + \sigma^2}\right) - \log_2\left(1 + \frac{P_0 g_2}{P_i h_{i,2} + \sigma^2}\right)\right]^+,$$

where $h_{i,1}$ denotes the channel gain of the link between the transmitter of D2D link i and the cellular user, and $h_{i,2}$ denotes the channel gain of the link between the transmitter of D2D link i and the eavesdropper. Let C_s denote the target secrecy rate of the cellular link. Then, the secrecy of the cellular link can be achieved if R'_s is above C_s. Otherwise, secrecy outage occurs in the cellular link, and the secrecy outage probability of the cellular link can be defined as $p_{\text{out}}^{(c)} = \mathbb{P}[R'_s < C_s]$. Assume the following secrecy constraint for the cellular link:

$$p_{\text{out}}^{(c)} \leq \zeta, \tag{10.1}$$

where $\zeta \in [0, 1]$ denotes the minimum secrecy requirement for the cellular link.

10.3.2   Optimal D2D link scheduling scheme

In the hybrid network, the aim of the D2D link scheduling scheme is to maximise D2D rates under the minimum secrecy outage probability constraint for the cellular link. Next, we study the optimal D2D link scheduling scheme, which determines which D2D link is allowed to underlay the cellular communication and how much transmission power it can use. In the following, we first derive the optimal transmission power of each D2D link under the constraint (10.1), and then select the D2D link that can achieve the maximum D2D data rate.

Assume D2D link i is allowed to underlay the cellular communication, then the optimal transmission power problem for D2D link i can be formulated by:

$$\max_{P_i} R_i, \quad \text{s.t.} \quad p_{\text{out}}^{(c)}(P_i) \leq \zeta, \quad 0 \leq P_i \leq P_{i,\max}, \tag{10.2}$$

where $P_{i,\max}$ denote the maximum transmission power of D2D link i. To solve this problem, we first calculate the value of $p_{\text{out}}^{(c)}(P_i)$. Assume g_2 and $h_{i,2}$ follow the exponential distributions with rate parameters α_2 and β_i, respectively, i.e. $g_2 \sim \exp(\alpha_2)$ and $h_{i,2} \sim \exp(\beta_i)$. Then, we have:

$$p_{\text{out}}^{(c)}(P_i) = 1 - \mathbb{P}\left[\log_2\left(1 + \frac{P_0 g_1}{P_i h_{i,1} + \sigma^2}\right) - \log_2\left(1 + \frac{P_0 g_2}{P_i h_{i,2} + \sigma^2}\right) \geq C_s \right]$$

$$= 1 - \mathbb{P}\left[\frac{P_0 g_2}{P_i h_{i,2} + \sigma^2} \leq 2^{-C_s}\left(1 + \frac{P_0 g_1}{P_i h_{i,1} + \sigma^2}\right) - 1 \right]$$

$$= 1 - \frac{P_i \beta_i \left[2^{-C_s}\left(1 + \frac{P_0 g_1}{P_i h_{i,1} + \sigma^2}\right) - 1 \right]}{P_0 \alpha_2 + P_i \beta_i \left[2^{-C_s}\left(1 + \frac{P_0 g_1}{P_i h_{i,1} + \sigma^2}\right) - 1 \right]} \exp\left(\frac{\sigma^2}{P_i \beta_i}\right).$$

It is easy to verify that $p_{\text{out}}^{(c)}(P_i)$ is a U-shape curve on P_i. Then, by solving $p_{\text{out}}^{(c)}(P_i) = \zeta$, we have $P_i = \underline{P_i}, \overline{P_i}$, where $\underline{P_i} \leq \overline{P_i}$. Therefore, $p_{\text{out}}^{(c)}(P_i) \leq \zeta$ yields $\underline{P_i} \leq P_i \leq \overline{P_i}$. Since $\log_2\left(1 + \frac{P_i h_{i,0}}{P_0 g_{i,0} + \sigma^2}\right)$ monotonically increases in P_i, the optimal solution of problem (10.2) is:

$$P_i^* = \min\{P_{i,\max}, \max\{0, \overline{P_i}\}\}.$$

Then, we can select the D2D link with maximum data rate to underlay the cellular communication, i.e.:

$$i^* = \arg\max_i \log_2\left(1 + \frac{P_i^* h_{i,0}}{P_0 g_{i,0} + \sigma^2}\right).$$

A numerical illustration of the network throughput improvement by employing the D2D link scheduling scheme is shown in Figure 10.3. Through the figure, we can see that as the secrecy outage probability requirement becomes larger, the improvement of network throughput increases. This is because when the network has a more stringent requirement for the secrecy of cellular communication, D2D links have more opportunities to communicate underlaying the cellular communication. This result shows that, if properly controlled, D2D communication underlaying cellular communication can improve the network throughput and at the same time guarantee the secrecy cellular communication.

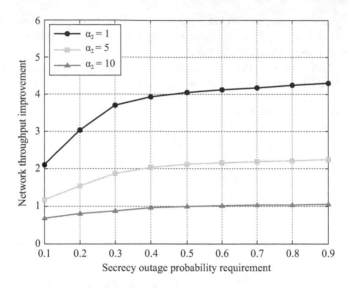

Figure 10.3 *Network throughput improvement by employing the D2D link scheduling scheme*

10.4 Secure transmission schemes for large-scale D2D-enabled cellular networks

10.4.1 Network model

Consider a hybrid network consisting of multiple cellular links, multiple D2D links, and a set of eavesdroppers that overhear the transmission of cellular links, as shown in Figure 10.4. Assume that the BSs are spatially distributed as a homogeneous Poisson point process (PPP) Φ_b of intensity λ_b, and the cellular users are located according to some independent stationary point process Φ_c. We focus on the downlink scenario for cellular communication, in which each cellular user connects with its strongest BS (i.e. the BS that offers the highest received SINR). The eavesdroppers are assumed to be spatially distributed as a homogeneous PPP Φ_e of intensity λ_e. Assume that each cellular link is exposed to all the eavesdroppers, and its secrecy rate is determined by the most detrimental eavesdropper (i.e. the eavesdropper with the highest received SINR of cellular signal). The locations of D2D transmitters are arranged according to a homogeneous PPP Φ_d of intensity λ_d, and for a given D2D transmitter, its associated receiver is assumed to be located at a fixed distance l away with isotropic direction.

Denote the transmission power of BSs and D2D transmitters by P_b and P_d, respectively. The channel model comprises path loss and Rayleigh fading for both cellular and D2D links: given a transmitter x_i with transmission power P, the received power at receiver x_j can be expressed as $Ph\|x_i - x_j\|^{-\alpha}$, where h is the fading factor following an exponential distribution with unit mean, i.e. $h \sim \exp(1)$, and $\alpha > 2$ is the path-loss

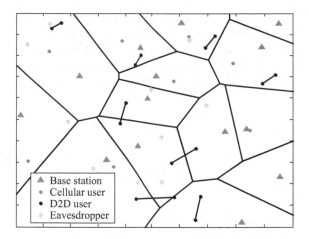

Figure 10.4   A large-scale D2D-enabled cellular network with eavesdroppers

exponent (for brevity of expressions, we use δ to denote $\frac{2}{\alpha}$). The noise power at the receiver is assumed to be additive and constant with value σ^2.

To fight against eavesdropping, we assume that each cellular transmitter (i.e. BS for downlink scenario) adopts Wyner code [21] to encode the data before transmission. Thus, the cellular transmitter needs to determine two kinds of rates – the rate of the transmitted codewords R_c and the rate of the transmitted message R_m ($R_c > R_m$), by considering both connection and secrecy of the cellular link: (*i*) *Connection*. If R_c is above the capacity of the cellular link, the received signal at the cellular receiver can be decoded with an arbitrarily small error, and thus perfect connection of the cellular link can be achieved. Otherwise, connection outage occurs in the cellular link. (ii) *Secrecy*. The rate redundancy $R_c - R_m$ is to provide secrecy. If the rate redundancy is above the capacity of the most detrimental eavesdropping link, the received signal at any eavesdropper provides no information about the transmitted message, and thus perfect secrecy of the cellular link can be achieved. Otherwise, secrecy outage occurs in the cellular link.

Next, we define the perfect transmission and (ϕ, ε)-perfect transmission for cellular links as follows:

- *Perfect transmission*. A cellular transmission is said to be perfect if $\mathsf{SINR}_c > T_\phi$ and $\mathsf{SINR}_e < T_\varepsilon$, where SINR_c, SINR_e denote the received SINRs at the cellular receiver and the most detrimental eavesdropper, respectively, and T_ϕ, T_ε are the corresponding threshold SINR values.
- (ϕ, ε)-*Perfect transmission*. A cellular transmission is said to be (ϕ, ε)-*perfect* if $\mathbb{P}(\mathsf{SINR}_c > T_\phi) \geq \phi$ and $\mathbb{P}(\mathsf{SINR}_e < T_\varepsilon) \geq \varepsilon$, where $0 \leq \phi$, $\varepsilon \leq 1$ denote the required minimum connection probability and minimum secrecy probability, respectively.

From the above definitions, we can see that the perfect transmission implies both perfect connection and perfect secrecy of the cellular link. However, due to time-varying radio environment, perfect transmission cannot be always guaranteed. Thus, in practical networks constraints on connection probability and secrecy probability can be pre-defined to control the network performance.

10.4.2 Secrecy transmission in large-scale D2D-enabled cellular networks

In this sub-section, we analyse the connection probability and secrecy probability of cellular links as well as the connection probability of D2D links in large-scale D2D-enabled cellular network.

10.4.2.1 Connection of cellular links

Without loss of generality, we conduct analysis on a typical cellular user located at the origin. Denote the distance between the BS located at x and the typical cellular user by r_x. Assume the fading factor of the link connecting BS x and the typical cellular user, which is denoted by g_x, is i.i.d. exponential, $g_x \sim \exp(1)$. Then, the received SINR at the typical cellular user from BS x can be expressed as:

$$\text{SINR}_c(x) = \frac{P_b g_x r_x^{-\alpha}}{\sigma^2 + I_c(x)},$$

where

$$I_c(x) = \sum_{x_i \in \Phi_b \backslash \{x\}} P_b g_i \|x_i\|^{-\alpha} + \sum_{y_i \in \Phi_d} P_d h_i \|y_i\|^{-\alpha}$$

is the cumulative interference from all other BSs that are located at x_i with fading factor g_i and D2D transmitters that are located at y_i with fading factor h_i.

A cellular user is connected to the network if its SINR from the strongest BS is above the threshold T_ϕ; otherwise, it is dropped from the network. We refer to the link connecting the typical cellular user and its strongest BS as the typical cellular link. Then, the connection probability of the typical cellular link can be defined as:

$$p_{\text{con}}^{(c)}(T_\phi) \triangleq \mathbb{P}\left[\max_{x \in \Phi_b} \text{SINR}_c(x) > T_\phi\right].$$

The following proposition provides an upper bound on $p_{\text{con}}^{(c)}(T_\phi)$.

Proposition 10.1. *The connection probability of the typical cellular link is upper-bounded by*

$$p_{\text{con}}^{(c)}(T_\phi) \leq 2\pi\lambda_b \int_0^\infty \exp\left(-P_b^{-1} T_\phi r_x^\alpha \sigma^2 - \pi r_x^2 T_\phi^\delta \mu\right) r_x \mathrm{d}r_x,$$

where $\mu = \lambda_b \left[1 + \frac{\lambda_d}{\lambda_b}\left(\frac{P_d}{P_b}\right)^\delta\right] \text{sinc}^{-1} \delta$. *The equality holds when* $T_\phi > 1 (0\ dB)$.

Proof. First, we derive the upper bound. The probability that the strongest SINR is above T_ϕ equals the probability that at least one SINR is above T_ϕ [22]. Therefore,

$$
p_{con}^{(c)}(T_\phi) \overset{\Delta}{=} \mathbb{P}\left[\max_{x\in\Phi_b}\mathsf{SINR}_c(x) > T_\phi\right] = \mathbb{E}\left[\mathbf{1}\left(\bigcup_{x\in\Phi_b}\mathsf{SINR}_c(x) > T_\phi\right)\right]
$$

$$
\overset{(a)}{\leq} \mathbb{E}\left[\sum_{x\in\Phi_b}\mathbf{1}\left(\mathsf{SINR}_c(x) > T_\phi\right)\right] = \mathbb{E}\left[\sum_{x\in\Phi_b}\mathbf{1}\left(\frac{P_b g_x r_x^{-\alpha}}{\sigma^2 + I_c(x)} > T_\phi\right)\right]
$$

$$
\overset{(b)}{=} \lambda_b\int_{\mathbb{R}^2}\mathbb{E}\left[\mathbf{1}\left(\frac{P_b g_x r_x^{-\alpha}}{\sigma^2 + I_c'} > T_\phi\right)\right]dx = \lambda_b\int_{\mathbb{R}^2}\mathbb{P}\left[\frac{P_b g_x r_x^{-\alpha}}{\sigma^2 + I_c'} > T_\phi\right]dx
$$

$$
\overset{(c)}{=} \lambda_b\int_{\mathbb{R}^2}e^{-P_b^{-1}T_\phi r_x^\alpha\sigma^2}\mathbb{E}_{I_c'}\left[e^{-P_b^{-1}T_\phi r_x^\alpha I_c'}\right]dx
$$

$$
\overset{(d)}{=} \lambda_b\int_{\mathbb{R}^2}e^{-P_b^{-1}T_\phi r_x^\alpha\sigma^2}\mathscr{L}_{I_c'}\left(P_b^{-1}T_\phi r_x^\alpha\right)dx
$$

$$
= 2\pi\lambda_b\int_0^\infty e^{-P_b^{-1}T_\phi r_x^\alpha\sigma^2}\mathscr{L}_{I_c'}\left(P_b^{-1}T_\phi r_x^\alpha\right)r_x\,dr_x. \tag{10.3}
$$

In the above equation, (a) follows from the property of union, and the equality holds if at most one BS in the network can provide a SINR above the threshold. In (b), $I_c' = \sum_{x_i\in\Phi_b}P_b g_i\|x_i\|^{-\alpha} + \sum_{y_i\in\Phi_d}P_d h_i\|y_i\|^{-\alpha}$ and the derivation of (b) follows from the Campbell–Mecke Theorem, (c) follows from the Rayleigh distribution assumption of channel fading. In (d), $\mathscr{L}_{I_c'}(\cdot)$ denotes the Laplace transform of I_c'.

Let $I_c' = I_{c-c}' + I_{c-d}'$, where $I_{c-c}' = \sum_{x_i\in\Phi_b}P_b g_i\|x_i\|^{-\alpha}$ and $I_{c-d}' = \sum_{y_i\in\Phi_d}P_d h_i\|y_i\|^{-\alpha}$ denote the interference from cellular links and D2D links, respectively. The Laplace transform of I_{c-c}' is given by:

$$
\mathscr{L}_{I_{c-c}'}(s) = \mathbb{E}_g\left[\mathbb{E}_{\Phi_b}\left[\prod_{x_i\in\Phi_b}\exp\left(-sP_b g_i\|x_i\|^{-\alpha}\right)\right]\right]
$$

$$
\overset{(e)}{=} \mathbb{E}_g\left[\exp\left(-\lambda_b\int_{\mathbb{R}^2}\left(1 - e^{-sP_b g_i\|x_i\|^{-\alpha}}\right)dx_i\right)\right]
$$

$$
\overset{(f)}{=} \mathbb{E}_g\left[\exp\left(-\lambda_b 2\pi\int_{r=0}^\infty\left(1 - e^{-sP_b gr^{-\alpha}}\right)r\,dr\right)\right]
$$

$$
\overset{(g)}{=} \mathbb{E}_g\left[\exp\left(-\lambda_b\pi\left(sP_b g\right)^\delta\Gamma(1-\delta)\right)\right] = \exp\left(-\frac{\pi\lambda_b P_b^\delta s^\delta}{\mathrm{sinc}\,\delta}\right).
$$

In the above equation, (e) follows from the probability generating functional (PGFL) of PPP. (f) follows from the double integral in polar coordinates. In (g),

$\Gamma(x) = \int_0^\infty t^{x-1}e^{-t}\,dt$ is the gamma function and $\delta = \frac{2}{\alpha}$. Similarly, $\mathscr{L}_{I'_{c-d}}(s) = \exp\left(-\frac{\pi \lambda_d P_d^\delta s^\delta}{\text{sinc}\,\delta}\right)$. Therefore, we have:

$$\mathscr{L}_{I'_c}(s) = \mathscr{L}_{I'_{c-c}}(s) \cdot \mathscr{L}_{I'_{c-d}}(s) = \exp\left(-\frac{\pi \lambda_b P_b^\delta s^\delta}{\text{sinc}\,\delta} - \frac{\pi \lambda_d P_d^\delta s^\delta}{\text{sinc}\,\delta}\right),$$

and thus:

$$\mathscr{L}_{I'_c}\left(P_b^{-1} T_\phi r_x^\alpha\right) = \exp\left(-\pi r_x^2 T_\phi^\delta \mu\right), \tag{10.4}$$

where $\mu = \lambda_b\left[1 + \frac{\lambda_d}{\lambda_b}\left(\frac{P_d}{P_b}\right)^\delta\right]\text{sinc}^{-1}\,\delta$. Then by plugging (10.4) into (10.3), we obtain the upper bound.

Next, we show that the equality holds when $T_\phi > 1(0\ dB)$. According to Lemma 1 in [22], at most one BS can provide a SINR above 1 if $T_\phi > 1$. Therefore, the equality in step (a) of (10.3) holds and the upper bound is achieved. $\square$

For tractability, in the following part of this chapter, we assume $T_\phi > 1$. Then, considering interference-limited networks, we have the following corollary.

Corollary 10.1. *In interference-limited D2D-enabled cellular networks, the connection probability of the typical cellular link with $T_\phi > 1$ is*

$$p_{\text{con}}^{(c)}\left(T_\phi\right) = \frac{\text{sinc}\,\delta}{\left[1 + \frac{\lambda_d}{\lambda_b}\left(\frac{P_d}{P_b}\right)^\delta\right] T_\phi^\delta}.$$

It is shown in Corollary 10.1 that $p_{\text{con}}^{(c)}\left(T_\phi\right)$ is inversely correlated with λ_d/λ_b and P_d/P_b. The result is intuitive since given λ_b and P_b, a larger density and transmission power of D2D links introduce more interference to cellular links.

10.4.2.2 Secrecy of cellular links

We consider a typical cellular link comprising a typical cellular user located at the origin and a typical BS located at x_0. For an eavesdropper located at z, its distance to the typical BS is denoted by r_z, and the fading factor of this eavesdropping link is denoted by g_z, $g_z \sim \exp(1)$. Then, the received SINR at eavesdropper z from the typical BS can be expressed as:

$$\text{SINR}_e(z) = \frac{P_b g_z r_z^{-\alpha}}{\sigma^2 + I_e(z)},$$

where

$$I_e(z) = \sum_{x_i \in \Phi_b \backslash \{x_0\}} P_b g_i \|x_i - z\|^{-\alpha} + \sum_{y_i \in \Phi_d} P_d h_i \|y_i - z\|^{-\alpha}$$

is the cumulative interference from all other BSs located at x_i (except the typical BS located at x_0) and D2D transmitters located at y_i.

Assume that each cellular link is exposed to all the eavesdroppers. The cellular transmission is not secure if there exists some eavesdropper z such that $\mathsf{SINR}_e(z)$ is above the threshold T_ε. Therefore, the secure transmission of a cellular link is determined by its most detrimental eavesdropper, and the secrecy probability of the typical cellular link can be defined as:

$$p_{\text{sec}}^{(c)}(T_\varepsilon) \triangleq \mathbb{P}\left[\max_{z\in\Phi_e}\mathsf{SINR}_e(z) < T_\varepsilon\right].$$

The following proposition gives the expression of $p_{\text{sec}}^{(c)}(T_\varepsilon)$.

Proposition 10.2. *The secrecy probability of the typical cellular link is*

$$p_{\text{sec}}^{(c)}(T_\varepsilon) = \exp\left(-2\pi\lambda_e\int_0^\infty e^{-P_b^{-1}T_\varepsilon r_z^\alpha\sigma^2 - \pi r_z^2 T_\varepsilon^\delta\mu}\, r_z\, dr_z\right),$$

where $\mu = \lambda_b\left[1 + \frac{\lambda_d}{\lambda_b}\left(\frac{P_d}{P_b}\right)^\delta\right]\text{sinc}^{-1}\delta.$

Proof. Since the probability that the most detrimental SINR (i.e. the maximum SINR) is below T_ε equals the probability that all the SINRs are below T_ε, the secrecy probability of the typical cellular link can be calculated as

$$p_{\text{sec}}^{(c)}(T_\varepsilon) \triangleq \mathbb{P}\left[\max_{z\in\Phi_e}\mathsf{SINR}_e(z) < T_\varepsilon\right] = \mathbb{P}\left[\bigcap_{z\in\Phi_e}\mathsf{SINR}_e(z) < T_\varepsilon\right]$$

$$= \mathbb{P}\left[\bigcap_{z\in\Phi_e}\mathsf{SINR}_e(z) < T_\varepsilon\right] = \mathbb{E}_{\Phi_e,\Phi_d}\left[1\left(\bigcap_{z\in\Phi_e}\mathsf{SINR}_e(z) < T_\varepsilon\right)\right]$$

$$\overset{(a)}{=} \mathbb{E}_{\Phi_e,\Phi_d}\left[\prod_{z\in\Phi_e}1(\mathsf{SINR}_e(z) < T_\varepsilon)\right] = \mathbb{E}_{\Phi_e}\left[\prod_{z\in\Phi_e}\mathbb{P}(\mathsf{SINR}_e(z) < T_\varepsilon \mid z)\right]$$

$$\overset{(b)}{=} \mathbb{E}_{\Phi_e}\left[\prod_{z\in\Phi_e}\left(1 - e^{-P_b^{-1}T_\varepsilon r_z^\alpha(\sigma^2 + I_e(z))}\right)\right]$$

$$\overset{(c)}{=} \exp\left(-2\pi\lambda_e\int_0^\infty e^{-P_b^{-1}T_\varepsilon r_z^\alpha\sigma^2}\,\mathscr{L}_{I_e(z)}\left(P_b^{-1}T_\varepsilon r_z^\alpha\right) r_z\, dr_z\right). \tag{10.5}$$

In the above equation, (a) follows from the property of intersection, (b) follows from the Rayleigh distribution assumption of channel fading, and (c) follows from the probability generating functional (PGFL) of PPP.

Next we calculate $\mathscr{L}_{I_e(z)}(s)$. By shifting the coordinates so that the eavesdropper z is located at the origin (note that translations do not change the distribution of PPP), $I_e(z)$ can be replaced by I_e, which equals $I_e(z = 0)$. Let $I_e = I_{e-c} + I_{e-d}$, where $I_{e-c} = \sum_{x_i\in\Phi_b\backslash\{x_0\}} P_b g_i\|x_i\|^{-\alpha}$ and $I_{e-d} = \sum_{y_i\in\Phi_d} P_d h_i\|y_i\|^{-\alpha}$ denote the interference from cellular links and D2D links, respectively. Following the proof of Proposition 10.1, we have

$$\mathscr{L}_{I_{d(z)}}\left(P_b^{-1}T_\varepsilon r_z^\alpha\right) = \exp\left(-\pi r_z^2 T_\varepsilon^\delta \mu\right), \tag{10.6}$$

where $\mu = \lambda_b\left[1 + \frac{\lambda_d}{\lambda_b}\left(\frac{P_d}{P_b}\right)^\delta\right]\mathrm{sinc}^{-1}\delta$. Then by plugging (10.6) into (10.5), we complete the proof. □

Considering interference-limited networks, we have the following corollary.

Corollary 10.2. *In interference-limited D2D-enabled cellular networks, the secrecy probability of the typical cellular link is*

$$p_{\mathrm{sec}}^{(c)}(T_\varepsilon) = \exp\left(-\frac{\lambda_e\,\mathrm{sinc}\,\delta}{\lambda_b\left[1 + \frac{\lambda_d}{\lambda_b}\left(\frac{P_d}{P_b}\right)^\delta\right]T_\varepsilon^\delta}\right).$$

It is shown in Corollary 10.2 that $p_{\mathrm{sec}}^{(c)}(T_\varepsilon)$ is negatively correlated with the eavesdropper intensity and is positively correlated with the D2D intensity and power. This is because, a larger population of eavesdroppers reduces the average distance of eavesdropping links, while a larger population of D2D users generate more interference to eavesdropping links.

10.4.2.3 Connection of D2D links

We conduct analysis on a typical D2D link comprising a typical D2D transmitter located at the origin and a typical D2D receiver located at distance l away. Denote the fading factor of the typical D2D link by h_0, $h_0 \sim \exp(1)$. The received SINR at the typical D2D receiver can be expressed as:

$$\mathrm{SINR}_d = \frac{P_d h_0 l^{-\alpha}}{\sigma^2 + I_d},$$

where

$$I_d = \sum_{x_i \in \Phi_b} P_b g_i \|x_i\|^{-\alpha} + \sum_{y_i \in \Phi_d \setminus \{y_0\}} P_d h_i \|y_i\|^{-\alpha}$$

is the cumulative interference from all the base stations located at x_i and other D2D transmitters located at y_i (except the typical D2D transmitter located at y_0).

The connection probability of the typical D2D link can be defined as:

$$p_{\mathrm{con}}^{(d)}(T_\sigma) \triangleq \mathbb{P}[\mathrm{SINR}_d > T_\sigma],$$

where T_σ is the SINR threshold. The following proposition gives the result of $p_{\mathrm{con}}^{(d)}(T_\sigma)$.

Proposition 10.3. *The connection probability of the typical D2D link is*

$$p_{\mathrm{con}}^{(d)}(T_\sigma) = \exp\left(-P_d^{-1}T_\sigma l^\alpha \sigma^2 - \pi l^2 T_\sigma^\delta v\right),$$

where $v = \lambda_d\left[1 + \frac{\lambda_b}{\lambda_d}\left(\frac{P_b}{P_d}\right)^\delta\right]\mathrm{sinc}^{-1}\delta.$

Proof. Given a fixed distance l, the connection probability can be calculated as

$$p_{\text{con}}^{(d)}(T_\sigma) \triangleq \mathbb{P}[\text{SINR}_d > T_\sigma] = e^{-P_d^{-1}T_\sigma l^\alpha \sigma^2} \mathscr{L}_{I_d}\big(P_d^{-1}T_\sigma l^\alpha\big). \tag{10.7}$$

Following the proofs of Propositions 10.1 and 10.2, we have

$$\mathscr{L}_{I_d}\big(P_d^{-1}T_\sigma l^\alpha\big) = \exp\big(-\pi l^2 T_\sigma^\delta \nu\big), \tag{10.8}$$

where $\nu = \lambda_d\left[1 + \frac{\lambda_b}{\lambda_d}\left(\frac{P_b}{P_d}\right)^\delta\right]\text{sinc}^{-1}\delta$. Then by plugging (10.8) into (10.7), we complete the proof. $\qquad\square$

Considering interference-limited networks, we have the following corollary.

Corollary 10.3. *In interference-limited D2D-enabled cellular networks, the connection probability of the typical D2D link is*

$$p_{\text{con}}^{(d)}(T_\sigma) = \exp\left(-\pi l^2 T_\sigma^\delta \lambda_d\left[1 + \frac{\lambda_b}{\lambda_d}\left(\frac{P_b}{P_d}\right)^\delta\right]\text{sinc}^{-1}\delta\right).$$

The results of Proposition 10.3 and Corollary 10.3 can be extended to cases where l is variable: Denote the probability density function of l as $f_l(l)$, then $p_{\text{con}}^{(d)}(T_\sigma)$ can be computed as $\int_0^\infty \mathbb{P}[\text{SINR}_d > T_\sigma \mid l] \cdot f_l(l) \, dl$. For example, l is Rayleigh distributed, $f_l(l) = 2\pi\lambda_d l e^{-\pi\lambda_d l^2}$, then $p_{\text{con}}^{(d)}(T_\sigma) = \lambda_d/\big(T_\sigma^\delta \nu + \lambda_d\big)$.

10.4.2.4 Performance guarantee criteria for cellular transmissions

According to Corollaries 10.2 and 10.3, $p_{\text{sec}}^{(c)}(T_\varepsilon)$ and $p_{\text{con}}^{(d)}(T_\sigma)$ can be enhanced by increasing λ_d and/or P_d. However, according to Corollary 10.1, increasing λ_d and/or P_d would reduce $p_{\text{con}}^{(c)}(T_\phi)$. Therefore, the scheduling parameters of D2D links (λ_d, P_d) should be carefully designed to guarantee the performance of cellular communication and meanwhile achieve a better performance of D2D communication.

To guarantee the performance of cellular communication, we propose the following two criteria.

- *Strong criterion for guaranteeing the cellular communication*: the probability of perfect cellular transmissions should not to be reduced after introducing D2D communication, i.e.:

$$p_{\text{con}}^{(c)}(T_\phi) p_{\text{sec}}^{(c)}(T_\varepsilon) \geq p_{\text{con}}^{(c)(0)}(T_\phi) p_{\text{sec}}^{(c)(0)}(T_\varepsilon), \tag{10.9}$$

where $p_{\text{con}}^{(c)(0)}(T_\phi)$ and $p_{\text{sec}}^{(c)(0)}(T_\varepsilon)$ denote the connection probability and secrecy probability of cellular links in the absence of D2D links, respectively.

- *Weak criterion for guaranteeing the cellular communication*: (ϕ, ε)-perfect cellular transmission should be guaranteed after introducing D2D communication, i.e.:

$$p_{\text{con}}^{(c)}(T_\phi) \geq \phi \quad \text{and} \quad p_{\text{sec}}^{(c)}(T_\varepsilon) \geq \varepsilon, \tag{10.10}$$

where $\phi\varepsilon \leq p_{\text{con}}^{(c)(0)}(T_\phi) p_{\text{sec}}^{(c)(0)}(T_\varepsilon)$.

The strong criterion requires that D2D communication should not degrade the performance of cellular communication, while the weak criterion can tolerate a certain level of performance degradation of cellular communication. In the following parts, we study the D2D link scheduling schemes under these two criteria, respectively.

10.4.3 Optimal D2D link scheduling schemes under the strong criterion

In this sub-section, we design D2D link scheduling schemes (including the intensity and power of D2D links) under the strong criterion. For purposes of mathematical tractability, we consider the interference-limited scenario.

10.4.3.1 Feasible region of D2D link scheduling parameters

Denote the average number of perfect cellular links per unit area by N_c, and thus:

$$N_c = \lambda_b p_{\text{con}}^{(c)}(T_\phi) p_{\text{sec}}^{(c)}(T_\varepsilon).$$

By Corollaries 10.1 and 10.2, we have:

$$N_c = \lambda_b a x \exp(-bx), \tag{10.11}$$

where $a = \frac{\text{sinc}\,\delta}{T_\phi^\delta} > 0$, $b = \frac{\lambda_e \text{sinc}\,\delta}{\lambda_b T_\varepsilon^\delta} > 0$, $x = \frac{1}{1 + \frac{\lambda_d}{\lambda_b}\left(\frac{P_d}{P_b}\right)^\delta} \in (0, 1]$. In the expression of N_c, $\frac{\lambda_d}{\lambda_b}\left(\frac{P_d}{P_b}\right)^\delta$ implies the impact of D2D communication on cellular communication. By letting $x = 1$, we get:

$$N_c^{(0)} = \lambda_b a \exp(-b), \tag{10.12}$$

where the superscript (0) indicates the case without D2D links. Therefore, the strong criterion in (10.9) is equivalent to $N_c \geq N_c^{(0)}$.

The following lemma shows the feasible region of D2D link scheduling parameters under the strong criterion.

Lemma 10.1. *The feasible region of D2D link scheduling parameters corresponding to the strong criterion is*

$$\mathscr{F}_{str} = \left\{ (\lambda_d, P_d) : \begin{cases} \lambda_d P_d^\delta \leq \left(-\frac{b}{W_p(-be^{-b})} - 1 \right) \lambda_b P_b^\delta, & \text{if } b > 1 \\ \lambda_d = P_d = 0, & \text{if } b \leq 1 \end{cases} \right\},$$

where $W_p(\cdot)$ is the real-valued principal branch of Lambert W-function.

Proof. First, we solve $N_c = N_c^{(0)}$. By (10.11) and (10.12), $N_c = N_c^{(0)}$ is equivalent to $e^{b(1-x)} = \frac{1}{x}$, yielding $-bxe^{-bx} = -be^{-b}$. By employing a change of variable $y = -bx$, we have $ye^y = -be^{-b}$, and thereby $y = W_p(-be^{-b}) \bigcup W_m(-be^{-b})$, where $W_p(\cdot)$ and $W_m(\cdot)$ are the real-valued principal branch and the other branch of Lambert W-function, respectively. Note that the solution of y has two branches due to $-be^{-b} \in \left(-\frac{1}{e}, 0\right)$. Then, by making an inverse variable change that $x = -\frac{1}{b}y$, we get $x = -\frac{1}{b}W_p(-be^{-b}) \bigcup -\frac{1}{b}W_m(-be^{-b})$. By examining the two branches of x, we find that $-\frac{1}{b}\mathrm{W_p}(-be^{-b}) \in (0,1]$ and $-\frac{1}{b}\mathrm{W_m}(-be^{-b}) \in [1,\infty)$. Considering $0 < x \le 1$, we reject the second branch and obtain the final solution to $N_c = N_c^{(0)}$: $x_0 = -\frac{1}{b}\mathrm{W_p}(-be^{-b})$.

Next, we evaluate $N_c \ge N_c^{(0)}$, which is equivalent to $xe^{-bx} \ge e^{-b}$. Let $f(x) = xe^{-bx} (0 < x \le 1)$. Then, $xe^{-bx} \ge e^{-b}$ is equivalent to $f(x) \ge f(1)$. Taking the derivative of $f(x)$ with respect to x, we get $f'(x) = (1 - bx) e^{-bx}$. Three cases are considered according to the value of b. (i) $b < 1$: In this case, $f'(x) > 0$ and thereby $f(x)$ monotonically increases in $0 < x \le 1$. Therefore, the solution of $f(x) \ge f(1)$ is $x = 1$, i.e. $\mathscr{F}_{str} = \{(\lambda_d, P_d) : \lambda_d = P_d = 0\}$. (ii) $b = 1$: It is easy to verify that the solution is also $x = 1$. (iii) $b > 1$: In this case, $f'(x) > 0$ when $x \in \left(0, \frac{1}{b}\right)$ and $f'(x) < 0$ when $x \in \left(\frac{1}{b}, 1\right]$. Thus $f(x)$ monotonically increases in $0 < x < \frac{1}{b}$, and monotonically decreases in $\frac{1}{b} < x \le 1$. Since $f(x) = f(1)$ holds at point $x_0 = -\frac{1}{b}\mathrm{W_p}(-be^{-b})$ and $x_1 = -\frac{1}{b}\mathrm{W_m}(-bx_0 e^{-bx_0}) = 1$, where $x_0 < x_1 = 1$, the solution of $f(x) \ge f(1)$ is $x \in [x_0, 1]$, i.e. $\mathscr{F}_{str} = \left\{(\lambda_d, P_d) : \lambda_d P_d^{\delta} \le \left(-\frac{b}{\mathrm{W_p}(-be^{-b})} - 1\right) \lambda_b P_b^{\delta}\right\}$. Combining the results of the three cases, we complete the proof. $\square$

Lemma 10.1 shows that, under the strong criterion, the performance of cellular communication is hampered by D2D links if $\lambda_e \le \frac{T_e^{\delta}}{\mathrm{sinc}\,\delta}\lambda_b$ and is enhanced otherwise. This is because, for D2D communication, its interfering effect on cellular links is critical when the eavesdropper intensity is small, while its jamming effect on eavesdropping links becomes dominant when the intensity is large.

10.4.3.2 D2D link scheduling schemes

Based on the feasible region of D2D link scheduling parameters, we study the D2D link scheduling problems. Since D2D links are blocked by the network when $b \le 1$, we focus only on the case that $b > 1$.

The first problem is how to obtain the maximum average number of perfect cellular links per unit area, i.e.:

$$\max_{(\lambda_d, P_d)} N_c, \quad \text{s.t. } (\lambda_d, P_d) \in \mathscr{F}_{str}. \tag{10.13}$$

Lemma 10.2. *The optimal solution of problem* (10.13) *is*

$$\mathscr{F}'_{str} = \left\{(\lambda_d, P_d) : \lambda_d P_d^{\delta} = (b - 1)\lambda_b P_b^{\delta}\right\}.$$

Proof. $f'(x) = 0$ implies $x = \frac{1}{b}$. Since $f''\left(\frac{1}{b}\right) < 0$, the optimal solution is $x = \frac{1}{b}$. $\square$

Lemma 10.2 shows that there exist a series of (λ_d, P_d) pairs that can achieve maximum N_c. Next, we further study which pair(s) among them can achieve the optimal D2D performance.

Scheme 1 (strong criterion)

Denoting the average number of perfect D2D links per unit area by N_d, we have $N_d = \lambda_d P_{con}^{(d)}(T_\sigma) = \lambda_d \exp\left(-\frac{c}{1-x}\lambda_d\right)$, where $c = \frac{\pi l^2 T_\sigma^{\delta}}{\mathrm{sinc}\,\delta} > 0$. Then, the optimisation problem can be formulated as

$$\text{P1}: \quad \max_{(\lambda_d, P_d)} N_d, \quad \text{s.t. } (\lambda_d, P_d) \in \mathscr{F}'_{str}.$$

Proposition 10.4. *The optimal solution of P1 is*

$$\mathscr{S}_1^* = \left\{ \left(\lambda_d^*, P_d^* \right) : \lambda_d^* = \frac{b-1}{bc}, \quad P_d^* = (bc\lambda_b)^{\frac{1}{\delta}} P_b \right\}.$$

Proof. The constraint of P1 is equivalent to $x = \frac{1}{b}$, and thereby $N_d = \lambda_d \exp\left(-\frac{bc}{b-1}\lambda_d\right)$, which depends only on λ_d. Since N_d monotonically increases in $\lambda_d \in \left(0, \frac{b-1}{bc}\right)$ and monotonically decreases in $\lambda_d \in \left(\frac{b-1}{bc}, \infty\right)$, the maximum value of N_d is obtained at point $\lambda_d^* = \frac{b-1}{bc}$. By Lemma 10.2, we have $P_d^* = (bc\lambda_b)^{\frac{1}{\delta}} P_b$. $\qquad\square$

Scheme 2 (strong criterion)

Relaxing the constraint of P1 to the strong criterion (10.9), we have

$$\textbf{P2}: \quad \max_{(\lambda_d, P_d)} N_d, \quad \text{s.t. } (\lambda_d, P_d) \in \mathscr{F}_{str}.$$

Proposition 10.5. *The optimal solution of P2 is*

$$\mathscr{S}_2^* = \left\{ \left(\lambda_d^*, P_d^* \right) : \lambda_d^* = \frac{b + W_p\left(-be^{-b}\right)}{bc}, P_d^* = \left(-\frac{bc\lambda_b}{W_p\left(-be^{-b}\right)} \right)^{\frac{1}{\delta}} P_b \right\}.$$

Proof. The constraint of P2 is equivalent to $x \in [x_0, 1]$. Then, $N_d = \lambda_d \exp\left(-\frac{c}{1-x}\lambda_d\right) \le \lambda_d \exp\left(-\frac{c}{1-x_0}\lambda_d\right)$, and the equality holds if $x = x_0$. Since the maximum value of the upper bound of N_d is obtained at point $\lambda_d^* = \frac{1-x_0}{c}$, by $x = x_0$, we have $P_d^* = \left(\frac{c\lambda_b}{x_0}\right)^{\frac{1}{\delta}} P_b$. $\qquad\square$

Propositions 10.4 and 10.5 give two D2D link scheduling schemes, $\mathscr{S}_1^*$ and $\mathscr{S}_2^*$, under the strong criterion. For scheme $\mathscr{S}_1^*$,

$$N_c^{(1)} = \frac{\lambda_b a}{b} e^{-1},$$

$$N_d^{(1)} = \frac{b-1}{bc} e^{-1},$$

and for scheme $\mathscr{S}_2^*$,

$$N_c^{(2)} = \lambda_b a e^{-b},$$

$$N_d^{(2)} = \frac{b + W_p\left(-be^{-b}\right)}{bc} e^{-1}.$$

Then, we can get that

$$N_c^{(1)} > N_c^{(2)} = N_c^{(0)},$$

$$N_d^{(1)} < N_d^{(2)}.$$

The results show that schemes $\mathscr{S}_1^*$ can achieve the optimal performance for cellular communication, while $\mathscr{S}_2^*$ can provide a better performance for D2D communication.

10.4.4 Optimal D2D link scheduling schemes under the weak criterion

In this sub-section, we design D2D link scheduling schemes for interference-limited networks under the weak criterion.

10.4.4.1 Feasible region of D2D link scheduling parameters

The weak criterion (10.10) requires the minimum connection probability ϕ and secrecy probability ε for cellular links. The following lemma gives the feasible region of D2D link scheduling parameters under this criterion.

Lemma 10.3. *The feasible region of D2D link scheduling parameters corresponding to the weak criterion is*

$$\mathscr{F}_{\text{weak}} = \left\{ (\lambda_d, P_d) : \left(\frac{b}{\ln \frac{1}{\varepsilon}} - 1 \right) \lambda_b P_b^\delta \leq \lambda_d P_d^\delta \leq \left(\frac{a}{\phi} - 1 \right) \lambda_b P_b^\delta \right\}.$$

Proof. By $p_{\text{con}}^{(c)}(T_\phi) \geq \phi$ and Corollary 10.1, we have $x \geq \frac{1}{a}\phi$, where $x = \frac{1}{1 + \frac{\lambda_d}{\lambda_b}\left(\frac{P_d}{P_b}\right)^\delta}$. By $p_{\text{sec}}^{(c)}(T_\varepsilon) \geq \varepsilon$ and Corollary 10.2, we have $x \leq \frac{1}{b} \ln \frac{1}{\varepsilon}$. Therefore, $\frac{1}{a}\phi \leq x \leq \frac{1}{b} \ln \frac{1}{\varepsilon}$. $\square$

According to Lemmas 10.1 and 10.3, $\mathscr{F}_{str}$ exists only for $b > 1$, while $\mathscr{F}_{\text{weak}}$ exists for any b. This is because, compared to the strong criterion, the weak criterion relaxes the secrecy constraint and hence provides more transmission opportunities for D2D links.

Before investigating the D2D link scheduling problems, we study the values of ϕ and ε. A reasonable range of (ϕ, ε) is

$$\mathscr{R} = \left\{ (\phi, \varepsilon) : 0 \leq \phi \leq a, \ e^{-b} \leq \varepsilon \leq 1, \ \phi\varepsilon \leq ae^{-b}, \ \frac{1}{\phi} \ln \frac{1}{\varepsilon} \geq \frac{b}{a} \right\},$$

in which the first and second conditions correspond to $0 \leq p_{\text{con}}^{(c)}(T_\phi) \leq a$ and $e^{-b} \leq p_{\text{sec}}^{(c)}(T_\varepsilon) \leq 1$, the third condition corresponds to the definition of the weak criterion $\phi\varepsilon \leq p_{\text{con}}^{(c)(0)}(T_\phi) p_{\text{sec}}^{(c)(0)}(T_\varepsilon)$, and the last one corresponds to the implied condition in Lemma 10.3 that $\frac{1}{a}\phi \leq \frac{1}{b} \ln \frac{1}{\varepsilon}$. In the following, we assume $(\phi, \varepsilon) \in \mathscr{R}$.

10.4.4.2 D2D link scheduling schemes

We first study how to obtain the maximum value of N_c, i.e.:

$$\max_{(\lambda_d, P_d)} N_c, \quad \text{s.t.} \quad (\lambda_d, P_d) \in \mathscr{F}_{\text{weak}}. \tag{10.14}$$

Lemma 10.4. *The optimal solution of problem* (10.14) *is*

$$
\mathscr{F}'_{\text{weak}} = \left\{ (\lambda_d, P_d) : \begin{cases} \lambda_d P_d^\delta = \left(\frac{b}{\ln \frac{1}{\varepsilon}} - 1 \right) \lambda_b P_b^\delta, & \text{if } b \le 1 \text{ or } b > 1, \varepsilon > e^{-1} \\ \lambda_d P_d^\delta = \left(\frac{a}{\phi} - 1 \right) \lambda_b P_b^\delta, & \text{if } b > 1, \phi > \frac{a}{b} \\ \lambda_d P_d^\delta = (b - 1) \lambda_b P_b^\delta, & \text{if } b > 1, \varepsilon \le e^{-1}, \phi \le \frac{a}{b} \end{cases} \right\}.
$$

Proof. Consider the following two cases. (i) $b \le 1$: Since N_c monotonically increases in $\frac{1}{a}\phi \le x \le \frac{1}{b} \ln \frac{1}{\varepsilon}$, the solution of (10.14) is $x = \frac{1}{b} \ln \frac{1}{\varepsilon}$. (ii) $b > 1$: If $\frac{1}{b} \ln \frac{1}{\varepsilon} < \frac{1}{b}$, i.e. $\varepsilon > e^{-1}$, then N_c monotonically increases in $\frac{1}{a}\phi \le x \le \frac{1}{b} \ln \frac{1}{\varepsilon}$, and thereby the solution is $x = \frac{1}{b} \ln \frac{1}{\varepsilon}$; if $\frac{1}{a}\phi > \frac{1}{b}$, i.e. $\phi > \frac{a}{b}$, then N_c monotonically decreases in $\frac{1}{a}\phi \le x \le \frac{1}{b} \ln \frac{1}{\varepsilon}$, and thereby the solution is $x = \frac{1}{a}\phi$; if $\frac{1}{a}\phi \le \frac{1}{b} \le \frac{1}{b} \ln \frac{1}{\varepsilon}$, i.e. $\varepsilon \le e^{-1}, \phi \le \frac{a}{b}$, then N_c monotonically increases in $\frac{1}{a}\phi < x < \frac{1}{b}$ and monotonically decreases in $\frac{1}{b} < x \le \frac{1}{b} \ln \frac{1}{\varepsilon}$, and thereby the solution is $x = \frac{1}{b}$. Combining the above results, we complete the proof. $\square$

Lemma 10.4 shows that there exist a series of (λ_d, P_d) pairs that can achieve maximum N_c. Next, we further study which pair(s) among them can achieve maximum N_d.

Scheme 3 (weak criterion)

$$
\mathbf{P3}: \quad \max_{(\lambda_d, P_d)} N_d, \quad \text{s.t. } (\lambda_d, P_d) \in \mathscr{F}'_{\text{weak}}.
$$

Proposition 10.6. *The optimal solution of P3 is*

$$
\mathscr{S}_3^* = \left\{ (\lambda_d^*, P_d^*) : \begin{cases} \lambda_d^* = \frac{b - \ln \frac{1}{\varepsilon}}{bc}, \; P_d^* = \left(\frac{bc}{\ln \frac{1}{\varepsilon}} \lambda_b \right)^{\frac{1}{\delta}} P_b, & \text{if } b \le 1 \text{ or } b > 1, \, \varepsilon > e^{-1} \\ \lambda_d^* = \frac{a - \phi}{ac}, \; P_d^* = \left(\frac{ac}{\phi} \lambda_b \right)^{\frac{1}{\delta}} P_b, & \text{if } b > 1, \, \phi > \frac{a}{b} \\ \lambda_d^* = \frac{b - 1}{bc}, \; P_d^* = (bc\lambda_b)^{\frac{1}{\delta}} P_b, & \text{if } b > 1, \, \varepsilon \le e^{-1}, \, \phi \le \frac{a}{b} \end{cases} \right\}.
$$

Proof. The maximum value of N_d can be obtained at point $\lambda_d^* = \frac{1 - x}{c}$, where x is the solution of problem (10.14) given in Lemma 10.4. $\square$

Scheme 4 (weak criterion)

Relaxing the constraint of P3 to the weak criterion (10.10), we have

$$
\mathbf{P4}: \quad \max_{(\lambda_d, P_d)} N_d, \quad \text{s.t. } (\lambda_d, P_d) \in \mathscr{F}_{\text{weak}}.
$$

Proposition 10.7. *The optimal solution of P4 is*

$$
\mathscr{S}_4^* = \left\{ (\lambda_d^*, P_d^*) : \lambda_d^* = \frac{a - \phi}{ac}, P_d^* = \left(\frac{ac}{\phi} \lambda_b \right)^{\frac{1}{\delta}} P_b \right\}.
$$

Proof. The constraint of P4 yields $x \in \left[\frac{1}{a}\phi, \frac{1}{b}\ln\frac{1}{\varepsilon}\right]$. Then, $N_d = \lambda_d \exp\left(-\frac{c}{1-x}\lambda_d\right) \leq \lambda_d \exp\left(-\frac{c}{1-\frac{1}{a}\phi}\lambda_d\right)$, and the equality holds if $x = \frac{1}{a}\phi$. Therefore, the maximum value

of N_d is obtained at point $\lambda_d^* = \frac{1-\frac{1}{a}\phi}{c}$. By $x = \frac{1}{a}\phi$, we have $P_d^* = \left(\frac{ac}{\phi}\lambda_b\right)^{\frac{1}{\delta}} P_b$. $\square$

Propositions 10.6 and 10.7 give two D2D link scheduling schemes, $\mathscr{S}_3^*$ and $\mathscr{S}_4^*$, under the weak criterion. For scheme $\mathscr{S}_3^*$,

$$
N_c^{(3)} = \begin{cases} \frac{\lambda_b a \varepsilon}{b}\ln\frac{1}{\varepsilon}, & \text{if } b \leq 1 \quad \text{or} \quad b > 1, \varepsilon > e^{-1} \\ \lambda_b \phi \exp\left(-\frac{b}{a}\phi\right), & \text{if } b > 1, \phi > \frac{a}{b} \\ \frac{\lambda_b a}{b}e^{-1}, & \text{if } b > 1, \varepsilon \leq e^{-1}, \phi \leq \frac{a}{b} \end{cases},
$$

$$
N_d^{(3)} = \begin{cases} \frac{b-\ln\frac{1}{\varepsilon}}{bc}e^{-1}, & \text{if } b \leq 1 \quad \text{or} \quad b > 1, \varepsilon > e^{-1} \\ \frac{a-\phi}{ac}e^{-1}, & \text{if } b > 1, \phi > \frac{a}{b} \\ \frac{b-1}{bc}e^{-1}, & \text{if } b > 1, \varepsilon \leq e^{-1}, \phi \leq \frac{a}{b} \end{cases},
$$

and for scheme $\mathscr{S}_4^*$,

$$
N_c^{(4)} = \lambda_b \phi \exp\left(-\frac{b}{a}\phi\right),
$$

$$
N_d^{(4)} = \frac{a-\phi}{ac}e^{-1}.
$$

Then, we can get that

$$
N_c^{(3)} \geq N_c^{(4)},
$$

$$
N_d^{(3)} \leq N_d^{(4)}.
$$

The results show that schemes $\mathscr{S}_3^*$ can achieve the optimal performance for cellular communication, while $\mathscr{S}_4^*$ can provide a better performance for D2D communication.

A summary of the proposed D2D link scheduling schemes is shown in Table 10.1.

Figure 10.5 compares N_c and N_d of the proposed schemes $\mathscr{S}_1^* - \mathscr{S}_4^*$. For scheme $\mathscr{S}_3^*, \mathscr{S}_4^*$, three cases of ϕ, ε are considered. Case 1: $\phi = e^{-b}$, $\varepsilon = 0.5$; Case 2: $\phi = \frac{1.5}{b}$; Case 3: $\phi = \frac{a}{2b}$, $\varepsilon = 0.2$. In Figure 10.5, in case 1, N_c of scheme $\mathscr{S}_3^*, \mathscr{S}_4^*$ are smaller than that of scheme $\mathscr{S}_1^*, \mathscr{S}_2^*$ respectively, while N_d of scheme $\mathscr{S}_3^*, \mathscr{S}_4^*$ are larger than that of scheme $\mathscr{S}_1^*, \mathscr{S}_2^*$ respectively; in cases 2 and 3, N_c, N_d of scheme $\mathscr{S}_3^*$ equal or approximately equal those of scheme $\mathscr{S}_1^*$, while N_c, N_d of scheme $\mathscr{S}_4^*$ are larger than those of scheme $\mathscr{S}_2^*$. This result suggests that by adjusting ϕ and ε, different performance levels of cellular and D2D links can be achieved. Moreover, for all these schemes, N_c is smaller when λ_e is larger, since a larger population of eavesdroppers lowers the probability of secrecy cellular transmissions. However, for

Table 10.1 *A summary of the proposed D2D link scheduling schemes. ($\mathscr{S}_1^*$, $\mathscr{S}_2^*$, $\mathscr{S}_3^*$, and $\mathscr{S}_4^*$ denote the four link scheduling schemes. For each scheme, the link scheduling parameters, λ_d and P_d, and the network performance metrics, N_c and N_d, are shown in the corresponding column.)*

	$\mathscr{S}_1^*$	$\mathscr{S}_2^*$	$\mathscr{S}_3^*$	$\mathscr{S}_4^*$
λ_d	$\dfrac{b-1}{bc}$	$\dfrac{b+\mathrm{W_p}(-be^{-b})}{bc}$	$\begin{cases} \dfrac{b-\ln\frac{1}{\varepsilon}}{bc}, & \text{if case 1} \\[2ex] \dfrac{a-\phi}{ac}, & \text{if case 2} \\[2ex] \dfrac{b-1}{bc}, & \text{if case 3} \end{cases}$	$\dfrac{a-\phi}{ac}$
P_d	$(bc\lambda_b)^{\frac{1}{\delta}}P_b$	$\left(-\dfrac{bc\lambda_b}{\mathrm{W_p}(-be^{-b})}\right)^{\frac{1}{\delta}}P_b$	$\begin{cases} \left(\dfrac{bc}{\ln\frac{1}{\varepsilon}}\lambda_b\right)^{\frac{1}{\delta}}P_b, & \text{case 1} \\[2ex] \left(\dfrac{ac}{\phi}\lambda_b\right)^{\frac{1}{\delta}}P_b, & \text{case 2} \\[2ex] (bc\lambda_b)^{\frac{1}{\delta}}P_b, & \text{case 3} \end{cases}$	$\left(\dfrac{ac}{\phi}\lambda_b\right)^{\frac{1}{\delta}}P_b$
N_c	$\dfrac{\lambda_b a}{b}e^{-1}$	$\lambda_b a e^{-b}$	$\begin{cases} \dfrac{\lambda_b a\varepsilon}{b}\ln\dfrac{1}{\varepsilon}, & \text{case 1} \\[2ex] \lambda_b\phi\exp\left(-\dfrac{b}{a}\phi\right), & \text{case 2} \\[2ex] \dfrac{\lambda_b a}{b}e^{-1}, & \text{case 3} \end{cases}$	$\lambda_b\phi e^{-\frac{b}{a}\phi}$
N_d	$\dfrac{b-1}{bc}e^{-1}$	$\dfrac{b+\mathrm{W_p}(-be^{-b})}{bc}e^{-1}$	$\begin{cases} \dfrac{b-\ln\frac{1}{\varepsilon}}{bc}e^{-1}, & \text{case 1} \\[2ex] \dfrac{a-\phi}{ac}e^{-1}, & \text{case 2} \\[2ex] \dfrac{b-1}{bc}e^{-1}, & \text{case 3} \end{cases}$	$\dfrac{a-\phi}{ac}e^{-1}$

(1) $a=\dfrac{\mathrm{sinc}\,\delta}{T_\phi^\delta}$, $b=\dfrac{\lambda_e\,\mathrm{sinc}\,\delta}{\lambda_b T_\varepsilon^\delta}$, $c=\dfrac{\pi l^2 T_\sigma^\delta}{\mathrm{sinc}\,\delta}$;

(2) case 1: $b\leq 1$ or $b>1, \varepsilon>e^{-1}$, case 2: $b>1, \phi>\dfrac{a}{b}$, case 3: $b>1, \varepsilon\leq e^{-1}, \phi\leq\dfrac{a}{b}$.

all these schemes, N_d is larger when λ_e is larger. This is because when the intensity of eavesdroppers increases, larger transmission power and intensity of D2D links are required to guarantee the performance of cellular transmissions, which creates more transmission opportunities for D2D links. In addition, it is noted that in the simulation b is above 1. For the scenario $b\leq 1$, the strong criterion blocks D2D communication,

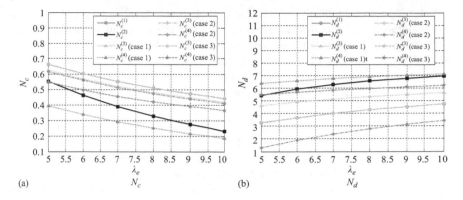

(a)

(b)

Figure 10.5 Performance comparison of schemes $\mathscr{S}_1^* - \mathscr{S}_4^*$

but the weak one admits some D2D links to the network and scheme $\mathscr{S}_3^*, \mathscr{S}_4^*$ are optimal link scheduling schemes. Therefore, scheme $\mathscr{S}_3^*, \mathscr{S}_4^*$ are applicable to a broader range of system parameters, compared with scheme $\mathscr{S}_1^*, \mathscr{S}_2^*$.

10.5 Summary

In this chapter, we investigated the physical layer security issues for D2D-enabled cellular networks. We first introduced the background of D2D communication and reviewed the state-of-the-art researches on physical layer security for D2D-enabled cellular networks. Then, we studied how D2D communication can affect the secrecy performance of cellular communication in both small-scale networks and large-scale networks. The results show that, with properly designed transmission schemes, the interference from D2D communication can be exploited to enhance physical layer security of cellular communication and meanwhile create extra transmission opportunities for D2D users.

References

[1] G. Fodor, E. Dahlman, G. Mildh, *et al.*, "Design aspects of network assisted device-to-device communications," *IEEE Communications Magazine*, vol. 50, no. 3, pp. 170–177, Mar. 2012.

[2] Y.-D. Lin and Y.-C. Hsu, "Multihop cellular: A new architecture for wireless communications," in *Proceedings of IEEE 2000 Annual Joint Conference of the Computer and Communications*, vol. 3, pp. 1273–1282, Mar. 2000.

[3] B. Kaufman and B. Aazhang, "Cellular networks with an overlaid device to device network," in *Proceedings of 2008 Asilomar Conference on Signals, Systems and Computers*, pp. 1537–1541, Oct. 2008.

[4] K. Doppler, M. Rinne, C. Wijting, C. Ribeiro, and K. Hugl, "Device-to-device communication as an underlay to LTE-advanced networks," *IEEE Communications Magazine*, vol. 47, no. 12, pp. 42–49, Dec. 2009.

[5] X. Wu, S. Tavildar, S. Shakkottai, *et al.*, "Flashlinq: A synchronous distributed scheduler for peer-to-peer ad hoc networks," in *Proceedings of 2010 Annual Allerton Conference on Communication, Control, and Computing*, pp. 514–521, Sep. 2010.

[6] 3GPP TR 36.843, "Technical specification group radio access network: Study on LTE device to device proximity services," Rel-12, 2014.

[7] IEEE 802.16n, "IEEE standard for air interface for broadband wireless access systems; Amendment 2: Higher reliability networks," 2013.

[8] R. Zhang, X. Cheng, and L. Yang, "Cooperation via spectrum sharing for physical layer security in device-to-device communications underlaying cellular networks," *IEEE Transactions on Wireless Communications*, vol. 15, no. 8, pp. 5651–5663, Aug. 2016.

[9] Z. Chu, K. Cumanan, M. Xu, and Z. Ding, "Robust secrecy rate optimisations for multiuser multiple-input-single-output channel with device-to-device communications," *IET Communications*, vol. 9, no. 3, pp. 396–403, Feb. 2015.

[10] Y. Wang, Z. Chen, Y. Yao, M. Shen, and B. Xia, "Secure communications of cellular users in device-to-device communication underlaying cellular networks," in *Proceedings of 2014 International Conference on Wireless Communications and Signal Processing (WCSP)*, pp. 1–6, Oct. 2014.

[11] H. Chen, Y. Cai, and D. Wu, "Joint spectrum and power allocation for green D2D communication with physical layer security consideration," *KSII Transactions on Internet and Information Systems*, pp. 1057–1073, Mar. 2015.

[12] L. Sun, Q. Du, P. Ren, and Y. Wang, "Two birds with one stone: Towards secure and interference-free D2D transmissions via constellation rotation," *IEEE Transactions on Vehicular Technology*, vol. 65, no. 10, pp. 8767–8774, Oct. 2016.

[13] L. Wang and H. Wu, "Jamming partner selection for maximising the worst D2D secrecy rate based on social trust," *Transactions on Emerging Telecommunications Technologies*, vol. 28, no. 2, pp. 1–11, Feb. 2017.

[14] J. Wang, C. Li, and J. Wu, "Physical layer security of D2D communications underlaying cellular networks," *Applied Mechanics and Materials. Trans Tech Publications*, pp. 951–954, 2014.

[15] Y. Luo, L. Cui, Y. Yang, and B. Gao, "Power control and channel access for physical-layer security of D2D underlay communication," in *Proceedings of 2015 International Conference on Wireless Communications and Signal Processing (WCSP)*, Oct. 2015, pp. 1–5.

[16] Y. Liu, L. Wang, S. A. R. Zaidi, M. Elkashlan, and T. Q. Duong, "Secure D2D communication in large-scale cognitive cellular networks: A wireless power transfer model," *IEEE Transactions on Communications*, vol. 64, no. 1, pp. 329–342, Jan. 2016.

[17] W. Saad, X. Zhou, B. Maham, T. Basar, and H. V. Poor, "Tree formation with physical layer security considerations in wireless multihop networks," *IEEE*

Transactions on Wireless Communications, vol. 11, no. 11, pp. 3980–3991, Nov. 2012.

[18] H. Zhang, T. Wang, L. Song, and Z. Han, "Radio resource allocation for physical-layer security in D2D underlay communications," in *Proceedings of 2014 IEEE International Conference on Communications (ICC)*, June 2014, pp. 2319–2324.

[19] D. Zhu, A. L. Swindlehurst, S. A. A. Fakoorian, W. Xu, and C. Zhao, "Device-to-device communications: The physical layer security advantage," in *Proceedings of 2014 IEEE International Conference on Acoustics, Speech and Signal Processing (ICASSP)*, May 2014, pp. 1606–1610.

[20] J. Barros and M. R. D. Rodrigues, "Secrecy Capacity of Wireless Channels," in *Proceedings of 2006 IEEE International Symposium on Information Theory (ISIT)*, July 2006, pp. 356–360.

[21] A. D. Wyner, "The wire-tap channel," *Bell System Technical Journal*, vol. 54, no. 8, pp. 1355–1387, Oct. 1975.

[22] H. Dhillon, R. Ganti, F. Baccelli, and J. Andrews, "Modeling and analysis of k-tier downlink heterogeneous cellular networks," *IEEE Journal on Selected Areas in Communications*, vol. 30, no. 3, pp. 550–560, Apr. 2012.

Chapter 11

Physical layer security for cognitive radio networks

Van-Dinh Nguyen[1], Trung Q. Duong[2], and Oh-Soon Shin[1]

11.1 Introduction

Radio frequency spectrum has recently become a scarce and expensive wireless resource due to the ever-increasing demand for multimedia services. Nevertheless, it has been reported in [1] that the majority of licenced users are idle at any given time and location. To significantly improve spectrum utilization, cognitive radio is widely considered a promising approach. There are two famous CR models, namely opportunistic spectrum access model and spectrum sharing model. In the former, the secondary users (SUs, unlicensed users) are allowed to use the frequency bands of the primary users (PUs, licenced users) only when these bands are not occupied [2–4]. In the latter, the PUs have prioritized access to the available radio spectrum, while the SUs have restricted access and need to avoid causing detrimental interference to the primary receivers [5–7]. Towards this end, several powerful techniques, such as spectrum sensing and beamforming design, have been used to protect the PUs from the interference from the SUs and to meet quality-of-service (QoS) requirements [8–10].

Physical layer (PHY) security has been a promising technique to tackle eavesdropping without upper layer data encryption. By exploiting random characteristics of the wireless channels, PHY-security aims to guarantee a positive secrecy rate of the legitimate user [11,12]. To this end, the quality of the legitimate channel is required to be better than the eavesdroppers' [13], yet in practice, this may not always be feasible. Conventionally, the secrecy rate can be improved by the use of multiple antennas at the transmitter for designing beamforming vectors [14–16], which will allow the concentration of the transmit signal towards the intended user's direction while reducing power leakage to the eavesdroppers. Further to this, the transmitter tries to debilitate the desired signal at the eavesdroppers.

Recently, jamming noise (JN), also known as artificial noise, has been proposed to be embedded at the transmitter. Specifically, JN will be transmitted simultaneously with the transmitter's own information signal to degrade the eavesdropping channel.

[1]School of Electronic Engineering and the Department of ICMC Convergence Technology, Soongsil University, Korea
[2]School of Electronics, Electrical Engineering and Computer Science, Queen's University Belfast, UK

Notably, the authors in [17] were the first to put forward the idea that the transmitter spends some of its available power to produce the JN in order to degrade the eavesdropping channels. Depending on how much the transmitter knows about the channel state information (CSI) of the eavesdropper, the use of JN for beamforming design can be categorized into passive and active eavesdropper cases. For the former, a system where the JN is designed to null out the interference to the legitimate user was considered in [18]. The authors in [19–22] then analysed and optimized the secrecy performance of this system by forcing JN beamforming into the null space of the legitimate channel, while fixing the beamforming for information transfer to maximum-ratio-transmission (MRT). For the latter case (i.e. active eavesdroppers), the authors in [23,24] investigated the secrecy rate maximization problem for a multiple-input–single-output (MISO) channel overheard by multiple eavesdroppers and showed that the rank relaxation is tight by applying a semi-definite program (SDP) relaxation. In addition, the imperfect knowledge of the eavesdropper's CSI at the legitimate transmitter has been an emerging topic for various research studies, e.g. in relation to the worst case robust transmit design [25,26] or the outage robust design [27].

On the other hand, cognitive radio networks (CRNs) are faced with serious security threats as a result of the openness of wireless media. However, it is only until recently that PHY-security of CRNs has been well investigated, e.g. in [14,28–31]. In [28], a beamforming vector at a secondary transmitter, which is equipped with multiple antennas, was designed to maximize the secrecy capacity of the secondary system and help improve that of the primary system. The authors of [29] studied the cooperative communication between the secondary and primary systems, aiming at both improving the secrecy capacity of the primary system and satisfying the quality of service (QoS) of the secondary system. Considering that the secondary transmitter does not have perfect CSI of all the channels and merely knows the uncertainty regions that contain the actual channels, the authors in [30] addressed the optimal robust design problem for secure MISO CRNs. In addition, the authors in [31] investigated the primary transmitter with a jamming beamforming that avoids interference to the primary receiver and derived a closed-form expression for the achievable rate. Though presenting in-depth results, most of the prior work in beamforming design for CRNs was based on the following impractical assumptions, particularly when the eavesdroppers are passive [32]: (i) either a single primary receiver or a single eavesdropper and (ii) that the CSI of the eavesdropping channel is perfectly known at the transmitter.

In this chapter, we attempt to provide key design issues and resource allocation problems in PHY-security for CRNs. In particular, we briefly present three popular system models, namely PHY-security for the primary system, secondary system, and cooperative CRNs. We then introduce a resource allocation framework for such networks and provide examples to show the effectiveness of the proposed methods and substantial performance improvements over existing approaches.

11.2 PHY-security of primary system

In this section, we consider a CRN in the presence of an eavesdropper that can overhear the transmissions from both the primary and secondary user, as depicted in

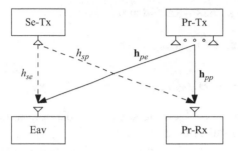

Figure 11.1   A cognitive radio network model with an eavesdropper

Figure 11.1. The primary system is assumed to adopt artificial noise and transmit beamforming as a security provisioning mechanism. We derive the secrecy capacity of the primary system in the presence of a secondary system. In particular, we derive exact closed-form expressions for the ergodic secrecy capacity as well as the probability density function (PDF) of the signal-to-interference-plus-noise ratio (SINR), in two extreme cases where the eavesdropping channel is very strong and very weak. Some numerical results will be presented to verify the analysis.

11.2.1   System model

11.2.1.1   Cognitive radio network

In this model, we assume that the primary transmitter (Pr-Tx) is equipped with M antennas, while the secondary transmitter (Se-Tx), primary receiver (Pr-Rx), and eavesdropper are equipped with a single antenna. The Se-Tx is assumed to employ an energy detector to sense the spectrum and decide on the status of the spectrum. Let $\mathscr{E}$ denote the test statistic for the energy detector. Then, the probability of false alarm for the energy detector is given as [33]:

$$\mathscr{P}_F = \Pr\{\mathscr{E} > \zeta | \mathscr{H}_0\} = \int_\zeta^\infty p_\mathscr{E}(\varepsilon | \mathscr{H}_0)d\varepsilon \tag{11.1}$$

where ζ denotes the decision threshold, $\mathscr{H}_0$ indicates the hypothesis that the primary system is inactive, and $p_\mathscr{E}(\varepsilon | \mathscr{H}_0)$ denotes the conditional PDF of $\mathscr{E}$ under the hypothesis $\mathscr{H}_0$. Similarly, the probability of detection can be computed as:

$$\mathscr{P}_D = \Pr\{\mathscr{E} > \zeta | \mathscr{H}_1\} = \int_\zeta^\infty p_\mathscr{E}(\varepsilon | \mathscr{H}_1)d\varepsilon \tag{11.2}$$

where $\mathscr{H}_1$ indicates the hypothesis that the primary system is active, and $p_\mathscr{E}(\varepsilon | \mathscr{H}_1)$ denotes the conditional PDF of $\mathscr{E}$ under the hypothesis $\mathscr{H}_1$. Let Θ denote the event that the secondary system is present ($\Theta = 1$) or absent ($\Theta = 0$) in the band, under the condition that the primary system is active (hypothesis $\mathscr{H}_1$). Then, Θ follows the Bernoulli distribution as:

$$\Theta = \begin{cases} 0 \text{ with probability } \mathscr{P}_0 = \mathscr{P}_D, \\ 1 \text{ with probability } \mathscr{P}_1 = 1 - \mathscr{P}_0. \end{cases} \tag{11.3}$$

The received signals at the Pr-Rx and at the Eav can be expressed as

$$y_p = \sqrt{\frac{\gamma_{pp}}{M}} \mathbf{h}_{pp} \mathbf{x}_p + \Theta \sqrt{\gamma_{sp}} h_{sp} x_s + n_p, \tag{11.4}$$

$$y_e = \sqrt{\frac{\gamma_{pe}}{M}} \mathbf{h}_{pe} \mathbf{x}_p + \Theta \sqrt{\gamma_{se}} h_{se} x_s + n_e \tag{11.5}$$

where $\mathbf{x}_p \in \mathbb{C}^{M \times 1}$ and $x_s \in \mathbb{C}$ are the signal transmitted from the Pr-Tx and Se-Tx, respectively, and they satisfy the power constraints $\frac{1}{M}\mathbb{E}\{\mathbf{x}_p^H \mathbf{x}_p\} = 1$ and $\mathbb{E}\{x_s^H x_s\} = 1$. $\mathbf{h}_{pp} \in \mathbb{C}^{1 \times M}$ and $\mathbf{h}_{pe} \in \mathbb{C}^{1 \times M}$ are fading channel gains from the Pr-Tx to the Pr-Rx and from the Pr-Tx to the Eav, respectively, such that $\mathbf{h}_{pp} \sim \mathscr{CN}(\mathbf{0}, \mathbf{I}_M)$ and $\mathbf{h}_{pe} \sim \mathscr{CN}(\mathbf{0}, \mathbf{I}_M)$.[1] Similarly, $h_{sp} \in \mathbb{C}$ and $h_{se} \in \mathbb{C}$ are fading channel gains from the Se-Tx to the Pr-Rx and from the Se-Tx to the Eav, respectively, such that $h_{sp} \sim \mathscr{CN}(0, 1)$ and $h_{se} \sim \mathscr{CN}(0, 1)$. $n_p \sim \mathscr{CN}(0, 1)$ and $n_e \sim \mathscr{CN}(0, 1)$ are additive white Gaussian noise (AWGN) at the Pr-Rx and at the Eav, respectively. γ_{pp} and γ_{pe} are the average SNR at the Pr-Rx and at the Eav, respectively, for the signal transmitted by the Pr-Tx. Similarly, γ_{sp} and γ_{se} are the average SNR at the Pr-Rx and at the Eav, respectively, for the signal transmitted by the Se-Tx.

11.2.1.2 Artificial noise

The use of artificial noise for secure communication has been proposed by Goel [17]. We assume that the Pr-Tx exploits artificial noise in combination with beamforming. The transmitter composes $\mathbf{x}_p$ as a weighted sum of information bearing signal $s_p \in \mathbb{C}$ and an artificial noise signal $\mathbf{w}_p \in \mathbb{C}^{(M-1) \times 1}$. Note that the power of s_p and $\mathbf{w}_p$ are normalized such that $\mathbb{E}\{|s_p|^2\} = 1$ and $\mathbf{w}_p \sim \mathscr{CN}(\mathbf{0}, \mathbf{I}_{M-1})$. Accordingly, $\mathbf{x}_p$ can be expressed as

$$\mathbf{x}_p = \sqrt{\phi} \mathbf{u}_p s_p + \sqrt{\frac{1-\phi}{M-1}} \mathbf{W}_p \mathbf{w}_p \tag{11.6}$$

where ϕ denotes the ratio of the power of information bearing signal to the total power $P : P = \sigma_s^2 + (M-1)\sigma_w^2$, where $\sigma_s^2 = \phi P$ and $\sigma_w^2 = (1-\phi)P/(M-1)$ with σ_s^2 and σ_w^2 denoting the signal power and the variance of each component of the artificial noise, respectively. The beamforming vector $\mathbf{u}_p$ in (11.6) is designed to maximize the power of the information bearing signal at the intended destination, such that $\mathbf{u}_p = \mathbf{h}_{pp}^H / \|\mathbf{h}_{pp}\|$, while the nulling matrix $\mathbf{W}_p \in \mathbb{C}^{M \times (M-1)}$ is chosen such that $\mathbf{h}_{pp}$ lies in the left-hand null space of $\mathbf{W}_p$, i.e. $\mathbf{h}_{pp} \mathbf{W}_p = \mathbf{0}$.

With $\mathbf{x}_p$ being defined in (11.6), the received signals in (11.4) and (11.5) can be re-written as

$$y_p = \sqrt{\frac{\phi \gamma_{pp}}{M}} \|\mathbf{h}_{pp}\| s_p + \Theta \sqrt{\gamma_{sp}} h_{sp} x_s + n_p, \tag{11.7}$$

$$y_e = \sqrt{\frac{\phi \gamma_{pe}}{M}} \psi_I s_p + \sqrt{\frac{1-\phi}{M-1} \frac{\gamma_{pe}}{M}} \psi_A \mathbf{w}_p + \Theta \sqrt{\gamma_{se}} h_{se} x_s + n_e \tag{11.8}$$

---

[1] $\mathscr{CN}(\mathbf{0}, \Sigma)$ denotes complex Gaussian distribution with zero mean and covariance matrix Σ.

where $\psi_I \triangleq \mathbf{h}_{pe}\mathbf{u}_p \in \mathbb{C}$ is associated with the information bearing signal, and $\psi_A \triangleq \mathbf{h}_{pe}\mathbf{W}_p \in \mathbb{C}^{1 \times (M-1)}$ is associated with the artificial noise. From (11.7), we can see that x_s, which is transmitted when spectrum sensing fails at the Se-Tx, causes interference to the primary system.

11.2.2 Ergodic secrecy capacity of the primary system

The secrecy capacity of primary system, denoted as C_s, is defined as [34]

$$C_s = \max\{C_p - C_e, 0\} = (C_p - C_e)^+ \tag{11.9}$$

where C_p is the ergodic capacity of the primary system, and C_e is the ergodic capacity of the eavesdropping channel, i.e. the channel between Pr-Tx and Eav. From (11.7), the ergodic capacity of primary system with perfect CSI is given as

$$C_p = \mathbb{E}_\Theta\{\log_2[1 + \mathrm{SINR}_p]\} \tag{11.10}$$

where SINR_p is the SINR at the Pr-Rx for decoding s_p, given as

$$\mathrm{SINR}_p = \frac{\left(\phi\frac{\gamma_{pp}}{M}\right)\|\mathbf{h}_{pp}\|^2}{1 + \gamma_{sp}\Theta\,|h_{sp}|^2}. \tag{11.11}$$

By substituting (11.11) into (11.10) and using (11.3), we obtain the ergodic capacity as

$$C_p = \mathscr{P}_0\log_2\left(1 + \phi\frac{\gamma_{pp}}{M}\|\mathbf{h}_{pp}\|^2\right) + \mathscr{P}_1\log_2\left(1 + \frac{\left(\phi\frac{\gamma_{pp}}{M}\right)\|\mathbf{h}_{pp}\|^2}{1 + \gamma_{sp}\,|h_{sp}|^2}\right). \tag{11.12}$$

On the other hand, from (11.8), the ergodic capacity of the primary-Eav channel with knowledge on the statistics of the eavesdropping channel is given by

$$\begin{aligned}
C_e &= \mathbb{E}_{\Theta, h_{se}, \psi_I, \psi_A}\{\log_2[1 + \mathrm{SINR}_e]\} \\
&= \mathscr{P}_0\mathbb{E}_{h_{se}, \psi_I, \psi_A}\{\log_2[1 + \mathrm{SINR}_{e|\Theta=0}]\} + \mathscr{P}_1\mathbb{E}_{h_{se}, \psi_I, \psi_A}\{\log_2[1 + \mathrm{SINR}_{e|\Theta=1}]\}
\end{aligned} \tag{11.13}$$

where SINR_e is the SINR at the Eav for decoding s_p, given as

$$\mathrm{SINR}_e = \frac{\left(\phi\frac{\gamma_{pe}}{M}\right)|\psi_I|^2}{1 + \gamma_{se}\Theta\,|h_{se}|^2 + \frac{1-\phi}{M-1}\frac{\gamma_{pe}}{M}\|\psi_A\|^2}. \tag{11.14}$$

11.2.2.1 Case of weak eavesdropping channel ($\gamma_{pe} \ll 1$)

Theorem 11.1. *In the case that the Pr-Tx is very far from the Eav ($\gamma_{pe} \ll 1$), the ergodic secrecy capacity of the primary system with perfect instantaneous CSI of the legitimate channel is given as*

$$
C_s = \left(\mathscr{P}_0 \log_2\left(1 + \phi \frac{\gamma_{pp}}{M} \|\mathbf{h}_{pp}\|^2\right) + \mathscr{P}_1 \log_2\left(1 + \frac{(\phi \frac{\gamma_{pp}}{M}) \|\mathbf{h}_{pp}\|^2}{1 + \gamma_{sp} |h_{sp}|^2}\right) \right.
$$
$$
\left. - \frac{\mathscr{P}_0}{\ln 2} e^{\frac{M}{\phi\gamma_{pe}}} E_1\left(\frac{M}{\phi\gamma_{pe}}\right) - \frac{\mathscr{P}_1}{\ln 2} \frac{1}{1-\alpha} \left\{ e^{\frac{M}{\phi\gamma_{pe}}} E_1\left(\frac{M}{\phi\gamma_{pe}}\right) - e^{\frac{1}{\gamma_{se}}} E_1\left(\frac{1}{\gamma_{se}}\right) \right\} \right)^+
$$

$$(11.15)$$

where $\alpha \triangleq \frac{M\gamma_{se}}{\phi\gamma_{pe}}$ and $E_1(u) \triangleq \int_1^\infty e^{-ut} t^{-1} dt$.

Proof. After dividing both the numerator and denominator of (11.14) by γ_{pe}, let γ_{pe} go to 0. Then, we can easily see that the third term in the denominator can be neglected, since $\|\psi_A\|^2$ associated with the artificial noise will be finite. Therefore, when $\gamma_{pe} \ll 1$, SINR_e in (11.14) is approximated as

$$
\text{SINR}_e \approx \mathscr{X} = \frac{(\phi \frac{\gamma_{pe}}{M}) |\psi_I|^2}{1 + \gamma_{se} \Theta |h_{se}|^2}.
$$

$$(11.16)$$

The PDF of $\mathscr{X}$ can be found as $f_{\mathscr{X}}(x) = \mathscr{P}_0 f_{\mathscr{X}}(x|\Theta = 0) + \mathscr{P}_1 f_{\mathscr{X}}(x|\Theta = 1)$ [31], where

$$
f_{\mathscr{X}}(x|\Theta = 0) = \frac{M}{\phi\gamma_{pe}} e^{-\frac{M}{\phi\gamma_{pe}}x},
$$
$$
f_{\mathscr{X}}(x|\Theta = 1) = \frac{M}{\phi\gamma_{pe}} e^{-\frac{M}{\phi\gamma_{pe}}x} \left\{ \left(1 + \frac{M\gamma_{se}}{\phi\gamma_{pe}}x\right)^{-1} + \gamma_{se}\left(1 + \frac{M\gamma_{se}}{\phi\gamma_{pe}}x\right)^{-2} \right\}.
$$

$$(11.17)$$

Using (11.17), C_e in (11.13) can be approximated as

$$
C_e = \mathscr{P}_0 C_{e|\Theta=0} + \mathscr{P}_1 C_{e|\Theta=1}
$$
$$
= \frac{\mathscr{P}_0}{\ln 2} e^{\frac{M}{\phi\gamma_{pe}}} E_1\left(\frac{M}{\phi\gamma_{pe}}\right) + \frac{\mathscr{P}_1}{\ln 2} \frac{1}{1-\alpha} \left\{ e^{\frac{M}{\phi\gamma_{pe}}} E_1\left(\frac{M}{\phi\gamma_{pe}}\right) - e^{\frac{1}{\gamma_{se}}} E_1\left(\frac{1}{\gamma_{se}}\right) \right\}
$$

$$(11.18)$$

where

$$C_{e|\Theta=0} = \int_0^\infty \log_2(1+x) f_{\mathscr{X}}(x|\Theta=0) dx = \frac{1}{\ln 2} e^{\frac{M}{\phi\gamma_{pe}}} E_1\left(\frac{M}{\phi\gamma_{pe}}\right),$$

$$C_{e|\Theta=1} = \int_0^\infty \log_2(1+x) f_{\mathscr{X}}(x|\Theta=1) dx$$

$$= \frac{1}{\ln 2}\frac{1}{1-\alpha}\left\{ e^{\frac{M}{\phi\gamma_{pe}}} E_1\left(\frac{M}{\phi\gamma_{pe}}\right) - e^{\frac{1}{\gamma_{se}}} E_1\left(\frac{1}{\gamma_{se}}\right)\right\}. \tag{11.19}$$

From (11.12) and (11.18), the ergodic secrecy capacity in (11.9) is computed as (11.15). □

Corollary 11.1. *From (11.18) and (11.19), we get*

$$\lim_{\gamma_{se}\to 0} C_e = \mathscr{P}_0 C_{e|\Theta=0} + \mathscr{P}_1 C_{e|\Theta=0} = C_{e|\Theta=0},$$

$$\lim_{\gamma_{se}\to\infty} C_e = \mathscr{P}_0 C_{e|\Theta=0} + \mathscr{P}_0 \cdot 0 = \mathscr{P}_0 C_{e|\Theta=0}. \tag{11.20}$$

According to Corollary 11.1, if γ_{se} changes from 0 to ∞, the C_s can be improved by at most the quantity $\max \Delta C_s = \max \Delta C_e = \mathscr{P}_1 C_{e|\Theta=0}$. Moreover, from (11.19), we can see that $C_{e|\Theta=0} \ll \frac{1}{\ln 2} e^2 E_1(2) \approx 0.521$, since $e^z E_1(z)$ decreases with $z \triangleq M/(\phi\gamma_{pe}) \gg M/\phi \geq 2$. Therefore, the quantity $\max \Delta C_s \ll 0.521 \mathscr{P}_1$ is not significant. In other words, even when the Se-Tx is very close to the Eav and location between Se-Tx and Pr-Rx is almost constant, its effect on the secrecy capacity of the primary system is slight.

Corollary 11.2. *Observing (11.18) and (11.19), we see that $C_{e|\Theta=0}$ and $C_{e|\Theta=1}$ increase with ϕ, as $e^{\frac{M}{\phi\gamma_{pe}}} E_1\left(\frac{M}{\phi\gamma_{pe}}\right)$ increases with ϕ.*

Corollary 11.2 implies that in low γ_{pe} regime, employing artificial noise is not effective to protect the primary user from eavesdropping. However, this will not be the case when there are many eavesdroppers, as discussed in [35], or when there are amplifying relays between the primary user and eavesdropper, as discussed in [17].

11.2.2.2   Case of strong eavesdropping channel ($\gamma_{pe} \gg 1$)

Theorem 11.2. *Assuming that the Pr-Tx is very close to the Eav ($\gamma_{pe} \gg 1$), the ergodic secrecy capacity of the primary system with perfect instantaneous CSI of the legitimate channel is given as*

$$C_s = \left(\mathscr{P}_0 \log_2\left(1 + \phi\frac{\gamma_{pp}}{M}\|\mathbf{h}_{pp}\|^2\right) + \mathscr{P}_1 \log_2\left(1 + \frac{\phi\frac{\gamma_{pp}}{M}\|\mathbf{h}_{pp}\|^2}{1+\gamma_{sp}|h_{sp}|^2}\right) \right.$$

$$\left. - \frac{\mathscr{P}_0}{\ln 2}\frac{1-\phi}{\phi}\mathscr{I}_M(\eta) - \frac{\mathscr{P}_1}{\ln 2}\left\{ \frac{A\ln\alpha}{\alpha(\alpha-1)} + \sum_{k=2}^M B_k \mathscr{I}_k(\eta) \right\} \right)^+ \tag{11.21}$$

where

$$\eta \triangleq \frac{1}{M-1}\frac{1-\phi}{\phi},$$

$$A \triangleq (\alpha - \eta)\left(1 - \frac{\eta}{\alpha}\right)^{-M}, \quad B_k \triangleq (1-k)(\alpha-\eta)\left(1 - \frac{\alpha}{\eta}\right)^{-2}\left(1 - \frac{\eta}{\alpha}\right)^{k-M},$$

$$\mathscr{I}_k(\eta) \triangleq \begin{cases} \dfrac{1}{(k-1)^2}, & \text{if } \eta = 1, \\ \dfrac{(1-\eta)^{1-k}}{(k-1)\eta}\left[-\ln \eta + \displaystyle\sum_{i=1}^{k-2}\frac{(k-2)!\,(-\eta)^i}{(k-i-2)!i!i}(\eta^{-i}-1)\right], & \text{otherwise.} \end{cases}$$

(11.22)

Proof. After dividing both the numerator and denominator of (11.14) by γ_{pe}, let γ_{pe} go to ∞. Then, we can easily see that the first term in the denominator can be neglected. Therefore, when $\gamma_{pe} \gg 1$, the SINR$_e$ in (11.14) can be approximated as

$$\text{SINR}_e \approx \mathscr{Y} = \frac{\phi\frac{\gamma_{pe}}{M}|\psi_I|^2}{\gamma_{se}\Theta|h_{se}|^2 + \frac{1-\phi}{M-1}\frac{\gamma_{pe}}{M}\|\psi_A\|^2}.$$

(11.23)

The PDF of $\mathscr{Y}$ can be expressed as $f_{\mathscr{Y}}(y) = \mathscr{P}_0 f_{\mathscr{Y}}(y|\Theta = 0) + \mathscr{P}_1 f_{\mathscr{Y}}(y|\Theta = 1)$ [31], where

$$f_{\mathscr{Y}}(y|\Theta = 0) = \frac{1-\phi}{\phi\left(1 + \frac{1-\phi}{\phi(M-1)}y\right)^M},$$

$$f_{\mathscr{Y}}(y|\Theta = 1) = \frac{\alpha + (M-1)\eta + M\alpha\eta y}{(1+\alpha y)^2(1+\eta y)^M}.$$

(11.24)

Using (11.24), C_e in (11.13) can be expressed as

$$C_e = \mathscr{P}_0 C_{e|\Theta=0} + \mathscr{P}_1 C_{e|\Theta=1}$$

$$= \frac{\mathscr{P}_0}{\ln 2}\frac{1-\phi}{\phi}\mathscr{I}_M(\eta) + \frac{\mathscr{P}_1}{\ln 2}\left(\frac{A\ln\alpha}{\alpha(\alpha-1)} + \sum_{k=2}^M B_k\mathscr{I}_k(\eta)\right)$$

(11.25)

where

$$C_{e|\Theta=0} = \int_0^\infty \log_2(1+y)f_{\mathscr{Y}}(y|\Theta = 0)dy = \frac{1}{\ln 2}\frac{1-\phi}{\phi}\mathscr{I}_M(\eta),$$

$$C_{e|\Theta=1} = \int_0^\infty \log_2(1+y)f_{\mathscr{Y}}(y|\Theta = 1)dy = \frac{1}{\ln 2}\left(\frac{A\ln\alpha}{\alpha(\alpha-1)} + \sum_{k=2}^M B_k\mathscr{I}_k(\eta)\right).$$

(11.26)

By substituting (11.12) and (11.25) into (11.9), we arrive at the desired result (11.21).

$\square$

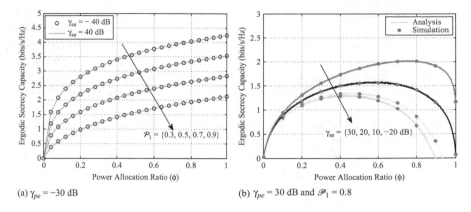

Figure 11.2   Ergodic secrecy capacity versus ϕ. ©IEEE 2015. Reprinted with permission from [31]

Corollary 11.3. *For $\gamma_{se} \ll \gamma_{pe}$, the term associated to γ_{se} in the denominator of (11.14) can be eliminated, which is equivalent to setting $\Theta = 0$.*

11.2.3   Numerical results

We now present numerical results to verify the analysis given in Section 11.2.2, and discuss the results in some specific scenarios of CRNs. In all the following figures, we set the number of antennas at the Pr-Tx to $M = 3$, unless otherwise specified. The average SNR γ_{pp} is set to 20 dB, and the average SNR γ_{sp} is set to 15 dB.

Figure 11.2(a) shows the ergodic secrecy capacity in (11.15) versus the ratio of the power ϕ for several values of $\mathscr{P}_1$, in the case that the Pr-Tx is very far from the Eav. We see that the ergodic secrecy capacity does not change as γ_{se} changes from -40 to 40 dB, which supports Corollary 11.1. It is also seen that the secrecy capacity has the largest value when $\phi = 1$, which corresponds to the case where artificial noise is not injected. This implies that the use of artificial noise is not effective when the Pr-Tx is very far from the Eav. Figure 11.2(b) shows the ergodic secrecy capacity in (11.21) versus the ratio of the power ϕ, in the case that the Pr-Tx is very close to the Eav. The value of γ_{pe} is set to 30 dB. The results show that in high γ_{pe} regime, the secrecy capacity changes significantly with γ_{se}. We can also find the optimal value of the power allocation ratio ϕ that maximizes the secrecy capacity. For instance, the optimal value of the power allocation ratio ϕ are 0.42, 0.45, 0.6, and 0.8 for $\gamma_{se} = -20$, 10, 20, and 30 dB, respectively.

11.3   PHY-security of secondary system

In this section, we consider the secure beamforming design for an underlay cognitive radio MISO broadcast channel in the presence of multiple passive eavesdroppers.

Our goal is to design a JN transmit strategy to maximize the secrecy rate of the secondary system. By utilizing the zero-forcing method to eliminate the interference caused by JN to the secondary user, we study the joint optimization of the information and JN beamforming for secrecy rate maximization of the secondary system while satisfying all the interference power constraints at the primary users as well as the per-antenna power constraint at the secondary transmitter. For an optimal beamforming design, the original problem is a non-convex program, which can be reformulated as a convex program by applying the rank relaxation method. To this end, we prove that the rank relaxation is tight and propose a barrier interior-point method to solve the resulting saddle point problem based on a duality result. To find the global optimal solution, we transform the considered problem into an unconstrained optimization problem. We then employ Broyden–Fletcher–Goldfarb–Shanno (BFGS) method to solve the resulting unconstrained problem which helps reduce the complexity significantly compared to conventional methods. Simulation results show the fast convergence of the proposed algorithm and substantial performance improvements over existing approaches.

11.3.1 System model and problem formulation

11.3.1.1 Signal model

We consider the PHY-security of a CRN consisting of one secondary transmitter (ST), one secondary receiver (SR), M primary users (PUs), and K eavesdroppers (Eves), as illustrated in Figure 11.3. The ST is equipped with N antennas, while the other

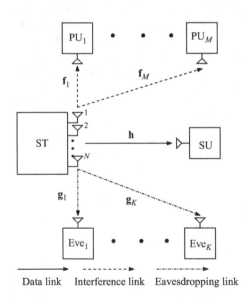

Figure 11.3   A CRN model with multiple eavesdroppers. © IEEE 2016. Reprinted with permission from [32]

nodes are equipped with a single antenna. In the secondary system, Eves intend to wiretap and decode confidential messages from the ST. All the channels are assumed to remain constant during a transmission block and change independently from one block to another.

The main objective of the design is to maximize the secrecy rate of the SU while satisfying a given interference power constraint I_m at the mth PU, for $m = 1, 2, \ldots, M$. We aim to design two beamforming vectors $\mathbf{w}$ and $\mathbf{u}$ at the ST, corresponding to the data and the JN as

$$\mathbf{x}_s = \mathbf{w}s_c + \mathbf{u} \tag{11.27}$$

where $s_c \in \mathbb{C}$ is the confidential message that the ST transmits to SU, with $\mathbb{E}\{|s_c|^2\} = 1$, which is weighted with the beamforming vector $\mathbf{w} \in \mathbb{C}^{N \times 1}$, and $\mathbf{u}$ is the JN vector whose elements are zero-mean complex Gaussian random variables with covariance matrix $\mathbf{U}$, i.e. $\mathbf{u} \sim \mathscr{C} \mathscr{N}(\mathbf{0}, \mathbf{U})$, where $\mathbf{U} \in \mathbb{H}^N$ and $\mathbf{U} \succeq \mathbf{0}$.

The SINR at the SU and at the kth Eve are, respectively, given by

$$\Gamma_s = \frac{|\mathbf{h}^H \mathbf{w}|^2}{\mathbf{h}^H \mathbf{U} \mathbf{h} + 1} \quad \text{and} \quad \Gamma_k = \frac{|\mathbf{g}_k^H \mathbf{w}|^2}{\mathbf{g}_k^H \mathbf{U} \mathbf{g}_k + 1}, \quad \forall k \in \mathscr{K} \tag{11.28}$$

where $\mathbf{h} \in \mathbb{C}^{N \times 1}$ and $\mathbf{g}_k \in \mathbb{C}^{N \times 1}$ are the baseband equivalent channels of the links from the ST to SU and to the kth Eve, respectively. $\mathscr{K}$ is defined as $\mathscr{K} \triangleq \{1, 2, \ldots, K\}$. Without loss of generality, the background thermal noise at each receiver is assumed to be a zero-mean and unit variance complex Gaussian random variable.

11.3.1.2   Problem formulation

The secrecy rate of the SU, R_s, is defined as [34]

$$R_s = \max\left(\log\left(1 + \Gamma_s\right) - \max_{k \in \mathscr{K}} \log\left(1 + \Gamma_k\right), 0\right). \tag{11.29}$$

If R_s is kept larger than zero while ensuring that the received interference at the mth PU is below the pre-determined threshold I_m, then the signal transmitted from the ST to the SU is 'undecodable', as indicated in [36].

The beamformer for the JN is designed to null out the interference to SU, such that

$$\Phi^H \mathbf{h} = \mathbf{0} \tag{11.30}$$

where we choose $\mathbf{U} = \Phi\Phi^H$ with $\Phi \in \mathbb{C}^{N \times (N-1)}$ [18,31]. Thus, the SINR of SU and the kth Eve can be re-written as

$$\Gamma_s = |\mathbf{h}^H \mathbf{w}|^2 \quad \text{and} \quad \Gamma_k = \frac{|\mathbf{g}_k^H \mathbf{w}|^2}{\|\mathbf{g}_k^H \Phi\|^2 + 1}, \quad \forall k \in \mathscr{K}. \tag{11.31}$$

The optimization problem can be formulated as

$$\textbf{P1} : \max_{\mathbf{w},\Phi} \quad \left\{ \log\left(1 + \Gamma_s\right) - \max_{k \in \mathcal{K}} \log\left(1 + \Gamma_k\right) \right\} \tag{11.32a}$$

subject to $\quad \Phi^H \mathbf{h} = \mathbf{0} \tag{11.32b}$

$$[\mathbf{w}\mathbf{w}^H]_{n,n} + [\Phi\Phi^H]_{n,n} \le P_n, \, \forall n \in \mathcal{N} \tag{11.32c}$$

$$|\mathbf{f}_m^H \mathbf{w}|^2 + \|\mathbf{f}_m^H \Phi\|^2 \le I_m, \, \forall m \in \mathcal{M} \tag{11.32d}$$

where $\mathcal{N} \triangleq \{1, 2, \ldots, N\}$ and $\mathcal{M} \triangleq \{1, 2, \ldots, M\}$. $\mathbf{f}_m \in \mathbb{C}^{N \times 1}$ is the baseband equivalent channel of the link from the ST to the mth PU. The constraint in (11.32c) represents the power constraint for the nth antenna at the ST. We note that each antenna is often equipped with its own power amplifier (PA). Thus, one may need to limit the per-antenna peak power to operate within the linear region of the PA [37]. The PAPCs in (11.32c) are different from the sum power constraint (SPC) considered in [24,29]; however, the proposed beamforming scheme in this paper can also be applied to the SPC with slight modifications. The constraint in (11.32d) is to protect the primary system, so that the interference power at the mth PU due to the ST is less than a given interference threshold $I_m, \forall m \in \mathcal{M}$.

By introducing an auxiliary variable Γ_{tol} and in the spirit of [38], **P1** has the same optimal solutions as the following new problem.

$$\textbf{P2} : \max_{\mathbf{w}, \Phi, \Gamma_{\text{tol}} > 0} \quad \log\left(1 + |\mathbf{h}^H \mathbf{w}|^2\right) - \log\left(1 + \Gamma_{\text{tol}}\right) \tag{11.33a}$$

subject to $\quad \displaystyle\max_{k \in \mathcal{K}} \frac{|\mathbf{g}_k^H \mathbf{w}|^2}{\|\mathbf{g}_k^H \Phi\|^2 + 1} \le \Gamma_{\text{tol}} \tag{11.33b}$

$$(11.32\text{b}), (11.32\text{c}), (11.32\text{d}) \tag{11.33c}$$

where $\Gamma_{\text{tol}} > 0$ is the maximum allowable SINR for Eves to wiretap the confidential messages from the ST. Intuitively, we have an equivalent problem with less difficulty by adjusting Γ_{tol}.

To further simplify **P2**, let $\bar{\mathbf{V}} \in \mathbb{C}^{N \times (N-1)}$ be the null space of $\mathbf{h}^H$. Then, we can write $\Phi = \bar{\mathbf{V}}\bar{\Phi}$, where $\bar{\Phi} \in \mathbb{C}^{(N-1) \times (N-1)}$ is the solution to the following problem.

$$\textbf{P3} : \max_{\mathbf{w}, \bar{\Phi}, \Gamma_{\text{tol}} > 0} \quad \log\left(1 + |\mathbf{h}^H \mathbf{w}|^2\right) - \log\left(1 + \Gamma_{\text{tol}}\right) \tag{11.34a}$$

subject to $\quad \displaystyle\max_{k \in \mathcal{K}} \frac{|\mathbf{g}_k^H \mathbf{w}|^2}{\mathbf{g}_k^H \bar{\mathbf{V}}\bar{\Phi}\bar{\Phi}^H \bar{\mathbf{V}}^H \mathbf{g}_k + 1} \le \Gamma_{\text{tol}} \tag{11.34b}$

$$[\mathbf{w}\mathbf{w}^H]_{n,n} + [\bar{\mathbf{V}}\bar{\Phi}\bar{\Phi}^H \bar{\mathbf{V}}^H]_{n,n} \le P_n, \, \forall n \in \mathcal{N} \tag{11.34c}$$

$$|\mathbf{f}_m^H \mathbf{w}|^2 + \|\mathbf{f}_m^H \bar{\mathbf{V}}\bar{\Phi}\|^2 \le I_m, \, \forall m \in \mathcal{M}. \tag{11.34d}$$

11.3.1.3 Channel state information

We consider the case that the CSI of $\mathbf{h}$ and $\mathbf{f}_m, \forall m$, is perfectly known at the ST [14,28,29], where the SU and PUs are active users. Explicitly, the ST sends pilot signals at the beginning of each scheduling slot to the SU and PUs. The channel vectors are estimated at the SU and PUs and then fed back to the ST using a dedicated control channel. We assume that Eves passively wiretap the confidential messages transmitted from the ST to SU without causing any interference to the SU and PUs. For the passive Eves, the entries of $\mathbf{g}_k$, $\forall k$, are modelled as independent and identically distributed (i.i.d.) Rayleigh fading channels, where the instantaneous information of these wiretap channels are not available at ST. These assumptions about the passive Eves are commonly used in the literature [18,31,39]. On the basis of the above settings, **P3** can be re-written as

P4 :
$$\max_{\mathbf{w}, \bar{\Phi}, \Gamma_{\text{tol}} > 0} \quad \log(1 + |\mathbf{h}^H \mathbf{w}|^2) - \log(1 + \Gamma_{\text{tol}}) \tag{11.35a}$$

$$\text{subject to} \quad \Pr\left(\max_{k \in \mathcal{K}} \frac{\mathbf{w}^H \mathbf{G}_k \mathbf{w}}{\text{tr}(\mathbf{G}_k \bar{\mathbf{V}} \bar{\Phi} \bar{\Phi}^H \bar{\mathbf{V}}^H) + 1} \leq \Gamma_{\text{tol}}\right) \geq \kappa \tag{11.35b}$$

$$[\mathbf{w}\mathbf{w}^H]_{n,n} + [\bar{\mathbf{V}} \bar{\Phi} \bar{\Phi}^H \bar{\mathbf{V}}^H]_{n,n} \leq P_n, \ \forall n \in \mathcal{N} \tag{11.35c}$$

$$|\mathbf{f}_m^H \mathbf{w}|^2 + \|\mathbf{f}_m^H \bar{\mathbf{V}} \bar{\Phi}\|^2 \leq I_m, \ \forall m \in \mathcal{M}. \tag{11.35d}$$

where $\mathbf{G}_k \triangleq \mathbf{g}_k \mathbf{g}_k^H$, and κ is a parameter for providing secure communication. In particular, the maximum received SINR at all Eves is required to be less than a given value Γ_{tol} with at least probability κ [39].

11.3.2 Optimization problem design

The global optimal Γ_{tol} of **P4** can be found from one dimensional search. Therefore, our main task in this section is to derive a convex optimization approach to **P4** with respect to $\mathbf{w}$ and $\bar{\Phi}$ for a fixed value of Γ_{tol}. A method to find the optimal solution of Γ_{tol} will be presented in the next section.

We note that **P4** is not a convex program. A standard way to solve (11.35) for a fixed Γ_{tol} is to consider the following problem.

P5 :
$$\max_{\mathbf{W} \succeq 0, \bar{\mathbf{U}} \succeq 0} \quad \log(1 + \mathbf{h}^H \mathbf{W} \mathbf{h}) \tag{11.36a}$$

$$\text{subject to} \quad \Pr\left(\max_{k \in \mathcal{K}} \frac{\text{tr}(\mathbf{G}_k \mathbf{W})}{\text{tr}(\mathbf{G}_k \bar{\mathbf{V}} \bar{\mathbf{U}} \bar{\mathbf{V}}^H) + 1} \leq \Gamma_{\text{tol}}\right) \geq \kappa \tag{11.36b}$$

$$[\mathbf{W}]_{n,n} + [\bar{\mathbf{V}} \bar{\mathbf{U}} \bar{\mathbf{V}}^H]_{n,n} \leq P_n, \ \forall n \in \mathcal{N} \tag{11.36c}$$

$$\text{tr}(\mathbf{F}_m \mathbf{W}) + \text{tr}(\bar{\mathbf{F}}_m \bar{\mathbf{U}}) \leq I_m, \ \forall m \in \mathcal{M} \tag{11.36d}$$

$$\text{rank}(\mathbf{W}) = 1 \tag{11.36e}$$

where $\mathbf{W} \triangleq \mathbf{w}\mathbf{w}^H$, $\bar{\mathbf{U}} \triangleq \bar{\Phi} \bar{\Phi}^H$, $\mathbf{F}_m \triangleq \mathbf{f}_m \mathbf{f}_m^H$, $\bar{\mathbf{F}}_m \triangleq \bar{\mathbf{f}}_m \bar{\mathbf{f}}_m^H$, and $\bar{\mathbf{f}}_m \triangleq \bar{\mathbf{V}}^H \mathbf{f}_m$. In addition, we have dropped $\log(1 + \Gamma_{\text{tol}})$ from the objective function in (11.36) to have a simpler problem without affecting optimality. We remark that the constraint $\text{rank}(\mathbf{W}) = 1$ must be satisfied to transmit the confidential message $s_c \in \mathbb{C}$. **P5** is still a non-convex program due to the non-convex constraints in (11.36b) and (11.36e).

To make **P5** a tractable problem, we first transform the constraint in (11.36b) into a linear matrix inequality and convex constraint according to the following lemma.

Lemma 11.1. *The constraint in (11.36b) can be transformed as*

$$\mathbf{W} - \Gamma_{\mathrm{tol}}\bar{\mathbf{V}}\bar{\mathbf{U}}\bar{\mathbf{V}}^H \preceq \mathbf{I}\xi \tag{11.37}$$

where $\xi = \Phi_N^{-1}(1 - \kappa^{1/K})\Gamma_{\mathrm{tol}}$, *with* $\Phi_N^{-1}(\cdot)$ *being the inverse cumulative distribution function of an inverse central chi-square random variable with* $2N$ *degrees of freedom.*

Proof. The probability in (11.36b) for the kth Eve link can be re-written as

$$\Pr\!\left(\mathrm{tr}(\mathbf{G}_k(\mathbf{W} - \Gamma_{\mathrm{tol}}\bar{\mathbf{V}}\bar{\mathbf{U}}\bar{\mathbf{V}}^H))\le \Gamma_{\mathrm{tol}}\right). \tag{11.38}$$

Let $\mathbf{Q} \triangleq \mathbf{W} - \Gamma_{\mathrm{tol}}\bar{\mathbf{V}}\bar{\mathbf{U}}\bar{\mathbf{V}}^H$. The probability in (11.38) cannot be computed directly unless specific properties of $\mathbf{Q}$ are satisfied. For $N \times N$ Hermitian matrices $\mathbf{G}_k$ and $\mathbf{Q}$, an inequality holds as [40]

$$\mathrm{tr}(\mathbf{G}_k\mathbf{Q}) \le \sum_{i=1}^{N} \lambda_i(\mathbf{G}_k)\lambda_i(\mathbf{Q}) \overset{(a)}{=} \lambda_{\max}(\mathbf{G}_k)\lambda_{\max}(\mathbf{Q}) \overset{(b)}{=} \mathrm{tr}(\mathbf{G}_k)\lambda_{\max}(\mathbf{Q}) \tag{11.39}$$

where $\lambda_i(\mathbf{X})$ denotes the ith eigenvalue of matrix $\mathbf{X} \in \mathbb{H}^{N \times N}$ and its orders are arranged as $\lambda_{\max}(\mathbf{X}) = \lambda_1(\mathbf{X}) \ge \lambda_2(\mathbf{X}) \ge \cdots \ge \lambda_N(\mathbf{X}) = \lambda_{\min}(\mathbf{X})$. In addition, the equalities (a) and (b) in (11.39) are obtained because $\mathbf{G}_k$ is a rank-one positive semi-definite matrix. Substituting (11.39) into (11.38), we have

$$\Pr\!\left(\mathrm{tr}(\mathbf{G}_k(\mathbf{W} - \Gamma_{\mathrm{tol}}\bar{\mathbf{V}}\bar{\mathbf{U}}\bar{\mathbf{V}}^H))\le \Gamma_{\mathrm{tol}}\right)\ge \Pr\!\left(\mathrm{tr}(\mathbf{G}_k)\lambda_{\max}(\mathbf{Q}) \le \Gamma_{\mathrm{tol}}\right). \tag{11.40}$$

Since the channel $\mathbf{g}_k$, $\forall k$, is modelled as i.i.d. Rayleigh fading, we have

$$\Pr\!\left(\max_{k\in\mathcal{K}} \frac{\mathrm{tr}(\mathbf{G}_k\mathbf{W})}{\mathrm{tr}(\mathbf{G}_k\bar{\mathbf{V}}\bar{\mathbf{U}}\bar{\mathbf{V}}^H) + 1} \le \Gamma_{\mathrm{tol}}\right) \ge \Pr\!\left(\mathrm{tr}(\mathbf{G})\lambda_{\max}(\mathbf{Q}) \le \Gamma_{\mathrm{tol}}\right)\ge \kappa^{1/K}$$

$$\Leftrightarrow \Pr\!\left(\frac{\lambda_{\max}(\mathbf{Q})}{\Gamma_{\mathrm{tol}}} \ge \frac{1}{\mathrm{tr}(\mathbf{G})}\right) \le 1 - \kappa^{1/K}$$

$$\overset{(c)}{\Leftrightarrow} \lambda_{\max}(\mathbf{Q}) \le \Phi_N^{-1}(1 - \kappa^{1/K})\Gamma_{\mathrm{tol}}$$

$$\Leftrightarrow \mathbf{Q} \preceq \mathbf{I}\!\left(\Phi_N^{-1}(1 - \kappa^{1/K})\Gamma_{\mathrm{tol}}\right) \tag{11.41}$$

where without loss of generality, we have removed the index of Eves and (c) is obtained similarly to the steps of Lemma 2 in [39]. $\Phi_N^{-1}(\cdot)$ denotes the inverse cumulative distribution function of an inverse central chi-square random variable with $2N$ degrees of freedom, and $\mathrm{tr}(\mathbf{G}) = \mathrm{tr}(|\mathbf{g}|^2)$ is the sum of the squares of N independent Gaussian random variables. This completes the proof. $\square$

Remark 11.1. *Note that the implication in (11.37) can be applied to any continuous channel distribution by replacing* $\Phi_N^{-1}(\cdot)$ *with the corresponding one. Consequently, the proposed solution introduced in this section also applies to other eavesdroppers' channel distributions with slight modification of the optimization problem.*

By replacing (11.36b) with (11.37), we obtain the following new problem.

$$\textbf{P6} : \max_{\textbf{W},\bar{\textbf{U}}} \quad \log\left(1 + \textbf{h}^H \textbf{W} \textbf{h}\right) \tag{11.42a}$$

$$\text{subject to} \quad \textbf{W} - \Gamma_{\text{tol}} \bar{\textbf{V}} \bar{\textbf{U}} \bar{\textbf{V}}^H \preceq \textbf{I}\xi \tag{11.42b}$$

$$[\textbf{W}]_{n,n} + [\bar{\textbf{V}} \bar{\textbf{U}} \bar{\textbf{V}}^H]_{n,n} \leq P_n, \ \forall n \in \mathcal{N} \tag{11.42c}$$

$$\text{tr}(\textbf{F}_m \textbf{W}) + \text{tr}(\bar{\textbf{F}}_m \bar{\textbf{U}}) \leq I_m, \ \forall m \in \mathcal{M} \tag{11.42d}$$

$$\textbf{W} \succeq \textbf{0}, \ \bar{\textbf{U}} \succeq \textbf{0}, \ \text{rank}(\textbf{W}) = 1. \tag{11.42e}$$

We note that the feasible solutions of (11.42) also satisfy (11.36) but not vice versa due to the inequality in (11.40). In other words, (11.37) is a relaxation of (11.36b) which yields a large feasible solution set for **P5**. Although (11.42) is a non-convex program, it can be efficiently solved with some numerical solvers by dropping the rank constraint in (11.42e); then, the considered problem **P6** becomes a so-called rank-relaxed problem. Importantly, we can prove that the rank relaxation is tight [32].

Remark 11.2. *We note that the optimization problem with a fixed Γ_{tol} in (11.42) is also applicable for the SINR-based design in [28,39]. More challengingly, the goal of this section is to provide a secrecy rate maximization for the secondary system, rather than a certain quality-of-service.*

Theorem 11.3. *Consider the following minimax problem.*

$$\textbf{P7} : \min_{\psi \geq 0, \textbf{D} \succeq 0} \ \max_{w \geq 0, \bar{\textbf{U}} \succeq 0} \quad \log\frac{|\Sigma + \textbf{h}w\textbf{h}^H|}{|\Sigma|}$$

$$\text{subject to} \quad w + \text{tr}(\Omega\bar{\textbf{U}}) \leq P \tag{11.43}$$

$$\xi\,\text{tr}(\textbf{D}) + \textbf{p}^T \psi \leq P$$

where $\Sigma = \textbf{D} + \text{diag}(\lambda) + \sum_{m=1}^{M} \mu_m \textbf{F}_m,\ \Omega = \bar{\textbf{V}}^H(-\Gamma_{\text{tol}}\textbf{D} + \text{diag}(\lambda) + \sum_{m=1}^{M} \mu_m \textbf{F}_m)\bar{\textbf{V}},$ *and* $\psi = [\lambda^T \ \ \mu^T]^T$. *Then, the optimal solution* $\textbf{W}^\star$ *of the relaxed problem* (11.42) *can be obtained from that of* (11.43) *as*

$$\textbf{W} = \frac{\Sigma^{-1}\textbf{h}w\textbf{h}^H\Sigma^{-1}}{\textbf{h}^H\Sigma^{-1}\textbf{h}}. \tag{11.44}$$

Proof. Here, we prove that **P7** is the dual problem of the relaxed problem of **P6**. In particular, we follow the same steps as in [41] while customizing them to our considered problem. The partial Lagrangian function of the relaxed version of **P6** can be defined as

$$\mathscr{L}(\textbf{W}, \bar{\textbf{U}}, \textbf{D}, \{\lambda_n\}, \{\mu_m\}) = \log(1 + \textbf{h}^H\textbf{W}\textbf{h}) - \text{tr}(\textbf{D}(\textbf{W} - \Gamma_{\text{tol}}\bar{\textbf{V}}\bar{\textbf{U}}\bar{\textbf{V}}^H - \textbf{I}\xi))$$

$$- \sum_{n=1}^{N} \lambda_n\left(\text{tr}(\textbf{W}\textbf{B}^{(n)}) + \text{tr}(\bar{\textbf{U}}\textbf{E}^{(n)}) - P_n\right) - \sum_{m=1}^{M} \mu_m\left(\text{tr}(\textbf{F}_m\textbf{W}) + \text{tr}(\bar{\textbf{F}}_m\bar{\textbf{U}}) - I_m\right)$$

$$\tag{11.45}$$

where $\mathbf{E}^{(n)} \triangleq \bar{\mathbf{T}}^H \bar{\mathbf{T}}$, $\bar{\mathbf{T}} = [\mathbf{0}_{n-1}^T \quad 1 \quad \mathbf{0}_{N-n}^T]\bar{\mathbf{V}}$, and $\bar{\mathbf{F}}_m \triangleq \bar{\mathbf{f}}_m \bar{\mathbf{f}}_m^H$. Next, the dual objective of **P6** is given by

$$\mathscr{D}(\mathbf{D}, \{\lambda_n\}, \{\mu_m\}) = \max_{\mathbf{W}, \bar{\mathbf{U}} \succeq 0} \mathscr{L}(\mathbf{W}, \bar{\mathbf{U}}, \mathbf{D}, \{\lambda_n\}, \{\mu_m\}). \tag{11.46}$$

For a given set $(\mathbf{D}, \{\lambda_n\}, \{\mu_m\})$, we first re-write the partial Lagrangian function as

$$\mathscr{L}(\mathbf{W}, \bar{\mathbf{U}}, \mathbf{D}, \lambda, \mu) = \log(1 + \mathbf{h}^H \mathbf{W} \mathbf{h}) - \operatorname{tr}(\Sigma \mathbf{W}) - \operatorname{tr}(\Omega \bar{\mathbf{U}}) + \xi \operatorname{tr}(\mathbf{D}) + \bar{\mathbf{p}}^T \lambda + \bar{I}^T \mu \tag{11.47}$$

where $\Omega \triangleq -\Gamma_{\text{tol}} \bar{\mathbf{V}}^H \mathbf{D} \bar{\mathbf{V}} + \sum_{n=1}^N \lambda_n \mathbf{E}^{(n)} + \sum_{m=1}^M \mu_m \bar{\mathbf{F}}_m$, $\bar{\mathbf{p}} = [P_1, P_2, \ldots, P_N]^T$, $\lambda = [\lambda_1, \lambda_2, \ldots, \lambda_N]^T$, $\bar{I} = [I_1, I_2, \ldots, I_M]^T$, and $\mu = [\mu_1, \mu_2, \ldots, \mu_M]^T$. Let $\bar{\mathbf{W}} = \Sigma^{1/2} \mathbf{W} \Sigma^{1/2}$. Since Σ is invertible, (11.47) can be re-written as

$$\mathscr{L}(\bar{\mathbf{W}}, \bar{\mathbf{U}}, \mathbf{D}, \lambda, \mu) = \log\left(1 + \mathbf{h}^H \Sigma^{-1/2} \bar{\mathbf{W}} \Sigma^{-1/2} \mathbf{h}\right) \\ - \operatorname{tr}(\bar{\mathbf{W}}) - \operatorname{tr}(\Omega \bar{\mathbf{U}}) + \xi \operatorname{tr}(\mathbf{D}) + \bar{\mathbf{p}}^T \lambda + \bar{I}^T \mu. \tag{11.48}$$

On the basis of the results in [42, Appendix A], the dual objective in (11.46) is equivalent to

$$\mathscr{D}(\mathbf{D}, \lambda, \mu) = \max_{w \geq 0, \bar{\mathbf{U}} \succeq 0} \log |\mathbf{I} + \Sigma^{-1/2} \mathbf{h} w \mathbf{h}^H \Sigma^{-1/2}| \\ - w - \operatorname{tr}(\Omega \bar{\mathbf{U}}) + \xi \operatorname{tr}(\mathbf{D}) + \bar{\mathbf{p}}^T \lambda + \bar{I}^T \mu \tag{11.49}$$

where the relationship between $\bar{\mathbf{W}}$ and w is given by

$$\bar{\mathbf{W}} = \frac{\Sigma^{-1/2} \mathbf{h} w \mathbf{h}^H \Sigma^{-1/2}}{\mathbf{h}^H \Sigma^{-1} \mathbf{h}}. \tag{11.50}$$

Next, we can write $\mathscr{D}(\mathbf{D}, \lambda, \mu)$ in a more compact form as

$$\mathscr{D}(\mathbf{D}, \psi) = \max_{w \geq 0, \bar{\mathbf{U}} \succeq 0} \log |\mathbf{I} + \Sigma^{-1/2} \mathbf{h} w \mathbf{h}^H \Sigma^{-1/2}| - w - \operatorname{tr}(\Omega \bar{\mathbf{U}}) + \xi \operatorname{tr}(\mathbf{D}) + \mathbf{p}^T \psi$$

$$= \max_{w \geq 0, \bar{\mathbf{U}} \succeq 0} \log \frac{|\Sigma + \mathbf{h} w \mathbf{h}^H|}{|\Sigma|} - w - \operatorname{tr}(\Omega \bar{\mathbf{U}}) + \xi \operatorname{tr}(\mathbf{D}) + \mathbf{p}^T \psi \tag{11.51}$$

where $\mathbf{p} = [\bar{\mathbf{p}}^T \quad \bar{I}^T]^T$ and $\psi = [\lambda^T \quad \mu^T]^T$. The dual problem is obtained by minimizing $\mathscr{D}(\mathbf{D}, \psi)$ as

$$\min_{\psi \geq 0, \mathbf{D} \succeq 0} \max_{w \geq 0, \bar{\mathbf{U}} \succeq 0} \log \frac{|\Sigma + \mathbf{h} w \mathbf{h}^H|}{|\Sigma|} - w - \operatorname{tr}(\Omega \bar{\mathbf{U}}) + \xi \operatorname{tr}(\mathbf{D}) + \mathbf{p}^T \psi. \tag{11.52}$$

To obtain an optimal point for the minimax problem in (11.52), we introduce another optimization variable $\varphi \geq 0$. Then, (11.52) can be re-written as

$$\min_{\psi \geq 0, \mathbf{D} \succeq 0} \max_{\varphi \geq 0, w \geq 0, \bar{\mathbf{U}} \succeq 0} \log \frac{|\Sigma + \mathbf{h} w \mathbf{h}^H|}{|\Sigma|} - \varphi P + \xi \operatorname{tr}(\mathbf{D}) + \mathbf{p}^T \psi$$

$$\text{subject to} \quad w + \operatorname{tr}(\Omega \bar{\mathbf{U}}) \leq \varphi P. \tag{11.53}$$

It is easy to see that (11.53) is equivalent to (11.52) since the inequality must hold with equality at the optimum; otherwise, we can scale down φ to achieve a strictly larger objective. Next, we make a change of variables as

$$\tilde{w} = w/\varphi, \quad \tilde{\psi} = \psi/\varphi, \quad \tilde{\mathbf{D}} = \mathbf{D}/\varphi, \quad \tilde{\mathbf{U}} = \bar{\mathbf{U}}/\varphi. \tag{11.54}$$

We now consider $\tilde{w}$, $\tilde{\psi}$, $\tilde{\mathbf{D}}$, and $\tilde{\mathbf{U}}$ as the new optimization variables. Then, (11.53) can be equivalently expressed as

$$\min_{\tilde{\psi} \geq 0, \tilde{\mathbf{D}} \geq 0} \max_{\varphi \geq 0, \tilde{w} \geq 0, \tilde{\mathbf{U}} \geq 0} \log \frac{|\tilde{\Sigma} + \mathbf{h}\tilde{w}\mathbf{h}^H|}{|\tilde{\Sigma}|} + \varphi(\xi \operatorname{tr}(\tilde{\mathbf{D}}) + \mathbf{p}^T \tilde{\psi} - P) \tag{11.55}$$
$$\text{subject to} \quad \tilde{w} + \operatorname{tr}(\Omega \tilde{\mathbf{U}}) \leq P$$

where $\tilde{\Sigma} = \Sigma/\varphi$. It is easy to see that the optimal dual variable $\varphi^\star$ can be obtained by considering the minimization of (11.55) over $\tilde{\psi}$ and $\tilde{\mathbf{D}}$. Hence, (11.55) is the dual of the following problem:

$$\min_{\tilde{\psi} \geq 0, \tilde{\mathbf{D}} \geq 0} \max_{\varphi \geq 0, \tilde{w} \geq 0, \tilde{\mathbf{U}} \geq 0} \log \frac{|\Sigma + \mathbf{h}\tilde{w}\mathbf{h}^H|}{|\Sigma|}$$
$$\text{subject to} \quad \tilde{w} + \operatorname{tr}(\Omega \tilde{\mathbf{U}}) \leq P \tag{11.56}$$
$$\xi \operatorname{tr}(\tilde{\mathbf{D}}) + \mathbf{p}^T \tilde{\psi} \leq P.$$

Finally, from the derivations in (11.50), (11.54), (11.56), and $\mathbf{W} = \Sigma^{-1/2} \bar{\mathbf{W}} \Sigma_k^{-1/2}$, the proof is finalized. □

11.3.3 Optimization over Γ_{tol}

P7 can be re-written by considering the optimization over Γ_{tol} as

$$\max_{\Gamma_{\text{tol}} > 0} f(\Gamma_{\text{tol}}) \tag{11.57}$$

where $f(\Gamma_{\text{tol}})$ is defined as

$$f(\Gamma_{\text{tol}}) = \min_{\psi \geq 0, \mathbf{D} \geq 0} \max_{w \geq 0, \tilde{\mathbf{U}} \geq 0} \log \frac{|\Sigma + \mathbf{h}w\mathbf{h}^H|}{|\Sigma|} - \log(1 + \Gamma_{\text{tol}})$$
$$\text{subject to} \quad w + \operatorname{tr}(\Omega \tilde{\mathbf{U}}) = P \tag{11.58}$$
$$\xi \operatorname{tr}(\mathbf{D}) + \mathbf{p}^T \psi = P.$$

Since the objective function is concave with respect to Γ_{tol}, a conventional method to find the optimal solution of Γ_{tol} is based on one dimensional search [23]. However, the major computational complexity comes from solving (11.58). Therefore, one-dimensional search method to seek a saddle point of Γ_{tol} may not be efficient since it often shows slow convergence. In this section, we propose an efficient method to find the global optimal $\Gamma_{\text{tol}}^\star$ which greatly reduces the complexity. To do this, we consider the following equivalent problem for a given set of $(w, \dot{\mathbf{U}}, \mathbf{D}, \lambda, \mu)$.

$$\min_{\Gamma_{\text{tol}} > 0} h(\Gamma_{\text{tol}}) \triangleq \log(1 + \Gamma_{\text{tol}})$$
$$\text{subject to} \quad -\operatorname{tr}(\bar{\mathbf{V}}^H \mathbf{D}\bar{\mathbf{V}}\bar{\mathbf{U}})\Gamma_{\text{tol}} = P - w - \operatorname{tr}(\tilde{\Omega}\tilde{\mathbf{U}}) \tag{11.59}$$
$$\tilde{\xi} \operatorname{tr}(\mathbf{D})\Gamma_{\text{tol}} = P - \mathbf{p}^T \psi$$

where $\tilde{\Omega} \triangleq \bar{V}^H(\text{diag}(\lambda) + \sum_{m=1}^{M} \mu_m F_m)\bar{V}$ and $\tilde{\xi} \triangleq \Phi_N^{-1}(1 - \kappa^{1/K})$. With implicit constraint $\Gamma_{\text{tol}} > 0$, we can derive the simpler form of the objective function in (11.59) without affecting optimality, as follows:

$$\tilde{h}(\Gamma_{\text{tol}}) \triangleq h(\Gamma_{\text{tol}} - 1) = \log(\Gamma_{\text{tol}}). \tag{11.60}$$

The Lagrangian function of (11.59) can be defined as [43]

$$\mathcal{L}(\Gamma_{\text{tol}}, \upsilon) = \tilde{h}(\Gamma_{\text{tol}}) + \upsilon^T a \Gamma_{\text{tol}} - \rho^T \upsilon \tag{11.61}$$

where $\rho \triangleq [P - w - \text{tr}(\tilde{\Omega}\bar{U}) \quad P - \mathbf{p}^T \psi]^T$, $a \triangleq [-\text{tr}(\bar{V}^H \mathbf{D}\bar{V}\bar{U}), \quad \tilde{\xi}\text{tr}(\mathbf{D})]^T$, and $\upsilon \triangleq [\upsilon_1 \quad \upsilon_2]^T$ with υ_1 and υ_2 being the dual variables related to the constraints in (11.59). Then, the solution of (11.59) can be found by solving the dual problem which is presented in the following theorem.

Theorem 11.4. *The dual problem of* (11.59) *is given by*

$$\max_{\upsilon} g(\upsilon) = -\rho^T \upsilon - \log(-a^T \upsilon) - 1 \tag{11.62}$$

with implicit constraint $a^T \upsilon < 0$.

Proof. The dual objective of (11.59) is

$$g(\upsilon) = \inf_{\Gamma_{\text{tol}}} \mathcal{L}(\Gamma_{\text{tol}}, \upsilon) = -\rho^T \upsilon + \inf_{\Gamma_{\text{tol}}} (\tilde{h}(\Gamma_{\text{tol}}) + \upsilon^T a \Gamma_{\text{tol}})$$

$$= -\rho^T \upsilon - \sup_{\Gamma_{\text{tol}}} ((-a^T \upsilon)^T \Gamma_{\text{tol}} - \tilde{h}(\Gamma_{\text{tol}})) \tag{11.63}$$

$$= -\rho^T \upsilon - \tilde{h}^*(-a^T \upsilon)$$

where $\tilde{h}^*(\cdot)$ is the convex conjugate of $\tilde{h}(\cdot)$. For the convex conjugate $\tilde{h}^*(-a^T \upsilon)$, we use the Legendre transform to a log function as $-\log(x) \to -(1 + \log(-x^*))$ [44]. Thus, the dual problem is

$$g(\upsilon) = -\rho^T \upsilon - 1 - \log(-a^T \upsilon) \tag{11.64}$$

and the proof is completed. $\square$

Corollary 11.4. *The optimal solution* Γ_{tol} *can be obtained from that of* (11.62) *as*

$$\Gamma_{\text{tol}} = -\frac{1}{\upsilon^T a}. \tag{11.65}$$

Proof. Once the optimal dual variable υ^* is found, we can obtain the optimal solution of Γ_{tol} from (11.59) by solving the Karush–Kuhn–Tucker (KKT) condition in (11.61) as

$$\nabla_{\Gamma_{\text{tol}}} \tilde{h}(\Gamma_{\text{tol}}) + \upsilon^T a = \Gamma_{\text{tol}}^{-1} + \upsilon^T a = 0. \tag{11.66}$$

This completes the proof. $\square$

Algorithm 1 Main algorithm to solve (11.57)

Initialization: $\upsilon := 1, \phi := -\nabla_\upsilon \tilde{g}(\upsilon), \phi' := 0, \mathbf{P} := I, s' := 1,$ and tolerance $\varepsilon' > 0$
1: **loop**
2:    Solve (11.58) to obtain $(w, \dot{\mathbf{U}}, \mathbf{D}, \lambda, \mu)$.
3:    Compute $\phi := \phi - \phi'$ and $\Delta \upsilon = \mathbf{P}\phi$.
4:    Stop, if $\|\phi\| < \varepsilon'$.
5:    Update $\upsilon, \phi',$ and $\mathbf{P}$ in the strict order:
        $\upsilon := \upsilon + s'\Delta \upsilon$
        $\phi' := \phi + \nabla_\upsilon \tilde{g}(\upsilon)$
        $\mathbf{P}$ as the result in (11.68).
6: **end loop**
7: **Ensure:** $(\upsilon, w, \dot{\mathbf{U}}, \mathbf{D}, \lambda, \mu)$

We now apply BFGS algorithm [45] to solve the unconstrained optimization problem (11.62). A standard method for solving the log concave function in (11.62) is to consider the following equivalent optimization problem.

$$\min_\upsilon \tilde{g}(\upsilon) = -g(\upsilon). \tag{11.67}$$

The BFGS algorithm can be described as

1. Compute $\Delta \upsilon$ as $\Delta \upsilon = -\mathbf{P}\nabla \tilde{g}(\upsilon)$, where $\nabla_\upsilon \tilde{g}(\upsilon) = \rho - a/(a^T \upsilon)$.
2. If $\|\nabla_\upsilon \tilde{g}(\upsilon + \Delta \upsilon)\| < \varepsilon'$ (tolerance), then stop.
3. Update the dual variables: $\upsilon := \upsilon + \Delta \upsilon$.
4. Compute $\phi' = \nabla_\upsilon \tilde{g}(\upsilon + \Delta \upsilon) - \nabla_\upsilon \tilde{g}(\upsilon)$.
5. Update $\mathbf{P}$ as [45]

$$\mathbf{P} := \mathbf{P} + \frac{(\Delta \upsilon^T \phi' + (\phi')^T \mathbf{P}\phi')(\Delta \upsilon \Delta \upsilon^T)}{(\Delta \upsilon^T \phi')^2} - \frac{\mathbf{P}\phi' \Delta \upsilon^T + \Delta \upsilon (\phi')^T \mathbf{P}}{\Delta \upsilon^T \phi'}. \tag{11.68}$$

6. Go back to Step 1.

The overall iterative algorithm which is based on the BFGS method and customized to our problem to solve (11.57) is given in Algorithm 1, and it can be summarized as follows. For a given υ, Γ_{tol} in (11.65) is computed, (11.58) (or equivalently (11.43)) is solved to obtain $(w, \dot{\mathbf{U}}, \mathbf{D}, \lambda, \mu)$; then update υ from line 5 until $\|\phi\|$ is below a specified accuracy level (i.e. the tolerance ε').

11.3.4 Numerical results

We provide numerical results to validate the performance of the proposed optimal approach. The entries of the channel vectors are all generated as independent circularly symmetric complex Gaussian (CSCG) random variables with zero-mean and unit variance. To guarantee secure communication, we set the probability $\kappa = 0.99$.

For simplicity, we assume that the interference thresholds at the PUs are equal, i.e. $I_m = I$ for all m, and the number of Eves is fixed to $K = 3$. The resulting power constraint for each antenna is $P_n = P/N$ for all n, where P is the total transmit power at the ST. We also compare the performance of the proposed scheme with existing schemes, namely the 'Isotropic JN scheme' [18,31] and 'No JN scheme' [14]. In the 'Isotropic JN scheme', the covariance matrix of the jamming beamforming is chosen as $\mathbf{U} = \frac{p_u \tilde{\mathbf{V}} \tilde{\mathbf{V}}^H}{N-1}$ where the variable p_u is used to control interference to the PUs and Eves. In the 'No JN scheme', the optimization problem with no JN is considered a benchmark, and the solution can be obtained from (11.57) by setting $\mathbf{U}$ to $\mathbf{0}$.

Figure 11.4(a) illustrates the average secrecy rate of the secondary system versus the transmit power at the ST. As seen, the curves overlap in low power regime. The reason is that in such a case, the ST mainly focuses on maximizing the secrecy rate, as the interference constraints are likely satisfied for all PUs. This indicates that JN may not be necessary in this regime. However, in high power regime, the schemes using JN outperform the scheme with no JN in terms of the secrecy rate. In addition, for the 'Optimal JN scheme', the ST is allowed to transmit with nearly full power, whereas the performance of the 'Isotropic JN scheme' and 'No JN scheme' tends to saturate. This is because the 'Optimal JN scheme' controls the interference to the PUs more efficiently than the other schemes thanks to the optimized transmission.

In this numerical example, we plot the average secrecy rate of the secondary system for the "Optimal JN scheme" under several different assumptions of sharing equally the resources, i.e., transmit power at the ST and interference thresholds at the PUs. In particular, for total transmit power at the ST, the information and JN beamforming are assumed to share 50% of the power resource, i.e., $[\mathbf{W}]_{n,n} \leq P_n/2$ and $[\tilde{\mathbf{V}}\mathbf{U}\tilde{\mathbf{V}}^H]_{n,n} \leq P_n/2$, which is referred to as equal transmit power (ETP). Likewise, the information and JN beamforming are assumed to share 50% of the interference threshold, i.e., $\mathrm{tr}(\mathbf{F}_m\mathbf{W}) \leq I_m/2$ and $\mathrm{tr}(\bar{\mathbf{F}}_m\mathbf{U}) \leq I_m/2$, which is referred to as equal interference threshold (EIT). Consequently, we compare the performance of

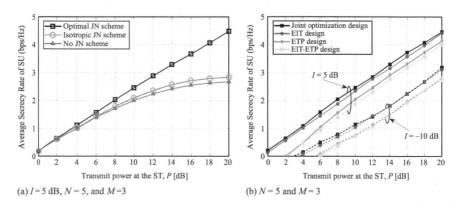

(a) $I = 5$ dB, $N = 5$, and $M = 3$

(b) $N = 5$ and $M = 3$

Figure 11.4   Average secrecy rate of the secondary system versus the total transmit power at the ST, P. (a) I = 5 dB, N = 5, and M = 3 and (b) N = 5 and M = 3. © IEEE 2016. Reprinted with permission from [32]

the proposed design with three other suboptimal methods, namely ETP, EIT, and EIT-ETP, and we present the results in Fig. 11.4(b). A general observation is that the joint optimization design outperforms the other designs in terms of the secrecy rate of the secondary system, especially when compared to the ETP design. In addition, decreasing the interference threshold I significantly degrades the secrecy rate of the secondary system. The performance gain achieved for higher interference threshold is due to the fact that more transmit power can be used when the interference threshold constraints are set to be high. Interestingly, the secrecy rate of the EIT design approaches the optimal one when the interference threshold is relatively small.

11.4 PHY-security of cooperative cognitive radio networks

In this section, we present a symbiotic approach for a secure primary network by allowing the secondary users to send the JN to degrade the wiretap ability of the eavesdropper. In particular, assuming that the global CSI is perfectly known at transceivers, we consider the case of the primary transmitter equipped with only one antenna, which implies that the primary transmitter does not have beamforming capability. As the reward of having access to the frequency spectrum which is licenced to the primary user, the secondary transmitter will assist the primary systems in terms of security by sending the JN to the eavesdropper. We propose an algorithm to find the optimal transmit power for maximizing the secrecy capacity of the primary system.

11.4.1 System model

We consider a PHY-security for a CRN model with a primary user (PU), a secondary user (SU), and an eavesdropper (Eve), as illustrated in Figure 11.5. The primary transmitter (PT) and secondary receiver (SR) are equipped with one antenna, while the secondary transmitter (ST), primary receiver (PR), and Eve are equipped with M, N, and K antennas, respectively. Eve intends to wiretap and decode confidential message from the PT.

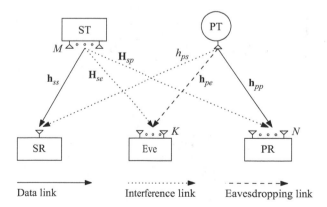

Figure 11.5 A cooperative cognitive radio network model with an eavesdropper

The received signals at the PR, SR, and Eve after the maximum-ratio-combining (MRC) processing can be expressed as

$$
y_{pr} = \frac{\mathbf{h}_{pp}^{H}}{\|\mathbf{h}_{pp}\|}\left(\sqrt{P_p}\mathbf{h}_{pp}x_p + \mathbf{H}_{sp}\mathbf{x}_s + \mathbf{n}_p\right),
\tag{11.69}
$$

$$
y_{sr} = \mathbf{h}_{ss}\mathbf{x}_s + \sqrt{P_p}h_{ps}x_p + n_s,
\tag{11.70}
$$

$$
y_e = \frac{\mathbf{h}_{pe}^{H}}{\|\mathbf{h}_{pe}\|}\left(\sqrt{P_p}\mathbf{h}_{pe}x_p + \mathbf{H}_{se}\mathbf{x}_s + \mathbf{n}_e\right)
\tag{11.71}
$$

where x_p and P_p are the unit power transmit signal from the PT and the corresponding transmit power, respectively. $\mathbf{x}_s$ is the beamforming vector of the ST. $\mathbf{h}_{pp} \in \mathbb{C}^{N \times 1}$, $\mathbf{h}_{pe} \in \mathbb{C}^{K \times 1}$, $h_{ps} \in \mathbb{C}$, $\mathbf{H}_{sp} \in \mathbb{C}^{N \times M}$, $\mathbf{H}_{se} \in \mathbb{C}^{K \times M}$, and $\mathbf{h}_{ss} \in \mathbb{C}^{1 \times M}$ are baseband equivalent channels of the links PT→PR, PT→Eve, PT→SR, ST→PR, ST→Eve, and ST→SR, respectively. $\mathbf{n}_p \sim \mathcal{CN}(0, \mathbf{I}_N)$, $\mathbf{n}_e \sim \mathcal{CN}(0, \mathbf{I}_K)$, and $n_s \sim \mathcal{CN}(0, 1)$ denote additive white Gaussian noise (AWGN) at the PR, Eve, and SR, respectively.

Let $\mathbf{x}_s = \mathbf{s} + \mathbf{q}$, where $\mathbf{s} \in \mathbb{C}^{M \times 1}$ and $\mathbf{q} \in \mathbb{C}^{M \times 1}$ are information vector and jamming vector, respectively. The covariance matrices of $\mathbf{s}$ and $\mathbf{q}$ are given as $\mathbb{E}[\mathbf{ss}^{H}] = \mathbf{S}$ and $\mathbb{E}[\mathbf{qq}^{H}] = \mathbf{W}$. As mentioned above, the beamforming vector $\mathbf{u}$ at the ST is designed to maximize the power of the information bearing signal at the intended destination SR, such that $\mathbf{S} = P_s\mathbf{uu}^{H}$ with $\mathbf{u} = \mathbf{h}_{ss}^{H}/\|\mathbf{h}_{ss}\|$, where P_s is a part of transmit power P_t at the ST. On the other hand, $\mathbf{W}$ should be designed to minimize the interference at the PR, to maximize the interference at Eve and to satisfy the capacity constraint R_{th} for the SR.

11.4.2 Optimization approach for beamforming of ST

Let C_{pr}, C_e, and C_{sr} be the capacity of the primary system, the eavesdropper, and the secondary system, respectively, which are given as

$$
C_{pr}(\mathbf{W}) = \log_2\left(1 + \frac{P_p\|\mathbf{h}_{pp}\|^2}{\dfrac{P_s\mathbf{h}_{pp}^{H}\mathbf{H}_{sp}\mathbf{h}_{ss}^{H}\mathbf{h}_{ss}\mathbf{H}_{sp}^{H}\mathbf{h}_{pp}}{\|\mathbf{h}_{ss}\|^2\|\mathbf{h}_{pp}\|^2} + \dfrac{\mathbf{h}_{pp}^{H}\mathbf{H}_{sp}\mathbf{W}\mathbf{H}_{sp}^{H}\mathbf{h}_{pp}}{\|\mathbf{h}_{pp}\|^2} + 1}\right),
\tag{11.72}
$$

$$
C_e(\mathbf{W}) = \log_2\left(1 + \frac{P_p\|\mathbf{h}_{pe}\|^2}{\dfrac{P_s\mathbf{h}_{pe}^{H}\mathbf{H}_{se}\mathbf{h}_{ss}^{H}\mathbf{h}_{ss}\mathbf{H}_{se}^{H}\mathbf{h}_{pe}}{\|\mathbf{h}_{ss}\|^2\|\mathbf{h}_{pe}\|^2} + \dfrac{\mathbf{h}_{pe}^{H}\mathbf{H}_{se}\mathbf{W}\mathbf{H}_{se}^{H}\mathbf{h}_{pe}}{\|\mathbf{h}_{pe}\|^2} + 1}\right),
\tag{11.73}
$$

$$
C_{sr}(\mathbf{W}) = \log_2\left(1 + \frac{P_s\|\mathbf{h}_{ss}\|^2}{\mathbf{h}_{ss}\mathbf{W}\mathbf{h}_{ss}^{H} + P_p|h_{ps}|^2 + 1}\right).
\tag{11.74}
$$

An optimization problem can be formulated as

$$\underset{P_s \geq 0, \mathbf{W} \succeq 0}{\text{maximize}} \quad C_{pr}(\mathbf{W}) - C_e(\mathbf{W}) \tag{11.75a}$$

$$\text{subject to} \quad P_s + \text{tr}(\mathbf{W}) \leq P_t, \tag{11.75b}$$

$$C_{sr}(\mathbf{W}) \geq R_{th} \tag{11.75c}$$

where $\text{tr}(\mathbf{W}) = P_w$ corresponds to the transmit power of JN, which is a part of the total transmit power P_t at the ST.

When $M > 1$, we aim to design the beamforming vector $\mathbf{W}$ to improve the secrecy capacity of the primary system. The optimization problem in (11.75) can be equivalently formulated as

$$\underset{\theta \geq 0}{\text{maximize}} \quad \Phi(\theta) \tag{11.76a}$$

where θ is an auxiliary variable, and $\Phi(\theta)$ is defined as

$$\Phi(\theta) = \underset{P_s \geq 0, \mathbf{W} \succeq 0}{\text{maximize}} \quad \log_2(1 + \theta) - C_e(\mathbf{W}) \tag{11.77a}$$

$$\text{subject to} \quad C_{pr} \leq \log_2(1 + \theta), \tag{11.77b}$$

$$P_s + \text{tr}(\mathbf{W}) \leq P_t, \tag{11.77c}$$

$$C_{sr}(\mathbf{W}) \geq R_{th}. \tag{11.77d}$$

The optimal solution θ^* to the problem (11.77) can be found through a one-dimensional search over θ. Next, we find the optimal solution $\mathbf{W}^*$. Note that (11.77) has the same optimal solution with the following problem.

$$\underset{P_s \geq 0, \mathbf{W} \succeq 0}{\text{maximize}} \quad \text{tr}(\bar{\mathbf{H}}_{se}\mathbf{W}) \tag{11.78a}$$

$$\text{subject to} \quad \xi_1 - \text{tr}(\bar{\mathbf{H}}_{sp}\mathbf{W}) \leq 0, \tag{11.78b}$$

$$\xi_2 + \text{tr}(\bar{\mathbf{H}}_{ss}\mathbf{W}) \leq 0, \tag{11.78c}$$

$$P_s + \text{tr}(\mathbf{W}) \leq P_t \tag{11.78d}$$

where

$$\xi_1 \triangleq \left[\frac{P_p \|\mathbf{h}_{pp}\|^2}{\theta} - (I_p + 1) \right] \|\mathbf{h}_{pp}\|^2, \quad \xi_2 \triangleq P_p |h_{ps}|^2 + 1 - \frac{P_s \|\mathbf{h}_{ss}\|^2}{2^{R_{th}} - 1}, \tag{11.79}$$

$$\bar{\mathbf{H}}_{sp} \triangleq \mathbf{H}_{sp}^H \mathbf{h}_{pp} \mathbf{h}_{pp}^H \mathbf{H}_{sp}, \quad \bar{\mathbf{H}}_{se} \triangleq \mathbf{H}_{se}^H \mathbf{h}_{pe} \mathbf{h}_{pe}^H \mathbf{H}_{se}, \quad \bar{\mathbf{H}}_{ss} \triangleq \mathbf{h}_{ss}^H \mathbf{h}_{ss}$$

with $I_p \triangleq \frac{P_s \mathbf{h}_{pp}^H \mathbf{H}_{sp} \mathbf{h}_{ss}^H \mathbf{h}_{ss} \mathbf{H}_{sp}^H \mathbf{h}_{pp}}{\|\mathbf{h}_{ss}\|^2 \|\mathbf{h}_{pp}\|^2}$. After some manipulations, it can be shown that the constraints in (11.78b) and (11.78c) are equivalent to the constraints in (11.77b) and (11.77d), respectively.

The objective function and the constraints of problem (11.78) are continuous and differentiable functions. However, the optimal solutions to (11.78) should satisfy KKT necessary conditions for optimality [47]. The Lagrangian for the optimization problem can be defined as

$$
\begin{aligned}
\mathcal{L}(\mathbf{W}, P_s, \alpha, \beta, \lambda) &= \mathrm{tr}(\bar{\mathbf{H}}_{se}\mathbf{W}) - \alpha(\xi_1 - \mathrm{tr}(\bar{\mathbf{H}}_{sp}\mathbf{W})) \\
&\quad - \beta(\xi_2 + \mathrm{tr}(\bar{\mathbf{H}}_{ss}\mathbf{W})) - \lambda(P_s + \mathrm{tr}(\mathbf{W}) - P_t) \\
&= \mathrm{tr}(\Lambda \mathbf{W}) - \alpha\xi_1 - \beta\xi_2 - \lambda P_s + \lambda P_t
\end{aligned}
\tag{11.80}
$$

where $\alpha \geq 0, \beta \geq 0$, and $\lambda \geq 0$ are the dual variables associated with the constraints from (11.78b) to (11.78d), respectively, and Λ is defined as

$$
\Lambda \triangleq \bar{\mathbf{H}}_{se} + \alpha\bar{\mathbf{H}}_{sp} - \beta\bar{\mathbf{H}}_{ss} - \lambda\mathbf{I}.
\tag{11.81}
$$

Denoting $\varphi = -\alpha\xi_1 - \beta\xi_2 - \lambda P_s + \lambda P_t$, the conditions for optimality of $\mathbf{W}^*$ can be formulated as

$$
\Lambda^* = \bar{\mathbf{H}}_{se} + \alpha^*\bar{\mathbf{H}}_{sp} - \beta^*\bar{\mathbf{H}}_{ss} - \lambda^*\mathbf{I},
\tag{11.82a}
$$

$$
\varphi^* = -\alpha^*\xi_1 - \beta^*\xi_2 - \lambda^*P_s + \lambda^*P_t,
\tag{11.82b}
$$

$$
\Lambda^*\mathbf{W}^* = 0
\tag{11.82c}
$$

where $\alpha^* \geq 0, \beta^* \geq 0$, and $\lambda^* \geq 0$ are the optimal dual variables associated to $\alpha \geq 0, \beta \geq 0$, and $\lambda \geq 0$, respectively.

Lemma 11.2. *Given that $\mathbf{h}_{ss}^H$ and $\mathbf{H}_{se}^H\mathbf{h}_{pe}$ are independent and randomly generated vectors, they should be linearly independent and the optimal solution to $\mathbf{W}^*$ and P_s^* satisfies $P_s^* + \mathrm{tr}(\mathbf{W}^*) = P_t$ [29].*

It should be noted that the optimal solution $\mathbf{W}^*$ of (11.77) should have rank $(\mathbf{W}^*) = 1$ [29]. Now, the optimal solution $\mathbf{W}^*$ to the maximization problem (11.77) can be found by using a standard convex tool as the ones in [43]. In order to get significant results, we consider a new equivalent problem. From Lemma 11.2, by introducing $\mathbf{W} = P_w\mathbf{w}\mathbf{w}^H$ with $\|\mathbf{w}\| = 1$, we have a new problem with an auxiliary variable z as

$$
\begin{aligned}
&\underset{\|\mathbf{w}\|=1}{\text{maximize}} \quad \mathbf{h}_{pe}^H\mathbf{H}_{se}\mathbf{w}\mathbf{w}^H\mathbf{H}_{se}^H\mathbf{h}_{pe} \\
&\text{subject to} \quad \mathbf{h}_{pp}^H\mathbf{H}_{sp}\mathbf{w}\mathbf{w}^H\mathbf{H}_{sp}^H\mathbf{h}_{pp} \triangleq z.
\end{aligned}
\tag{11.83}
$$

Let $f(z)$ denote the objective function in (11.83). From Lemma 11.2, by introducing $P_s = \phi P_t$ and $P_w = (1 - \phi)P_t$, where ϕ $(0 \leq \phi \leq 1)$ denotes the ratio of the power of information bearing signal to the total power, the maximization problem (11.75) can then be modified to

$$
\underset{0 \le \phi \le 1, z \ge 0}{\text{maximize}} \quad \chi(\phi, z) \triangleq \frac{1 + \dfrac{P_p \left\| \mathbf{h}_{pp} \right\|^2}{\phi P_t \mathscr{I}_p + (1 - \phi) P_t \dfrac{z}{\left\| \mathbf{h}_{pp} \right\|^2} + 1}}{1 + \dfrac{P_p \left\| \mathbf{h}_{pe} \right\|^2}{\phi P_t \mathscr{I}_e + (1 - \phi) P_t \dfrac{f(z)}{\left\| \mathbf{h}_{pe} \right\|^2} + 1}} \tag{11.84a}
$$

$$
\text{subject to} \quad \phi \ge \frac{P_t \mathbf{h}_{ss} \mathbf{w} \mathbf{w}^H \mathbf{h}_{ss}^H + P_p \left| h_{ps} \right|^2 + 1}{P_t \left(\dfrac{\left\| \mathbf{h}_{ss} \right\|^2}{2^{R_{th}} - 1} + \mathbf{h}_{ss} \mathbf{w} \mathbf{w}^H \mathbf{h}_{ss}^H \right)} \tag{11.84b}
$$

where $\mathscr{I}_p \triangleq \frac{\mathbf{h}_{pp}^H \mathbf{H}_{sp} \mathbf{h}_{ss}^H \mathbf{h}_{ss} \mathbf{H}_{sp}^H \mathbf{h}_{pp}}{\left\| \mathbf{h}_{ss} \right\|^2 \left\| \mathbf{h}_{pp} \right\|^2}$ and $\mathscr{I}_e \triangleq \frac{\mathbf{h}_{pe}^H \mathbf{H}_{se} \mathbf{h}_{ss}^H \mathbf{h}_{ss} \mathbf{H}_{se}^H \mathbf{h}_{pe}}{\left\| \mathbf{h}_{ss} \right\|^2 \left\| \mathbf{h}_{pe} \right\|^2}$ are constant in each transmission block. The constraint in (11.84b) is obtained by separating ϕ from $C_{sr}(\mathbf{W}) \ge R_{th}$.

Lemma 11.3. *Given ϕ, let $a = \left| \frac{\mathbf{h}_{pp}^H \mathbf{H}_{sp} \mathbf{H}_{se}^H \mathbf{h}_{pe}}{\left\| \mathbf{H}_{se}^H \mathbf{h}_{pe} \right\| \left\| \mathbf{H}_{sp}^H \mathbf{h}_{pp} \right\|} \right|$. Then, $f(z) = 1 - \left(a\sqrt{1 - z} - \sqrt{(1 - a^2)z} \right)^2$ is a concave function of z.*

Proof. The derivation of $f(z)$ is obtained by following the same steps as [46]. It is simple to check $d^2 f(z)/dz^2 < 0$. $\qquad\square$

From Theorem 1 in [46], for a fixed ϕ, $\chi(z)$ is quasi-concave in z. Its optimal solution can be found through a one-dimensional search over z.

Lemma 11.4. *Given an optimal solution $\mathbf{w}^*$ or z^*, the optimal solution ϕ^* is given as*

$$
\phi^* = \min \left(\frac{P_t \mathbf{h}_{ss} \mathbf{w}^* \mathbf{w}^{*,H} \mathbf{h}_{ss}^H + P_p \left| h_{ps} \right|^2 + 1}{P_t \left(\dfrac{\left\| \mathbf{h}_{ss} \right\|^2}{2^{R_{th}} - 1} + \mathbf{h}_{ss} \mathbf{w}^* \mathbf{w}^{*,H} \mathbf{h}_{ss}^H \right)}, \ 1 \right). \tag{11.85}
$$

Proof. In (11.84), if we ignore the constraint in (11.84b) and consider the optimal solution $\mathbf{w}^*$, by taking the first derivative with respect to ϕ, it is easy to show that $\chi(\phi)$ is a decreasing function with respect to ϕ, i.e. $\chi(\phi)$ is maximum at $\phi^* = 0$. However, the constraint in (11.84b) should be satisfied with the condition $R_{th} < \log_2 \left(1 + \frac{P_t \| \mathbf{h}_{ss} \|^2}{P_p | h_{ps} |^2 + 1} \right)$. This completes the proof. $\qquad\square$

11.4.3   Optimization with transmit power of PT

From the results of Section 11.4.2, we know that the secrecy capacity of the primary system $\bar{C}_{pr}$ increases as the optimal ϕ^* decreases. Furthermore, ϕ^* is a function of R_{th} in Lemma 11.4. If R_{th} increases, which makes ϕ^* increase, then the secrecy capacity of the primary system decreases. In the worst case of $\phi^* = 1$ due to large R_{th}, no JN can be transmitted. To solve this problem and improve $\bar{C}_{pr}$, we incorporate P_p into the maximization problem to reduce ϕ^* as

$$\underset{P_p}{\text{maximize}}\quad \bar{C}_{pr}(P_p) = \log_2 \left(\frac{1 + \dfrac{P_p \left\| \mathbf{h}_{pp} \right\|^2}{\phi P_t \mathscr{I}_p + (1 - \phi)P_t \dfrac{z}{\left\| \mathbf{h}_{pp} \right\|^2} + 1}}{1 + \dfrac{P_p \left\| \mathbf{h}_{pe} \right\|^2}{\phi P_t \mathscr{I}_e + (1 - \phi)P_t \dfrac{f(z)}{\left\| \mathbf{h}_{pe} \right\|^2} + 1}} \right) \tag{11.86a}$$

$$\text{subject to}\quad P_p \left| h_{ps} \right|^2 \le \bar{\gamma}(\phi), \tag{11.86b}$$

where $\bar{\gamma}(\phi) \triangleq \phi P_t \left(\frac{\|\mathbf{h}_{ss}\|^2}{2^{R_{th}} - 1} + \mathbf{h}_{ss} \mathbf{w} \mathbf{w}^H \mathbf{h}_{ss}^H \right) - P_t \mathbf{h}_{ss} \mathbf{w} \mathbf{w}^H \mathbf{h}_{ss}^H - 1$. Note that (11.86b) can be derived by separating $P_p \left| h_{ps} \right|^2$ from $C_{sr}(\mathbf{W}) \ge R_{th}$. In order to solve (11.86), we introduce a classical augmented Lagrangian method using a succession of unconstrained optimization programs as [48]. At each step, the augmented Lagrangian function

$$\mathscr{L}(P_p, b, \mu) = \bar{C}_{pr}(P_p) + \frac{b}{2} \left(\left(\max\{0, \mu + \frac{1}{b}(P_p \left| h_{ps} \right|^2 - \bar{\gamma}(\phi))\} \right)^2 - \mu^2 \right) \tag{11.87}$$

is maximized with respect to P_p, where μ is a Lagrangian dual variable, and b is an adjustable penalty parameter. The Lagrangian dual variable is updated as

$$\mu^+ = \max \left\{ 0, \mu + \frac{1}{b} \left(P_p^+ \left| h_{ps} \right|^2 - \bar{\gamma}(\phi) \right) \right\}, \tag{11.88}$$

where μ^+ and P_p^+ are the values of μ and P_p at the next stage of the iterative algorithm.

   We can develop an efficient iterative algorithm to solve (11.86) by updating μ and b until the convergence criterion is met. Moreover, convergence may be obtained without driving the penalty parameter b to 0. The algorithm that shows the steps of the power control at the PT is summarized in Algorithm 2.

11.4.4   Numerical results

Figure 11.6(a) depicts the secrecy capacity of the primary system versus the transmit power P_t at the ST. We can see that the 'BF and transmit power optimization' scheme outperforms both of the 'BF optimization' and 'No cooperation' schemes, and the performance of the 'BF optimization' scheme is superior to that of 'No cooperation' scheme in all cases. The secrecy capacity of 'No cooperation' scheme is almost zero

---

Algorithm 2 The optimal power control of P_p

Input : $\mathbf{h}_{ss}, \mathbf{H}_{se}, \mathbf{H}_{sp}, \mathbf{h}_{pp}, \mathbf{h}_{pe}, h_{ps}, P_t, P_p^0, b^0,$
$\quad\quad \mu^0, \varepsilon$, and $n = 0$
Output : P_p^n
repeat
Perform the classical augmented Lagrangian function $\mathscr{L}(P_p^n, b^n, \mu^n)$ in (11.87) to obtain P_p^n
Update: $P_p^{n+1} := P_p^n$, μ^{n+1} in (11.88), and $b^{n+1} := b^n$
Set $n := n + 1$
until $\left| \mu^{n+1} - \mu^n \right| \leq \varepsilon$

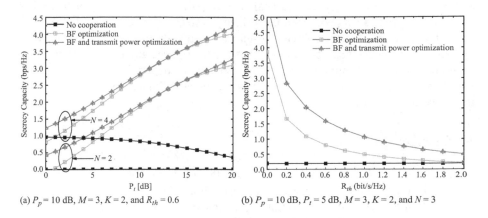

(a) $P_p = 10$ dB, $M = 3$, $K = 2$, and $R_{th} = 0.6$ (b) $P_p = 10$ dB, $P_t = 5$ dB, $M = 3$, $K = 2$, and $N = 3$

Figure 11.6   Secrecy capacity versus P_t in (a) and R_{th} in (b)

due to the setting $N = K = 2$, since the PT cannot protect the primary system from Eve. It can be seen that the secrecy capacity of these schemes increases with increasing the number of antennas at the ST. Furthermore, we can see that the difference between the secrecy capacity of 'BF and transmit power optimization' and that of 'BF optimization' scheme becomes significant when P_t is smaller than 10 dB. This is because a large P_t will decrease ϕ^*. Finally, we observe that the system is not interference-limited as Lemma 11.2.

In Figure 11.6(b), we investigate the effect of the capacity constraint R_{th} of the secondary system on the secrecy capacity of the primary system. The 'No cooperation' does not suffer from the capacity constraint R_{th}; therefore, the secrecy capacity remains constant. As expected, the secrecy capacities of 'BF and transmit power optimization' and 'BF optimization' schemes decrease as R_{th} increases. All cooperation schemes outperform the 'No cooperation' scheme, even when R_{th} is as high as

2 (bps/Hz). It can also be seen that the performance gap between 'No cooperation' scheme and 'BF optimization' scheme becomes smaller for higher R_{th}. This is due to the fact that as R_{th} increases, ϕ^* will increase to allocate more energy to its own signal.

11.5   Conclusions

In this chapter, we have presented physical layer (PHY)-security of cognitive radio network (CRN). Specifically, we have described fundamental PHY-security in CRN and pointed out some recent enhanced protocols available in the literature. Typical applications of artificial noise for the primary system, secondary system, and cooperative CRN have been investigated. We have also presented a powerful technique such as beamforming design on resource allocation problems for such schemes. For the primary system, based on the derived capacity formula, the impact of the secondary system on the secrecy capacity is analysed. In particular, we point out that when the eavesdropper is very far from the primary system, the use of artificial noise is not effective to protect the primary system from eavesdropping. For the secondary system, the proposed approach offers a better performance and is quite robust when compared to the existing approaches. In addition, a cooperative CRN is also presented to improve PHY-security of the primary system. Simulation results are shown to verify the theoretical developments.

References

[1]   Spectrum Policy Task Force, Report of the spectrum efficiency working group Fed. Commun. Commiss., 2002, Tech. Rep. ET Docket-135.

[2]   J. Mitola III and G. Q. Maguire, "Cognitive radios: Making software radios more personal," *IEEE Personal Commun.*, vol. 6, no. 4, pp. 13–18, Aug. 1999.

[3]   E. C. Y. Peh, Y.-C. Liang, Y. L. Guan, and Y. Zeng, "Optimization of cooperative sensing in cognitive radio networks: A sensing-throughput tradeoff view," *IEEE Trans. Veh. Technol.*, vol. 58, no. 9, pp. 5294–5299, Nov. 2009.

[4]   D. He, Y. Lin, C. He, and L. Jiang, "A novel spectrum-sensing technique in cognitive radio based on stochastic resonance," *IEEE Trans. Veh. Technol.*, vol. 59, no. 4, pp. 1680–1688, May 2010.

[5]   J. T. Wang, "Maximum–minimum throughput for MIMO system in cognitive radio network," *IEEE Trans. Veh. Technol.*, vol. 63, no. 1, pp. 217–224, Jan. 2014.

[6]   S. Hua, H. Liu, X. Zhuo, M. Wu, and S. S. Panwar, "Exploiting multiple antennas in cooperative cognitive radio networks," *IEEE Trans. Veh. Technol.*, vol. 63, no. 7, pp. 3318–3330, Sep. 2014.

[7]   L. Zhang, Y. Liang, and Y. Xin, "Joint beamforming and power allocation for multiple access channels in cognitive radio networks," *IEEE J. Select. Areas Commun.*, vol. 26, no. 1, pp. 38–51, Jan. 2008.

[8] F. A. Khan, C. Masouros, and T. Ratnarajah, "Interference-driven linear precoding in multiuser MISO downlink cognitive radio network," *IEEE Trans. Veh. Technol.*, vol. 61, no. 6, pp. 2531–2543, Jul. 2012.

[9] E. A. Gharavol, Y.-C. Liang, and K. Mouthaan, "Robust downlink beamforming in multiuser MISO cognitive radio networks with imperfect channel-state information," *IEEE Trans. Veh. Technol.*, vol. 59, no. 6, pp. 2852–2860, Jul. 2010.

[10] V.-D. Nguyen, L.-N. Tran, T. Q. Duong, O.-S. Shin, and R. Farrell, "An efficient precoder design for multiuser MIMO cognitive radio networks with interference constraints," *IEEE Trans. Veh. Technol.*, vol. 66, no. 5, pp. 3991–4004, May 2017.

[11] A. D. Wyner, "The wire-tap channel," *Bell System Tech. J.*, vol. 54, no. 8, pp. 1355–1387, Oct. 1975.

[12] M. Bloch and J. Barros, *Physical-Layer Security*. Cambridge: Cambridge University Press, 2011.

[13] F. Oggier and B. Hassibi, "The secrecy capacity of the MIMO wiretap channel," *IEEE Trans. Inform. Theory*, vol. 57, no. 8, pp. 4691–4972, Aug. 2011.

[14] Y. Pei, Y.-C. Liang, L. Zhang, K. C. Teh, and K. H. Li, "Secure communication over MISO cognitive radio channels," *IEEE Trans. Wireless Commun.*, vol. 9, no. 4, pp. 1494–1502, Apr. 2010.

[15] A. Khisti and G. W. Wornell, "Secure transmission with multiple antennas-Part I: The MISOME wiretap channel," *IEEE Trans. Inform. Theory*, vol. 56, no. 7, pp. 3088–3104, Jul. 2010.

[16] A. Khisti and G. W. Wornell, "Secure transmission with multiple antennas-Part II: The MIMOME wiretap channel," *IEEE Trans. Inform. Theory*, vol. 56, no. 11, pp. 5515–5532, Nov. 2010.

[17] S. Goel and R. Negi, "Guaranteeing secrecy using artificial noise," *IEEE Trans. Wireless Commun.*, vol. 7, no. 6, pp. 2180–2189, Jun. 2008.

[18] X. Zhou and M. R. McKay, "Secure transmission with artificial noise over fading channels: Achievable rate and optimal power allocation," *IEEE Trans. Veh. Technol.*, vol. 59, no. 8, pp. 3831–3842, Oct. 2010.

[19] T. V. Nguyen and H. Shin, "Power allocation and achievable secrecy rates in MISOME wiretap channels," *IEEE Commun. Lett.*, vol. 15, no. 11, pp. 1196–1198, Nov. 2011.

[20] X. Zhang, X. Zhou, and M. R. McKay, "On the design of artificial-noise-aided secure multi-antenna transmission in slow fading channels," *IEEE Trans. Veh. Technol.*, vol. 62, no. 5, pp. 2170–2181, Jun. 2013.

[21] N. Yang, S. Yan, J. Yuan, R. Malaney, R. Subramanian, and I. Land, "Artificial noise: Transmission optimization in multi-input single-output wiretap channels," *IEEE Trans. Commun.*, vol. 63, no. 5, pp. 1771–1783, May 2015.

[22] H.-M. Wang, T. Zheng, and X.-G. Xia, "Secure MISO wiretap channels with multiantenna passive eavesdropper: Artificial noise vs. artificial fast fading," *IEEE Trans. Wireless Commun.* vol. 14, no. 1, pp. 94–106, Jan. 2015.

[23]   F. Zhu, F. Gao, M. Yao, and H. Zou, "Joint information and jamming beam-forming for physical layer security with full duplex base station," *IEEE Trans. Signal Process.*, vol. 62, no. 24, pp. 6391–6401, Dec. 2014.

[24]   Q. Li and W. K. Ma, "Spatially selective artificial-noise aided transmit optimization for MISO multi-eves secrecy rate maximization," *IEEE Trans. Signal Process.*, vol. 61, no. 10, pp. 2704–2717, May 2013.

[25]   J. Huang and A. L. Swindlehurst, "Robust secure transmission in MISO channels based on worst-case optimization," *IEEE Trans. Signal Process.*, vol. 60, no. 4, pp. 1696–1707, Apr. 2012.

[26]   Q. Li and W.-K. Ma, "Optimal and robust transmit designs for MISO channel secrecy by semidefinite programming," *IEEE Trans. Signal Process.*, vol. 59, no. 8, pp. 3799–3812, Aug. 2011.

[27]   S. Gerbrach, C. Scheunert, and E. A. Jorswieck, "Secrecy outage in MISO systems with partial channel information," *IEEE Trans. Inform. Forensics & Security*, vol. 7, no. 2, pp. 704–716, Apr. 2012.

[28]   F. Zhu and M. Yao, "Improving physical layer security for CRNs using SINR-based cooperative beamforming," *IEEE Trans. Veh. Technol.*, vol. 65, no. 3, pp. 1835–1841, Mar. 2016.

[29]   V.-D. Nguyen, T. Q. Duong, and O.-S. Shin, "Physical layer security for primary system: A symbiotic approach in cooperative cognitive radio networks," in *Proc. IEEE Global Commun. Conf. 2015 (GLOBECOM 2015)*, San Diego, CA, Dec. 2015.

[30]   Y. Pei, Y.-C. Liang, L. Zhang, K. C. Teh, and K. H. Li, "Secure communication in multiantenna cognitive radio networks with imperfect channel state information," *IEEE Trans. Signal Process.*, vol. 59, no. 4, pp. 1683–1693, Apr. 2011.

[31]   V.-D. Nguyen, T. M. Hoang, and O.-S. Shin, "Secrecy capacity of the primary system in a cognitive radio network," *IEEE Trans. Veh. Technol.*, vol. 64, no. 8, pp. 3834–3843, Aug. 2015.

[32]   V.-D. Nguyen, T. Q. Duong, O. A. Dobre, and O.-S. Shin, "Joint information and jamming beamforming for secrecy rate maximization in cognitive radio networks," *IEEE Trans. Inform. Forensics Security*, vol. 11, no. 11, pp. 2609–2623, Nov. 2016.

[33]   Y. Liang, Y. Zeng, E. Peh, and A. Hoang, "Sensing-throughput tradeoff for cognitive radio networks," *IEEE Trans. Wireless Commun.*, vol. 7, no. 4, pp. 1326–1337, Apr. 2008.

[34]   P. Gopala, L. Lai, and H. Gamal, "On the secrecy capacity of fading channels," *IEEE Trans. Inform. Theory*, vol. 54, no. 10, pp. 4687–4698, Oct. 2008.

[35]   Z. Shu, Y. Yang, Y. Qian, and R. Q. Hu, "Impact of interference on secrecy capacity in a cognitive radio network," in *Proc. IEEE Global Commun. Conf. 2011 (GLOBECOM 2011)*, Houston, TX, pp. 1–6, Dec. 2011.

[36]   E. Tekin and A. Yener, "The general Gaussian multiple-access and two-way wiretap channels: Achievable rates and cooperative jamming," *IEEE Trans. Inform. Theory*, vol. 54, no. 6, pp. 2735–2751, Jun. 2008.

[37] W. Yu and T. Lan, "Transmitter optimization for the multi-antenna downlink with per-antenna power constraints," *IEEE Trans. Signal Process.*, vol. 55, no. 6, pp. 2646–2660, Jun. 2007.

[38] W.-C. Liao, T.-H. Chang, W.-K. Ma, and C.-Y. Chi, "QoS-based transmit beamforming in the presence of eavesdroppers: an optimized artificial noise-aided approach," *IEEE Trans. Signal Process.*, vol. 59, no. 3, pp. 1202–1216, Mar. 2011.

[39] D. W. K. Ng, E. S. Lo, and R. Schober, "Robust beamforming for secure communication in systems with wireless information and power transfer," *IEEE Trans. Wireless Commun.*, vol. 13, no. 8, pp. 4599–4615, Aug. 2014.

[40] J. B. Lasserre, "A trace inequality for matrix product," *IEEE Trans. Autom. Control,* vol. 40, no. 8, pp. 1500–1501, Aug. 1995.

[41] L.-N. Tran, M. Juntti, M. Bengtsson, and B. Ottersten, "Beamformer designs for MISO broadcast channels with zero-forcing dirty paper coding," *IEEE Trans. Wireless Commun.*, vol. 12, no. 3, pp. 1173–1185, Mar. 2013.

[42] S. Vishwanath, N. Jindal, and A. Goldsmith, "Duality, achievable rates and sum rate capacity of Gaussian MIMO broadcast channels," *IEEE Trans. Inform. Theory*, vol. 49, no. 10, pp. 2658–2668, Oct. 2003.

[43] S. Boyd and L. Vandenberghe, *Convex Optimization.* Cambridge, UK: Cambridge University Press, 2007.

[44] W. Fenchel, "On conjugate convex functions," *Canadi. J. Math.*, vol. 1, pp. 73–77, 1949.

[45] A. M. Nezhad, R. A. Shandiz, and A. E. Jahromi, "A particle swarm-BFGS algorithm for nonlinear programming problems," *Comput. Oper. Res.*, vol. 40, no. 4, pp. 963–972, Apr. 2013.

[46] G. Zheng, I. Krikidis, J. Li, A. P. Petropulu, and B. Ottersten, "Improving physical layer secrecy using full-duplex jamming receivers," *IEEE Trans. Signal Process.*, vol. 61, no. 20, pp. 4962–4974, Oct. 2013.

[47] D. P. Bertsekas, *Nonlinear Programming.* Belmont, MA: Athena Scientific, 1999.

[48] J. Nocedal and S. J. Wright, *Numerical Optimization.* New York: Springer Verlag, 2006.

Chapter 12

Physical layer security in mmWave cellular networks

Hui-Ming Wang[1]

Recent research has shown that millimetre wave (mmWave) communications can offer orders of magnitude increases in the cellular capacity. However, the physical layer secrecy performance of a mmWave cellular network has not been investigated so far. Leveraging the new path-loss and blockage models for mmWave channels, which are significantly different from conventional microwave channels, this chapter studies the network-wide physical layer security performance of the downlink transmission in a mmWave cellular network under a stochastic geometry framework. We first study the secure connectivity probability and the average number of perfect communication links per unit area in a mmWave network in the presence of non-colluding eavesdroppers. Then, the case of colluding eavesdroppers is studied. Numerical results demonstrate the network-wide secrecy performance, and provide interesting insights into how the secrecy performance is influenced by network parameters.

12.1 Introduction

Within the next 20 years, wireless data traffic is anticipated to skyrocket by 10,000 folds, spurred by the popularity of various intelligent devices, smart phones and tablets. However, today's radio spectrum of below 10 GHz, which could be utilised more for wireless services, has become congested due to the widespread deployment of various wireless communications systems, such as cellular and wireless local area network (WLAN) microwave systems. This challenge inspires a wave of research on exploiting a new frontier of spectrum recently, i.e. millimetre wave (mmWave) communications. The frequencies of mmWave span from 30 to 300 GHz, with the wave lengths ranging from 10 to 1 mm. Recent research shows that over 20 GHz of spectrum is waiting to be used for wireless services in the 28, 38, and 72 GHz alone, and hundreds of GHz more could be used at frequencies above 100 GHz. Given the large amount of spectrum available, mmWave communications are believed to be a strong candidate for the fifth-generation (5G) cellular communications

[1]School of Electronic and Information Engineering, Xi'an Jiaotong University, P. R. China

systems [1]. In this chapter, we will discuss the physical layer security in mmWave cellular networks.

Compared with conventional microwave networks in band below 6 GHz [1–3], recent field measurements have shown the significantly differences of the mmWave networks on account of small wavelength. In particular, the mmWave cellular network has the following distinctive characteristics: propagation sensitivity to blockages, variable propagation laws, being equipped with a large number of antennas, etc. [4]. For example, mmWave signals are more sensitive to blockage effects. Based on the real-world measurements in [3], spatial statistical models of the mmWave channel have been built in [5], which reveals the different path-loss characteristics of the line-of-sight (LOS) and non-line-of-sight (NLOS) links.

Various mmWave channel models have been proposed in [4–11] to characterise the blockage effects of mmWave signals. In [6], an exponential blockage model has been proposed, and such a model has been approximated as a LOS-ball-based blockage model for the coverage analysis in [4,7]. In [8], the authors adopted the exponential blockage model to perform the coverage and capacity analysis for mmWave ad hoc networks. In [10], the authors proposed a ball based blockage model which is validated by using field measurements in New York and Chicago. Taking the outage state emerging in the mmWave communications into consideration, in [11], the authors have proposed a two-ball approximate blockage model for the analysis of the coverage and average rate of multi-tier mmWave cellular networks.

Taking into consideration these new characteristics of mmWave channels, the secrecy performance of a mmWave cellular network will be significantly different from the conventional microwave network, which should be re-evaluated. The efficiency of traditional physical layer security techniques should be re-checked as well. Recently, the secrecy performance of a point-to-point mmWave communication has been studied in [12], which has shown that mmWave systems can enable significant secrecy improvement compared with conventional microwave systems. However, the network-wide secrecy performance of the mmWave cellular communication is still unknown, which is the main topic of this chapter.

In this chapter, using the stochastic geometry framework and the blockage model [10], we propose a systematic secrecy performance analysis approach for the mmWave cellular communication, by modelling the random locations of the base stations (BSs) and eavesdroppers as two independent homogeneous Poisson point processes (PPPs). We characterise the secrecy performance of a noise-limited mmWave cellular network that is applicable to medium/sparse network deployments, where each BS adopts only the directional beamforming to transmit the confidential information. Considering two cases of non-colluding/colluding eavesdroppers, we derive the analysis result of the secure connectivity probability and the cumulative distribution function (CDF) of the received SNR at the typical receiver and eavesdropper, respectively. The secure connectivity probability facilitates the evaluation of the probability of the existence of secure connections from a typical transmitter to its intended receiver. With the CDF of the received SNR at the typical receiver and eavesdropper, we can characterise the average number of perfect communication links per unit area statistically in the random network. We show that the high gain narrow beam antenna is very important for enhancing the secrecy performance of mmWave networks.

Chapter Outline: In Section 12.2, the system model and mmWave channel characteristics are introduced. In Section 12.3, we characterise the secure connectivity probability and average number of perfect communication links per unit area. Numerical results are provided in Section 12.4, and the chapter is concluded in Section 12.5.

Notation: $x \sim \text{gamma}(k, m)$ denotes the gamma-distributed random variable with shape k and scale m, $\gamma(x, y)$ is the lower incomplete gamma function [13, 8.350.1], $\Gamma(x)$ is the gamma function [13, Eq. (8.310)], and $\Gamma(a, x)$ is the upper incomplete function [13, 8.350.2]. $b(o, D)$ denotes a ball whose centre is origin and radius is D. The factorial of a non-negative integer n is denoted by $n!$, $\mathbf{x} \sim \mathscr{CN}(\Lambda, \Delta)$ denotes the circular symmetric complex Gaussian vector with mean vector Λ and covariance matrix Δ, $\binom{n}{k} = \frac{n!}{(n-k)!k!}$. $\mathscr{L}_X(s)$ denotes the Laplace transform of X, i.e. $\mathbb{E}(e^{-sX})$. $_2F_1(\alpha, \beta; \gamma, z)$ is the Gauss hypergeometric function [13, Eq. (9.100)].

12.2 System model and problem formulation

We consider the secure downlink communication in a mmWave cellular network, where multiple spatially distributed BSs transmit the confidential information to authorised users in the presence of multiple malicious eavesdroppers. In the following sub-sections, we first introduce the system model and channel characteristics, which have been validated in [4,8,10]. We then give some important results on probability theory which will be used in the performance analysis.

12.2.1 mmWave cellular system

12.2.1.1 BS and eavesdropper layout

The locations of the BSs are modelled by a homogeneous PPP Φ_B of intensity λ_B. Using PPP for modelling the irregular BS locations has been shown to be mathematically tractable and able to characterise the downlink performance tendency of the cellular network [14–16]. The locations of multiple eavesdroppers are modelled as an independent homogeneous PPP, Φ_E, of intensity λ_E. Such random PPP model is well motivated by the random and unpredictable eavesdroppers' locations, for example, similar assumption on the locations of eavesdroppers has been adopted in the analysis of ad hoc networks [17], cellular networks [18,19] and D-2-D networks [20]. Furthermore, we consider the **worst case** scenario by facilitating the eavesdroppers' multi-user decodability [17,21–23], i.e. the eavesdroppers can perform successive interference cancellation [24] to eliminate the interference due to the information signals from other interfering BSs. The total transmit power of each BS is P_t.

12.2.1.2 Directional beamforming

For compensating the significant path loss at mmWave frequencies, highly directional beamforming antenna arrays are deployed at the BSs to perform the directional beamforming. For mathematical tractability and similar to [4,8,10,11], the antenna pattern is approximated by a sectored antenna model as in [25]. In particular:

$$G_b(\theta) = \begin{cases} M_s, & \text{if} \quad |\theta| \leq \theta_b \\ m_s, & \text{otherwise,} \end{cases} \tag{12.1}$$

where θ_b is the beam width of the main lobe, M_s and m_s are the array gains of main and sidelobes, respectively. We assume that each BS can get the perfect channel state information (CSI) estimation, including angles of arrivals and fading, so that they can then adjust their antenna steering orientation array for adjusting the boresight direction of antennas to their intended receivers and maximising the directivity gains. In the following, we denote the boresight direction of the antennas as $0°$. Therefore, the directivity gain of the intended link is M_s. For each interfering link, the angle θ is independently and uniformly distributed in $[-\pi, \pi]$, which results in a random directivity gain $G_b(\theta)$. For simplifying the performance analysis, the authorised users and malicious eavesdroppers are both assumed to be equipped with a single omnidirectional antenna [10,12].[1]

12.2.1.3   Small-scale fading

The small-scale fading of each link follows independent Nakagami fading [4,8], and the Nakagami fading parameter of the LOS (NLOS) link is N_L (N_N). For simplicity, N_L and N_N are both assumed to be positive integers. Measurement results in [1] showed that highly directional antennas alleviate the effect of small-scale fading, especially for the LOS link. Such scenario can be approximated by using a large Nakagami fading parameter [4,8]. In the following, the small-scale channel gain from the BS at $x \in \mathbb{R}^2$ to the authorised user (eavesdropper) at $y \in \mathbb{R}^2$ is expressed as h_{xy} (g_{xy}).

12.2.1.4   Blockage model

The blockage model proposed in [10] is adopted, which can be regarded as an approximation of the statistical blockage model in [5, Eq. (8)], [11], and incorporates the LOS ball model proposed in [4,7] as a special case. As shown by [9,10], the blockage model proposed in [10] is simple yet flexible enough to capture blockage statistics, coverage and rate trends in mmWave cellular networks. In particular, defining $q_L(r)$ as the probability that a link of length r is LOS:

$$q_L(r) = \begin{cases} C, & \text{if} \quad r \leq D, \\ 0, & \text{otherwise,} \end{cases} \tag{12.2}$$

for some $0 \leq C \leq 1$. The parameter C can be interpreted as the average LOS area in the spherical region around a typical user. The empirical (C, D) for Chicago and Manhattan are $(0.081, 250)$ and $(0.117, 200)$, respectively [10], which are adopted in the simulation results. With such blockage model, the BS process in $b(o, D)$ can be divided into two independent PPPs: the LOS BS process Φ_L with intensity $C\lambda_B$ and

---

[1]This assumption is just for simplifying the performance analysis. However, the obtained analysis methods can be extended to the multiple antennas case directly by modelling the array pattern at authorised users and malicious eavesdroppers in a similar way as (12.1).

NLOS BS process with intensity $(1 - C)\lambda_B$ [14, Proposition 1.3.5]. Outside $b(o, D)$, only the NLOS BS process exists with intensity λ_B. We denote the whole NLOS BS process as Φ_N.

12.2.1.5 Path-loss model

In mmWave propagation, different path-loss laws are applied to LOS and NLOS links [4,10]. In particular, given a link from $x \in \mathbb{R}^2$ to $y \in \mathbb{R}^2$, its path loss $L(x,y)$ can be calculated by

$$L(x,y) = \begin{cases} C_L||x - y||^{-\alpha_L}, & \text{if link } x \to y \text{ is LOS link,} \\ C_N||x - y||^{-\alpha_N}, & \text{if link } x \to y \text{ is NLOS link,} \end{cases} \quad (12.3)$$

where α_L and α_N are the LOS and NLOS path-loss exponents, and $C_L \triangleq 10^{-\frac{\beta_L}{10}}$ and $C_N \triangleq 10^{-\frac{\beta_N}{10}}$ can be regarded as the path-loss intercepts of LOS and NLOS links at the reference distance. Typical α_j and β_j for $j \in \{L, N\}$ are defined in [5, Table I]. For example, for 28 GHz bands, $\beta_L = 61.4$, $\alpha_L = 2$, and $\beta_N = 72$, $\alpha_N = 2.92$. From the measured values of C_j and $\alpha_j, j \in \{L, N\}$ in [5, Table I], we know that it satisfies $C_L > C_N$ and $\alpha_L < \alpha_N$.

12.2.1.6 User association

For maximising the receiving quality of the authorised users, one authorised user is assumed to be associated with the BS that offers the lowest path loss to him, since the network under consideration is homogeneous [4,10]. Thus, for the typical authorised user at the origin, its serving BS is located at $x^* \triangleq \arg\max_{x \in \Phi_B} L(x, o)$. We denote the distance from the typical authorised user to the nearest BS in Φ_j as d_j^* for $j = \{L, N\}$. The following Lemma 1 provides their probability distribution functions (pdf), and the obtained statistics hold for a generic authorised user, according to Slivnyak's theorem [14].

Lemma 12.1. *Given that the typical authorised user observes at least one LOS BS, the pdf of d_L^* is*

$$f_{d_L^*}(r) = \frac{2\pi C \lambda_B r \exp(-\pi C \lambda_B r^2)}{1 - \exp(-\pi C \lambda_B D^2)}, \text{ for } r \in [0, D]. \quad (12.4)$$

On the other hand, the pdf of d_N^* is given by

$$f_{d_N^*}(r) = 2\pi(1 - C)\lambda_B r e^{-\pi(1-C)\lambda_B r^2} \mathbb{I}(r \le D)$$

$$+ 2\lambda_B \pi r e^{-\lambda_B \pi (r^2 - D^2)} e^{-\pi(1-C)\lambda_B D^2} \mathbb{I}(r > D), \quad (12.5)$$

where $\mathbb{I}(.)$ is the indicator function.

Proof. The proof is given in Appendix A. $\square$

Then, the following lemma gives the probability that the typical authorised user is associated with a LOS or NLOS BS.

Lemma 12.2. *The probability that the authorised user is associated with a NLOS BS,* A_N, *is given by*

$$
A_N = \int_0^{\mu} \left(\left(e^{-\pi C \lambda_B \left(\frac{C_L}{C_N} \right)^{\frac{2}{\alpha_L}} x^{\frac{2\alpha_N}{\alpha_L}}} - e^{-\pi C \lambda_B D^2} \right) \right.
$$

$$
\left. 2\pi (1 - C) \lambda_B x e^{-\pi (1-C) \lambda_B x^2} \, dx \right) + e^{-\pi C \lambda_B D^2}, \tag{12.6}
$$

where $\mu \triangleq \left(\frac{C_L}{C_N} \right)^{-\frac{1}{\alpha_N}} D^{\frac{\alpha_L}{\alpha_N}}$. *The probability that the typical authorised user is associated with a LOS BS is given by* $A_L = 1 - A_N$.

Proof. The proof is given in Appendix B. □

With the smallest path-loss association rule, the typical authorised user would be associated with the nearest LOS BS in Φ_L or the nearest NLOS BS in Φ_N. The following lemma gives the pdf of the distance between the typical authorised user and its serving BS in Φ_j, i.e. r_j, $\forall j \in \{L, N\}$.

Lemma 12.3. *On the condition that the serving BS is in* Φ_L, *the pdf of the distance from the typical authorised user to its serving BS in* Φ_L *is given in (12.7).*

$$
f_{r_L}(r) = \frac{\exp\left(-(1 - C)\lambda_B \pi \left(\frac{C_N}{C_L} \right)^{\frac{2}{\alpha_N}} r^{\frac{2\alpha_L}{\alpha_N}} \right) 2\pi C \lambda_B r \exp(-C \lambda_B \pi r^2)}{A_L}, \ r \in [0, D].
$$

$$
\tag{12.7}
$$

On the condition that the serving BS is in Φ_N, *the pdf of the distance from the typical authorised user to its serving BS in* Φ_N *is given in (12.8)*

$$
f_{r_N}(r) = \frac{2\pi \lambda_B r \exp(-\pi \lambda_B r^2)\left((1 - C) \exp(\pi C \lambda_B (r^2 - D^2)) \, \mathbb{I}(r \le D) + \mathbb{I}(r \ge D) \right)}{A_N}
$$

$$
+ \frac{2\pi (1 - C)\lambda_B r e^{-(1-C)\lambda_B \pi r^2} \left(e^{-C \lambda_B \pi \left(\frac{C_L}{C_N} \right)^{\frac{2}{\alpha_L}} r^{\frac{2\alpha_N}{\alpha_L}}} - e^{-C \lambda_B \pi D^2} \right)}{A_N}
$$

$$
\times \mathbb{I}\left(r \le \left(\frac{C_N}{C_L} \right)^{\frac{1}{\alpha_N}} D^{\frac{\alpha_L}{\alpha_N}} \right). \tag{12.8}
$$

Proof. The proof is given in Appendix C. □

12.2.2 Secrecy performance metrics

We assume that the channels are all quasi-static fading channels. The legitimate receivers and eavesdroppers can obtain their own CSI, but mmWave BSs do not know the instantaneous CSI of eavesdroppers. To protect the confidential information from wiretapping, each BS encodes the confidential data by the Wyner code [26]. Then, two code rates, namely the rate of the transmitted codewords R_b and the rate of the confidential information R_s, should be determined before transmission, and $R_b - R_s$ is the cost for securing that information. The details of the code construction can be found in [26,27]. Here, we adopt the fixed rate transmission, where R_b and R_s are fixed during the information transmission [17,27,29]. For the secrecy transmissions over quasi-static fading channels, the perfect secrecy cannot always be guaranteed. Therefore, we adopt outage-based secrecy performance metrics [27–33]. We analyse the secrecy performance of the mmWave communication by considering both the secure connectivity probability and average number of perfect communication links per unit area.

1. *Secure connectivity probability.* Secure connectivity probability is defined as the probability that the secrecy rate is non-negative [28]. Using the secure connectivity probability, we aim to statistically characterise the existence of secure connection between any randomly chosen BS and its intended authorised user in the presence of multiple eavesdroppers.
2. *Average number of perfect communication links per unit area [29].* When R_b and R_s are given, we define the links that have perfect connection and secrecy as perfect communication links. Then, the mathematical definition of the average number of perfect communication links per unit area is given as follows.
   - *Connection probability.* When R_b is below the capacity of legitimate links, authorised users can decode signals with an arbitrary small error, and thus perfect connection can be assured. Otherwise, connection outage would occur. The connection probability is denoted as p_{con}.
   - *Secrecy probability.* When the wiretapping capacity of eavesdroppers is below the rate redundancy $R_e \triangleq R_b - R_s$, there will be no information leakage to potential eavesdroppers, and thus perfect secrecy of the link can be assured [26]. Otherwise, secrecy outage would occur. The secrecy probability is denoted as p_{sec}.

   Following [29, Eq. (29)], the average number of perfect communication links per unit area is

$$N_p = \lambda_B p_{con} p_{sec}. \tag{12.9}$$

Remark 12.1. *With the given R_b and R_s, the average achievable secrecy throughput per unit area ω can be calculated by $\omega = N_p R_s$.*

12.3 Secrecy performance of millimetre wave cellular networks

In this section, we evaluate the secrecy performance of the direct transmission for the noise-limited mmWave communication. As pointed out by [3,5,10,11], highly directional transmissions used in mmWave systems combined with short cell radius results in links that are noise-dominated, especially for densely blocked settings (e.g. urban settings) and medium/sparse network deployments [10,11]. This distinguishes from current dense cellular deployments where links are overwhelmingly interference-dominated. Therefore, we study the secrecy performance of the noise-limited mmWave communication without considering the effect of inter-cell interference.

The received SNR by the typical authorised user at origin and the eavesdropper at z with respect to the serving BS can be expressed as $\text{SNR}_U = \frac{P_t M_s L(x^*, o) h_{x^* o}}{N_0}$ and $\text{SNR}_{E_z} = \frac{P_t G_b(\theta) L(x^*, z) g_{x^* z}}{N_0}$. N_0 is the noise power in the form of $N_0 = 10^{\frac{N_0(dB)}{10}}$, where $N_0(dB) = -174 + 10\log_{10}(\text{BW}) + \mathscr{F}_{dB}$, BW is the transmission bandwidth, and $\mathscr{F}_{dB}$ is the noise figure [11]. With the array pattern in (12.1), $G_b(\theta)$ seen by the eavesdropper is a Bernoulli random variable whose probability mass function (PMF) is given by

$$G_b(\theta) = \begin{cases} M_s, & \Pr_{G_b}(M_s) \triangleq \Pr(G_b(\theta) = M_s) = \frac{\theta_b}{180}, \\ m_s, & \Pr_{G_b}(m_s) \triangleq \Pr(G_b(\theta) = m_s) = \frac{180 - \theta_b}{180}. \end{cases} \quad (12.10)$$

12.3.1 Non-colluding eavesdroppers

In this sub-section, assuming that the random distributed eavesdroppers are **non-colluding**, we evaluate the secrecy performance of the mmWave cellular network.

12.3.1.1 Secure connectivity probability

We first study the secure connectivity probability, τ_n, of the mmWave communication in the presence of multiple non-colluding eavesdroppers. A secure connection is possible, if the condition $\frac{M_s L(x^*, o) h_{x^* o}}{\max_{z \in \Phi_E} G_b(\theta) L(x^*, z) g_{x^* z}} \geq 1$ holds [28], and the secure connectivity probability can be calculated by $\tau_n = \Pr\left(\frac{M_s L(x^*, o) h_{x^* o}}{\max_{z \in \Phi_E} G_b(\theta) L(x^*, z) g_{x^* z}} \geq 1\right)$. We can see that the wiretapping capability of multiple eavesdroppers is determined by the path-loss process $G_b(\theta) L(x^*, z) g_{x^* z}$. Thus, for facilitating the performance evaluation, the following process is introduced.

Definition 12.1. *The path-loss process with fading (PLPF), denoted as $\mathscr{N}_E$, is the point process on $\mathbb{R}^+$ mapped from Φ_E, where $\mathscr{N}_E \triangleq \left\{ \varsigma_z = \frac{1}{G_b(\theta) g_{xz} L(x, z)}, z \in \Phi_E \right\}$ and x denotes the location of the wiretapped BS. We sort the elements of $\mathscr{N}_E$ in ascending order and denote the sorted elements of $\mathscr{N}_E$ as $\{\xi_i, i = 1, \ldots\}$. The index is introduced such that $\xi_i \leq \xi_j$ for $\forall i < j$.*

Note that $\mathscr{N}_E$ involves both the impact of small fading and spatial distribution of eavesdroppers, which is an ordered process. Consequently, $\mathscr{N}_E$ determines the wiretapping capability of eavesdroppers. We then have the following lemma.

Lemma 12.4. *The PLPF $\mathscr{N}_E$ is an one-dimensional non-homogeneous PPP with the intensity measure*

$$\Lambda_E(0, t) = 2\pi \lambda_E \left(\sum_{j \in \{L, N\}} q_j \left(\Omega_{j, in}(M_s, t) + \Omega_{j, in}(m_s, t) \right) \right.$$

$$\left. + \Omega_{N, out}(M_s, t) + \Omega_{N, out}(m_s, t) \right), \quad (12.11)$$

where $q_L \triangleq C$, $q_N \triangleq 1 - C$, and $\Omega_{j,in}(V,t) \triangleq Pr_{G_b}(V)\frac{(VC_j t)^{\frac{2}{\alpha_j}}}{\alpha_j} \sum_{m=0}^{N_j-1} \frac{\gamma\left(m+\frac{2}{\alpha_j}, \frac{D^{\alpha_j}}{VC_j i}\right)}{m!}$,

$\Omega_{j,out}(V,t) \triangleq Pr_{G_b}(V)\frac{(VC_j t)^{\frac{2}{\alpha_j}}}{\alpha_j} \sum_{m=0}^{N_j-1} \frac{\Gamma\left(m+\frac{2}{\alpha_j}, \frac{D^{\alpha_j}}{VC_j i}\right)}{m!}$, with $V \in \{M_s, m_s\}$.

Proof. The proof is given in Appendix D. $\qquad\qquad\qquad\qquad\qquad\qquad\qquad\square$

With the lemma, the following theorem gives the analytical result of the secure connectivity probability in the presence of non-colluding eavesdroppers.

Theorem 12.1. *In the case of non-colluding eavesdroppers, the secure connectivity probability is*

$$\tau_n = \sum_{j\in\{L,N\}} A_j \left(\int_0^{+\infty} f_{r_j}(r)dr \int_0^{+\infty} \frac{e^{-\Lambda_E\left(0, \frac{r^{\alpha_j}}{M_s w C_j}\right)} w^{N_j-1} e^{-w}}{\Gamma(N_j)} dw. \right) \qquad (12.12)$$

Proof. The detailed derivations are given as follows.

$$\Pr\left(\frac{M_s L(x^*, o) h_{x^* o}}{\max_{z\in\Phi_E} G_b(\theta) L(x^*, z) g_{x^* z}} \geq 1 \right)$$

$$= \Pr\left(\min_{z\in\Phi_E} \frac{1}{G_b(\theta) L(x^*, z) g_{x^* z}} \geq \frac{1}{M_s L(x^*, o) h_{x^* o}} \right)$$

$$\overset{(e)}{=} \Pr\left(\xi_1 \geq \frac{1}{M_s L(x^*, o) h_{x^* o}} \right) \overset{f}{=} \mathbb{E}_{L(x^*, o), h_{x^* o}} \left(\exp\left(-\Lambda_E\left(0, \frac{1}{M_s L(x^*, o) h_{x^* o}} \right) \right) \right)$$

$$\overset{(g)}{=} A_L \mathbb{E}_{r_L, h_{x^* o}} \left(\exp\left(-\Lambda_E\left(0, \frac{r_L^{\alpha_L}}{M_s C_L h_{x^* o}} \right) \right) |\text{Serving BS is a LOS BS} \right)$$

$$+ A_N \mathbb{E}_{r_N, h_{x^* o}} \left(\exp\left(-\Lambda_E\left(0, \frac{r_N^{\alpha_N}}{M_s C_N h_{x^* o}} \right) \right) |\text{Serving BS is a NLOS BS} \right).$$

$$(12.13)$$

where step (e) is due to Definition 12.1, step (f) follows the PPP's void probability [14], and step (g) is due to the law of total probability. When the serving BS is a LOS BS, $h_{x^* o} \sim \text{gamma}(N_L, 1)$ and the pdf of r_L is given by (12.7), and when the serving BS is a NLOS BS, $h_{x^* o} \sim \text{gamma}(N_N, 1)$ and the pdf of r_N is given by (12.8). Finally, substituting the pdf of $h_{x^* o}$, r_L, and r_N into (12.13), τ_n can be obtained. $\qquad\square$

Remark 12.2. *Although a closed-form expression of the analytical result of (12.12) is difficult to obtain, the double integral term involved in (12.12) can be evaluated with the iterative numerical method in [34], which facilitates the evaluation of the secure connectivity of mmWave communication in the presence of multiple non-colluding eavesdroppers.*

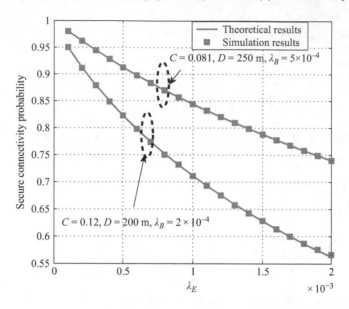

Figure 12.1   Secure connectivity probability of the mmWave communication in the presence of multiple non-colluding eavesdroppers versus λ_E with $BW = 2$ GHz, $P_t = 30$ dB, $\mathscr{F}_{dB} = 10$, $\theta_b = 9°$, $M_s = 15$ dB, and $m_s = -3$ dB. © IEEE 2016. Reprinted with permission from Wang C. and Wang H.-M. Physical Layer Security in Millimetre Wave Cellular Networks. IEEE Trans. Wireless Commun. 2016 Aug; 15(8):5569–85

We plot the secure connectivity probability τ_n versus λ_E in Figure 12.1. For all the simulations, 100,000 trials are used. From Figure 12.1, we can find that theoretical curves obtained by calculating (12.12) coincide with the simulation ones well, which validates the theoretical result in Theorem 12.1.

12.3.1.2   Average number of perfect communication links per unit area

In the following, we study the average number of perfect communication links per unit area, N_p, of the mmWave communication in the presence of non-colluding eavesdroppers. First, we should derive the analytical result of the connection probability and secrecy probability of a mmWave communication link, given by

$$p_{con} \triangleq \Pr(\text{SNR}_U \geq T_c)$$

$$p_{sec,n} \triangleq \Pr\left(\max_{z \in \Phi_E} \text{SNR}_{E_z} \leq T_e\right), \tag{12.14}$$

respectively, where $T_c \triangleq 2^{R_c} - 1$ and $T_e \triangleq 2^{R_e} - 1$. Then, N_p can be calculated by (12.9). We have the following theorem.

Theorem 12.2. *For the non-colluding eavesdroppers case, the analytical result of* p_{con} *is given by*

$$p_{con} = \int_0^D \frac{\Gamma\left(N_L, \frac{N_0 T_c r^{\alpha_L}}{P_t M_S C_L}\right)}{\Gamma(N_L)} f_{r_L}(r) \, dr A_L + \int_0^{+\infty} \frac{\Gamma\left(N_N, \frac{N_0 T_c r^{\alpha_N}}{P_t M_S C_N}\right)}{\Gamma(N_N)} f_{r_N}(r) \, dr A_N, \quad (12.15)$$

and the analytical result of $p_{sec,n}$ *is given by*

$$p_{sec,n} = \exp\left(-\Lambda_E\left(0, \frac{1}{T e N_0}\right)\right). \tag{12.16}$$

Proof. p_{con} can be derived as follows:

$$p_{con} = \Pr\left(h_{x^*o} \geq \frac{N_0 T_c}{P_t M_S L(x^*, o)} \middle| \text{Serving BS is a LOS BS}\right) A_L$$

$$+ \Pr\left(h_{x^*o} \geq \frac{N_0 T_c}{P_t M_S L(x^*, o)} \middle| \text{Serving BS is a NLOS BS}\right) A_N$$

$$= \int_0^D \frac{\Gamma\left(N_L, \frac{N_0 T_c r^{\alpha_L}}{P_t M_S C_L}\right)}{\Gamma(N_L)} f_{r_L}(r) \, dr A_L + \int_0^{+\infty} \frac{\Gamma\left(N_N, \frac{N_0 T_c r^{\alpha_N}}{P_t M_S C_N}\right)}{\Gamma(N_N)} f_{r_N}(r) \, dr A_N.$$

$$\tag{12.17}$$

$p_{sec,n}$ can be derived as follows:

$$p_{sec,n} = \Pr\left\{\frac{\max_{z\phi_{E_z}} G_b(\theta) L(x^*, z) g_{x^*,z}}{N_0} \leq T_e\right\}$$

$$\overset{(g)}{=} \Pr\left\{\frac{1}{\xi_1 N_0} \leq T_e\right\} \overset{(h)}{=} \exp\left(-\Lambda_E\left(0, \frac{1}{T_e N_0}\right)\right), \tag{12.18}$$

where step (g) is due to Definition 12.1, and step (h) is due to the PPP's void probability [14]. □

With (12.15) and (12.16), the average number of perfect communication links per unit area, N_p, can be calculated by (12.9).

12.3.2 Colluding eavesdroppers

In this sub-section, we study the secrecy performance of the mmWave communication by considering the worst case: **colluding eavesdroppers**, where geographically dispersed eavesdroppers adopt the maximal ratio combining to process the wiretapped confidential information.

12.3.2.1 Secure connectivity probability

The secure connectivity probability τ_c in the presence of multiple colluding eavesdroppers can be calculated by

$$\tau_c = \Pr\left(\frac{M_S L(x^*, o) h_{x^*o}}{I_E} \geq 1\right), \tag{12.19}$$

where $I_E \triangleq \sum_{z \in \Phi_E} G_b(\theta) L(x^*, z) g_{x^*z}$. We have the following theorem.

Theorem 12.3. *In the case of colluding eavesdroppers, τ_c can be calculated by*

$$\tau_c = \sum_{j \in \{L,N\}} \mathbb{E}_{r_j} \left[\sum_{m=0}^{N_j-1} \left(\frac{r_j}{M_S C_j} \right)^{\alpha_j} \frac{A_j}{\Gamma(m+1)} (-1)^m \mathscr{L}_{I_E}^{(m)} \left(\frac{r_j^{\alpha_j}}{M_S C_j} \right) \right], \qquad (12.20)$$

where $\mathscr{L}_{I_E}(s) \triangleq \exp(\Xi(s))$, $\mathscr{L}_{I_E}^{(m)}(s) \triangleq \frac{d^m \mathscr{L}_{I_E}(s)}{ds^m}$, and $\Xi(s)$ is given as follows:

$$\Xi(s) \triangleq$$

$$-s \left(2\pi \lambda_E \sum_{j \in \{L,N\}} q_j \left(\sum_{V \in \{M_s, m_s\}} Pr_{G_b}(V) \frac{(VC_j)^{\frac{2}{\alpha_j}}}{\alpha_j} \sum_{m=0}^{N_j-1} \frac{\left(\frac{D^{\alpha_j}}{VC_j} \right)^{m+\frac{2}{\alpha_j}}}{\left(m + \frac{2}{\alpha_j} \right) \left(s + \frac{D^{\alpha_j}}{VC_j} \right)^{m+1}} \right. \right.$$

$$\left. {}_2F_1 \left(1, m+1; m+\frac{2}{\alpha_j}+1; \frac{\frac{D^{\alpha_j}}{VC_j}}{\frac{D^{\alpha_j}}{VC_j}+s} \right) \right) + 2\pi \lambda_E \sum_{V \in \{M_s, m_s\}} Pr_{G_b}(V) \frac{(VC_N)^{\frac{2}{\alpha_N}}}{\alpha_N}$$

$$\sum_{m=0}^{N_N-1} \frac{\left(\frac{D^{\alpha_N}}{VC_N} \right)^{m+\frac{2}{\alpha_N}}}{\left(1 - \frac{2}{\alpha_N} \right) \left(s + \frac{D^{\alpha_N}}{VC_N} \right)^{m+1}} {}_2F_1 \left(1, m+1; 2 - \frac{2}{\alpha_N}; \frac{s}{s + \frac{D^{\alpha_N}}{VC_N}} \right) \right). \qquad (12.21)$$

Proof. The proof is given in Appendix E. □

Although the analytical result given in Theorem 12.3 is general and exact, it is rather unwieldy, motivating the interest in acquiring a more compact expression. Exploring the tight lower bound of the CDF of the gamma random variable in [35], a tight upper bound of τ_c can be calculated as follows.

Theorem 12.4. *τ_c can be tightly upper bounded by*

$$\tau_c \lesssim \sum_{j \in \{L,N\}} \sum_{n=1}^{N_j} \binom{N_j}{n} (-1)^{n+1} \int_0^{+\infty} f_{r_j}(r) \mathscr{L}_{I_E} \left(\frac{a_j n r^{\alpha_j}}{M_S C_j} \right) dr, \qquad (12.22)$$

where $a_L \triangleq (N_L)^{-\frac{1}{N_L}}$ and $a_N \triangleq (N_N)^{-\frac{1}{N_N}}$.

Proof. We leverage the tight lower bound of the CDF of a normalised gamma random variable, g with N degrees of freedom as $\Pr(x \leq y) \gtrsim \left(1 - e^{-\kappa y} \right)^N$ [35], where

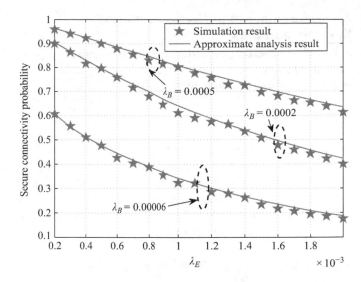

Figure 12.2 Secure connectivity probability of mmWave communication in the presence of multiple colluding eavesdroppers versus λ_E. The system parameters are $P_t = 30\,dB$, $\beta_L = 61.4\,dB$, $\alpha_L = 2$, $\beta_N = 72\,dB$, $\alpha_N = 2.92$, $BW = 2\,GHz$, $\mathscr{F}_{dB} = 10$, $\lambda_B = 0.0005, 0.0002, 0.00006$, $C = 0.12$, $D = 200\,m$, $\theta_b = 9°$, $M_s = 15\,dB$, and $m_s = -3\,dB$. © IEEE 2016. Reprinted with permission from Wang C. and Wang H.-M. Physical Layer Security in Millimetre Wave Cellular Networks. IEEE Trans. Wireless Commun. 2016 Aug; 15(8):5569–85

$\kappa = (N!)^{-\frac{1}{N}}$. Since $h_{x^*,o}$ is a normalised gamma random variable, we have:

$$\tau_c \lesssim 1 - \sum_{j \in \{L,N\}} \mathbb{E}_{I_E, r_j}\left(\left(1 - \exp\left(-\frac{na_j I_E r_j^{\alpha_j}}{M_S C_j} \right) \right)^{N_j} \right). \tag{12.23}$$

Using the binomial expansion, we can obtain (12.24) as follows:

$$\tau_c \lesssim \sum_{j \in \{L,N\}} \sum_{n=1}^{N_j} \binom{N_j}{n} (-1)^{n+1} \mathbb{E}_{r_j}\left(\mathbb{E}_{I_E}\left(\exp\left(-\frac{a_j n r_j^{\alpha_j} I_E}{M_S C_j} \right) \right) \right)$$

$$= \sum_{j \in \{L,N\}} \sum_{n=1}^{N_j} \binom{N_j}{n} (-1)^{n+1} \int_0^{+\infty} f_{r_j}(r) \mathscr{L}_{I_E}\left(\frac{a_j n r^{\alpha_j}}{M_S C_j} \right). \qquad \Box \tag{12.24}$$

The bounds in Theorem 12.4 are validated in Figure 12.2, where we plot the secure connectivity probability τ_c versus λ_E. From Figure 12.2, we can find that theoretical curves coincide with simulation ones well, which show that the upper bound given in Theorem 12.4 is tight.

12.3.2.2   Average number of perfect communication links per unit area

In the case of colluding eavesdroppers, the connection probability p_{con} of the typical authorised user can be still calculated by (12.15) in Theorem 12.2, and the achievable secrecy probability p_{sec} can be calculated by

$$p_{sec,c} = \Pr\left\{\frac{P_t I_E}{N_0} \leq T_e\right\}. \tag{12.25}$$

To get the analytical result of $p_{sec,c}$ in (12.25), the CDF of I_E should be available. Although the CDF of I_E can be obtained from its Laplace transform $\mathscr{L}_{I_E}(s)$ by using the inverse Laplace transform calculation [36], it could get computationally intensive in certain cases and may render the analysis intractable. As an alternative, we resort to an approximation method widely adopted in [4,8,10] for getting an approximation of $p_{sec,c}$ which is given in the following theorem.

Theorem 12.5. *In the case of multiple colluding eavesdroppers, the approximation of $p_{sec,c}$ is given by*

$$p_{sec,c} \lesssim \sum_{n=1}^{N} (-1)^{n+1} \mathscr{L}_{I_E}\left(-\frac{an}{n_0 T_e}\right), \tag{12.26}$$

where $\mathscr{L}_{I_E}(s)$ is given in Theorem 12.3, $a \triangleq (N!)^{\frac{1}{N}}$, and N is the number of terms used in approximation.

Proof.

$$p_{sec,c} = \Pr\left\{\frac{P_t I_E}{N_0 T_e} \leq 1\right\} \stackrel{(i)}{\approx} \Pr\left\{\frac{P_t I_E}{N_0 T_e} \leq w\right\}. \tag{12.27}$$

In (*i*), w is a normalised gamma random variable with a shape parameter, N, and the approximation in (*i*) is due to the fact that a normalised gamma random variable converges to identity when its shape parameter goes to infinity [4,8].

Then, using the tight lower bound of the CDF of a normalised gamma random variable in [35], p_{sec} can be tightly upper bounded by

$$p_{sec,c} \lesssim 1 - \left(1 - \exp\left(-\frac{aP_t I_E}{N_0 T_e}\right)\right)^N. \tag{12.28}$$

Finally, using the binomial expansion, (12.28) can be further re-written as (12.26).   □

The approximate analysis result in Theorem 12.5 is validated in Figure 12.3. From Figure 12.3, we can find that when $N = 5$, (12.26) can give an accurate approximation. Then, in the following simulations, we set $N = 5$ to calculate $p_{sec,c}$, approximately.

Finally, the number of perfect links in the network, N_p, can be calculated with (12.15) and (12.26), approximately.

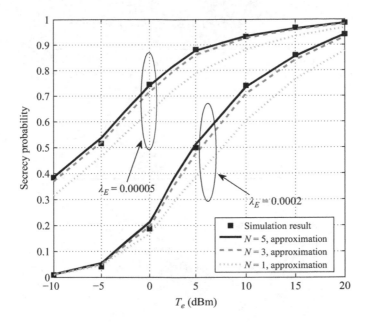

Figure 12.3 *Secrecy probability of the mmWave communication in the presence of multiple colluding eavesdroppers versus T_e. The system parameters are $P_t = 30$ dB, $\theta_b = 9°$, $M_s = 15$ dB, $m_a = -3$ dB, $\beta_L = 61.4$ dB, $\alpha_L = 2$, $\beta_N = 72$ dB, $\alpha_N = 2.92$, BW $= 2$ GHz, $\mathscr{F}_{dB} = 10$, $\lambda_B = 0.0005$, $C = 0.081$, and $D = 250$ m. © IEEE 2016. Reprinted with permission from Wang C. and Wang H.-M. Physical Layer Security in Millimetre Wave Cellular Networks. IEEE Trans. Wireless Commun. 2016 Aug; 15(8):5569–85*

12.4 Simulation result

In this section, more simulation results are provided to characterise the secrecy performance of mmWave networks and the effect of different network parameters.

Part of the simulation parameters are summarised in Table 12.1, and the others are separately specified in each figure. Since the theoretical analysis results obtained have been validated by the simulation results in Figures 12.1–12.3, all of the simulation results in this section are theoretical analysis results. Employing analytical results, we illustrate the secrecy performance of mmWave networks in the presence of non-colluding and colluding eavesdroppers.

Figure 12.4 plots the secrecy connectivity probability of the mmWave communication in the presence of multiple non-colluding and colluding eavesdroppers versus λ_E. Obviously, the wiretapping capability of the colluding eavesdroppers is larger than the non-colluding case. Therefore, compared with the non-colluding eavesdroppers, the secrecy connectivity probability for the colluding case deteriorates.

Table 12.1   Part of the simulation parameters

Parameter	Value
P_t	30 dBm
$N_L(N_N)$	3(2)
F_c	28 GHz
BW	2 GHz
$\mathscr{F}_{dB}$	10 dB
$\alpha_L(\alpha_N)$	2(2.92)
$\beta_L(\beta_N)$	61.4(72)

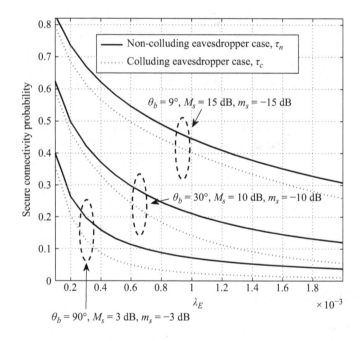

Figure 12.4   Secrecy connectivity probability of the mmWave communication in the presence of multiple eavesdroppers versus λ_E. The system parameters: $\lambda_B = 0.00005$, $C = 0.081$, and $D = 250$ m. © IEEE 2016. Reprinted with permission from Wang C. and Wang H.-M. Physical Layer Security in Millimetre Wave Cellular Networks. IEEE Trans. Wireless Commun. 2016 Aug; 15(8):5569–85

With the increasing λ_E, the wiretapping capability of eavesdroppers increases and the secrecy connectivity probability decreases. Furthermore, the secrecy performance would be enhanced with the improving directionality of the beamforming of each BS. This can be explained by the fact that the high gain narrow beam antenna decreases

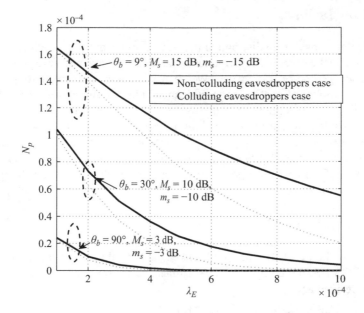

Figure 12.5 Average number of perfect communication links per unit, N_p of mmWave communication in the presence of multiple eavesdroppers versus λ_E. The system parameters: $T_c = 10dB$, $T_e = 0\ dB$, $\lambda_B = 0.0002$, $C = 0.12$, and $D = 200\ m$. © IEEE 2016. Reprinted with permission from Wang C. and Wang H.-M. Physical Layer Security in Millimetre Wave Cellular Networks. IEEE Trans. Wireless Commun. 2016 Aug; 15(8):5569–85

the information leakage, improves the receive performance of the authorised user, and increases the secure connectivity probability.

Figure 12.5 plots N_p versus λ_E. Compared with the case of non-colluding eavesdroppers, the performance deterioration of the colluding case increases with the increasing λ_E, especially for the BS equipped with highly directional antenna arrays. Furthermore, the simulation results show that the directional beamforming is very important for the secrecy communication. For example, with non-colluding eavesdroppers, when $\theta_b = 9°$, $M_s = 15$ dB, $M_a = 3$ dB, $\lambda_E = 4 \times 10^{-4}$, $N_p \approx 1.1 \times 10^{-4}$, and more than half of communication links are perfect, on average. However, for other two cases of array patterns, N_p reduces greatly due to the increasing beam width of the main lobe and the decreasing array gains of the intended sector.

The simulation results above show that the directional beamforming of BSs is very important for the secrecy performance of mmWave networks. Therefore, in practice, BSs should perform the highly directional beamforming.

12.5 Conclusions

In this chapter, considering distinguishing features of the mmWave cellular networks, we characterise the secrecy performance of the noise-limited mmWave network using the stochastic geometry framework. We analyse the secure connectivity probability and average number of perfect communication links per unit area for colluding and non-colluding eavesdroppers. Simulation results validated our analysis. From the simulation results, we find that the array pattern has a great influence on the system secrecy performance. When the highly directional beamforming is utilised, high secure connectivity probability can be achieved. Another important system parameter is the density of the eavesdroppers, the increase of which will cause a sharp decline in secrecy performance.

A.1 Appendix A

We first show the derivation of $f_{d_L^*}(r)$. Given the typical authorised user observes at least one LOS BS, the complementary cumulative distribution function (CCDF) of d_L^* can be derived as

$$
\Pr\left(d_L^* \geq r\right) \triangleq \Pr\left(\Phi_{B_L}(B(o,r)) = 0 | \Phi_{B_L}(B(o,D)) \neq 0\right)
$$

$$
= \frac{e^{-C\lambda_B \pi r^2}\left(1 - e^{-C\lambda_B \pi (D^2 - r^2)}\right)}{1 - e^{-C\lambda_B \pi D^2}}, \tag{A.1}
$$

Then, with (A.1), the pdf $f_{d_L^*}(r) = -\frac{d\Pr(d_L^* \geq r)}{dr}$ that can be derived as (12.4).

Second, invoking the PPP's void probability [14], the CCDF $\Pr\left(d_N^* \geq r\right)$ can be derived as follows:

$$
\Pr\left(d_N^* \geq r\right) = \Pr\left(\Phi_{B_N}(B(o,r)) = 0\right) \mathbb{I}(r \leq D) + \Pr\left(\Phi_{B_N}(B(o,D)) = 0,\right.
$$

$$
\times \Phi_{B_N}(B(o,r)/B(o,D)) = 0\right) \mathbb{I}(r > D)
$$

$$
= \exp\left((1 - C)\lambda_B \pi D^2\right)\exp\left(-\lambda_B \pi \left(r^2 - D^2\right)\right)\mathbb{I}(r > D)
$$

$$
+ \exp\left((1 - C)\lambda_B \pi r^2\right)\mathbb{I}(r \leq D). \tag{A.2}
$$

Finally, calculating $-\frac{d\Pr(d_N^* \geq r)}{dr}$, the pdf $f_{d_N^*}(r)$ can be derived as (12.5).

B.2 Appendix B

The analysis result of A_N can be derived with procedures as follows:

$$
A_N = \Pr\left(C_L(d_L^*)^{-\alpha_L} \leq C_N(d_N^*)^{-\alpha_N}\right)\Pr\left(\Phi_{B_L}(B(o,D)) \neq 0\right) + \Pr\left(\Phi_{B_L}(B(o,D)) = 0\right)
$$

$$
= \Pr\left(\left(\frac{C_L}{C_N}\right)^{\frac{1}{\alpha_L}}(d_N^*)^{\frac{\alpha_N}{\alpha_L}} \leq d_L^*\right)\left(1 - e^{-\pi C\lambda_B D^2}\right) + e^{-\pi C\lambda_B D^2}
$$

$$
= \mathbb{E}_{d_N^* \leq \left(\frac{C_L}{C_N}\right)^{-\frac{1}{\alpha_N}} D^{\frac{\alpha_L}{\alpha_N}}}\left(e^{-\pi C\lambda_B \left(\frac{C_L}{C_N}\right)^{\frac{2}{\alpha_L}}(d_N^*)^{\frac{2\alpha_N}{\alpha_L}}} - e^{-\pi C\lambda_B D^2}\right) + e^{-\pi C\lambda_B D^2}. \tag{B.1}
$$

Finally, (12.6) can be obtained by employing the pdf $f_{d_N^*}(r)$ in (12.5).

Accordingly, the analysis result of A_L can be derived with the procedures in (B.2) as follows:

$$A_L = \Pr\left(C_L(d_L^*)^{-\alpha_L} \geq C_N(d_N^*)^{-\alpha_N}\right)\Pr\left(\Phi_{B_L}(B(o,D)) \neq 0\right)$$

$$= \left(\Pr\left(D \geq \left(\frac{C_L}{C_N}\right)^{\frac{1}{\alpha_L}}(d_N^*)^{\frac{\alpha_N}{\alpha_L}} \geq d_L^*\right)\right.$$

$$\left. + \Pr\left(D \leq \left(\frac{C_L}{C_N}\right)^{\frac{1}{\alpha_L}}(d_N^*)^{\frac{\alpha_N}{\alpha_L}}\right)\right)\Pr\left(\Phi_{B_L}(B(o,D)) \neq 0\right)$$

$$= \left(\int_0^\mu \left(\Pr\left(\left(\frac{C_L}{C_N}\right)^{\frac{1}{\alpha_L}} r^{\frac{\alpha_N}{\alpha_L}} \geq d_L^*\right)\right)f_{d_N^*}(r)dr + \Pr\left(d_N^* \geq \mu\right)\right)$$

$$\times \Pr\left(\Phi_{B_L}(B(o,D)) \neq 0\right). \tag{B.2}$$

Since $\Pr(d_L^* \leq r) = 1 - \Pr(d_L^* \geq r)$ and $\Pr(d_L^* \geq r)$ has been defined in (A.1), substituting the analytical expression of $\Pr(d_L^* \leq r)$ and $f_{d_N^*}(r)$ in (12.5) into (B.2), we obtain $A_L = 1 - A_N$.

C.3 Appendix C

We first show the derivation of $f_{r_L}(r)$. The CCDF $\Pr(r_L \geq r)$ can be derived with the procedures in (C.1) as follows:

$$\Pr(r_L \geq r) \triangleq \Pr\left(\Phi_{B_L}(B(o,r)) = 0 \mid C_L(d_L^*)^{-\alpha_L} \geq C_N(d_N^*)^{-\alpha_N}\right)$$

$$= \frac{\Pr\left(\Phi_{B_L}(B(o,D)) \neq 0\right)\Pr\left(r \leq d_L^*, C_L(d_L^*)^{-\alpha_L} \geq C_N(d_N^*)^{-\alpha_N}\right)}{\Pr\left(C_L(d_L^*)^{-\alpha_L} \geq C_N(d_N^*)^{-\alpha_N}\right)}$$

$$\overset{(a)}{=} \frac{1}{A_L}\int_r^D \Pr\left(\Phi_N \cap b\left(o, \left(\frac{C_N}{C_L}\right)^{\frac{1}{\alpha_N}} y^{\frac{\alpha_L}{\alpha_N}}\right) = \emptyset\right)2\pi C\lambda_B y\exp\left(-\pi C\lambda_B y^2\right)dy. \tag{C.1}$$

Step (a) can be derived by the PPP's void probability and $f_{d_L^*}(r)$. Then, calculating $f_{r_L}(r) = -\frac{d\Pr(r_L \geq r)}{dr}$, the pdf $f_{r_L}(r)$ can be derived as (12.7).

We show the derivation of $f_{r_N}(r)$. The CCDF $\Pr(r_N \geq r)$ is equivalent to the conditional CCDF in (C.2) as follows:

$$\Pr(r_N \geq r) \triangleq \Pr\left(d_N^* \geq r \mid C_L(d_L^*)^{-\alpha_L} \leq C_N(d_N^*)^{-\alpha_N}\right)$$

$$= \frac{\Pr\left(r \leq d_N^*, C_L(d_L^*)^{-\alpha_L} \leq C_N(d_N^*)^{-\alpha_N}\right)\Pr\left(\Phi_{B_L}(B(o,D)) \neq 0\right)}{A_N}$$

$$+ \frac{\Pr\left(r \leq d_N^*, \Phi_{B_L}(B(o,D)) = 0\right)}{A_N}. \tag{C.2}$$

We firstly derive the first term in (C.2) as follows:

$$\frac{\mathbb{E}_{d_N^* \ge r}\left(\Pr\left(\left(\frac{C_L}{C_N}\right)^{\frac{1}{\alpha_L}}(d_N^*)^{\frac{\alpha_N}{\alpha_L}} \le d_L^*\right)\right)\Pr\left(\Phi_{B_L}(B(o,D)) \neq 0\right)}{A_N}$$

$$\stackrel{(b)}{=} \frac{\mathbb{E}_{d_N^* \ge r}\left(\exp\left(-C\lambda_B\pi\left(\frac{C_L}{C_N}\right)^{\frac{2}{\alpha_L}}(d_N^*)^{\frac{2\alpha_N}{\alpha_L}}\right) - e^{-C\lambda_B\pi D^2}\right)}{A_N}\mathbb{I}\left(r \le \left(\frac{C_N}{C_L}\right)^{\frac{1}{\alpha_N}}D^{\frac{\alpha_L}{\alpha_N}}\right),$$

$$\text{(C.3)}$$

where step (b) is obtained according to the PPP's void probability.

Finally, we derive the second term in (C.2). Since the LOS BS process and NLOS BS process are two independent PPPs, we have:

$$\frac{\Pr\left(r \le d_N^*, \Phi_{B_L}(B(o,D)) = 0\right)}{A_N} = \frac{\exp\left(-(1-C)\lambda_B\pi r^2 - C\lambda_B\pi D^2\right)}{A_N}\mathbb{I}(r \le D)$$

$$+ \frac{e^{-\lambda_B\pi r^2}}{A_N}\mathbb{I}(r \ge D). \quad \text{(C.4)}$$

Substituting (C.4) and (C.3) into (C.2), $f_{r_N}(r) = -\frac{d\Pr(r_N \ge r)}{dr}$, which can be derived as (12.8).

D.4   Appendix D

The point process $\mathcal{N}_E$ can be regarded as a transformation of the point process Φ_E by the probability kernel $p(z,A) = \Pr\left(\frac{1}{G_b(\theta)g_{xz}L(x,z)} \in A\right), z \in \mathbb{R}^2, A \in \mathcal{B}(\mathbb{R}^+)$. According to the displacement theorem [14], $\mathcal{N}_E$ is a PPP on $\mathbb{R}^+$ with the intensity measure $\Lambda_E(0,t)$ given by

$$\Lambda_E(0,t) = \lambda_E \int_{\mathbb{R}^2} \Pr\left(\frac{1}{G_b(\theta)g_{xz}L(x,z)} \in [0,t]\right)dz. \quad \text{(D.1)}$$

From the blockage model in Section 12.2.1, we know that Φ_E is divided into two independent point processes, i.e. the LOS and NLOS eavesdropper processes. Furthermore, the directivity gains received at the eavesdroppers in the main and sidelobes are different. Therefore, changing to a polar co-ordinate system, $\Lambda_E(0,t)$ in (D.1) can be further derived as (D.2) given by

$$\Lambda_E(0,t) = 2\pi\lambda_E \sum_{V \in \{M_s, m_s\}} \Pr_{G_b}(V) \sum_{j \in \{L,N\}} q_j \int_0^D \Pr\left(\frac{r^{\alpha_j}}{G_b(\theta)g_r C_j} \le t | G_b(\theta) = V\right)rdr$$

$$+ 2\pi\lambda_E \sum_{V \in \{M_s, m_s\}} \Pr_{G_b}(V) \int_D^{+\infty} \Pr\left(\frac{r^{\alpha_N}}{G_b(\theta)g_r C_N} \le t | G_b(\theta) = V\right)rdr,$$

$$\text{(D.2)}$$

where g_r denotes the small-scale fading of eavesdropper which has distance r from the target BS at x. $g_r \sim$ gamma $(N_L, 1)$ if the link between the eavesdropper and the target BS is LOS, otherwise, $g_r \sim$ gamma$(N_N, 1)$.

For getting the analytical result $\Lambda_E(0, t)$, the integral formulas in (D.2) should be derived. First, the integral $\int_0^D \Pr\left(\frac{r^{\alpha_j}}{G_b(\theta)g_r C_j} \leq t | G_b(\theta) = V\right) r dr$ can be derived with the following procedures:

$$\int_0^D \Pr\left(g_r \geq \frac{r^{\alpha_j}}{VC_j t}\right) r dr \overset{(a)}{=} \int_0^D \left(1 - \frac{\gamma\left(N_j, \frac{r^{\alpha_j}}{VC_j t}\right)}{\Gamma(N_j)}\right) r dr \overset{(b)}{=} \int_0^D \frac{\Gamma\left(N_j, \frac{r^{\alpha_j}}{VC_j t}\right)}{\Gamma(N_j)} r dr$$

$$\overset{(c)}{=} \int_0^D e^{-\frac{r^{\alpha_j}}{VC_j t}} \sum_{m=0}^{N_j-1} \left(\frac{r^{\alpha_j}}{VC_j t}\right)^m \frac{1}{m!} r dr \overset{(d)}{=} \frac{(VC_j t)^{\frac{2}{\alpha_j}}}{\alpha_j} \sum_{m=0}^{N_j-1} \frac{\gamma\left(m + \frac{2}{\alpha_j}, \frac{D^{\alpha_j}}{VC_j t}\right)}{m!}, \qquad \text{(D.3)}$$

where step (a) is due to $g_r \sim$ gamma $(N_j, 1)$, step (b) is due to [13, Eq. (8.356.3)], step (c) is due to [13, Eq. (8.352.2)], and step (d) is due to [13, Eq. (3.381.1)].

With a similar procedure, the integral $\int_D^{+\infty} \Pr\left(\frac{r^{\alpha_j}}{G_b(\theta)g_r C_j} \leq t | G_b(\theta) = V\right) r dr$ can be derived as

$$\int_D^{+\infty} \Pr\left(\frac{r^{\alpha_j}}{G_b(\theta)g_r C_j} \leq t | G_b(\theta) = V\right) r dr = \frac{(VC_j t)^{\frac{2}{\alpha_j}}}{\alpha_j} \sum_{m=0}^{N_j-1} \frac{\Gamma\left(m + \frac{2}{\alpha_j}, \frac{D^{\alpha_j}}{VC_j t}\right)}{m!}.$$

$$\text{(D.4)}$$

Finally, substituting (D.3) and (D.4) into (D.2), the proof can be completed.

E.5 Appendix E

The achievable secure connectivity probability, τ_c can be calculated as (E.1) given by

$$\tau_c \overset{(g)}{=} \sum_{j\in\{L,N\}} \mathbb{E}_{r_j}\left[\mathbb{E}_{I_E}\left[e^{-\frac{I_E r_j^{\alpha_j}}{M_s C_j}} \sum_{m=0}^{N_j-1}\left(\frac{I_E r_j^{\alpha_j}}{M_s C_j}\right)^m \frac{A_j}{\Gamma(m+1)}\right]\right]$$

$$\overset{(h)}{=} \sum_{j\in\{L,N\}} \mathbb{E}_{r_j}\left[\sum_{m=0}^{N_j-1}\left(\frac{r_j^{\alpha_j}}{M_s C_j}\right)^m \frac{A_j}{\Gamma(m+1)}(-1)^m \mathscr{L}_{I_E}^{(m)}\left(\frac{r_j^{\alpha_j}}{M_s C_j}\right)\right], \qquad \text{(E.1)}$$

where step (g) holds since the serving BS at x^* can be a LOS or NLOS BS, and step (h) is due to the Laplace transform property $t^n f(t) \overset{\mathscr{L}}{\leftrightarrow} (-1)^n \frac{d^n}{ds^n} \mathscr{L}_{f(t)}(s)$.

In the following, we derive the analysis result of $\mathscr{L}_{I_E}(s)$:

$$\mathscr{L}_{I_E}(s) = \mathbb{E}\left(e^{-s\sum_{z\in\Phi_E} G_b(\theta)L(x^*,z)g_{x^*z}}\right) \overset{(i)}{=} \exp\left(\int_0^{+\infty}\left(e^{-\frac{s}{x}}-1\right)\Lambda_E(0,dx)\right)$$

$$\overset{(k)}{=} \exp\left(-\int_0^{+\infty}\Lambda_E(0,x)\frac{s}{x^2}e^{-\frac{s}{x}}dx\right) \overset{(v)}{=} \exp\left(\underbrace{-\int_0^{+\infty}\Lambda_E\left(0,\frac{1}{z}\right)se^{-sz}dz}_{T}\right),$$

$$\text{(E.2)}$$

where step (i) is obtained by using the probability generating functional (PGFL), step (k) is obtained by using integration by parts, and step (v) is obtained by the variable replacing $z = \frac{1}{x}$. Then, we concentrate on deriving the analysis result of T in (E.2). Substituting $\Lambda_E(0,\frac{1}{z})$ in (12.11) into T, T can be re-written as follows:

$$T = -s\left(2\pi\lambda_E \sum_{j\in\{L,N\}} q_j \left(\sum_{V\in\{M_s,m_s\}} \rho_j \sum_{m=0}^{N_j-1}\frac{1}{m!}\underbrace{\int_0^\infty \frac{\gamma\left(m+\frac{2}{\alpha_j},\frac{D^{\alpha_j}z}{VC_j}\right)}{z^{\frac{2}{\alpha_j}}}e^{-sz}dz}_{H_1}\right)\right.$$

$$\left.+ \sum_{V\in\{M_s,m_s\}}\Pr\nolimits_{G_b}(V)\rho_N\sum_{m=0}^{N_N-1}\frac{1}{m!}\underbrace{\int_0^\infty\frac{\Gamma\left(m+\frac{2}{\alpha_N},\frac{D^{\alpha_N}z}{VC_N}\right)}{z^{\frac{2}{\alpha_N}}}e^{-sz}dz}_{H_2}\right), \quad \text{(E.3)}$$

where $\rho_j \triangleq \frac{\Pr_{G_b}(V)(VC_j)^{\frac{2}{\alpha_j}}}{\alpha_j}$.

Using [13, Eq. (6.455.1)] and [13, Eq. (6.455.2)], the integral terms H_1 and H_2 can be calculated as

$$H_1 = \frac{(D^{\alpha_j}/(VC_j))^{m+\frac{2}{\alpha_j}}\,\Gamma(m+1)}{\left(m+\frac{2}{\alpha_j}\right)(s+D^{\alpha_j}/(VC_j))^{m+1}}{}_2F_1\left(1,m+1;m+\frac{2}{\alpha_j}+1;\frac{D^{\alpha_j}/(VC_j)}{s+D^{\alpha_j}/(VC_j)}\right),$$

$$\text{(E.4)}$$

$$H_2 = \frac{(D^{\alpha_N}/(VC_N))^{m+\frac{2}{\alpha_N}}\,\Gamma(m+1)}{\left(1-\frac{2}{\alpha_N}\right)(s+D^{\alpha_N}/(VC_j))^{m+1}}{}_2F_1\left(1,m+1;2-\frac{2}{\alpha_N}+1;\frac{s}{s+D^{\alpha_j}/(VC_j)}\right).$$

$$\text{(E.5)}$$

Finally, substituting (E.4) and (E.5) into (E.3), the closed-form result of $\mathscr{L}_{I_E}(s)$ can be obtained.

References

[1] T. S. Rappaport, S. Sun, R. Mayzus *et al.*, "Millimeter wave mobile communications for 5G cellular: It will work!" *IEEE Access*, vol. 1, pp. 335–349, 2013.

[2] T. A. Thomas and F.W. Vook, "System level modeling and performance of an outdoormm Wave local area access system", in *Proc. IEEE Int. Symp. Personal Indoor and Mobile Radio Commun.*, Washington, USA, Sep. 2014.

[3] S. Rangan, T. S. Rappaport, and E. Erkip "Millimeter-wave cellular wireless networks: Potentials and challenges," *Proc. IEEE*, vol. 102, no. 3, pp. 365–385, March 2014.

[4] T. Bai and R. W. Heath, Jr., "Coverage and rate analysis for millimeter wave cellular networks," *IEEE Trans. Wireless Commun.* vol. 14, no. 2, pp. 1100–1114, Feb. 2015.

[5] M. R. Akdeniz, Y. Liu, M. K. Samimi *et al.*, "Millimeter wave channel modeling and cellular capacity evaluation," *IEEE J. Sel. Areas Comm.*, vol. 32, no. 6, pp. 1164–1179, Jun. 2014.

[6] T. Bai, R. Vaze, and R. W. Heath, Jr., "Analysis of blockage effects on urban cellular networks," *IEEE Trans. Wireless Commun.*, vol. 13, no. 9, pp. 5070–5083, Sep. 2014.

[7] T. Bai, A. Alkhateeb, and R. W. Heath, Jr., "Coverage and capacity of millimeter-wave cellular networks," *IEEE Commun. Mag.* vol. 52, no. 9, pp. 70–77, Sep. 2014.

[8] A. Thornburg, T. Bai and R. W. Heath, "Performance analysis of outdoor mmWave Ad Hoc networks," *IEEE Trans. Signal Process.*, vol. 64, no. 15, pp. 4065-4079, Aug.1, 1 2016.

[9] M. N. Kulkarni, S. Singh and J. G. Andrews, "Coverage and rate trends in dense urban mmWave cellular networks," in *Proc. IEEE Global Communications Conference (GLOBECOM)*, Austin, USA, Dec. 2014.

[10] S. Singh, M. N. Kulkarni, A. Ghosh, and J. G. Andrews, "Tractable model for rate in self-backhauled millimeter wave cellular networks", *IEEE J. Sel. Areas Commun.*, vol. 33, no. 10, pp. 2196–2211, Sep. 2015.

[11] M. Di Renzo, "Stochastic geometry modeling and analysis of multi-tier millimeterwave cellular networks," *IEEE Trans. Wireless Commun.*, vol. 14, no. 9, pp. 5038–5057, Sep. 2015.

[12] L. Wang, M. Elkashlan, T. Q. Duong, and R. W. Heath, Jr, "Secure communication in cellular networks: The benefits of millimeter wave mobile broadband," in *Proc. IEEE Signal Processing Advances in Wireless Communications (SPAWC)*, Toronto, Canada, Jun. 2014.

[13] I. S. Gradshteyn and I. M. Ryzhik, *Table of Integrals, Series, and Products*, 7th ed. NewYork: Academic, 2007.

[14] M. Haenggi, J. G. Andrews, F. Baccelli, O. Dousse, and M. Franceschetti, "Stochastic geometry and random graphs for the analysis and design of wireless networks," *IEEE J. Sel. Areas Commun.*, vol. 27, no. 7, pp. 1029–1046, Sep. 2009.

[15] B. Blaszczyszyn, M. K. Karray, and H.-P. Keeler, "Using Poisson processes to model lattice cellular networks," in *Proc. IEEE Intl. Conf. on Comp. Comm. (INFOCOM)*, Apr. 2013.

[16] H. Wang and M. C. Reed, "Tractable model for heterogeneous cellular networks with directional antennas", in *Proc. Australian Communications Theory Workshop (AusCTW)*, Wellington, New Zealand, Jan. 2012.

[17] X. Zhang, X. Zhou, and M. R. McKay, "Enhancing secrecy with multi-antenna transmission in wireless ad hoc networks," *IEEE Trans. Infor. Forensics Sec.*, vol. 8, no. 11, pp. 1802–1814, Nov. 2013.

[18] T.-X. Zheng, H.-M. Wang, J. Yuan, D. Towsley, and M. H. Lee, "Multi-antenna transmission with artificial noise against randomly distributed eavesdroppers," *IEEE Trans. Commun.*, vol. 63, no. 11, pp. 4347–4362, Nov. 2015.

[19] T.-X. Zheng and H.-M. Wang, "Optimal power allocation for artificial noise under imperfect CSI against spatially random eavesdroppers," *IEEE Trans. Veh. Technol.*, vol. 65, no. 10, pp. 8812–8817, Oct. 2016.

[20] C. Ma, J. Liu, X. Tian, H. Yu, Y. Cui, and X. Wang, "Interference exploitation in D2D-enabled cellular networks: A secrecy perspective," *IEEE Trans. Commun.*, vol. 63, no. 1, pp. 229–242, Jan. 2015

[21] O. O. Koyluoglu, C. E. Koksal, and H. E. Gamal, "On secrecy capacity scaling in wireless networks," in *Proc. Inf. Theory Applicat. Workshop*, La Jolla, CA, USA, Feb. 2010, pp. 1–4.

[22] S. Vasudevan, D. Goeckel, and D. Towsley, "Security-capacity trade-off in large wireless networks using keyless secrecy," in *Proc. ACM Int. Symp. Mobile Ad Hoc Network Comput.*, Chicago, IL, USA, 2010, pp. 210–230.

[23] X. Zhou, M. Tao, and R. A. Kennedy, "Cooperative jamming for secrecy in decentralized wireless networks," in *Proc. IEEE Int. Conf. Commun.*, Ottawa, Canada, Jun. 2012, pp. 2339–2344.

[24] D. Tse and P. Viswanath, *Fundamentals of wireless communication*, Cambridge University Press 2005.

[25] J. Wang, L. Kong, and M.-Y. Wu, "Capacity of wireless ad hoc networks using practical directional antennas," in *Proc. IEEE Wireless Communications and Networking Conference (WCNC)*, Sydney, Australia, Apr. 2010

[26] A. D. Wyner, "The wire-tap channel," *Bell Sys. Tech. J.*, vol. 54, pp. 1355–1387, 1975.

[27] B. He and X. Zhou, "New physical layer security measures for wireless transmissions over fading channels," in *Proc. IEEE Global Communications Conference (GLOBECOM'14)*, Austin, USA, Dec. 2014.

[28] X. Zhou, R. K. Ganti, and J. G. Andrews, "Secure wireless network connectivity with multi-antenna transmission," *IEEE Trans. Wireless Commun.*, vol. 10, no. 2, pp. 425–430, Feb. 2011.

[29] C. Ma, J. Liu, X. Tian, H. Yu, Y. Cui, and X. Wang, "Interference exploitation in D2D-enabled cellular networks: A secrecy perspective," *IEEE Trans. Commun.*, vol. 63, no. 1, pp. 229–242, Jan. 2015

[30] H.-M. Wang, T.-X. Zheng, J. Yuan, D. Towsley, and M. H. Lee, "Physical layer security in heterogeneous cellular networks," *IEEE Trans. Commun.*, vol. 64, no. 3, pp. 1204–1219, Mar. 2016.

[31] H.-M. Wang, C. Wang, D. W. Kwan Ng, M. H. Lee, and J. Xiao, "Artificial noise assisted secure transmission for distributed antenna systems," *IEEE Trans. Signal Process.*, to appear, 2016.

[32] G. Geraci, H. S. Dhillon, J. G. Andrews, J. Yuan, and I. B. Collings, "Physical layer security in downlink multi-antenna cellular networks," *IEEE Trans. Commun.*, vol. 62, no. 6, pp. 2006–2021, Jun. 2014.

[33] T. Zheng, H.-M. Wang, and Q. Yin, "On transmission secrecy outage of multiantenna system with randomly located eavesdroppers," *IEEE Commun. Lett.*, vol. 18, no. 8, pp. 1299–1302, Aug. 2014.

[34] Justin Krometis, "Multidimensional Numerical Integration," 2015 [Online]. Available: http://www.arc.vt.edu/about/personnel/krometis/krometisc_cuba ture_2013.pdf

[35] H. Alzer, "On some inequalities for the incomplete gamma function," *Mathematics of Computation*, vol. 66, no. 218, pp. 771–778, 1997. [Online]. Available: *http://www.jstor.org/stable/2153894*.

[36] J. Abate and W. Whitt, "Numerical inversion of Laplace transforms of probability distributions," *ORSA J. Comput.*, vol. 7, no. 1, pp. 36–43, 1995.

Part IV

Physical layer security with emerging modulation technologies

Chapter 13

Directional-modulation-enabled physical-layer wireless security

Yuan Ding[1] and Vincent Fusco[1]

Directional modulation (DM), as a promising keyless physical-layer security technique, has rapidly developed within the last decade. This technique is able to directly secure wireless communications in the physical layer by virtue of the property of its direction-dependent signal-modulation-formatted transmission. This chapter reviews the development in DM technology over recent years and provides some recommendations for future studies.

13.1 Directional modulation (DM) concept

Directional modulation (DM) concept was first proposed in the antenna and propagation community [1–3]. Normally, signal modulations are conducted at digital baseband in conventional communication systems. At that time, researchers found that when signal modulations were performed at radio frequency (RF) stages, e.g. antenna array radiation structures [1,2] and phases of RF carriers feeding into antenna arrays [3], the signal waveforms radiated along different spatial directions in free space were differently combined. In other words, signal-modulation formats were direction dependent. This is due to the fact that the far-field radiation patterns produced by each array element are summed differently along different directions, i.e. spatially dependent. When carefully designed, signal waveforms containing required information can be generated along an a-priori-selected secured communication direction. In this way, only receivers located along the pre-specified direction are able to capture the correct signal signature, leading to successful data recovery, while distorted signal waveforms radiated along other directions make eavesdropping difficult. In order to clearly illustrate the DM concept, take as an example a DM system modulated for quadrature phase-shift keying (QPSK), as shown in Figure 13.1. Here, the standard-formatted QPSK constellation patterns, i.e. central-symmetric square in in-phase and quadrature (IQ) space, is preserved only along an a-priori-defined observation direction θ_0, while along all other spatial directions, the signal formats are distorted.

[1]The Institute of Electronics, Communications and Information Technology (ECIT) Queen's University of Belfast, UK

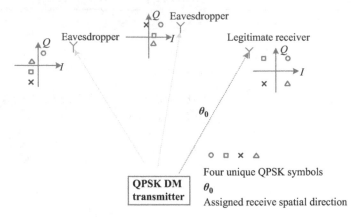

Figure 13.1   Illustration of the major properties of a QPSK DM system. Along a pre-specified direction θ_0 a usable constellation is formed. Away from the desired direction the constellations are scrambled [4]. © 2015 Cambridge University Press and the European Microwave Association. Reprinted with permission from Ding Y. and Fusco V. "A review of directional modulation technology". International Journal of Microwave and Wireless Technologies. 2015;1–13

In such a fashion, potential eavesdroppers positioned along all other directions suffer a significantly reduced possibility of successful interception.

This DM feature is quite distinct from the general broadcast nature normally associated with a conventional wireless transmission system where identical copies, subject to amplitude differences, of signal waveforms are projected into the entire radiation coverage space.

At first, it was wrongly believed that the property of the direction dependent modulation transmission was obtained because the signal modulation occurred at the RF stages [1,2]. In fact, in some reported DM transmitters, signal modulation still takes place at digital baseband [5]. It was discovered that

- from the antenna and propagation perspective, the essence of the DM technique requires updating antenna array far-field radiation patterns at transmission symbol rates. In contrast, in conventional wireless transmission systems, the far-field radiation patterns of antenna or antenna arrays are updated at channel block fading rates.
- from the signal processing perspective, the essence of the DM technique is injecting artificial interference that is orthogonal, in the spatial domain, to the information signal conveyed to the desired receiver [6,7]. In contrast, in conventional wireless transmission systems no orthogonal artificial interference is used.

The remaining sections in this chapter are organised as follows: in Section 13.2, types of DM transmitter physical architecture are presented and categorised, while

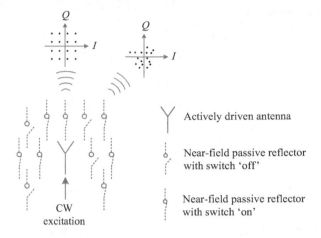

Figure 13.2   Concept of NFDAM DM structure

a general mathematical model is established in Section 13.3. Section 13.4 describes DM synthesis approaches associated with different types of DM architectures, and the metrics for assessing secrecy performance of DM systems are briefly presented and summarised in Section 13.5. The extension of the DM technique for multi-beam and multi-path applications is investigated in Section 13.6. DM demonstrators reported to date are discussed in Section 13.7. Finally, the recommendations for future DM studies are provided in Section 13.8.

13.2   DM transmitter architectures

In this section, we firstly describe the reported DM transmitter architecture in chronological order, and then categorise them into three groups in order to facilitate discussions of their respect properties.

13.2.1   Near-field direct antenna modulation

The first type of DM transmitter, named by the authors as near-field direct antenna modulation (NFDAM), consists of a large number of re-configurable passive reflectors coupled in the near field of a centre-driven antenna [1,8]. The concept of the NFDAM DM transmitter is shown in Figure 13.2. Each set of switch state combinations on the near-field passive reflectors contributes to a unique far-field radiation pattern, which can be translated into constellation points in IQ space detected along each spatial direction in free space. After measurement of a multiplicity of patterns, with regard to a large number of possible switch combinations, has been done, a usable constellation pattern detected along the desired direction may be selected. Subsequently the corresponding switch settings for secure transmission of each symbol is memorised in order to reproduce this usable state. Since the far-field patterns

Baseband information data
controlled RF attenuators
and phase shifters

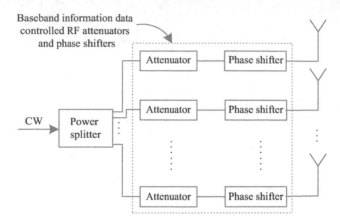

Figure 13.3   *Generic excitation-re-configurable DM transmitter array [4].*
© 2015 Cambridge University Press and the European Microwave
Association. Reprinted with permission from Ding Y. and Fusco V.
"A review of directional modulation technology". International
Journal of Microwave and Wireless Technologies. 2015;1–13

associated with each selected switch combination are different functions of the spatial direction, the detected signal constellation patterns along undesired communication directions are distorted.

13.2.2   DM using re-configurable antennas in an array configuration

Similar to the NFDAM, by replacing each antenna element, both the central active-driven antenna and the surrounding passive reflectors, with an active-driven pattern re-configurable antenna element, a DM transmitter can also be constructed [2]. When the array elements are well separated, e.g. half wavelength spaced, the current on antenna elements can be actively manipulated using the re-configurable components associated with each of the antenna radiation structures. Performance optimisation of the DM transmitter proceeds using optimisation algorithms [3,9].

13.2.3   DM using phased antenna array

In [3], the authors built a DM transmitter using a phased antenna array, see Figure 13.3. This type of architecture exploits individual array element excitation re-configurability, instead of radiation structure re-configurability, in order to achieve direction dependent modulation transmission. This excitation-re-configurable DM structure has then been heavily investigated under various hardware constraints [9–13] mainly because it is synthesis-friendly and it can be readily implemented.

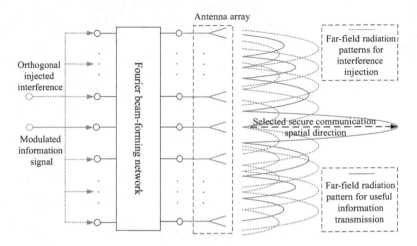

Figure 13.4 Fourier beamforming-network-enabled DM transmitters

13.2.4 DM using Fourier beamforming networks

In [14,15], DM transmitters were constructed using Fourier beamforming networks, e.g. Butler matrix and Fourier Rotman lens, whose orthogonality property in beam space helps project uncontaminated signals only along the selected spatial direction that corresponds to the excited information beam port, Figure 13.4. This type of DM arrangement reduces the number of required RF chains to two, and it was experimentally verified for 10-GHz operation for real-time data transmission with both analogue and digital modulations in [16] and [7], respectively.

13.2.5 DM using switched antenna arrays

Researchers in [17,18] revealed that by inserting a switch array before antennas to randomly select a subset of elements in an antenna array for signal transmission on a per transmitted symbol basis, DM transmitters could have been constructed. Actually, this concept had been proposed long before in [19]. This architecture, named as antenna subset modulation (ASM) in [17] or 4-D antenna arrays in [20], requires only one RF chain the expense of reduced beamforming gain.

13.2.6 DM using digital baseband

The DM transmitter in Figure 13.3 uses RF components to re-configure antenna array excitations. It is natural to replace these RF components with more flexible and more precise digital baseband arrangements [6,21]. This approach facilitates the DM technique being applied in modern digital wireless communication systems [5].

The above-mentioned types of DM architecture can be sorted into three categories, namely, radiation structure re-configurable DM (*A* and *B*), excitation

Table 13.1   Summary of characteristics of different types of DM architecture

	Radiation structure re-configurable DM		Excitation re-configurable DM		Synthesis-free DM	
	A	*B*	*C*	*F*	*D*	*E*
Synthesis	Difficult		Medium	Easy	Not required	
Optimisation[a]	Difficult		Medium	Easy	Medium	
Multi-beam[b]	Difficult		Medium		Easy	Difficult
Multi-path	Difficult		Medium	Easy	Difficult	
System complexity	High		High	Medium	Low	Medium
Cost	High		High	Medium	Low	

[a] It refers to the enhancement of the system secrecy performance.
[b] It refers to secure independent transmission along multiple spatial directions simultaneously.

re-configurable DM (*C* and *F*), and synthesis-free DM (*D* and *E*, as well as retrodirective DM in [22]). Their individual characteristics are summarised in Table 13.1. Since the design of radiation structure re-configurable DM is greatly dependent on the selected antenna structure, no effective universal synthesis methods are available. While the synthesis-free DM can be considered as hardware realisation of some excitation re-configurable DM. As a consequence, only the details for the excitation re-configurable DM are provided later in this chapter.

13.3   Mathematical model for DM

(© 2015 Cambridge University Press and the European Microwave Association. Reprinted with permission from Ding Y. and Fusco V. "A review of directional modulation technology". International Journal of Microwave and Wireless Technologies. 2015;1–13).

In order to facilitate the synthesis of DM transmitters, a mathematical model was established in [6,23]. This model is the key to understanding the essence of the DM technique presented in Section 13.1, and thus is elaborated as below.

The superimposed radiation from a series of radiating antenna elements in free space can be formed for N elements, with the summed far-field radiation pattern $\boldsymbol{P}$ at some distant observation point in free space, given as

$$\boldsymbol{P}(\theta) = \frac{e^{-j\vec{k}\cdot\vec{r}}}{|\vec{r}|} \begin{bmatrix} \boldsymbol{R}_1(\theta) \\ \boldsymbol{R}_2(\theta) \\ \vdots \\ \boldsymbol{R}_N(\theta) \end{bmatrix} \cdot \begin{bmatrix} A_1 e^{-j\vec{k}\cdot\vec{x}_1} \\ A_2 e^{-j\vec{k}\cdot\vec{x}_2} \\ \vdots \\ A_N e^{-j\vec{k}\cdot\vec{x}_N} \end{bmatrix}, \tag{13.1}$$

where R_n ($n \in (1, 2, \ldots, N)$) refers to the far-field active element pattern of the nth array element, which is a function of spatial direction θ. A_n is the excitation applied at the nth antenna port. $\vec{k}$ is the wavenumber vector along each spatial direction θ. $\vec{r}$ and $\vec{x}_n$, respectively, represent the location vectors of the receiver and the nth array element relative to the array phase centre.

In a conventional wireless transmitter beamforming array, when the communication direction is known, denoted as θ_0, the antenna excitation A_n is designed such that $R_n(\theta_0)A_n e^{-j\vec{k}\cdot\vec{x}_n}$ for each n are in-phase. As a result, the main radiation beam is steered towards the direction θ_0. The information data D, which is a complex number corresponding to a constellation point in IQ space, can then be identical applied onto each A_n, i.e. DA_n, for wireless transmission with the signal magnitude spatial distribution governed by $|P(\theta)|$. During the entire data transmission, $P(\theta)$ is unchanged.

In a DM transmitter array, the information data is wirelessly delivered using more degrees of freedom within the array. The far-field radiation pattern $P(\theta)$ in (13.1) does not remain constant during the data stream transmission as occurs in the conventional beamforming system. Instead, the pattern $P(\theta)$ is differently synthesised for each symbol transmitted. $P(\theta_0)$ can be directly designed as D_i which is the ith transmitted symbol, or it can be designed to be a constant value with the required data stream applied afterwards. Both approaches are equivalent. As can be seen in (13.1), there are two ways to alter the far-field pattern $P(\theta)$; one is to re-configure the radiation patterns R_n of each antenna element, as adopted in the radiation structure re-configurable DM, and the other is to update the antenna excitations A_n, as exploited in the excitation re-configurable DM, in both cases at the transmission symbol rate. The former method, i.e. altering R_n, either passively or actively, is complicated because there is no closed-form link between the geometry of the antenna radiation structure and its far-field radiation pattern. Thus, the focus of the DM mathematical model described in this section is on the excitation re-configurable DM structure, though the model can, in principle, be utilised to analyse the radiation structure re-configurable DM and the synthesis-free DM.

In order to facilitate discussion and simplify mathematical expressions, isotropic antenna active element patterns, i.e. $R_n(\theta) = 1$, and uniform one-dimensional (1D) array elements with spacing $|\vec{x}_{n+1} - \vec{x}_n|$ of one half wavelength ($\lambda/2$) are assumed for the establishment of the DM model. This model can be readily extended for general DM transmitter scenarios, e.g. arbitrary antenna active element patterns and other array arrangements. Under these assumptions the far-field pattern $P(\theta)$ is solely determined by the array element excitations A_n:

$$P(\theta) = \sum_{n=1}^{N} \left(A_n e^{j\pi \left(n - \frac{N+1}{2}\right)\cos\theta} \right) \tag{13.2}$$

The term $e^{-j\vec{k}\vec{r}}/|\vec{r}|$ is dropped, because it is a constant complex scaling factor for the pattern P in all directions.

Since there are analogue and digital means for updating A_n as discussed in Section 13.2, a description technique which is architecture independent and which lends itself to both analysis and synthesis of any class of excitation re-configurable DM structures is required. Thus, the requirements that array excitations A_n need to satisfy for a DM transmitter are analysed, regardless of the means of generating A_n. For clarity, we add an additional subscript 'i' to the relevant notations to refers to their updated values in the ith transmitted symbol slot.

As previously stated, the two key properties of a DM transmitter are

- preservation of the transmitted signal format (standard constellation pattern in IQ space) along a pre-specified communication direction θ_0;
- distortion of constellation patterns along all other communication directions.

From a signal processing perspective, $P(\theta)$ in (13.2) can be regarded as a detected constellation point in IQ space at receiver sides, denoted as $D_i(\theta)$ for the ith symbol transmitted, (13.3):

$$D_i(\theta) = \sum_{n=1}^{N} \underbrace{A_{ni} \cdot e^{j\pi\left(n-\frac{N+1}{2}\right)\cos\theta}}_{B_{ni}(\theta)}. \tag{13.3}$$

For each symbol transmitted, the vector summation of B_{ni} has to yield the standard constellation point D_i^{st} in IQ space along, and only along, the direction θ_0. This statement, in mathematical description, can be expressed in (13.4):

$$\sum_{n=1}^{N} B_{ni}(\theta_0) = D_i^{st}. \tag{13.4}$$

From (13.3), by scanning the observation angle θ, the constellation track in IQ space, $D_i(\theta)$, of the ith symbol can be obtained.

General properties of constellation tracks can be observed from (13.3) and are summarised as follows:

(a) For an array with an odd number of elements, constellation tracks are closed loci. In some extreme cases, these loci collapse to line segments. The starting point ($\theta = 0°$) always overlaps with end point ($\theta = 180°$). For the case of even number elements, $D_i(0°) = -D_i(180°)$. When the array element spacing is changed, the start and end angle will change accordingly.

(b) Changing the desired communication direction θ_0 does not affect the shape of the constellation track pattern in IQ space. It determines only where the tracks start ($\theta = 0°$) and end ($\theta = 180°$).

(c) Different vector paths (trajectories of the vector summation, $\sum_{n=1}^{N} B_{ni}(\theta_0)$) inevitably lead to different constellation tracks. This is guaranteed by the orthogonality property of $e^{jn\pi\cos\theta}$ for different n within the spatial range from 0° to 180°.

(d) When $B_{np}(\theta_0)$ are the scaled $B_{nq}(\theta_0)$ with the same scaling factor K for each n, the corresponding constellation track $D_p(\theta)$ is also the scaled $D_q(\theta)$ with the same scaling factor K.

Properties (c) and (d) indicate that constellation distortion at other directions can be guaranteed when vector $[\boldsymbol{B}_{1p}(\theta_0)\ \boldsymbol{B}_{2p}(\theta_0)\ \ldots\ \boldsymbol{B}_{Np}(\theta_0)]$ and the other vector $[\boldsymbol{B}_{1q}(\theta_0)\ \boldsymbol{B}_{2q}(\theta_0)\ \ldots\ \boldsymbol{B}_{Nq}(\theta_0)]$ are linearly independent, see (13.5). Here the pth and the qth symbols in the data stream are different modulated symbols:

$$\left[\boldsymbol{B}_{1p}(\theta_0)\ \boldsymbol{B}_{2p}(\theta_0)\ \ldots\ \boldsymbol{B}_{Np}(\theta_0)\right] \neq \boldsymbol{K}\left[\boldsymbol{B}_{1q}(\theta_0)\ \boldsymbol{B}_{2q}(\theta_0)\ \ldots\ \boldsymbol{B}_{Nq}(\theta_0)\right] \tag{13.5}$$

The combination of (13.4) and (13.5) forms the necessary conditions for DM transmitter arrays. Thus, the requirement for the DM array excitations A_{ni} can then be obtained using (13.3).

With the DM vector model described above the static and dynamic DM transmitters are defined below [4]. This classification facilitates the discussions of DM synthesis approaches that are presented in Sections 13.4.

Definition: If along the DM secure communication direction, the vector path in IQ space reaching each unique constellation point is independent and fixed, which results in a distorted, but static with respect to time, constellation pattern along other spatial directions, the transmitters are termed 'static DM'; if the vector paths are randomly re-selected, on a per transmitted symbol basis in order to achieve the same constellation symbol in the desired direction, then the symbol transmitted at the different time slots in the data stream along spatial directions other than the prescribed direction would be scrambled dynamically and randomly. This we call the 'dynamic DM' strategy.

With the above definition, the DM transmitters reported in [1–3,8–12] fall into the static DM category, while the DM systems reported in [5–7,14,15,17,20], involving time as a variable to update system settings, can be labelled as dynamic.

It is noted that here, we assume the potential eavesdroppers located along directions other than the intended secure communication direction do not collude. For the colluding eavesdroppers, especially when the number of colluding eavesdroppers are greater than the number of DM transmit array elements, the successful estimation of the useful information conveyed to the desired receivers is possible. Thus a different system design rules should be adopted. Interested readers can find more details on this aspect in [21], where DM system is viewed and analysed from a signal processing perspective.

13.4 Synthesis approaches for DM transmitters

As discussed in Section 13.2, the design of the radiation structure re-configurable DM is complicated due to the lack of closed-form link between the geometry of the antenna radiation structure and its far-field radiation pattern. Seeking DM synthesis approaches for the excitation re-configurable DM transmitter architecture is the most active area in DM research recently. Among all synthesis approaches, the orthogonal vector method developed in [6,24], which shares a similar idea with the artificial noise concept [25,26] studied in the information theory community, was found to furnish a fundamental and universal DM synthesis strategy. It also provides the key to understand the synthesis-free DM architecture. Thus, the orthogonal vector synthesis approach is next elaborated in this section.

13.4.1   Orthogonal vector approach for DM synthesis

(© 2015 Cambridge University Press and the European Microwave Association. Reprinted with permission from Ding Y. and Fusco V. "A review of directional modulation technology". International Journal of Microwave and Wireless Technologies. 2015;1–13.)

In Section 13.3, it is showed that when the same constellation symbol detected along the desired communication direction is reached via different vector paths in IQ space, the resulting constellation tracks are altered accordingly. This leads to the constellation pattern distortion along spatial directions other than the prescribed direction, which is the key property of DM systems. In other words, it is the difference between each two vector paths selected to achieve a same standard constellation point in IQ space that enables DM behaviour. This difference is defined as the orthogonal vector, since it is always orthogonal to the conjugated channel vector along the intended spatial direction.

A one-dimensional (1D) five-element array with half wavelength spacing is taken as an example below in order to explain the orthogonal vector concept. It is assumed that each antenna element has an identical isotropic far-field radiation pattern.

The channel vector for this system along the desired communication direction θ_0 in free space can be written as

$$\vec{H}(\theta_0) = \left[\underbrace{e^{j2\pi \cos \theta_0}}_{H_1} \ \underbrace{e^{j\pi \cos \theta_0}}_{H_2} \ \underbrace{e^{j0}}_{H_3} \ \underbrace{e^{-j\pi \cos \theta_0}}_{H_4} \ \underbrace{e^{-j2\pi \cos \theta_0}}_{H_5} \right]^T, \tag{13.6}$$

where '$[\cdot]^T$' refers to the transpose operator.

When the excitation signal vectors $[A_{1i}(\theta_0) A_{2i}(\theta_0) \ldots A_{Ni}(\theta_0)]^T$ at the array input for the same symbol transmitted at the uth and vth time slots chosen to be

$$\vec{S}_u = \left[\underbrace{e^{j\left(-\frac{\pi}{4}+2\pi \cos \theta_0\right)}}_{S_{1u}} \ \underbrace{e^{j\pi \cos \theta_0}}_{S_{2u}} \ \underbrace{e^{j0}}_{S_{3u}} \ \underbrace{e^{-j\pi \cos \theta_0}}_{S_{4u}} \ \underbrace{e^{-j2\pi \cos \theta_0}}_{S_{5u}} \right]^T \tag{13.7}$$

and

$$\vec{S}_v = \left[\underbrace{e^{j\left(\frac{\pi}{4}+2\pi \cos \theta_0\right)}}_{S_{1v}} \ \underbrace{e^{j\left(\frac{\pi}{4}+\pi \cos \theta_0\right)}}_{S_{2v}} \ \underbrace{e^{-j\frac{\pi}{4}}}_{S_{3v}} \ \underbrace{e^{j\left(\frac{3\pi}{4}-\pi \cos \theta_0\right)}}_{S_{4v}} \ \underbrace{e^{j\left(\frac{\pi}{4}-2\pi \cos \theta_0\right)}}_{S_{5v}} \right]^T, \tag{13.8}$$

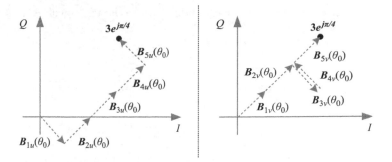

Figure 13.5   Illustration example of two vector paths derived from two different excitation settings [4]. © 2015 Cambridge University Press and the European Microwave Association. Reprinted with permission from Ding Y. and Fusco V. "A review of directional modulation technology". International Journal of Microwave and Wireless Technologies. 2015;1–13

the resulting received vector paths in IQ space along this spatial direction θ_0 are

$$\vec{B}_u = \vec{H}^*(\theta_0) \circ \vec{S}_u$$

$$= \left[\underbrace{H_1^* \cdot S_{1u}}_{B_{1u}} \ \underbrace{H_2^* \cdot S_{2u}}_{B_{2u}} \ \underbrace{H_3^* \cdot S_{3u}}_{B_{3u}} \ \underbrace{H_4^* \cdot S_{4u}}_{B_{4u}} \ \underbrace{H_5^* \cdot S_{5u}}_{B_{5u}} \right]^T$$

$$= \left[e^{-j\frac{\pi}{4}} \ e^{j\frac{\pi}{4}} \ e^{j\frac{\pi}{4}} \ e^{j\frac{\pi}{4}} \ e^{j\frac{3\pi}{4}} \right]^T \tag{13.9}$$

and

$$\vec{B}_v = \vec{H}^*(\theta_0) \circ \vec{S}_v$$

$$= \left[\underbrace{H_1^* \cdot S_{1v}}_{B_{1v}} \ \underbrace{H_2^* \cdot S_{2v}}_{B_{2v}} \ \underbrace{H_3^* \cdot S_{3v}}_{B_{3v}} \ \underbrace{H_4^* \cdot S_{4v}}_{B_{4v}} \ \underbrace{H_5^* \cdot S_{5v}}_{B_{5v}} \right]^T$$

$$= \left[e^{j\frac{\pi}{4}} \ e^{j\frac{\pi}{4}} \ e^{-j\frac{\pi}{4}} \ e^{j\frac{3\pi}{4}} \ e^{j\frac{\pi}{4}} \right]^T, \tag{13.10}$$

where '$[\cdot]^*$' returns the conjugation of the enclosed term, and operator '∘' denotes Hahamard product of two vectors. These vector paths described in (13.9) and (13.10) are plotted in Figure 13.5.

The vector summations of $\sum_{n=1}^{5} B_{nu}$ and $\sum_{n=1}^{5} B_{nv}$ can be calculated by $\vec{H}^\dagger(\theta_0)\vec{S}_u$ and $\vec{H}^\dagger(\theta_0)\vec{S}_v$, both reading position $3 \cdot e^{j\pi/4}$ in IQ space. '$(\cdot)^\dagger$' denotes complex conjugate transpose (Hermitian) operator.

The difference between the corresponding two excitation signal vectors is the orthogonal vector:

$$\Delta \vec{S} = \vec{S}_v - \vec{S}_u$$

$$= \sqrt{2} \left[e^{j\left(\frac{\pi}{2} + 2\pi \cos \theta_0\right)} \quad 0 \quad e^{-j\frac{\pi}{2}} \quad e^{j\pi(1 - \cos \theta_0)} \quad e^{-j2\pi \cos \theta_0} \right]^T, \qquad (13.11)$$

which is orthogonal to the conjugated channel vector $\vec{H}^*(\theta_0)$ since $\vec{H}^*(\theta_0) \cdot \Delta \vec{S} = 0$.

With the help of the orthogonal vector concept, a generalised DM synthesis approach was developed in [6] and further refined in [27]. The synthesised DM transmitter array excitation vector $\vec{S}_{ov}$ takes the form:

$$\vec{S}_{ov} = \vec{\Lambda} + \vec{W}_{ov}, \qquad (13.12)$$

where $\vec{\Lambda}$ is a vector with each entry of array excitations before injecting the orthogonal vector $\vec{W}_{ov}$ that is orthogonal to the conjugated channel vector $\vec{H}^*(\theta_0)$. Denote $\vec{Q}_p$ $(p = 1, 2, \ldots, N - 1)$ to be the orthonormal basis in the null space of $\vec{H}^*(\theta_0)$, then $\vec{W}_{ov} = 1/(N - 1) \sum_{p=1}^{N-1} \left(\vec{Q}_p \cdot v_p \right)$. v_p is a random viable used to control the power of the injected orthogonal vector $\vec{W}_{ov}$. Here, $\vec{\Lambda}$ is defined as the excitations of non-DM beam steering arrays, selected according to the system requirements, e.g. uniform magnitude excitations used in conventional beam steering phased arrays result in narrower secure transmission spatial region, while other magnitude-tapered excitations can reduce the amount of leaked information through sidelobe directions at the cost of slightly widened secure transmission spatial region. This excitation contains the information symbol D_i to be transmitted, e.g. $e^{j\pi/4}$ corresponds to the QPSK symbol 11.

During the synthesis of DM arrays from the non-DM beam steering arrays two parameters which are crucial to system performance, need to be discussed. One is the length of each excitation vector. From a practical implementation perspective, the square of the excitation vector should fall into the linear range of each power amplifier located within each RF path. The other is the extra power that is required to be injected into the non-DM array in order for signal distortion along other directions.

In order to describe this extra power, the DM power efficiency (PE_{DM}) is defined:

$$PE_{DM} = \frac{\sum_{i=1}^{I} \left(\sum_{n=1}^{N} |\Lambda_{ni}|^2 \right)}{\sum_{i=1}^{I} \left(\sum_{n=1}^{N} |S_{ov\_ni}|^2 \right)} \times 100\% = \frac{\sum_{i=1}^{I} \left(\sum_{n=1}^{N} |B_{ni\_non-DM}|^2 \right)}{\sum_{i=1}^{I} \left(\sum_{n=1}^{N} |B_{ni\_DM}|^2 \right)} \times 100\%,$$

$$(13.13)$$

where I is, for static DM, the number of modulation states, e.g. 4 for QPSK, or, for dynamic DM, the symbol number T in a data stream. Λ_{ni} and S_{ov_ni} are the nth array element excitation for the ith symbol in the non-DM array and in the synthesised

DM array, respectively. In noiseless free space, their modulus equal to the normalised modulus of corresponding B_{ni_non-DM} and B_{ni_DM} from the receiver side perspective. Generally, the larger the allowable range of the excitation vector lengths are, and the lower PE_{DM} is, the better the DM system secrecy performance that can be achieved. PE_{DM} can also be expressed as a function of v_p as in (13.14) and (13.15):

$$PE_{DM} = \frac{1}{I} \cdot \sum_{i=1}^{I} PE_{DM\_}i, \tag{13.14}$$

$$PE_{DM\_}i = \frac{1}{1 + \sum_{p=1}^{N-1} \left(\frac{1}{N-1} \cdot v_p \right)^2} \times 100\%. \tag{13.15}$$

13.4.2 Other DM synthesis approaches

In the last sub-section, the universal DM synthesis method, i.e. the orthogonal vector approach, is presented. However, in some application scenarios, some other requirements on DM system properties may need to be considered. These requirements, or constraints, which have been investigated, include the bit error rate (BER) spatial distribution, the array far-field radiation characteristics, and the interference spatial distribution. All these synthesis methods can be viewed as seeking a subset of orthogonal vectors that satisfy prescribed DM system requirements.

The BER-driven [28] and the constrained array far-field radiation pattern [29–31] DM synthesis approaches share a similar idea, i.e. via the iterative transformations between the array excitations and the required DM properties, namely the BER spatial distributions and the array far-field radiation patterns, the constraints on DM characteristics can be imposed. Since iteration processes are involved, these two methods are not suitable for dynamic DM synthesis.

Another DM array far-field pattern separation synthesis approach was developed in [32,33]. Here by virtue of the far-field null steering approach, the DM array far-field radiation patterns can be separated into information patterns, which describe information energy projected along each spatial direction, and interference patterns, which represent disturbance on genuine information. Through this separation methodology, we can identify the spatial distribution of information transmission and hence focus interference energy into the most vulnerable directions with regard to interception, i.e. information sidelobes, and in doing so submerge leaked information along unwanted directions. This method is closely linked to the orthogonal vector approach. In fact, the separated interference patterns can be considered as far-field patterns generated by the injected orthogonal vectors. However, it is more convenient to apply constraints, such as interference spatial distribution, with the pattern separation approach. This approach is compatible to both static and dynamic DM systems.

The relationships among the four DM synthesis approaches are illustrated in Figure 13.6.

Through the various DM synthesis approaches presented above and their associated examples, it is found that DM functionality is always enabled by projecting

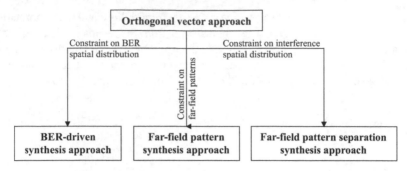

Figure 13.6   Relationships among the four DM synthesis approaches [4]. © 2015
Cambridge University Press and the European Microwave
Association. Reprinted with permission from Ding Y. and Fusco V.
"A review of directional modulation technology". International
Journal of Microwave and Wireless Technologies. 2015;1–13

extra energy into undesired communication directions in free space. This extra energy, which can be either static or dynamic with respect to time, corresponding to static and dynamic DM systems, acts as interference which scrambles constellation symbol relationships along the unselected directions. Intuitively, the larger the interference energy projected, the more enhanced the DM system secrecy performance that can be achieved. It is concluded that the essence of a functional DM system synthesis approaches lies in generating artificial interference energy that is orthogonal to the directions where the intended receivers locate.

13.4.3   A note on synthesis-free DM transmitters

Two types of synthesis-free DM transmitters have been reported to date. They are the Fourier beamforming network assisted DM [7] and the ASM [17], as described in Section 13.2. Their 'synthesis-free' property is enabled by the adopted hardware, i.e. Fourier beamforming network and antenna subset selection switches, which has the capability of generating orthogonal vectors without additional calculation for interference projection along all undesired spatial directions.

In addition to the above two types of synthesis-free DM transmitters, another promising synthesis-free DM architecture that is based on the retrodirective arrays (RDAs) [34] has yet been reported. This RDA-enabled DM transmitter is able to overcome the weaknesses inherent in the Fourier beamforming network and the ASM DM systems, which are

- transmitters need to acquire receiver's direction in advance;
- in the Fourier beamforming network assisted DM systems, receivers can only locate at some discrete spatial directions in order to maintain orthogonality between information and artificial interference. The number of these spatial

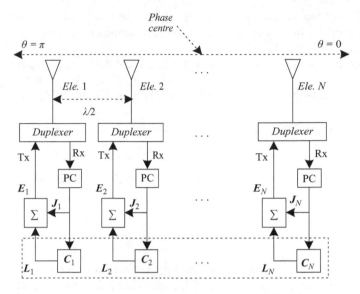

Figure 13.7 Architecture of synthesis-free RDA DM transmitter

directions and their angular spacings are determined by the number of array elements;

- in the ASM DM systems, the available transmission gain is greatly reduced because only a small subset of antennas in the array are utilised for beamforming.

An RDA is capable of re-transmitting signals back along the spatial direction along which the array was interrogated by the incoming signals without requiring a-priori knowledge of its direction of arrival. In order to achieve retrodirective re-transmit functionality, a phase conjugator (PC) is required [35]. Next we discuss how a classical RDA is altered to form a synthesis-free DM transmitter. The system block diagram associated with this is illustrated in Figure 13.7.

In the receive (Rx) mode, the pilot tone signal from a legitimate receiver along θ_0 in free space impinged on the antenna array with N elements is phase conjugated. The phase conjugated signal can be expressed as a vector $\vec{J}$, of which the nth entry is $J_n = H_n^*(\theta_0)$, seen in Figure 13.7. $H_n(\theta_0)$ is the channel coefficient between the distant receiver and the nth RDA element, which for the RDA arrangement in Figure 13.7 is $e^{-j\pi(n-\frac{N+1}{2})\cos\theta_0}$ after the path loss is normalised.

In a classical RDA the phase conjugated $\vec{J}$ is directly used for re-transmission, i.e. $\vec{E} = \vec{J}$. Here $\vec{E}$, as well as $\vec{H}$, $\vec{C}$, and $\vec{L}$ that will be used later in this sub-section, is defined similarly as $\vec{J}$. In this case, the distant pilot source node receives:

$$F = \vec{H}(\theta_0) \cdot \vec{E} = \vec{H}(\theta_0) \cdot \vec{J} = \vec{H}(\theta_0) \cdot \vec{H}^*(\theta_0) = N, \tag{13.16}$$

which indicates the perfect beamforming towards the legitimate receiver. However, as was pointed out in Section 13.1, this classical beamforming preserves the signal formats in all radiation directions.

By adding an extra block, enclosed within the dotted box in Figure 13.7, the synthesis-free DM can be constructed. Its operation is elaborated as below.

With the additional enclosed block, the received signal along (θ_0) is

$$F = \vec{H}(\theta_0) \cdot \vec{E} = \vec{H}(\theta_0) \cdot (\vec{J} + \vec{L}) = N + \vec{H}(\theta_0) \cdot (\vec{C} \circ \vec{J}) = N + \vec{C} \cdot (\vec{H}(\theta_0) \circ \vec{J}). \tag{13.17}$$

Since $\vec{H}(\theta_0) \circ \vec{J} = \vec{H}(\theta_0) \circ \vec{H}^*(\theta_0)$ is a vector with all N elements of unity, (13.17) can be written as

$$F = N + \sum_{n=1}^{N} C_n. \tag{13.18}$$

From (13.18), it can be seen that when $\sum_{n=1}^{N} C_n$ is kept as a constant during the entire data transmission, the formats of re-transmitted information signal D, applied onto $\vec{E}$ afterwards, can be well preserved along θ_0, i.e. FD. However, along all other directions, the signal formats are scrambled due to the randomly updated $\vec{C}(\vec{H}(\theta) \circ \vec{J})$, $(\theta \neq \theta_0)$. As a consequence, a synthesis-free DM transmitter can be successfully realised.

13.5 Assessment metrics for DM systems

In order to evaluate the performance of DM systems in a way that is consistent and which allows direct comparison between different systems, assessment metrics were systematically investigated in [36]. It was shown that for static DM systems BER, calculated from either closed-form equations or random data streams, as well as secrecy rate were applicable for system performance evaluation, whereas error-vector-magnitude-like (EVM-like) metrics did not perform well. For dynamic DM systems under the scenarios of zero-mean Gaussian distributed orthogonal interference, EVM-like metrics, BER, and secrecy rate were equivalent and can be converted between each other. For other interference distributions no closed-form BER and secrecy rate equations were found.

In order to provide readers with a clear picture on metrics for assessing performance of DM systems, all of the findings presented in [36] are summarised in Table 13.2.

13.6 Extensions to the DM technique

DM technology was first proposed for securing wireless information transmission along *one* pre-specified direction only in *free space*. It is natural to consider

Table 13.2 Summaries of Metrics for DM System Performance Assessment [37].
© 2015 Cambridge University Press and the European Microwave
Association. Reprinted with permission from Ding Y. and Fusco V.
"A review of directional modulation technology". International
Journal of Microwave and Wireless Technologies. 2015;1–13

		EVM	BER		Secrecy rate	
			Closed form equation	Data stream simulation	Numerical calculus	Bit-wise [38]
	Static DM	−	•	+	+	−
Dynamic DM	Zero-mean Gaussian orthogonal vectors	+	•	+	+	+
	Zero-mean non-Gaussian orthogonal vectors	−	−	+	−	−
	Non-zero-mean orthogonal vectors	−	−	+	−	−
	Calculation complexity	Low	Low	Medium	Medium	High

'+': Metric works.
'•': Metric works for QPSK, but not for higher order modulations.
'−': Metric cannot be calculated or does not work.

developing multi-beam DM systems and also extending DM application to multi-path scenarios. These two aspects are addressed separately below.

13.6.1 Multi-beam DM

Multi-beam DM transmitters have the capability of projecting multiple independent information data streams into different spatial directions, while simultaneously distorting information signal formats along all other unselected directions. The first multi-beam DM synthesis attempt was based on analogue excitation re-configurable DM transmitter architecture with 2-bit phase shifters [39]. Here, the limited states of the phase shifters reduce the number of independent users that can be supported in the DM systems. Further in [40,41], the orthogonal vector approach was successfully adapted for general-case multi-beam DM synthesis. By realising that the orthogonal vectors generate far-field radiation patterns, termed as interference patterns, which have nulls along all desired secure directions, the far-field pattern separation approach was developed [33]. All of the above methods are equivalent.

It is worth noting that a recent study in [42] revealed that the DM system can be regarded as a kind of multiple-input and multiple-output (MIMO) system. It is well known that under multi-path rich wireless channel conditions systems which deploy both multiple transmit and multiple receive antennas provide an additional spatial dimension for communication and yield degree-of-freedom gain. The additional degree-of-freedom can be exploited by spatially multiplexing several parallel independent data streams onto the MIMO channel leading to an increase in channel capacity [43]. In order to retain MIMO spatial multiplexing in free space, the multiple receive antennas have to be separately placed along different spatial directions [43].

In MIMO systems, firstly, the singular value decomposition (SVD) is performed on the channel matrix $[H]$ which can be obtained through channel training beforehand, i.e.:

$$[H] = [U][\Sigma][V]^{\dagger}. \tag{13.19}$$

Here, $[U]$ and $[V]$ are unitary matrices, and $[\Sigma]$ is a matrix whose diagonal elements are non-negative real numbers and whose off-diagonal elements are zero. Then by designing networks $[V]$ and $[U]^{\dagger}$ and inserting them into transmit and receive sides, respectively, the MIMO multiplexing channels whose gains are diagonal entries in $[\Sigma]$, are created.

In contrast in DM systems, the receivers do not have knowledge about the channel matrix $[H]$, and also they cannot collaborate for signal processing. Under these prerequisites, the channel SVD is not applicable. Instead, the channel matrix has to be decomposed as

$$[H] = [Z]^{-1}[Q], \tag{13.20}$$

where $[Z]$ is the DM-enabling matrix (network) at transmit side, and $[Q]$ is a diagonal matrix, whose diagonal entries, similar to those in $[\Sigma]$, represent the gains of the multiplexing channels associated with each independent receivers. The decomposition in (13.20) is not unique, which should be determined according to the system requirements, such as the gain for each receiver and the hardware constraints at transmit side.

The establishment of a link between the DM and MIMO technologies is of great importance, since it may open a way for further DM development, e.g. DM operation in a multi-path environment, as discussed in the following sub-section.

13.6.2 DM in a multi-path environment

'Spatial direction' only makes sense for free space communication, in terms of where receivers locate. In a multi-path environment, a more relevant concept is that of a channel that determines the response each receiver detects. Thus, the extension of the DM technology for multi-path application can be readily achieved by replacing the transmission coefficients in free space, which are functions of spatial directions, with the channel responses in multi-path environment, which are functions of spatial positions. In [44,45], examples were provided for the extension of the orthogonal vector approach for multi-path environment with the realisation facilitated by RDAs [34] that have the ability to obtain the required channel response automatically. Other DM synthesis methods could have equally been adapted in a similar way for multi-path applications.

13.7   DM demonstrators

Up to date there have been only a few DM demonstrators built for real data transmission.

The first demonstrator was constructed based on the passive DM architecture in 2008 [8]. Since there are no effective synthesis methods associated with this type of DM structure, as was discussed in Section 13.2, no further developments in this branch has been subsequently reported.

Instead of radiation structure re-configurable DM, the excitation re-configurable DM array demonstrator was built based on the analogue approach in [10]. Since the iterative BER-driven synthesis approach was adopted, only the static DM transmitter was realised.

A 7-element digital DM demonstrator for 2.4-GHz operation [5] was realised with the help of the Wireless Open-Access Research Platform (WARP) [46]. This digital DM architecture is compatible with any of the synthesis methods presented in Sections 13.4.

The other DM demonstrator [7] that can be considered as hardware realisations of the orthogonal vector DM synthesis approach, utilised the beam orthogonality property possessed by Fourier beamforming networks to orthogonally inject information and interference along the desired secure communication direction. This structure avoids the use of analogue re-configurable RF devices, thus leads to an effective step towards practical field applications. Several system level experiments based on a 13-by-13 Fourier Rotman lens for 10-GHz operation were conducted in an anechoic chamber. One is for amplitude modulation (AM), the video of which can be found in [16]. Another Fourier Rotman lens DM experiment employed WARP boards, which allowed the digital modulation to be adopted and BER to be measured [7]. The experimental setup and the results for both received constellation patterns and BER spatial distributions can be found in [7].

The properties of the four DM demonstrators are summarised in Table 13.3.

13.8   Conclusions and recommendations for future studies on DM

This chapter reviewed the development in the DM technique up to the present time. Specifically, Section 13.1 described the DM concept. Section 13.2 enumerated the reported DM physical architecture, whose mathematical model, synthesis approaches, and assessment metrics were, respectively, addressed in Sections 13.3–13.5. Section 13.6 was devoted to the extension of the DM technology for multi-beam application and multi-path environment.

Table 13.3   *Summary of four DM demonstrators for real-time data transmission [4].*
© 2015 Cambridge University Press and the European Microwave
Association. Reprinted with permission from Ding Y. and Fusco V.
"A review of directional modulation technology". International Journal
of Microwave and Wireless Technologies. 2015;1–13

Article DM architecture	[8] NFDAM	[10] Antenna- re-configurable	[5] Digital	[7] Fourier lens
No. of array elements	900	4	7	13
Synthesis method	Trial and error	BER-driven	Orthogonal vector	Not required
Static or dynamic DM	Static	Static	Dynamic	Dynamic
Signal modulation	Non-standard	QPSK	DQPSK	DQPSK DBPSK
Operating frequency	60 GHz	7 GHz	2.4 GHz	10 GHz
Bit rate	Not specified	200 kbps	5 Mbps	5 Mbps
Realisation complexity	High	Medium	Medium	Low
Complexity of steering θ_0	High	High	Low	Low
Complexity for higher order modulation	High	Medium	Low	Low
Complexity for multi-beam DM	High	High	Medium	Low
Complexity for DM in multi-path	High	High	Medium	High

Although rapid development in the DM technology has been achieved in recent years, the field is still not mature, and needs to be perfected in the following aspects;

- The vector model of the DM technology was established based on a static signal constellation pattern in IQ space. This makes the model not usable for certain modulation schemes, e.g. frequency modulation (FM) and frequency-shift keying (FSK), which exhibit trajectories with respect to time when represented in IQ space.
- In [15], it was mentioned that it was the beam orthogonality property possessed by the Fourier transform beamforming networks that enabled the DM functionality. However, in terms of successfully constructing DM transmitters, it is required that only beam orthogonality along the desired communication direction occurs, i.e. the far-field patterns excited by the interference applied at the relevant beam ports have nulls along the direction where the main information beam projects, whereas the main beam projected by the interference signal applied at each beam port lies at the null of the information pattern. Alignment of the nulls in this way is not strictly required as all that is needed for DM operation is that interference occurs everywhere except along the projected information direction. This means that the strict Fourier constraint may potentially be relaxed. The extent of relaxation that can be applied could be investigated.

- Current DM technology was developed based on the assumption of narrow band signals. Wideband transmissions, such as CDMA and OFDM signals, would require new mathematical models and associated synthesis methods.
- Assessment metrics and power efficiency concepts may need to be revisited when recently emerged multi-beam DM systems are under consideration.
- More synthesis-free DM transmitters that can function in multi-beam mode within multi-path environment are to be expected. Furthermore associated physical implementations and experiments on real-time data transmission are of works of interest.
- More suitable candidates for multi-port antenna structures, acting as DM transmitter array elements, are required to further reduce the DM system complexity and physical profile. One preliminary work on this topic can be found in [47].

References

[1] A. Babakhani, D. B. Rutledge, and A. Hajimiri, "Near-field direct antenna modulation," *IEEE Microwave Magazine*, vol. 10, no. 1, pp. 36–46, 2009.

[2] M. P. Daly and J. T. Bernhard, "Beamsteering in pattern reconfigurable arrays using directional modulation," *Antennas and Propagation, IEEE Transactions on*, vol. 58, no. 7, pp. 2259–2265, 2010.

[3] M. P. Daly and J. T. Bernhard, "Directional modulation technique for phased arrays," *Antennas and Propagation, IEEE Transactions on*, vol. 57, no. 9, pp. 2633–2640, 2009.

[4] Y. Ding and V. Fusco, "A review of directional modulation technology," *International Journal of Microwave and Wireless Technologies*, vol. 8, no. 7, pp. 981–993, 2016.

[5] Y. Ding and V. Fusco, "Experiment of digital directional modulation transmitters," *Forum for Electromagn. Research Methods and Application Technol. (FERMAT)*, vol. 11, 2015.

[6] Y. Ding and V. F. Fusco, "A vector approach for the analysis and synthesis of directional modulation transmitters," *IEEE Transactions on Antennas and Propagation*, vol. 62, no. 1, pp. 361–370, Jan. 2014.

[7] Y. Ding, Y. Zhang, and V. Fusco, "Fourier Rotman lens enabled directional modulation transmitter," *International Journal of Antennas and Propagation*, vol. 2015, pp. 1–13, 2015.

[8] A. Babakhani, D. B. Rutledge, and A. Hajimiri, "Transmitter architectures based on near-field direct antenna modulation," *IEEE Journal of Solid-State Circuits*, vol. 43, no. 12, pp. 2674–2692, 2008.

[9] Y. Ding and V. Fusco, "Directional modulation transmitter synthesis using particle swarm optimization," in *Proc. Loughborough Antennas and Propag. Conf. (LAPC)*, Loughborough, UK, Nov. 2013, pp. 500–503.

[10] M. P. Daly, E. L. Daly, and J. T. Bernhard, "Demonstration of directional modulation using a phased array," *IEEE Transactions on Antennas and Propagation*, vol. 58, no. 5, pp. 1545–1550, May 2010.

[11]　H. Shi and T. Alan, "Direction dependent antenna modulation using a two element array," in *Antennas and Propagation (EUCAP), Proceedings of the Fifth European Conference on*, Rome, Italy, Apr. 2011, pp. 812–815.

[12]　H. Shi and A. Tennant, "An experimental two element array configured for directional antenna modulation," in *Antennas and Propagation (EUCAP), 2012 6th European Conference on*, Prague, Czech, Mar. 2012, pp. 1624–1626.

[13]　H. Shi and A. Tennant, "Covert communication using a directly modulated array transmitter," in *2014 Eighth European Conference on Antennas and Propagation (EuCAP)*, The Hague, Netherlands, Apr. 2014, pp. 352–354.

[14]　Y. Ding and V. Fusco, "Sidelobe manipulation using butler matrix for 60 GHz physical layer secure wireless communication," in *Proc. Loughborough Antennas and Propag. Conf. (LAPC)*, Loughborough, UK, Nov. 2013, pp. 11–12.

[15]　Y. Zhang, Y. Ding, and V. Fusco, "Sidelobe modulation scrambling transmitter using Fourier Rotman lens," *IEEE Transactions on Antennas and Propagation*, vol. 61, no. 7, pp. 3900–3904, Jul. 2013.

[16]　Y. Ding. (2014, Jun.) Fourier Rotman lens directional modulation demonstrator experiment. [Online]. Available: www.youtube.com/watch?v=FsmCcxo-TPE.

[17]　N. Valliappan, A. Lozano, and R. W. Heath, "Antenna subset modulation for secure millimeter-wave wireless communication," *IEEE Transactions on Communications*, vol. 61, no. 8, pp. 3231–3245, Aug. 2013.

[18]　N. N. Alotaibi and K. A. Hamdi, "Switched phased-array transmission architecture for secure millimeter-wave wireless communication," *IEEE Transactions on Communications*, vol. 64, no. 3, pp. 1303–1312, Mar. 2016.

[19]　E. J. Baghdady, "Directional signal modulation by means of switched spaced antennas," *IEEE Transactions on Communications*, vol. 38, no. 4, pp. 399–403, Apr. 1990.

[20]　Q. Zhu, S. Yang, R. Yao, and Z. Nie, "Directional modulation based on 4-D antenna arrays," *IEEE Transactions on Antennas and Propagation*, vol. 62, no. 2, pp. 621–628, Feb. 2014.

[21]　A. Kalantari, M. Soltanalian, S. Maleki, S. Chatzinotas, and B. Ottersten, "Directional modulation via symbol-level precoding: a way to enhance security," *IEEE Journal of Selected Topics in Signal Processing*, vol. 10, no. 8, pp. 1478–1493, Dec. 2016.

[22]　Y. Ding and V. Fusco, "A synthesis-free directional modulation transmitter using retrodirective array," *IEEE Journal of Selected Topics in Signal Processing*, vol. 11, no. 2, pp. 428–441, Mar. 2017.

[23]　Y. Ding and V. Fusco, "Vector representation of directional modulation transmitters," in *Antennas and Propagation (EuCAP), 2014 Eighth European Conference on*, Hague, Netherlands, Apr. 2014, pp. 367–371.

[24]　J. Hu, F. Shu, and J. Li, "Robust synthesis method for secure directional modulation with imperfect direction angle," *IEEE Communication Letters*, vol. 20, no. 6, pp. 1084–1087, Jun. 2016.

[25]　R. Negi and S. Goel, "Secret communication using artificial noise," in *Proceedings on IEEE Vehicular Technology Conference*, vol. 62, no. 3, pp. 1906–1910, Sep. 2005.

[26] S. Goel and R. Negi, "Guaranteeing secrecy using artificial noise," *IEEE Transactions on Wireless Communications*, vol. 7, no. 6, pp. 2180–2189, Jun. 2008.

[27] J. Hu, F. Shu, and J. Li, "Robust synthesis method for secure directional modulation with imperfect direction angle," *IEEE Communications Letters*, vol. 20, no. 6, pp. 1084–1087, June 2016.

[28] Y. Ding and V. Fusco, "BER-driven synthesis for directional modulation secured wireless communication," *International Journal of Microwave and Wireless Technologies*, vol. 6, no. 2, pp. 139–149, Apr. 2014.

[29] Y. Ding and V. Fusco, "Directional modulation transmitter radiation pattern considerations," *IET Microwaves, Antennas & Propagation*, vol. 7, no. 15, pp. 1201–1206, Dec. 2013.

[30] Y. Ding and V. F. Fusco, "Constraining directional modulation transmitter radiation patterns," *IET Microwaves, Antennas & Propagation*, vol. 8, no. 15, pp. 1408–1415, Dec. 2014.

[31] B. Zhang, W. Liu, and X. Gou, "Compressive sensing based sparse antenna array design for directional modulation," *IET Microwaves, Antennas & Propagation*, vol. 11, no. 5, pp. 634–641, Apr. 2017.

[32] Y. Ding and V. Fusco, "A far-field pattern separation approach for the synthesis of directional modulation transmitter arrays," in *General Assembly and Scientific Symposium (URSI GASS), 2014 XXXIth URSI*, Aug. 2014, pp. 1–4.

[33] Y. Ding and V. F. Fusco, "Directional modulation far-field pattern separation synthesis approach," *Microwaves, Antennas & Propagation, IET*, vol. 9, no. 1, pp. 41–48, Jan. 2015.

[34] V. Fusco and N. Buchanan, "Developments in retrodirective array technology," *Microwaves, Antennas & Propagation, IET*, vol. 7, no. 2, pp. 131–140, Jan. 2013.

[35] L. Chen, Y. C. Guo, X. W. Shi, and T. L. Zhang, "Overview on the phase conjugation techniques of the retrodirective array," *International Journal of Antennas and Propagation*, vol. 2010, pp. 1–10, 2010.

[36] Y. Ding and V. F. Fusco, "Establishing metrics for assessing the performance of directional modulation systems," *IEEE Transactions on Antennas and Propagation*, vol. 62, no. 5, pp. 2745–2755, May 2014.

[37] Y. Ding and V. F. Fusco, "Developments in directional modulation technology," *Forum for Electromagn. Research Methods and Application Technol. (FERMAT)*, vol. 13, 2016.

[38] S. Ten Brink, "Convergence behavior of iteratively decoded parallel concatenated codes," *IEEE Transactions on Communications*, vol. 49, no. 10, pp. 1727–1737, 2001.

[39] H. Shi and A. Tennant, "Simultaneous, multichannel, spatially directive data transmission using direct antenna modulation," *IEEE Transactions on Antennas and Propagation*, vol. 62, no. 1, pp. 403–410, Jan. 2014.

[40] Y. Ding and V. Fusco, "Orthogonal vector approach for synthesis of multibeam directional modulation transmitters," *IEEE Antennas and Wireless Propagation Letters*, vol. 14, pp. 1330–1333, Feb. 2015.

[41] M. Hafez and H. Arslan, "On directional modulation: An analysis of transmission scheme with multiple directions," in *Communication Workshop (ICCW), 2015 IEEE International Conference on*. Piscataway, NJ: IEEE, 2015, pp. 459–463.

[42] Y. Ding and V. Fusco, "MIMO inspired synthesis of multi-beam directional modulation systems," *IEEE Antennas and Wireless Propagation Letters*, vol. 15, pp. 580–584, Mar. 2016.

[43] D. Tse and P. Viswanath, *Fundamentals of wireless communication*. New York: Cambridge University Press, 2005.

[44] Y. Ding and V. Fusco, "Improved physical layer secure wireless communications using a directional modulation enhanced retrodirective array," in *General Assembly and Scientific Symposium (URSI GASS), 2014 XXXIth URSI*, Beijing, China, Aug. 2014, pp. 1–4.

[45] Y. Ding and V. Fusco, "Directional modulation-enhanced retrodirective array," *Electronics Letters*, vol. 51, no. 1, pp. 118–120, Jan. 2015.

[46] Warp project. [Online]. Available: http://www. warpproject.org. [Accessed: 14-Jun-2014].

[47] A. Narbudowicz, D. Heberling, and M. J. Ammann, "Low-cost directional modulation for small wireless sensor nodes," in *2016 10th European Conference on Antennas and Propagation (EuCAP)*. Piscataway, NJ: IEEE, 2016, pp. 1–3.

Chapter 14

Secure waveforms for 5G systems

Stefano Tomasin[1]

The fifth generation (5G) of cellular systems addresses challenging objectives in terms of connection speed, latency, energy consumption, traffic type, number of served users and their type (not just humans but mostly devices or even herds of animals). In this plethora of requests, a proper design of the waveforms is seen by many in the field as a key factor for the success of the new standard. This chapter will look at the possible waveform candidates from a physical layer security (PLS) point of view, providing an overview of on-going research activity on how they can be exploited to either secretly transmit a message on a wireless channel or extract a secret key between a couple of users.

The focus will be on the most promising waveforms candidates, i.e. orthogonal frequency division multiplexing (OFDM), filterbank modulation (FBMC), single-carrier frequency division multiple access (SC-FDMA), universal filter multi-carrier (UFMC) and generalized frequency division multiplexing (GFDM). These waveforms have been compared with respect to a number of issues that are not related to security, such as spectral containment, impulse duration, spectral efficiency, complexity of the transmitter/receivers, ease of integration with multiple-input–multiple-output (MIMO) systems, etc. [1–5]. However, their comparison in terms of suitability for PLS is not yet available. The chapter will present the comparison addressing some issues common to all approaches, but also considering the peculiarities of each system. It turns out that all proposed waveforms convert a wideband channel into a set of parallel narrowband channels; therefore, we will provide an in-depth analysis of this channel model from a security standpoint. Then, we will also consider how Eve can exploit the waveforms peculiarities [in particular the cyclic prefix (CP) used by most waveforms] to improve its eavesdropping capabilities, deriving performance and possible countermeasures.

The first part of this chapter will deal with PLS over parallel narrowband channels. This channel model is actually common to all proposed 5G waveforms. We will consider both secret message transmission and secret key agreement over the parallel channels. For secret message transmission, resource allocation and coding strategies with various objectives (e.g. maximisation of either the secrecy rate or the secrecy outage rate) will be discussed. Since 5G systems will be multi-user, the design of

---

[1]Department of Information Engineering, University of Padova, Italy

secure systems for multiple users will be considered, including the transmission of a common unsecure message to all users, that can be particularly useful in an Internet of Things (IoT) or sensor network context. For secret key agreement, we will analyse the achievable rate and the resource allocation, the key can then be used for various security purposes and since this approach is quite simple, it may find application in 5G systems (including low-power, low-complexity IoT devices) sooner than solutions based on wiretap coding.

The second part of the chapter will deal with the peculiarities of each waveform choice, in order to see their impact on PLS. We will see how CP and the transmit filters can be exploited by an eavesdropper to know more about the secret message and which countermeasures can be taken. We will also see specific security solutions proposed for the various waveforms.

Notation: Vectors and matrices are denoted in boldface, x^T, x^* and x^H denote the transpose, conjugate and Hermitian operator of x, respectively. det (A) and trace (A) denote the determinant and trace of matrix A, respectively. $\mathbb{E}[\cdot]$ denotes the expectation. $\mathbb{P}[\cdot]$ denotes the probability. $\log x$ denotes the base-2 logarithm. $\ln x$ denotes the natural-base logarithm. $[x]^+ = \max\{0, x\}$. Set $\mathscr{S}$ has cardinality $|\mathscr{S}|$. $I(X; Y)$ denotes the mutual information between X and Y. $a \otimes b$ denotes the Kroneker product between matrices a and b. $\mathbf{0}_{n \times k}$ is a $n \times k$ matrix with all entries equal to zero. I_n is the square $n \times n$ identity matrix.

14.1　Secret transmission over parallel channels

As we will see in more details in the second part of the chapter, all the proposed 5G waveforms convert the wideband channel into a set of parallel narrowband *sub-channels*. Therefore, we first consider the problem of secure transmission and secrecy key extraction over a set of parallel narrowband channels.

14.1.1　Single user case

We start considering here the case of a point-to-point transmission shown in Figure 14.1. In particular, we consider a system where Alice aims at transmitting a message $\mathscr{M}$ with spectral efficiency R_S to Bob, ensuring that eavesdropper Eve does not obtain any information on $\mathscr{M}$. The transmission occurs on a set of N parallel additive white Gaussian noise (AWGN) flat sub-channels.

Message $\mathscr{M}$ is encoded with a wiretap code into N signals, x_n, $n = 1, \ldots, N$ (block ENC in Figure 14.1). Let $x = [x_1, \ldots, x_N]^T$ be the signal transmitted on the N sub-channels, where the power of each signal is $P_n = \mathbb{E}[|x_n|^2]$, $n = 1, \ldots, N$. We assume a power constraint at Alice, i.e.:

$$\sum_{n=1}^{N} P_n \leq \bar{P}. \tag{14.1}$$

The received signal at Bob and Eve are $y = [y_1, \ldots, y_N]^T$ and $z = [z_1, \ldots, z_N]^T$ with:

$$y_n = h_n x_n + v_n, \quad z_n = g_n x_n + w_n, \quad n = 1, \ldots, N, \tag{14.2}$$

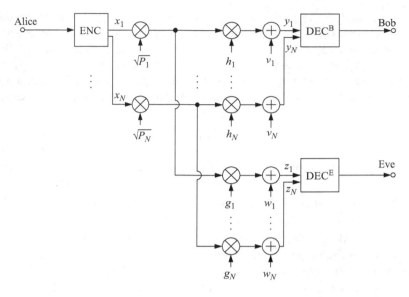

Figure 14.1   Model for a wiretap transmission over parallel channels

with v_n and w_n are AWGN terms that will be assumed zero-mean and with unitary variance, and $\boldsymbol{h} = [h_1, \ldots, h_N]^T$ and $\boldsymbol{g} = [g_1, \ldots, g_N]^T$ are the legitimate and eavesdropper channels, respectively. We also indicate the sub-channels' power gains as $\boldsymbol{H} = [H_1, \ldots, H_N]^T = [|h_1|^2, \ldots, |h_N|^2]^T$ and $\boldsymbol{G} = [G_1, \ldots, G_N]^T = [|g_1|^2, \ldots, |g_N|^2]^T$.

Bob and Eve attempt to decode the received signals by decoders DEC^B and DEC^E as shown in Figure 14.1.

For the transmission of a secret message, two strategies may be considered, namely coding per sub-channel (CPS) and coding across sub-channels (CAS). With coding across sub-channels (CAS) the message, $\mathcal{M}$ is encoded with a single wiretap code, mapped into constellation symbols, each of which is transmitted on a different sub-channel. With CPS instead, the message $\mathcal{M}$ is first split into N sub-messages with rates R_n, $n = 1, \ldots, N$, each separately encoded with a wiretap code, mapped into constellation symbols and transmitted on a single sub-channel. The sum rate of the message is also in this case $R_S = \sum_{n=1}^{N} R_n$.

14.1.1.1   Full channel-state information at the transmitter (CSIT) and Gaussian input

We will first consider the case in which Alice knows both the legitimate and eavesdropper channel (perfect and complete CSIT). This scenario may apply when we assume that the eavesdropper is honest but curious, so it will reveal its channel conditions to Alice while still trying to get information on $\mathcal{M}$. This situation may occur because Eve either has some other advantages in being honest or is at least partially under the control of Alice. In any case, the study of this case is useful before considering more general (and often more realistic) scenarios in which Alice has only a partial CSIT.

We will now analyse the secrecy capacity of this system focusing in particular on the problem of optimising the power split among the sub-channels. With perfect and full CSIT, the secrecy capacity[1] is the same for both CAS and CPS schemes [6–9], i.e. $I(y;x) - I(z;x)$, and can be achieved by a Gaussian channel input.

The resource allocation problem is typically stated as the choice of the power levels at each subchannel P_n (power allocation) that maximises the secrecy capacity under the total power constraint (14.1), i.e.:

$$R_S^* = \max_{\{P_n\}} I(y;x) - I(z;x), \quad \text{s.t. (14.1)}, \tag{14.3}$$

which for Gaussian signals on AWGN channels becomes:

$$R_S^* = \max_{\{P_n\}} \sum_{n=1}^{N} \log(1 + P_n H_n) - \log(1 + P_n G_n), \quad \text{s.t. (14.1)}. \tag{14.4}$$

If the eavesdropper's channel has a higher gain than the legitimates one ($G_n > H_n$), no secrecy can be achieved; therefore, the optimum choice is to set $P_n = 0$ for that channel. When $H_n > G_n$ we note that the sum in (14.4) is concave in P_n and therefore we have a closed-form solution for the optimal power allocation, yielding:

$$P_n = \left[-\frac{H_n + G_n}{2H_n G_n} + \frac{\sqrt{(H_n - G_n)^2 + 4H_n G_n(H_n - G_n)/\mu}}{2G_n H_n} \right]^+, \tag{14.5}$$

where $\mu > 0$ must be chosen in order to satisfy the power constraint (14.1).

14.1.1.2   Full CSIT and finite constellations

For finite constellations, (14.4) does not hold anymore, and the achievable secrecy rate can be written as

$$R_S^* = \max_{\{P_n\}} \sum_{n=1}^{N} \bar{I}(P_n H_n) - \bar{I}(P_n G_n) \tag{14.6}$$

where $\bar{I}(P_n H_n)$ and $\bar{I}(P_n G_n)$ are the mutual information for finite constellation transmissions over sub-channel n for the channel to Bob and Eve, respectively. It turns out that the function to be maximised in (14.6) is non-concave in P_n when $H_n > G_n$ [10], and the optimization problem does not yield trivial solutions. In [10], a sub-optimal efficient dual optimization algorithm is proposed, having a duality gap that decreases with N in the order of $\mathcal{O}(1/\sqrt{N})$, provided that the channel has a finite delay spread. In [11], the authors have exploited the result of [12], i.e. linking the mutual information with the minimum mean square error (MMSE), i.e.:

$$\frac{d\bar{I}(\rho)}{d\rho} = \text{MMSE}(\rho) \tag{14.7}$$

---

[1]The secrecy capacity is the maximum rate of message $\mathcal{M}$ that, for infinite length codewords, can be decoded with vanishing error probability by Bob while providing zero mutual information between $\mathcal{M}$ and z, i.e. no information leakage, to Eve.

where MMSE(ρ) has been computed in [13] for various constellations. From this, a dual decomposition method has been applied where the optimization problem is split into two sub-problems involving a single variable linked together by a master problem solved by a gradient method over the variable.

14.1.1.3 Partial CSIT and Gaussian input

As we mentioned before, having the complete CSIT may not be realistic in many cases, for example when the eavesdropper is completely passive and is not inserted into the legitimate network, or when it does not behave honestly, thus reporting a different channel gain with respect to what she actually experiences. Therefore, it is useful to consider the case in which Alice has a partial CSIT, in particular, we only assume the knowledge of the statistics of G and in particular G_n are independent and identically distributed (iid) with exponential distribution having average α. In this case, we cannot always ensure perfect secrecy, but there is the possibility that the secrecy capacity for a given power allocation is below the transmit secret spectral efficiency, i.e. $C_S < R_S$, where $C_S = I(y;x) - I(z;x)$. In this case will have a *secrecy outage event*, that occurs with probability:

$$p_{\text{out}} = \mathbb{P}[C_S \leq R_S]. \tag{14.8}$$

Note that here we assume that Alice and Bob still have perfect information on the legitimate channel h; therefore, we can ensure that Bob always decodes the secret message.

The two coding strategies CAS and CPS are not any longer equivalent in terms of secrecy outage probability. For the CAS scheme, the outage event jointly involves to all sub-channels. For CPS, the outage event occurs if on *any* sub-channel n (with transmit secrecy code rate R_n) is in outage, i.e.:

$$p_{\text{out}} = \mathbb{P}\left[\cup_1^N \{I(x_n;y_n) - I(x_n;y_n) \leq R_n\}\right]. \tag{14.9}$$

Assuming independent channel realisations for each sub-channel, and for the AWGN channels, for CPS we obtain:

$$p_{\text{out}} = \prod_{n=1}^{N} \mathbb{P}\left[\log(1 + H_n P_n) - \log(1 + G_n P_n) \leq R_n\right] = \prod_{n=1}^{N} p_{\text{out},n}. \tag{14.10}$$

The resource allocation problem in the case of partial CSIT can be formulated as the choice of the power allocation that maximises the sum rate $R_S = \sum_{n=1}^{N} R_n$ for a given outage probability $p_{\text{out}} = \varepsilon$. For CAS, we do not have any closed-form solution of the power allocation. However, a closed-form expression (involving Gamma functions and generalised Fox H-functions) of the secrecy outage probability has been derived in [9].

For CPS instead, we can formulate the total secrecy outage probability as a function of the secrecy outage probability on each channel, $p_{\text{out},n}$ as

$$p_{\text{out}} = 1 - \prod_{n=1}^{N} (1 - p_{\text{out},n}). \tag{14.11}$$

We can impose a constraint on the secrecy outage probability of each sub-channel, i.e. $p_{\text{out},n} \leq \varepsilon_n$, choosing the values of ε_n such that $p_{\text{out}} = \varepsilon$. Imposing the constraint on a per-subchannel basis defines an equivalent Eve channel, whose gains $\bar{G}_n$ are such that the probability that the effective Eve channel gain G_n is higher than $\bar{G}_n$ is equal to ε_n, $\mathbb{P}[G_n > \bar{G}_n] = \varepsilon_n$. For iid Rayleigh fading channels, we have for example $\bar{G}_n = -\alpha \ln \varepsilon_n$. Then, the sub-channel and power can be allocated as for the full CSIT case.

In [14], another approach has been adopted, where the AWGN-affecting Eve has been neglected to consider the worst case scenario, but an artificial noise (AN) is transmitted by Alice. Also in this case, the optimisation problem of maximising the secrecy outage rate is non-convex.

14.1.1.4   Multiple eavesdroppers

Until now, we have considered the presence of a single eavesdropper in the system. However, it may be possible that multiple eavesdroppers are present. For example, we can assume that any other user in the system is a potential eavesdropper. In this chapter, we will focus on non-colluding eavesdroppers, i.e. they do not exchange information or collaborate to obtain the secret message but act individually. When $M > 1$ eavesdroppers are considered, the resource allocation problems for CAS and CPS become significantly different. Let $G_n^{(m)}$ be the channel gain of eavesdropper $m = 1, \ldots, M$ on sub-channel n. For CPS, we may consider an equivalent eavesdropper having channel gains $\bar{G}_n = \max_m G_n^{(m)}$ and then perform the resource allocation as described for the single eavesdropper, where the channel gains are those of the equivalent eavesdropper. For CAS instead, we must solve the following max–min problem (over Gaussian channels):

$$\max_{\{P_n\}} \min_m \sum_{n=1}^{N} \log(1 + P_n H_n) - \log(1 + P_n G_n^{(m)}), \quad \text{s.t. (14.1)}, \tag{14.12}$$

which is a NP-hard problem.

14.1.2   Multiple users case

When K users are considered into the system, as typical of 5G systems, we may want to make sure that each user receives a message that remains unknown to possibly multiple eavesdroppers. Let $R_{\text{S},k}$ be the secrecy rate achieved for user $k = 1, \ldots, K$. We consider as objective function the maximisation of the weighted sum rate:

$$R_{\text{S}} = \sum_{k=1}^{K} \omega_k R_{\text{S},k} \tag{14.13}$$

where $\omega_k > 0$ are weights chosen in order to ensure some kind of fairness among users. The special case $\omega_k = 1$ for all k provides the sum-rate among all users.

We note that in a multi-user scenario, also legitimate users may behave as eavesdroppers, in addition to possible external eavesdroppers. Therefore, each user may have a different set of eavesdroppers. In the following, we suppose that there are M external eavesdroppers and that all other users operate as eavesdroppers as well, for a total of $K + M$ eavesdroppers.

In the following, we focus on the Gaussian input.

14.1.2.1 Downlink

For downlink transmission, the base station (BS) is transmitting to the mobile terminals (MTs), under total power constraint. Let $H_{k,n}$ be the channel gain of the BS-MT k link on sub-channel n. Under the assumption of perfect CSIT, the resource allocation consists in deciding the power to be allocated on each sub-channel and the user to be served on each sub-channel. We denote as $P_{k,n}$ the power allocated for user k on sub-channel n. Considering single-user receivers (i.e. each user does not attempt to process interference due to transmissions to other users), each user rate is maximised if we use each sub-channel to serve a single user. The power constraint becomes in this case:

$$\sum_{n=1}^{N}\sum_{k=1}^{K}P_{k,n} \leq \bar{P}. \tag{14.14}$$

We first assume that there is only **one external eavesdropper** with channel gains G_n, and no curious users. Aiming at maximising the secrecy sum-rate ($\omega_k = 1$), we note that for any power allocation, the maximum rate is achieved by assigning sub-channel n to user k_n^* having the maximum secrecy rate on that sub-channel [15], i.e. $k_n^* = \operatorname{argmax}_k H_{k,n}$. Then, we have an equivalent Bob user having channel gains $H_{k_n^*,n}$, $n = 1, \ldots, N$, and we can apply the power optimization techniques described before for the single-user case, to achieve the secrecy multi-user capacity (where CAS and CPS are equivalent). User-k rate $R_{S,k}$ will be obtained by adding the secrecy rates of sub-channels assigned to him. Note that if we use CPS and we do not have full CSIT, we can still define equivalent channel gains such that higher gains are taken with probability ε_n and optimise the sub-channel and power allocation on these equivalent gains.

For the maximum weighted secrecy sum-rate problem (i.e. ω_k are different for each user), we observe that for a given power allocation, the following equations must be satisfied:

$$\left(\frac{1+H_{k_n^*,n}P_n}{1+G_nP_n}\right)^{\omega_{k_n^*}} \geq \left(\frac{1+H_{k,n}P_n}{1+G_nP_n}\right)^{\omega_k} \quad \forall k \neq k_n^*. \tag{14.15}$$

Therefore, the sub-channel allocation to the users depends on the allocated power. Indeed, the resulting problem requires to check all possible sub-channel allocations, having a complexity that grows exponentially with N. A possible sub-optimal solution to the problem is the solution of the dual problem (see also [16,17]). We also denote as $\mathscr{P}$ the set of powers $P_{k,n}$ that satisfy the constraint that for each sub-channel only one power is positive, while all other users' powers are zero. Then, the Lagrange dual function of the optimization problem is

$$g(\mu) = \max_{\{P_{k,n}\}\in\mathscr{P}} \sum_{k=1}^{K}\sum_{n=1}^{N}\omega_k\left[\log(1+P_{k,n}H_{k,n}) - \log(1+P_{k,n}G_n)\right]$$

$$+ \mu\left(\bar{P} - \sum_{n=1}^{N}\sum_{k=1}^{K}P_{k,n}\right),$$

where $\mu > 0$ is the Lagrange multiplier to be chosen to satisfy the power constraint. The dual problem is:

$$\min_{\mu} g(\mu), \quad \mu > 0. \tag{14.16}$$

Now, $g(\mu)$ can be decomposed into N independent sub-functions as:

$$g(\mu) = \sum_{n=1}^{N} g_n(\mu) + \mu \bar{P}, \tag{14.17}$$

$$g_n(\mu) = \max_{\{P_{k,n}\} \in \mathscr{P}} \sum_{k=1}^{K} \omega_k \big[\log(1 + P_{k,n} H_{k,n}) - \log(1 + P_{k,n} G_n)\big] - \mu \sum_{k=1}^{K} P_{k,n}, \tag{14.18}$$

We note that for a given μ, the maximisation problem in (14.18) provides as power allocation $P_{k,n} = 0$, if $H_{k,n} < G_n$ and

$$P_{k,n}^* = \left[-\frac{H_{k,n} + G_n}{2H_{k,n}G_n} + \frac{\sqrt{(H_{k,n} - G_n)^2 + 4H_{k,n}G_n(H_{k,n} - G_n)\omega_k/\mu}}{2G_n H_{k,n}} \right]^+. \tag{14.19}$$

if $H_{k,n} \geq G_n$. The choice of μ is not trivial, since we cannot use the gradient descent algorithm, since $g(\mu)$ is not differentiable due to the maximisation over the discrete set. We instead resort to the sub-gradient:

$$\delta\mu = \bar{P} - \sum_{n=1}^{N} \sum_{k=1}^{K} P_{k,n}, \tag{14.20}$$

that can be used to optimise μ. Note that this is a sub-optimal solution, however, in similar problems (such as multi-user power allocation in orthogonal frequency division multiplexing (OFDM) systems [16,17]), it has been shown to have a negligible duality gap.

When **users are eavesdroppers and multiple external eavesdroppers** are also present, again we must distinguish between CPS and CAS. For CAS, we already have seen that even with a single user the problem becomes very complex, and we do not discuss it here further. For CPS instead, again we have zero secrecy rate if we allocate a sub-channel to a user not having the highest gain; therefore, user allocation is straightforwardly solved, obtaining k_n^*. Let us indicate with $G_{k,n}^{(m)}$ the channel gain of eavesdropper m for user k on sub-channel n. We have that for each sub-channel, we can consider an equivalent eavesdropper having channel gain:

$$\bar{G}_n = \max_m G_{k_n^*,n}^{(m)}, \tag{14.21}$$

and then perform the resource allocation as described for the single eavesdropper. For maximum weighted secrecy sum-rate problem, we must in any case allocate the sub-channel to the user with the best channel gain (since all others achieve no secrecy rate). Therefore, we can solve the problem of power optimisation similarly to what is obtained for the single-user single-eavesdropper case: we allocate power only when $H_{k_n^*,n} > \bar{G}_n$, and in this case, the optimum power is as in (14.19) with $k = k_n^*$ and G_n replaced by $\bar{G}_n$.

14.1.2.2 Uplink

For the uplink, we must consider K separate power constraints, i.e.:

$$\sum_{n=1}^{N} P_{k,n} \leq \bar{P}, \quad k = 1, \ldots, K. \tag{14.22}$$

Also in this case, we cannot de-couple the problems of sub-channel and power allocations.

If there is **one external eavesdropper** with full CSIT (or also CPS with partial CSIT), we can still address the dual problem now with K Lagrange multipliers μ_k, with $k = 1, \ldots, K$, one for each power constraint, obtaining the dual function:

$$g(\mu) = \sum_{n=1}^{N} g_n(\mu) + \sum_{k=1}^{K} \mu_k \bar{P}, \tag{14.23}$$

power allocation $P_{k,n} = 0$ if $H_{k,n} < G_n$ and (14.19), if $H_{k,n} \geq G_n$ and K sub-gradients:

$$\delta \mu_k = \bar{P} - \sum_{n=1}^{N} P_{k,n}. \tag{14.24}$$

When **users are eavesdroppers**, we can proceed as for the downlink, focusing on CPS and considering only the highest channel gain for each user and sub-channel $\bar{G}_{k,n} = \max_m G_{k,n}^{(m)}$, and we have a power allocation as in (14.19) with $\bar{G}_{k,n}$ replacing G_n. Note however that in this scenario, we must know the channels between any user couple, and this information is not typically available. Indeed, typically only the knowledge of the MT-BS channel is available at the terminals.

Alternative approaches may include greedy iterative algorithms, where at each iteration one sub-carrier is assigned to the user that provides the highest secrecy rate increase, similarly to what has been proposed for non-secure orthogonal frequency division multiple access (OFDMA) [18].

14.1.3 Downlink with common message

Another scenario that might be relevant for 5G system is the downlink transmission of confidential messages to K users, together with the transmission of a common non-secure message to all users. This common message may be either a control signal targeting all users or a content that must be shared among all users and does not need to be secured, at least at the physical layer. It is multiplexed with the secret messages and jointly encoded across the sub-channels. Its rate is constrained by the minimum among all rates that can be achieved with each user. At the receiver, the common message is decoded first (thus being subject to the interference of the secret messages) and then subtracted from the received signal. Therefore, the achievable rate of the common message that can be received by user k is

$$R_{0k} = \sum_{n=1}^{N} \log\left(1 + H_{k,n} P_{0,n} + H_{k,n} \sum_{i=1}^{K} P_{i,n}\right) - \log\left(1 + G_{k,n} \sum_{i=1}^{K} P_{i,n}\right), \tag{14.25}$$

where $P_{0,n}$ is the power used on sub-channel n to transmit the common message. Since the common message must be decodable by all users, its maximum rate is

$$R_{S,0} = \min_k R_{0k}.$$ (14.26)

Note that we still use 'S' in $R_{S,0}$ although this message is not secret.

When we have a common message, we may consider as optimization objective the weighted sum-rate:

$$R_S = \sum_{k=0}^{K} \omega_k R_{S,k},$$ (14.27)

where ω_0 is the weight of the common message. The total power constraint (14.14) still holds and is applied also to the common message; thus, the sum over n now starts at $n = 0$. The maximisation of (14.27) under the total power constraint has been addressed in [19].

14.1.3.1 Power allocation for per-user encoding

The maximisation of R_S by proper sub-channel and power allocation has a high complexity, due to the involved max–min problem, that grows super-exponential versus the number of users. In [19], a sub-optimal approach has been proposed where (a) the sub-channel allocation is separated from power allocation and (b) the total power is fixed a-priori for groups of sub-channels.

About sub-channel allocation, the set of sub-channels allocated to user k is

$$\mathcal{L}_k = \left\{ n : H_{k,n} \geq H_{j,n} \, \forall j \in \{1, \ldots, K\}, j \neq k \right\}, \quad k = 1, \ldots, K$$ (14.28)

$$\mathcal{L}_0 = \{1, \ldots, N\} \setminus \cup_{k=1,\ldots,K} \mathcal{L}_k,$$ (14.29)

where $\mathcal{L}_0$ is the set of sub-channels allocated exclusively for the transmission of the common message. Still, other sub-channels are also used for the transmission of the common message, that will suffer from interference of secret messages.

About power allocation for user k, we assign a fixed total power proportional to the number of sub-channels in $\mathcal{L}_k$, i.e.:

$$\sum_{n \in \mathcal{L}_k} P_{k,n} \leq |\mathcal{L}_k| \bar{P}/N, \quad k = 0, \ldots, K.$$ (14.30)

The power allocation problem to maximise the weighted sum-rate R_S under the power constraints (14.30) can be split into $K + 1$ independent sub-problems, one for each user (including the virtual user transmitting the common message) within the set $\mathcal{L}_k$, thus reducing the complexity of the original max–min problem.

14.1.3.2 Power allocation for per-channel encoding

Another simplification of the power and sub-channel allocation problem is obtained by considering a CPS approach, where both the common message and the confidential messages are split into sub-messages, and a separate encoder is used on each channel. The secret rate for user $k \geq 1$ is:

$$R_{S,k} = \sum_{n \in \mathscr{L}_k} \log(1 + H_{k,n}P_{k,n}) - \log(1 + \bar{G}_n P_{k,n}), \tag{14.31}$$

where $\bar{G}_n$ is given by (14.21). For the common rate, we define $M_n = \min_j H_{j,n}$, and from (14.25), we have:

$$R_{S,0} = \sum_{n=1}^{N} \left[\log\left(1 + M_n P_{0,n} + M_n \sum_{i=1}^{K} P_{i,n}\right) - \log\left(1 + M_n \sum_{i=1}^{K} P_{i,n}\right) \right]. \tag{14.32}$$

The problem of the maximisation of the weighted sum-rate with common message and CPS has a closed-form solution. In particular, we define:

$$\Lambda_{k,n}(\lambda) = \frac{1}{2}\left[\left(\frac{1}{G_n} - \frac{1}{H_{k,n}}\right)\left(\frac{1}{G_n} - \frac{1}{H_{k,n}} + \frac{4\omega_k}{\lambda \ln 2}\right)\right]^{1/2} - \frac{1}{2}\left(\frac{1}{G_n} + \frac{1}{H_{k,n}}\right) \tag{14.33a}$$

$$\Psi_n(\lambda) = \frac{\omega_0}{\lambda \ln 2} - \frac{1}{M_n} \tag{14.33b}$$

$$\Delta_{k,n} = \left(\frac{1}{\bar{G}_n} - \frac{1}{H_{k,n}}\right)^2\left[\left(\frac{\omega_k}{\omega_0}\right)^2 + 2\frac{\omega_k}{\omega_0}\frac{\frac{2}{M_n} - \frac{1}{H_{k,n}} - \frac{1}{\bar{G}_n}}{\frac{1}{\bar{G}_n} - \frac{1}{H_{k,n}}} + 1\right] \tag{14.33c}$$

$$\Theta_{k,n} = \frac{1}{2}\left[\frac{\omega_k}{\omega_0}\left(\frac{1}{\bar{G}_n} - \frac{1}{H_{k,n}}\right) - \left(\frac{1}{\bar{G}_n} + \frac{1}{H_{k,n}}\right) + \sqrt{\Delta_{k,n}}\right]. \tag{14.33d}$$

The solution of power allocation problem [19] is for $n \in \mathscr{L}_k$:

$$P_{0,n} = \begin{cases} \left[\Psi_\ell(\lambda) - \Theta_{k,\ell}\right]^+, & H_{k,n} - G_n > \frac{M_n \omega_0}{\omega_k} \\ \left[\Psi_n(\lambda)\right]^+ & \text{otherwise} \end{cases} \tag{14.34}$$

while for $k = 1, \ldots, K$, we have

$$P_{k,n} = \begin{cases} \left[\min\{\Lambda_{k,n}(\lambda); \Theta_{k,n}\}\right]^+ & H_{k,n} - G_n > \frac{M_n \omega_0}{\omega_k} \\ 0 & \text{otherwise,} \end{cases} \tag{14.35}$$

where λ is chosen to satisfy the total power constraints on each $\mathscr{L}_k$ with equality. Note that this solution is fully determined apart for the choice of λ and its complexity scales only linearly with the number of users.

14.2   Secret key agreement

With a secret key agreement (SKA) procedure, Alice and Bob aim at sharing a sequence of random bits that are not known to Eve. The sequence of random bits (key) can then be used to encrypt data, provide authentication or other security features in forthcoming transmissions. Typically, once the key has been used for these purposes, it cannot be used again without letting Eve infer something on the key and therefore break the security. Therefore, we need efficient methods to renew the key, or generate another one. Here, we focus on SKA over parallel channels, that we have seen is a typical scenario for the waveforms transmissions likely to be used in 5G systems.

SKA needs a source of randomness for the generation of the random bits of the key. Depending on the place in which the randomness source is available, we have two main models for SKA. In the *source model* SKA, the randomness is provided directly by the physical channel connecting Alice to Bob, which typically has some randomness (in amplitude, phase, multi-path components...). In the *channel model* instead, Alice needs a random bit generator and the channel still plays some role as a source of randomness, as will be discussed later. We will now apply these two models to the parallel channel transmission, investigating the design of a SKA procedure and the resource allocation strategies over the sub-channels.

14.2.1   *Channel-model SKA over parallel channels*

The channel model provides the following steps for SKA [20]:

1. Alice transmits a random bit sequence to Bob.
2. Bob decodes (*distils*) a set of bits from the received signals of step 1. The decoded bits are in general affected by errors due to the noise; therefore, they do not coincide with the transmitted bit sequence. Bob transmits on a public channel the indices of the distilled bits.
3. Some redundancy on the distilled bits is shared between Alice and Bob in order to correct the errors.
4. Now that Alice and Bob share the same bit sequence, a new shorter sequence is obtained from the original one, by some hashing function.

Note that the randomness of the key comes from the second step, where noise corrupts the transmitted signal in a independent fashion for Bob and Eve. Therefore, bits that are moderately affected by noise for Bob have a good chance of being more corrupted for Eve. In step 2, Bob selects the *best* bits, i.e. those that are most likely to be detected correctly. Eve does not have the possibility to pick her best bits because Bob decides which bits are to be picked for the key generation. This provides an advantage to Bob over Eve. Note that Alice needs to transmit a random bit sequence to prevent Eve to know directly the bits; however, the true randomness that distinguishes Eve from Bob comes from the noise (which impacts the distillation process). As the bits distilled by Bob may still be corrupted by noise, step 3 corrects residual errors. However, the correction procedure is done on the public wireless channel, thus is overheard also by Eve, who can exploit it to correct some the her own received bits. At the end of step 3, Alice and Bob share the same sequence, and Eve knows correctly some of the bits, but not all. With the last step 4 (*privacy amplification*), both Alice and Bob apply a deterministic hash function to the shared sequence in order to obtain a smaller sequence. The point here is that many long sequences are mapped to the same short sequence, and that small differences in the long sequence leads to completely different short sequences. Therefore, even if Eve knows the long sequence with only a few errors, with high probability the short sequence that she will obtain will be significantly different from that obtained by Alice and Bob.

In [21], this system has been investigated for a transmission over parallel channels. With reference to model (14.2), the signal transmitted by Alice is binary phase shift keying (BPSK) modulated ($x_n = \pm\sqrt{P_n}$), and in step 2, Bob distils the received samples on sub-channel n having a log likelihood ratio (LLR) larger than a threshold A_n. In particular, letting the LLR be

$$\Lambda_n = \ln \frac{\mathbb{P}[x_n = 1 | y_n]}{\mathbb{P}[x_n = -1 | y_n]} = 4h_n y_n, \tag{14.36}$$

the bit is distilled, if $\Lambda_n > A_n$. We assume here that Alice has full CSIT of both Bob's and Eve's channels. In the third step, Alice transmits to Bob the syndrome column vector obtained from the distilled bits using a suitably selected error-correcting code. Bob will correct the sign of the LLRs by maximum likelihood decoding.

A performance metric for this system is the secret key throughput (SKT), defined as the ratio between the number of *secret key bits* and the number of channel uses (i.e. the number of transmitted symbols) needed to obtain them. Considering a BPSK transmission, the SKT can also be computed as the ratio between the average number of secret bits per distilled bit and the average number of random bits that Alice needs to transmit to distil one bit.

In particular, we let $\sigma(\boldsymbol{P}, \boldsymbol{A})$ be the average number of secret key bits per distilled bit for power allocation $\boldsymbol{P} = [P_1, P_2, \ldots, P_N]$ and threshold choice $\boldsymbol{A} = [A_1, A_2, \ldots, A_N]$. Note that $\sigma(\boldsymbol{P}, \boldsymbol{A})$ is a random variable that depends on the fading channels signal-to-noise ratios (SNRs). We also let $\tau(\boldsymbol{P}, \boldsymbol{A})$ be the average number of channel uses needed to distil one bit. Again note that (a) $\tau(\boldsymbol{P}, \boldsymbol{A})$ depends on the thresholds and the allocated powers and (b) $\tau(\boldsymbol{P}, \boldsymbol{A})$ is a random variable. The SKT is then defined as

$$\eta(\boldsymbol{P}, \boldsymbol{A}) = \frac{\mathbb{E}[\sigma(\boldsymbol{P}, \boldsymbol{A})]}{\mathbb{E}[\tau(\boldsymbol{P}, \boldsymbol{A})]}. \tag{14.37}$$

A lower bound on the SKT has been derived in [21]. In particular, we first define $\pi(P_n, A_n)$ as the probability that a bit is distilled on sub-channel n, which depends on the Alice–Bob signal to noise ratio (SNR) and threshold value, i.e.:

$$\pi(P_n, A_n) = \mathbb{P}[|\Lambda_n| \geq A_n]. \tag{14.38}$$

Then, defining

$$\mu_n(P_n, A_n) = H(x_n | z_n) - 1 + I(x_n; y_n) \tag{14.39}$$

we obtain the bound

$$\eta(\boldsymbol{P}, \boldsymbol{A}) \geq \left[\sum_{n=1}^{N} \pi(P_n, A_n)\mu_n(P_n, A_n) \right]^{+} = \bar{\eta}(\boldsymbol{P}, \boldsymbol{A}). \tag{14.40}$$

14.2.1.1 Resource allocation problem for full CSIT

The resource allocation problem for SKA aims at optimising both the LLR thresholds $\boldsymbol{A}$ and the allocated powers $\boldsymbol{P}$ with the purpose of maximising the SKT lower bound (14.40), under the total power constraint (14.1). In formulas, the resource allocation problem can be written as:

$$\eta_{\text{opt}} = \max_{P,A} \bar{\eta}(P,A), \quad \text{s.t. (14.1).} \tag{14.41}$$

Since computing $\bar{\eta}(P,A)$ requires the evaluation of integrals depending on the optimisation variables [21], solving this problem is not trivial.

It is possible first to cascade the two problems of power and threshold optimisation, having as inner problem the maximisation with respect to the thresholds and as outer problem the maximisation with respect to the powers, i.e.:

$$\eta_{\text{opt}} = \max_{P:\sum P_n \leq \bar{P}} \max_{A} \bar{\eta}(P,A). \tag{14.42}$$

Then, from (14.40), we note that the SKT is the sum of N functions, each depending on an individual threshold A_n, on channel n. Moreover, constraint (14.1) is only on powers. Therefore, for a given power allocation, the threshold optimization can be independently performed for each channel. From (14.40), we define the threshold-optimised SKT for channel n as:

$$\phi(H_n P_n, \gamma_n) = \max_{A_n} \pi(P_n, A_n)\mu_n(P_n, A_n), \tag{14.43}$$

where $\gamma_n = G_n/H_n$, which highlights the dependence on Eve's channel as well. Using (14.43), the maximisation problem (14.42) becomes:

$$\eta_{\text{opt}} = \max_{P} \sum_{n=1}^{N} [\phi(H_n P_n, \gamma_n)]^+, \quad \text{s.t. (14.1).} \tag{14.44}$$

In [21], it has been shown that a good fitting of $\phi(P, \gamma)$ is provided by

$$\tilde{\phi}(P, \gamma) = \tilde{a}(\gamma) + \tilde{b}(\gamma) \exp[\tilde{c}(\gamma)P] \quad \text{for } P \leq \tilde{P}(\gamma). \tag{14.45}$$

where

$$\tilde{a}(\gamma) = a_3\gamma^3 + a_2\gamma^2 + a_1\gamma + a_0, \quad \tilde{b}(\gamma) = b_3\gamma^3 + b_2\gamma^2 + b_1\gamma + b_0 \tag{14.46}$$

$$\tilde{c}(\gamma) = c_0\gamma^2 + c_1 + c_2 \exp[c_3\gamma], \quad \tilde{P}(\gamma) \approx d_1 + d_2 \exp[d_3\gamma], \tag{14.47}$$

where parameters a_i, b_i, c_i and d_i, $i = 1, 2, 3$, are chosen to provide the best fit.

Now, using the approximation, our resource allocation problem becomes:

$$\tilde{\eta}_{\text{opt}} = \max_{P} \frac{1}{N} \sum_{n=1}^{N} \tilde{\phi}(H_n P_n, \gamma_n), \quad \text{s.t. (14.1)} \tag{14.48}$$

With this modification, (14.48) is now a convex problem. Its Lagrangian function is

$$\mathcal{L}(\lambda, P) = \frac{1}{N} \sum_{n=1}^{N} \tilde{\phi}(H_n P_n, \gamma_n) - \lambda \left[\sum_{n=1}^{N} P_n - \bar{P} \right], \tag{14.49}$$

where $\lambda \geq 0$ is the Lagrange multiplier.

By maximising the Lagrangian function for a given value of the Lagrange multiplier λ, we have the following optimal power allocation:

$$P_n^\star(\lambda) = \min\left\{ \frac{\tilde{P}(\gamma_n)}{H_n}, \left[\frac{1}{\tilde{c}(\gamma_n)H_n} \ln \frac{\lambda}{N\tilde{b}(\gamma_n)\tilde{c}(\gamma_n)H_n} \right]^+ \right\}. \tag{14.50}$$

Since the target function is strictly increasing with respect to P_n, within its domain $P_n \leq \tilde{P}(\gamma_n)/H_n$, we have that the power constraint is satisfied with equality in correspondence of the solution, thus the value of λ corresponding to the solution of (14.48) satisfies:

$$\sum_{n=1}^{N} P_n^\star(\lambda) - P_{\max}' = 0, \tag{14.51}$$

where

$$P_{\max}' = \min\left\{ \bar{P}, \sum_{n=1}^{N} \frac{\tilde{P}(\gamma_n)}{H_n} \right\}. \tag{14.52}$$

Finally, observing that $\sum_{n=1}^{N} P_n^\star(\lambda)$ is a decreasing function of λ, a simple dichotomic search over λ solves (14.51).

To summarise, it is only needed to store coefficients a_i, b_i, c_i, with $i = 0, \ldots, 3$, and d_i, with $i = 1, 2, 3$, for the computation of the SKT, and the optimal power allocation is obtained by a dichotomic search over λ until (14.51) is satisfied. Note that from the original problem (14.41) with $2N$ variables a problem with a single real variable is obtained.

From [21], we consider a scenario with $N = 16$ parallel channels, with Rayleigh statistics. The mean SNR of the Alice–Eve channel is $\Gamma^{(E)} = 0$ dB. Figure 14.2 shows the SKT η_{opt} averaged over the channel realisations as a function of the total power constraint $\bar{P}$, here normalised by the number of channels N, for various values of $\Gamma^{(B)}$. We observe a higher SKT both for a higher transmit power and a lower eavesdropper average channel gain. Moreover, we observe that as $\bar{P}$ goes to infinity, the SKT achieves a maximum value, as both Bob and Eve benefit from the power increases, thus preventing an unbounded growth of the SKT.

14.2.2 Source-model SKA over parallel channels

With the source-model SKA, the reciprocity of the wireless channel (i.e. the fact that the Alice–Bob channel is the same as the Bob–Alice one) is exploited, so that the randomness source shared by Alice and Bob (and not known to Eve) is the channel itself. In particular, SKA is performed in three phases:

1. Alice transmits a training block of T symbols on the N sub-channels, which are collected into the $N \times T$ matrix x. The training block is known to Alice, Bob and Eve. Bob receives the matrix signal y, while Eve receives the matrix signal z.
2. Bob transmits a training block of T symbols on the N sub-channels, which are collected into matrix $\tilde{x}$. The training block is known to Alice, Bob and Eve. Alice receives the matrix signal $\tilde{y}$, while Eve receives the matrices signal $\tilde{z}$.
3. Alice and Bob perform, respectively, on an estimate of their channels, distillation (see e.g. [22]), information reconciliation and privacy amplification [20, sec. 4.3].

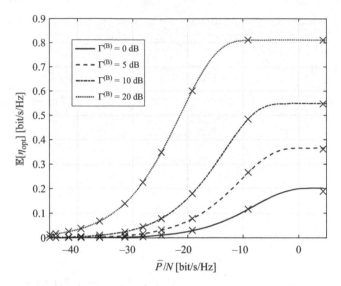

*Figure 14.2   Average SKT versus $\bar{P}/N$ for different values of $\Gamma^{(B)}$. © IEEE.
From [21]*

Eve listens to on-going transmissions and attempts to infer the Alice–Bob channel
from estimates of her channels to Alice and Bob. By this procedure, Eve attempts
to extract the same key of Alice and Bob. Indeed, Eve can get some information
on the secret key if her channels to Alice and Bob are correlated to the Alice–Bob
channel. For example, if the Alice–Bob channel is the same of the Alice–Eve channel
(and Eve knows this property), Eve can fully extract the key (which is not anymore
secret). Therefore, we must define the correlation among the various channels. At the
beginning of the chapter, we have already introduced the (parallel) channels $\boldsymbol{h}$ and
$\boldsymbol{g}$ between Alice and Bob/Eve. We now also introduce the N-size parallel channel
$\boldsymbol{f} = [f_1, \ldots, f_N]^T$ between Bob and Eve.

According to the Rayleigh fading assumption, all channel matrices are zero-mean
Gaussian random variables. The channel correlations are:

$$\mathbb{E}[\boldsymbol{hg}^H] = \boldsymbol{G}^{\mathrm{BA-AE}}, \quad \mathbb{E}[\boldsymbol{gf}^H] = \boldsymbol{G}^{\mathrm{AE-BE}}, \quad \mathbb{E}[\boldsymbol{hf}^H] = \boldsymbol{G}^{\mathrm{BA-BE}}, \tag{14.53}$$

while the auto-correlation matrices of the Bob–Alice, Alice–Eve, Bob–Eve
channels are:

$$\mathbb{E}[\boldsymbol{hh}^H] = \boldsymbol{G}^{\mathrm{BA}}, \quad \mathbb{E}[\boldsymbol{gg}^H] = \boldsymbol{G}^{\mathrm{AE}}, \quad \mathbb{E}[\boldsymbol{ff}^H] = \boldsymbol{G}^{\mathrm{BE}}. \tag{14.54}$$

Note that in general the correlation matrices are full. This is due to the fact that sub-
channels are correlated. As it will be seen later in this chapter, for OFDM systems,
sub-channels correspond to different frequency of a broadband channel, and typically
the number of sub-channels is larger than the number of independent channel taps
in the time domain. Moreover, transmit and receive filters introduce correlation among
the channel taps. Therefore, for the performance analysis of source-model SKA, we

can refer to [23] where a lower bound on the secret key-rate over correlated MIMO multiple-Eve (MIMOME) Rayleigh fading channels has been derived.

In particular, using [20, Proposition 5.4], a lower bound on the number of bits that can be securely extracted from the channel is provided by:

$$C_{\text{lower}}(x, \tilde{x}) = \text{I}(\tilde{y}; y) - \min\{\text{I}(\tilde{y}; z), \text{I}(y; z)\}. \tag{14.55}$$

For an explicit computation of the bound, first observe that h can be obtained from h^H by a suitable permutation, i.e. $h = \bar{\Pi} h^H$, with $\bar{\Pi}$ being the $N^2 \times N^2$ permutation matrix with $[\bar{\Pi}]_{n+(m-1)N, m+(n-1)N} = 1, n, m = 1, \ldots, N$, and all other entries equal to zero. Then, defining:

$$L = \begin{bmatrix} I_N \otimes x & 0 \\ 0 & I_N \otimes x \end{bmatrix}, \tag{14.56}$$

$$R_{yy} = \mathbb{E}[yy^H] = x\bar{\Pi}G^{\text{BA}}\bar{\Pi}^H x^H + I_{NT}, \quad R_{xx} = \tilde{x}G^{\text{BA}}\tilde{x}^H + I_{NT}, \tag{14.57}$$

$$R_{xy} = \mathbb{E}[xy^H] = \tilde{x}G^{\text{BA}}\bar{\Pi}^H x^H, \quad R_{zz} = L\begin{bmatrix} G^{\text{AE}} & G^{\text{AE−BE}} \\ G^{\text{AE−BE}H} & G^{\text{BE}} \end{bmatrix}L^H + I_{2NT}, \tag{14.58}$$

$$R_{xz} = \mathbb{E}[xz^H] = \tilde{x}[G^{\text{BA−AE}} \ G^{\text{BA−BE}}]L^H, \quad R_{yz} = x\bar{\Pi}[G^{\text{BA−AE}} \ G^{\text{BA−BE}}]L^H, \tag{14.59}$$

and

$$U_{xy} = \begin{bmatrix} R_{xx} & R_{xy} \\ R_{xy}^H & R_{yy} \end{bmatrix}, \quad U_{xz} = \begin{bmatrix} R_{xx} & R_{xz} \\ R_{xz}^H & R_{zz} \end{bmatrix} \quad U_{yz} = \begin{bmatrix} R_{yy} & R_{yz} \\ R_{yz}^H & R_{zz} \end{bmatrix}, \tag{14.60}$$

we obtain:

$$\text{I}(x; y) = \log \frac{\det R_{yy} \det R_{xx}}{\det U_{xy}}, \quad \text{I}(x; z) = \log \frac{\det R_{zz} \det R_{xx}}{\det U_{xz}}, \tag{14.61}$$

$$\text{I}(y; z) = \log \frac{\det R_{zz} \det R_{yy}}{\det U_{yz}}. \tag{14.62}$$

14.3 Waveforms peculiarities

We will now consider various waveform candidates for 5G systems and assess their security performance. In particular, we will establish links between the waveforms choices and the parallel channel model described in the previous section. We will also highlight further attack strategies by the eavesdropper.

14.3.1 OFDM

An OFDM transceiver processes groups of N complex data symbols, e.g. quadrature amplitude modulation (QAM) symbols. We focus here on the transmission of a single block of symbols and refer to the transceiver model of Figure 14.3. Let x_n be the symbol $n = 1, \ldots, N$ of the group. At the transmitter, Alice first applies an N-size inverse discrete Fourier transform (IDFT) on vector $x = [x_1, \ldots, x_N]$ to obtain vector:

$$\tilde{x} = F^{-1}x, \tag{14.63}$$

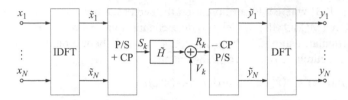

Figure 14.3   OFDM transceiver with dispersive AWGN channel

where $\boldsymbol{F}^{-1}$ is an $N \times N$ matrix with entries $[\boldsymbol{F}^{-1}]_{p,q} = 1/\sqrt{N}e^{2\pi j\frac{(p-1)(q-1)}{M}}$, $p, q = 1, \ldots, N$. Then, a CP of size L is added to each block, i.e. a vector $\boldsymbol{S} = [S_0, \ldots, S_{L+N}]$ of size $N + L$ is formed, where the first L samples are the last L samples of $\tilde{\boldsymbol{x}}$, while the remaining entries are $\tilde{\boldsymbol{x}}$. Vector $\boldsymbol{S}$ is then sequentially transmitted on the channel represented by a filter with impulse response $\tilde{H}_\ell$, $\ell = 0, \ldots, L$ so that the received signal is:

$$R_k = \sum_{\ell=0}^{L} \tilde{H}_{k-\ell} S_\ell + V_k \tag{14.64}$$

where V_k is the AWGN term with zero-mean and unitary variance. Note that the convolution between $\tilde{H}_\ell$ and S_k will generate $N + 2L - 1$ terms, and $L - 1$ of them will interfere with the next transmitted group of symbols, in correspondence of the next CP. Transmission is performed under the power constraint (14.1).

At the receiver, Bob discards the first L samples corresponding to the CP (which is corrupted by the interference of the previous group transmission), and the next N samples are collected into the vector $\boldsymbol{R} = [R_{L+1}, \ldots, R_{N+L-1}]$. Then, a discrete Fourier transform (DFT) is applied on $\boldsymbol{R}$ to obtain:

$$\boldsymbol{y} = \boldsymbol{FR}, \tag{14.65}$$

where $\boldsymbol{F}$ is an $N \times N$ matrix with entries $[\boldsymbol{F}]_{p,q} = 1/\sqrt{N}e^{-2\pi j\frac{(p-1)(q-1)}{M}}$, $p, q = 1, \ldots, N$. Due to the circulant nature of S (see the presence of the CP), the convolution with $\tilde{H}$ corresponds in the discrete dual domain of the DFT to the product of the DFTs of the transmitted signal and the channel impulse response. That is, defining $\boldsymbol{h} = \boldsymbol{F}[\tilde{H}_0, \ldots, \tilde{H}_L, 0, \ldots, 0]^T$, we obtain:

$$y_n = h_n x_n + v_n, \quad n = 1, \ldots, N, \tag{14.66}$$

where v_n is a zero-mean unit variance AWGN term, DFT of $\boldsymbol{W}$. Clearly, if Eve too adopts the same receiver, denoting with $\boldsymbol{g}$ the N-size DFT of the channel to Eve (still assumed to be of length less than L), we obtain a similar model. In particular, we observe that we obtain the model (14.2) with N parallel AWGN channels. Therefore, all considerations on the secrecy capacity and the resource allocation presented in Section 14.1 can be applied to the OFDM system.

14.3.1.1 Advanced Eve's receiver

We may wonder if adopting a conventional OFDM receiver based on a DFT is the best choice for Eve in order to get as much information as possible on the secret message. Indeed, we note that discarding the portion corresponding to the CP implies also discarding some useful information contained in the CP. This will come at the cost of increased complexity, which however should not be considered an issue for the eavesdropper. Therefore, a better processing for Eve should involve also the CP. We now derive the secrecy capacity of a system where Alice transmits an OFDM signal, Bob has a conventional OFDM receiver and Eve can use any receiver, thus not being constrained to use an OFDM receiver.

In order to derive the secrecy capacity in this context and then discuss resource allocation strategies, consider the equivalent block model where:

$$T = \begin{bmatrix} \mathbf{0}_{L \times (N-L)} & I_L \\ I_N \end{bmatrix} F^{-1}, \quad S = Tx, \tag{14.67}$$

represent the IDFT and the insertion of the cyclic prefix, respectively. The received block at Bob can be written as:

$$R = G_B S + W, \tag{14.68}$$

where G_B and G_E are $(N + L) \times (N + L)$ Toeplitz matrices with the first columns $[\tilde{H}_0, \ldots, \tilde{H}_L, 0, \ldots, 0]^T$ and $[G_0, \ldots, G_L, 0, \ldots, 0]^T$, respectively, and Z is the $N + L$ column vector of the signal received by Eve. Then:

$$\bar{R} = F[\mathbf{0}_{N \times L} I_N], \quad y = \bar{R}R, \tag{14.69}$$

represent the CP discard and the DFT. Considering the operations of the OFDM transmitter and receiver, we can write the input–output relation with the following block model:

$$y = \tilde{h}x + v, \quad Z = Gx + W, \tag{14.70}$$

where $\tilde{h} = \text{diag}(h)$ is a diagonal matrix having on the main diagonal the vector h, W is an AWGN zero-mean unit variance column vector and

$$G = G_E T. \tag{14.71}$$

Note that this model does not consider the interference due to the transmission of the previous OFDM symbol, which is perfectly cancelled if an OFDM receiver is used but will be present on Z, if Eve does not use it.

The model (14.70) describes a flat MIMOME channel with N antennas for Alice and Bob and $N + L$ antennas for Eve. The power constraint (14.1) can be re-written as a trace constraint on the correlation matrix of x, $K_x = \mathbb{E}[xx^H]$, i.e.:

$$\text{trace}(K_x) \leq \bar{P}. \tag{14.72}$$

Hence, the secrecy capacity of this system under the trace constraint can be written as:

$$C_{\text{OFDM}} = \max_{K_x : \text{trace}(K_x) \leq \bar{P}} [\log \det (I_N + \tilde{h}K_x\tilde{h}^H) + \log \det (I_{N+L} - GK_xG^H)]^+. \tag{14.73}$$

However, considering the specific structure of the channels due in this system, a simplified expression can be obtained, as derived in [24], as:

$$C_{\text{OFDM}} = \max_{K : \text{trace}(K) \leq P} \left[\log \det (I + \tilde{H}_B K\tilde{H}_B^H) - \log \det (I + \tilde{H}_E K\tilde{H}_E^H) \right] \tag{14.74}$$

where

$$\tilde{H}_{\mathrm{B}} = \tilde{H}DF, \quad \tilde{H}_{\mathrm{E}} = G_{\mathrm{E}}DF, \quad D = \begin{bmatrix} I_{N-L} & 0 \\ 0 & \frac{1}{\sqrt{2}}I_L \end{bmatrix} \tag{14.75}$$

and the corresponding input covariance is:

$$K_x = FDKDF. \tag{14.76}$$

From (14.76), we conclude that when Eve does not use a conventional OFDM receiver, it is advantageous to transmit correlated symbols on the OFDM sub-channels rather than independent symbols. A general solution for the maximisation of (14.74) is not available; however, for the high SNR regime ($\bar{P} \to \infty$), the optimum matrix K under constraint (14.76) can be obtained from the generalised singular value decomposition (GSVD) of $(\tilde{H}_{\mathrm{B}}, \tilde{H}_{\mathrm{E}})$. In particular, the GSVD provides unitary matrices U_{B} and U_{E} and a non-singular matrix Ω such that

$$U_{\mathrm{B}}^H \tilde{H}_{\mathrm{B}} \Omega = D_{\mathrm{B}}, \quad U_{\mathrm{E}}^H \tilde{H}_{\mathrm{E}} \Omega = D_{\mathrm{E}}, \tag{14.77}$$

and $D_{\mathrm{B}} = \mathrm{diag}(d_{\mathrm{R},1}, \ldots, d_{\mathrm{R},N})$ and $D_{\mathrm{E}} = \mathrm{diag}(d_{\mathrm{E},1}, \ldots, d_{\mathrm{E},N})$. The optimum matrix K is [25]:

$$K = \bar{P} \frac{\Omega_\xi \Omega_\xi^H}{\mathrm{trace}(\Omega_\xi \Omega_\xi^H)}, \tag{14.78}$$

where Ω_ξ gathers the last ξ columns of Ω, and ξ is the number of generalised singular values $d_{\mathrm{R},i}/d_{\mathrm{R},i}$ that are greater than one.

The advanced Eves' receiver provides advantages also when channel-model SKA is considered. By using the MIMOME model (14.70), the secret key capacity can be written as [26]:

$$C_{\mathrm{SKA}} = \log \det \left[I_N + K_x^{\frac{1}{2}}(\tilde{H}^H \tilde{H} + G^H G)K_x^{\frac{1}{2}} \right] - \log \det \left[I + K_x^{\frac{1}{2}} G^H G K_x^{\frac{1}{2}} \right]. \tag{14.79}$$

About the resource allocation in this scenario, in [27], it has been derived a result for the low-power regime, i.e. when $\bar{P}$ is close to zero. In this case, it turns out that it is convenient to concentrate all the power along the eigenspace of the legitimate channel $\tilde{H}$ corresponding to the maximum eigenvalue, regardless of the eavesdropper's channel. In other words, Alice will transmit only on the sub-channel having the maximum gain, which has index $n^* = \mathrm{argmax}_n H_n$, and the secret key rate can be written as:

$$C_{\mathrm{SKA}} = \log \frac{1 + \Lambda_{\mathrm{B}} + \Lambda_{\mathrm{E}}}{1 + \Lambda_{\mathrm{E}}}, \tag{14.80}$$

where $\Lambda_{\mathrm{B}} = \frac{\bar{P}}{1+\rho} H_{n^*}$, $\Lambda_{\mathrm{E}} = \frac{\bar{P}}{1+\rho} \|H_{\mathrm{E}:,n^*}\|^2$, are the Bob and Eve's SNRs and $H_{\mathrm{E}:,n^*}$ is the n^*th column of $\tilde{H}_{\mathrm{E}}$ and $\rho = L/(N+L)$. Note that (14.80) depends on the Alice–Bob highest sub-channel gain and on the resulting Alice–Eve channel SNR Λ_{E}. When CSIT is available, we can determine the secret key rate. As discussed also for the secret message transmission, the gain of the eavesdropper channel may not be available, and in this case, we must consider the secret-key outage probability. To this end, we need the cumulative distribution function (CDF) of Λ_E, which has been derived in [26] for the case of independent Rayleigh fading channels.

14.3.2 SC-FDMA

The SC-FDMA system [28,29] can be seen as a modified OFDM system where the IDFT at the transmitter (14.63) is not performed, i.e. $\tilde{x} = x$, and the CP is inserted on the single-carrier signal. By discarding the CP portion at the receiver and applying a DFT, we have that (14.70) is replaced by:

$$\hat{y} = \hat{H}x + v, \quad Z = \hat{G}x + W, \tag{14.81}$$

where $\hat{H} = \tilde{H}F$ and $\hat{G} = G_E TF$. Therefore, for the evaluation of the secrecy capacity, we can resort to the MIMOME results (14.73) with $\tilde{H}$ replaced by $\hat{H}$ and G replaced by $\hat{G}$.

Bob typically applies a zero-forcing (ZF) receiver, i.e. he computes:

$$\tilde{y}_n = \frac{\hat{y}_n}{h_n}, \quad n = 1, \ldots, N. \tag{14.82}$$

Then, an IDFT is applied, and we obtain:

$$y_n = x_n + v_n, \tag{14.83}$$

with v_n being zero-mean Gaussian with power $\mathbb{E}[|v_n|^2] = \frac{1}{N} \sum_{k=1}^{N} \frac{1}{|h_k|^2}$. If Eve applies the same receiver (equalising now her channel g), she gets:

$$z_n = x_n + w_n, \tag{14.84}$$

with w_n being zero-mean Gaussian with power $\mathbb{E}[|w_n|^2] = \frac{1}{N} \sum_{k=1}^{N} \frac{1}{|g_k|^2}$. Although v_n are correlated for various n, and w_n are also correlated for various n, a typical receiver will ignore this correlation. Therefore, the channel seen by Bob and Eve will be two parallel channels, and we can apply all results obtained in the previous sections for the parallel channel model.

14.3.3 GFDM

In GFDM, instead of using a simple IDFT to generate the modulated signal as in OFDM, more elaborate filters and linear transformations are used. As described in [30], the transmitted signal can be modelled as a linear transformation of the data block x, i.e. (14.63) is replaced now by:

$$\tilde{x} = Ax, \tag{14.85}$$

where A is an $N \times N$ matrix that can be designed according to various criteria, in particular, to enable pulse shaping on a per-subchannel basis. Then, a CP is inserted which is removed at the receiver, and a ZF receiver implemented in the frequency domain is used, as for SC-FDMA. The resulting signal is:

$$Q = Ax + \tilde{V}, \tag{14.86}$$

(noise vector $\tilde{V}$ entries are correlated) after which typically a ZF demodulator is applied, i.e.:

$$y = A^{-1}Q = x + A^{-1}\tilde{V}. \tag{14.87}$$

Again, the correlation among noise terms is ignored, and the obtained model is a parallel channel model. For Eve, we can either consider a similar receiver, and use all the results for parallel channels, or we can observe that she receives:

$$Z = G_E TAx + W, \tag{14.88}$$

and we can use the results on MIMOME channels seen for the OFDM system.

14.3.4   UFMC

UFMC generalises the OFDM system by replacing the CP with the ramp-up and ramp-down of suitably designed filters. In particular, for UFMC, the transmitted vector S of size $N + L$ can be written as in (14.67) where now $T = VF^{-1}$, and V is a Toeplitz matrix composed of the UFMC filter impulse response, which can be designed according to various criteria. At the receiver, a $2N$-size DFT is applied on a zero-padded version of Y, and the even sub-carriers are retained, while the odd sub-carriers are discarded. By a proper design of the transmit filters, the resulting N-size vector can be written as the parallel channel model (14.2), where the noise signal v_n has power:

$$\frac{N + L - 1}{N}, \tag{14.89}$$

and the channel h_n is replaced by $h_n u_n$, where u_n, $n = 1, \ldots, N$ is the N-size DFT of the transmit shaping filter impulse response. Therefore, we can apply the results on parallel channel models.

Also in this case, we can apply results on more advanced Eve's receivers as for OFDM by resorting to (14.73) with modified MIMOME channels.

14.3.5   FBMC

Also for FBMC, data symbols are collected into groups of N symbols where the nth symbol of group k is $x_{k,n}$, $n = 1, \ldots, N$. FBMC is obtained as the combination of a filter bank with offset QAM (OQAM). Indeed, the complex symbols are split into $2N$ *real* symbols $\tilde{x}_{k,q}$, $q = 1, \ldots, 2N$ where the even symbols are the real part of $x_{k,n}$, i.e. $\tilde{x}_{k,2n} = \text{Re}\{x_{k,n}\}$, while the odd symbols are the imaginary part of $x_{k,n}$, i.e. $\tilde{x}_{k,2n+1} = \text{Im}\{x_{k,n}\}$. The imaginary part will be delayed by half of a symbol period with respect to the real part. Moreover, the $2N$ symbols will be transmitted using a filtered multi-carrier system, where the impulse response of the prototype filter is ϕ_ℓ, $\ell = 0, \ldots, L_\phi - 1$ and each filter is shifted by $1/(2N)$ of the bandwidth. Assuming a low-pass prototype filter with passband of $1/4N$, the resulting spectrum of the transmitted signal will see the alternation of real and imaginary components, where in correspondence of $1/N$ of the bandwidth (approximately corresponding to one

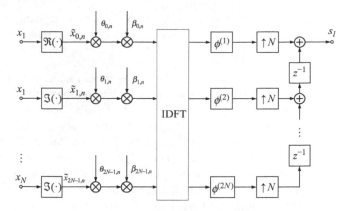

Figure 14.4 FBMC transmitter

OFDM sub-carrier), we have two adjacent signals, one pure real and the other pure complex, carrying the real and imaginary part of one QAM symbol.

The motivation of OQAM is that for transmission over a non-distorting channel, we can apply a bank of matched filter (to the transmit filters) at the receiver, obtaining $2N$ signals, each corresponding to the real and imaginary parts of the transmitted symbols. If the prototype filter has a passband slightly larger than $1/4N$ (or it has some transition band), interference will arise between these adjacent symbols. However, if the overlap of the prototype filter is limited to two (to the left and right in the spectrum) filter responses, the interference can be removed by taking the real and imaginary parts. However, in general a complex dispersive channel will introduce interference among adjacent real/imaginary transmissions that cannot be removed with this technique, and we need an equaliser to compensate partially for the effects of the channel, and then we can take the real/imaginary part.

In formulas, the transmitted signal can be written as

$$s_\ell = \sum_{n=1}^{2N} \sum_{k=-\infty}^{\infty} \tilde{x}_{k,n} \theta_{k,n} \beta_{k,n} \phi_{m-nN} e^{j\frac{\pi}{N}k(n-1)}, \tag{14.90}$$

where $\theta_{k,n} = j^{j+n}$, $\beta_{k,n} = e^{-j\frac{\pi k}{N}\left(\frac{L_\phi - 1}{2}\right)}$. An efficient implementation of FBMC transmitter is shows in Figure 14.4, where $\phi^{(k)}$, $k = 1, \ldots, 2N$ are the polyphase filter components of $\{\phi_n\}$.

At the receiver, a bank of filters with the same prototype filter is applied, followed by a partial equalisation of the channel and OQAM demodulation. A second equalisation on each real/imaginary branch is possible to compensate for channel distortions within the sub-channel filter band. As a result of these operations, the input/output relation can be seen as a set of equivalent AWGN parallel channels, i.e. the wiretap channel model described in Section 14.1 and all considerations on the secrecy capacity and the resource allocation presented in that section can be applied to the FBMC system. Note that in this system no CP is present; thus, Eve cannot exploit additional redundancy to get information on the secret message.

14.3.5.1 Filter hopping FBMC

In [31], a specific solution for confidential message transmission using FBMC has been proposed. In particular, it is proposed to change the prototype filter in a pseudo-random fashion, e.g. by changing the roll-off factor of a raised cosine filter (so that all filters have the same bandwidth). Letting Alice and Bob share a secret sequence of roll-off factors allows Bob to adapt the receive filters and being able to decode the transmitted signal. On the other hand, Eve will not know this sequence and will use a mismatched receive filter that will reduce her ability to decode the message. Although this scheme exploits the peculiarities of FBMC, it suffers from numerous drawbacks. The main one is that Eve can use more sophisticated receivers, for example a bank of receivers, each using a filter matched to one of the possible roll-off factors. Then, a maximum likelihood receiver is applied that picks the roll-off factor and the corresponding sequence of decoded symbols that is most probable to have been sent. Moreover, a shared secret (the sequence of used roll-off factors) between Alice and Bob is assumed in this scheme, which makes it weak and open questions on how to share the secret without being intercepted by Eve.

14.3.6 Performance comparison

Figure 14.5 shows the average capacity for the various considered waveforms, as a function of the average eavesdropper channel gain $\Gamma^{(E)}$, when $\Gamma^{(B)} = 0$ dB. The performance of a pure parallel channel system is also reported, which accounts for

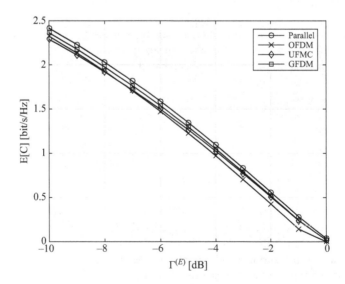

Figure 14.5   Average secrecy capacity for various transmission systems using advanced receivers versus $\Gamma^{(E)}$

the case when also Eve uses a conventional receiver, not exploiting the CP, or for the case in which FBMC is used. We note that the parallel channel system outperforms all other configurations, as Eve gets an extra advantage in using the CP. When comparing the various waveforms, we observe that they all perform quite closely to the parallel channel conditions, with UFMC and GFDM providing a slightly higher average secrecy capacity than OFDM.

References

[1]   Aminjavaheri A, Farhang A, RezazadehReyhani A, and Farhang-Boroujeny B. Impact of timing and frequency offsets on multicarrier waveform candidates for 5G. In: Signal Processing and Signal Processing Education Workshop (SP/SPE), 2015 IEEE; 2015. p. 178–183.

[2]   Schaich F, Wild T, and Chen Y. Waveform contenders for 5G – suitability for short packet and low latency transmissions. In: 2014 IEEE 79th Vehicular Technology Conference (VTC Spring); 2014. p. 1–5.

[3]   Banelli P, Buzzi S, Colavolpe G, Modenini A, Rusek F, and Ugolini A. Modulation Formats and Waveforms for 5G Networks: Who Will Be the Heir of OFDM?: An overview of alternative modulation schemes for improved spectral efficiency. IEEE Signal Processing Magazine. 2014 Nov;31(6):80–93.

[4]   Ibars C, Kumar U, Niu H, Jung H, and Pawar S. A comparison of waveform candidates for 5G millimeter wave systems. In: 2015 49th Asilomar Conference on Signals, Systems and Computers; 2015. p. 1747–1751.

[5]   Sahin A, Guvenc I, and Arslan H. A Survey on Multicarrier Communications: Prototype Filters, Lattice Structures, and Implementation Aspects. IEEE Communications Surveys Tutorials. 2014 Third;16(3):1312–1338.

[6]   Li Z, Yates R, and Trappe W. Secrecy Capacity of Independent Parallel Channels. In: Liu R, Trappe W, editors. Securing Wireless Communications at the Physical Layer. Boston, MA: Springer US; 2010. p. 1–18. Available from: http://dx.doi.org/10.1007/978-1-4419-1385-2_1.

[7]   Liang Y, Poor HV, and Shamai S. Secure Communication Over Fading Channels. IEEE Transactions on Information Theory. 2008 June;54(6):2470–2492.

[8]   Jorswieck EA, and Wolf A. Resource allocation for the wire-tap multi-carrier broadcast channel. In: Telecommunications, 2008. ICT 2008. International Conference on Telecommunications; 2008. p. 1–6.

[9]   Baldi M, Chiaraluce F, Laurenti N, Tomasin S, and Renna F. Secrecy Transmission on Parallel Channels: Theoretical Limits and Performance of Practical Codes. IEEE Transactions on Information Forensics and Security. 2014 Nov;9(11):1765–1779.

[10] Qin H, Sun Y, Chang TH, *et al.* Power Allocation and Time-Domain Artificial Noise Design for Wiretap OFDM with Discrete Inputs. IEEE Transactions on Wireless Communications. 2013 June;12(6):2717–2729.

[11] Bashar S, Ding Z, and Xiao C. On Secrecy Rate Analysis of MIMO Wiretap Channels Driven by Finite-Alphabet Input. IEEE Transactions on Communications. 2012 December;60(12):3816–3825.

[12] Guo D, Shamai S, and Verdu S. Mutual information and minimum mean-square error in Gaussian channels. IEEE Transactions on Information Theory. 2005 April;51(4):1261–1282.

[13] Lozano A, Tulino AM, and Verdu S. Optimum power allocation for parallel Gaussian channels with arbitrary input distributions. IEEE Transactions on Information Theory. 2006 July;52(7):3033–3051.

[14] Ng DWK, Lo ES, and Schober R. Energy-Efficient Resource Allocation for Secure OFDMA Systems. IEEE Transactions on Vehicular Technology. 2012 July;61(6):2572–2585.

[15] Khisti A, Tchamkerten A, and Wornell GW. Secure Broadcasting Over Fading Channels. IEEE Transactions on Information Theory. 2008 June;54(6): 2453–2469.

[16] Wang X, Tao M, Mo J, and Xu Y. Power and Subcarrier Allocation for Physical-Layer Security in OFDMA-Based Broadband Wireless Networks. IEEE Transactions on Information Forensics and Security. 2011 Sept;6(3):693–702.

[17] Seong K, Mohseni M, and Cioffi JM. Optimal Resource Allocation for OFDMA Downlink Systems. In: 2006 IEEE International Symposium on Information Theory; 2006. p. 1394–1398.

[18] Kim K, Han Y, and Kim SL. Joint subcarrier and power allocation in uplink OFDMA systems. IEEE Communications Letters. 2005 Jun;9(6):526–528.

[19] Benfarah A, Tomasin S, and Laurenti N. Power Allocation in Multiuser Parallel Gaussian Broadcast Channels with Common and Confidential Messages. IEEE Transactions on Communications. 2016; 64(6):2326–2339.

[20] Bloch MB, and Barros J. Physical-Layer Security: From Information Theory to Security Engineering. Cambridge: Cambridge University Press; 2011.

[21] Tomasin S, and Dall'Arche A. Resource Allocation for Secret Key Agreement Over Parallel Channels With Full and Partial Eavesdropper CSI. IEEE Transactions on Information Forensics and Security. 2015 Nov;10(11):2314–2324.

[22] Tomasin S, Trentini F, and Laurenti N. Secret Key Agreement by LLR Thresholding and Syndrome Feedback over AWGN Channel. IEEE Communications Letters. 2014 January;18(1):26–29.

[23] Tomasin S, and Jorswieck E. Pilot-based secret key agreement for reciprocal correlated MIMOME block fading channels. In: 2014 IEEE Globecom Workshops (GC Wkshps); 2014. p. 1343–1348.

[24] Renna F, Laurenti N, and Poor HV. Physical-Layer Secrecy for OFDM Transmissions Over Fading Channels. IEEE Transactions on Information Forensics and Security. 2012 Aug;7(4):1354–1367.

[25] Khisti A, Wornell G, Wiesel A, and Eldar Y. On the Gaussian MIMO Wiretap Channel. In: 2007 IEEE International Symposium on Information Theory; 2007. p. 2471–2475.

[26] Renna F, Laurenti N, Tomasin S, et al. Low-power secret-key agreement over OFDM. In: Proceedings of the 2Nd ACM Workshop on Hot Topics on Wireless Network Security and Privacy. HotWiSec '13. New York, NY, USA: ACM; 2013. p. 43–48. Available from: http://doi.acm.org/10.1145/2463183.2463194.

[27] Renna F, Bloch MR, and Laurenti N. Semi-Blind Key-Agreement over MIMO Fading Channels. IEEE Transactions on Communications. 2013 February;61(2):620–627.

[28] Myung HG, Lim J, and Goodman DJ. Single carrier FDMA for uplink wireless transmission. IEEE Vehicular Technology Magazine. 2006 Sept;1(3):30–38.

[29] Benvenuto N, Dinis R, Falconer D, and Tomasin S. Single Carrier Modulation With Nonlinear Frequency Domain Equalization: An Idea Whose Time Has Come–Again. Proceedings of the IEEE. 2010 Jan;98(1):69–96.

[30] Michailow N, Matthé M, Gaspar IS *et al.* Generalized Frequency Division Multiplexing for 5th Generation Cellular Networks. IEEE Transactions on Communications. 2014 Sept;62(9): 3045–3061.

[31] Lücken V, Singh T, Cepheli, Kurt GK, Ascheid G, and Dartmann G. Filter hopping: Physical layer secrecy based on FBMC. In: 2015 IEEE Wireless Communications and Networking Conference (WCNC); 2015. p. 568–573.

Chapter 15

Physical layer security in non-orthogonal multiple access

Hui-Ming Wang[1], Yi Zhang[1], and Zhiguo Ding[2]

Non-orthogonal multiple access (NOMA) has been widely considered as a promising technology to enable high-efficient wireless transmissions in future 5G communication systems. In this chapter, we investigate single-input single-output (SISO) NOMA systems from the perspective of physical layer security. There into, two different SISO NOMA systems are studied in sequence so as to explore security issues in NOMA. First, we attempt the physical layer security technique in a SISO NOMA system which consists of a transmitter, multiple legitimate users and an eavesdropper who aims to wiretap the messages intended for all legitimate users. The objective is to maximise sum of secrecy rates subject to an individual quality of service constraint for each legitimate user, respectively. The investigations in this system will provide a preliminary analysis of the secure performance of SISO NOMA systems. Second, on the basis of the SISO NOMA system previously studied, a multi-antenna jammer is additionally equipped to enhance secure transmissions of the system. The joint optimisation of power allocation and beamforming design is considered. The efforts in this second system aim to propose an effective solution for realising secure transmissions for each legitimate user.

15.1 Introduction

Non-orthogonal multiple access (NOMA) has been recently recognised as a promising technology for 5G communication systems because of its remarkable spectral efficiency (SE) [1]. Different from conventional orthogonal multiple access (OMA) such as time-division multiple access (TDMA), NOMA exploits the power domain to realise simultaneous transmissions of multiple data streams at different power levels [2]. At a transmitter, the superposition of different data streams is broadcasted, and at the receivers, the successive interference cancellation (SIC) technique is generally used to realise multi-user detection [3]. Previous works on NOMA have mainly focused on the improvement of SE. In [4,5], different cooperative NOMA schemes

[1]School of Electronic and Information Engineering, Xi'an Jiaotong University, P.R. China
[2]School of Computing and Communications, Lancaster University, UK

have been proposed for improving the performance of spectrum utilisation. In [6], the authors have studied the downlink sum rate maximisation problem in a multiple-input single-output (MISO) NOMA system. Furthermore, applications of multiple-input multiple-output (MIMO) technologies have been proposed in [7,8] to further improve the SE of NOMA. Moreover, NOMA has also been studied from many other interesting viewpoints. For example, in [9], the power allocation problem has been studied by taking the user fairness into account in a single-input single-output (SISO) NOMA system. In [10], a cognitive radio inspired NOMA scheme has been initially proposed to guide the power allocation in two-user NOMA systems. In addition, some recent works on NOMA have involved the energy efficiency (EE) issues [11,12].

Besides SE and EE issues, security is another important aspect for future 5G communication systems. Naturally, this new concept of security, i.e. physical layer security, can be also applied to NOMA for serving as a complement to traditional cryptography techniques, making a more integrated and robust secure transmission. A wealth of relevant research has achieved a significant success in the conventional secure communication systems [13–17] and is potential to be used as references in strengthening the security of NOMA systems. One of the possible approaches is to send the artificially generated noise to interfere the eavesdropper. The concept of applying artificial noise (AN) to enhance the secure transmission has been firstly proposed in [18], and it has been further studied in [19]. Isotropic AN and spatially selective AN can be implemented according to whether the channel state information (CSI) is available at the transmitter. Therefore, it is promising to apply the existing physical layer security techniques into NOMA systems for realising a high-rate transmission with a certain secure guarantee.

However, the secrecy performance and the design of secure transmissions in NOMA are still unclear. Only a few works have studied security issues in NOMA systems. In [20], security issue has been initially considered in a SISO NOMA system. In [21], the physical layer security of applying NOMA in large-scale networks has been investigated. All the aforementioned observations greatly motivate us to study secure transmission designs in NOMA systems.

The objectives of this chapter are to firstly present a preliminary analysis of the secure performance of SISO NOMA systems and to subsequently propose an effective solution for realising secure transmissions in SISO NOMA systems. In this chapter, two different SISO NOMA systems are studied in sequence, which are illustrated as follows:

1. First, we investigate the power allocation among users for a SISO NOMA system consisting of a transmitter, multiple legitimate users, and an eavesdropper who tries to wiretap the messages intended for all the legitimate users. There into, our goal is to maximise sum of secrecy rates (SSR) of the system subject to pre-defined quality of service (QoS) requirements for all the legitimate users. To this end, we firstly identify the feasible range of the transmit power that satisfies all the users' QoS requirements. Then, we derive the closed-form expression for an optimal power allocation policy that maximises the SSR. A preliminary analysis of the secure performance of the SISO NOMA system is provided by studying this primitive NOMA system.

2. Second, on the basis of the SISO NOMA system that is previously studied, a multi-antenna jammer is additionally equipped at the same position of the transmitter to further enhance the secure transmission of the system. We then focus on the joint optimisation of the power allocation among users at the transmitter and the AN design at the jammer, subject to an individual secrecy rate constraint for each legitimate user. An efficient algorithm is proposed, based on the sequential convex approximation (SCA) method, to solve the corresponding optimisation problem therein.

Simulation results show that NOMA is superior to conventional OMA in terms of SSR, and the performance gain achieved by NOMA is more significant as user number increases. In addition, the highly demanding QoS requirements will degrade the SSR performance of the system. Furthermore, it is defective that the secure transmission for the users whose channel gains are smaller than that of the eavesdropper cannot be ensured in that primitive SISO NOMA system. Nevertheless, our theoretical analysis and simulation results further demonstrate that secure transmissions can be well guaranteed for all the users by exploiting a multi-antenna jammer. In particular, the system performance increases with the growth of the antenna number of the jammer.

Chapter Outline: The rest of this chapter is organised as follows. In Section 15.2, we study the first SISO NOMA system where no jammer is equipped at the system. From this primitive SISO NOMA system, we will obtain a preliminary analysis of the secure performance of SISO NOMA systems. In Section 15.3, we further investigate the benefit of adding a multi-antenna jammer in guaranteeing the secure transmission for each legitimate user. Conclusions and open issues are provided in Section 15.4.

Notations: Boldface uppercase letters and boldface lowercase letters are used to denote matrices and column vectors, respectively. $(\cdot)^T$, $(\cdot)^H$, $\text{Tr}(\cdot)$, and $\mathbb{E}(\cdot)$ are the transpose, Hermitian transpose, trace, and expectation operators, respectively. $\mathbf{x} \sim CN(\mathbf{0}, \mathbf{I}_M)$ indicates that $\mathbf{x}$ is a circular symmetric complex Gaussian random vector whose mean vector is $\mathbf{0}$ and covariance matrix is $\mathbf{I}_M$. $\supseteq$ and $\cap$ represent the superset and the intersection in set theory, respectively.

15.2   Preliminary analysis of the secure performance of SISO NOMA systems

15.2.1   System model

Consider the downlink transmission of a system which consists of a single-antenna transmitter, M single-antenna legitimate users and a single-antenna eavesdropper, which is shown as Figure 15.1. The channel gain from the transmitter to the mth legitimate user is denoted by $h_m = d_m^{-\frac{\alpha}{2}} g_m$, where g_m is the Rayleigh fading channel gain, d_m is the distance between the transmitter and the mth user, and α is the path loss exponent. In a similar way, the channel gain from the transmitter to the eavesdropper can be modelled as $h_e = d_e^{-\frac{\alpha}{2}} g_e$. In this work, the instantaneous CSI of each user and of the eavesdropper are supposed to be available at the transmitter. Without loss of generality, we suppose that the channel gains are sorted as follows:

$$0 < |h_1|^2 \leq |h_2|^2 \leq \ldots |h_{M_e}|^2 \leq |h_e|^2 < |h_{M_e+1}|^2 \ldots \leq |h_M|^2, \qquad (15.1)$$

where M_e represents the number of the legitimate users whose channel gains are not greater than that of the eavesdropper.

According to the principle of the NOMA scheme [2,3], the transmitter broadcasts a linear combination of M signals to its M users. The transmitted superposition signal can be given as $\sum_{m=1}^{M} \sqrt{\gamma_m P} s_m$, where s_m is the message intended for the mth legitimate user, P represents the total power available at the transmitter, and γ_m denotes the power allocation coefficient, i.e. the ratio of the transmit power for the mth user's message to the total power P. Meanwhile, the eavesdropper tries to intercept all the legitimate users' messages.[1]

On the basis of the signal model and the channel model given above, the received signal at the mth user, denoted by y_m, and at the eavesdropper, denoted by y_e, can be respectively given as follows:

$$y_m = h_m \sum_{i=1}^{M} \sqrt{\gamma_i P} s_i + n_m, \quad 1 \leq m \leq M, \qquad (15.2)$$

$$y_e = h_e \sum_{i=1}^{M} \sqrt{\gamma_i P} s_i + n_e, \qquad (15.3)$$

where s_i with $\mathbb{E}(|s_i|^2) = 1$ denotes the message for ith user, n_m and n_e are both the additive Gaussian noise with zero mean and variance σ^2 for all receivers including the eavesdropper, respectively.

15.2.1.1 Achievable rates of legitimate users

As mentioned before, the users often apply the SIC technique to decode their own messages [3]. In SISO NOMA systems, the decoding order is generally determined by the sort of channel gains given in (15.1) without regarding to the power allocations among the users, which means that a user will firstly decode the messages intended for the users whose channel gains are smaller than its, in a successive way. To be specific, in the order $i = 1, 2, \ldots, m - 1$, the mth user firstly decodes the ith user's message for $i < m$ and then eliminates this message from its received mixture; the messages for the ith user for $i > m$ are treated as noise. Therefore, the achievable rate

---

[1] In fact, the eavesdropper may be interested only in a specific user's message. But in this work, a more conservative assumption is made that the eavesdropper intercepts all users' messages, because the transmitter does not know which user the eavesdropper wants to wiretap. It is worth pointing out that for the case that the eavesdropper wiretaps a specific user's message, the problem formulations and analysis can be obtained in a similar way.

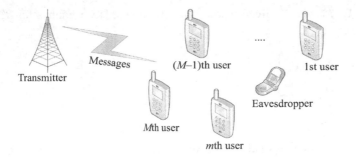

Figure 15.1   System model

of the mth legitimate user in the SISO NOMA system for $1 \leq m \leq M$ can be given by [2]:

$$R_b^m = \log_2\left(1 + \frac{P|h_m|^2 \gamma_m}{P|h_m|^2 \sum_{i=m+1}^{M} \gamma_i + \sigma^2}\right). \tag{15.4}$$

In particular, given (15.1), it can be easily verified that the mth user can always successfully decode the message intended for the ith user for $i < m$, which ensures the correctness of (15.4) [2].

15.2.1.2   Sum of secrecy rates of the SISO NOMA system

Regarding the eavesdropper, we use R_e^m to represent the achievable rate of the eavesdropper when it detects the message intended for the mth legitimate user by using SIC. Further, R_s^m and R_s are used to denote the secrecy rate of the mth user and the SSR of the system, respectively. Since the eavesdropper uses the SIC technique, R_e^m can be given as follows:

$$R_e^m = \begin{cases} \log_2\left(1 + \dfrac{P|h_e|^2 \gamma_m}{P|h_e|^2 \sum_{i=m+1}^{M} \gamma_i + \sigma^2}\right), & 1 \leq m \leq M_e, \\[4mm] \log_2\left(1 + \dfrac{P|h_e|^2 \gamma_m}{P|h_e|^2 \sum_{i=M_e+1, i\neq m}^{M} \gamma_i + \sigma^2}\right), & M_e + 1 \leq m \leq M, \end{cases} \tag{15.5}$$

In this chapter, the first $m - 1$ users' messages are assumed to have already been correctly decoded before the eavesdropper decodes the mth user's message, which overestimates the decoding capability of the eavesdropper that employs SIC. Therefore, $\tilde{R}_e^m$ provided as below can serve as an upper bound of the decoding capability of the eavesdropper that uses SIC:

$$\tilde{R}_e^m = \log_2\left(1 + \frac{P|h_e|^2 \gamma_m}{P|h_e|^2 \sum_{i=m+1}^{M} \gamma_i + \sigma^2}\right), \quad 1 \leq m \leq M. \tag{15.6}$$

Accordingly, $\tilde{R}_{\text{s}}^m$ and $\tilde{R}_{\text{s}}$ given below can serve as the lower bounds for R_{s}^m and the SSR, respectively:

$$\tilde{R}_{\text{s}}^m = \left[R_{\text{b}}^m - \tilde{R}_{\text{e}}^m \right]^+ , \tag{15.7}$$

$$\tilde{R}_{\text{s}} = \sum_{m=1}^{M} \tilde{R}_{\text{s}}^m , \tag{15.8}$$

where $[\cdot]^+ \triangleq \max(0, \cdot)$. In addition, we can verify that $R_{\text{b}}^m \leq \tilde{R}_{\text{e}}^m$ when $|h_m|^2 \leq |h_{\text{e}}|^2$ [2], which makes $\tilde{R}_{\text{s}}^m$ be zero when $1 \leq m \leq M_{\text{e}}$. Then, $\tilde{R}_{\text{s}}$ can be rewritten as

$$\tilde{R}_{\text{s}} = \sum_{m=M_{\text{e}}+1}^{M} \left(R_{\text{b}}^m - \tilde{R}_{\text{e}}^m \right). \tag{15.9}$$

As a result, in the system considered, the secure transmission for the mth user when $1 \leq m \leq M_{\text{e}}$ cannot be guaranteed with physical layer security. This drawback will be further discussed and addressed in another SISO NOMA system where the multi-antenna technique is exploited to enhance secure transmissions in SISO NOMA systems. In the following text, we use SSR to refer to $\tilde{R}_{\text{s}}$.[2]

15.2.2 Maximisation of the sum of secrecy rates

In this primitive SISO NOMA system, the guarantees of the QoS requirements of all the legitimate users are considered, which demands the transmitter to send the message to each user with a minimum data rate, respectively.

In this sub-section, we propose a power allocation policy which maximises the SSR of the system, i.e. $\tilde{R}_{\text{s}}$, subject to all the users' QoS requirements. To be specific, the closed-form expressions for the optimal power allocation coefficients $\{\gamma_m^{\text{Opt}}\}_{m=1}^{M}$, which maximise the SSR, are derived analytically.

Let $\bar{R}_{\text{b}}^m$ represent the minimum data rate required by the mth user and then the QoS constraints can be characterised as

$$R_{\text{b}}^m \geq \bar{R}_{\text{b}}^m, \quad 1 \leq m \leq M. \tag{15.10}$$

By substituting (15.4) into (15.10), the QoS constraints are transformed into

$$\gamma_m \geq A_m \left(P|h_m|^2 \sum_{i=m+1}^{M} \gamma_i + \sigma^2 \right), \quad 1 \leq m \leq M, \tag{15.11}$$

---

[2]In fact, $\tilde{R}_{\text{s}}$ can be verified to equal to the accurate secrecy sum rate of the SISO NOMA system according to [22,23], since the eavesdropper can employ some unknown but more advanced decoding strategies rather than SIC. This indicates $\tilde{R}_{\text{s}}$ can well serve as a secure metric for our preliminary analysis.

where $A_m \triangleq \frac{2^{\tilde{R}_b^m}-1}{P|h_m|^2}$. As a result, the SSR maximisation problem is formulated as

$$\max_{\gamma_m, 1 \leq m \leq M} \quad \tilde{R}_s \tag{15.12a}$$

$$\text{s.t.} \qquad \sum_{m=1}^{M} \gamma_m \leq 1 \quad \text{and} \quad (15.11). \tag{15.12b}$$

Due to the QoS requirements, there must exist a minimum transmit power, denoted by $P_{\min}$, that can satisfy all the users' QoS requirements. Then, problem (15.12) is feasible only under the condition that $P \geq P_{\min}$. As a result, it is important to firstly identify the feasible range of the transmit power before dealing with problem (15.12).

15.2.2.1 Minimum transmit power that satisfies QoS requirements

Let P_m represent the power of the mth user's signal, then the problem of figuring out $P_{\min}$ is formulated as:

$$P_{\min} \triangleq \min_{P_m, 1 \leq m \leq M} \sum_{m=1}^{M} P_m \tag{15.13a}$$

$$\text{s.t.} \qquad P_m \geq B_m \left(|h_m|^2 \sum_{i=m+1}^{M} P_i + \sigma^2 \right), \quad 1 \leq m \leq M, \tag{15.13b}$$

where $B_m \triangleq \frac{2^{\tilde{R}_b^m}-1}{|h_m|^2}$ and (15.13b) comes from the QoS constraints in (15.10). Problem (15.13) is solved by the following theorem.

Theorem 15.1. *The objective function in (15.13a) is minimised when all the constraints in (15.13b) are active. The optimal solution to problem (15.13), denoted by* $\{P_m^{\text{Min}}\}_{m=1}^{M}$, *is given as*

$$P_m^{\text{Min}} = B_m \left(|h_m|^2 \sum_{i=m+1}^{M} P_i^{\text{Min}} + \sigma^2 \right), \quad 1 \leq m \leq M. \tag{15.14}$$

Proof. This theorem is going to be proved by contradiction as follows:

Assume that $\{P_m^*\}_{m=1}^{M}$ is the optimal solution to problem (15.13) with at least one constraint in (15.13b) inactive. Suppose that the nth constraint in (15.13b) is inactive, i.e.:

$$P_n^* > B_n \left(|h_n|^2 \sum_{i=n+1}^{M} P_i^* + \sigma^2 \right). \tag{15.15}$$

Then, we define a new set $\{P_m^{**}\}_{m=1}^{M}$ by setting $P_m^{**} = P_m^*$ for $m \neq n$ and P_n^{**} to the right-hand side (RHS) of (15.15). It is obvious that the newly defined set $\{P_m^{**}\}_{m=1}^{M}$ satisfies the constraint in (15.13b) for $m = n$.

We further verify that $\{P_m^{**}\}_{m=1}^M$ still satisfies the constraint in (15.13b) for $m \neq n$. Observing the special structure of (15.13b), one can see that for an arbitrary m from 1 to M, $\{P_i\}_{i=1}^{m-1}$ does not appear in (15.13b), which indicates that the setting of P_n^{**} has no impact on the constraints in (15.13b) for $n+1 \leq m \leq M$. Thereby, the newly defined set $\{P_m^{**}\}_{m=1}^M$ satisfies the constraint in (15.13b) for $n+1 \leq m \leq M$.

As for $1 \leq m \leq n-1$, we have:

$$P_m^{**} = P_m^* \geq B_m\left(|h_m|^2 \sum_{i=m+1}^M P_i^* + \sigma^2\right) > B_m\left(|h_m|^2 \sum_{i=m+1}^M P_i^{**} + \sigma^2\right), \quad (15.16)$$

which indicates that $\{P_m^{**}\}_{m=1}^M$ can also ensure that the constraints in (15.13b) hold for $1 \leq m \leq n-1$.

As a result, this newly defined set $\{P_m^{**}\}_{m=1}^M$ ensures that all the constraints in (15.13b) hold. However, according to (15.15), we have $\sum_{m=1}^M P_m^{**} < \sum_{m=1}^M P_m^*$, which contradicts the initial assumption that $\{P_m^*\}_{m=1}^M$ is the optimal solution to problem (15.13). Therefore, the assumption that problem (15.13) is solved with at least one constraint in (15.13b) being inactive must be false. As a result, all the constraints in (15.13b) must be active when the objective function in (15.13a) is minimised, which provides the optimal solution to problem (15.13) that is given by (15.14). Furthermore, it is worth pointing out that $\{P_i^{\text{Min}}\}_{i=1}^M$ can be calculated sequentially in the order $M, M-1, \ldots, 1$, since P_m^{Min} can be calculated with $\{P_i^{\text{Min}}\}_{i=m+1}^M$. By now, we have obtained the closed-form solution to problem (15.13) and thus complete the proof. $\qquad\qquad\Box$

According to Theorem 15.1, with the instantaneous CSI of all the legitimate users, $\{P_m^{\text{Min}}\}_{m=1}^M$ can be calculated sequentially in the order $m = M, M-1, \ldots, 1$ with (15.14) since P_m^{Min} is determined by $\{P_i^{\text{Min}}\}_{i=m+1}^M$ and P_M^{Min} is a known quantity, i.e. $B_M \sigma^2$. Therefore, we have $P_{\min} = \sum_{m=1}^M P_m^{\text{Min}}$ and the feasible range of the transmit power is $P \geq P_{\min}$. In addition, P_{Min} can be used as a threshold to verify whether P is large enough to meet all the users' QoS requirements.

15.2.2.2 Optimal power allocation policy

After obtaining $P_{\min}$, we are now going to deal with the SSR maximisation problem in (15.12) under the condition that $P \geq P_{\min}$.

By substituting (15.4) and (15.6) into (15.9), we firstly reformulate $\tilde{R}_s$ as follows:

$$\tilde{R}_s = \log_2\left(P|h_{M_e+1}|^2 \sum_{i=M_e+1}^M \gamma_i + \sigma^2\right) - \log_2\left(P|h_e|^2 \sum_{i=M_e+1}^M \gamma_i + \sigma^2\right)$$

$$+ \sum_{m=M_e+1}^{M-1}\left[\log_2\left(P|h_{m+1}|^2 \sum_{i=m+1}^M \gamma_i + \sigma^2\right) - \log_2\left(P|h_m|^2 \sum_{i=m+1}^M \gamma_i + \sigma^2\right)\right].$$

$$(15.17)$$

For notational simplicity, we further define:

$$C_m \triangleq \begin{cases} P\,|h_e|^2, & m = M_e, \\ P\,|h_m|^2, & M_e + 1 \leq m \leq M, \end{cases} \tag{15.18a}$$

$$t_m \triangleq \sum_{i=m+1}^{M} \gamma_i, \quad M_e \leq m \leq M - 1, \tag{15.18b}$$

$$J_m(t_m) \triangleq \log_2\left(C_{m+1}t_m + \sigma^2\right) - \log_2\left(C_m t_m + \sigma^2\right). \tag{15.18c}$$

Then $\tilde{R}_s$ in (15.17) can be recast as:

$$\tilde{R}_s = \sum_{m=M_e}^{M-1} J_m(t_m). \tag{15.19}$$

By observing (15.18) and (15.19), we can find that problem (15.12) has two important properties: (1) the objective function $\tilde{R}_s$ is the sum of $M - M_e$ non-convex sub-functions which shares similar forms; (2) the arguments $\{\gamma_m\}_{m=1}^{M}$ are coupled with each other in the constraints in (15.11) by a complicated way.

On the basis of the observed properties, we propose an optimisation algorithm to solve problem (15.12), which can be elaborated in two steps illustrated as follows:

Step 1: We individually solve the maximisation problem of each sub-function $J_m(t_m)$ for $M_e \leq m \leq M - 1$ subject to the constraints in (15.12b).

Step 2: We demonstrate that the optimal solution set to each maximisation problem possesses a *unique common solution*.

In other words, we can find a unique solution that simultaneously maximises $J_m(t_m)$ for each m from M_e to $M - 1$ with all the constraints in (15.12b) being satisfied. Thereby, this unique solution must be the optimal solution to problem (15.12). Mathematically, let Φ_m represent the optimal solution set that maximises $J_m(t_m)$ subject to all the constraints in (15.12b), where $M_e \leq m \leq M - 1$, then our goal is to prove:

$$\Phi_{M_e} \cap \Phi_{M_e+1} \cap \cdots \cap \Phi_{M-1} = \left\{ \{\gamma_m^{\text{Opt}}\}_{m=1}^{M} \right\}, \tag{15.20}$$

where $\{\gamma_m^{\text{Opt}}\}_{m=1}^{M}$ is the unique common solution of the $M - M_e$ optimisation problems.

We now deal with these $M - M_e$ optimisation problems. First of all, we transform the original problem of maximising $J_m(t_m)$ by using its monotonicity. The first-order derivative of $J_m(t_m)$ is given by

$$\frac{dJ_m(t_m)}{dt_m} = \frac{(C_{m+1} - C_m)\sigma^2}{\ln 2\left(C_{m+1}t_m + \sigma^2\right)\left(C_m t_m + \sigma^2\right)} \geq 0, \tag{15.21}$$

which indicates that $J_m(t_m)$ is a monotonically increasing function of t_m. Therefore, maximising J_m is equivalent to maximising t_m. As a result, these $M - M_e$ optimisation problems can be uniformly formulated as:

$$\max_{\gamma_i, 1 \leq i \leq M} t_m \tag{15.22a}$$

$$\text{s.t.} \quad \gamma_i \geq A_i \left(P|h_i|^2 \sum_{j=i+1}^{M} \gamma_j + \sigma^2 \right), \quad 1 \leq i \leq M, \tag{15.22b}$$

$$\sum_{i=1}^{M} \gamma_i \leq 1. \tag{15.22c}$$

Problem (15.22) is solved by the following proposition.

Proposition 15.1. *The necessary and sufficient condition for the optimal solution to problem (15.22) is that both the constraints in (15.22b) for $1 \leq i \leq m$ and the constraint in (15.22c) are active. Moreover, the closed-form solution to problem (15.22) is given by*

$$\gamma_i = \frac{A_i \left[P|h_i|^2 \left(1 - \sum_{j=1}^{i-1} \gamma_j \right) + \sigma^2 \right]}{2\varrho_i}, \quad 1 \leq i \leq m. \tag{15.23a}$$

$$t_m = 1 - \sum_{i=1}^{m} \gamma_i. \tag{15.23b}$$

Proof. It is obvious that problem (15.22) is convex and thereby the following Karush–Kuhn–Tucker (KKT) conditions are necessary and sufficient for the optimal solution to problem (15.22):

$$\lambda = \begin{cases} \mu_k - \sum_{i=1}^{k-1} \mu_i A_i P |h_i|^2, & 1 \leq k \leq m, \\ \mu_k - \sum_{i=1}^{k-1} \mu_i A_i P |h_i|^2 + 1, & m < k \leq M, \end{cases} \tag{15.24a}$$

$$\mu_i \left[A_i \left(P|h_i|^2 \sum_{j=i+1}^{M} \gamma_j + \sigma^2 \right) - \gamma_i \right] = 0, \quad 1 \leq i \leq M, \tag{15.24b}$$

$$\mu_i \geq 0, \quad 1 \leq i \leq M, \tag{15.24c}$$

$$\lambda \left(\sum_{i=1}^{M} \gamma_i - 1 \right) = 0, \tag{15.24d}$$

$$\lambda \geq 0, \tag{15.24e}$$

where $\{\mu_i\}_{i=1}^{M}$ and λ are the Lagrange multipliers for the inequality constraints in (15.22b) and (15.22c), respectively. For sake of proving that the constraints in (15.22b) are active when $1 \leq i \leq m$ and that the constraint in (15.22c) is active, it is necessary and sufficient to demonstrate $\mu_i \neq 0$ for $1 \leq i \leq m$ and $\lambda \neq 0$, respectively. To this end, we firstly proof $\mu_1 \neq 0$ by contradiction:

Supposing $\mu_1 = 0$ and setting $k = 1$ in (15.24a), we obtain:

$$\lambda = \mu_1 = 0. \tag{15.25}$$

By substituting (15.25) into (15.24a), we can obtain:

$$\mu_k = \sum_{i=1}^{k-1} \mu_i A_i P |h_i|^2, \quad 1 \leq k \leq m. \tag{15.26}$$

Apparently, (15.26) demonstrates that $\mu_k = 0$ for $1 \leq k \leq m$, because $\mu_1 = 0$ and μ_k can be calculated sequentially in the order $2, 3, \ldots, k$. However, under the condition that $\mu_k = 0$, $1 \leq k \leq m$, we set $k = m + 1$ in (15.24a) and then obtain $\lambda = \mu_{m+1} + 1 > 0$, which contradicts (15.25) that is obtained from the assumption $\mu_1 = 0$. Therefore, we can conclude that $\mu_1 \neq 0$ and

$$\lambda = \mu_1 \neq 0, \tag{15.27}$$

which indicates the inequality constraint in (15.22c) must be active.

We then substitute (15.27) into (15.24a) for $1 \leq k \leq m$ and further obtain:

$$\mu_k = \sum_{i=1}^{k-1} \mu_i A_i P |h_i|^2 + \lambda, \quad 1 \leq k \leq m, \tag{15.28}$$

which proves that $\mu_k > 0$ for $1 \leq k \leq m$, since λ is a positive number according to (15.24e) and (15.27). Therefore, the constraints in (15.22b) are active for $1 \leq k \leq m$.

To derive the closed-form solution of problem (15.22), we replace $\sum_{j=i+1}^{M} \gamma_j$ by $1 - \sum_{j=1}^{i} \gamma_j$ in (15.22b) since (15.22c) has been proved to be active. By setting the constraints in (15.22b) active for $1 \leq i \leq m$, we have the expressions of $\{\gamma_i\}_{i=1}^{m}$ and of t_m given by (15.23a) and (15.23b), respectively. According to (15.23a), γ_i can be obtained by sequential calculation in the order $1, 2, \ldots, m$. Therefore, the closed-form expression for the optimal solution, which maximises t_m, can be obtained when both the constraints in (15.22b) for $1 \leq i \leq m$ and the constraint in (15.22c) are active. By now, we have completed the proof. $\square$

On the basis of Proposition 15.1, which solves problem (15.22) with the closed-form solution given in (15.23), the following theorem further provides the unique solution that maximises $\tilde{R}_s$, namely the optimal solution to problem (15.12).

Theorem 15.2. *The unique optimal power allocation coefficients* $\{\gamma_m^{\text{Opt}}\}_{m=1}^{M}$ *that maximise* $\tilde{R}_s$, *are given by*

$$\gamma_m^{\text{Opt}} = \begin{cases} \dfrac{A_m \left[P |h_m|^2 \left(1 - \sum_{i=1}^{m-1} \gamma_i^{\text{Opt}} \right) + \sigma^2 \right]}{2^{\tilde{R}_b^m}}, & 1 \leq m < M, \\ 1 - \sum_{i=1}^{M-1} \gamma_i^{\text{Opt}}, & m = M. \end{cases} \tag{15.29}$$

Proof. According to Proposition 15.1, when both the constraints in (15.22b) for $1 \leq i \leq m$ and the constraint in (15.22c) are active, the arguments $\{\gamma_i\}_{i=1}^m$ are uniquely determined with the formulas in (15.23a). This implies that more power allocation coefficients can be uniquely determined with the growth of m. In other words, the size of the optimal solution set to problem (15.22), i.e. Φ_m, becomes smaller as m increases, which can be further mathematically characterised as follows:

$$\Phi_{M_e} \supseteq \Phi_{M_e+1} \supseteq \cdots \supseteq \Phi_{M-1}, \tag{15.30a}$$

$$\Phi_{M_e} \cap \Phi_{M_e+1} \cap \cdots \cap \Phi_{M-1} = \Phi_{M-1}. \tag{15.30b}$$

As a result, Φ_{M-1} is the optimal solution set that simultaneously maximises $J_m(t_m)$ for $M_e \leq m \leq M-1$, thereby becoming the optimal solution set to problem (15.12). By setting $m = M-1$ in the closed-form solution provided by (15.23) in Proposition 15.1, the first $M-1$ arguments $\{\gamma_i^{\mathrm{Opt}}\}_{i=1}^{M-1}$ are uniquely determined with (15.23a) in the order $1, 2, \ldots, M-1$ and the last argument γ_M^{Opt} is also uniquely determined with $\gamma_M^{\mathrm{Opt}} = 1 - \sum_{i=1}^{M-1} \gamma_i^{\mathrm{Opt}}$ since the constraint in (15.22c) has been proved to be satisfied at equality. Therefore, we have found the optimal solution to problem (15.12) and thus complete the proof. ☐

The analysis and derivations above have demonstrated that under the condition that $P \geq P_{\min}$, the optimal power allocation policy that maximises the SSR of the SISO NOMA system is to use the extra power $(P - P_{\min})$ only for increasing the Mth user's secrecy rate. This is because the Mth user has the largest channel gain, and it achieves the highest secrecy rate among all users with the same amount of power. In other words, the Mth user can use power more effectively than the other users do. As a result, the nature of maximising the SSR of the system is to enlarge the secrecy rate of the user with the largest channel gain as much as possible. However, this does not signify that the extra power $(P - P_{\min})$ should be totally allocated to the Mth user, since its signal also interferes with the other $M-1$ users. In addition, this nature can also well explain a mathematical phenomenon that the terms h_e and M_e do not appear in the closed-form solution given by (15.29), i.e. the proposed power allocation policy does not need the eavesdropper' CSI.

15.2.3   Simulation results

In this section, numerical results are provided to show the secure performance of the SISO NOMA system in terms of SSR, whereby our proposed power allocation policy is applied. Besides, a TDMA system is used for benchmarking, in which the time slots with equal duration are allocated to users individually. In each time slot, total power is allocated to the user for maximising its secrecy rate. 50,000 channel realisations are randomly generated with parameters $g_m, g_e \sim CN(0, 1)$, $1 \leq m \leq M$, $\alpha = 3$, $d_m = d_e = 80$ m, $1 \leq m \leq M$, and $\sigma^2 = -70$ dBm. In particular, when the available transmit power P is not in the feasible range, the transmitter will not send the messages. In other words, when P is not large enough to satisfy all users' QoS requirements, the SSR of the system is set to zero.

Figure 15.2 depicts the average SSR of the system versus the available transmit power P. It can be seen that NOMA has a superior secure performance than

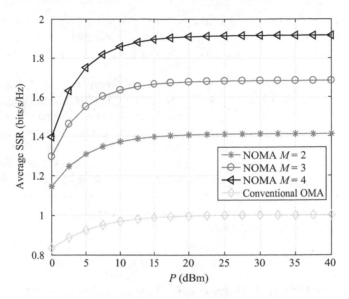

Figure 15.2   *Average SSR (bits/s/Hz) versus the available transmit power P (dBm) for different user numbers with parameters $\bar{R}_b^m = 1$ bits/s/Hz, $1 \le m \le M$. © [2016] IEEE. Reprinted, with permission, from [20]*

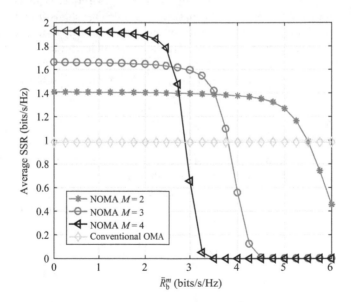

Figure 15.3   *Average SSR (bits/s/Hz) versus $\bar{R}_b^m$ (bits/s/Hz) for different numbers of users. The transmit power P is set to 20 dBm. © [2016] IEEE. Reprinted, with permission, from [20]*

conventional OMA and the performance gain obtained by NOMA becomes more significant as M increases. This is because, a higher diversity gain is offered when M is large, and higher spectral efficiency can be achieved when more users are simultaneously served.

Figure 15.3 demonstrates the effect of the minimum required QoS $\bar{R}_b^m$ on the SSR of the SISO NOMA system. One can see that the SSR decreases as $\bar{R}_b^m$ increases. This is because, the increase of $\bar{R}_b^m$ requires the transmitter to use the extra power for improving the data rate of the users with poor channel conditions, which obviously degrades the SSR of the system. Moreover, as $\bar{R}_b^m$ becomes very large, the SSR approaches to zero, since P is not large enough to satisfy all users' QoS requirements when $\bar{R}_b^m$ is too large and then the transmitter does not send the messages to the users, which makes the SSR be zero. This implies that NOMA is more suitable for low-rate communications and less robust for the increase of data rate requirements in comparison with conventional OMA.

15.3 Secure transmissions realised by a multi-antenna jammer

From our previous investigations on the primitive SISO NOMA system, we can conclude that the secure transmission cannot be provided for those legitimate users whose channel gains are smaller than that of the eavesdropper. This is one of the biggest natural defects of physical layer security in SISO systems. To overcome this drawback, we are going to equip a multi-antenna jammer at the same position of the transmitter in order to provide secure transmissions for all legitimate users by sending AN to interfere the eavesdropper.[3] With the well-designed AN, secure transmissions can be guaranteed with a great possibility for the legitimate users whose channel conditions are worse than that of the eavesdropper.

15.3.1 System model

We consider that a jammer with N antennas is additionally equipped at the same position of the transmitter for the system studied previously. The model of this new system is given by Figure 15.4. Denote $\mathbf{h}_{Jm}$ for $1 \le m \le M$ as the $N \times 1$ channel vector from the jammer to the mth legitimate user and then it can be modelled as:

$$\mathbf{h}_{Jm} = \mathbf{g}_{Jm} d_m^{-\frac{\alpha}{2}}, \quad 1 \le m \le M, \tag{15.31}$$

---

[3]In fact, the jammer could be placed at any other positions and our proposed algorithm will work similarly. In this chapter, we equip the jammer at the same position of the transmitter. This setting helps to avoid defining the position of the jammer, which facilitates illustrations of the following algorithm.

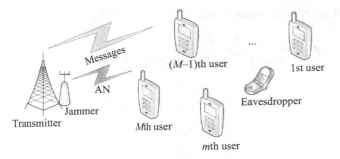

Figure 15.4   System model

where $\mathbf{g}_{Jm} \sim CN(\mathbf{0}, \mathbf{I}_N)$ is the small-scale fading coefficient, d_m is the distance from the jammer (i.e. the transmitter) to the mth legitimate user. Similarly, the channel from the jammer to the eavesdropper $\mathbf{h}_{Je}$ is modelled as:

$$\mathbf{h}_{Je} = \mathbf{g}_{Je} d_e^{-\frac{\alpha}{2}}. \tag{15.32}$$

In this new system, we use R_{Jb}^m, R_{Je}^m, and R_{Js}^m to represent, respectively, the achievable rate of the mth legitimate user, the achievable rate of the eavesdropper in detecting the mth user's message, and the secrecy rate of the mth legitimate user. Their own analytical expressions will be discussed and given in the following.

15.3.1.1   Achievable rate of the mth legitimate user R_{Jb}^m

As mentioned before, the users often apply the SIC technique to realise multi-user detection in NOMA systems. In this new system, the decoding order is the same to the one used in Section 15.2. However, due to the AN broadcasted by the multi-antenna jammer, the feasibility of the implementation of the SIC cannot be naturally guaranteed under the pre-defined order, i.e. the aforementioned ascending order of the channel gains given in (15.1). Therefore, it is important to firstly analyse the used decoding order and then add some necessary constraints to ensure its feasibility.

Let SINR_m^i represent the SINR of the mth user in detecting the ith user's message where $i < m$, then SINR_m^i can be given by:

$$\text{SINR}_m^i = \frac{P_i |h_m|^2}{|h_m|^2 \sum_{k=i+1}^{M} P_k + \mathbf{h}_{Jm}^H \Lambda \mathbf{h}_{Jm} + \sigma^2}, \tag{15.33}$$

where Λ represents the covariance matrix of the AN broadcasted by the jammer. Similarly, the SINR of the ith user in detecting its own message, i.e. SINR_i^i, can be given by:

$$\text{SINR}_i^i = \frac{P_i |h_i|^2}{|h_i|^2 \sum_{k=i+1}^{M} P_k + \mathbf{h}_{Ji}^H \Lambda \mathbf{h}_{Ji} + \sigma^2}. \tag{15.34}$$

Then the difference between the two SINRs are given by:

$$
\begin{aligned}
\text{SINR}_m^i - \text{SINR}_i^i &= \frac{P_i |h_m|^2}{|h_m|^2 \sum_{k=i+1}^{M} P_k + \mathbf{h}_{Jm}^H \Lambda \mathbf{h}_{Jm} + \sigma^2} \\
&\quad - \frac{P_i |h_i|^2}{|h_i|^2 \sum_{k=i+1}^{M} P_k + \mathbf{h}_{Ji}^H \Lambda \mathbf{h}_{Ji} + \sigma^2} \\
&= \frac{P_i \left(\frac{\mathbf{h}_{Ji}^H \Lambda \mathbf{h}_{Ji} + \sigma^2}{|h_i|^2} - \frac{\mathbf{h}_{Jm}^H \Lambda \mathbf{h}_{Jm} + \sigma^2}{|h_m|^2} \right)}{\left(\sum_{k=i+1}^{M} P_i + \frac{\mathbf{h}_{Jm}^H \Lambda \mathbf{h}_{Jm} + \sigma^2}{|h_m|^2} \right) \left(\sum_{k=i+1}^{M} P_i + \frac{\mathbf{h}_{Ji}^H \Lambda \mathbf{h}_{Ji} + \sigma^2}{|h_i|^2} \right)}.
\end{aligned} \tag{15.35}
$$

According to (15.35), we know that when the jammer does not broadcast the AN, i.e. $\Lambda = \mathbf{0}$, $\text{SINR}_m^i \geq \text{SINR}_i^i$ always holds since $|h_i|^2 \leq |h_m|^2$, which indicates that the mth user can always successfully decode the ith user's message for $i < m$ without regard to their power allocation coefficients. This also validates (15.4) in Section 15.2. Therefore, the following constraints must be satisfied so as to ensure $\text{SINR}_m^i \geq \text{SINR}_i^i$, which guarantees the feasibility of the SIC, even though the jammer broadcasts the AN to interfere the eavesdropper:

$$
\frac{\mathbf{h}_{Ji}^H \Lambda \mathbf{h}_{Ji} + \sigma^2}{|h_i|^2} \geq \frac{\mathbf{h}_{Jm}^H \Lambda \mathbf{h}_{Jm} + \sigma^2}{|h_m|^2}, \quad 1 \leq m \leq M, \quad 1 \leq i \leq m-1. \tag{15.36}
$$

For convenience, we refer to (15.36) as the SIC constraints. With the SIC constraints satisfied, the achievable rate of the mth legitimate user R_{Jb}^m can be given by:

$$
R_{Jb}^m = \log_2 \left(1 + \frac{P_m |h_m|^2}{|h_m|^2 \sum_{i=m+1}^{M} P_i + \mathbf{h}_{Jm}^H \Lambda \mathbf{h}_{Jm} + \sigma^2} \right), \quad 1 \leq m \leq M. \tag{15.37}
$$

15.3.1.2   Discussions on the capability of the eavesdropper

First of all, the secrecy rate of the mth legitimate user R_{Js}^m can be given by:

$$
R_{Js}^m = \left[R_{Jb}^m - R_{Je}^m \right]^+, \tag{15.38}
$$

where R_{Je}^m denotes the achievable rate of the eavesdropper in detecting the mth user's message, of which the expression will be provided later. Before giving the definition of R_{Je}, we firstly introduce a preliminary constraint on R_{Js}^m in the following:

Our design is based on providing a secure transmission for each legitimate user. To this end, we denote $\bar{R}_{Js}^m$ as the minimum secrecy rate required by the mth legitimate user and then the secrecy rate constraints can be characterised as:

$$
R_{Js}^m \geq \bar{R}_{Js}^m, \quad 1 \leq m \leq M. \tag{15.39}
$$

With (15.39), the decoding ability of the eavesdropper is greatly degraded. As a result, the eavesdropper has to treat all the other users' messages as noise when it is trying to decode the message of a certain user, which means that the application of the SIC technique is unavailable for the eavesdropper. Therefore, the achievable rate of the eavesdropper in detecting the mth user's message is given as follows:

$$R_{\text{Je}}^m = \log_2\left(1 + \frac{P_m|h_e|^2}{|h_e|^2 \sum_{i\neq m}^M P_i + \mathbf{h}_{\text{Je}}^H \Lambda \mathbf{h}_{\text{Je}} + \sigma^2}\right), \qquad 1 \le m \le M. \qquad (15.40)$$

15.3.2 Secure transmissions based on secrecy rate guarantees

In this section, we propose a secure transmission design which protects each user's message from being intercepted by the eavesdropper. To be specific, our design is subject to the individual secrecy rate constraint for each user, which means that each legitimate user can achieve a pre-defined positive secrecy rate. In contrast with the power allocation policy in Section 15.2, this newly proposed design can guarantee each user's secure transmission.

15.3.2.1 Secrecy rate constraints

The decoding capability of the eavesdropper is greatly degraded by a well-designed AN at the jammer. If the jammer does not send the AN, it is obvious that the eavesdropper can apply the SIC technique to remove partial inter-user interference. However, in this new system, the AN will be carefully designed in order to protect each user's message from being wiretapped. The secrecy rate constraints has already been given in (15.39), which ensure the correctness of (15.40).

15.3.2.2 Problem formulations and approximations

In the following, we consider one of the most basic problems in physical layer security, which is to minimise the transmit power including the power allocated to users' signal and the power for radiating the designed AN, subject to the individual secrecy rate constraint for each user. On the basis of the signal models established above, this optimisation problem can be formulated as:

$$P_{\text{Tot}} \triangleq \min_{\{\gamma_m\}_{m=1}^M, \Lambda} \quad \sum_{m=1}^M P_m + \text{Tr}(\Lambda) \qquad\qquad (15.41\text{a})$$

$$\text{s.t. } (15.36) \quad \text{and} \quad (15.39), \qquad\qquad (15.41\text{b})$$

where P_{Tot} denotes the minimum required total transmit power, the constraints in (15.36) are the SIC constraints which ensure the feasibility of the SIC given the pre-defined decoding order and the constraints in (15.39) are the secrecy rate constraints which guarantee the secure transmission for each legitimate user.

According to the formulation of problem (15.41), both its objective function in (15.41a) and the constraints in (15.36) are convex, which makes the main difficulty of solving problem (15.41) lie in the non-convex constraints in (15.39). Generally, it is difficult to obtain a global optimal solution to this sort of non-convex problems, which leads us to turn to certain efficient algorithms helping to obtain a local optimal solution to a non-convex problem. In the following, an efficient algorithm is proposed to solve problem (15.41) based on the SCA method.

The basic idea of the SCA method is to approximate a non-convex problem by a sequence of convex problems iteratively. In each iteration, every non-convex constraint is replaced by its appropriate inner convex approximation. A local optimal solution is guaranteed by the SCA method [24]. For using SCA, we are going to firstly analyse the non-convexity of the constraints in (15.39) and then seek their appropriate inner convex approximations.

First of all, the non-convex constraints in (15.39) can be rewritten as:

$$R_{\mathrm{Js}}^m = R_{\mathrm{Jb}}^m - R_{\mathrm{Je}}^m$$

$$= \log_2\left(1 + \frac{P_m|h_m|^2}{|h_m|^2 \sum_{i=m+1}^M P_i + \mathbf{h}_{\mathrm{Jm}}^H \Lambda \mathbf{h}_{\mathrm{Jm}} + \sigma^2}\right)$$

$$\quad - \log_2\left(1 + \frac{P_m|h_e|^2}{|h_e|^2 \sum_{i\neq m}^M P_i + \mathbf{h}_{\mathrm{Je}}^H \Lambda \mathbf{h}_{\mathrm{Je}} + \sigma^2}\right) \tag{15.42}$$

$$= \log_2\left(|h_m|^2 \sum_{i=m}^M P_i + \mathbf{h}_{\mathrm{Jm}}^H \Lambda \mathbf{h}_{\mathrm{Jm}} + \sigma^2\right) - \log_2\left(|h_m|^2 \sum_{i=m+1}^M P_i + \mathbf{h}_{\mathrm{Jm}}^H \Lambda \mathbf{h}_{\mathrm{Jm}} + \sigma^2\right)$$

$$\quad - \log_2\left(|h_e|^2 \sum_{i=1}^M P_i + \mathbf{h}_{\mathrm{Je}}^H \Lambda \mathbf{h}_{\mathrm{Je}} + \sigma^2\right) + \log_2\left(|h_e|^2 \sum_{i\neq m}^M P_i + \mathbf{h}_{\mathrm{Je}}^H \Lambda \mathbf{h}_{\mathrm{Je}} + \sigma^2\right)$$

$$\geq \bar{R}_{\mathrm{Js}}^m.$$

where the operator $[\cdot]^+$ in (15.39) is eliminated since $\bar{R}_{\mathrm{Js}}^m$ is a positive number. According to (15.42), we know that its non-convexity is due to the term $-\log_2(x)$, which is not concave but in the left-hand side of the greater than or equal to sign ($\geq$). In fact, an inner convex approximation for $-\log_2(x)$ can be obtained as follows:

$$-\log_2(x) \geq -\log_2(x_0) - \frac{(x - x_0)}{x_0}, \tag{15.43}$$

where $-\log_2(x_0) - \frac{(x-x_0)}{x_0}$ is the first-order Taylor approximation of $-\log_2(x)$ around x_0. With (15.43), the non-convex constraint in (15.42) can be approximated by a more stringent but convex constraint given by:

$$T_1 + T_2 + T_3 + T_4 \geq \bar{R}_{\mathrm{Js}}^m, \tag{15.44}$$

where the definition of T_1, T_2, T_3, and T_4 are given as follows:

$$T_1 \triangleq \log_2\left(|h_m|^2 \sum_{i=m}^M P_i + \mathbf{h}_{\mathrm{Jm}}^H \Lambda \mathbf{h}_{\mathrm{Jm}} + \sigma^2\right), \tag{15.45a}$$

$$T_2 \triangleq -\log_2\left(|h_m|^2 \sum_{i=m+1}^M P_i^{(n-1)} + \mathbf{h}_{\mathrm{Jm}}^H \Lambda^{(n-1)} \mathbf{h}_{\mathrm{Jm}} + \sigma^2\right)$$

$$\quad - \frac{|h_m|^2 \sum_{i=m+1}^M \left(P_i - P_i^{(n-1)}\right) + \mathbf{h}_{\mathrm{Jm}}^H \left(\Lambda - \Lambda^{(n-1)}\right) \mathbf{h}_{\mathrm{Jm}}}{|h_m|^2 \sum_{i=m+1}^M P_i^{(n-1)} + \mathbf{h}_{\mathrm{Jm}}^H \Lambda^{(n-1)} \mathbf{h}_{\mathrm{Jm}} + \sigma^2}, \tag{15.45b}$$

$$T_3 \triangleq -\log_2\left(|h_e|^2 \sum_{i=1}^{M} P_i + \mathbf{h}_{Je}^{H}\Lambda\mathbf{h}_{Je} + \sigma^2\right)$$

$$- \frac{|h_e|^2 \sum_{i=1}^{M}\left(P_i - P_i^{(n-1)}\right) + \mathbf{h}_{Je}^{H}\left(\Lambda - \Lambda^{(n-1)}\right)\mathbf{h}_{Je}}{|h_e|^2 \sum_{i=1}^{M} P_i^{(n-1)} + \mathbf{h}_{Je}^{H}\Lambda\mathbf{h}_{Je} + \sigma^2}, \tag{15.45c}$$

$$T_4 \triangleq \log_2\left(|h_e|^2 \sum_{i\neq m}^{M} P_i + \mathbf{h}_{Je}^{H}\Lambda\mathbf{h}_{Je} + \sigma^2\right), \tag{15.45d}$$

where $\left\{P_i^{(n-1)}\right\}_{i=1}^{M}$ and $\Lambda^{(n-1)}$ refers to, respectively, the optimal power allocated to the users and the optimal covariance matrix of AN that are obtained in the $(n-1)$th iteration. Therefore, the original problem (15.41) becomes an iterative convex program. To be specific, the nth iteration is to solve the following convex optimisation problem:

$$\min_{\{\gamma_m\}_{m=1}^{M},\ \Lambda} \quad \sum_{m=1}^{M} P_m + \mathrm{Tr}(\Lambda) \tag{15.46a}$$

$$\text{s.t.} \quad (15.36) \quad \text{and} \quad (15.44). \tag{15.46b}$$

Herein, we recall that $\left\{\left\{P_i^{(n)}\right\}_{i=1}^{M},\ \Lambda^{(n)}\right\}$ denotes the optimal solution to problem (15.46). Furthermore, the SCA-based algorithm for solving problem (15.41) is outlined in Algorithm 1. Moreover, the used approach to generate the initial feasible solution $\left\{\left\{P_i^{(0)}\right\}_{i=1}^{M},\ \Lambda^{(0)}\right\}$ and the convergence analysis of our proposed iterative algorithm will be discussed in the following.

Algorithm 1 Transmit power minimisation subject to secrecy rate constraints

---

Input: Initial feasible solution $\left\{\left\{P_i^{(0)}\right\}_{i=1}^{M},\ \Lambda^{(0)}\right\}$, $\{h_m\}_{m=1}^{M}$, h_e, $\{\mathbf{h}_{Jm}\}_{m=1}^{M}$, $\mathbf{h}_{Je}$, σ^2;

1: $n = 1$;
2: Initialise $\left\{P_i^{(0)}\right\}_{i=1}^{M}$ and $\Lambda^{(0)}$.
3: **repeat**
4:     Solve problem (15.46);
5:     Update $n = n + 1$;
6:     Update $\left\{P_i^{(n)}\right\}_{i=1}^{M}$ and $\Lambda^{(n)}$;
7: **until** Convergence or limitation of the number of iterations;

15.3.2.3   Generation of initial feasible solution

We herein propose an efficient approach to find an initial feasible solution for the proposed SCA-based algorithm. In the following, we show that problem (15.41) can be approximated by a convex program without any iterations.

We firstly rewrite the non-convex constraints in (15.39) as below by adding an auxiliary variable t:

$$R_{Jb}^m \geq t + \bar{R}_{Js}^m, \tag{15.47a}$$

$$R_{Je}^m \leq t, \tag{15.47b}$$

whereby t can be interpreted as the maximum achievable rate of the eavesdropper in detecting an arbitrary user's message. Then, (15.47) can be further transformed into:

$$|h_m|^2 \sum_{i=m}^M P_i + \mathbf{h}_{Jm}^H \Lambda \mathbf{h}_{Jm} + \sigma^2 \geq 2^{(t+\bar{R}_{Js}^m)} \left(|h_m|^2 \sum_{i=m+1}^M P_i + \mathbf{h}_{Jm}^H \Lambda \mathbf{h}_{Jm} + \sigma^2 \right), \tag{15.48a}$$

$$|h_e|^2 \sum_{i=1}^M P_i + \mathbf{h}_{Je}^H \Lambda \mathbf{h}_{Je} + \sigma^2 \leq 2^t \left(|h_e|^2 \sum_{i \neq m}^M P_i + \mathbf{h}_{Je}^H \Lambda \mathbf{h}_{Je} + \sigma^2 \right). \tag{15.48b}$$

Note that if the auxiliary variable t is fixed as a constant, the constraints in (15.48) become convex, which helps to create the following convex program:

$$\min_{\{\gamma_m\}_{m=1}^M, \Lambda} \quad \sum_{m=1}^M P_m + \mathrm{Tr}(\Lambda) \tag{15.49a}$$

$$\text{s.t.} \quad (15.36) \quad \text{and} \quad (15.48). \tag{15.49b}$$

In fact, problem (15.49) is created by reducing the feasible range of problem (15.41) and can be easily solved by general interior-point methods [25]. Let us denote the optimal solution to problem (15.49) as $\left\{ \left\{ P_i^{(0)} \right\}_{i=1}^M, \Lambda^{(0)} \right\}$ and set it as the initial feasible solution for Algorithm 1, then our proposed SCA-based algorithm can solve problem (15.41) in the vicinity of this initial solution by finding a better solution. Concerning the setting of t, it can be randomly and repeatedly generated for ensuring the feasibility of problem (15.49).

15.3.2.4   Convergence analysis

In fact, the convergence of Algorithm 1 has already been studied in [24]. From [24, Theorem 1], we can easily verify that the first-order Taylor approximations in (15.43) satisfies the constraints for ensuring the convergence of the SCA method. Thereby, it is guaranteed that the solution returned by Algorithm 1 converges to a KKT solution of the original problem (15.46).

15.3.3 Simulation results

In this section, numerical results are provided to show the secure performance of the jammer assisted SISO NOMA system. The original optimisation problem is solved for 300 times by using CVX package [26] with randomly generated channel realisations. The channels $\{\mathbf{h}_{Jm}\}_{m=1}^{M}$, $\mathbf{h}_{Je}$, $\{\mathbf{h}_{m}\}_{m=1}^{M}$ and $\mathbf{h}_{e}$ are generated with $\{\mathbf{g}_{Jm}\}_{m=1}^{M}, \mathbf{g}_{Je} \sim CN(\mathbf{0}, \mathbf{I}_M)$, $g_m, g_e \sim CN(0, 1)$ and $\{d_m\}_{m=1}^{M}$, d_e are fixed. Thus, the average in simulation is taken over the small-scale fading components of the channel vectors and the channel coefficients from the transmitter to the receivers including the eavesdropper. Besides, the variance of the additive Gaussian noise is $\sigma^2 = -70$ dBm. In particular, all legitimate users' minimum required secrecy rates $\{\bar{R}_{Js}^{m}\}_{m=1}^{M}$ are set to the same value, which is denoted by $\bar{R}_{Js}$.

Figure 15.5 shows the minimum required total transmit power P_{Tot} versus the minimum required secrecy rate $\bar{R}_{Js}$ for different numbers of legitimate users. First of all, it is obviously that P_{Tot} increases as $\bar{R}_{Js}$ increases. Furthermore, as the legitimate user number M increases, it is natural that more power is required so as to satisfy all legitimate users' secrecy rate requirements.

Figure 15.6 depicts the minimum required total transmit power P_{Tot} versus the minimum required secrecy rate $\bar{R}_{Js}$ for different numbers of the antennas of the jammer N. It can be seen that as N increases, less total power is required to attain the target secrecy rates for all legitimate users thanks to the array gains provided by

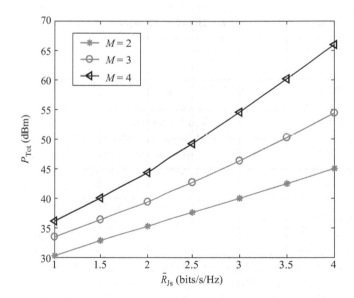

Figure 15.5 Minimum required total transmit power P_{Tot} (dBm) versus the minimum required secrecy rate $\bar{R}_{Js}$ (bits/s/Hz) for different numbers of users

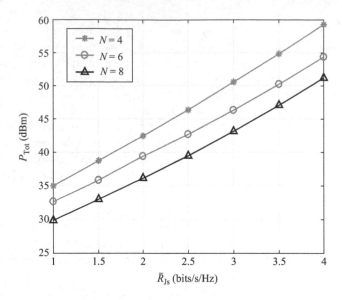

Figure 15.6   Minimum required total transmit power P_{Tot} (dBm) versus the minimum required secrecy rate $\bar{R}_{Js}$ (bits/s/Hz) for different antenna numbers of the jammer N

adding more antennas at the jammer. Besides, the increases of N brings more degrees of freedom which helps to facilitate the design of AN.

15.4   Conclusions and open issues

In this chapter, we have studied physical layer security in SISO NOMA systems. We have initially attempted to provide a preliminary analysis of the secure performance of a SISO NOMA system which consists of a transmitter, multiple legitimate users and an eavesdropper. In this considered system, the SSR of the system has been max-imised subject to the pre-defined QoS requirements for all legitimate users. Numerical results have shown that NOMA has the superior SSR performance compared with conventional OMA and the performance gain achieved by NOMA is more signifi-cant as the user number increases. In addition, highly demanding QoS requirements will degrade the SSR performance. However, a great defect of this SISO NOMA system is that the secure transmission cannot be assured for the users whose channel gains are smaller than that of the eavesdropper. To address this serious problem, we have further exploited AN to interfere the eavesdropper by equipping a multi-antenna jammer at the same position of the transmitter. There into, a joint optimisation of the power allocation and the AN design has been solved, subject to an individual secrecy rate constraint for each legitimate user, respectively. Our theoretical analysis and simulation results have demonstrated that the secure transmission can be well guaranteed for all users by exploiting the multi-antenna jammer. In particular, the system performance increases with the growth of the antenna number of the jammer.

One promising future direction for physical layer security in NOMA is to adopt multiple antenna settings at both the transmitter and the legitimate users to further enhance secure transmissions. The current works only focus on the single-antenna transceivers. However, the problem of establishing the SIC decoding order in MIMO NOMA systems will be a meaningful problem but with great difficulties. Besides, the designs of SIC decoding order, transmission rates, and power allocation can be also jointly investigated. Further, secrecy outage probability can be adopted as a secrecy metric.

Another possible research direction is to consider the uplink secure transmission of NOMA systems. Early works mainly focus on downlink NOMA systems, and very few contributions investigate uplink NOMA. In [27], an uplink power control scheme is proposed to achieve diverse arrived power in uplink NOMA systems. Accordingly, it is interesting to study physical layer security for uplink NOMA systems.

References

[1] Y. Saito, A. Benjebbour, Y. Kishiyama, and T. Nakamura, "System level performance evaluation of downlink non-orthogonal multiple access (NOMA)," in *Proc. IEEE Annu. Symp. Personal, Indoor and Mobile Radio Commun. (PIMRC)*, London, UK, Sep. 2013, pp. 611–615.

[2] Z. Ding, Z. Yang, P. Fan, and H. V. Poor, "On the performance of non-orthogonal multiple access in 5G systems with randomly deployed users," *IEEE Signal Process. Lett.*, vol. 21, no. 12, pp. 1501–1505, Dec. 2014.

[3] L. Dai, B. Wang, Y. Yuan, S. Han, C.-L. I, and Z. Wang, "Non-orthogonal multiple access for 5G: Solutions, challenges, opportunities, and future research trends," *IEEE Commun. Mag.*, vol. 53, no. 9, pp. 74–81, Sep. 2015.

[4] J. Choi, "Non-orthogonal multiple access in downlink coordinated two point systems," *IEEE Commun. Lett.*, vol. 18, no. 2, pp. 313–316, Feb. 2014.

[5] Z. Ding, M. Peng, and H. V. Poor, "Cooperative non-orthogonal multiple access in 5G systems," *IEEE Commun. Lett.*, vol. 19, no. 8, pp. 1462–1465, Aug. 2015.

[6] M. F. Hanif, Z. Ding, T. Ratnarajah, and G. K. Karagiannidis, "A minorization–maximization method for optimizing sum rate in the downlink of non-orthogonal multiple access systems," *IEEE Trans. Signal Process.*, vol. 64, no. 1, pp. 76–88, Jan. 2016.

[7] Z. Ding, F. Adachi, and H. V. Poor, "The application of MIMO to non-orthogonal multiple access," *IEEE Trans. Wireless Commun.*, vol. 15, no. 1, pp. 537–552, Jan. 2016.

[8] Z. Ding, R. Schober, and H. V. Poor, "A general MIMO framework for NOMA downlink and uplink transmission based on signal alignment," *IEEE Trans. Wireless Commun.*, vol. 15, no. 6, pp. 4438–4454, Jun. 2016.

 [9] S. Timotheou and I. Krikidis, "Fairness for non-orthogonal multiple access in 5G systems," *IEEE Signal Process. Lett.*, vol. 22, no. 10, pp. 1647–1651, Oct. 2015.

[10] Z. Ding, P. Fan, and H. V. Poor, "Impact of user pairing on 5G non-orthogonal multiple access," *IEEE Trans. Veh. Technol.*, vol. 65, no. 8, pp. 6010–6023, Aug. 2016.

[11] Q. Sun, S. Han, C.-L. I, and Z. Pan, "Energy efficiency optimization for fading MIMO non-orthogonal multiple access systems," in *Proc. IEEE Int. Conf. Commun. (ICC)*, London, U.K., Jun. 2015, pp. 2668–2673.

[12] Y. Zhang, H.-M. Wang, T.-X. Zheng, and Q. Yang, "Energy-efficient transmission design in non-orthogonal multiple access," *IEEE Trans. Veh. Technol.*, vol. 66, no. 3, pp. 2852–2857, Mar. 2017.

[13] G. Geraci, S. Singh, J. G. Andrews, J. Yuan, and I. B. Collings, "Secrecy rates in broadcast channels with confidential messages and external eavesdroppers," *IEEE Trans Wireless Commun.*, vol. 13, no. 5, pp. 2931–2943, May 2014.

[14] X. He, A. Khisti, and A. Yener, "MIMO broadcast channel with an unknown eavesdropper: Secrecy degrees of freedom," *IEEE Trans. Commun.*, vol. 62, no. 1, pp. 246–255, Jan. 2014.

[15] M. Bloch and J. Barros, *Physical layer security: From information theory to security engineering*. Cambridge, UK: Cambridge University Press, 2011.

[16] H.-M. Wang and X.-G. Xia, "Enhancing wireless secrecy via cooperation: signal design and optimization," *IEEE Commun. Mag.*, vol. 53, no. 12, pp. 47–53, Dec. 2015.

[17] S. Yang, M. Kobayashi, P. Piantanida, and S. Shamai (Shitz), "Secrecy degrees of freedom of MIMO broadcast channels with delayed CSIT," *IEEE Trans. Inf. Theory*, vol. 59, no. 9, pp. 5244–5256, Sep. 2013.

[18] R. Negi and S. Goel, "Secret communication using artificial noise," in *Proc. IEEE Veh. Technol. Conf. (VTC)*, pp. 1906–1910, Sep. 2005.

[19] Q. Li and W.-K. Ma, "Spatially selective artificial-noise aided transmit optimization for MISO multi-eves secrecy rate maximization," *IEEE Trans. Signal Process.*, vol. 61, no. 10, pp. 2704–2717, May 2013.

[20] Y. Zhang, H.-M. Wang, Q. Yang, and Z. Ding, "Secrecy sum rate maximization in non-orthogonal multiple access," *IEEE Commun. Lett.*, vol. 20. no. 5, pp. 930–933, May 2016.

[21] Z. Qin, Y. Liu, Z. Ding, Y. Gao, and M. Elkashlan, "Physical layer security for 5G non-orthogonal multiple access in large-scale networks," in *Proc. IEEE Int. Conf. Commun. (ICC)*, Kuala Lumpur, Malaysia, May 2016.

[22] G. Bagherikaram, A. S. Motahari, and A. K. Khandani, "Secrecy capacity region of Gaussian broadcast channel," in *Proc. 43rd Annu. Conf. Inf. Sci. Syst. (CISS)*, pp. 152–157, Mar. 2009.

[23] E. Ekrem and S. Ulukus, "Secrecy capacity of a class of broadcast channels with an eavesdropper," *EURASIP J. Wireless Commun. Netw.*, vol. 2009, no. 1, pp. 1–29, 2009.

[24] B. R. Marks and G. P. Wright, "A general inner approximation algorithm for nonconvex mathematical programs," *Operat. Res.*, vol. 26, no. 4, pp. 681–683, Jul.–Aug. 1977.

[25] S. Boyd and L. Vandenberghe, *Convex Optimization*. Cambridge, UK: Cambridge University Press, 2004.

[26] M. Grant and S. Boyd, "CVX: Matlab software for disciplined convex programming, version 2.1," http://cvxr.com/cvx/, Jun. 2015.

[27] N. Zhang, J. Wang, G. Kang, and Y. Liu, "Uplink non-orthogonal multiple access in 5G systems," *IEEE Commun. Lett.*, vol. 20, no. 3, pp. 458–461, Mar. 2016.

Chapter 16

Physical layer security for MIMOME-OFDM systems: spatial versus temporal artificial noise

Ahmed El Shafie[1], Zhiguo Ding[2], and Naofal Al-Dhahir[1]

16.1 Introduction

Information secrecy is crucial to a wireless communication system due to the broadcast nature of radio-frequency transmissions. Traditionally, secrecy has been provided by designing upper layer sophisticated protocols. More specifically, the problem of securing the information from malicious eavesdropping nodes has been tackled by relying on the conventional encryption mechanism. To improve the system security, physical (PHY) layer security has been recently recognised as a valuable tool to guarantee information-theoretic confidentiality of messages transmitted over the wireless medium.

Perfect secrecy [17] was introduced as the statistical independence between the information bearing message and the eavesdropper's observations. The seminal works of Wyner [20] and Csiszár and Korner [2] characterised the secrecy capacity for a wiretap channel as the maximal rate at which information can be transmitted over the wireless channel while guaranteeing a vanishing mutual information leakage per channel use. On the basis of these works, the secrecy capacity has been investigated for different channel models and network settings. For more details, the interested readers are referred to [10,11].

In recent years, orthogonal frequency-division multiplexing (OFDM) has been widely adopted in the PHY layer of most wireless and wired communication standards thanks to its efficiency in converting frequency-selective fading channels to frequency flat fading sub-channels, resulting in high performance at practical complexity. Hence, it is important to investigate PHY-layer security for OFDM systems. In the information theoretic security literature, OFDM has usually been modelled as a set of parallel Gaussian channels. In [9], the authors derived the secrecy capacity and developed the corresponding power-allocation schemes for OFDM systems. In [16], the system security was defined in terms of minimum mean-squared error (MMSE) at the eavesdropper. In [15], the OFDM wiretap channel was treated as

[1]Department of Electrical Engineering, University of Texas at Dallas, USA
[2]School of Computing and Communications, Lancaster University, UK

a special instance of the MIMO wiretap channel and its secrecy rates were studied by using both Gaussian inputs and quadrature amplitude modulation (QAM) signal constellations through asymptotic high and low signal-to-noise ratio (SNR) analyses.

Recently, many papers studied precoding for PHY-layer security. For example, linear precoding was studied in [5,18,19]. In [5], with the full channel state information (CSI) knowledge at the legitimate transmitter, the authors proposed a Vandermonde precoding scheme that enables the legitimate transmitter to transmit the information signal in the null space of the equivalent eavesdropper multiple-input multiple-output (MIMO) channel matrix. In [19], the authors investigated the optimal power-allocation scheme for artificial noise (AN) secure precoding systems. It was assumed that the legitimate transmitter has perfect CSI of the link to its legitimate receiver and knows only the statistics of the potential eavesdropper's CSI. The author of [18] studied the MIMO wiretap problem in which three multiple-antenna nodes share the channel. The author proposed using enough power to guarantee a certain quality of service (QoS) for the legitimate receiver measured by a predefined signal-to-interference-plus-noise ratio (SINR) for successful decoding, and then allocating the remaining power to generate AN to confuse the eavesdropper. Note that, as explained in [11], using the constraints of the bit error rate (BER), mean-squared error (MSE), or SINR at eavesdropping nodes does not satisfy either weak or strong secrecy requirements, but it often simplifies system design and analysis.

In [13], the authors proposed temporal AN injection for the single-input single-output single-antenna-eavesdropper (SISOSE) OFDM system in which a time domain AN signal is added to the data signal before transmission. The temporal AN signal is designed to be cancelled at the legitimate receiver prior to data decoding. The authors demonstrated that injecting the AN signal with the data in the frequency domain is not beneficial since it degrades the secrecy rate. In [1], the authors proposed a temporal AN injection scheme in the time domain for the multiple-input multiple-output multiple-antenna-eavesdropper (MIMOME) OFDM system. The authors assumed that the precoders are used over the available sub-carriers, which couples the sub-carriers and complicates both encoding and decoding processes at the legitimate nodes.

While spatial AN-aided (in a non-OFDM system) [18,19] and temporal AN-aided [1,13] PHY-layer security schemes have been proposed in the literature, we are not aware of any existing work which compares both approaches in terms of average secrecy rates and implementation feasibility and complexity. Our goal in this chapter is to answer the following two questions:

- Under what scenarios is spatial AN preferable over temporal AN and vice versa?
- Under a total AN average power constraint, is there any advantage in the average secrecy rate for a hybrid spatio-temporal AN scheme over a purely temporal or spatial AN scheme?

To answer these questions, we propose a novel hybrid spatio-temporal AN-aided PHY-layer security scheme for MIMOME-OFDM channels which is parameterised by the fraction of the total AN power allocated to each type of AN signals. In addition, we analyse the average secrecy rate achieved by this hybrid scheme and derive tight asymptotic bounds on the average secrecy rate as the number of transmit antennas at the legitimate transmitter becomes large.

In this chapter, we consider a general scenario with a hybrid time and spatial AN-injection scheme. In particular, we consider the MIMOME-OFDM wiretap channel, where each node is equipped with multiple antennas and the legitimate transmitter (Alice) has perfect knowledge of the CSI for the wireless links to her legitimate receiver (Bob) only, while the eavesdropper (Eve) has perfect CSI knowledge of all the links in the network. This global CSI assumption at Eve represents the best case scenario for Eve (the worst case scenario for Alice/Bob) since she knows all the channels between Alice and Bob as well as the used data and AN precoders at Alice. We exploit the spatial and temporal degrees of freedom provided by the available antennas and by the cyclic prefix (CP) structure of OFDM blocks, respectively, to confuse Eve and, hence, increase the secrecy rate of the legitimate system.

Notation: Unless otherwise stated, lower and upper case bold letters denote vectors and matrices, respectively. Lower and upper-case letters denote time domain and frequency domain signals, respectively. $\mathbf{I}_N$ and $\mathbf{F}$ denote, respectively, the identity matrix whose size is $N \times N$ and the fast Fourier transform (FFT) matrix. $\mathbb{C}^{M \times N}$ denotes the set of all complex matrices of size $M \times N$. $(\cdot)^T$ and $(\cdot)^*$ denote transpose and Hermitian (i.e. complex-conjugate transpose) operations, respectively. $\mathbb{R}^{M \times M}$ denotes the set of real matrices of size $M \times M$. $\|\cdot\|$ denotes the Euclidean norm of a vector. $[\cdot]_{k,l}$ denotes the (k, l)th entry of a matrix, $[\cdot]_k$ denotes the kth entry of a vector. The function $\min(\cdot, \cdot)$ $(\max(\cdot, \cdot))$ returns the minimum (maximum) among the values enclosed in brackets. $\text{blkdiag}(\mathbf{A}_1, \ldots, \mathbf{A}_j, \ldots, \mathbf{A}_M)$ denotes a block diagonal matrix where the enclosed elements are the diagonal blocks. $\mathbb{E}[\cdot]$ denotes statistical expectation. $(\cdot)^{-1}$ is the inverse of the matrix in brackets. $\mathbf{0}$ denotes the all-zero matrix and its size is understood from the context. $\otimes$ is the Kronecker product. $\text{Trace}\{\cdot\}$ denotes the sum of the diagonal entries of enclosed matrix in brackets.

16.2 Preliminary

In this section, we describe the basic principles of the spatial and temporal AN injection. We start with the spatial AN scheme. Afterwards, we review the temporal AN scheme.

16.2.1 Spatial AN

The great interest in MIMO systems led to the realisation that the available spatial dimensions could enhance the secrecy capabilities of wireless channels [11]. Consider a fading MIMO channel where Alice, Bob, and Eve are equipped with N_A, N_B, and N_E antennas, respectively. Assuming Alice wishes to transmit $N_s \leq \min\{N_A, N_B\}$ data streams to Bob, a general representation for the signals received by Bob and Eve are

$$\mathbf{y}^B = \mathbf{H}_{A-B}(\mathbf{P}_{\text{data}}\mathbf{x} + \mathbf{P}_{\text{AN}}\mathbf{d}^s) + \mathbf{z}^B, \tag{16.1}$$

$$\mathbf{y}^E = \mathbf{H}_{A-E}(\mathbf{P}_{\text{data}}\mathbf{x} + \mathbf{P}_{\text{AN}}\mathbf{d}^s) + \mathbf{z}^E, \tag{16.2}$$

where $\mathbf{P}_{\text{data}} \in \mathbb{C}^{N_A \times N_s}$ is the data precoding matrix at Alice, $\mathbf{x} \in \mathbb{C}^{N_s \times 1}$ is the transmit data vector by Alice, $\mathbf{P}_{\text{AN}} \in \mathbb{C}^{N_A \times (N_A - N_s)}$ is the spatial-AN precoding matrix at Alice, $\mathbf{d}^s \in \mathbb{C}^{(N_A - N_s) \times 1}$ is the spatial-AN vector, $\mathbf{H}_{A-B} \in \mathbb{C}^{N_B \times N_A}$ is the MIMO complex

Gaussian channel matrix from Alice to Bob, $\mathbf{H}_{A-E} \in \mathbb{C}^{N_E \times N_A}$ is the complex Gaussian channel matrix from Alice to Eve, and $\mathbf{z}^B \in \mathbb{C}^{N_B \times 1}$, $\mathbf{z}^E \in \mathbb{C}^{N_E \times 1}$ are zero-mean complex AWGN vectors.

The key idea is to design $\mathbf{P}_{data}$ and $\mathbf{P}_{AN}$ to guarantee that the received data and AN signals at Bob lie in orthogonal sub-spaces by forming them from the right-singular vectors of $\mathbf{H}_{A-B}$. That is, Alice uses the singular value decomposition (SVD) of the channel matrix $\mathbf{H}_{A-B}$ to calculate the precoding matrices $\mathbf{P}_{data}$ and $\mathbf{P}_{AN}$, where $\mathbf{P}_{data}$ is the N_s columns of right singular-vectors matrix of $\mathbf{H}_{A-B}$ corresponding to the N_s largest non-zero singular values. To decode the data, Bob filters the received vector $\mathbf{y}^B$ using a filtering matrix $\mathbf{C}_B^*$. The columns of $\mathbf{C}_B$ are the N_s columns of left singular-vectors matrix of $\mathbf{H}_{A-B}$ corresponding to the N_s largest non-zero singular values. The condition to cancel the spatial-AN signal at Bob is:

$$\mathbf{C}_B^* \mathbf{H}_{A-B} \mathbf{P}_{AN} = \mathbf{0}. \tag{16.3}$$

16.2.2 Temporal AN

In this sub-section, we describe the temporal-AN scheme for the SISOSE case where $N_A = N_B = N_E = 1$ to explain the basic idea [13]. Temporal AN has been proposed recently to enhance the PHY-layer security of multi-carrier systems [13]. In an OFDM system, the frequency domain signal vector $\mathbf{x}$ is transformed to the time domain using the inverse fast Fourier transform (IFFT) and then the CP is inserted to mitigate the inter-symbol interference. After that, an AN signal is added to this time-domain signal before transmission. At the receiving end, the CP is discarded the resulting signal is transformed to the frequency domain using the FFT operation.

Let N_{cp} denote the length of CP, N denote the number of OFDM sub-carriers, and $N_O = N + N_{cp}$ denote the total size of an OFDM block. The matrices for CP insertion and removal are represented by $\mathbf{T}^{cp}$ and $\mathbf{R}^{cp}$, respectively. Let $\nu \leq N_{cp}$ denote the maximum delay spread. The received signals at the legitimate receiver (Bob) and the eavesdropping node (Eve) can be expressed as:

$$\mathbf{y}^B = \mathbf{F} \mathbf{R}^{cp} \tilde{\mathbf{H}} (\mathbf{T}^{cp} \mathbf{F}^* \mathbf{x} + \mathbf{Q} \mathbf{d}^t) + \mathbf{z}^B, \tag{16.4}$$

$$\mathbf{y}^E = \mathbf{F} \mathbf{R}^{cp} \tilde{\mathbf{G}} (\mathbf{T}^{cp} \mathbf{F}^* \mathbf{x} + \mathbf{Q} \mathbf{d}^t) + \mathbf{z}^E, \tag{16.5}$$

where $\mathbf{Q}$ is the temporal-AN precoding matrix, $\mathbf{d}^t \in \mathbb{C}^{N_{cp} \times 1}$ is the temporal-AN vector which is modelled as a zero-mean complex Gaussian random vector, $\tilde{\mathbf{H}} \in \mathbb{C}^{N_O \times N_O}$ is the channel impulse response (CIR) matrix of the Alice–Bob link, $\tilde{\mathbf{G}} \in \mathbb{C}^{N_O \times N_O}$ is CIR matrix of the Alice–Eve link, and $\mathbf{z}^B \in \mathbb{C}^{N \times 1}$, $\mathbf{z}^E \in \mathbb{C}^{N \times 1}$, are the AWGN vectors at Bob and Eve, respectively.

In order to cancel the interference caused by the AN signal at the legitimate receiver, the AN signal should be designed to lie in the null space of the matrix $\mathbf{F} \mathbf{R}^{cp} \tilde{\mathbf{H}}$. Hence, we have the following design condition to cancel the temporal-AN signal at Bob:

$$\mathbf{R}^{cp} \tilde{\mathbf{H}} \mathbf{Q} = \mathbf{0}, \ \mathbf{Q}^* \mathbf{Q} = \mathbf{I}_{N_{cp}}, \tag{16.6}$$

where the condition $\mathbf{Q}^*\mathbf{Q} = \mathbf{I}_{N_{cp}}$ is necessary to ensure the orthonormality of matrix $\mathbf{Q}$ and that the transmit power does increase after precoding the temporal-AN symbols.

We have now described the main principles of the spatial-only and temporal-only schemes. In the following sections, we will explain the adopted system model in this chapter and the proposed hybrid spatio-temporal AN scheme.

16.3 System model and artificial noise design

In this section, we describe the system model adopted in this chapter and explain our proposed AN-aided scheme.

16.3.1 System model and assumptions

The investigated transmission scenario assumes one legitimate transmitter (Alice), one legitimate receiver (Bob), and one passive eavesdropper (Eve). For each OFDM block, Alice transmits over N orthogonal sub-carriers by sending N_s streams per sub-carrier. We assume that the channel matrices remain constant during the coherence time. Alice converts the frequency-domain signals to time-domain signals using an N-point IFFT and adds a CP of N_{cp} samples to the beginning of each OFDM block. We assume that the CP length is longer than the delay spreads of all the channels between Alice and Bob to eliminate inter-block interference at Bob. Moreover, we assume that the delay spreads of the Alice–Eve channels are shorter than or equal to the CP length.[1] To simplify the analysis in Section 16.4, we assume that all channels have the same delay spread, denoted by ν. All the channel coefficients are assumed to be independent and identically distributed (i.i.d.) zero-mean circularly symmetric complex Gaussian random variables with variance σ_{A-B}^2 for the Alice–Bob links and σ_{A-E}^2 for the Alice–Eve links, respectively. The channel coefficients between Alice and Bob are assumed to be known at Alice, Bob, and Eve, and the channel coefficients between Alice and Eve are assumed known to Eve only. The number of antennas at Node $\ell \in \{A, B, E\}$ is N_ℓ. The thermal noise samples at receiver ℓ are modelled as zero-mean complex circularly symmetric Gaussian random variables with variance κ_ℓ Watts/Hz, $\ell \in \{B, E\}$. A description of the key variables and their dimensions is given in Table 16.1.

16.3.2 Proposed hybrid spatio-temporal AN-aided scheme

To enhance the security of her transmissions, Alice exploits both the temporal and spatial dimensions provided by the CP and multiple antennas at both Alice and Bob to generate and transmit spatial and temporal AN symbols, in order to degrade the effective channels between Alice and the passive eavesdropper.

---

[1]This is a best case assumption for Eve; otherwise, her rate will be degraded due to inter-block and intra-block interference.

Table 16.1   List of key variables and their dimensions. © 2016 IEEE. Reference [3]

Symbol	Description	Symbol	Description
N and N_{cp}	# Sub-carriers and CP length	$N_O = N + N_{cp}$	Size of an OFDM block
N_s	# Data streams per a sub-carrier	N_ℓ	# Antennas at Node $\ell \in \{A, B, E\}$
P	Average transmit power budget	κ_ℓ	AWGN variance at Node ℓ
$\Gamma_\ell = P/\kappa_\ell$	Ratio between P and noise variance	$\bar{x}$ and $\tilde{x}$	$1 - x$ and $1 + x$, respectively, when x is scalar
θ	Data power fraction	$1 - \theta$	AN power fraction
α	Spatial-AN power fraction	$1 - \alpha$	Temporal-AN power fraction
σ^2_{A-B}	Variance of Alice–Bob channels	σ^2_{A-E}	Variance of Alice–Eve channels
$\mathbf{y}^\ell \in \mathbb{C}^{N_\ell N \times 1}$	Received signal vector at Node ℓ	$\mathbf{P}_{N_\ell} \in \mathbb{C}^{N_\ell N \times N_\ell N}$	Permutation matrix
$\mathbf{R}^{cp}_{N_\ell} \in \mathbb{C}^{N_\ell N \times N_\ell N_O}$	CP removal matrix at Node ℓ	$\mathbf{T}^{cp}_n \in \mathbb{C}^{N_\ell N_O \times N_\ell N}$	CP insertion matrix
$\tilde{\mathbf{H}} \in \mathbb{C}^{N_B N_O \times N_A N_O}$	CIR matrix of Alice–Bob link	$\mathbf{H} \in \mathbb{C}^{N_B N \times N_A N}$	Frequency-domain matrix of Alice–Bob link
$\tilde{\mathbf{G}} \in \mathbb{C}^{N_E N_O \times N_A N_O}$	CIR matrix of Alice–Eve link	$\mathbf{G} \in \mathbb{C}^{N_E N \times N_A N}$	Frequency-domain matrix of Alice–Eve link
$\mathbf{B} \in \mathbb{C}^{N_A N \times (N_A - N_s)N}$	Overall spatial-AN precoding matrix	$\mathbf{B}_k \in \mathbb{C}^{N_A \times (N_A - N_s)}$	Spatial-AN precoding matrix at sub-carrier k
$\mathbf{Q} \in \mathbb{C}^{N_A N_O \times (N(N_A - N_s) + N_{cp} N_A)}$	Temporal-AN precoding matrix	$\mathbf{A} \in \mathbb{C}^{N_A N \times N_s N}$	Overall data precoding matrix
$\mathbf{A}_k \in \mathbb{C}^{N_A \times N_s}$	Data precoding matrix at sub-carrier k	$\mathbf{C}^*_B \in \mathbb{C}^{N_s N \times N_B N}$	Overall receive filter matrix at Bob
$\mathbf{C}^*_k \in \mathbb{C}^{N_s \times N_B}$	Receive filter matrix of Bob	$\mathbf{z}^\ell$	AWGN vector at Node ℓ

Define the *data symbols* permutation matrix $\mathbf{P}_{N_\ell} \in \mathbb{R}^{N_\ell N \times N_\ell N}$ which re-arranges the pre-coded data symbols transmitted over the multiple antennas into OFDM blocks. Moreover, define the FFT operation at a node with N_ℓ antennas as $\mathbf{F}_{N_\ell} = \mathbf{I}_{N_\ell} \otimes \mathbf{F} \in \mathbb{C}^{N_\ell N \times N_\ell N}$. In addition, let $\mathbf{d}^s = (\mathbf{d}_1^{s\top}, \mathbf{d}_2^{s\top}, \ldots, \mathbf{d}_N^{s\top})^\top$ be the overall spatial-AN vector and $\mathbf{d}_k^s$ be the injected spatial-AN vector over sub-carrier k. Moreover, let $\mathbf{d}^t$ be the temporal-AN vector. The AN vectors $\mathbf{d}^s$ and $\mathbf{d}^t$ are modelled as complex Gaussian random vectors. As explained in Sub-section 16.3.5, the spatial-AN vector is of dimension $N_A - N_s$ and the temporal-AN vector is of dimension $N(N_A - N_s) + N_{cp} N_A$.

Alice transmits the following vector:

$$s_A = T_{N_A}^{cp} F_{N_A}^* P_{N_A} (Ax + Bd^s) + Qd^t, \tag{16.7}$$

where $x = (x_1^\top, x_2^\top, \ldots, x_N^\top)^\top \in \mathbb{C}^{N_s N \times 1}$ is the data vector, $x_k \in \mathbb{C}^{N_s \times 1}$ is the data vector transmitted over sub-carrier k, $P_{N_A} \in \mathbb{R}^{N_A N \times N_A N}$ is a permutation matrix that re-arranges the pre-coded Sub-carriers,[2] $T_{N_A}^{cp} \in \mathbb{C}^{N_A N_O \times N_A N}$ is the CP insertion matrix, $Q \in \mathbb{C}^{N_A N_O \times (N(N_A - N_s) + N_{cp} N_A)}$ is the temporal-AN precoder matrix, and $A \in \mathbb{C}^{N_A N \times N_s N}$ and $B \in \mathbb{C}^{N_A N \times (N_A - N_s)N}$ are the data and spatial-AN precoders, respectively.

The average transmit power constraint is $\mathbb{E}\{(T_{N_A}^{cp} F_{N_A}^* P_{N_A} Bd^s)^* (T_{N_A}^{cp} F_{N_A}^* P_{N_A} Bd^s)\} + \mathbb{E}\{d^{t*} d^t\} + \mathbb{E}\{(T_{N_A}^{cp} F_{N_A}^* P_{N_A} Ax)^* (T_{N_A}^{cp} F_{N_A}^* P_{N_A} Ax)\} = P$, where P is the average transmit power budget at Alice in Watts/Hz. Define θP to be the data transmission power, then we have $\mathbb{E}\{(T_{N_A}^{cp} F_{N_A}^* P_{N_A} Ax)^* (T_{N_A}^{cp} F_{N_A}^* P_{N_A} Ax)\} = \theta P$ and $\mathbb{E}\{(T_{N_A}^{cp} F_{N_A}^* P_{N_A} Bd^s)^* (T_{N_A}^{cp} F_{N_A}^* P_{N_A} Bd^s)\} + \mathbb{E}\{d^{t*} d^t\} = \bar{\theta} P$, where $\bar{\theta} = 1 - \theta$. Letting α be the fraction of $\bar{\theta} P$ allocated to spatial AN, then the power allocated to temporal AN is $\bar{\alpha}\bar{\theta} P$, where $\bar{\alpha} = 1 - \alpha$. We assume that the power of the N_O samples of an OFDM block is distributed equally and, hence, the fraction of power allocated for data transmission over Sub-carrier k is $1/N_O$. Since Eve's instantaneous CSI is unknown at Alice, Alice distributes the power of the spatial and temporal AN uniformly over their precoders' columns. Hence, the power of a spatial AN symbol is $\alpha\bar{\theta} P/(N_O(N_A - N_s))$, while the power of a temporal AN symbol is $\bar{\alpha}\bar{\theta} P/(N(N_A - N_s) + N_{cp} N_A)$.

Our proposed hybrid AN-aided transmission scheme is summarised as follows (see Figure 16.1):

- Alice computes the SVD of the channel matrix at each sub-carrier. Then, she uses the N_s columns of the right singular vectors of the sub-carrier channel matrix corresponding to the N_s largest non-zero singular values to precode the data. The remaining right singular vectors are used to precode the spatial-AN symbols.
- Alice performs the IFFT on the data-plus-spatial AN vector and adds the CP.
- Alice then adds the pre-coded temporal AN to the data-and-spatial-AN time-domain vector.

As explained in the next sub-section, the spatio-temporal AN precoders are designed to ensure that spatial and temporal AN are transparent to Bob.

16.3.3 Received signal vector at Bob

After applying the linear filter $C_B^* \in \mathbb{C}^{N_s N \times N_B N}$, the filtered output vector at Bob is:

$$C_B^* y^B = C_B^* P_{N_B}^\top F_{N_B} R_{N_B}^{cp} \tilde{H} s_A + C_B^* z^B, \tag{16.8}$$

---

[2]This matrix is used because Alice re-arranges the data as a block of N_s pre-coded symbols per sub-carrier. The multiplication of the permutation matrix and the pre-coded data re-arranges the data into OFDM blocks where each OFDM block consists of N sub-carriers. For instance, OFDM block j spans entries $N(j - 1) + 1$ to Nj.

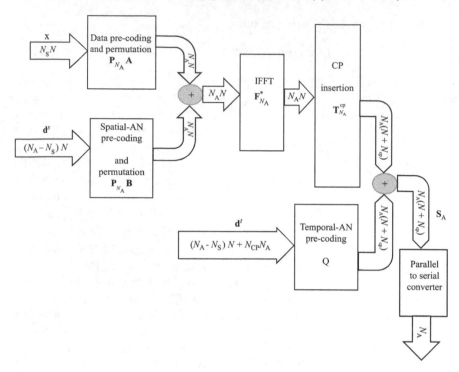

Figure 16.1   Block diagram of our proposed hybrid spatial and temporal AN-aided scheme. © 2016 IEEE. Reference [3]

where $\mathbf{y}^{\mathrm{B}} \in \mathbb{C}^{NN_{\mathrm{B}} \times 1}$ is the received signal vector at Bob, and $\mathbf{P}_{N_{\mathrm{B}}}^{\top}$ is a permutation matrix used by Bob to re-arrange the received sub-carriers. In addition, $\mathbf{F}_{N_{\mathrm{B}}} = \mathrm{blkdiag}(\mathbf{F}, \mathbf{F}, \ldots, \mathbf{F}) \in \mathbb{C}^{N_{\mathrm{B}}N \times N_{\mathrm{B}}N}$ is the FFT matrix, and $\mathbf{R}_{N_{\mathrm{B}}}^{\mathrm{cp}} \in \mathbb{C}^{N_{\mathrm{B}}N \times N_{\mathrm{B}}N_{\mathrm{O}}}$ is the CP removal matrix at Bob. Finally, $\tilde{\mathbf{H}} \in \mathbb{C}^{N_{\mathrm{B}}N_{\mathrm{O}} \times N_{\mathrm{A}}N_{\mathrm{O}}}$ is the CIR matrix of the Alice–Bob link, and $\mathbf{z}^{\mathrm{B}} \in \mathbb{C}^{N_{\mathrm{B}}N \times 1}$ is the AWGN vector at Bob.

The matrix $\mathbf{H} = \mathbf{P}_{N_{\mathrm{B}}}^{\top} \mathbf{F}_{N_{\mathrm{B}}} \mathbf{R}_{N_{\mathrm{B}}}^{\mathrm{cp}} \tilde{\mathbf{H}} \mathbf{T}_{N_{\mathrm{A}}}^{\mathrm{cp}} \mathbf{F}_{N_{\mathrm{A}}}^{*} \mathbf{P}_{N_{\mathrm{A}}}$, whose size is $N_{\mathrm{B}}N \times N_{\mathrm{A}}N$, is block-diagonal, i.e. $\mathbf{H} = \mathrm{blkdiag}(\mathbf{H}_1, \mathbf{H}_2, \ldots, \mathbf{H}_N)$ where $\mathbf{H}_k \in \mathbb{C}^{N_{\mathrm{B}} \times N_{\mathrm{A}}}$ is the frequency-domain channel matrix of the Alice–Bob link at Sub-carrier k. The data precoding matrix of the kth sub-carrier at Alice is denoted by $\mathbf{A}_k \in \mathbb{C}^{N_{\mathrm{A}} \times N_{\mathrm{s}}}$ where $[\mathbf{A}_k]_{i,j}$, for all j, represents the weights multiplied by the data symbol $X_{i,k}$. Hence, the overall data precoding matrix is denoted by $\mathbf{A} = \mathrm{blkdiag}(\mathbf{A}_1, \mathbf{A}_2, \ldots, \mathbf{A}_N) \in \mathbb{C}^{N_{\mathrm{A}}N \times N_{\mathrm{s}}N}$.

The spatial-AN precoder matrix of Sub-carrier k at Alice is $\mathbf{B}_k \in \mathbb{C}^{N_{\mathrm{A}} \times (N_{\mathrm{A}} - N_{\mathrm{s}})}$ where $[\mathbf{B}_k]_{i,j}$, for all j, represents the weights multiplied by the spatial AN symbol $[\mathbf{d}_k^{\mathrm{s}}]_i$. The overall spatial-AN precoding matrix is $\mathbf{B} = \mathrm{blkdiag}(\mathbf{B}_1, \mathbf{B}_2, \ldots, \mathbf{B}_N) \in \mathbb{C}^{N_{\mathrm{A}}N \times (N_{\mathrm{A}} - N_{\mathrm{s}})N}$. Moreover, the receive filter matrix applied at Bob over Sub-carrier k is denoted by $\mathbf{C}_k^{*} \in \mathbb{C}^{N_{\mathrm{s}} \times N_{\mathrm{B}}}$. The overall filtering matrix at Bob, denoted by $\mathbf{C}_{\mathrm{B}}^{*} \in \mathbb{C}^{N_{\mathrm{s}}N \times N_{\mathrm{B}}N}$, is given by $\mathbf{C}_{\mathrm{B}}^{*} = \mathrm{blkdiag}(\mathbf{C}_1^{*}, \mathbf{C}_2^{*}, \ldots, \mathbf{C}_N^{*})$.

16.3.4 Design of Alice's data precoder matrix and Bob's receive filter matrix

The received signal vector at Bob after cancelling the temporal and spatial AN vectors is given by:

$$
\mathbf{C}_\mathrm{B}^* \mathbf{y}^\mathrm{B} = \begin{pmatrix} \mathbf{C}_1^* \mathbf{y}_1^\mathrm{B} \\ \mathbf{C}_2^* \mathbf{y}_2^\mathrm{B} \\ \vdots \\ \mathbf{C}_N^* \mathbf{y}_N^\mathrm{B} \end{pmatrix} = \mathbf{C}_\mathrm{B}^* \mathbf{P}_{N_\mathrm{B}}^\top \mathbf{F}_{N_\mathrm{B}} \mathbf{R}_{N_\mathrm{B}}^{\mathrm{cp}} \tilde{\mathbf{H}} \mathbf{T}_{N_\mathrm{A}}^{\mathrm{cp}} \mathbf{F}_{N_\mathrm{A}}^* \mathbf{P}_{N_\mathrm{A}} \mathbf{A} \mathbf{x} + \mathbf{C}_\mathrm{B}^* \mathbf{z}^\mathrm{B} \tag{16.9}
$$

$$
= \mathbf{C}_\mathrm{B}^* \mathbf{H} \mathbf{A} \mathbf{x} + \mathbf{C}_\mathrm{B}^* \mathbf{z}^\mathrm{B}, \tag{16.10}
$$

where $\mathbf{y}_k^\mathrm{B} \in \mathbb{C}^{N_\mathrm{B} \times 1}$ is the received vector at Sub-carrier k and $\mathbf{C}_k^* \mathbf{y}_k^\mathrm{B} = \mathbf{C}_k^* \mathbf{H}_k \mathbf{A}_k \mathbf{x}_k + \mathbf{C}_k^* \mathbf{z}_k^\mathrm{B}$.

To maximise the SINR per symbol at Bob, both Alice and Bob use the SVD of the per-sub-carrier channel matrix, $\mathbf{H}_k = \mathbf{U}_k \mathbf{\Sigma}_k \mathbf{V}_k^*$, where $\mathbf{\Sigma}_k$ is a diagonal matrix containing the singular values of $\mathbf{H}_k$, the columns of $\mathbf{U}_k$ are the left singular vectors of $\mathbf{H}_k$, and the columns of $\mathbf{V}_k$ are the right singular vectors of $\mathbf{H}_k$. Hence, Alice selects the data precoding matrix $\mathbf{A}_k$ to be the N_s columns of $\mathbf{V}_k$ and $\mathbf{C}_k$ to be the N_s columns of $\mathbf{U}_k$ which correspond to the N_s largest non-zero singular values of $\mathbf{H}_k$.

16.3.5 Design of Alice's temporal and spatial AN precoders

Our aim is to cancel the spatial and temporal AN at Bob. From (16.8), the condition to cancel the spatial-AN vector at Bob is:

$$
\mathbf{C}_\mathrm{B}^* \mathbf{H} \mathbf{B} = 0 \Leftrightarrow \mathbf{C}_k^* \mathbf{H}_k \mathbf{B}_k = 0, \quad \forall k \in \{1, \ldots, N\}. \tag{16.11}
$$

Given $\mathbf{C}_k^*$, Alice designs the spatial-AN precoding matrix $\mathbf{B}_k$ to lie in the null space of $\mathbf{C}_k^* \mathbf{H}_k \in \mathbb{C}^{N_\mathrm{s} \times N_\mathrm{A}}$ which is non-trivial only if $N_\mathrm{A} > N_\mathrm{s}$. If this condition is not satisfied (i.e. $N_\mathrm{A} = N_\mathrm{s}$), Alice will not be able to inject any spatial AN.

The spatial-AN precoding matrix is designed to make the AN lie in the subspace spanned by the remaining vectors of $\mathbf{V}_k$. That is, Alice combines the remaining $N_\mathrm{A} - N_\mathrm{s}$ columns of $\mathbf{V}_k$ using Gaussian random variables each with zero mean and variance $\alpha \bar{\theta} P / (N_\mathrm{O}(N_\mathrm{A} - N_\mathrm{s}))$.

After designing the data and spatial AN linear precoders and receive filters, Alice designs the temporal-AN precoder matrix $\mathbf{Q}$, based on the knowledge of the receive filter at Bob, according to the following condition:

$$
\mathbf{C}_\mathrm{B}^* \mathbf{P}_{N_\mathrm{B}}^\top \mathbf{F}_{N_\mathrm{B}} \mathbf{R}_{N_\mathrm{B}}^{\mathrm{cp}} \tilde{\mathbf{H}} \mathbf{Q} = 0, \tag{16.12}
$$

where $\mathbf{Q}$ lies in the null space of $\mathbf{C}_\mathrm{B}^* \mathbf{P}_{N_\mathrm{B}}^\top \mathbf{F}_{N_\mathrm{B}} \mathbf{R}_{N_\mathrm{B}}^{\mathrm{cp}} \tilde{\mathbf{H}}$ which has a dimension of $N_\mathrm{s} N \times N_\mathrm{A} N_\mathrm{O}$. The condition to cancel the temporal-AN vector at Bob is:

$$
N_\mathrm{A} N_\mathrm{O} > N_\mathrm{s} N \Rightarrow \left(1 + \frac{N_\mathrm{cp}}{N} \right) > \frac{N_\mathrm{s}}{N_\mathrm{A}}. \tag{16.13}
$$

This condition is always satisfied since $N_\mathrm{A} \geq N_\mathrm{s}$.

16.3.6 Received signal vector at Eve

Since Eve has N_E receive antennas, her received signal vector is given by:

$$\mathbf{y}^E = \mathbf{P}_{N_E}^\top \mathbf{F}_{N_E} \mathbf{R}_{N_E}^{cp} \tilde{\mathbf{G}} \mathbf{s}_A + \mathbf{z}^E, \tag{16.14}$$

where $\mathbf{y}^E \in \mathbb{C}^{N_E N \times 1}$ is the received signal vector at Eve, $\mathbf{F}_{N_E} \in \mathbb{C}^{N_E N \times N_E N}$ is the FFT operation performed at Eve, $\mathbf{R}_{N_E}^{cp} \in \mathbb{C}^{N_E N \times N_E N_O}$ is the CP removal matrix used at Eve, $\tilde{\mathbf{G}} \in \mathbb{C}^{N_E N_O \times N_A N_O}$ is the CIR matrix of the Alice–Eve link, and $\mathbf{z}^E \in \mathbb{C}^{N_E N \times 1}$ is the AWGN vector at Eve's receiver.

The per-sub-carrier received signal vector at Eve, denoted by $\mathbf{y}_k^E \in \mathbb{C}^{N_E \times 1}$, is:

$$\mathbf{y}_k^E = \mathbf{G}_k \left(\mathbf{A}_k \mathbf{x}_k + \mathbf{B}_k \mathbf{d}_k^s \right) + \mathbf{E}_k \mathbf{d}^t + \mathbf{z}_k^E, \tag{16.15}$$

where $\mathbf{G}_k \in \mathbb{C}^{N_B \times N_A}$ is the frequency-domain channel matrix at sub-carrier k of the Alice–Eve link, $\mathbf{z}_k^E$ are rows $(k-1)N_E + 1$ to kN_E of the AWGN vector $\mathbf{z}^E$, and $\mathbf{E}_k$ consists of the N_E rows from $(k-1)N_E + 1$ to kN_E of matrix $\mathbf{E} = \mathbf{P}_{N_E}^\top \mathbf{F}_{N_E} \mathbf{R}_{N_E}^{cp} \tilde{\mathbf{G}} \mathbf{Q}$. We define the frequency-domain block-diagonal matrix of the Alice–Eve link as $\mathbf{G} = \mathbf{P}_{N_E}^\top \mathbf{F}_{N_E} \mathbf{R}_{N_E}^{cp} \tilde{\mathbf{G}} \mathbf{T}_{N_A}^{cp} \mathbf{F}_{N_A}^* \mathbf{P}_{N_A} = \text{blkdiag}(\mathbf{G}_1, \mathbf{G}_2, \ldots, \mathbf{G}_N)$.

16.4 Average secrecy rate

Our goal in this section is to derive closed-form expressions for the average secrecy rate. Assume that the received signal vector at Node ℓ_2 due to a transmission from Node ℓ_1, where $\ell_1, \ell_2 \in \{A, B, E\}$, is $\mathbf{r} + \mathbf{j} + \mathbf{z}^{\ell_2}$, where $\mathbf{r}$ is the received data vector, $\mathbf{z}^{\ell_2}$ is the AWGN vector at Node ℓ_2, and $\mathbf{j}$ is a Gaussian interference vector. The instantaneous rate of the $\ell_1 - \ell_2$ link is [12,19]:

$$R_{\ell_1 - \ell_2} = \log_2 \det \left(\mathbb{E}\{\mathbf{r}\mathbf{r}^*\} \left[\mathbb{E}\{ (\mathbf{j} + \mathbf{z}^{\ell_2})(\mathbf{j} + \mathbf{z}^{\ell_2})^* \} \right]^{-1} + \mathbf{I}_{N_{\ell_2}} \right), \tag{16.16}$$

where $\mathbb{E}\{\mathbf{r}\mathbf{r}^*\}$ is the data covariance matrix, and $\mathbb{E}\{ (\mathbf{j} + \mathbf{z}^{\ell_2})(\mathbf{j} + \mathbf{z}^{\ell_2})^* \}$ is the noise-plus-interference covariance matrix.

Define the input SNRs of the Alice–Bob and Alice–Eve links as $\Gamma_B = P/\kappa_B$ and $\Gamma_E = P/\kappa_E$, respectively. Assuming that the transmit power is allocated equally among the N_s independent data symbols, and using (16.9) and (16.16), the rate of the Alice–Bob link, denoted by R_{A-B}, is given by:

$$R_{A-B} = \log_2 \det \left(\theta \Gamma_B \mathbf{C}_B^* \mathbf{H} \mathbf{A} \Sigma_\mathbf{x} \left(\mathbf{C}_B^* \mathbf{H} \mathbf{A} \right)^* + \mathbf{I}_{N_s N} \right) \tag{16.17}$$

$$= \sum_{k=1}^N \log_2 \det \left(\theta \Gamma_B \mathbf{C}_k^* \mathbf{H}_k \mathbf{A}_k \Sigma_{\mathbf{x}_k} \left(\mathbf{C}_k^* \mathbf{H}_k \mathbf{A}_k \right)^* + \mathbf{I}_{N_s} \right), \tag{16.18}$$

where $\theta P \Sigma_\mathbf{x}$ is the data covariance matrix with $\Sigma_\mathbf{x} = \frac{1}{N_s N_O} \mathbf{I}_{N_s N}$ and $\Sigma_{\mathbf{x}_k} = \frac{1}{N_s N_O} \mathbf{I}_{N_s}$.

We consider and compare two decoding approaches at Eve. In the first approach, she decodes the signals on all the sub-carriers jointly by taking into account the correlation among the sub-carriers due to the temporal AN. In the second approach, Eve decodes the signals at each sub-carrier separately assuming that each sub-carrier is a MIMO channel. If Eve performs joint processing across all of her sub-carriers, using (16.14), the rate of the Alice–Eve link, denoted by $R_{\text{A--E}}$, is given by:[3]

$$R_{\text{A--E}} = \log_2 \det\left(\theta\Gamma_{\text{E}}\mathbf{GA}\mathbf{\Sigma}_{\mathbf{x}}(\mathbf{GA})^*\left(\mathbf{\Sigma}_{\text{AN}} + \mathbf{I}_{N_E N}\right)^{-1} + \mathbf{I}_{N_E N}\right), \tag{16.19}$$

where

$$\mathbf{\Sigma}_{\text{AN}} = \overline{\theta}\Gamma_{\text{E}}\left(\frac{\alpha\mathbf{GBB}^*\mathbf{G}^*}{N_0(N_A - N_s)} + \frac{\overline{\alpha}\mathbf{EE}^*}{N(N_A - N_s) + N_{\text{cp}}N_A}\right). \tag{16.20}$$

The eavesdropper's rate expression in (16.19) assumes that Eve knows the channels between Alice and Bob and, hence, she knows the used data and AN precoders by Alice. This assumption represents a best case scenario for Eve.

Lemma 16.1. *When* $1 - \frac{N_s}{N_A} \gg \frac{N_{\text{cp}}}{N}$, *Eve's rate is independent of* α.

Proof. See Appendix A.1.1. □

If Eve performs per-sub-carrier processing due to its reduced complexity, her rate is given by

$$R_{\text{A--E}} = \sum_{k=1}^{N}\left[\log_2 \det\left(\theta\Gamma_{\text{E}}\mathbf{G}_k\mathbf{A}_k\mathbf{\Sigma}_{\mathbf{x}_k}(\mathbf{G}_k\mathbf{A}_k)^*\left(\mathbf{\Sigma}_{\text{AN},k} + \mathbf{I}_{N_E}\right)^{-1} + \mathbf{I}_{N_E}\right)\right], \tag{16.21}$$

where

$$\mathbf{\Sigma}_{\text{AN},k} = \overline{\theta}\Gamma_{\text{E}}\left(\frac{\alpha}{N_0(N_A - N_s)}\mathbf{G}_k\mathbf{B}_k\mathbf{B}_k^*\mathbf{G}_k^* + \frac{\overline{\alpha}}{N(N_A - N_s) + N_{\text{cp}}N_A}\mathbf{E}_k\mathbf{E}_k^*\right). \tag{16.22}$$

The secrecy rate of the legitimate system is given by [7,8,11,19]:

$$R_{\text{sec}} = (R_{\text{A--B}} - R_{\text{A--E}})^+, \tag{16.23}$$

where $(\cdot)^+ = \max(\cdot, 0)$. The average secrecy rate of the legitimate system is approximately given by [7,8,11,19]:

$$\mathbb{E}\{R_{\text{sec}}\} = \mathbb{E}\{R_{\text{A--B}}\} - \mathbb{E}\{R_{\text{A--E}}\}. \tag{16.24}$$

The average secrecy rate loss due to eavesdropping, which represents the reduction in the average secrecy rate due to the presence of an eavesdropper (i.e. the difference between the achievable rate of the Alice–Bob link without eavesdropping and the secrecy rate of the Alice–Bob link), denoted by $\mathscr{S}_{\text{Loss}}$, is given by

$$\mathscr{S}_{\text{Loss}} = R_{\text{A--B,noEve}} - R_{\text{sec}}, \tag{16.25}$$

where $R_{\text{A--B,noEve}}$ is the achievable rate at Bob when there is no Eve.

The following relation is useful in our secrecy rate derivations:

$$\mathbf{G}_k\mathbf{G}_k^* = \mathbf{G}_k\mathbf{A}_k(\mathbf{G}_k\mathbf{A}_k)^* + \mathbf{G}_k\mathbf{B}_k(\mathbf{G}_k\mathbf{B}_k)^*, \tag{16.26}$$

---

[3]This expression represents the best rate for Eve which will be reduced if Eve applies a linear filter such as MMSE or zero-forcing filters as in [14].

where the columns of $\mathbf{A}_k$ are the N_s right singular vectors corresponding to the largest non-zero singular values of $\mathbf{H}_k$ and the columns of $\mathbf{B}_k$ are the remaining $N_A - N_s$ right singular vectors of $\mathbf{H}_k$. Equation (16.26) can be easily obtained by first noting that $\mathbf{A}_k\mathbf{A}_k^* + \mathbf{B}_k\mathbf{B}_k^* = \mathbf{I}_{N_A}$, since the columns of $\mathbf{A}_k$ and $\mathbf{B}_k$ form orthonormal bases. Then, (16.26) is obtained by left and right multiplication by $\mathbf{G}_k$ and $\mathbf{G}_k^*$, respectively.

In the following sub-section, we investigate the asymptotic average secrecy rate of MIMOME-OFDM systems as $N_A \to \infty$. We derive a lower bound on the average secrecy rate as well as the average secrecy rate loss, which represents the reduction in the average secrecy rate due to the presence of an eavesdropper. Moreover, we carry out the asymptotic analysis at low and high SNRs with a sufficiently large N_A.

16.4.1 Asymptotic average rates in MIMOME-OFDM channels

In this sub-section, we investigate the asymptotic case of $N_A \to \infty$. Wireless base stations with large numbers of transmit antennas, also known as massive MIMO systems, have attracted considerable research attention recently, e.g. see [4,6] and the references therein. It is important to point out that massive MIMO is a key-enabling technology for emerging 5G communication networks.

For MIMOME-OFDM systems, the rate of the Alice–Eve link, when N_A is very large, is given by:

$$R_{A-E} \approx \log_2 \det(\theta \Gamma_E \mathbf{GA}\Sigma_x(\mathbf{GA})^*(\Sigma_{AN} + \mathbf{I}_{N_EN})^{-1} + \mathbf{I}_{N_EN}), \tag{16.27}$$

where

$$\Sigma_{AN} \approx \overline{\theta}\Gamma_E\left(\frac{\alpha}{N_O(N_A - N_s)}\mathbf{GG}^* + \frac{\overline{\alpha}}{N(N_A - N_s) + N_{cp}N_A}\mathbf{EE}^*\right). \tag{16.28}$$

Note that we used the fact that $\mathbf{BB}^* \approx \mathbf{I}_{N(N_A-N_s)}$ and $\mathbf{QQ}^* \approx \mathbf{I}_{N_AN_O}$ as $N_A \to \infty$. To analyse the average secrecy rate, we need the statistics of $\mathbf{P}_{N_E}^{\top}\mathbf{F}_{N_E}\mathbf{R}_{N_E}^{cp}\tilde{\mathbf{G}}$, $\mathbf{G}_k\mathbf{A}_k\mathbf{A}_k^*\mathbf{G}_k^*$, and $\mathbf{H}_k\mathbf{A}_k\mathbf{A}_k^*\mathbf{H}_k^*$ which are derived in Appendices A.1.2 and A.1.3.

Lemma 16.2. *As $N_A \to \infty$ and when $1 - \frac{N_s}{N_A} \gg \frac{N_{cp}}{N}$, Eve's rate is independent of α and is given by*

$$R_{A-E} \approx \log_2 \det\left(\theta\Gamma_E\mathbf{GA}\Sigma_x(\mathbf{GA})^*\left(\frac{\overline{\theta}\Gamma_E}{N(N_A - N_s)}\mathbf{GG}^* + \mathbf{I}_{N_EN}\right)^{-1} + \mathbf{I}_{N_EN}\right). \tag{16.29}$$

In addition, the secrecy rate is given by

$$R_{sec} \approx \sum_{k=1}^{N} \log_2 \det\left(\theta\Gamma_B\mathbf{C}_k^*\mathbf{H}_k\mathbf{A}_k\Sigma_{x_k}\left(\mathbf{C}_k^*\mathbf{H}_k\mathbf{A}_k\right)^* + \mathbf{I}_{N_s}\right) \tag{16.30}$$

$$- \log_2 \det\left(\theta\Gamma_E\mathbf{GA}\Sigma_x(\mathbf{GA})^*\left(\frac{\overline{\theta}\Gamma_E}{N(N_A - N_s)}\mathbf{GG}^* + \mathbf{I}_{N_EN}\right)^{-1} + \mathbf{I}_{N_EN}\right).$$

Proof. See Appendix A.1.4.   □

Lemma 16.3. *As $N_A \to \infty$ and when $1 - \frac{N_s}{N_A} \gg \frac{N_{cp}}{N}$, per-sub-carrier processing at Eve achieves the same rate as that achieved with joint sub-carrier processing. Hence, the average secrecy rate is given by*

$$R_{\text{sec}} \approx \sum_{k=1}^{N} \left(\log_2 \det(\theta \Gamma_B \mathbf{C}_k^* \mathbf{H}_k \mathbf{A}_k \mathbf{\Sigma}_{\mathbf{x}_k} (\mathbf{C}_k^* \mathbf{H}_k \mathbf{A}_k)^* + \mathbf{I}_{N_s}) \right. \tag{16.31}$$

$$\left. - \log_2 \det \left(\theta \Gamma_E \mathbf{G}_k \mathbf{A}_k \mathbf{\Sigma}_{\mathbf{x}_k} (\mathbf{G}_k \mathbf{A}_k)^* \left(\frac{\overline{\theta}\Gamma_E}{N(N_A - N_s)} \mathbf{G}_k \mathbf{G}_k^* + \mathbf{I}_{N_E} \right)^{-1} + \mathbf{I}_{N_E} \right) \right).$$

Proof. See Appendix A.1.5.   □

Lemma 16.4. *As $N_A \to \infty$, the average secrecy rate for the MIMOME-OFDM system in bits/sec/Hz is lower bounded as follows:*

$$\mathbb{E}\{R_{\text{sec}}\} \gtrsim \frac{N_s N \log_2\left(\frac{\theta \Gamma_B}{N_s N_O} N_A \tilde{\nu}\sigma_{A-B}^2 + 1\right) - N_E N \log_2\left(\frac{\frac{\theta \Gamma_E}{N_O}\tilde{\nu}\sigma_{A-E}^2}{\frac{\overline{\theta}\Gamma_E}{N_O}\tilde{\nu}\sigma_{A-E}^2 + 1} + 1\right)}{N_O}, \tag{16.32}$$

where $\tilde{\nu} = \nu + 1$.

Proof. See Appendix A.1.6.   □

The average secrecy rate in (16.32) is independent of α, which implies that α can take any value from 0 to 1 without reducing the average secrecy rate.

Remark 16.1. *The lower bound in (16.32) holds only when N_A is sufficiently large so that our use of the law of large numbers is accurate. When N_A is not sufficiently large, the bound in (16.32) should be considered an approximation.*

Lemma 16.5. *As $N_A \to \infty$, Eve's average rate when she performs per-sub-carrier processing is upper bounded as follows:*

$$\mathbb{E}\{R_{A-E,k}\} \lesssim N_E \log_2 \left(\frac{\frac{\theta \Gamma_E}{N_O}\tilde{\nu}\sigma_{A-E}^2}{\frac{\overline{\theta}\Gamma_E}{N_O}\tilde{\nu}\sigma_{A-E}^2 + 1} + 1 \right). \tag{16.33}$$

In addition, the average secrecy rate of the legitimate system is lower bounded as follows:

$$\mathbb{E}\{R_{\text{sec}}\} \gtrsim \frac{N_s N \log_2\left(\frac{\theta \Gamma_B}{N_s N_O} N_A \tilde{\nu}\sigma_{A-B}^2 + 1\right) - N_E N \log_2\left(\frac{\frac{\theta \Gamma_E}{N_O}\tilde{\nu}\sigma_{A-E}^2}{\frac{\overline{\theta}\Gamma_E}{N_O}\tilde{\nu}\sigma_{A-E}^2 + 1} + 1\right)}{N_O}. \tag{16.34}$$

Proof. Following the same steps as those used to prove Lemma 16.4, we can prove Lemma 16.5.   □

Lemma 16.6. *When $N_A \to \infty$ and for high $\Gamma_B \sigma_{A-B}^2$ and $\Gamma_E \sigma_{A-E}^2$ with $\Gamma_B \sigma_{A-B}^2 \geq \Gamma_E \sigma_{A-E}^2$, the average secrecy rate for MIMOME-OFDM systems can be lower bounded as follows:*

$$\mathbb{E}\{R_{sec}\} \gtrsim \frac{N_s N \log_2\left(\frac{\Gamma_B}{N_s N_O} N_A \tilde{v} \sigma_{A-B}^2\right) + N_E N \log_2\left(\left(\frac{N_E}{N_E+N_s}\right)^{\frac{N_s}{N_E}} \left(\frac{N_s}{N_E+N_s}\right)\right)}{N_O}, \quad (16.35)$$

when Alice uses the power allocation

$$\theta = \theta^\star = \frac{N_E}{N_E + N_s}. \quad (16.36)$$

Proof. See Appendix A.1.7. ☐

Lemma 16.6 suggests that when the number of transmit antennas at Alice, N_A, is large and the SNR is high, the optimal power allocated for data transmission to maximise the average secrecy rate is a function of the number of antennas at Eve and the number of data streams at Alice.

Corollary 16.1. *As $N_A \to \infty$ and for high $\Gamma_B \sigma_{A-B}^2$ and $\Gamma_E \sigma_{A-E}^2$ with $\Gamma_B \sigma_{A-B}^2 \geq \Gamma_E \sigma_{A-E}^2$, the average rate loss due to eavesdropping in MIMOME-OFDM systems can be upper bounded by*

$$\mathbb{E}\{\mathscr{S}_{Loss}\} \lesssim \frac{N_s N}{N_O} \log_2\left(\frac{N_E + N_s}{N_E}\right) + \frac{N_E N}{N_O} \log_2\left(\frac{N_E + N_s}{N_s}\right), \quad (16.37)$$

when Alice uses the power allocation

$$\theta = \theta^\star = \frac{N_E}{N_E + N_s}. \quad (16.38)$$

Proof. See Appendix A.1.7. ☐

Remark 16.2. *From (16.36), in the high-SNR regime, the optimal power allocation policy between the data and the AN depends only on the number of receive antennas at Eve and the number of transmitted data streams at Alice. Hence, Alice can optimise θ which maximises her average secrecy rate without the need to know the CSI of Eve. If $N_E = N_s$, the optimal power allocation policy is to set θ to $1/2$. This implies that the optimal power-allocation policy for Alice is to split her transmit power equally between the data symbols and the AN symbols. If $N_E \gg N_s$, $\theta^\star = 1$ which means that all of Alice's transmit power should be assigned for data transmission to maximise the average secrecy rate, and there is no need to waste power on transmitting AN.*

Remark 16.3. *If $N_E \gg N_s$, the first term of the right hand side in (16.37) is approximately zero. Hence,*

$$\mathbb{E}\{\mathscr{S}_{Loss}\} \lesssim \frac{N_E N}{N_O} \log_2\left(\frac{N_E + N_s}{N_s}\right) \approx \frac{N_E N}{N_O} \log_2\left(\frac{N_E}{N_s}\right). \quad (16.39)$$

Remark 16.4. *For $N_E = N_s$, $N_A \to \infty$ and for high $\Gamma_B \sigma_{A-B}^2 \geq \Gamma_E \sigma_{A-E}^2$ and high $\Gamma_E \sigma_{A-E}^2$, the upper-bound on the average rate loss due to eavesdropping in MIMOME-OFDM systems is given by:*

$$\mathbb{E}\{\mathscr{S}_{\text{Loss}}\} \lesssim \frac{2 N_E N}{N_O} \tag{16.40}$$

which is obtained by substituting with $N_s = N_E$ into (16.37). □

Lemma 16.7. *As $N_A \to \infty$, $\Gamma_B \sigma_{A-B}^2 \geq \Gamma_E \sigma_{A-E}^2$, and for low $\Gamma_E \sigma_{A-E}^2$, the average secrecy rate loss due to eavesdropping in MIMOME-OFDM systems is upper bounded as follows:*

$$\mathbb{E}\{\mathscr{S}_{\text{Loss}}\} \lesssim \frac{N_E N}{N_O} \log_2\left(\frac{\Gamma_E}{N_O} \tilde{v} \sigma_{A-E}^2 + 1\right), \tag{16.41}$$

when Alice uses the power allocation policy $\theta = \theta^\star = 1$.

Proof. See Appendix A.1.8. □

Lemma 16.7 suggests that when the number of transmit antennas at Alice is large and the SNR is low, AN injection slightly affects the average secrecy rate. Thus, Alice should allocate all of her power to data transmission (i.e. $\theta = 1$).

Remark 16.5. *From (A.49) in Appendix A.1.8, when $N_E = N_s = N_B$ and for low input SNR case, the average secrecy rate is lower bounded as follows:*

$$\mathbb{E}\{R_{\text{sec}}\} \gtrsim \frac{N_s N}{N_O}\left(\log_2\left(\frac{\Gamma_B}{N_s N_O} N_A \tilde{v} \sigma_{A-B}^2 + 1\right) - \log_2\left(\frac{\Gamma_E}{N_O} \tilde{v} \sigma_{A-E}^2 + 1\right)\right). \tag{16.42}$$

Since $N_A > N_s$ and $N_A \to \infty$, we have

$$\log_2\left(\frac{\Gamma_B}{N_s N_O} N_A \tilde{v} \sigma_{A-B}^2 + 1\right) \gg \log_2\left(\frac{\Gamma_E}{N_O} \tilde{v} \sigma_{A-E}^2 + 1\right). \tag{16.43}$$

Consequently, (16.42) is approximately given by

$$\mathbb{E}\{R_{\text{sec}}\} \gtrsim \frac{N_s N}{N_O} \log_2\left(\frac{\Gamma_B}{N_s N_O} N_A \tilde{v} \sigma_{A-B}^2 + 1\right) = \mathbb{E}\{R_{A-B,\text{noEve}}\}. \tag{16.44}$$

From (16.44) and given that $\mathbb{E}\{R_{A-B,\text{noEve}}\} \geq \mathbb{E}\{R_{\text{sec}}\}$, we conclude that at low input power levels and for a large number of transmit antennas at Alice, the average secrecy rate loss is negligible.

16.4.2 Temporal AN versus spatial AN

We conclude this section by comparing the pros and cons of spatial and temporal AN for MIMOME-OFDM systems.

- Since Alice encodes the data on each sub-carrier individually, and the received signals at different sub-carriers at Bob are independent, Bob can perform per-sub-carrier processing of the received signals without loss of optimality. On the other hand, due to the presence of the temporal AN, which couples the signals at different sub-carriers (i.e. results in a correlated noise vector at Eve's receiver), Eve's optimal detection strategy should be based on joint processing on the received sub-carriers. In our numerical results, Eve's rate loss due to per-sub-carrier processing is not significant for the considered scenarios, as shown in Figure 16.3. However, for the SISO case, the per-sub-carrier processing can be much worse than the joint sub-carrier processing and may lead to a significant rate loss for Eve. This demonstrates the benefits of injecting the temporal AN to increase the decoding complexity at Eve.

- When $N_A = N_s$, the dimension of the AN vector is zero. Hence, spatial AN cannot be applied, and Alice can only apply temporal AN to confuse Eve.
- The implementation complexity of spatial AN is lower than that of temporal AN. More specifically, the spatial AN precoder is implemented using the SVD of the N per-sub-carrier matrices where the dimension of each sub-carrier matrix is $N_B \times N_A$. Hence, the complexity of computing the overall spatial-AN precoder matrix, **B**, is linear in N and cubic in N_B and N_A. On the other hand, the temporal-AN design requires the computation of the null space of a large matrix whose size is $NN_s \times N_0N_A$, which is cubic in N.
- Temporal AN injection increases the encoding complexity at Alice as well as the decoding complexity at Eve. On the other hand, spatial AN injection decreases complexity at both nodes. However, since Eve does not know exactly which AN Alice is injecting in her transmissions, Alice might use the spatial-AN scheme while Eve performs joint sub-carrier processing.
- Given that both schemes provide comparable average secrecy rates as proved in Proposition 16.1 and demonstrated numerically in Figure 16.4, Alice can inject the spatial-only AN to reduce the system design complexity. However, spatial AN *cannot* be implemented when $N_A = N_s$. We emphasise here that the proposed null space computation technique in Appendix A of [3], which is used to compute the temporal-AN precoding matrix, reduces the null space computation complexity and can increase the randomness at Eve since Alice can use random matrices in her design. Thus, we conjecture that temporal AN injection can further increase the secrecy rate using a randomised temporal-AN precoder matrix.

16.5 Simulation results

In this section, we evaluate the average secrecy rate of our proposed hybrid spatio-temporal AN-aided scheme. We use the following common parameters: $v = N_{cp} = 16$, $N = 64$, $\Gamma_B = \Gamma_E = 20$ dB, and $\sigma^2_{A-B} = \sigma^2_{A-E} = 1$. Figure 16.2 shows the impact of the number of antennas at Eve on the average secrecy rate. The parameters used to generate the figure are the common parameters and $\theta = \alpha = 0.5$. As N_E increases, Eve can better mitigate the impact of the AN and, hence, the average secrecy rate decreases. When N_E is large enough, Eve can always decode Alice's data reliably and, hence, the secrecy rate is *zero*. This figure also quantifies the impact of N_A on the average secrecy rate. For example, when $N_E = 4$, the average secrecy rate is increased from 0 to 2 bits/sec/Hz as N_A increases from 2 to 4.

Figure 16.3 quantifies two important aspects: (1) the impact of the power fraction assigned to information-bearing signal, θ, on the average secrecy rate; (2) the accuracy

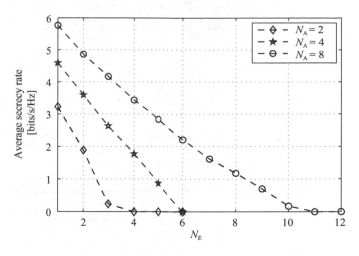

Figure 16.2 Average secrecy rate in bits/sec/Hz versus N_E for different values of N_A. © 2016 IEEE. Reference [3]

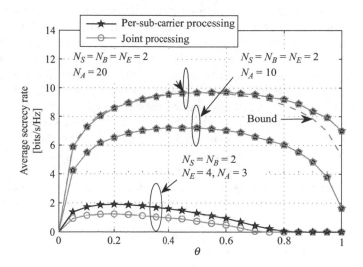

Figure 16.3 Average secrecy rate in bits/sec/Hz versus θ for different antenna configurations. © 2016 IEEE. Reference [3]

of our proposed approximations and bounds. The parameters used to generate the figure are the common parameters and $\alpha = 0.5$. As the power fraction increases, the average secrecy rate increases until it achieves a maximum, then it decreases until it achieves the zero secrecy rate since there is no sufficient power to transmit information-bearing signal. If Eve's instantaneous CSI is known at Alice, she can optimise over θ for each OFDM block to maximise the instantaneous secrecy rate. If the statistics of the Alice–Eve links are known at Alice, she can optimise θ to

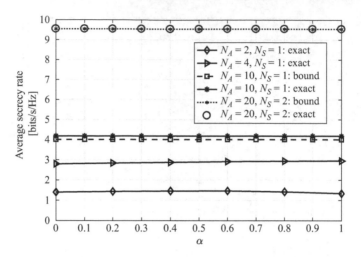

Figure 16.4   Average secrecy rate in bits/sec/Hz versus α. [3] © 2016 IEEE

maximise the average secrecy rate. However, if Alice does not have any information about Eve's channels, she chooses a value for θ and operates under this assumption.

When N_A is very large, as shown in Section 16.4.1, the optimal θ depends on the number of transmit data streams at Alice and the number of receive antennas at Eve. Hence, Alice just needs to know the number of receive antennas at Eve to adjust θ such that the average secrecy rate is maximised. For $N_s = N_B = N_E = 2$ and $N_A = 10$, per-sub-carrier processing achieves the same average secrecy rate as that of joint sub-carrier processing. This means that Eve can reduce the decoding complexity at her receiver by employing per-sub-carrier processing. On the other hand, for the scenario of $N_s = N_B = 2$, $N_E = 4$, and $N_A = 3$, joint processing increases Eve's rate and, hence, decreases the secrecy rate of the legitimate system. We emphasise here that if Alice does not know the statistics of the Alice–Eve links, she cannot optimise θ to maximise her own average secrecy rate. However, a good approximation to the optimal θ is given by (A.40), e.g. $\theta^\star = N_E/(N_E + N_s) = 1/2$ is optimal for $N_A = 10$ and $N_A = 20$. Eve can still decide to perform per-sub-carrier processing even if she achieves a lower rate to reduce the decoding complexity since the average secrecy rate loss is not high in many scenarios. Furthermore, since we assume the worst case scenario for Alice/Bob in which Eve knows all channels and precoding matrices, Eve can choose the best decoding strategy for her. When $N_A = 20$, our proposed lower bound on the average secrecy rate is tight for low θ levels. When θ is near unity, our proposed lower bound is close to the exact average secrecy rate. Figure 16.3 reveals that if Alice allocates all her power to the data (i.e. $\theta = 1$), the average secrecy rate may approach zero as in the scenario of $N_s = N_B = 2$, $N_E = 4$, and $N_A = 3$. This demonstrates the benefits of AN injection.

Figure 16.4 demonstrates the impact of the transmit power distribution between the spatial and temporal AN on the average secrecy rate. This figure is generated using the common parameters, $\theta = 0.5$, $N_B = N_s$, and $N_E = 2$. We plot the average secrecy rate by varying α, which is the fraction of the AN power allocated to the spatial AN. This figure also shows the increase in the average secrecy rate with increasing the numbers of transmit antennas at Alice and receive antennas at Bob. The impact of α is relatively small and the secrecy rate versus α is almost flat, which corroborates our discussions in the previous sections. The benefit of increasing the number of transmit antennas is evident since the average secrecy rate is almost doubled when N_A is doubled. The figure also demonstrates the tightness of our proposed lower bound. The bound completely coincides with the exact value when $N_A = 20$ and $N_s = 2$. For $N_A = 10$ and $N_s = 2$, the bound is very close to the exact value and the gap is less than 5%.

16.6   Conclusions

In this chapter, we proposed a hybrid spatio-temporal AN-aided scheme to secure a MIMOME-OFDM system in the presence of a passive eavesdropper. We assumed that Eve's instantaneous CSI is unknown at the legitimate system. We analysed the nodes' rates, system's secrecy rate, and system's average secrecy rate. In addition, we derived lower bounds on the secrecy rate of the MIMOME-OFDM systems when the number of transmit antennas at Alice, N_A, is very large. When N_A is sufficiently large, closed-form solutions for the optimal power allocation between the AN and data and between the temporal and spatial AN were obtained from these bounds at high and low average SNRs.

In the low SNR regime, Alice should not allocate power to AN and, hence, all Alice's transmit power is allocated to data transmission. The corresponding average secrecy rate loss due to the presence of an eavesdropper is negligible. In the high SNR regime, Alice should allocate $\theta^\star = N_E/(N_E + N_s)$ of her total power for information transmission and the remaining power for AN transmission. If $N_E = N_s$, in the high average SNR regime, Alice should allocate *half* of her total power to AN transmission and the other *half* to data transmission. The corresponding average secrecy rate loss due to the presence of an eavesdropper is $2/N_O$ bits/sec/Hz, which is linearly decreasing with the OFDM block size N. In addition, we numerically demonstrated that this power allocation solution, $\theta^\star = N_E/(N_E + N_s)$, is also a good approximation for the case with a small N_A and can be used without knowledge of the Alice–Eve links CSI at Alice. Furthermore, we showed that the average secrecy rate is independent of α for $N_{cp}/N \ll 1 - N_s/N_A$. In addition, for the considered numerical scenario, we showed that Eve can perform per-sub-carrier decoding instead of joint decoding without significant reduction in her rate. However, for the SISO case, the per-sub-carrier processing can significantly degrade Eve's rate.

A.1   Appendices

A.1.1   Proof of Lemma 16.1

In this appendix, we prove that Eve's rate is independent of $0 \le \alpha \le 1$ by showing that when $N_{cp}/N \ll 1 - N_s/N_A$, the eigenvalues of the interference-plus-noise covariance matrix, denoted by $\boldsymbol{\Sigma}_{\mathrm{AN}}$ in (16.20), do not depend on α. Starting from (16.11) and (16.12), we define $\mathscr{X}_{\tilde{\mathbf{H}}} = \mathbf{C}_{\mathrm{B}}^* \mathbf{P}_{N_{\mathrm{B}}}^{\mathsf{T}} \mathbf{F}_{N_{\mathrm{B}}} \mathbf{R}_{N_{\mathrm{B}}}^{\mathrm{cp}} \tilde{\mathbf{H}}$. Hence, we can re-write the terms in (16.11) and (16.12) as $\mathbf{C}_{\mathrm{B}}^* \mathbf{H} \mathbf{B} = \mathscr{X}_{\tilde{\mathbf{H}}} \mathbf{T}_{N_{\mathrm{A}}}^{\mathrm{cp}} \mathbf{F}_{N_{\mathrm{A}}}^* \mathbf{P}_{N_{\mathrm{A}}} \mathbf{B}$ and $\mathbf{C}_{\mathrm{B}}^* \mathbf{P}_{N_{\mathrm{B}}}^{\mathsf{T}} \mathbf{F}_{N_{\mathrm{B}}} \mathbf{R}_{N_{\mathrm{B}}}^{\mathrm{cp}} \tilde{\mathbf{H}} \mathbf{Q} = \mathscr{X}_{\tilde{\mathbf{H}}} \mathbf{Q}$. When $N_{\mathrm{cp}}/N \ll 1$, the CP insertion and removal matrices are approximately identity matrices. Hence, the spatial AN cancellation condition in (16.11) can be re-written as:

$$\mathscr{X}_{\tilde{\mathbf{H}}} \hat{\mathbf{B}} = \mathbf{0}, \tag{A.1}$$

where $\hat{\mathbf{B}} = \mathbf{T}_{N_{\mathrm{A}}}^{\mathrm{cp}} \mathbf{F}_{N_{\mathrm{A}}}^* \mathbf{P}_{N_{\mathrm{A}}} \mathbf{B}$ is an orthonormal-column matrix. The temporal AN cancellation condition in (16.12) can be re-written as:

$$\mathscr{X}_{\tilde{\mathbf{H}}} \mathbf{Q} = \mathbf{0}. \tag{A.2}$$

Since the columns of $\mathbf{Q}$ form an orthonormal basis for the null space of $\mathscr{X}_{\tilde{\mathbf{H}}}$, and the columns of $\hat{\mathbf{B}}$ are orthonormal, and they lie in the null space of $\mathscr{X}_{\tilde{\mathbf{H}}}$, every column in $\hat{\mathbf{B}}$ can be written as a linear combination of the columns of $\mathbf{Q}$. Hence:

$$\mathbf{Q} \mathbf{L} = \hat{\mathbf{B}}, \tag{A.3}$$

where $\mathbf{L} \in \mathbb{C}^{(N(N_{\mathrm{A}}-N_{\mathrm{s}})+N_{\mathrm{cp}}N_{\mathrm{A}}) \times N(N_{\mathrm{A}}-N_{\mathrm{s}})}$ is a linear transformation matrix. Since $\mathbf{Q}^* \mathbf{Q} = \mathbf{I}_{N(N_{\mathrm{A}}-N_{\mathrm{s}})+N_{\mathrm{cp}}N_{\mathrm{A}}}$ and $\hat{\mathbf{B}}^* \hat{\mathbf{B}} = \mathbf{I}_{N(N_{\mathrm{A}}-N_{\mathrm{s}})}$, we have:

$$\mathbf{L}^* \mathbf{Q}^* \mathbf{Q} \mathbf{L} = \hat{\mathbf{B}}^* \hat{\mathbf{B}} = \mathbf{I}_{N(N_{\mathrm{A}}-N_{\mathrm{s}})}. \tag{A.4}$$

Consequently,

$$\mathbf{L}^* \mathbf{L} = \mathbf{I}_{N(N_{\mathrm{A}}-N_{\mathrm{s}})}. \tag{A.5}$$

This implies that $\mathbf{L}$ is an orthonormal-column matrix. Substituting from (A.3) into (16.20), we have:

$$\boldsymbol{\Sigma}_{\mathrm{AN}} = \overline{\theta} \Gamma_{\mathrm{E}} \mathscr{X}_{\tilde{\mathbf{G}}} \left(\frac{\alpha \mathbf{Q} \mathbf{L} \mathbf{L}^* \mathbf{Q}^*}{N_{\mathrm{O}}(N_{\mathrm{A}}-N_{\mathrm{s}})} + \frac{\overline{\alpha} \mathbf{Q} \mathbf{Q}^*}{N(N_{\mathrm{A}}-N_{\mathrm{s}})+N_{\mathrm{cp}}N_{\mathrm{A}}} \right) \mathscr{X}_{\tilde{\mathbf{G}}}^* \tag{A.6}$$

$$= \overline{\theta} \Gamma_{\mathrm{E}} \mathscr{X}_{\tilde{\mathbf{G}}} \mathbf{Q} \left(\frac{\alpha \mathbf{L} \mathbf{L}^*}{N_{\mathrm{O}}(N_{\mathrm{A}}-N_{\mathrm{s}})} + \frac{\overline{\alpha} \mathbf{I}_{N(N_{\mathrm{A}}-N_{\mathrm{s}})+N_{\mathrm{cp}}N_{\mathrm{A}}}}{N(N_{\mathrm{A}}-N_{\mathrm{s}})+N_{\mathrm{cp}}N_{\mathrm{A}}} \right) \mathbf{Q}^* \mathscr{X}_{\tilde{\mathbf{G}}}^*, \tag{A.7}$$

where $\mathscr{X}_{\tilde{\mathbf{G}}} = \mathbf{P}_{N_{\mathrm{E}}}^{\mathsf{T}} \mathbf{F}_{N_{\mathrm{E}}} \mathbf{R}_{N_{\mathrm{E}}}^{\mathrm{cp}} \tilde{\mathbf{G}}$. Since $N_{\mathrm{cp}}/N \ll 1$, and assuming that $N_{\mathrm{cp}}/N \ll 1 - N_{\mathrm{s}}/N_{\mathrm{A}}$, $\boldsymbol{\Sigma}_{\mathrm{AN}}$ is re-written as:

$$\boldsymbol{\Sigma}_{\mathrm{AN}} = \frac{\overline{\theta} \Gamma_{\mathrm{E}}}{N(N_{\mathrm{A}}-N_{\mathrm{s}})} \mathscr{X}_{\tilde{\mathbf{G}}} \mathbf{Q} (\alpha \mathbf{L} \mathbf{L}^* + \overline{\alpha} \mathbf{I}_{N(N_{\mathrm{A}}-N_{\mathrm{s}})+N_{\mathrm{cp}}N_{\mathrm{A}}}) \mathbf{Q}^* \mathscr{X}_{\tilde{\mathbf{G}}}^*. \tag{A.8}$$

Define the SVD of $\mathbf{L}$ as $\mathbf{L} = \mathbf{U}_{\mathbf{L}} \boldsymbol{\Lambda}_{\mathbf{L}} \mathbf{V}_{\mathbf{L}}^*$, where the columns of $\mathbf{U}_{\mathbf{L}}$ are the left singular vectors of $\mathbf{L}$, $\boldsymbol{\Lambda}_{\mathbf{L}} = \mathrm{blkdiag}(\mathbf{I}_{N(N_{\mathrm{A}}-N_{\mathrm{s}})}, \mathbf{0}) \in \mathbb{C}^{(N(N_{\mathrm{A}}-N_{\mathrm{s}})+N_{\mathrm{cp}}N_{\mathrm{A}}) \times (N(N_{\mathrm{A}}-N_{\mathrm{s}})+N_{\mathrm{cp}}N_{\mathrm{A}})}$ is the diagonal matrix containing the singular values, and $\mathbf{V}_{\mathbf{L}}$ are the right singular vectors of $\mathbf{L}$. Thus:

$$\boldsymbol{\Sigma}_{\mathrm{AN}} = \frac{\overline{\theta} \Gamma_{\mathrm{E}}}{N(N_{\mathrm{A}}-N_{\mathrm{s}})} \mathscr{X}_{\tilde{\mathbf{G}}} \mathbf{Q} (\alpha \mathbf{U}_{\mathbf{L}} \boldsymbol{\Lambda}_{\mathbf{L}}^2 \mathbf{U}_{\mathbf{L}}^* + \overline{\alpha} \mathbf{I}_{N(N_{\mathrm{A}}-N_{\mathrm{s}})+N_{\mathrm{cp}}N_{\mathrm{A}}}) \mathbf{Q}^* \mathscr{X}_{\tilde{\mathbf{G}}}^*. \tag{A.9}$$

By replacing $\mathbf{I}_{N(N_A-N_s)+N_{cp}N_A}$ in (A.9) with $\mathbf{U_L U_L^*}$, and since $\mathbf{\Lambda_L^2} = \mathbf{\Lambda_L}$, (A.9) is re-written as:

$$\mathbf{\Sigma}_{AN} = \frac{\overline{\theta}\Gamma_E}{N(N_A - N_s)} \mathscr{X}_{\tilde{\mathbf{G}}} \mathbf{Q}(\alpha\mathbf{U_L \Lambda_L U_L^*} + \overline{\alpha}\mathbf{U_L U_L^*}) \mathbf{Q}^* \mathscr{X}_{\tilde{\mathbf{G}}}^* \qquad (A.10)$$

$$= \frac{\overline{\theta}\Gamma_E}{N(N_A - N_s)} \mathscr{X}_{\tilde{\mathbf{G}}} \mathbf{Q U_L}(\alpha\mathbf{\Lambda_L} + \overline{\alpha}\mathbf{I}_{N(N_A-N_s)+N_{cp}N_A}) \mathbf{U_L^* Q^*} \mathscr{X}_{\tilde{\mathbf{G}}}^* \qquad (A.11)$$

$$= \overline{\theta}\Gamma_E \mathscr{X}_{\tilde{\mathbf{G}}} \mathbf{Q U_L} \frac{\mathbf{M}}{N(N_A - N_s)} \mathbf{U_L^* Q^*} \mathscr{X}_{\tilde{\mathbf{G}}}^*, \qquad (A.12)$$

where $\mathbf{M} = \alpha\mathbf{\Lambda_L} + \overline{\alpha}\mathbf{I}_{N(N_A-N_s)+N_{cp}N_A} = \text{blkdiag}(\mathbf{I}_{N(N_A-N_s)}, \overline{\alpha}\mathbf{I}_{N_{cp}N_A})$. When the dimension of the matrix $\mathbf{I}_{N(N_A-N_s)}$ is much greater than the dimension of the matrix $\overline{\alpha}\mathbf{I}_{N_{cp}N_A}$, we can ignore the matrix $\overline{\alpha}\mathbf{I}_{N_{cp}N_A}$. The condition is thus given by $N_{cp}/N \ll 1 - N_s/N_A$. In this case, the variation of $\mathbf{\Sigma}_{AN}$ with α is negligible which makes Eve's rate almost flat with α.

A.1.2   Distribution of $\mathbf{F}_{N_E}\tilde{\mathbf{G}}_{toep}$

In this appendix, we derive the distribution of $\mathbf{f}_k\tilde{\mathbf{G}}_{toep}$ and $\|\mathbf{f}_k\tilde{\mathbf{G}}_{toep}\|^2$, where $\tilde{\mathbf{G}}_{toep} = \mathbf{R}_{N_E}^{cp}\tilde{\mathbf{G}}$ and $\mathbf{f}_k$ is the kth row of $\mathbf{F}_{N_E}$. Since the matrix $\tilde{\mathbf{G}}_{toep}$ is a block Toeplitz matrix whose (i,j) block, denoted by $\tilde{\mathbf{G}}_{i,j}$, is the CIR of the $i-j$ link and is an upper-triangular Teoplitz matrix with first row equal to $(\tilde{a}, \tilde{b}, \ldots, 0, \ldots, 0)$. This matrix $\tilde{\mathbf{G}}_{ij} \in \mathbb{C}^{N_O \times N_O}$ has the following properties:

- Each row has $(\nu + 1)$ non-zero circularly symmetric complex Gaussian random variables.
- The ℓth column, $\ell \leq \nu$, has ℓ non-zero complex Gaussian entries. The non-zero entries are i.i.d. circuitry-symmetric complex Gaussian.
- Each column from $(\nu + 1)$ to N has $\nu + 1$ non-zero complex Gaussian entries.
- Column $(N + k)$ has $(\nu - k + 1)$ non-zero entries. Finally, Column $(N + \nu)$ has only 1 non-zero complex Gaussian entry, and the other entries are zero.

The above structure is repeated each $N + \nu$ columns and rows of $\tilde{\mathbf{G}}_{toep}$. The random vector $\mathbf{f}_k\tilde{\mathbf{G}}_{toep}$ is composed of entries which follow complex Gaussian distribution with different means and variances based on the structure of $\tilde{\mathbf{G}}_{toep}$. Assuming that $N_s = N_B = N_E = 1$ as in the MISOSE-OFDM case, the random vector $\mathbf{f}_k\tilde{\mathbf{G}}_{toep}$ is given by:

$$[\mathbf{f}_k\tilde{\mathbf{G}}_{toep}]_{1,j} = \sum_{i=1}^{N}[\mathbf{f}_k]_{1,i}\left[\tilde{\mathbf{G}}_{toep}\right]_{i,j}, \qquad (A.13)$$

with $\left|[\mathbf{f}_k]_{1,i}\right|^2 = 1/N$ for all i. The jth entry of $\mathbf{f}_k\tilde{\mathbf{G}}_{\text{toep}}$, $[\mathbf{f}_k\tilde{\mathbf{G}}_{\text{toep}}]_{1,j}$, is a complex Gaussian distributed random variable with zero mean and variance:

$$\sigma_j^2 = \begin{cases} \frac{\sigma_{\text{A}-\text{E}}^2}{N}j, \text{if } j \leq v \\ \frac{\sigma_{\text{A}-\text{E}}^2}{N}(v+1), \text{if } v+1 \leq j \leq N \\ \frac{\sigma_{\text{A}-\text{E}}^2}{N}(v-k+1), \text{if } j \geq N+k, \forall 1 \leq k \leq v, \end{cases} \tag{A.14}$$

where $j \leq N + v$. Using (A.14), the random variable $\|\mathbf{f}_k\tilde{\mathbf{G}}_{\text{toep}}\|^2$ has a mean of:

$$\begin{aligned}
\mathbb{E}\{\|\mathbf{f}_k\tilde{\mathbf{G}}_{\text{toep}}\|^2\} &= \mathbb{E}\left\{\sum_{j=1}^{NN_A}\left|\sum_{i=1}^{N}[\mathbf{f}_k]_{1,i}\left[\tilde{\mathbf{G}}_{\text{toep}}\right]_{i,j}\right|^2\right\} \\
&= \left(2\frac{\sigma_{\text{A}-\text{E}}^2}{N} + 2(2)\frac{\sigma_{\text{A}-\text{E}}^2}{N} + 2(3)\frac{\sigma_{\text{A}-\text{E}}^2}{N}\right. \\
&\qquad \left. + \cdots + 2(v)\frac{\sigma_{\text{A}-\text{E}}^2}{N} + (N-v)\tilde{v}\frac{\sigma_{\text{A}-\text{E}}^2}{N}\right)N_A \\
&= \left(2\frac{\sigma_{\text{A}-\text{E}}^2}{N}\sum_{m=1}^{v}m + (N-v)\tilde{v}\frac{\sigma_{\text{A}-\text{E}}^2}{N}\right)N_A \\
&= \left(2\frac{\sigma_{\text{A}-\text{E}}^2}{N}\frac{v\tilde{v}}{2} + (N-v)\tilde{v}\frac{\sigma_{\text{A}-\text{E}}^2}{N}\right)N_A = \sigma_{\text{A}-\text{E}}^2\tilde{v}N_A,
\end{aligned} \tag{A.15}$$

since $\sum_{m=1}^{v}m = \frac{v(v+1)}{2} = \frac{v\tilde{v}}{2}$. Using the law of large numbers when $N_A \to \infty$, $\|\mathbf{f}_k\tilde{\mathbf{G}}_{\text{toep}}\|^2$ is approximately given by:

$$\|\mathbf{f}_k\tilde{\mathbf{G}}_{\text{toep}}\|^2 \approx \sigma_{\text{A}-\text{E}}^2\tilde{v}N_A. \tag{A.16}$$

A.1.3   Distributions of $\mathbf{G}_k\mathbf{A}_k\mathbf{A}_k^*\mathbf{G}_k^*$ and $\mathbf{H}_k\mathbf{A}_k\mathbf{A}_k^*\mathbf{H}_k^*$

In this appendix, we derive an analytic expression for the distribution of $\|[\mathbf{G}_k\mathbf{A}_k]_{i,j}\|^2$. The (i,j)th element of $\mathbf{G}_k$, $[\mathbf{G}_k]_{i,j}$, is given by:

$$[\mathbf{G}_k]_{i,j} = g_{0,i,j} + \sum_{\ell=1}^{v}g_{\ell,i,j}\omega^{\ell k}, \tag{A.17}$$

where $\omega = \exp(-2\pi\sqrt{-1}/N)$, $g_{\ell,i,j}$ is the ℓth tap of the CIR of the (i,j)th Alice–Eve channel. Since $g_{\ell,i,j}$ follows an i.i.d. Gaussian distribution, $g_{0,i,j} + \sum_{\ell=1}^{v}g_{\ell,i,j}\omega^{\ell k}$ is also Gaussian distributed with zero mean and variance:

$$\begin{aligned}
\mathbb{E}\{[\mathbf{G}_k]_{i,j}[\mathbf{G}_k]_{i,j}^*\} &= \mathbb{E}\left\{\left(g_{0,i,j} + \sum_{\ell=1}^{v}g_{\ell,i,j}\omega^{\ell k}\right)\left(g_{0,i,j} + \sum_{r=1}^{v}g_{r,i,j}\omega^{rk}\right)^*\right\} \\
&= \tilde{v}\sigma_{\text{A}-\text{E}}^2.
\end{aligned} \tag{A.18}$$

Since the channels are i.i.d., the expected values of the cross terms are zero. Each of the diagonal elements of $\mathbf{G}_k\mathbf{G}_k^*$ is the sum of magnitude of squares of N_A Gaussian random variables. Hence, each entry is Chi-square distributed. The off-diagonal entries are the sums of the products of independent complex Gaussian random variables, hence, their means are zero. As $N_A \to \infty$, the matrix $\mathbf{G}_k\mathbf{G}_k^*$ is approximately given by:

$$\mathbf{G}_k\mathbf{G}_k^* \approx \sigma_{A-E}^2 \tilde{\nu} N_A \mathbf{I}_{N_E}. \tag{A.19}$$

The random variable $[\mathbf{G}_k\mathbf{A}_k\mathbf{A}_k^*\mathbf{G}_k^*]_{i,i}/(\frac{1}{2}\sigma_{A-E}^2\tilde{\nu})$ is Chi-square distributed with $2N_s$ degrees of freedom since $[\mathbf{G}_k\mathbf{A}_k\mathbf{A}_k^*\mathbf{G}_k^*]_{i,i}$ is a complex Gaussian variable with zero mean and variance $\tilde{\nu}\sigma_{A-E}^2$. Hence, the expected value of $[\mathbf{G}_k\mathbf{A}_k\mathbf{A}_k^*\mathbf{G}_k^*]_{i,i}$ is $\tilde{\nu}\sigma_{A-E}^2 N_s$. The expected values of the off-diagonal entries of $\mathbf{G}_k\mathbf{A}_k\mathbf{A}_k^*\mathbf{G}_k^*$, $[\mathbf{G}_k\mathbf{A}_k\mathbf{A}_k^*\mathbf{G}_k^*]_{i,j}$, are zeros. In a similar fashion, we can derive the distribution of the random matrix $\mathbf{H}_k\mathbf{A}_k\mathbf{A}_k^*\mathbf{H}_k^*$.

A.1.4 Proof of Lemma 16.2

When $1 - \frac{N_s}{N_A} \gg \frac{N_{cp}}{N}$, the CP insertion matrix can be approximated by an identity matrix since the CP size is negligible. Moreover, as $N_A \to \infty$, $\mathbf{BB}^* \approx \mathbf{I}_{N(N_A-N_s)}$ and $\mathbf{QQ}^* \approx \mathbf{I}_{N_AN_O}$. Hence, $\mathbf{GG}^* \approx \mathbf{F}_{N_E}\mathbf{R}_{N_E}^{cp}\tilde{\mathbf{G}}(\mathbf{F}_{N_E}\mathbf{R}_{N_E}^{cp}\tilde{\mathbf{G}})^*$ and the covariance matrix of the interference at Eve in (16.28) becomes:

$$\begin{aligned}
\mathbf{\Sigma}_{AN} &= \bar{\theta}\Gamma_E\left(\frac{\alpha}{N_O(N_A-N_s)}\mathbf{GG}^* + \frac{\bar{\alpha}}{N(N_A-N_s)+N_{cp}N_A}\mathbf{GG}^*\right) \\
&= \bar{\theta}\Gamma_E\mathbf{GG}^*\left(\frac{\alpha}{N_O(N_A-N_s)} + \frac{\bar{\alpha}}{N(N_A-N_s)+N_{cp}N_A}\right) \\
&\approx \frac{\bar{\theta}\Gamma_E}{N(N_A-N_s)}\mathbf{GG}^*,
\end{aligned} \tag{A.20}$$

where the last equality holds when $1 - \frac{N_s}{N_A} \gg \frac{N_{cp}}{N}$. Substituting with $\mathbf{\Sigma}_{AN}$ in (A.20) into (16.27) leads to (16.29). Moreover, substituting with R_{A-B} in (16.17) and R_{A-E} in (16.29) into (16.23), we obtain the secrecy rate expression in (16.30), which is independent of α.

A.1.5 Proof of Lemma 16.3

Since $\mathbf{GA}\mathbf{\Sigma}_x(\mathbf{GA})^*$ and $\mathbf{GG}^*$ in (16.29) are block-diagonal matrices, Eve's rate in (16.29) can be re-written as

$$R_{A-E} = \sum_{k=1}^{N}\left[\log_2\det\left(\theta\Gamma_E\mathbf{G}_k\mathbf{A}_k\mathbf{\Sigma}_{x_k}(\mathbf{G}_k\mathbf{A}_k)^*\left(\frac{\bar{\theta}\Gamma_E\mathbf{G}_k\mathbf{G}_k^*}{N(N_A-N_s)} + \mathbf{I}_{N_E}\right)^{-1} + \mathbf{I}_{N_E}\right)\right], \tag{A.21}$$

which is equivalent to processing each sub-carrier individually. This completes the proof.

A.1.6 Proof of Lemma 16.4

Using Appendices A.1.2 and A.1.3, we approximate the matrices in (16.27) and (16.28) when N_A is very large as follows:

- From Appendix A.1.3, we showed that $\|[\mathbf{G}_k\mathbf{A}_k]_{i,i}\|^2/(\frac{1}{2}\sigma_{A-E}^2\tilde{\nu})$ follows the Chi-square distribution with $2N_s$ degrees of freedom, and the mean of $\|[\mathbf{G}_k\mathbf{A}_k]_{i,i}\|^2$ is $\mathbb{E}\{\|[\mathbf{G}_k\mathbf{A}_k]^2]_{i,i}\} = \tilde{\nu}\sigma_{A-E}^2 N_s$.
- The matrix $\mathbf{GA} = \text{blkdiag}(\mathbf{G}_1\mathbf{A}_1,\ldots,\mathbf{G}_N\mathbf{A}_N)$ has diagonal entries whose variances are $N_s\tilde{\nu}\sigma_{A-E}^2$. The matrix $\mathbf{GA(GA)}^*$ has off-diagonal elements $[\mathbf{GA(GA)}^*]_{i,j} = \mathbf{g}_i\mathbf{A}_k\mathbf{A}_k^*\mathbf{g}_j^*$ which represent the sum of N_s i.i.d. complex Gaussian products, where $\mathbf{g}_i$ is the ith row of the matrix $\mathbf{G}_k$. The ith diagonal element of $\mathbf{GA(GA)}^*$ is $[\mathbf{GA(GA)}^*]_{i,i} = \mathbf{g}_i\mathbf{A}_k\mathbf{A}_k^*\mathbf{g}_i^*$ which follows a Chi-square distribution with mean $\sigma_{A-E}^2\tilde{\nu}N_s$.
- Consider the matrix $\mathbf{GG}^*$. The matrix $\mathbf{G}$ is a block-diagonal matrix and, hence, $\mathbf{GG}^*$ is also a block-diagonal matrix. The expected value of each of the diagonal entries of $\mathbf{GG}^*$ is equal to $N_A\tilde{\nu}\sigma_{A-E}^2$. The off-diagonal entries are the sum of the product of independent complex Gaussian random variables, hence, their means are zero. Hence, $\mathbf{GG}^*$ is approximated as $N_A\tilde{\nu}\sigma_{A-E}^2\mathbf{I}_{N_EN}$.
- Each row of $\mathbf{R}_{N_E}^{\text{cp}}\tilde{\mathbf{G}}$ is composed of $\nu + 1$ i.i.d. complex Gaussian random variables. Hence, the expected values of the off-diagonal elements of the matrix $\mathbf{R}_{N_E}^{\text{cp}}\tilde{\mathbf{G}}\left(\mathbf{R}_{N_E}^{\text{cp}}\tilde{\mathbf{G}}\right)^* = \mathbf{R}_{N_E}^{\text{cp}}\tilde{\mathbf{G}}\tilde{\mathbf{G}}^*\mathbf{R}_{N_E}^{\text{cp}*}$ are zero. As $N_A \to \infty$, we can approximate this matrix by $N_A\tilde{\nu}\sigma_{A-E}^2\mathbf{I}_{N_EN}$. Hence, $\mathbf{P}_{N_E}^{\top}\mathbf{F}_{N_E}\mathbf{R}_{N_E}^{\text{cp}}\tilde{\mathbf{G}}\tilde{\mathbf{G}}^*\mathbf{R}_{N_E}^{\text{cp}*}\mathbf{F}_{N_E}^*\mathbf{P}_{N_E} \approx N_A\tilde{\nu}\sigma_{A-E}^2\mathbf{I}_{N_EN}$ as $N_A \to \infty$.

On the basis of our earlier discussions, we have the following relations:

$$\mathbb{E}\left\{\left[\mathbf{GA}\boldsymbol{\Sigma}_\mathbf{x}(\mathbf{GA})^*\right]_{i,i}\right\} = \frac{\tilde{\nu}\sigma_{A-E}^2}{N_O}, \tag{A.22}$$

where $\boldsymbol{\Sigma}_\mathbf{x} = \frac{1}{N_sN_O}\mathbf{I}_{N_sN}$, and

$$\boldsymbol{\Sigma}_{AN} = \bar{\theta}\Gamma_E N_A\tilde{\nu}\sigma_{A-E}^2\left(\frac{\alpha}{N_O(N_A - N_s)} + \frac{\bar{\alpha}}{N(N_A - N_s) + N_{\text{cp}}N_A}\right)\mathbf{I}_{N_EN}. \tag{A.23}$$

Hence,

$$\boldsymbol{\Sigma}_{AN} + \mathbf{I}_{N_EN} = p(\alpha)\mathbf{I}_{N_EN}, \tag{A.24}$$

where the scalar $p(\alpha)$ is defined as

$$p(\alpha) = \bar{\theta}\Gamma_E N_A\tilde{\nu}\sigma_{A-E}^2\left(\frac{\alpha}{N_O(N_A - N_s)} + \frac{\bar{\alpha}}{N(N_A - N_s) + N_{\text{cp}}N_A}\right) + 1. \tag{A.25}$$

The inverse of $\boldsymbol{\Sigma}_{AN} + \mathbf{I}_{N_EN}$ is $\frac{1}{p(\alpha)}\mathbf{I}_{N_EN}$. Hence,

$$\theta\Gamma_E\mathbf{GA}\boldsymbol{\Sigma}_\mathbf{x}(\mathbf{GA})^*\left(\boldsymbol{\Sigma}_{AN} + \mathbf{I}_{N_EN}\right)^{-1} + \mathbf{I}_{N_EN} = \frac{\theta\Gamma_E}{p(\alpha)}\mathbf{GA}\boldsymbol{\Sigma}_\mathbf{x}(\mathbf{GA})^* + \mathbf{I}_{N_EN}. \tag{A.26}$$

Using the Hadamard inequality [21], we can upper bound Eve's rate in (16.27) as follows:

$$\log_2 \det\left(\theta\Gamma_E \mathbf{GA}\mathbf{\Sigma}_x(\mathbf{GA})^*\left(\mathbf{\Sigma}_{AN} + \mathbf{I}_{N_E N}\right)^{-1} + \mathbf{I}_{N_E N}\right)$$

$$\leq \log_2 \prod_{i=1}^{N_E N}\left(\frac{\theta\Gamma_E}{p(\alpha)}[\mathbf{GA}\mathbf{\Sigma}_x(\mathbf{GA})^*]_{i,i} + 1\right). \tag{A.27}$$

Using the properties of the logarithmic function,

$$R_{A-E} \leq \sum_{i=1}^{N_E N}\log_2\left(\frac{\theta\Gamma_E}{p(\alpha)}[\mathbf{GA}\mathbf{\Sigma}_x(\mathbf{GA})^*]_{i,i} + 1\right). \tag{A.28}$$

The upper bound in (A.28) is the sum of $N_E N$ concave functions, hence, by using Jensen's inequality, Eve's average rate is upper bounded by

$$\mathbb{E}\{R_{A-E}\} \leq \sum_{i=1}^{N_E N}\log_2\left(\frac{\theta\Gamma_E}{p(\alpha)}\mathbb{E}\left\{[\mathbf{GA}\mathbf{\Sigma}_x(\mathbf{GA})^*]_{i,i}\right\} + 1\right). \tag{A.29}$$

Using (A.22), we get

$$\mathbb{E}\{R_{A-E}\} \lesssim \sum_{i=1}^{N_E N}\log_2\left(\frac{\theta\Gamma_E}{p(\alpha)}\frac{\tilde{\nu}\sigma_{A-E}^2}{N_0} + 1\right) = N_E N\log_2\left(\frac{\theta\Gamma_E}{p(\alpha)}\frac{\tilde{\nu}\sigma_{A-E}^2}{N_0} + 1\right). \tag{A.30}$$

Substituting for $p(\alpha)$ from (A.25) into (A.30) and using the fact that $N_A \gg N_s$, the upper bound on Eve's average rate is given by

$$\mathbb{E}\{R_{A-E}\} \lesssim N_E N\log_2\left(\frac{\frac{\theta\Gamma_E}{N_0}\tilde{\nu}\sigma_{A-E}^2}{\overline{\theta}\Gamma_E\tilde{\nu}\sigma_{A-E}^2\left(\frac{\alpha}{N_0} + \frac{\overline{\alpha}}{N_0}\right) + 1} + 1\right). \tag{A.31}$$

Since $\frac{\alpha}{N_0} + \frac{\overline{\alpha}}{N_0} = \frac{1}{N_0}$, the right hand side of (A.31) is independent of α, which implies that the upper bound on Eve's average rate is not a function of α. The upper bound on the average rate of the Alice–Eve link is thus given by

$$\mathbb{E}\{R_{A-E}\} \lesssim N_E N\log_2\left(\frac{\frac{\theta\Gamma_E}{N_0}\tilde{\nu}\sigma_{A-E}^2}{\frac{\overline{\theta}\Gamma_E}{N_0}\tilde{\nu}\sigma_{A-E}^2 + 1} + 1\right). \tag{A.32}$$

The rate of the Alice–Bob link is given by

$$R_{A-B} = \log_2\det\left(\theta\Gamma_B\mathbf{HA}\mathbf{\Sigma}_x(\mathbf{HA})^* + \mathbf{I}_{N_B N}\right). \tag{A.33}$$

The matrix $\mathbf{HA\Sigma_x(HA)^*}$ can be approximated by its diagonal elements as a weighted identity matrix. Hence, $\mathbf{HA\Sigma_x(HA)^*} \approx \frac{1}{N_sN_O}N_A\tilde{v}\sigma_{A-B}^2\mathbf{I}_{N_sN}$. Consequently, Bob's average rate can be approximated by

$$
\begin{aligned}
\mathbb{E}\{R_{A-B}\} &\approx \log_2 \det\left(\frac{\theta\Gamma_B}{N_sN_O}N_A\tilde{v}\sigma_{A-B}^2\mathbf{I}_{N_sN} + \mathbf{I}_{N_sN}\right) \\
&= \log_2\left(\frac{\theta\Gamma_B}{N_sN_O}N_A\tilde{v}\sigma_{A-B}^2 + 1\right)^{N_sN} \\
&= N_sN \log_2\left(\frac{\theta\Gamma_B}{N_sN_O}N_A\tilde{v}\sigma_{A-B}^2 + 1\right).
\end{aligned}
\tag{A.34}
$$

Using (A.32) and (A.34), and since $\mathbb{E}\{R_{sec}\} = \mathbb{E}\{R_{A-B}\} - \mathbb{E}\{R_{A-E}\}$, we get the result in (16.32).

A.1.7 Proof of Lemma 16.6

When $\Gamma_B\sigma_{A-B}^2$ and $\Gamma_E\sigma_{A-E}^2$ are sufficiently high, we have

$$
\mathbb{E}\{R_{sec}\} \gtrsim N_sN \log_2\left(\frac{\theta\Gamma_B}{N_sN_O}N_A\tilde{v}\sigma_{A-B}^2\right)
\tag{A.35}
$$

$$
- N_EN \log_2\left(\frac{\frac{\theta}{N_O}}{\overline{\theta}\left(\frac{\alpha}{N_O} + \frac{\overline{\alpha}}{N_O}\right)} + 1\right).
\tag{A.36}
$$

Hence, we can lower bound the average secrecy rate as follows:

$$
\mathbb{E}\{R_{sec}\} \gtrsim N_sN \log_2\left(\frac{\theta\Gamma_B}{N_sN_O}N_A\tilde{v}\sigma_{A-B}^2\right) - N_EN \log_2\left(\frac{\theta}{\overline{\theta}} + 1\right).
\tag{A.37}
$$

Thus,

$$
\begin{aligned}
\mathbb{E}\{R_{sec}\} &\gtrsim N_sN \log_2\left(\frac{\Gamma_B}{N_sN_O}N_A\tilde{v}\sigma_{A-B}^2\right) + N_EN \log_2\left(\theta^{\frac{N_s}{N_E}}\right) + N_EN \log_2(\overline{\theta}) \\
&= N_sN \log_2\left(\frac{\Gamma_B}{N_sN_O}N_A\tilde{v}\sigma_{A-B}^2\right) + N_EN \log_2\left(\theta^{\frac{N_s}{N_E}}\overline{\theta}\right).
\end{aligned}
\tag{A.38}
$$

The first term is independent of θ, hence to increase the lower bound, Alice needs to choose the value of θ which maximises the second term only. This is equivalent to maximise the term inside the logarithmic function. Letting $f(\theta) = \theta^{\frac{N_s}{N_E}}\overline{\theta}$, the first derivative of $f(\theta)$ is given by

$$
\frac{\delta f(\theta)}{\delta\theta} = \frac{N_s}{N_E}\theta^{\frac{N_s}{N_E}-1}\overline{\theta} - \theta^{\frac{N_s}{N_E}}.
\tag{A.39}
$$

The root of the first derivative is given by

$$
\theta^\star = \frac{1}{1 + \frac{N_s}{N_E}} = \frac{N_E}{N_E + N_s}.
\tag{A.40}
$$

The second derivative is given by

$$\frac{\delta^2 f(\theta)}{\delta \theta^2} = \frac{N_s}{N_E} \left(\frac{N_s}{N_E} - 1 \right) \theta^{\frac{N_s}{N_E} - 2} - \left(\frac{N_s}{N_E} + 1 \right) \frac{N_s}{N_E} \theta^{\frac{N_s}{N_E} - 1} \leq 0, \tag{A.41}$$

since $N_E \geq N_s$, which is a reasonable assumption to enable Eve to decode the data.

Substituting with the optimal value of θ in (A.40) into (A.38), the average secrecy rate is lower bounded as follows:

$$\mathbb{E}\{R_{sec}\} \gtrsim N_s N \log_2 \left(\frac{\Gamma_B}{N_s N_O} N_A \tilde{\nu} \sigma_{A-B}^2 \right) \tag{A.42}$$

$$+ N_E N \log_2 \left(\left(\frac{N_E}{N_E + N_s} \right)^{\frac{N_s}{N_E}} \left(1 - \frac{N_E}{N_E + N_s} \right) \right).$$

Re-arranging the expression in (A.42), we obtain the result in (16.35).

The average secrecy rate loss due to the presence of Eve is thus upper bounded as follows:

$$\mathbb{E}\{\mathscr{S}_{Loss}\} = \mathbb{E}\{R_{A-B,noEve}\} - \mathbb{E}\{R_{sec}\}$$

$$\lesssim N_s N \log_2 \left(\frac{\Gamma_B}{N_s N_O} N_A \tilde{\nu} \sigma_{A-B}^2 \right) \tag{A.43}$$

$$- N_s N \log_2 \left(\frac{\Gamma_B}{N_s N_O} N_A \tilde{\nu} \sigma_{A-B}^2 \right)$$

$$- N_E N \log_2 \left(\left(\frac{N_E}{N_E + N_s} \right)^{\frac{N_s}{N_E}} \left(1 - \frac{N_E}{N_E + N_s} \right) \right).$$

Re-arranging (A.43), we get

$$\mathbb{E}\{\mathscr{S}_{Loss}\} \lesssim N_s N \log_2 \left(\frac{N_E + N_s}{N_E} \right) + N_E N \log_2 \left(\frac{N_E + N_s}{N_s} \right). \tag{A.44}$$

Hence, the average secrecy rate loss in bits/sec/Hz is given by the expression in (16.37).

A.1.8 Proof of Lemma 16.7

In the low $\Gamma_E \sigma_{A-E}^2 \ll 1$ regime, $\bar{\theta} \frac{\Gamma_E}{N_O} \tilde{\nu} \sigma_{A-E}^2 \ll 1$. Hence, the upper bound on Eve's average rate in (A.32) is approximated as follows:

$$\mathbb{E}\{R_{A-E}\} \lesssim N_E N \log_2 \left(\frac{\frac{\theta \Gamma_E}{N_O} \tilde{\nu} \sigma_{A-E}^2}{\frac{\bar{\theta} \Gamma_E}{N_O} \tilde{\nu} \sigma_{A-E}^2 + 1} + 1 \right) \tag{A.45}$$

$$= N_E N \log_2 \left(\frac{\frac{\Gamma_E}{N_O} \tilde{\nu} \sigma_{A-E}^2 + 1}{\frac{\bar{\theta} \Gamma_E}{N_O} \tilde{\nu} \sigma_{A-E}^2 + 1} \right) \tag{A.46}$$

$$\approx N_E N \log_2 \left(\frac{\Gamma_E}{N_O} \tilde{\nu} \sigma_{A-E}^2 + 1 \right). \tag{A.47}$$

Consequently:

$$\mathbb{E}\{R_{\text{sec}}\} \gtrapprox N_s N \log_2\left(\frac{\theta \Gamma_B}{N_s N_O} N_A \tilde{v} \sigma_{A-B}^2 + 1\right)$$
$$- N_E N \log_2\left(\frac{\Gamma_E}{N_O} \tilde{v} \sigma_{A-E}^2 + 1\right). \tag{A.48}$$

From (A.48), the lower bound on the average secrecy rate is maximised when $\theta = 1$. Hence,

$$\mathbb{E}\{R_{\text{sec}}\} \gtrapprox N_s N \log_2\left(\frac{\Gamma_B}{N_s N_O} N_A \tilde{v} \sigma_{A-B}^2 + 1\right)$$
$$- N_E N \log_2\left(\frac{\Gamma_E}{N_O} \tilde{v} \sigma_{A-E}^2 + 1\right). \tag{A.49}$$

When there is no eavesdropper in the network, the average secrecy rate (i.e. average rate of the Alice–Bob link) is

$$\mathbb{E}\{R_{A-B,\text{noEve}}\} = N_s N \log_2\left(\frac{\Gamma_B}{N_s N_O} N_A \tilde{v} \sigma_{A-B}^2 + 1\right). \tag{A.50}$$

Accordingly, in the low $\Gamma_E \sigma_{A-E}^2 \leq \Gamma_B \sigma_{A-B}^2$ regime, the average secrecy rate loss is given by

$$\mathbb{E}\{\mathscr{S}_{\text{Loss}}\} = \mathbb{E}\{R_{A-B,\text{noEve}}\} - \mathbb{E}\{R_{\text{sec}}\} \lessapprox N_E N \log_2\left(\frac{\Gamma_E}{N_O} \tilde{v} \sigma_{A-E}^2 + 1\right). \tag{A.51}$$

The average secrecy rate loss in bits/sec/Hz is given by

$$\mathbb{E}\{\mathscr{S}_{\text{Loss}}\} \lessapprox \frac{N_E N}{N_O} \log_2\left(\frac{\Gamma_E}{N_O} \tilde{v} \sigma_{A-E}^2 + 1\right). \tag{A.52}$$

References

[1] T. Akitaya, S. Asano, and T. Saba, "Time-domain artificial noise generation technique using time-domain and frequency-domain processing for physical layer security in MIMO-OFDM systems," in *IEEE International Conference on Communications Workshops (ICC)*, 2014, pp. 807–812.

[2] I. Csiszár and J. Korner, "Broadcast channels with confidential messages," *IEEE Transactions on Information Theory*, vol. 24, no. 3, pp. 339–348, 1978.

[3] A. El Shafie, Z. Ding, and N. Al-Dhahir, "Hybrid spatio-temporal artificial noise design for secure MIMOME-OFDM systems," *IEEE Transactions on Vehicular Technology*, vol. 66, no. 5, pp. 3871–3886, May 2017.

[4] J. Hoydis, S. Ten Brink, and M. Debbah, "Massive MIMO in the UL/DL of cellular networks: How many antennas do we need?" *IEEE Journal on Selected Areas in Communications*, vol. 31, no. 2, pp. 160–171, 2013.

[5] M. Kobayashi, M. Debbah, and S. Shamai, "Secured communication over frequency-selective fading channels: A practical vandermonde precoding," *EURASIP Journal on Wireless Communications and Networking*, vol. 2009, p. 2, 2009.

[6] E. Larsson, O. Edfors, F. Tufvesson, and T. Marzetta, "Massive MIMO for next generation wireless systems," *IEEE Communications Magazine*, vol. 52, no. 2, pp. 186–195, 2014.

[7] J. Li and A. Petropulu, "Ergodic secrecy rate for multiple-antenna wiretap channels with Rician fading," *IEEE Transactions on Information Forensics and Security*, vol. 6, no. 3, pp. 861–867, Sept 2011.

[8] J. Li and A. Petropulu, "On ergodic secrecy rate for Gaussian MISO wiretap channels," *IEEE Transactions on Wireless Communications*, vol. 10, no. 4, pp. 1176–1187, April 2011.

[9] Z. Li, R. Yates, and W. Trappe, "Secrecy capacity of independent parallel channels," in *Securing Wireless Communications at the Physical Layer*. Springer, 2010, pp. 1–18.

[10] Y. Liang and H. V. Poor, "Information theoretic security," *Foundations and Trends in Communications and Information Theory*, vol. 5, no. 4–5, pp. 355–580, 2009.

[11] A. Mukherjee, S. Fakoorian, J. Huang, and A. Swindlehurst, "Principles of physical layer security in multiuser wireless networks: A survey," *IEEE Communications Surveys Tutorials*, vol. 16, no. 3, pp. 1550–1573, 2014.

[12] F. Negro, S. P. Shenoy, I. Ghauri, and D. Slock, "On the MIMO interference channel," in *Information Theory and Applications Workshop (ITA)*, 2010, pp. 1–9.

[13] H. Qin, Y. Sun, T.-H. Chang *et al.*, "Power allocation and time-domain artificial noise design for wiretap OFDM with discrete inputs," *IEEE Transactions on Wireless Communications*, vol. 12, no. 6, pp. 2717–2729, June 2013.

[14] H. Reboredo, J. Xavier, and M. Rodrigues, "Filter design with secrecy constraints: The MIMO Gaussian wiretap channel," *IEEE Transactions on Signal Processing*, vol. 61, no. 15, pp. 3799–3814, Aug 2013.

[15] F. Renna, N. Laurenti, and H. Poor, "Physical-layer secrecy for OFDM transmissions over fading channels," *IEEE Transactions on Information Forensics and Security*, vol. 7, no. 4, pp. 1354–1367, Aug 2012.

[16] M. R. Rodrigues and P. D. Almeida, "Filter design with secrecy constraints: The degraded parallel Gaussian wiretap channel," in *Global Telecommunications Conference (Globecom)*. IEEE, 2008, pp. 1–5.

[17] C. E. Shannon, "Communication theory of secrecy systems," *Bell System Technical Journal*, vol. 28, no. 4, pp. 656–715, 1949.

[18]   A. Swindlehurst, "Fixed SINR solutions for the MIMO wiretap channel," in *IEEE International Conference on Acoustics, Speech and Signal Processing*, April 2009, pp. 2437–2440.

[19]   S.-H. Tsai and H. Poor, "Power allocation for artificial-noise secure MIMO precoding systems," *IEEE Transactions on Signal Processing*, vol. 62, no. 13, pp. 3479–3493, July 2014.

[20]   A. D. Wyner, "The wire-tap channel," *Bell System Technical Journal*, vol. 54, no. 8, pp. 1355–1387, 1975.

[21]   D. Zwillinger, *Table of Integrals, Series, and Products*.   Elsevier, 2014.

Part V

Applications of physical layer security

Chapter 17

Physical layer security for real-world applications: use cases, results and open challenges

Stephan Ludwig[1], René Guillaume[1],
and Andreas Müller[1]

In the past, physical layer security has been a topic of primarily academic interest since the technology has not been matured enough yet for widespread practical deployments. Existing implementations are usually restricted to proof-of-concepts of selected aspects, and possible use cases as well as associated requirements for real-world applications have not been fully understood yet. In this chapter, we therefore discuss in more detail what role physical layer security may play in practical systems in future, what should be considered in this respect and what performance may be expected in realistic scenarios. To this end, extensive investigations and experiments have been carried out, the results of which are summarized hereafter. Due to space constraints and the relative broad scope of this chapter, however, it is unfortunately not possible to present all aspects and analyses in every single detail.

17.1 Introduction

In this section, we briefly present why physical layer security may become an important building block for dealing with some novel security challenges, especially in the context of the Internet of Things (IoT). In addition, we discuss the suitability of different flavours of physical layer security from a practical point of view, we outline some promising use cases and associated requirements, and we finally provide a brief comparison of physical layer security with alternative existing approaches.

17.1.1 Why physical layer security?

Communications security has been a highly active research field for more than half a century, and as an outcome of these efforts, a wide variety of different technologies have been developed for inter-connecting different devices in a secure manner.

[1]Robert Bosch GmbH, Corporate Sector Research and Advance Engineering, Germany

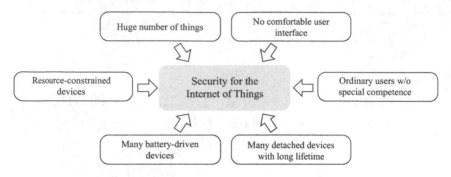

Figure 17.1   Overview of various IoT-related security challenges

However, not just technology improves over time, but also requirements and relevant use cases are subject to a permanent change. That is why existing security solutions permanently have to be refined in order to meet latest needs and expectations. Traditionally, security research has mainly focused on classic IT and telecommunication systems, where a reasonable number of devices, such as computers, mobile phones or tablet PCs, have to be securely connected to a certain network or server. For these kinds of problems, many well-established and proven technologies exist, including a wide range of different ciphers and key management systems. However, with various recent developments, such as the emerging IoT, the requirements and boundary conditions for establishing a secure communication change radically. Some examples for these novel challenges are depicted in Figure 17.1.

First, the IoT will be made up of a huge number of different devices, which often even do not have any comfortable user interface, such as a display or keyboard. Therefore, a manual distribution of cryptographic keys, e.g. by entering a password or some other form of pre-shared key, is not manageable in many cases. Furthermore, IoT devices will often have only very limited computational, memory and energy resources. For that reason, the use of complex asymmetric cryptographic schemes may turn out to be overly demanding for practical scenarios. In addition, IoT devices will often be detached (if you think of connected sensors in a Smart City, for example) with a rather long lifetime, therefore resulting in a need for easy, scalable yet secure re-keying. Finally, ordinary users without any special engineering expertise should be able to set up secure connections in a plug-and-play manner, thus requiring new levels of usability for security solutions.

Taking all these aspects into account, we can conclude that symmetric ciphers might be better suited for many IoT devices than asymmetric ones from a complexity and energy consumption point of view, but then the distribution of the symmetric keys required for that purpose becomes a major challenge due to the other mentioned constraints. Therefore, any scheme that may contribute to a simplification of the key management procedure or that completely supersedes the need for keys represents a promising candidate for many IoT applications. This is exactly where physical layer security may come into play. In this regard, two different flavours have to be distinguished, which will be briefly discussed in the next section.

17.1.2   Flavours of physical layer security

In general, two main flavours of physical layer security can be distinguished [1–5]:

1.   *Channel-based key generation (CBKG)*, where the wireless channel between two nodes is considered a joint source of randomness, out of which the nodes may generate symmetric cryptographic keys. These keys can then be used with established symmetric ciphers, such as the Advanced Encryption Standard (AES), for securing the communication between both nodes.
2.   *Secrecy coding*, where the sender encodes a message using a special secrecy code in such a way that the intended recipient is able to successfully decode the message, whereas a potential attacker eavesdropping on the channel is not. This may be done with or without feedback from the receiver to the sender, thus resulting in different solutions. Classic ciphers and keys are not needed anymore in this case.

Even though secrecy coding seems to be rather attractive since no classic ciphers are required, thus opening the door for highly efficient and low-cost implementations, there are also some major challenges involved. These challenges restrict – at least for the time being – the use in practical systems. On the one hand, this is because the design of powerful yet low complexity codes requires still additional research work. On the other hand and even more importantly, secrecy coding may only be used in a reasonable manner if some assumptions can be made regarding the channel from the sender to the eavesdropper, e.g. that the signal-to-noise ratio (SNR) of this channel is worse than the SNR between the sender and the legitimate receiver. This may be hard to guarantee in many real-world deployments.

Channel-based key generation, in contrast, is considered to be more attractive for practical applications in the short-run since the approach is already more mature. Besides, it enables a straightforward co-existence between classic key establishment schemes and channel-based approaches. This is because the only difference would be how keys are established while – once corresponding keys are available – the same security primitives may be used. For these reasons, we will mainly focus on channel-based key generation throughout this article. For further details on secrecy coding, the reader is referred to the literature, see for example [1–3,6].

17.1.3   Use cases and major requirements

As already outlined in Section 17.1.1, physical layer security represents a promising approach for many different IoT applications. Especially with CBKG, symmetric ciphers of low complexity and therefore with rather low resource requirements may be used, while the key management may be largely automated. In the following, three potential use cases will be outlined in more detail, but it should be noted that this list is far away from being exhaustive. It should just provide a general idea of potentially promising application scenarios.

Smart homes: Future smart homes will provide an unprecedented degree of automation, comfort and control. This will be enabled by inter-connecting all kinds of devices – be it kitchen appliances, heating and air conditioning systems, the light system, etc. – and making them accessible to standard user interfaces, such as smart-phones or tablet PCs. This access is often realized via a dedicated home controller, which in many cases coincides with a wireless access point. Wireless connectivity itself represents a natural choice in this case since on the one hand many smart home devices are portable and on the other hand it enables easy retrofit solutions, even for non-portable devices. The key challenge here would be to take the burden of setting up secure connections away from the typically non-qualified user and to maintain a solid level of security by periodically refreshing keys in an automated manner.

Smart cities and precision farming: Smart city and precision farming applications are usually characterized by a large number of typically inexpensive sensors and/or actuators, which in many cases are connected using wireless links due to the absence of a fixed network infrastructure. In precision farming, for example, large sensor networks deployed in a field may measure the temperature, humidity, fertility and other parameters in order to optimize the cultivation and thus yield. Likewise, in smart cities, parking sensors integrated into the ground may detect if a parking lot is available and thus provide valuable input for parking guidance systems, for example. The sheer number of devices to be securely inter-connected in these cases along with the resource-constrained nature of many of these sensors or actuators make physical layer security in general and CBKG in particular an attractive alternative compared to other solutions. Also, the devices may remain in use for a relatively long period of time, therefore requiring a convenient way to regularly update keys again.

Healthcare: In the healthcare domain, small sensors may be used to monitor and track the well-being and fitness of people, for example by measuring relevant vital signs or their activity levels. To this end, wearable devices or sensors directly attached to the body are required, which often transmit the collected data first of all to a smartphone or a similar standard user interface, from where it then may be forwarded to a remote server for analysis or storage. Security is of utmost importance in this case since the involved data may be quite sensitive, but wearables and sensors often lack a sophisticated user interface for entering passwords or the like. Furthermore, they are also highly resource-constrained and should be usable by ordinary users. Therefore, again physical layer security certainly may help.

The exact set of requirements on a CBKG scheme is certainly specific to the use case and depends on the concrete scenario and boundary conditions. In general, however, some of the core requirements that should be relevant in most cases of practical interest include the following:

- Arbitrary lengths of the keys to be generated should be supported.
- The generated keys should have full entropy. In particular, they should not exhibit any statistical pattern that may be exploited by a potential attacker.
- The key generation mechanism should not contribute significantly to the memory, computing, storage or energy consumption of a device.
- The (initial) key generation procedure should be fast enough such that a user is willing to wait until it has been completed. For a typical smart home use case,

Table 17.1   General comparison of three major key establishment schemes

Criterion	CBKG	ECDH	Passwords
Computational complexity	Moderate	High	Low
Bandwidth requirements	Moderate	Moderate	None
Implementation costs	Moderate	High	Low
Scalability with no. of nodes	Good	Fair	Not given
Ease of use	High	High	Low
Post-quantum ready	Yes	No	Yes
Efficient re-keying support	Yes	Partially	No

the tolerable worst case setup time is considered to be in the order of 10 s, with a preferred setup time of at most 2 s.
- The residual probability that two generated keys do not match should be lower than 1% and the system should be able to detect such a mis-alignment.

17.1.4   Comparison to alternative key establishment schemes

Before making use of any key establishment scheme, it should be benchmarked against possible alternative solutions in order to make sure that the best scheme for a particular use case is selected. In this section, we provide a rather general comparison of three major approaches for that purpose, namely CBKG, the elliptic-curve Diffie-Hellman key exchange (ECDH) protocol, which is one of the most widely used asymmetric key agreement protocols, and the manual distribution of keys by means of suitable passwords. The actual comparison is given in Table 17.1.

Clearly, a password-based approach excels in terms of computational complexity, bandwidth requirements (i.e. the amount of data that has to be transmitted for establishing a secure connection) and implementation costs but is still not really favourable due to lack of scalability, low ease of use and lack of efficient re-keying support. In this respect, re-keying refers to the complete or partial replacement of a cryptographic key in order to limit the time during which a certain key is used, thus leading to a higher security level. Moreover, it should be noted that if people themselves choose suitable passwords, the obtained keys usually do not have full entropy. ECDH, in contrast, has comparatively high requirements on the computational complexity and may be hard to use on a very lightweight 8-bit micro-controller, for example. In return, it is easy to use and provides at least to some extent re-keying support, even if this most probably would involve the generation of a complete full-length key. However, another major drawback of ECDH is the fact that it cannot be considered secure in the post-quantum age, thus coming along with the risk that with the advent of sufficiently powerful quantum computers a system may no longer be secure from one day to the other [7]. Depending on the concrete realization, CBKG may have moderate computational and bandwidth requirements, especially if the channel is estimated by the involved nodes anyway (e.g. along with actual data transmissions). It is highly flexible, provides a high ease of use and very efficient re-keying support, as with very little effort some new secret bits may be extracted out of the channel, thus yielding refreshed keys if combined with the old ones in an appropriate way. Also, it can be considered to be secure even in the post-quantum age as the secrecy comes from suitable physical properties rather than from computationally hard mathematical problems. Hence, especially compared to the other two alternatives, CBKG is certainly a promising solution that should be more and more put into consideration for practical use in future systems.

17.2 Fundamentals

The following sections summarize some fundamental aspects of key generation from reciprocal channel properties. In Section 17.2.1, we provide a compact summary of the information-theoretic foundations of CBKG, whereas Section 17.2.2 gives a general overview of the key generation principle and introduces a system architecture suitable for practical implementations. Finally, in Section 17.2.3, suitable key performance indicators are introduced and discussed.

17.2.1 Information-theoretic foundation

The problem of generating secret keys from common randomness was first studied in [5,8], allowing unlimited public feedback ('discussion') from one legitimate party, Bob, to the other one, Alice. Concise introductions into information-theoretic physical layer security are given in [1–3]. As reasoned above, herein we solely consider the case where each legitimate party obtains a set of size n of possibly noisy, memoryless channel observations as source of common randomness, which is termed the *source model*. This common knowledge of Alice, Bob and a passive eavesdropper Eve is modelled as three random variables, X^n, Y^n and Z^m, which follow a publicly known joint distribution. For the ease of this introduction, we assume that the m channel observations Z^m of an eavesdropper Eve are independent from those of Alice and Bob, which could be guaranteed in practice by having a sufficiently large distance between Eve and Alice/Bob. Alice and Bob are allowed to exchange k symbols of information (Ψ^k and Φ^k) with each other over an error-free channel. These messages are assumed to be known to Eve, hence the term public discussion. From this knowledge, each legitimate party generates a key $K_A = K_A(U_A, X^n, \Psi^k)$ and $K_B = K_B(U_B, Y^n, \Phi^k)$, respectively, where $U_{A/B}$ are two independent random variables generated by the respective terminals for initial randomization.

Then, a secret key rate R_S (in bit per observation/channel use) is achievable [2], if for every $\varepsilon > 0$ and sufficiently large n, there exists a private communication strategy such that (i) both generated keys are almost surely identical, $\Pr(K_A \neq K_B) < \varepsilon$; (ii) the 'leaked' mutual information $I(\Psi^k, \Phi^k; K_A)/n < e$ between the public discussion and the key is negligible; (iii) the key has an entropy rate $H(K_A)/n > R_S - \varepsilon$ close to the secret key rate; and (iv) the key has 'full' entropy (rate), i.e. $\log |\mathcal{K}|/n < H(K_A)/n + \varepsilon$, where $\mathcal{K}$ is the finite range of the random variable K_A representing the

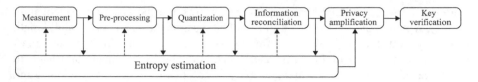

Figure 17.2   Generic system architecture of channel-based key generation

secret key. The largest achievable secret key rate (among all distributions of exchanged public information) is termed secret key capacity C_{SK}. For the source model $C_{SK} = I(X^n; Y^n)$, and it is achievable with a single forward or backward transmission [1].

17.2.2   General system architecture

The purpose of this section is to give a general overview of the system architecture for key generation. Practical aspects that occurred while implementing such a scheme will be addressed in Section 17.3. We consider Alice and Bob, who want to extract a symmetric cryptographic key from the wireless fading channel between them, and a passive attacker, Eve, eavesdropping on the communication of Alice and Bob and trying to determine exactly the same key. Under the given constraints, the key genera-tion process can be described as a modular sequence of various building blocks which generate binary strings, and thus eventually keys, from correlated measurements of reciprocal channel properties. A general representation of this system architecture is illustrated in Figure 17.2. In general, this process is conducted by both nodes, Alice and Bob, respectively.

In the first block *Channel Measurement*, the communicating nodes probe the wireless channel in order to measure suitable channel characteristics $c(t)$. Typically, this is either channel state information (CSI), e.g. the coefficients of the transfer function or its impulse response or the received signal strength indicator (RSSI), as will be discussed in more detail in Section 17.3.1.1. Let $c_{BA}(t_i)$ be the measurement obtained by Alice at time t_i and $c_{AB}(t_i + \Delta t_{AB})$ the measurement obtained by Bob at time $t_i + \Delta t_{AB}$. Then, Δt_{AB} needs to be considerably smaller than the coherence time T_c of the channel in order to retain the reciprocity of the channel at least in the noise and interference free case. After sometime $\Delta t_m = t_{i+1} - t_i$, Alice and Bob probe the channel again. They repeat this process until they have obtained a sequence of channel measurements of a length required to proceed with the next module.

The next block *Pre-processing* prepares the channel data for further processing, and especially and foremost compensates for differing measurement times and for correlations between subsequent measurements. A summary of suitable schemes and corresponding analyses can be found in e.g. [9,10].

The block *Quantization* then maps the fine-granular measurement values onto binary bit strings. Basically, lossless and lossy quantization can be distinguished. While lossless quantization (e.g. [11]) finds a corresponding binary symbol for every input value, lossy quantization (e.g. [12]) suppresses unreliable measurements in order to increase robustness against bit disagreement. In fact, there is always a general trade-off between the number of quantization bins and the bit disagreement ratio (BDR), which is the ratio of the number of bit disagreements and the overall length of the bit sequences. An increased number of quantization bins allows to retain most of the information carried by a measurement, but choosing too many bins leads to an increased BDR due to the quantization scheme's increased sensitivity to noise and interference. The output sequence of the quantization is called *Initial Key*.

Due to imperfect channel reciprocity, the initial keys of Alice and Bob are likely to mismatch. For coping with these mismatches, the key generation process includes the *information reconciliation (IR)* module. In this block, the communicating nodes run a dedicated protocol to exchange, for instance, parity information, which allows them to identify and correct possible disagreements between the initial keys with high probability. An extensive survey on different approaches for IR can be found in [13]. Although we call the bit string resulting from IR the *Aligned Key*, it should be noted that there is still some non-zero probability of a residual mismatch. IR relies on public information exchange between Alice and Bob. Hence, the attacker may have eavesdropped on this communication and may also have partial knowledge of the channel measurements of Alice and Bob through correlated channel observations. Consequently, there is a need for an *Entropy Estimation* module. In Figure 17.2, this is illustrated as a functional block parallel to the previously introduced processing sequence. It allows to gather results from each processing step, perform some statistical analysis and generate control signals for adaptivity purposes. Taking potential knowledge of the attacker about the secret key into account allows to estimate the residual security level of the derived and aligned bit string.

Having this knowledge, in the following block *Privacy Amplification*, the effective entropy of the aligned key can be condensed into a shorter key sequence by applying for example cryptographic hash functions. Finally, since there is still a residual probability of mismatching keys, in the *Key Validation* phase the nodes follow a protocol for again checking the keys to be identical.

17.2.3   Major metrics for performance evaluation

A basic requirement for efficient CBKG is a high degree of channel reciprocity. As a first order approximation, reciprocity can be measured by the Pearson cross-correlation coefficient between Alice's and Bob's channel measurements. We note that using cross-correlation itself is not a suitable description of reciprocity, because for random processes with non-zero mean, their mean value usually dominates the cross-correlation but, by nature, is not part of the reciprocal randomness. In contrast, the cross-correlation coefficient is the cross-covariance normalized by both standard deviations, which is better suited for that purpose. However, this coefficient does not sufficiently describe the distributions if they are non-Gaussian. Furthermore, the correlation coefficient does not give any information about the amount of randomness contained in the channel. The original measure for the amount of usable randomness offered by the channel is the mutual information between both measurements. It is also a suitable description of reciprocity for non-Gaussian channels and it closely relates to the cross-correlation coefficient for normally distributed channels.

Quite many publications (e.g. [10–12]) provide key generation rates that the respective authors have achieved in practical systems. However, care should be taken when comparing these results with each other since they might have been determined in different ways. Some authors (e.g. [10]), for example, give a key generation rate directly after quantization without requiring the bits to be properly reconciled. However, when relaxing IR requirements, arbitrarily high key generation rates can be obtained with any quantization scheme. Some other authors (e.g. [11,12]) compare the key generation rate after IR but do not consider the entropy contained in these bits. As an example, by many measurements within the coherence time, one could have obtained highly correlated samples. Proper IR might then be possible with few parity bits even if using a multi-level quantization, because of the correlation of the bits. Hence, the key generation rate after IR will be high, but the entropy contained in the bits will be low or even zero due to the correlation. For a fair comparison, the entropy of reconciled bits or, equivalently, the effective key bit generation rate (KBGR), which is the output bit rate of the privacy amplification, should be used. Naturally, this is then equivalent to the remaining (after subtracting the leaked parity information) mutual information between the reconciled bits of Alice and Bob, because the entropy of the bits given the other party's bits vanishes after proper IR.

17.3   Channel-based key generation in practice

CBKG uses the wireless fading channel between two legitimate parties as a common source of randomness. How such schemes can be implemented under the constraints of practical systems is described in this section. It shortly summarizes the bottom line of experiences we made with our practical implementations.

17.3.1   Channel characterization

In this section, we review typical characteristics of wireless fading channels that can be used for key generation in practice. Furthermore, two possible types of raw data, namely RSSI and CSI, are compared with respect to their suitability for practical key generation. Finally, various effects occurring while measuring the channel and resulting consequences for practical key generation are explained.

17.3.1.1   Wireless channel model

Scattering, reflection and diffraction cause signal fading effects, which are attributed to the wireless channel[1] [14]. We focus on the effects being relevant for CBKG in an indoor/office scenario and consider fading on three different types of scales [15]: (1) Short-term: During the transmission of a single packet, the channel is approximately static, but between two packet transmissions the channel evolves such that subsequent channel observations will be correlated. (2) Mid-term: Small-scale fading describes the evolution of the channel between packet transmissions in terms of

---

[1]All systems and signals are considered in the equivalent, complex-valued low-pass domain.

temporal correlation. It is restricted to spatial variations smaller than approximately ten carrier wave lengths λ_c. Here, the channel is typically wide-sense stationary and uncorrelated scattering (WSSUS) as will be explained in the sequel. (3) Long-term: Large-scale fading is caused by movements along distances being $> 10\lambda_c$. Shadowing and path-loss effects cause the channel to become non-wide-sense stationary.

With broadband transmission schemes like orthogonal frequency division multiplex (OFDM) or frequency-hopping spread spectrum (FHSS), frequency diversity can be exploited for key generation in addition to the eminent temporal diversity. Using linear time-variant (LTV) system theory, the time-variant (TV) impulse response $h(t; \tau)$, which is the response of the system to a Dirac impulse issued at the input the lag time τ seconds ago, determines the channel. The system's TV transfer function is given by the Fourier transform with respect to lag time $\tau \circ\!\!-\!\!\bullet f$. Its spreading function is the Fourier transform with respect to time $t \circ\!\!-\!\!\bullet \nu$. If the system is time-invariant, then $h(t; \tau) = h(\tau)$. Due to causality, $h(t; \tau) = 0 \, \forall \tau < 0$ holds.

Although the physics of the wave propagation, and hence the channel, are well-defined, exact and deterministic with perfect knowledge of the environment (physical dimensions, material constants, surface roughness, etc.), all these details cannot be captured in practice. Therefore, these effects can be assumed to be random to some extent. The small-scale fading effects typically yield a Rician fading channel, where the samples of the impulse response form a circular complex Gaussian random vector[2] (over time and lag). The large-scale fading effects influence mean and variance of the Rician channel and are hence arbitrary but fixed. If there exists a specular component, e.g. a line-of-sight (LOS) component or a specular reflection on a flat (with respect to λ_c) surface, then the corresponding contribution to the channel will be deterministic[3] [15]. The other, diffuse parts of the channel are of zero mean such that the second order statistics of the TV impulse response are given as

$$\mathbb{E}\left\{\tilde{h}(t_1; \tau_1)\tilde{h}^*(t_2; \tau_2)\right\} = \bar{h}(\tau_1)\bar{h}^*(\tau_2) + \sigma_h^2 r(\Delta t; \tau_1)\delta(\tau_2 - \tau_1), \qquad (17.1)$$

where $\mathbb{E}$ denotes the ensemble average, σ_h^2 the total variance and $r(\Delta t; \tau_1)$ is the time-lag correlation function [16] normalized to unit variance (see below). For simplicity, let's assume here that the channel has a (deterministic) time-invariant mean $\bar{h}(\tau)$. For spatial variations (of transmitter, scatterers or receiver) within a range of approximately $10\lambda_c$ [15], the channel typically is WSSUS [16], which is already included in the right hand side of (17.1): With wide-sense stationary (WSS), at least the first and second order moments are stationary (for Gaussian processes this implies strict stationarity), i.e. they only depend on the time difference $\Delta t = t_2 - t_1$. Scatterers located at such different locations that their paths observe different delay times, typically cause uncorrelated contributions to the channel. Hence, the random variables at different lags τ_1 and τ_2 are uncorrelated.

---

[2]In fact, if the signal is continuous in time, its source is a random *process*, but the distinction between random variable, random vector and random process will be neglected in the sequel.

[3]By deterministic we here mean that these values are constant over spatial ranges of some $10\lambda_c$.

In general, large-scale fading will only influence mean and variance of $h(t; \tau)$, which depend on the location of Alice and Bob and hence will be TV. Furthermore, movements cause a Doppler shift of the specular components, which results in a TV mean rotating in the complex plane with its Doppler frequency. Hence, the WSS property will be relaxed accordingly in Section 17.3.1.3.

The time-lag correlation function $r(\Delta t; \tau_1)$ describes the temporal auto-correlation of each (uncorrelated) channel tap. It extends the classic temporal correlation function towards lag time and hence is essential for temporal de-correlation of the channel measurements. Its Fourier transform with respect to $\Delta t \circ\!\!-\!\!\bullet\, \nu_1$, the scattering function $C(\nu_1; \tau_1)$, shows the power, or variance, of each uncorrelated pair of Doppler frequency ν_1 and lag τ_1. Therefore, it can be used for calculating the mutual information between Alice's and Bob's channel measurements as detailed in Section 17.4.1. It can be estimated from measured spreading functions using the Wiener–Khinchin theorem, which also holds for the TV case under some mild conditions [16].

17.3.1.2 RSSI versus CSI

CBKG requires that an eavesdropper's channel measurements quickly become independent to those of the legitimate parties within small distances. Hence, in practice, only small-scale fading characteristics of the channel should be used for CBKG, and the choice of an appropriate type of channel measurement plays a vital role for the quality of the raw material. Typically, RSSI and CSI are used. As CSI, we consider the TV channel impulse response (CIR) or any of its Fourier transforms. For these, there exists a well-understood analytic model (cf. Section 17.3.1.1), which matches practice and still is mostly analytically traceable due to the Gaussian distribution such that optimum signal processing like de-correlation over time, and frequency is easier to obtain. Theoretical limits are also available for comparison, and CSI values exhibit large temporal/local dynamics. Hence, they offer a large amount of entropy for key generation, although it is challenging to extract reciprocal randomness from the vivid phase of the complex-valued CSI in practice. With its circular-complex Gaussian random variables, the channel's real and imaginary part are independent from each other, and this implies that magnitude and phase of this random variable are correlated with each other. Hence, they cannot be used as separate input to two parallel key generation processes. From a practical point of view, CSI values are seldom accessible from higher layers. Some wireless schemes like IEEE 802.11n or LTE require CSI values internally for equalization or beamforming, while others like IEEE 802.15.4 get along without explicit knowledge of CSI. Thus, using CSI values requires deep access into the integrated circuit (IC) for baseband signal processing.

In contrast, RSSI values are easy to obtain as they are almost certainly provided to the user space by the used hardware driver. However, they vary slowly compared to CSI values due to the RSSI mostly being part of the (slow) automatic gain control (AGC) loop. In addition, RSSI values are just real numbers compared to complex-valued CSI. Both of these properties cause a decreased key generation rates. Furthermore, usually frequency selectivity is hard to exploit with RSSI, because one RSSI value is obtained from a broadband energy measurement. Exemptions are FHSS schemes in primal IEEE 802.11 (WiFi) or in IEEE 802.15.1/4 (Bluetooth/ZigBee physical layer) standards, where some baseband ICs provide RSSI values per used frequency hopping channel. Since the distribution of the RSSI values is typically found to be non-Gaussian, signal processing like de-correlation becomes very challenging. Even theoretical limits are hard to determine. Because the RSSI are determined near the beginning of the receive chain by nature, the necessarily broadband measurement method offers the possibility of manipulations by an active attacker: Narrowband, high power signals transmitted near the edges of the transmission band can dominate the measured RSSI values and hence can force measurements of the legitimate parties to take values known by the attacker. This manipulating signal does not necessarily disturb the in-band data transmission significantly such that the legitimate parties may not detect the manipulation.

In conclusion, RSSI values are easy to obtain but only provide rather small key generation rates and suffer from an increased vulnerability to active attacks. It is thus preferable to use CSI values for their larger key generation rate and their larger robustness against the RSSI attack, although they are more complex to obtain. In order to reduce this effort, it is advisable to integrate CBKG into the baseband signal processing IC. Alternatively, the IC manufacturer could provide application programming interfaces which offer frequent CSI value updates to the user space.

17.3.1.3 Practical channel model and implications

When considering practical scenarios, Alice and Bob transmit training sequences to each other in order to let the other party measure the channel. Then, two synchronization imperfections need to be considered in addition to the characteristics described in Section 17.3.1.1: The carrier frequency offset (CFO) is due to a Doppler shift of the carrier frequency and non-synchronized carrier oscillators at transmitter and receiver. It is considered being sufficiently small (maybe after applying a synchronization scheme) such that the sub-carriers of an OFDM scheme remain free of inter-carrier interference (ICI) during the transmission of one packet. However, the residual CFO after synchronization will let subsequent measurements undergo a phase rotation between two channel measurements of several multiples of 2π such that the bare measurements cannot be assumed having a reciprocal carrier phase.

Second, a timing phase offset (TPO) of a fraction of a sample duration will not cause inter-symbol interference (ISI) but a linear non-reciprocal phase progression of the transfer function over frequency. A timing frequency offset is practically not significant during one symbol transmission such that its effect can be neglected.

In addition, because both legitimate parties have asynchronous local oscillators, which cause non-reciprocal carrier phases (often further influenced by carrier tracking loops), only differences with respect to a phase reference can be used for CBKG. If the spatial variation becomes larger than $10\lambda_c$, the channel becomes non-stationary. The large-scale channel magnitude will be proportional to path loss and shadowing. The overall model of the measured channel $\hat{h}(t; \tau)$ then becomes:

$$\hat{h}(t; \tau) = L(\bar{t})e^{j2\pi \Delta vt + j\Phi_0(t)}h(t; \tau - \tau_0(t)) + n(t; \tau), \tag{17.2}$$

where path-loss and shadowing effects are combined into the amplitude factor $L(\bar{t})$. The term further factors into CFO and unknown carrier phase, into the actual WSSUS channel $h(t; \tau - \tau_0(t))$ with lag-shift by $\tau_0(t)$ caused by TPO and adds additive white Gaussian noise (AWGN) $n(t, \tau)$. While a dependency on t denotes a variation with every measurement, a dependency on $\bar{t}$ denotes long-term effects. Furthermore, the measured CIR will not be causal, because the transmit signal and hence the measurement of the channel's transfer function is bandlimited.

As already noted, large-scale fading effects should not be exploited for key generation because of their large area of high spatial correlation. This influences the usable SNR for quantization. Let γ_{rx} denote the SNR of a signal received over a Rician channel with Rice factor $K = \bar{h}^2/\sigma_h^2$. Then, only the share of variance σ_h^2 is usable such that the SNR γ_Q for quantization becomes:

$$\gamma_Q = \frac{\gamma_{\text{rx}}}{1 + K}. \tag{17.3}$$

If the K factor is e.g. 20 dB, then the SNR will be reduced by approximately 20 dB.

17.3.2 Pre-processing

Pre-processing includes *Interpolation*, *Reciprocity Enhancement* and *De-correlation*. Interpolation addresses possible disagreements between Alice's and Bob's measurements, which are introduced if Alice and Bob cannot probe exactly the same channel instance. This may be due to channel usage at different times or frequencies, which, for instance, occurs when following time division duplex (TDD) or frequency division duplex (FDD) measurement schemes, respectively. Reciprocity enhancement is dedicated to mismatches introduced by imperfect channel reciprocity, which is typically induced by heterogeneous hardware, interference and noise. These defects can be mitigated, for instance, by curve fitting techniques like regression or Savitzky–Golay filtering. Finally, de-correlation aggregates the entropy of the measured sequence and allows to quantize the de-correlated samples separately from each other. In time domain, a straightforward way to de-correlate the measurement data could be to estimate the coherence time T_c on basis of the estimated auto-correlation function and to adjust the measurement rate to be larger than $1/T_c$. This will result in a very energy efficient CBKG scheme albeit with low KBGR, because every uncorrelated measurement value contains a maximum of entropy. In practice, we observed uncorrelated measurements, when taking them at two to four times $1/T_c$. Alternatively, appropriate de-correlation transforms such as the Karhunen–Loéve transform (KLT) can be applied, which diagonalizes the components' covariance matrix. For stationary processes as in a WSSUS channels, the KLT is given by the discrete Fourier transform (DFT). This transform has the practical advantage that the transform is independent from the actual channel's statistics and hence does not require an alignment of the transform matrix. In [9], it was shown that applying identical de-correlation transform matrices at both, Alice and Bob, is detrimental in practical systems.

17.3.3 Quantization

A wide variety of different quantization schemes has been proposed in the literature. Hershey *et al.* were the first ones to publish a scheme for CBKG in [17], while practically more feasible implementations were proposed in [18]. They are similar to the quantization schemes described in [11,19], which considered RSSI values as the channel metric. While the scheme in [11] divides the range between the minimum to the maximum measurement value into a given number of equally sized bins, the method in [19] estimates the mean and standard deviation of the measured sequence in order to determine multiple thresholds. [10] describes a scheme resulting in equally likely quantization symbols based on the channel's empirical probability density function (PDF). This increases the entropy of the generated symbol sequence. However, at moderate and high SNR values, which are those where CBKG is only practically feasible at the moment, we observed from theoretical considerations that the gain in entropy is negligible compared to uniform quantization, which requires less effort to implement. Several more key generation schemes have been proposed (e.g. [12,20–23]). Comparisons of selected quantization techniques in the field of CBKG can be found in [24]. From a practical point of view, also complexity, security and adaptivity are crucial. A PDF-based approach may have beneficial properties at first sight, but it can be rather complex and inaccurate to compute thresholds based on few observations. Furthermore, some quantization schemes have been shown to be vulnerable to active attacks [25]. Particularly in TV environments, quantization needs to be sufficiently adaptive to varying channel conditions. In practice, we had good experience with (lossy) quantization schemes using guard intervals, which adapted their number of quantization bins depending on the BDR estimated from the number of corrected errors in IR as well as adapted their thresholds to the estimated standard deviation of the measurements.

17.3.4 Information reconciliation

Two types of protocols can be distinguished: *source coding with side information* and *interactive error correction protocols*. In the first case, Alice encodes her sequence X to $a(X)$, which is sent to Bob. If X is sufficiently correlated to Bob's sequence Y, Bob should be able to reconstruct X from $a(X)$ and Y, while the amount of entropy leaked to Eve by sharing $a(X)$ should be as small as possible. Corresponding schemes for reducing entropy leakage include *Code-Offset Construction* and *Syndrome Construction* [26], which employ error correcting codes to derive $a(X)$, such as Turbo or LDPC codes. If using interactive error correction protocols, Alice and Bob find disagreements through iterative bi-section and public discussion. Well-known protocols are Cascade [27] and the Winnow [28] protocol, for example. An extensive survey on different approaches for IR can be found in [13]. In practice, we prefer the former schemes (in particular code-offset construction) to interactive protocols, although the latter being inherently adaptive to changing channel conditions. The reason is that interactive protocols typically exchange more packets and hence are less energy efficient as a main share of the overall energy effort is caused by packet transmissions.

Furthermore, interactive protocols tend to leak slightly more information than code-based ones given a certain input error rate. Code-offset construction in particular can be used with practically any error correction code and is hence very flexible. For an adaptive system, we choose from a codebook of available codes based on feedback of the number of corrected errors in previous blocks.

17.3.5 Entropy estimation

By applying entropy estimation tests, e.g. as proposed by the National Institute of Standards and Technology (NIST) in [29], the entropy per bit of the aligned key can be estimated. Taking potential knowledge of the attacker about the secret key into account allows to estimate the residual security level of the derived and aligned bit string. Since the estimation is performed during the actual key generation procedure, only rather short data sequences are available. However, the statistical analysis of short random sequences implies a high degree of uncertainty, so appropriate schemes need to be chosen carefully. Several methods have been proposed for estimation of entropy based on observations [30–32]. However, these results only apply to asymptotically long sequences, which may not be appropriate here. Pessimistic estimates cause overhead and decrease the efficiency of the system. Optimistic estimates reduce the overall security of the system. In practice, we found that decomposed context tree weighting (DCTW) [33] and Lempel–Ziv compression [32] might be promising candidates. Apart from that, in practice, a surveillance module should be implemented, which detects critical systems states, e.g. arising from static channels or interfered channel measurements. A proper means of detecting static channel is e.g. to count the zero-crossing points within a time period. If too few zero-crossings are within the signal, the channel is considered being static, and measurements within this interval are excluded from key generation.

17.3.6 Privacy amplification and key verification

A practical scheme for applying privacy amplification is to buffer reconciled bits until their total estimated entropy is larger than the required level of secrecy. Then, the privacy amplification is applied onto all data using a cryptographic hash function in order to obtain a secret key. For security reasons, we chose the future-proof SHA-3 algorithm. In practice, it is essential to verify the key afterwards [34]. To this end, Alice may use her secret key K_A to encrypt a random number n and transmit the result to Bob afterwards, who de-crypts it and adds 1. After encrypting the result with his secret key K_B, Bob transmits it to Alice, who again de-crypts and checks if the random number is indeed $n + 1$. If so, the keys K_A and K_B must be identical.

17.3.7 Security considerations and energy consumption of CBKG

Apart from the already considered passive eavesdropping, also active attacks are possible, e.g. jamming/denial of service (DoS) or interference with receive signals. Depending on an assessment of the power of an adversary (technical expertise, knowledge about the scheme, window of opportunity, equipment, etc.), we found the following attacks to be immediately threatening: (i) classic man-in-the-middle attacks. By impersonating Alice or Bob and causing both nodes to re-pair, the attacking node may become able to relay and replay exchanged data. While relaying of the RF signal is hard to detect or prevent in practice, man-in-the-middle attacks can be addressed by proper authentication and time-stamp mechanisms (cf. Section 17.5.2). Implementing a proper surveillance module (cf. previous section), which detects an implausible change of channel conditions, can be another suitable countermeasure. (ii) the manipulation of RSSI values as described in Section 17.3.1.2. Besides the mentioned countermeasures, key verification (cf. previous section) is an essential building block for detecting such attacks. (iii) jamming and a classic DoS attack like flooding one party with key exchange requests/initiations. Basically, it is hard to secure a scheme against jamming or DoS attacks under the given constraints.

In general, CBKG is considered a feasible alternative to well-established schemes like ECDH, which may require complex arithmetic computations and put a high strain onto the device's battery power. Although a major advantage of CBKG is its low computational complexity, little effort has been put into analysing CBKG's energy consumption on embedded devices. While well-optimized implementations of ECDH are available, CBKG is still to be analysed in modern security research. In [35], a direct comparison of these two key agreement schemes is provided, taking the respective energy consumption into account. As results show in line with [36], the overall energy consumption of CBKG is dominated by the energy required for the transmission and reception of data (including the estimation of the channel). For the considered systems, ECDH consumes less energy than CBKG. However, the energy required by CBKG for generating a single key significantly depends on the utilized quantization and IR schemes. It is shown that CBKG may consume less energy than ECDH if the number of required packets can be limited to a certain amount, e.g. by further optimizing the used algorithms. Also the required channel estimates could be taken from regular data exchange or from communication during the reconciliation phase, which relaxes the need for dedicated probing transmissions. Furthermore, sensors may be used to detect movements of the device. This way channel probing can be triggered whenever dynamic channels are expected.

17.4 Experimental results

In the previous sections, the general process of CBKG and typical channel characteristics have been described and a link to practical issues was established. In the sequel, two practical implementations are presented and evaluated by experiments. First, a setup for obtaining real-world CSI measurements is described in Section 17.4.1. Subsequently, another off-the-shelf setup is presented, which limits the access of channel properties to RSSI values. Although results of experiments on CBKG have been presented in the literature already, the evaluation methods often have been rather individual by considering different building blocks, parameters and performance metrics. Most of these experiments were carried out using RSSI values, e.g. [10,11]. As already stated in Section 17.2.3, some of the publications gave results which are hard to compare with each other in a fair way. For instance, [10] provides key generation rates of un-reconciled bit strings. [11,12] do not consider the entropy contained in these bits. Based on the system architecture introduced in Figure 17.2 and previously defined performance metrics, a clear and reproducible evaluation framework can be designed. Another extension to existing work is the consideration of online entropy estimation as introduced in Section 17.3.5. Although this cannot guarantee asymptotic entropy estimation, it prevents the results from being too optimistic and supports to compare the results in a fair manner. The results obtained for BDR and KBGR emphasize the individual key generation performance depending on the considered scenario and the way of evaluation.

17.4.1 CSI-based experiments

Few experiments exist, which use CSI measurements obtained by means of wireless transmission standards (in our case IEEE 802.11). Ye *et al.* [37] apply a level-crossing algorithm to the first tap of a CIR of an IEEE 802.11a system. Mathur *et al.* [12] only use the magnitude of the strongest tap of the CIR measured by a 802.11a format compliant signal. However, neither frequency selectivity nor the channel phase have been used. References [38–40] use an IEEE 802.11n system for measuring CSI and give practical key generation rates for very specific systems. However, neither of the previously cited publications calculate a theoretical limit on base of practical measurements. Reference [23] uses a channel sounder for measuring eight frequency points of the channel's transfer function in a 80 MHz wide band at 2.5 GHz and calculate the mutual information between Alice's and Bob's measurements, also given Eve's measurements. However, the authors do not used a standard-compliant system. Our experiment using CSI as raw data is implemented on a software-defined radio (SDR) as a flexible prototyping platform since arbitrary parameters of a wireless transmission can be accessed and adjusted. We use this experiment in order to find out how close a practical CBKG scheme may get to the theoretic optimum for a specific measured channel. Using GNU Radio with Ettus USRP N210 hardware, we implemented a channel measurement and data transmission scheme according to the IEEE 802.11n WiFi standard. On the application layer, Alice and Bob exchange OFDM packets every 10 ms in a time-division duplex scheme with a sequence number for the alignment of measurements as payload, which is saved in a file together with the channel estimates for offline signal processing in MATLAB. There, Alice's and Bob's measurements are aligned by the saved sequence number and pre-processed as described below.

The perturbations of the channel measurements, as described in Section 17.3.1.3, need to be compensated in order to obtain a reciprocal and WSS TV channel estimate. The procedure implemented for this purpose is as follows: The linear phase of the channel transfer function caused by the TPO is estimated using linear regression of the phase and then compensated by multiplying the sub-carriers of the transfer function with the conjugate complex of that estimate. In order to compensate for the CFO and an unknown carrier phase between measurements, all sub-carriers are de-rotated by the phase of the reference sub-carrier. In order to compensate for the large-scale fading effects and to obtain a WSS channel measurement, the slowly TV power of the overall received signal (all sub-carriers) is estimated. Then, the channel measurements are normalized such that they have unit variance. For channels with uncorrelated scattering in time domain, discrete sub-carriers of the channel transfer function are de-correlated by applying the inverse discrete Fourier transform (DFT). From thereon, each tap is handled independently. Now each tap might have a specular component with independent Doppler frequency resulting in a slowly TV mean. Furthermore, some taps might gain influence over time, while others diminish. Hence, mean and variance are estimated and tracked over time. The measurements are compensated for the mean and variance, which results in a TV variance of the noise and hence TV SNR. By applying the extension of the Wiener–Khinchin theorem and averaging over multiple measurements, the scattering function is obtained.

As of the time writing, the experiment was still work-in-progress such that they were only implemented up to the pre-processing block of Figure 17.2. Further results from implementing the remaining signal processing chain are yet to come. For these steps of the key generation process only those taps are used in those time intervals, where they exhibit a SNR being sufficiently large for successful quantization and IR. This of course depends on the used schemes in these steps.

An exemplary measurement result of the experiment is shown in Figure 17.3 for an office scenario, where Alice and Bob were at fixed positions 10 m apart from each other, but people crossed the LOS path. For a measurement bandwidth of 40 MHz, the TV channel transfer function is shown with magnitude and phase after the previously described signal processing in Figure 17.3(a) and (b), respectively. It can be observed visually that Alice and Bob obtain quite similar measurements, both in amplitude and in phase. In this office scenario, the channel is only moderately frequency selective, as the magnitude varies by only some 15 dB. Apart from the zero crossing for the non-occupied zero sub-carrier its phase changes only moderately, either. This is supported by the power delay profile in Figure 17.3(c), where only few taps show significant power contributions. The Doppler power spectral density (PSD) in Figure 17.3(d) does not exhibit a U-shape typical for the Jakes PSD, but rather a bell-shaped one with centre around zero Doppler frequency. This implies that the scattering points are not equally distributed on a circle around the receiver as in the Jakes model, but rather off the LOS path and to its sides.

In order to obtain an upper bound R_S (given in bit/s) to the maximum number of (secret) bits, which can be extracted from a particular set of measurements, we extend [41, Sec. II.A–C] to WSSUS TV frequency-selective channels. We calculate the mutual information $I(\mathbf{h}_A; \mathbf{h}_B)$ between the vector of Alice's and Bob's measurement samples of the TV CIR, $\mathbf{h}_A$ and $\mathbf{h}_B$, respectively, obtained over a period T_m. Under the empirically backed assumption of equal albeit TV SNR at Alice and Bob, the mutual information (as an upper bound) is calculated assuming that the measurements follow a joint Gaussian distribution, characterized by the discrete scattering function $C(\nu_l; \tau_l)$. For a time-discrete WSSUS channel, the (inverse) DFT is the solution to the Eigen problem of the KLT and hence de-correlates samples if applied to both dimensions, time and (transfer) frequency. Hence, the scattering function of such a channel corresponds to their variance, and for AWGN thus to the SNR γ_i,

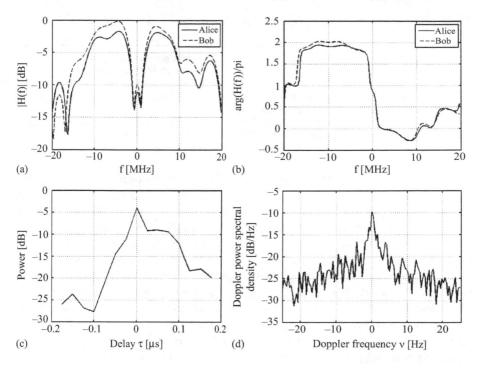

Figure 17.3 *Exemplary measurement results for CSI in an office scenario: (a) magnitude and (b) phase of the TV channel transfer function at one time instant; (c) power delay profile and (d) Doppler power spectral density*

of each uncorrelated random variable x_i on the Doppler frequency–lag grid. Then, the upper bound to the secret key rate is easily derived for Gaussian distributions as [41, Sec. II.C]:

$$R_S = \frac{1}{T_m} I(\mathbf{h}_A; \mathbf{h}_B) = \frac{1}{T_m} \sum_{i:(\tau_1, v_1)} \log_2 \left(1 + \frac{\gamma_i^2}{1 + 2\gamma_i} \right), \tag{17.4}$$

where the sum is taken over all sample tuples of $(v_1; \tau_1)$.

In order to give some coarse impression on the achievable secret KBGR, Table 17.2 shows summarized and averaged results for the upper bound on the secret key rate calculated from a channel measurement campaign, which was conducted in our open-plan office in the 2.4 GHz ISM/WiFi band at 40 MHz bandwidth in different scenarios. In the first kind of scenarios, Alice and Bob were static on a desk approximately 1 m apart from each other with LOS connection. It becomes obvious that the completely static environment exhibits almost no usable randomness for CBKG. In the second kind of scenarios, Alice was static while Bob was moved on the same desk, still maintaining a strong LOS connection at approximately 1 m distance. Short distances between Alice and Bob result in channels with large Rician K-factors, where

Table 17.2   Secret key rates of real-world channels for different scenarios

Approx. distance Alice–Bob (m)	Scenario	Upper bound R_S [bit/s] in 40 MHz bandwidth		
1	All static	0.01	...	0.4
1	Mobile environment	0.2	...	30
10	Mobile, strong LOS	70	...	100
10	Mobile, non-LOS	120	...	240

the unusable specular components predominate the diffuse ones usable for CBKG. The same is true considering larger distances, where a strong LOS link exists, e.g. when Alice is placed static on a desk and Bob is moved around within the office at pedestrian velocities. The largest secret key rate is offered by mobile channels with weak specular components, which is consistent with the considerations made in Section 17.3.1.1. This kind of scenarios occurred when Alice was placed static on the desk and Bob was moving through metal/glass doors into rooms with metal-shielded walls at distances of about 10 m. Similar results were reported in [23] for a LOS link of up to 50 m, which find the mutual information to be within 40 Bits to 80 Bits per channel. However, [23] measured only on eight sub-carriers but over 80 MHz of bandwidth (at 2.5 GHz carrier frequency) and at a measurement rate of $\frac{1}{3\,\text{ms}}$. As mutual information strongly and non-linearly depends on the number of measurements and their correlation (increasing the measurement rate monotonically increases mutual information), it is hard to compare both results with each other. At least, [38–40] achieve a secret KBGR of up to 60 to 90 bits/s in mobile scenarios with IEEE 802.11 systems, which is smaller than the theoretical limits given in Table 17.2 but of the same order of magnitude.

From the evaluation of a measurement campaign in office scenarios, the following conclusion can be drawn: In scenarios, which are considered difficult for wireless transmissions like non-LOS scenarios, large distances or in-band interference, one can still obtain reciprocal channel measurements after complex signal conditioning. Even without any movement of Alice and Bob, the resulting channel still provides highly reciprocal TV randomness, mainly due to not necessarily noticeable movement of scattering points around the devices. However, interference from other devices on the same WiFi channel cause some non-reciprocity. And even with a non-LOS connection there can be strong specular components by reflections of flat (with respect to λ_c) surfaces, e.g. metal coated window glass or metal frames in walls, which cause non-zero mean taps. Non-LOS scenarios show shorter coherence times of the channel, i.e. a higher secret key rate per unit time. Furthermore, the channel exhibit only moderate frequency-selectivity even across a bandwidth of 40 MHz. Approximately, seven of the 128 taps have such a large SNR that they significantly contribute to the secret key rate. Using these taps, let the secret key rate increase by a factor of 2.5 to 3 compared to the secret key rate for the first tap taken alone.

Since the channel is moderately frequency-selective within the considered transmission band, the phase between the reference and any sub-carrier of the band can be highly correlated. In these cases, only little randomness can be extracted from the channel phase. In general measurements exist, where the distribution of the CSI fits well a Gaussian distribution, but there are also measurements, where it is clearly non-Gaussian. However, a complex Gaussian distribution of the first channel tap was found in every case. We never observed a Jakes PSD, but rather a bell-shaped one. This implies that less entropy is offered by the channel than with Jakes PSD for a given channel coherence time or maximum occurring Doppler frequency.

17.4.2 RSSI-based experiments

For running RSSI-based experiments, another system has been implemented using off-the-shelf hardware. The setup consists of a laptop PC, which serves as a monitoring device, and three Raspberry Pi devices (Alice, Bob and Eve), which can be equipped with different WiFi adapters in order to realize heterogeneous communication. The WiFi adapters are configured in 802.11n monitoring mode, which allows to observe the channel and to capture every packet in the air. The channel is probed periodically by injecting WiFi management frames. Clearly, these packets may also be observed by Eve eavesdropping on the channel. Every time a packet is received, a RSSI value can be determined, thus yielding a sequence of RSSI values over time. For evaluation purposes, the measurements are provided to the monitoring system, which processes the data and mimics a distributed key generation procedure of Alice, Bob and Eve, as summarized in Table 17.3. The procedure considered in the given setup is based on the architecture known from Figure 17.2 [42] as follows [42]:

The pre-processing schemes summarized in Section 17.2.2 are optional and omitted here. Instead, a moving average filter is applied, which mitigates the effects of path loss and shadowing. This becomes essential particularly in case of high mobility scenarios with significantly varying distance between transmitter and receiver. The resulting data is then quantized by applying a lossless 2-bit scheme according to [11]. This scheme detects each sample in one of multiple quantization bins and assigns a binary symbol of certain bit length. The RSSI sequence is divided into sub-blocks of 250 measurements in order to allow individual computations of corresponding quantization bounds. Then, the range from the minimum to the maximum RSSI value in each sub-block is divided into quantization bins of equal size. Afterwards, we use a syndrome-based IR scheme [26] with a (255,47,85) BCH code, where Alice computes a syndrome $S_A = \text{syn}(X)$ of her RSSI sequence X and transmits it to Bob. Upon receiving S_A, Bob equivalently computes $S_B = \text{syn}(Y)$ of his sequence Y and derives $\text{syn}(X - Y) = \text{syn}(X) - \text{syn}(Y)$. By decoding $\text{syn}(X - Y)$, Bob obtains a sequence Z such that $X = Y + Z$.

Different to Figure 17.2, entropy estimation is not performed after each modular step, but only after IR. This is done by utilizing DCTW [33], a lossless compression scheme utilizing variable order Markov models to derive paths of a context-tree as the probabilities of symbol occurrence. Although compression cannot be expected to allow exhaustive entropy estimation, still it is a suitable way to address redundancy as one statistical defect. Additionally, the entropy loss due to disclosure during IR is taken into account. These considerations allow improved approximation of the actual

Table 17.3   Considered practical system realization

Building block	Scheme
Pre-processing	Moving average filter
Quantization	Lossless multi-level quantization [11]
Information reconciliation	Syndrome-based information reconciliation [26]
Entropy estimation	Decomposed context tree weighting (DCTW) [33]
Privacy amplification	SHA-3

secret information commonly known to Alice and Bob. Finally, once a sufficient amount of entropy has been gathered, for privacy amplification a SHA-3 function is used to compress the obtained bit sequence to a key string of required length. Direct comparison of Alice's and Bob's keys validates successful key agreement.

The previously described setup has been used to perform measurements in different environments and mobility levels. We considered outdoor (O) and indoor office environments (I), as well as static (S) and mobile (M) nodes, and homogeneous (Hom) and heterogeneous (Het) setups. This results in eight different scenarios with significantly different results, as will be shown later. The measurements are performed for a fixed duration of 10 min each, while Bob is either moving along a circular trajectory in mobile scenarios or staying at a fixed position without LOS in static setups. The attacker Eve is always considered to stay at a fixed location with a constant distance of 15 cm to Alice, while the distance between Alice and Bob is 7.5 m in the static indoor scenario and around 20 m in the static outdoor scenario. Due to the circular movement trajectory, in the mobile setups the distance between Alice and Bob varies in a range from 5 to 12 m (indoor) and 10 to 30 m (outdoor). Although the channel probing rate is set to 1 kHz, due to packet loss the effective probing rate varies for each scenario. For instance, in case of the homogeneous and mobile indoor scenario (I-M-Hom), the effective probing rate is 170 probes/s, while it drops to 122 probes/s in case of the heterogeneous setup I-M-Het.

Figure 17.4(a) gives an impression of the RSSI measurements gathered by Alice ($RSSI_{BA}$), Bob ($RSSI_{AB}$) and Eve ($RSSI_{BE}$) in case of the static indoor scenario I-S-Hom. Apparently, the average signal strength is almost constant over time. Although there are random variations in a small range of discrete RSSI levels, these are clearly different for Alice and Bob and, hence, are not reciprocal. This leads to the conclusion that the given randomness is primarily induced through additive noise and not through reciprocal channel fading. Contrarily, in Figure 17.4(b), an excerpt of measurement sequences from the mobile scenario I-M-Hom is plotted versus the probe indices. As can be noticed, the sequences from Alice and Bob are highly correlated, indicating a high degree of channel reciprocity. In addition, the sequence $RSSI_{BE}$ barely correlates with $RSSI_{BA}$ and $RSSI_{AB}$. Hence, Eve's observations do not grant her significant insight into the random source of Alice and Bob.

The recorded RSSI traces are quantitatively evaluated by giving the respective correlation coefficient ρ_{pp}, BDR, the raw KBGR, compression ratio ν_c and the effective KBGR for every considered scenario. The results are summarized in Table 17.4.

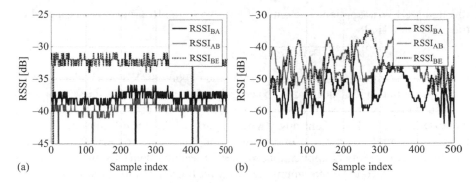

Figure 17.4   Measured RSSI sequences in a static (a) and mobile (b) indoor scenario

Table 17.4   Experimental results for different measurement scenarios

Scenario	ρ_{pp}		BDR		KBGR$_{raw}$ [bit/s]	ν_{DCTW}	KBGR$_{eff}$ [bit/s]
	AB-BA	**AB-BE**	**AB-BA**	**AB-BE**			
I-S-Hom	0.203	0.072	0.469	0.499	0.84	0.308	0.05
I-S-Het	0.077	0.055	0.498	0.498	0.41	0.196	0.01
I-M-Hom	0.835	0.373	0.180	0.414	158.59	0.529	15.45
I-M-Het	0.840	0.315	0.187	0.430	124.63	0.582	13.37
O-S-Hom	0.093	0.077	0.470	0.487	2.53	0.360	0.17
O-S-Het	0.075	0.060	0.495	0.488	0.64	0.295	0.04
O-M-Hom	0.815	0.449	0.221	0.381	85.47	0.502	7.90
O-M-Het	0.671	0.347	0.301	0.422	25.06	0.449	2.08

It can be noticed that particularly the node movement in mobile scenarios has significant impact on the reciprocity that can be utilized. In mobile scenarios, the correlation coefficient ρ_{pp} for Alice and Bob (AB-BA) can be derived to be around 80%, resulting in a BDR of approximately 20%. In static scenarios however, ρ_{pp} decreases to less than 10%, leading to bit mismatches of nearly 50%. Consequently, secret key generation becomes very inefficient. The results derived from Eve's observations (AB-BE) give insight into her capabilities to compute the secret key of Alice and Bob. As Table 17.4 shows, Eve experiences a BDR of around 40% in case of mobile setups. Guessing the secret key of Alice and Bob just from her correlated channel observations therefore is rather inefficient.

For the purpose of measuring the rate of key bit generation, the raw KBGR is defined. It comprises the number of extracted bits per second, which are successfully reconciled and commonly known to Alice and Bob. Nevertheless, as the channel probing rate is not adjusted to the instantaneous coherence time of the channel, potential over-sampling has to be taken into account (cf. Section 17.2.3) and compression through DCTW is applied. The entropy loss due to the IR process is considered by the effective KBGR

$$KBGR_{eff} = \nu_{DCTW} \cdot (KBGR_{raw} - \Phi), \tag{17.5}$$

where ν_{DCTW} denotes the compression ratio when applying DCTW and Φ covers the entropy loss due to IR. Equation (17.5) considers Φ to be affected by the same amount of redundancy as the rest of the sequence. Hence, every data bit exchanged for IR discloses $\nu_{DCTW} \leq 1$ bit of entropy, where $\nu_{DCTW}\Phi$ denotes the effective entropy loss. In fact, this might be a rather optimistic assumption. As a conservative approach, e.g. in context of critical applications with strict security demands, the entropy loss should be assumed to be maximum, i.e. every bit transmitted for IR also discloses 1 bit of entropy. From Table 17.4, it can be noticed that the compression ratio ν_{DCTW} indeed is significantly higher in mobile setups than in static setups, which can be explained by the decreased coherence time of mobile channels. Thus, the effective KBGR can be up to 15 bit/s in mobile cases, while static scenarios remain challenging with respect to efficient key generation. Here, the effective KBGR is significantly lower than 1 bit/s.

Obviously, the previously presented scheme is of rather basic nature as it is implemented with off-the-shelf hardware and does not consider any adaptivity or system optimization. However, the analysed results give valuable insights into the impact of individual mobility scenarios and environment. Moreover, especially when comparing the results of $KBGR_{raw}$ and $KBGR_{eff}$, the significant difference between bit generation rate and entropy rate becomes apparent.

17.5 Further aspects

Physical layer security – no matter whether it is achieved by means of channel-based key generation or secrecy coding – has the potential to establish a confidential radio link between two devices. Yet, there are still major challenges to be addressed, including a more detailed analysis of different attacker models [43,44]. However, considered on its own it is generally not able to address all security and system requirements of practical systems. Therefore, additional mechanisms are required in order to open doors for real-world deployments. This will be briefly addressed in this section, together with some elaborations on a possible physical layer security scheme for wireline systems. So far, physical layer security for wireline systems has not really been addressed in the literature yet but offers quite some potential and interesting applications for various practical applications.

17.5.1 Missing building blocks

Apart from establishing a confidential radio link, which physical layer security can achieve, additional building blocks are required for setting up a comprehensive security framework. A major challenge in this respect is to preserve the unique benefits that physical layer security may offer, especially the high ease-of-use, while combining it with other approaches in an appropriate way. Some selected aspects will be outlined in the following, but it should be noted that this overview is certainly far away from being complete. The goal here is just to sensitise to additional challenges that have to be overcome before physical layer security can really become a widespread success in practical systems.

Initial authentication: Initial (entity) authentication is required in order to make sure that a node communicating with another one is really the one he pretends to be. Without a suitable initial authentication scheme, an attacker Eve could simply pretend to be Bob, for example, and then generate a symmetric key together with Alice, who thinks that she is talking to Bob. Such a scheme is basically only required for the initial authentication. Once a node has been successfully authenticated and a symmetric key pair has been established, the (follow-up) authentication can be traced back to the knowledge of the generated symmetric keys.

In current systems, this problem is often tackled by means of certificates, which, however, involve rather high resource requirements again as well as the availability of a complex and costly public key infrastructure [34]. Therefore, a certificate-based solution would largely compromise many of the benefits that physical layer security can offer. Likewise, pre-shared keys automatically can achieve initial authentication as well, as it is assumed that only legitimate nodes can be in possession of a pre-shared key. However, if pre-shared keys are used in conjunction with physical layer security, one has to question oneself why physical layer security is needed at all and why the pre-shared keys are not used also as the basis for any cipher. Hence, there is clearly the need for an authentication scheme, which exhibits similar properties in terms of complexity, usability, etc. as physical layer security and which may be combined with physical layer security in an appropriate way. One possible example pointing into this direction will be presented in Section 17.5.2.

End-to-end security: For many if not most use cases of practical interest, it is required to secure the communication between a certain end device and another device (e.g. a server or cloud) connected to a different network. Hence, we essentially have to establish a secure connection on an end-to-end basis. Physical layer security on its own, however, can only secure a wireless link or possibly multiple wireless hops if implemented in the right way, but as of now cannot include a wired backbone network, such as the Internet. One solution could be to achieve end-to-end security on a higher layer, e.g. by using transport layer security (TLS) or something similar. However, this would compromise the unique advantages of physical layer security again, as this would imply that the device is capable of using asymmetric cryptography anyway. Another argument in favour of such a combination could be to establish a multi-stage solution in order to improve the overall security level, but it seems more than questionable that this would justify the additional effort. Therefore, innovative ideas are required concerning how to embed a wireless physical layer security scheme in an extended framework such that end-to-end security can be achieved. One rather obvious alternative could be to realize a two-stage approach: First, the radio link from end devices to a wireless access point acting as a gateway is secured using physical layer security and then the connection from the gateway to any other server in the Internet may be secured using conventional mechanisms, such as public key cryptography. This requires, of course, that the gateway can be trusted, but for smart home or healthcare applications, for example, this assumption may be reasonable. Nevertheless, additional ideas for tackling this problem may give physical layer security a strong push forward.

Consistent performance: Another pre-requisite for a widespread adoption of physical layer security schemes in practice is a consistent performance in many different scenarios. A major challenge here is that the actual propagation conditions, which are crucial for CBKG, may differ significantly, depending on where such schemes should be used. However, the end user always expects a comparable performance, e.g. with respect to the required key setup time. This may be particularly critical in case of static environments since in such a case the wireless channel may not contain enough entropy in order to generate secure keys in any reasonable time. For that reason, on the one hand adaptive approaches are required, which always can get the most out of a particular propagation environment, and on the other hand one has to come up with innovative ideas for dealing with static environments. One possible direction to think of is to artificially introduce some randomness in the channel, e.g. by means of some kind of helper device [45].

Standardisation: Last but not least, another necessity for paving the way for physical layer security in practice is standardisation. In many relevant use case scenarios, including the ones outlined in Section 17.1.3, the vendor of an end device (e.g. a body sensor or smart thermostat) is not the one who sells also the corresponding wireless access point or smartphone to the end customer. Though pure software-based solutions that may run over-the-top on a multitude of different devices may be possible in principle, this would require additional effort and expertise at the end user's side. It may further lead to a questionable (and potentially insecure) performance, as – at least with today's chipsets – only rather coarse RSSI values may be accessed from the outside world. Since inter-operability is a key pre-requisite for the success of the IoT in general, corresponding physical layer security schemes definitely should be standardised in relevant standardisation bodies.

17.5.2　Sensor-assisted authentication

As has been outlined in Section 17.5.1, for an overall usable security solution it is generally not sufficient to just generate symmetric cryptographic keys between two devices, but in addition to that initial entity authentication is required in order to make sure that a device is really the one it pretends to be. One possible way to do that is to exploit sensors contained in a device – as it should be the case for many IoT applications – along with a trusted device, such as a personal smartphone or tablet PC. Such a device may be securely integrated into a certain network using conventional approaches (e.g. using pre-shared keys, certificates, etc.) as the special constraints listed in Figure 17.1 are not given here. The integration of a new IoT device into an existing wireless network could then be realized as follows:

First of all, the IoT device establishes symmetric cryptographic keys with the wireless access point using CBKG. Afterwards, they have an encrypted but not yet authenticated connection. The wireless access point then sends an authentication challenge to the user integrating the IoT device via its trusted device. This authentication challenge tells the user how to interact with the IoT device (or to be more precise: the sensors contained in the IoT device) in order to prove that it is really a legitimate one. Such an interaction could be a gesture that should be performed with the IoT device (for example, in case that the IoT device contains inertial sensors), a sound/noise pattern to be generated, e.g. by humming a certain melody (in case of microphones) or a light intensity pattern to be created. In general, numerous possibilities exist, and the best approach certainly depends on the available sensors as well as the concrete IoT device to be integrated. In fact, the wireless access point or a dedicated authentication server accessed by the wireless access point may always select a suitable challenge on-the-fly. Once the user has interacted with the IoT device as requested by the wireless access point, the sensor values obtained during this period are transmitted via the already encrypted connection back to the wireless access point, and there a simple pattern recognition is performed for checking if the action conducted by the user corresponds to the actually requested action. If this is the case, the (initial) authentication procedure is complete and thus an encrypted and authenticated link has been successfully set up.

Clearly, with such an approach, the main benefits of CBKG (cf. Table 17.1) can be preserved while complementing it with the initial entity authentication required for most scenarios of practical interest. Also, it is quite obvious that numerous variants of the procedure described above are possible. For example, the sensors contained in an IoT device may not only be excited by the user interacting with the device in an appropriate way but also by the trusted device itself, e.g. by generating a certain vibration pattern. Besides, the IoT device may already perform a pattern recognition based on the sensed values during the excitement and then signal only the index in a codebook back to the wireless access point.

Sensor-assisted authentication certainly represents just one possible approach for plug-and-play-like authentication tailored to the specific constraints of the IoT, but in general, it is just important to keep the need for such methods in mind when designing appropriate physical layer security schemes, as physical layer security alone may not be able to satisfy all security requirements in real-world deployments.

17.5.3 *Physical layer security for wireline systems*

Even though more and more devices are connected wirelessly, especially in the IoT, there is a still a huge number of devices which also in future will rely on wireline technologies. However, CBKG is usually not feasible in that case since wire-bound channels tend to be time-invariant. Likewise, secrecy coding over a wireline channel (e.g. a linear bus) seems to be intractable in practice as well. Nevertheless, it might be possible to make use of some kind of physical layer security also for wireline channels if we rethink the way it is done. In fact, recently in [46], the authors have proposed a novel approach for establishing symmetric cryptographic keys between different devices connected to a controller area network (CAN) by exploiting special properties of the CAN physical layer. CAN is a widely used serial bus system, which can be found in in-vehicle networks as well as industrial and building automation systems, for example. The base topology is a linear bus with multiple devices connected to it. A special property of the CAN PHY is that there are dominant ('0') and recessive ('1') bits. If two (or more) devices simultaneously transmit a certain bit, a dominant bit will always overwrite recessive bits. This property is conventionally exploited for bus arbitration and turns a CAN bus essentially into a wired AND function. It is interesting to note that exactly the same behaviour can be found in other bus systems as well, such as I2C or LIN, to name just two of them. Therefore, the proposed scheme may be readily applied to any of these bus systems as well. In this case, it is possible to establish symmetric keys between the devices connected to the bus in a very low-complexity and low-cost manner – similar to CBKG for wireless systems. In the following, we briefly outline the basic idea of this approach, but for more details we refer to [46].

We consider two nodes, Alice and Bob, and assume that they are connected to the same bus resembling a wired AND function, such as CAN. In a first step, both nodes generate independently of each other random bit strings of a certain length N, which will be denoted as S_A and S_B in the following. Thereupon, these random bit strings are extended in such a way that after each bit the corresponding inverse bit is inserted, thus yielding the extended bit strings S'_A and S'_B of length $2N$. The extended bit strings are then *simultaneously* transmitted by Alice and Bob over the CAN bus. In consequence, we obtain a super-position of the individual bits, where the effective bit string on the bus corresponds to the logical AND function applied to S'_A and S'_B, i.e. $S_{\text{eff}} = S'_A$ AND S'_B. This effective bit string, which can be read back by Alice and Bob, is made up of different tuples, where each tuple corresponds to one bit in the original bit strings S_A and S_B. In general, two cases have to be distinguished:

1. If there is a '1' in the tuple, i.e. a recessive bit, this implies that both nodes must have transmitted a '1'. Since the other bit in a tuple is always the corresponding inverse bit, consequently both nodes must have transmitted a '0' at this position. Clearly, a passive eavesdropper may draw exactly the same conclusion and therefore these bits are of no value for us. For that reason, Alice and Bob simply discard all bits in their original bit strings S_A and S_B which correspond to a tuple in the effective bit string containing a '1'.

2. If a tuple of the effective bit string S_{eff} corresponds to '00', i.e. two dominant bits, we can directly conclude that Alice and Bob must have transmitted inverse tuples, i.e. either Alice has transmitted '01' and Bob '10' or vice versa. A passive eavesdropper can conclude exactly the same, but she cannot tell who has transmitted what. Alice and Bob, in contrast, do know what they have transmitted themselves, they also observe the '00' on the bus, and therefore they readily know what the respective other node has transmitted as well. Thus, they have an advantage over a passive eavesdropper and can use that in order to generate a shared secret. This shared secret actually can be obtained by keeping all bits in the original bit strings S_A and S_B that correspond to a tuple '00' in the effective bit

string S_{eff}, where the resulting shortened sequence of Bob is exactly the inverse of the corresponding shortened sequence of Alice.

Hence, Alice and Bob can agree on a shared secret (and thus eventually on a secret key) by simply transmitting and receiving CAN messages and by interpreting the resulting effective bit string on the bus in an appropriate way. Due to its simplicity and low complexity, this approach has been denoted as *plug-and-secure communication for CAN*. It has been outlined in [46] that under certain assumptions neither a passive eavesdropper nor an active attacker are able to successfully attack the system, except for possible DoS attacks. This, however, is always possible in CAN networks if a node simply floods the bus with high priority messages. In summary, *plug-and-secure communication for CAN* hence has the potential to become a major building block for future secure CAN networks, and the approach shows that new notions of physical layer security are possible, even for wireline systems.

17.6   Summary and outlook

Physical layer security may play an important role for establishing security in future communication systems in a plug-and-play manner. This particularly holds for the IoT, where special constraints, such as the sheer number of things to be securely inter-connected or the resource-constrained nature of many devices, make it hard to directly make use of existing approaches. From a practical point of view, the generation of symmetric cryptographic keys based on suitable channel properties seems to be more viable for real-world deployments, since with secrecy coding certain assumptions regarding the location and capabilities of a potential attacker have to be made, which might be hard to guarantee in practice. We have performed extensive analyses and experiments for evaluating the practical use of CBKG, the essence of which has been put together herein, omitting many details due to space constraints. In summary, however, we could verify that it is feasible to generate such keys in relevant propagation environments fast enough, at least under certain conditions. In general, approaches based on detailed CSI are to be preferred compared to RSSI-based schemes due to a better performance as well as better security properties. Before putting physical layer security into practice, however, certain open challenges still have to be addressed, which include the following:

1.   It should be possible to generate keys in a reasonable time also in rather static environments. In any case, it should at least be possible to accurately estimate how much entropy is contained in a certain key.
2.   Aside from the generation of cryptographic keys, also other aspects need to be addressed. Examples are the initial authentication of devices that should be securely connected as well as possible ways to establish a secure connection on an end-to-end basis rather than just for a radio link.
3.   Potential attacks (esp. active ones) should be investigated in more detail and corresponding countermeasures should be developed.

4. The performance should be further improved, such that a user does not have to wait for a long time until a key pair has been established.
5. The key generation schemes should be deeply integrated into future wireless modules, with direct access to detailed channel state information.

Once all these challenges have been overcome, nothing should be in the way anymore for a widespread adoption and success of physical layer security in practice.

References

[1] Y. Liang, V. H. Poor, and S. Shamai (Shitz), "Information theoretic security," *Foundations and Trends in Communications and Information Theory*, vol. 5, no. 45, pp. 355–580, 2009.

[2] X. Zhou, L. Song, and Y. Zhang, *Physical Layer Security in Wireless Communications*. Boca Raton, FL, USA: CRC Press, 2014.

[3] M. Bloch and J. Barros, *Physical-Layer Security: From Information Theory to Security Engineering*. Cambridge, UK: Cambridge University Press, 2011.

[4] A. D. Wyner, "The wire-tap channel," *Bell System Technical Journal*, vol. 54, no. 8, pp. 1355–1387, 1975.

[5] R. Ahlswede and I. Csiszár, "Common randomness in information theory—Part I: Secret sharing," *IEEE Transactions on Information Theory*, vol. 39, no. 4, pp. 1121–1132, 1993.

[6] W. K. Harrison, J. Almeida, M. R. Bloch, S. W. McLaughlin, and J. Barros, "Coding for secrecy an overview of error control coding techniques for physical layer security," *IEEE Signal Processing Magazine*, vol. 30, no. 5, pp. 41–50, 2013.

[7] A. Montanaro, "Quantum algorithms: An overview," *npj Quantum Information*, vol. 2, no. 15023, 2016.

[8] U. Maurer, "Secret key agreement by public discussion from common information," *IEEE Transaction on Information Theory*, vol. 39, no. 3, pp. 733–742, 1993.

[9] S. Gopinath, R. Guillaume, P. Duplys, and A. Czylwik, "Reciprocity enhancement and decorrelation schemes for PHY-based key generation," in *IEEE Globecom Workshop on Trusted Communications with Physical Layer Security*, 2014.

[10] N. Patwari, J. Croft, S. Jana, and S. Kasera, "High-rate uncorrelated bit extraction for shared secret key generation from channel measurements," *IEEE Transaction on Mobile Computing*, vol. 9, no. 1, pp. 17–30, 2010.

[11] S. Jana, S. Premnath, M. Clark, S. Kasera, N. Patwari, and S. Krishnamurthy, "On the effectiveness of secret key extraction from wireless signal strength in real environments," in *International Conference on Mobile Computing and Networking*, 2009.

[12] S. Mathur, W. Trappe, N. Mandayam, C. Ye, and A. Reznik, "Radio-telepathy: Extracting a secret key from an unauthenticated wireless channel," in *International Conference on Mobile Computing and Networking*, 2008.

[13]   C. Huth, R. Guillaume, T. Strohm, P. Duplys, I. A. Samuel, and T. Güneysu, "Information reconciliation schemes in physical-layer security: A survey," *Computer Networks – Special Issue on Recent Advances in Physical-Layer Security*, vol. 109, no. 1, pp. 84–104, 2016.

[14]   J. G. Proakis and M. Salehi, *Digital Communications*, 5th ed. New York: McGraw-Hill, 2008.

[15]   W. C. Jakes, *Microwave Mobile Communications*. New York: Wiley, 1974.

[16]   F. Hlawatsch and G. Matz, *Wireless Communications Over Rapidly Time-Varying Channels*. Burlington, MA, USA: Academic Press, 2011.

[17]   J. Hershey, A. Hassan, and R. Yarlagadda, "Unconventional cryptographic keying variable management," *IEEE Transaction on Communications*, vol. 43, no. 1, pp. 3–6, 1995.

[18]   H. Koorapaty, A. Hassan, and S. Chennakeshu, "Secure information transmission for mobile radio," *IEEE Communications Letters*, vol. 4, no. 2, pp. 52–55, 2000.

[19]   A. Ambekar, M. Hassan, and H. D. Schotten, "Improving channel reciprocity for effective key management systems," in *International Conference on Signals, Systems and Electronics*, 2012.

[20]   M. A. Tope and J. C. McEachen, "Unconditionally secure communications over fading channels," in *IEEE Military Communications Conference Communications for Network-Centric Operations: Creating the Information Force*, vol. 1, 2001.

[21]   T. Aono, K. Higuchi, T. Ohira, B. Komiyama, and H. Sasaoka, "Wireless secret key generation exploiting reactance-domain scalar response of multipath fading channels," *IEEE Transaction on Antennas and Propagation*, vol. 53, no. 11, pp. 3776–3784, 2005.

[22]   A. Sayeed and A. Perrig, "Secure wireless communications: Secret keys through multipath," in *IEEE Int. Conf. on Acoustics, Speech and Signal Processing*, 2008.

[23]   J. W. Wallace and R. K. Sharma, "Automatic secret keys from reciprocal mimo wireless channels: Measurement and analysis," *IEEE Transaction on Information Forensics and Security*, vol. 5, no. 3, pp. 381–392, 2010.

[24]   C. Zenger, J. Zimmer, and C. Paar, "Security analysis of quantization schemes for channel-based key extraction," in *Workshop on Wireless Communication Security at the Physical Layer*, 2015.

[25]   S. Eberz, M. Strohmeier, M. Wilhelm, and I. Martinovic, "A practical man-in-the-middle attack on signal-based key generation protocols," in *European Symp. on Research in Computer Security*, 2012.

[26]   Y. Dodis, R. Ostrovsky, L. Reyzin, and A. Smith, "Fuzzy extractors: How to generate strong keys from biometrics and other noisy data," *SIAM Journal on Computing*, vol. 38, no. 1, pp. 97–139, 2008.

[27]   G. Brassard and L. Salvail, "Secret-key reconciliation by public discussion," in *Int. Conf. on the Theory and Applications of Cryptographic Techniques*, 1994, pp. 410–423.

[28] W. T. Buttler, S. K. Lamoreaux, J. R. Torgerson, G. H. Nickel, C. H. Donahue, and C. G. Peterson, "Fast, efficient error reconciliation for quantum cryptography," *Physical Review A*, vol. 67, p. 052303, 2003.

[29] E. Barker and J. Kelsey, "Recommendation for the entropy sources used for random bit generation," *NIST, SP 800-90B*, 2012.

[30] J. Beirlant, E. J. Dudewicz, L. Győrfi, and E. C. Van Der Meulen, "Non-parametric entropy estimation: An overview," *International Journal of Mathematical and Statistical Sciences*, vol. 6, no. 1, pp. 17–39, 1997.

[31] C. Caferov, B. Kaya, R. O'Donnell, and A. C. C. Say, "Optimal bounds for estimating entropy with PMF queries," in *International Symposium on Mathematical Foundations of Computer Science, Part II*, 2015.

[32] J. Ziv and A. Lempel, "Compression of individual sequences via variable-rate coding," *IEEE Transaction on Information Theory*, vol. 24, no. 5, pp. 530–536, 2006.

[33] P. A. J. Volf, *Weighting Techniques in Data Compression: Theory and Algorithms*. Eindhoven, The Netherlands: Technische Universiteit Eindhoven, 2002.

[34] A. J. Menezes, S. A. Vanstone, and P. C. V. Oorschot, *Handbook of Applied Cryptography*. Boca Raton, FL, USA: CRC Press, 1996.

[35] C. Huth, R. Guillaume, P. Duplys, K. Velmurugan, and T. Güneysu, "On the energy cost of channel based key agreement," in *International Workshop on Trustworthy Embedded Devices*, 2016.

[36] G. de Meulenaer, F. Gosset, O.-X. Standaert, and O. Pereira, "On the energy cost of communication and cryptography in wireless sensor networks," in *IEEE Int. Conf. on Wireless and Mobile Computing*, 2008.

[37] C. Ye, S. Mathur, A. Reznik, Y. Shah, W. Trappe, and N. B. Mandayam, "Information-theoretically secret key generation for fading wireless channels," *IEEE Transaction on Information Forensics and Security*, vol. 5, no. 2, pp. 240–254, 2010.

[38] W. Xi, X. Li, C. Qian, *et al.*, "KEEP: Fast secret key extraction protocol for D2D communication," in *IEEE International Symposium of Quality of Service*, 2014.

[39] Z. Wang, J. Han, W. Xi, and J. Zhao, "Efficient and secure key extraction using channel state information," *The Journal of Supercomputing*, vol. 70, no. 3, pp. 1537–1554, 2014.

[40] H. Liu, Y. Wang, J. Yang, and Y. Chen, "Fast and practical secret key extraction by exploiting channel response," in *IEEE International Conference on Computer Communications*, 2013.

[41] C. Ye, A. Reznik, G. Sternburg, and Y. Shah, "On the secrecy capabilities of ITU channels," in *IEEE Vehicular Technology Conference*, 2007.

[42] R. Guillaume, F. Winzer, A. Czylwik, C. T. Zenger, and C. Paar, "Bringing PHY-based key generation into the field: An evaluation for practical scenarios," in *IEEE Vehicular Technology Conference*, 2015.

[43] W. Trappe, "The challenges facing physical layer security," *IEEE Communications Magazine*, vol. 35, no. 6, pp. 16–20, 2015.

[44] K. Zeng, "Physical layer key generation in wireless networks: challenges and opportunities," *IEEE Communications Magazine*, vol. 35, no. 6, pp. 33–39, 2015.

[45] R. Guillaume, S. Ludwig, A. Müller, and A. Czylwik, "Secret key generation from static channels with untrusted relays," in *IEEE International Conference on Wireless and Mobile Computing, Networking and Communications*, Abu Dhabi, United Arab Emirates, 2015.

[46] A. Müller and T. Lothspeich, "Plug-and-secure communication for CAN," in *International CAN Conference*, 2015, pp. 04-1-04-8.

Chapter 18

Key generation from wireless channels: a survey and practical implementation

Junqing Zhang[1], Trung Q. Duong[1], Roger Woods[1], and Alan Marshall[2]

18.1 Introduction

The broadcast nature of the wireless medium allows all the users within communication range to receive and possibly decode a signal, thus making this form of wireless communications vulnerable to eavesdropping. Data confidentiality is usually implemented by cryptographic primitives, consisting of symmetric encryption and public key cryptography (PKC) [1]. The former is used to encrypt the data with a common key, which is then usually distributed between legitimate users by the latter. PKC relies on the computational hardness of some mathematical problems, e.g. discrete logarithm. In addition, it requires a public key infrastructure (PKI) to distribute the public keys among users. Due to the computationally expensive feature and the requirement for PKI, PKC is not applicable to many networks that contain low-cost devices and have a de-centralized topology.

Key generation from wireless channels has emerged as a promising technique to establish cryptographic keys [2]. Two legitimate users, Alice and Bob, generate keys from the randomness of their common channel. Because key generation leverages the unpredictable characteristics of the wireless channel, it is information-theoretical secure. This technique is lightweight as it does not employ computationally complex operations [3]. Furthermore, it does not require any aid from a third party. Due to the above reasons, key generation is a promising alternative to PKC in many key distribution applications.

The rest of this chapter is organized as follows. In Section 18.2, we review the key generation technique by discussing the principles, evaluation metrics, procedure, and applications. In Section 18.3, we carry out a case study by implementing a key generation system using a customized hardware platform and studying key generation principles. Section 18.4 concludes the chapter.

[1]School of Electronics, Electrical Engineering and Computer Science, Queen's University Belfast, UK
[2]Department of Electrical Engineering and Electronics, University of Liverpool, UK

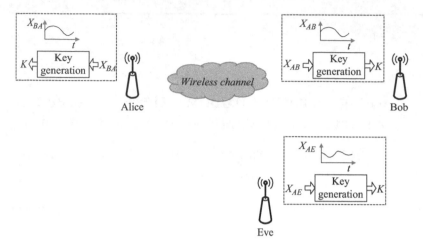

Figure 18.1   Key generation source model

18.2 A survey of wireless key generation

As shown in Figure 18.1, in a key generation source model, Alice and Bob try to establish the same key by exploiting the randomness from their common channel. A passive eavesdropper (Eve) is observing all the transmissions but does not initiate active attacks such as jamming. In this section, we review the key generation principles, evaluation metrics, procedure, and channel parameters.

18.2.1 Principles

Key generation is based on three principles: temporal variation, channel reciprocity, and spatial de-correlation.

Temporal variation guarantees the randomness of the keys. In a dynamic channel, the randomness can be introduced by the movement of users and/or objects in the environment. The faster the channel changes, the more randomness there is. It has been proven in indoor and outdoor experiments that temporal variation is an ideal random source for key generation [4–6]. The temporal variation can be quantified by the auto-correlation function (ACF) of the channel measurement, which can be written as

$$R_{X_{uv}}(t, \Delta t) = \frac{E\{(X_{uv}(t) - \mu_{X_{uv}})(X_{uv}(t + \Delta t) - \mu_{X_{uv}})\}}{E\{|X_{uv}(t) - \mu_{X_{uv}}|^2\}}, \tag{18.1}$$

where $E\{\cdot\}$ denotes the expectation calculation, $X_{uv}(t)$ represents the channel measurement, $\mu_{X_{uv}}$ is the mean value of $X_{uv}(t)$, u and v denote transmitter and receiver, respectively.

Channel reciprocity indicates that the channel features of a link, e.g. phase, delay, and attenuation, are identical at each end of the link. Most current key generation systems use half-duplex hardware which can become a problem whenever the channel changes too fast, as the non-simultaneous measurements will affect the cross-correlation of the signals between Alice and Bob. Therefore, most of the key generation protocols are only applicable in slow-fading channels. Channel reciprocity is also impacted by the independent hardware noise. It has been theoretically proven in [7] and experimentally verified in [8] that the channel reciprocity is more vulnerable to noise in a slow fading channel. The signal similarity can be quantified by the cross-correlation relationship between the channel measurements, which can be given as

$$\rho_{uv,u'v'} = \frac{E\{X_{uv}X_{u'v'}\} - E\{X_{uv}\}E\{X_{u'v'}\}}{\sigma_{X_{uv}}\sigma_{X_{u'v'}}}, \tag{18.2}$$

where $\sigma_{X_{uv}}$ is the standard deviation of $X_{uv}(t)$.

Spatial de-correlation implies that when the eavesdroppers are located more than half-wavelength (0.5 λ) away from either Alice or Bob, the eavesdropping channels are uncorrelated from the legitimate channels, therefore, the keys generated by legitimate users will be different from those generated by the eavesdroppers. This assumption is based on a rich scattering Rayleigh environment. When the number of scatterers goes to infinity, the correlation relationship is a Jakes model [9], and the correlation coefficient decreases to 0 when the distance is 0.4 λ. This property has been verified by experiments, e.g. in ultra-wideband (UWB) systems [10–13] and IEEE 802.11g systems [14]. However, the channel conditions such as multipath level may not always be satisfied and spatial de-correlation will not hold [15–18]. In this case, key generation is vulnerable to passive eavesdropping and requires special design consideration. Spatial de-correlation can also be quantified by the cross-correlation between the measurements of legitimate users and eavesdroppers, which is defined in (18.2).

18.2.2 Evaluation metrics

The keys generated are used for cryptographic applications, e.g. authentication and encryption. For practical applications, there are requirements in terms of randomness, the key generation rate (KGR), and the key disagreement rate (KDR), which are discussed in detail in this section.

18.2.2.1 Randomness

The randomness of the keys is the most essential requirement of cryptographic applications. A non-random key will significantly decrease the search space by brute force attacks, which makes the cryptosystems vulnerable. There is a statistical tool provided by National Institute of Standards and Technology (NIST) [19] to test the randomness feature of the random number generator (RNG) and pseudo-random number generator (PRNG). Key generation is a RNG by generating keys from the randomness of wireless channels, and therefore, NIST test suite has been widely applied in key generation applications [4–7,20–23].

There are 15 tests in total provided by the NIST random test suite, each evaluating a specific randomness feature, e.g. the proportion between 0 s and 1 s, periodic feature, and approximate entropy. The NIST provides a C implementation of the test suite and the source code is free to download [19].

18.2.2.2 Key generation rate

The KGR describes the number of key bits generated in 1 second/measurement. Cryptographic applications need a certain length of key. For example, the key length for the advanced encryption standard (AES) is 128, 192, or 256-bits [24]. In order to guarantee the security of the cryptosystem, the keys should be refreshed regularly.

18.2.2.3 Key disagreement rate

The KDR quantifies the ratio of the mismatched bits between users after quantization, which is given as

$$
\text{KDR}_{uv,u'v'} = \frac{\sum_{i=1}^{N} |K_{uv}(i) - K_{u'v'}(i)|}{N}, \tag{18.3}
$$

where K_{uv} and $K_{u'v'}$ are the keys quantized from X_{uv} and $X_{u'v'}$, respectively, and N is the length of the keys.

18.2.3 Key generation procedure

Key generation usually consists of four stages, i.e. channel probing, quantization, information reconciliation, and privacy amplification, as shown in Figure 18.2, which will be described in detail in this section. Without loss of generality, Alice is selected as the initiator of the entire process.

18.2.3.1 Channel probing

Alice and Bob probe the common wireless channel alternately and measure its randomness.

In the ith sampling round, Alice sends a public pilot signal to Bob at time $t_{i,A}$. Bob will measure a channel parameter, e.g. received signal strength (RSS), channel state information (CSI), and store it in $X'_{BA}(i)$. The channel parameters suitable for key generation are introduced separately in Section 18.2.4. Bob then also sends a public pilot signal to Alice at $t_{i,B}$ and Alice will measure the same parameter and put it in $X'_{AB}(i)$. Alice and Bob will keep doing this until they collect enough measurements.

Although there are research efforts on using full-duplex hardware for key generation [25–27], most of the commercial hardware platforms work in half-duplex mode and they cannot sample the channel simultaneously. The $\Delta t_{AB} = |t_{i,A} - t_{i,B}| > 0$ always holds for half-duplex hardware. In addition, channel measurements are also impacted by hardware noises, which reside independently in different platforms. Therefore, the cross-correlation relationship between the measurements of Alice and Bob, i.e. X'_{BA} and X'_{AB}, are affected by the non-simultaneous measurements and noises. These effects can be mitigated by signal pre-processing algorithms, e.g. interpolation [5,28] and filtering [7,21,29–32].

As the sampling interval $|t_{i+1,A} - t_{i,A}|$ is usually very small, there is correlation between the adjacent samples. The channel measurements are re-sampled at a probing rate, T_p, and a subset of the measurements, X^n, are selected.

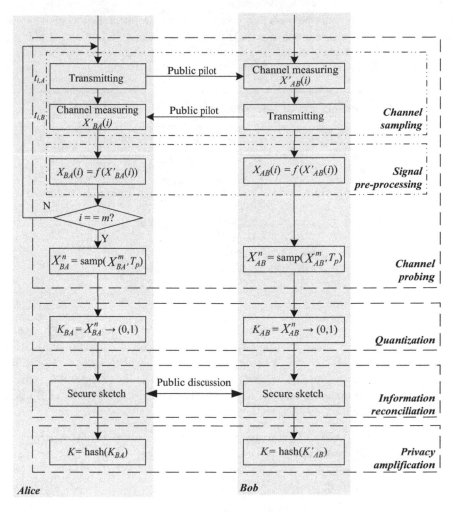

Figure 18.2   Key generation procedure

18.2.3.2   Quantization

The quantization scheme maps the analogue channel measurements to binary values. A quantizer has two factors, i.e. quantization level (QL) and thresholds. QL is the number of bits that each measurement is converted to and is usually determined by the signal-to-noise ratio (SNR) of the channel. Gray code is adopted in the multi-bit quantization to reduce the bit disagreement. Thresholds are the reference levels and can be set according to the mean and standard deviation, or by using cumulative distribution function (CDF).

The mean and standard-deviation-based quantizer [4] is given in Algorithm 1, which can be configured as lossy or lossless quantization by a parameter α [4]. When

$\alpha \neq 0$, the measurements between η_+ and η_- will be dropped. The samples above η_+/below η_- will be converted to 1/0. This quantizer is easy to implement but may generate unbalanced proportions of 1 s and 0 s. For example, when there are spikes in the signal, the data will not distribute evenly along the thresholds.

Algorithm 1 Mean and standard-deviation-based quantization algorithm

---

  1: $\eta_+ = \mu_{X_{uv}} + \alpha \times \sigma_{X_{uv}}$
  2: $\eta_- = \mu_{X_{uv}} - \alpha \times \sigma_{X_{uv}}$
  3: **for** $i \leftarrow 1$ **to** n **do**
  4:     **if** $X_{uv}(i) > \eta_+$ **then**
  5:         $K_{uv}(i) = 1$
  6:     **else if** $X_{uv}(i) \leq \eta_-$ **then**
  7:         $K_{uv}(i) = 0$
  8:     **else**
  9:         $X_{uv}(i)$ dropped
10:     **end if**
11: **end for**

A CDF-based quantizer calculates the thresholds based on the distribution of the measurements [28,33], as shown in Algorithm 2. The thresholds can be set evenly according to the distribution and therefore the same amount of 1 s and 0 s can be quantized. In addition, it is simple to adjust the QL, thus the scheme can be configured as a multi-bit quantizer.

Algorithm 2 CDF-based quantization algorithm

  1: $F(x) = \Pr(X_{uv} < x)$
  2: $\eta_0 = -\infty$
  3: **for** $j \leftarrow 1$ **to** $2^{QL} - 1$ **do**
  4:     $\eta_j = F^{-1}(\frac{j}{2^{QL}})$
  5: **end for**
  6: $\eta_{2^{QL}} = \infty$
  7: Construct Gray code b_j and assign them to different intervals $[\eta_{j-1}, \eta_j]$
  8: **for** $i \leftarrow 1$ **to** n **do**
  9:     **if** $\eta_{j-1} \leq X_{uv}(i) < \eta_j$ **then**
10:         $K_{uv}(i, QL) = b_j$
11:     **end if**
12: **end for**

Performance comparisons between different quantization schemes can be found in [4,14,34].

18.2.3.3 Information reconciliation

As mentioned in Section 18.2.3.1, the cross-correlation relationship between the measurements of Alice and Bob are impacted. The binary bits quantized thereafter usually will not match. Information reconciliation is thus designed to make the users agree on the same keys by using protocols such as Cascade [4,31,35,36] or error correcting code (ECC) like low-density parity-check (LDPC) [37–39], BCH code [40,41], Reed-Solomon code [42], Golay code [5,6,43], and Turbo code [44].

Secure sketch is a popular information reconciliation technique [40]. Alice selects a codeword c from the ECC C, calculates the syndrome s by XOR operation, i.e. $s = XOR(K_{BA}, c)$, and sends the syndrome to Bob via the public channel. Bob calculates the codeword by $c'' = XOR(K_{AB}, s)$ and decodes c' from c''. Bob then obtains the new K'_{AB} by XOR operation, i.e. $K'_{AB} = XOR(s, c')$. When the hamming distance between c and c'' is smaller than the correction capacity of the used ECC, $c' = c$, and then Bob can get the same keys as Alice. The agreement can be confirmed by cyclic redundancy check (CRC). The correction capacity is determined by the adopted ECCs or protocols. For example, BCH code can correct up to 25% disagreement [7]. Whenever the disagreement exceeds the correction capacity, key generation fails.

An extensive survey on the information reconciliation techniques applied in key generation can be found in [45].

18.2.3.4 Privacy amplification

There is information exchange via the public channel in the information reconciliation stage, which reveals information of the keys to eavesdroppers. Privacy amplification is adopted to remove the information leakage [46]. This can be implemented by extractor [47], or universal hashing functions, such as leftover hash lemma [4,48], cryptographic hash functions [42,44], and Merkle-Damgard hash function [36].

18.2.4 Channel parameters

RSS is the most popular parameter used for key generation, especially for practical implementation, because it is available in almost all wireless systems, and provided by many off-the-shelf network interface cards (NICs). IEEE 802.11 [49] is widely used in our daily life, such as laptops, smartphones, tablets. IEEE 802.15.4 [50] is the standard used for wireless sensor networks (WSNs). MICAz [51] and TelosB [52] are two popular sensor motes platform supporting this standard, and many key generation applications have been reported. RSS-based key generation has also been applied to Bluetooth systems.

CSI is less popular for practical key generation because it is not currently available in most of the commercial NICs, with the exception of Intel WiFi 5300 NIC [53]. Customized hardware platforms are also able to provide CSI, such as universal software radio peripheral (USRP) and wireless open-access research platform (WARP). CSI can be extracted by IEEE 802.11a/g/n/ac systems and UWB systems, etc. Compared to RSS, CSI is a finer-grained channel parameter which contains detailed channel information such as amplitude and phase of the channel gains [7,37].

Table 18.1   *Key generation applications in wireless networks*

Technique		Modulation	Parameter	Testbed	Representative References
IEEE 802.11	n	MIMO OFDM	RSS, CSI	RSS: all NICs; CSI: Intel 5300 NIC, and customized hardware platforms, such as USRP [54] and WARP [55]	RSS-based: [56] CSI-based: [21,22]
	a	OFDM	RSS, CSI		RSS-based:
	g	OFDM, DSSS	RSS, CSI		[4,20,48,57]
	b	DSSS	RSS		CSI-based: [58]
IEEE 802.15.4		DSSS	RSS	MICAz [51], TelosB [52]	[5,6,28,30,59,60]
Bluetooth		FHSS	RSS	Smartphones	[23]
UWB		Pulse	CIR	Constructed by oscilloscope, waveform generator, etc.	[10–13,61–63]
LTE		MIMO OFDM	RSS, CSI	Smartphones	[64,65]

Many RSS-based and CSI-based key generation systems have been reported and some examples are shown in Table 18.1.

18.3   Case study: practical implementation of an RSS-based key generation system

In this section, we implement a key generation system using a customized field-programmable gate array (FPGA)-based hardware platform, namely WARP, in order to test the key generation principles. We first introduce the background information such as IEEE 802.11 protocol and WARP hardware. We then describe our measurement system including the testbed and scenarios. Finally, we present the test results extracted from real indoor experiments and study the key generation principles.

18.3.1   Preliminary

18.3.1.1   IEEE 802.11 protocol

IEEE 802.11, more commonly known as Wi-Fi, is the most popular wireless local area network (WLAN) standard [49]. The standard defines both the physical (PHY) layer and media access control (MAC) layer protocols.

The PHY layer techniques of IEEE 802.11 are listed in Table 18.2. Direct sequence spread spectrum (DSSS) spreads the signal to a larger bandwidth which makes the transmission robust to the noise and interference. Orthogonal frequency-division multiplexing (OFDM) technique modulates the signals to multiple orthogonal subcarriers, which can significantly improve the bandwidth efficiency and transmission rate.

Table 18.2   PHY layer techniques of IEEE 802.11

Amendment	Release date	Frequency/GHz	Modulation
b	1999	2.4	DSSS
a	1999	5	OFDM
g	2003	2.4	OFDM
n	2009	2.4/5	MIMO OFDM
ac	2013	5	MIMO OFDM

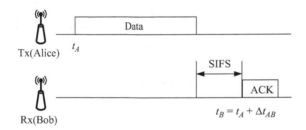

Figure 18.3   Timing between data and ACK packets in DCF protocol

Multiple-input and multiple-output (MIMO) OFDM modulation further explores the spatial diversity and the system performance is greatly enhanced. In all these wireless modulation techniques, the received signal power can be measured and reported as RSS.

Distributed coordination function (DCF) MAC protocol is used in IEEE 802.11. As shown in Figure 18.3, at time t_A, the transmitter sends a data packet. If the receiver decodes it successfully, at time t_B, the receiver will transmit a positive acknowledgement (ACK) frame back to the transmitter after waiting short interframe space (SIFS). The successful data packet and ACK packet always appear in pair, and can be employed to measure the channel.

18.3.1.2   WARP hardware

WARP is a scalable and extensible programmable wireless platform and allows fast prototype of physical layer algorithms [55]. WARP v3 hardware integrates an FPGA, two programmable radio frequency (RF) interfaces and a variety of peripherals. The Virtex-6 FPGA LX240T serves as the centre control system running the PHY and MAC layer codes. The RF interface is consisted of power amplifier (PA) and transceiver MAX2829 [66], which supports dual-band IEEE 802.11 a/b/g.

WARP 802.11 reference design is a real-time FPGA implementation of the IEEE 802.11 OFDM PHY and DCF MAC for WARP v3 hardware. In order to control the behaviour of the PHY and MAC without interfering with the real-time operation of the wireless interfaces, a Python experiments framework has been developed

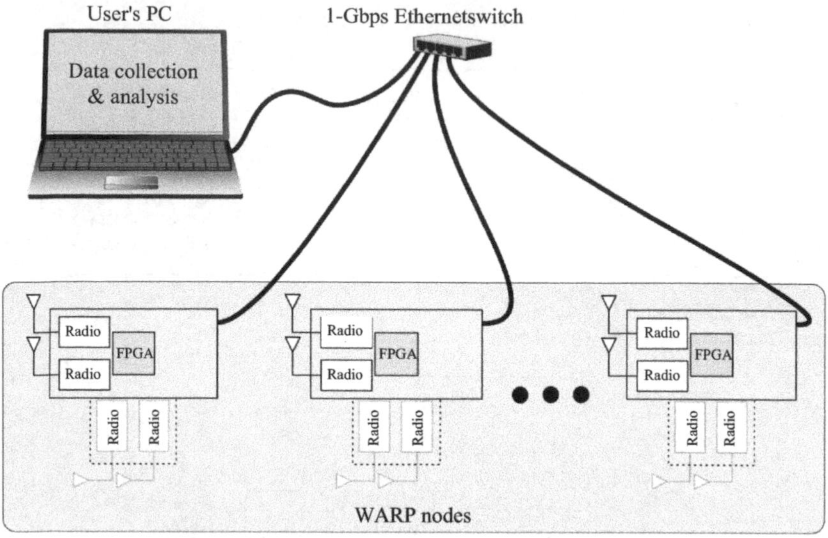

*Figure 18.4   Configuration of WARP 802.11 reference design experiments
framework*

to log the transmission parameters, such as timestamp, rate, transmission power,
received signal power, and channel estimation. The system configuration is shown in
Figure 18.4, where WARP nodes are running 802.11 reference design and the PC
is operating Python experiments framework. The WARP and PC are connected by a
1-Gbps Ethernet switch so that the logged data can be transferred to the PC for further
processing.

18.3.2   Measurement system and test scenario

Due to the limited number of WARP platforms, we used eight WARP boards to
construct the measurement system, with one Alice, one Bob, and six eavesdroppers,
as shown in Figure 18.5. All the users were running WARP 802.11 reference design
and operating at carrier frequency of 2.412 GHz. Alice and Bob were the legitimate
users wishing to establish a secure key between them. They were configured as access
point (AP) and station (STA), respectively, and formed an infrastructure basic service
set (BSS). Eavesdroppers were not associated to Alice but could overhear and record
all the transmissions in the network and did not attempt to initiate active attacks such
as disrupting the transmissions by jamming.

Alice sent data packets every 0.96 ms, through which Bob could measure the RSS
$P_{AB}(t)$. The successful data packets were acknowledged by Bob, and Alice could also
measure the RSS of the ACK packet, $P_{BA}(t)$. The time difference Δt_{AB} was configured
as 0.06 ms [58], which was quite short compared to the channel variation and can
guarantee a high cross-correlation of the channel measurements.

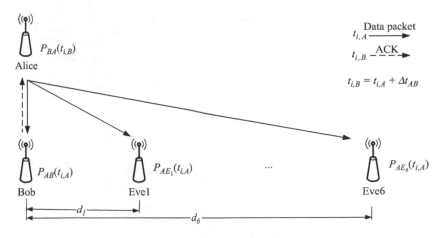

Figure 18.5   Experiment setup

The experiments were carried out in an office room which was a typical indoor environment with chairs, cupboards, desks, etc. Three scenarios were considered.

- Static: All the users were stationary and no object was moving in the room.
- Object moving: All the users were stationary and an object (a person) was moving around in the room with a speed of about 1 m/s.
- Mobile: Bob and eavesdroppers were stationary and Alice was placed on a trolley and moved by a person in the room with a speed of about 1 m/s.

We used the channel measurements of Alice and Bob from one experiment to study temporal variation and channel reciprocity. Regarding spatial de-correlation, we carried out several experiments with different distance configurations (by changing the distances between Bob and eavesdroppers) and the results were put together. All the experiments ran for 60 s and about 60,000 packets were collected. The keys were then quantized by single-bit CDF-based quantization scheme, by setting $QL = 1$ in Algorithm 2.

18.3.3   Experiment results

We used the collected channel measurements to study the key generation principles, i.e. temporal variation, channel reciprocity, and spatial de-correlation.

18.3.3.1   Temporal variation

Temporal variation is an ideal random source for key generation. Whenever the wireless channel changes at a constant rate, it becomes a wide sense stationary (WSS) random process [7]. The mean value is a constant and the ACF only depends on the time difference Δt but is irrelevant of the observation time t in a WSS random process. This property can simplify the channel-probing design for key generation, as a fixed probing rate will suffice.

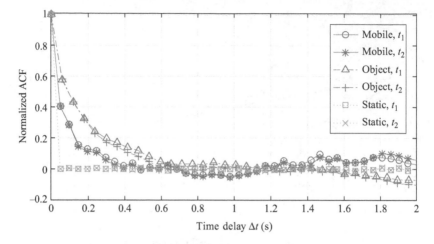

Figure 18.6   Normalized ACF $R_{P_{AB}}(\Delta t)$ observed at time t_1 and t_2 with mobile, object moving, and static scenarios. $t_2 = t_1 + 10$ s

The ACFs can be calculated via (18.1) by substituting $P_{AB}(t)$ and shown in Figure 18.6. In the static scenario, the only variation was contributed by the hardware noise, which is temporally independent. Therefore, there is no auto-correlation unless when the sequences are aligned, i.e. $\Delta t = 0$. This would appear to be beneficial for key generation, as the probing rate can be very small. However, as discussed in Section 18.3.3.2, key generation cannot work in the static scenario due to the low cross-correlation between the channel measurements of Alice and Bob. In the dynamic scenarios, i.e. object moving and mobile scenario, the received power is temporally correlated. The ACF only depends on the Δt, which indicates $P_{AB}(t)$ is a WSS random process. $R_{P_{AB}}(\Delta t)$ in mobile scenario decreased faster than that of object moving scenario, because the channel was changing more dynamically.

18.3.3.2   Channel reciprocity

The cross-correlation coefficients and KDR can be calculated via (18.2) and (18.3), respectively, and given in Table 18.3. In the static scenario, although the channel remained the same, there was interference from other networks, such as commercial Wi-Fi systems. Therefore, the correlation coefficients were not exactly zero. In the mobile scenario, $\rho_{AB,BA}$ was much higher than that of object moving scenario. This is because when one user was moving, the channel was changing more significantly than when only one object was moving. When we had two objects moving in the room, the correlation was almost as high as that of the mobile scenario, which is intuitive because more randomness was introduced. It is worth noting that in all the dynamic scenarios, the disagreement can be corrected by information reconciliation techniques.

Table 18.3   *Cross-correlation coefficients, $\rho_{AB,BA}$, and KDRs, $KDR_{AB,BA}$, with static, object moving, and mobile scenarios*

	Static	**Object moving**	**Two objects moving**	**Mobile**
CorrCoeff	0.1152	0.7152	0.929	0.935
KDR	0.3523	0.2148	0.1037	0.087

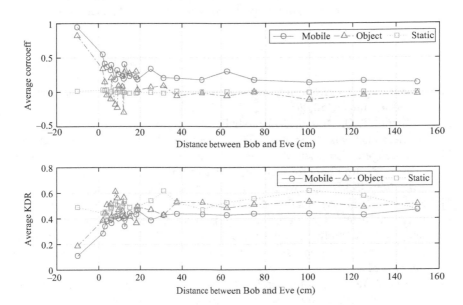

Figure 18.7   *Cross-correlation coefficients, ρ_{AB,AE_j}, and KDRs, KDR_{AB,AE_j}, with static, object moving, and mobile scenarios. Eavesdroppers were in linear placement. $\lambda = 12.44\,cm$. The points with distance smaller than 0 are $\rho_{AB,BA}$ and $KDR_{AB,BA}$ and shown for comparison*

18.3.3.3   Spatial de-correlation

The cross-correlation coefficients and KDR can be calculated via (18.2) and (18.3), respectively, and shown in Figure 18.7. The points with distance smaller than 0 are $\rho_{AB,BA}$ and $KDR_{AB,BA}$ and shown for comparison.

When the channel was static, the eavesdropping channels were not correlated to the legitimate channels, and the eavesdroppers cannot get any useful information. KDR_{AB,AE_j} is about 0.5, which is no better than random guess. In the dynamic scenarios, ρ_{AB,AE_j} decreases very fast with the distance. Even when the eavesdropper is very close to Bob, ρ_{AB,AE_j} is not high. This property is quite essential for the security of key generation applications.

18.4   Conclusion

This chapter reviewed the key generation from wireless channels and presented a case study by implementing an RSS-based key generation system. We introduced the key generation principles, evaluation metrics, procedure, and channel parameters. We then implemented a key generation system by using WARP hardware, which is a customized FPGA-based platform. We carried out several experiments in the indoor environment and tested the key generation principles, i.e. temporal variation, channel reciprocity, and spatial de-correlation. We concluded that key generation is workable in dynamic environment but cannot operate properly in static channels.

References

[1]   W. Stallings, *Cryptography and Network Security: Principles and Practice*, 6th ed. Prentice Hall, 2013.

[2]   J. Zhang, T. Q. Duong, A. Marshall, and R. Woods, "Key generation from wireless channels: A review," *IEEE Access*, vol. 4, pp. 614–626, Mar. 2016.

[3]   C. T. Zenger, J. Zimmer, M. Pietersz, J.-F. Posielek, and C. Paar, "Exploiting the physical environment for securing the internet of things," in *Proc. New Security Paradigms Workshop*, Twente, The Netherlands, Sep. 2015, pp. 44–58.

[4]   S. Jana, S. N. Premnath, M. Clark, S. K. Kasera, N. Patwari, and S. V. Krishnamurthy, "On the effectiveness of secret key extraction from wireless signal strength in real environments," in *Proc. 15th Annu. Int. Conf. Mobile Computing and Networking (MobiCom)*, Beijing, China, Sep. 2009, pp. 321–332.

[5]   H. Liu, J. Yang, Y. Wang, and Y. Chen, "Collaborative secret key extraction leveraging received signal strength in mobile wireless networks," in *Proc. 31st IEEE Int. Conf. Comput. Commun. (INFOCOM)*, Orlando, Florida, USA, Mar. 2012, pp. 927–935.

[6]   H. Liu, J. Yang, Y. Wang, Y. Chen, and C. Koksal, "Group secret key generation via received signal strength: Protocols, achievable rates, and implementation," *IEEE Trans. Mobile Comput.*, vol. 13, no. 12, pp. 2820–2835, 2014.

[7]   J. Zhang, A. Marshall, R. Woods, and T. Q. Duong, "Efficient key generation by exploiting randomness from channel responses of individual OFDM subcarriers," *IEEE Trans. Commun.*, vol. 64, no. 6, pp. 2578–2588, 2016.

[8]   J. Zhang, R. Woods, T. Q. Duong, A. Marshall, and Y. Ding, "Experimental study on channel reciprocity in wireless key generation," in *Proc. IEEE Int. Workshop on Signal Processing Advances in Wireless Communications*, Edinburgh, UK, Jul. 2016, pp. 1–5.

[9]   A. Goldsmith, *Wireless Communications*. Cambridge University Press, 2005.

[10] M. G. Madiseh, S. He, M. L. McGuire, S. W. Neville, and X. Dong, "Verification of secret key generation from UWB channel observations," in *Proc. IEEE Int. Conf. Commun. (ICC)*, Dresden, Germany, Jun. 2009, pp. 1–5.

[11] S. T.-B. Hamida, J.-B. Pierrot, and C. Castelluccia, "An adaptive quantization algorithm for secret key generation using radio channel measurements," in *Proc. 3rd Int. Conf. New Technologies, Mobility and Security (NTMS)*, Cairo, Egypt, Dec. 2009, pp. 1–5.

[12] S. T.-B. Hamida, J.-B. Pierrot, and C. Castelluccia, "Empirical analysis of UWB channel characteristics for secret key generation in indoor environments," in *Proc. 21st IEEE Int. Symp. Personal Indoor and Mobile Radio Commun. (PIMRC)*, Instanbul, Turkey, Sep. 2010, pp. 1984–1989.

[13] F. Marino, E. Paolini, and M. Chiani, "Secret key extraction from a UWB channel: Analysis in a real environment," in *Proc. IEEE Int. Conf. Ultra-WideBand (ICUWB)*, Paris, France, Sep. 2014, pp. 80–85.

[14] C. T. Zenger, J. Zimmer, and C. Paar, "Security analysis of quantization schemes for channel-based key extraction," in *Workshop Wireless Commun. Security at the Physical Layer*, Coimbra, Portugal, Jul. 2015, pp. 1–6.

[15] M. Edman, A. Kiayias, and B. Yener, "On passive inference attacks against physical-layer key extraction," in *Proc. 4th European Workshop System Security*, Salzburg, Austria, Apr. 2011, pp. 8:1–8:6.

[16] X. He, H. Dai, W. Shen, and P. Ning, "Is link signature dependable for wireless security?" in *Proc. 32nd IEEE Int. Conf. Comput. Commun. (INFOCOM)*, Turin, Italy, Apr. 2013, pp. 200–204.

[17] X. He, H. Dai, Y. Huang, D. Wang, W. Shen, and P. Ning, "The security of link signature: A view from channel models," in *IEEE Conf. Commun. and Network Security (CNS)*, San Francisco, California, USA, Oct. 2014, pp. 103–108.

[18] X. He, H. Dai, W. Shen, P. Ning, and R. Dutta, "Toward proper guard zones for link signature," *IEEE Trans. Wireless Commun.*, vol. 15, no. 3, pp. 2104–2117, Mar. 2016.

[19] A. Rukhin, J. Soto, J. Nechvatal, *et al.*, "A statistical test suite for random and pseudorandom number generators for cryptographic applications," National Institute of Standards and Technology, Tech. Rep. Special Publication 800-22 Revision 1a, Apr. 2010.

[20] S. Mathur, W. Trappe, N. Mandayam, C. Ye, and A. Reznik, "Radio-telepathy: Extracting a secret key from an unauthenticated wireless channel," in *Proc. 14th Annu. Int. Conf. Mobile Computing and Networking (MobiCom)*, San Francisco, California, USA, Sep. 2008, pp. 128–139.

[21] H. Liu, Y. Wang, J. Yang, and Y. Chen, "Fast and practical secret key extraction by exploiting channel response," in *Proc. 32nd IEEE Int. Conf. Comput. Commun. (INFOCOM)*, Turin, Italy, Apr. 2013, pp. 3048–3056.

[22] W. Xi, X. Li, C. Qian, *et al.*, "KEEP: Fast secret key extraction protocol for D2D communication," in *Proc. 22nd IEEE Int. Symp. of Quality of Service (IWQoS)*, Hong Kong, May 2014, pp. 350–359.

[23] S. N. Premnath, P. L. Gowda, S. K. Kasera, N. Patwari, and R. Ricci, "Secret key extraction using Bluetooth wireless signal strength measurements," in *Proc. 11th Annu. IEEE Int. Conf. Sensing, Commun., and Networking (SECON)*, Singapore, Jun. 2014, pp. 293–301.

[24] *Advanced Encryption Standard*, Federal Information Processing Standards Publication Std. FIPS PUB 197, 2001. [Online]. Available: http://csrc.nist.gov/publications/fips/fips197/fips-197.pdf

[25] H. Vogt and A. Sezgin, "Full-duplex vs. half-duplex secret-key generation," in *Proc. IEEE Int. Workshop Information Forensics Security (WIFS)*, Rome, Italy, Nov. 2015, pp. 1–6.

[26] H. Vogt, K. Ramm, and A. Sezgin, "Practical secret-key generation by full-duplex nodes with residual self-interference," in *Proc. 20th Int. ITG Workshop Smart Antennas*, Munich, Germany, Mar. 2016, pp. 1–5.

[27] A. Sadeghi, M. Zorzi, and F. Lahouti, "Analysis of key generation rate from wireless channel in in-band full-duplex communications," *arXiv preprint arXiv:1605.09715*, 2016.

[28] N. Patwari, J. Croft, S. Jana, and S. K. Kasera, "High-rate uncorrelated bit extraction for shared secret key generation from channel measurements," *IEEE Trans. Mobile Comput.*, vol. 9, no. 1, pp. 17–30, 2010.

[29] B. Azimi-Sadjadi, A. Kiayias, A. Mercado, and B. Yener, "Robust key generation from signal envelopes in wireless networks," in *Proc. 14th ACM Conf. Comput. Commun. Security (CCS)*, Alexandria, USA, Oct. 2007, pp. 401–410.

[30] S. Ali, V. Sivaraman, and D. Ostry, "Eliminating reconciliation cost in secret key generation for body-worn health monitoring devices," *IEEE Trans. Mobile Comput.*, vol. 13, no. 12, pp. 2763–2776, Dec. 2014.

[31] X. Zhu, F. Xu, E. Novak, C. C. Tan, Q. Li, and G. Chen, "Extracting secret key from wireless link dynamics in vehicular environments," in *Proc. 32nd IEEE Int. Conf. Comput. Commun. (INFOCOM)*, Turin, Italy, Apr. 2013, pp. 2283–2291.

[32] J. Zhang, R. Woods, A. Marshall, and T. Q. Duong, "An effective key generation system using improved channel reciprocity," in *Proc. 40th IEEE Int. Conf. Acoustics, Speech and Signal Process. (ICASSP)*, Brisbane, Australia, Apr. 2015, pp. 1727–1731.

[33] C. Chen and M. A. Jensen, "Secret key establishment using temporally and spatially correlated wireless channel coefficients," *IEEE Trans. Mobile Comput.*, vol. 10, no. 2, pp. 205–215, 2011.

[34] R. Guillaume, A. Mueller, C. T. Zenger, C. Paar, and A. Czylwik, "Fair comparison and evaluation of quantization schemes for PHY-based key generation," in *Proc. 18th Int. OFDM Workshop (InOWo'14)*, Essen, Germany, Aug. 2014, pp. 1–5.

[35] G. Brassard and L. Salvail, "Secret-key reconciliation by public discussion," in *Advances in Cryptology-EUROCRYPT*, 1994, pp. 410–423.

[36] Y. Wei, K. Zeng, and P. Mohapatra, "Adaptive wireless channel probing for shared key generation based on PID controller," *IEEE Trans. Mobile Comput.*, vol. 12, no. 9, pp. 1842–1852, 2013.

[37] Y. Liu, S. C. Draper, and A. M. Sayeed, "Exploiting channel diversity in secret key generation from multipath fading randomness," *IEEE Trans. Inf. Forensics Security*, vol. 7, no. 5, pp. 1484–1497, 2012.

[38] M. Bloch, J. Barros, M. R. Rodrigues, and S. W. McLaughlin, "Wireless information-theoretic security," *IEEE Trans. Inf. Theory*, vol. 54, no. 6, pp. 2515–2534, 2008.

[39] C. Ye, S. Mathur, A. Reznik, Y. Shah, W. Trappe, and N. B. Mandayam, "Information-theoretically secret key generation for fading wireless channels," *IEEE Trans. Inf. Forensics Security*, vol. 5, no. 2, pp. 240–254, 2010.

[40] Y. Dodis, R. Ostrovsky, L. Reyzin, and A. Smith, "Fuzzy extractors: How to generate strong keys from biometrics and other noisy data," *SIAM J. Comput.*, vol. 38, no. 1, pp. 97–139, 2008.

[41] D. Chen, Z. Qin, X. Mao, P. Yang, Z. Qin, and R. Wang, "Smokegrenade: An efficient key generation protocol with artificial interference," *IEEE Trans. Inf. Forensics Security*, vol. 8, no. 11, pp. 1731–1745, 2013.

[42] J. Zhang, S. K. Kasera, and N. Patwari, "Mobility assisted secret key generation using wireless link signatures," in *Proc. 32nd IEEE Int. Conf. Comput. Commun. (INFOCOM)*, San Diego, California, USA, Mar. 2010, pp. 1–5.

[43] S. Mathur, R. Miller, A. Varshavsky, W. Trappe, and N. Mandayam, "Proximate: Proximity-based secure pairing using ambient wireless signals," in *Proc. 9th Int. Conf. Mobile Systems, Applications, and Services (MobiSys)*, Washington, DC, USA, Jul. 2011, pp. 211–224.

[44] A. Ambekar, M. Hassan, and H. D. Schotten, "Improving channel reciprocity for effective key management systems," in *Proc. Int. Symp. Signals, Syst., Electron. (ISSSE)*, Potsdam, Germany, Oct. 2012, pp. 1–4.

[45] C. Huth, R. Guillaume, T. Strohm, P. Duplys, I. A. Samuel, and T. Güneysu, "Information reconciliation schemes in physical-layer security: A survey," *Computer Networks*, vol. 109, pp. 84–104, 2016.

[46] C. H. Bennett, G. Brassard, C. Crépeau, and U. M. Maurer, "Generalized privacy amplification," *IEEE Trans. Inf. Theory*, vol. 41, no. 6, pp. 1915–1923, 1995.

[47] Q. Wang, H. Su, K. Ren, and K. Kim, "Fast and scalable secret key generation exploiting channel phase randomness in wireless networks," in *Proc. 30th IEEE Int. Conf. Comput. Commun. (INFOCOM)*, Shanghai, China, Apr. 2011, pp. 1422–1430.

[48] S. N. Premnath, S. Jana, J. Croft, *et al.*, "Secret key extraction from wireless signal strength in real environments," *IEEE Trans. Mobile Comput.*, vol. 12, no. 5, pp. 917–930, 2013.

[49] *Wireless LAN Medium Access Control (MAC) and Physical Layer (PHY) Specification*, IEEE Std. 802.11, 2012.

[50] *Low-Rate Wireless Personal Area Networks (LR-WPANs)*, IEEE Std. 802.15.4, 2011.

[51] MICAz wireless measurement system. [Online]. Available: http://www.memsic.com/userfiles/files/Datasheets/WSN/micaz_datasheet-t.pdf.

[52] Crossbow TelosB mote platform. [Online]. Available: http://www.willow.co.uk/TelosB_Datasheet.pdf.

[53]   D. Halperin, W. Hu, A. Sheth, and D. Wetherall, "Tool release: gathering 802.11n traces with channel state information," *ACM SIGCOMM Comput. Commun. Review*, vol. 41, no. 1, pp. 53–53, 2011.

[54]   Ettus research. [Online]. Available: http://www.ettus.com. Accessed on 10 July 2017.

[55]   WARP project. [Online]. Available: http://warpproject.org. Accessed on 10 July 2017.

[56]   K. Zeng, D. Wu, A. Chan, and P. Mohapatra, "Exploiting multiple-antenna diversity for shared secret key generation in wireless networks," in *Proc. 29th IEEE Int. Conf. Comput. Commun. (INFOCOM)*, San Diego, California, USA, Mar. 2010, pp. 1–9.

[57]   R. Guillaume, F. Winzer, and A. Czylwik, "Bringing PHY-based key generation into the field: An evaluation for practical scenarios," in *Proc. 82nd IEEE Veh. Technology Conf. (VTC Fall)*, Boston, USA, Sep. 2015, pp. 1–5.

[58]   J. Zhang, R. Woods, A. Marshall, and T. Q. Duong, "Verification of key generation from individual OFDM subcarrier's channel response," in *Proc. IEEE GLOBECOM Workshop Trusted Commun. with Physical Layer Security (TCPLS)*, San Diego, California, USA, Dec. 2015, pp. 1–6.

[59]   T. Aono, K. Higuchi, T. Ohira, B. Komiyama, and H. Sasaoka, "Wireless secret key generation exploiting reactance-domain scalar response of multipath fading channels," *IEEE Trans. Antennas Propag.*, vol. 53, no. 11, pp. 3776–3784, 2005.

[60]   M. Wilhelm, I. Martinovic, and J. B. Schmitt, "Secure key generation in sensor networks based on frequency-selective channels," *IEEE J. Sel. Areas Commun.*, vol. 31, no. 9, pp. 1779–1790, 2013.

[61]   R. Wilson, D. Tse, and R. Scholtz, "Channel identification: Secret sharing using reciprocity in ultrawideband channels," *IEEE Trans. Inf. Forensics Security*, vol. 2, no. 3, pp. 364–375, 2007.

[62]   M. G. Madiseh, M. L. McGuire, S. S. Neville, L. Cai, and M. Horie, "Secret key generation and agreement in UWB communication channels," in *Proc. IEEE GLOBECOM*, New Orleans, Louisiana, USA, Nov. 2008, pp. 1–5.

[63]   J. Huang and T. Jiang, "Dynamic secret key generation exploiting ultra-wideband wireless channel characteristics," in *Proc. IEEE Wireless Commun. and Networking Conf. (WCNC)*, New Orleans, Los Angeles, USA, Mar. 2015, pp. 1701–1706.

[64]   D. Wang, A. Hu, and L. Peng, "A novel secret key generation method in OFDM system for physical layer security," *Int. J. Interdisciplinary Telecommunications and Networking (IJITN)*, vol. 8, no. 1, pp. 21–34, 2016.

[65]   K. Chen, B. Natarajan, and S. Shattil, "Secret key generation rate with power allocation in relay-based LTE-A networks," *IEEE Trans. Inf. Forensics Security*, vol. 10, no. 11, pp. 2424–2434, 2015.

[66]   MAX2828/MX2829 Single-/Dual-Band 802.11 a/b/g World-Band Transceiver ICs. 2004. [Online]. Available: http://datasheets.maximintegrated.com/en/ds/MAX2828-MAX2829.pdf. Accessed on 23 August 2017.

Chapter 19

Application cases of secret key generation in communication nodes and terminals

Christiane Kameni Ngassa[1], Taghrid Mazloum[2],
François Delaveau[1], Sandrine Boumard[3], Nir Shapira[4],
Renaud Molière[1], Alain Sibille[2], Adrian Kotelba[3],
and Jani Suomalainen[3]

19.1 Introduction

The main objective of this chapter is to study explicit key extraction techniques and algorithms for the security of radio communication. After some recalls on the main processing steps (Figure 19.1(a)) and on theoretical results relevant to the radio wiretap model (Figure 19.1(b)), we detail recent experimental results on randomness properties of real field radio channels. Furthermore, we detail a practical implantation of secret key generation (SKG) schemes, based on the Channel Quantization Alternate (CQA) algorithm helped with channel decorrelation techniques, into modern public networks such as WiFi and radio-cells of fourth generation (LTE, long-term evolution). Finally, through realistic simulations and real field experiments of radio links, we analyze the security performance of the implemented SKG schemes, and highlight their significant practical results and perspectives for future implantations into existing and next-generation radio standards.

The chapter is organized as follows.

Section 19.2 introduces the usage of a shared source of randomness under a reciprocity assumption, explains the particular interest of radio propagation for achieving confidentiality of wireless links in the wiretap channel model, provides security metrics suited to wireless links and analyzes the impact of radio-channel properties from simulated samples.

Section 19.3 provides a complete implementation of a SKG scheme, adapted to orthogonal frequency division multiplexing (OFDM) signals such as encountered in modern digital Wireless Access Networks (802.11n/ac and LTE). Then, simulations

[1]Thales Communications and Security, USA
[2]Institut Mines Telecom – Telecom ParisTech, France
[3]VTT Technical Research Centre of Finland, Finland
[4]Celeno Communications

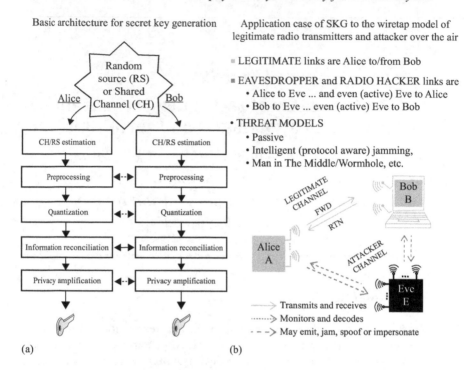

*Figure 19.1   Principle of secret key generation—application case to the wiretap
radio channel. (a) basic architecture for secret key generation;
(b) application case of SKG to the wiretap model of legitimate radio
transmitters and attacker over the air*

of SKG into realistic LTE links and practical over-the-air implementation of SKG
into WiFi chipsets assess the feasibility proof and provide the practical performance,
while computation of entropy and mutual information complete the security analysis.

Section 19.4 points out the potential advantages of these techniques for radio
standards. After some recalls on existing vulnerabilities of public radio networks,
many potential application cases of SKG are highlighted for enhancing the privacy
of subscribers and confidentiality of data streams into existing and future wireless
standards. Especially, practical tracks are proposed for securing the early stages of
radio access protocols where identification and authentication are performed.

19.2   Fundamental aspects of secret key generation

Given a common or highly correlated source of randomness and a public error-free
authenticated channel, legitimate users may generate a shared secret key about which
an eavesdropper that has an uncorrelated access to the random source would have
negligible information [1]. The users extract correlated random sequences from their
correlated source of randomness and subsequently agree on the same random key
through public communications. Messages exchanged between legitimate users over

the public channel do not carry enough information to allow the eavesdropper to recover the same cryptographic key. This SKG approach was introduced in 1993 [2,3] and provides information-theoretic security as the eavesdropper is assumed to have unlimited computing power.

19.2.1 Channel-based random bit generators

While the first implementations of SKG concerned quantum physics, the propagation channel has also captured very much attention in establishing on-the-fly secure key bits [4–8]. Indeed, it is possible to create a random bit generator based on the characteristics of the propagation channel, owing to the stochastic nature of the multipath propagation and to the intrinsic reciprocity of the electromagnetic (EM) transmission media.

19.2.1.1 A shared source owing to reciprocity

The reciprocity law states that the multipath properties are the same on both directions of a link since EM waves undergo the same physical interactions in both senses of propagation. Such a reciprocity especially holds in time division duplex (TDD) systems, e.g., IEEE 802.11, LTE and next generation (5G) of wireless standards, where both uplink and downlink use the same frequency band. Accordingly and without any feedback, any two entities (e.g., legitimate ones) may share common information extracted from the reciprocal channel, from which identical key bits may be generated.

However, some practical issues disrupt reciprocity. On the one hand, in TDD systems, the channel must be estimated over a duration smaller than the coherence time, in order to reduce the discrepancies between Alice and Bob. On the other, hardware calibration should be performed in order to account for the asymmetric properties of certain electronic components in the transmit and receive communication chains. Briefly speaking, both the non-reciprocity sources and the channel noise limit the number of shared bits that legitimate parties may reliably share [9].

19.2.1.2 Randomness owing to multipath propagation

Both length and randomness are fundamental properties of a robust cryptographic key. The former is required to avoid any brute force attack while the latter increases the eavesdropper uncertainty. A long key may result from the concatenation of several subkeys where each one, of a limited number of reliable bits, comes from a single channel sample [9,10]. Moreover, the channel samples must be statistically decorrelated as much as possible in order to enhance the randomness versus a single sample.

Owing to multipath propagation, the radio channel is subject to intrinsic stochastic variations in the time (e.g., through terminal movement or people motion in the surrounding area), space (e.g., using multiple antennas systems) or frequency (e.g., using an OFDM scheme) domains. Indeed, it is shown in [11] that, for multiple-input–multiple-output systems, the richer the channel in multipaths, the richer in randomness. Consistently, the authors in [12] proved by the mean of a RT tool that the diffuse scattering components of the channel play an essential role in improving the performance of SKG.

Nonetheless, the challenge resides on how to sample the propagation channel in order to select sufficiently decorrelated realizations. Intuitively, this relies on the transmission medium itself from the point of view of its richness in scattered obstacles as well as on the source domain of randomness (e.g., time, space or frequency domains). As shown in [10,13], if we consider SKG by investigating either the space, the frequency or the time domains, the performance can be respectively assessed according to the coherence distance, the coherence bandwidth (BW) or the coherence time. For better performance, it is therefore proposed in [10] to jointly exploit both the space and the frequency propagation channel variability and in [13] to jointly exploit both the space and time propagation channel variability.

19.2.1.3 Confidentiality

The purpose of an information-theoretic framework is to achieve a configuration where the eavesdropper does not have enough information to collapse the key difference between the computation of legitimates and its own computations.

Owing to spatial channel decorrelation of the multipath propagation, Eve may measure decorrelated radio channel with respect to Alice–Bob channel. The key confidentiality is ensured when the correlation coefficient does not exceed a certain threshold that depends on the implemented SKG scheme. The efficiency of such a scheme, described in Section 19.3.1, relies on providing independent keys even from highly correlated channels, however, without disturbing the reliability between Alice and Bob (i.e., obtaining exactly the same key from very highly correlated channels).

Furthermore, the channel correlation, and subsequently the key confidentiality, is impacted by the relative position of Eve to at least one of the legitimate users (e.g., Bob) as well as to the channel characteristics (e.g., the coherence distance when considering space diversity).

The worst scenario occurs when Eve and Bob are collocated; they measure the same channel except independent noise which may lead to some discrepancies between the extracted keys. These discrepancies increase when Eve moves away from Bob as the correlation coefficient decreases. Even that, the correlation is still relatively high as they are in the same stationary region where they share the same multipath components. In this case, Eve may employ tools such as RT, while exploiting further information (e.g., the location of Bob and the environment characteristics), in order to improve her learning about the key. Nevertheless, the correlation is very low when Eve and Bob are not in the same stationary area, where they do not share the same multipath components. In this latter case, the security is ensured.

Experimental results in [13] show that special decorrelation occurs after one-half wave length distance in most of dispersive radio-channel configurations encountered in Non Line Of Sight (NLOS) geometry, while a distance of several wavelength (up to 4 wavelengths) is usually required to achieve space decorrelation in more stationary radio channels, such as encountered in Line Of Sight (LOS) geometry.

19.2.2 *Metrics for secret key generation assessment*

As already shown, the robustness of the key relies on its length, randomness and confidentiality. All these features may be assessed, theoretically, by the secret key rate through mutual information computations [2,11,14]. On the other hand, from the practical point of view, the quality of the key resulting from the implementation of the SKG protocol may be assessed in terms of both reliability and confidentiality through the KBER [6,9] and also in terms of randomness through statistical tests [6,10].

Secret key rate: In an information-theoretic framework, the maximum amount of random information reliably shared between Alice and Bob is measured by the mutual information between their legitimate channels (i.e., $I_K = I(\hat{\mathbf{h}}_a, \hat{\mathbf{h}}_b)$). Such an amount is entirely secure if Eve experiences channels that are statistically independent from those measured by the legitimate terminals (e.g., Eve is sufficiently far from both Alice and Bob). Otherwise and more generally, the secret key rate is the mutual information between Alice and Bob's channels, given Eve's observations. We note that the mutual information is expressed by the covariance matrices of channel observations if the latter are both marginally and jointly Gaussian distributed [11,15]. Otherwise, there is no closed-form expression for the mutual information.

Key bit error rate (KBER): More practically, the SKG performance can be assessed after applying the protocol in the whole or part of it. Thus, it is crucial to compare the keys through the evaluation of the KBER[1] by computing the ratio of the number of differing bits to the key length. Obviously, when comparing the keys generated by legitimate terminals, the KBER should approach 0 while for the comparison versus the key computed by Eve, it should approach 1/2 [9].

Statistical tests for randomness: The randomness of the key may be assessed by specific statistical tests, among which are notably the National Institute of Standard and Technology (NIST) test suite [16] and the Intel Health Check test [17]. The NIST test suite is composed of 16 tests, where each attempts to detect if the key bits follow a certain deterministic behavior resulting from imperfect randomness in the key generation process. Among the 16 tests, selected examples below have used both the "mono-bit frequency" test and the "runs" test. The former investigates the occurrence of bit 0 on the entire key bits, while the latter checks whether the transition between bits 0 and 1 is too fast or too slow. Since NIST tests are complex to embed into nodes and terminals, we also employ the Intel Health Check test, which checks the entropy of the generated keys by evaluating the occurrence of six different bit patterns in a 256-bit sequence.

19.2.3 Impact of channel characteristics

In this section, we address some concrete aspects of SKG in relation with the characteristics of the radio channel. We particularly show here the behavior of I_K when increasing the number of investigated frequency-varying subchannels. Thus, we build a key from N_u consecutive subcarriers, even correlated, selected among N_f total subcarriers in a given BW of an OFDM system. We assume that the power spectral density (PSD) of both the subcarrier and the noise is constant, with a signal-to-noise ratio (SNR) of 15 dB.

---

[1]We choose to evaluate the BER instead of the symbol error rate in order to account for the reconciliation phase which operates at the bit level.

For the sake of comparison, we consider the following different types of channel data:

1. **A power decay profile model**: We consider a simple dispersive channel model, based on periodic independent multipaths in the delay domain, with an exponentially decreasing mean power. Each path complex amplitude is circularly-symmetric Gaussian distributed.

2. **A ray-tracing outdoor data**: We exploit deterministic simulations using a commercial ray tracing (RT) tool [18], specially extended by incorporating diffuse scattering, implemented according to the effective roughness approach [19]. We consider an outdoor suburban environment located in Paris (i.e., the "carrousel du Louvre"). Alice is represented by a fixed base station at 48 m above the ground level, whereas Bob takes several positions covering the considered area at 1.5-m antenna height. I_K is statistically computed for each position of Bob by considering a small scale stationarity region around each RT computed Bob location, where the variability stems from the varying phase for each antenna position and for each path.

3. **Measured indoor channels**: Channel coefficients have been recorded in the 2–6-GHz frequency band using a vector network analyzer, in the premises of Telecom ParisTech [9,10,13]. For each position of Bob in either a room, a corridor or a hall, the transmitter representing Alice is spatially scanned over a 11×11 square grid confined to a small area, which allows the evaluation of I_K.

I_K is assessed in the frequency domain by stacking N_u subcarriers within a single vector. Since the channel model and the RT tool produce channel impulse responses (CIRs), we perform a discrete Fourier transform in order to obtain N_f channel transfer functions within a given BW.

Figure 19.2 shows the behavior of I_K with respect to N_u, thus for the three aforementioned channel models. The dashed curves concern the extreme case where the N_u subcarriers are independent and identically distributed (i.i.d.). The set of solid curves corresponds to a set of various positions of Bob, resulting in different channel features. Obviously, increasing N_u yields an increase in the amount of randomness, which depends on the correlation of the subcarriers. In fact, a set of N_u subcarriers is equivalent, in the delay domain, to N_u resolved paths of a gain obtained by using a cardinal sinus filter. When the resolved paths are independent, no more randomness is obtained by increasing N_u, which should yield to a saturation behavior. However, as shown in Figure 19.2(a), beyond the saturation point, I_K still increases with N_u, although slowly. This is explained by the improvement on the SNR per resolved path, owing to the increase in the total transmitted power as N_u increases. Actually, it is shown in [14] that there is an optimal BW that maximizes I_K if the total transmitted power is fixed, rather than the PSD.

Moreover, in Figure 19.2(b) and (c), the curves show a sublinear behavior, which is more pronounced for the RT data. This is explained by the fact that the rays computed by the RT tool are not evenly distributed over all the delay bins, while measured CIRs turn out to be very dense in multipaths. This requires a much higher BW than 160 MHz in order to exploit the rich degrees of freedom of the channel. Furthermore, we notice

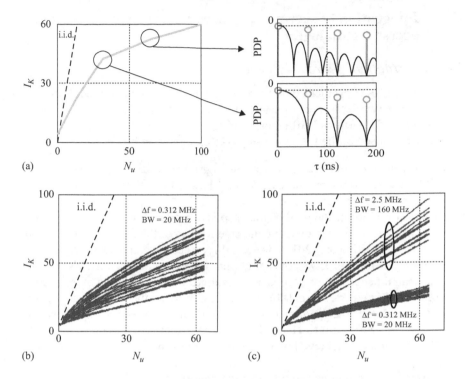

*Figure 19.2 I_K with respect to the number of used subcarriers N_u: (a) power decay
profile model, (b) RT data model and (c) measured channel data*

that, for a given N_u subcarriers, I_K values change from one measured location to
another one since the channel coherence BW changes according to the richness in
multipaths as well as on their spread in the delay domain [12].

As a conclusion of practical interest, as shown in [12], it appears that indoor
environments are less prone to SKG from frequency-variant channels, since such
environments have naturally large coherence BWs, as opposed to outdoor environments where delay spreads are usually longer that implies smaller coherence BW and
higher frequency diversity.

When considering spatial-variant channels,

- increasing randomness can be captured into indoor radio-channels, especially
  with large BW channel sounding, relevant coherence distance is low (half
  wave length) to medium (usually less than ten wavelengths) depending on the
  NLOS/LOS configuration.
- Outdoor radio-channel, usually more dispersive than indoor radio-channels, provide native randomness at limited BW channel sounding, relevant coherence
  distance is usually low (half wavelength), and very low especially in dense urban
  NLOS geometries.

19.3   Integration of secret key generation into existing radio access technologies

19.3.1   Practical secret key generation scheme

The proposed SKG protocol targets existing and future radio-communication standards, such as Wireless Local Access Networks (WLAN) 802.11n/ac and radio-cells of second, third, fourth and fifth generations.

It is composed of the following steps (as mentioned in Figure 19.1).

Instantaneous channel estimation at the signal frame level: this first step of the SKG scheme computes the CIR or the Channel Frequency Response (CFR) for the signal frame.

Spatial decorrelation: the goal of this step is to reduce spatial correlation between channel measurements by using the eigenvectors of the full covariance matrix of the channel estimates [20]. However, this process is computationally expensive and only helpful in LOS configurations. Therefore, unless otherwise specified, the spatial decorrelation step is not performed in the remaining of the chapter.

Channel coefficient decorrelation over several frames: this second step optimizes the selection of the randomness material in stationary environments. We apply a selection algorithm to channel coefficients issued from the previous CIRs or CFR, in order to achieve low crosscorrelation of these coefficients over several frames.

Quantization: this step uses the CQA algorithm introduced by Wallace to quantize selected channel coefficients [11], that minimizes key mismatch between the legitimate users Alice and Bob.

Information reconciliation: this step corrects the remaining mismatch between Alice and Bob keys. Secure sketches and error correcting codes are employed to allow Bob to recover the same key than Alice. To do so, Alice has to send the secure sketch over the public channel, what possibly leaks a controlled amount of information to the eavesdropper Eve.

Privacy amplification: this step improves the randomness of the secret key and removes the information leaked during the information reconciliation step. To do so, hash functions are used and the key length is reduced if necessary. This final step guarantees that the generated secret key is fully decorrelated from the key computed by the eavesdropper.

Note: in the following, practical implementation of SKG scheme inside communication devices is studied. Hence robust and simple algorithms are considered. For example, simple algebraic Forward Error Correction (FEC) code are employed to reconcile Alice and Bob keys and a classical family of 2-universal hash function is chosen in the privacy amplification step [21].

19.3.1.1   Channel estimation—application case to OFDM signals

When considering an OFDM (such as defined in WiFi and LTE), the component of the CFR H_f in the frequency domain quantifies the fading applying on each subcarrier.

In a sampled system, considering a finite response and band, the kth frequency component f_k of the CFR can be calculated as follows:

$$H_f(k) = \frac{Y(f_k)}{X(f_k)} \tag{19.1}$$

where Y is the received signal and X is the emitted signal also referred to as the reference signal (RS). In the time domain, an equivalent CIR estimation can be deduced from the CFR by IFFT, as follows:

$$H_{\text{IFFT}} = \text{IFFT}(H_f) \tag{19.2}$$

When considering now Time Division Multiple Access of Code Division Multiple Access waveforms defined in 2G and 3G Radio Access Technologies (RAT), CIR can be computed directly in the time domain by applying filter estimations techniques to RS X.

19.3.1.2 Channel decorrelation

Secret key bits should be completely random to keep them unpredictable by Eve. Therefore any deterministic component in the radio propagation channel should be removed. To obtain key bits with equal probability, the quantization algorithm should tamper with channel coefficients as random and decorrelated as possible. The goal of this step is thus to decrease the negative effect of channel correlation by a careful selection of the channel coefficient to be quantized. Channel correlation can be observed in time and frequency domains.

Time correlation is decreased between channel coefficients. To do so, channel coefficients are recorded during a given acquisition time, constituting a frame. Then crosscorrelation coefficients are computed between two consecutive frames and finally only frames with low crosscorrelation coefficient (above a given threshold T_t) are selected.

Similarly, the frequency correlation is decreased by first computing crosscorrelation coefficients between two consecutive frequency carriers. Only frequency carriers for which the crosscorrelation coefficient is above a given threshold T_f are selected. In addition, lowest and highest frequency carriers are dropped. Finally, Alice sends to Bob the position of the channel coefficients over the public channel. Hence, Eve also knows which coefficients were dropped and which ones were selected but she does not have any additional information on their value. Therefore, there is no information leakage during the channel decorrelation step.

19.3.1.3 Quantization

After measuring the radio channel, Alice and Bob jointly employ an algorithm to quantize the channel taps that they have estimated in order to generate a common sequence of key bits, under reciprocity assumption.

However, due to noise and channel estimation errors, Alice and Bob may disagree on some key bits. Several quantization algorithms employing censoring schemes have been developed to limit this mismatch between Alice and Bob keys.

A typical censoring algorithm defines guard band intervals and discards any channel measurement falling into them [11]; leading to an inefficient exploitation of channel measurements and to a lower number of generated key bits.

Other schemes employ different quantization maps where each one is adapted to the channel observations, e.g., CQA algorithm [11]. The principle consists in choosing the adaptive quantization map where the current observation is less sensitive to mismatch. Moreover this method keeps higher number of key bits. Consequently, CQA algorithm is applied on complex channel coefficients to generate secret key bits in remaining of the chapter.

19.3.1.4 Information reconciliation

This step suppresses remaining mismatches between Alice and Bob keys by using secure sketch based on error-correcting codes [22]. The key computed by Alice is considered as the reference secret key that Bob wants to obtain using the key K_b extracted from his channel measurements.

Alice first selects a random codeword c from an error-correcting code $\mathscr{C}$. She then computes the secure sketch $s = K_a \oplus c$ and sends s to Bob over the public channel. Bob subtracts s from its computed key $K_b : c_b = K_b \oplus s$ ($=K_b \oplus K_a \oplus c$), decodes c_b to recover c and gets $\hat{c}$. Finally Bob computes K_a by shifting back and gets $(K_a) = \hat{c} \oplus \hat{s}$.

Perfect reconciliation is achieved when Bob perfectly retrieves the random codeword chosen by Alice, meaning that $c = \hat{c}$. As a result, there is no mismatch between Alice and Bob keys ($K_a = K_b$).

The secure sketch s, sent over the public channel, allows the exact recovery of the secret key without revealing the exact value of the key. However, s might leak some information on the secret key over the public channel as Eve can also use the secure sketch to retrieve the secret key K_a. Thus, a final step is then necessary to suppress the leaked information and to improve the quality of the secret key.

19.3.1.5 Privacy amplification

The objective of the privacy amplification step is to erase the information leaked to Eve on the secret key during the information reconciliation step and to improve the randomness of the key. We interpret the secret key K as an element of the Galois Field $GF(2^n)$ and we choose the following two-universal family of hash functions [23] where n is the number of bits of the key K.

For $1 \leq r \leq n$ and for $a \in GF(2^n)$, the functions $\{0, 1\}^n \rightarrow \{0, 1\}^r$ assigning to the key K the first r bits of key $aK \in GF(2^n)$ define a two-universal family of hash functions. r is the final length of the secret key. In practice, at each new key computation, the parameter a is randomly chosen by Alice who sends it to Bob over the public channel. Alice and Bob then compute the product $aK \in GF(2^n)$.

The hash mechanism spreads any bit error all over the final key $(aK)_{(r \text{ bits})}$ (first r bits of aK), thus when Eve tries to recover the initial key K (at the reconciliation step), any error on K will make the final key $(aK)_{(r \text{ bits})}$ unusable for her. Nevertheless, Bob has to perfectly recover the initial key K (i.e., reconciliation should be perfectly achieved) in order to get the usage of the final key $(aK)_{(r \text{ bits})}$.

(a) Propagation of legitimate and attacker radio channel

(b) Acquisition of real field LTE/WiFi signals
with a static antenna geometry
in an empty tennis court

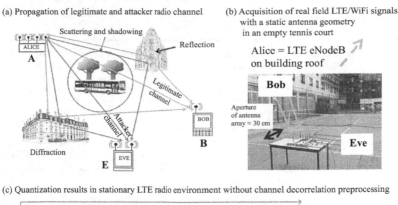

Alice = LTE eNodeB
on building roof

(c) Quantization results in stationary LTE radio environment without channel decorrelation preprocessing

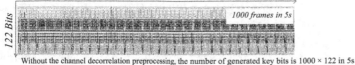

Without the channel decorrelation preprocessing, the number of generated key bits is 1000 × 122 in 5s
=> High time correlation and stationary patterns in the quantized bits that can be exploited by Eve

(d) Quantization results in stationary LTE radio environment with channel decorrelation preprocessing

With the channel decorrelation preprocessing, the number of generated key bits decreases to 200×36 in 5s
=> Less stationary pattern in the quantized bits

(e) SKG results in several LTE and WLAN/802/11n radio-environments

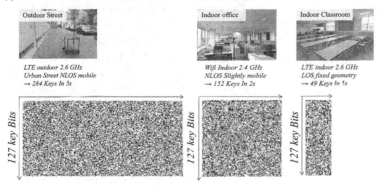

Figure 19.3   SKG principle and results

19.3.2   Simulation results from single sense recorded signals

In this section, we generate keys from real LTE and WiFi signals acquired using the PHYLAWS test bed presented in Figure 19.3 and described in [13]. The test bed emulates Bob and Eve. Bob is considered to have 2 antennas spaced out of 33 cm corresponding to 3λ where λ is the signal's wavelength. Eve is considered to have 4 antennas spaced out of 11 cm corresponding to λ.

Table 19.1   NIST test results

NIST test	LTE indoor 2.6 GHz		LTE outdoor 2.6 GHz		WiFi LOS 2.4 GHz		WiFi NLOS 2.4 GHz	
	Frequency	Runs	Frequency	Runs	Frequency	Runs	Frequency	Runs
Quantization	98% (48/49)	27% (13/49)	99% (281/284)	80% (228/284)	87% (132/152)	84% (128/152)	100% (171/171)	99% (169/171)
Amplification	100% (49/49)	100% (49/49)	100% (284/284)	100% (284/284)	99% (151/152)	98% (149/152)	100% (171/171)	99% (170/171)

19.3.2.1 Impact of the SKG preprocessing step on the randomness of generated keys

Figure 19.3(c) shows the output of the CQA quantization algorithm applied on LTE signals which were recorded during 5 s in a very stationary propagation environment at 2627.5 MHz with a total BW of 10 MHz. CFR were computed from the Primary Synchronization Signal that occupies only 1.4 MHz.

When using 4 quantization regions, the CQA produces 1,000 frames detections and 122 secret bits per frames. However, we can notice a repetitive pattern on the generated keys meaning that CFR coefficients are highly correlated in time and frequency. This high correlation represents a major vulnerability as the generated secret key bits will not be random enough.

Figure 19.3(d) shows key bits obtained when the channel coefficient decorrelation processing is applied on the same record. The algorithm manages to extract the useful information and remove most of the repetitions of the bit pattern. Hence the number of key bits is reduced at the output of the quantization algorithm but the correlation between obtained bits is significantly decreased both in time and frequency.

19.3.2.2 Evaluation of the randomness of the keys using NIST statistical tests

In this section, the quality of keys generated from previous records of LTE and WiFi signals is evaluated with two randomness tests defined in the NIST statistical test suite [16].

NIST frequency mono-bit test: the goal of this test is to determine whether the numbers of 0s and 1s in the key are approximately the same, as expected for a truly random sequence. Table 19.1 provides the percentage of keys that passed the frequency mono-bit test for the previous LTE and WiFi signals. As expected, almost all the keys pass the test after quantization since the CQA algorithm intrinsically distributes 0s and 1s in a uniform way.

NIST runs test: the goal of this test is to determine whether the oscillation between 0s and 1s is too fast or too slow compared to what expected for a truly random sequence. The results in Table 19.1 shows that after quantization, only 27% of keys generated in the stationary LTE indoor environment pass the runs test whereas a high percentage ($\geq 80\%$) is achieved with keys generated in other environments which are much more dispersive.

The runs test better captures the randomness of a sequence.

As expected, near 100% of generated keys pass NIST tests after privacy amplification step, even in the static indoor environment. This final step of the SKG scheme is therefore crucial for practical application in low dispersive radio environments and narrow band signals.

19.3.2.3 Entropy estimation and analysis

The aim of this section is to evaluate the percentage of entropy bits extractable from the radio channel in realistic radio environment. To do so, we estimate the min-entropy of channels, first between Alice and Bob, then between Alice and Eve, at the output of the quantization step of the SKG scheme (without applying the channel decorrelation). The computation uses NIST tests for estimating the min-entropy of non-IID sources described in [24]. We also estimate the joint entropy and the maximum mutual information between pairs of antennas in order to evaluate the percentage of information shared by two distinct antennas. Finally, for a given pair of antennas, the entropy and the mutual information can provide us an experimental insight on the percentage of secure entropy bits.

Table 19.2 provides the entropy estimates for the six antennas of the PHYLAWS test bed shown in Figure 19.3(b). The results are provided for two extreme propagation environments. The first one, very stationary, is an empty tennis indoor court surrounded by building on the top of which is an LTE e-nodeB. The configuration is fixed and LOS. The second one, much less stationary is an indoor office where antennas are slightly moving. WiFi signals come from NLOS access point. The results show that there are at least 20% of entropy bits in the first (worse) case and around 70% of entropy bits in second (better) case. In addition, the computed maximum value of the mutual information between pairs of antennas reveals that one antenna on Eve's array shares only few information (around 20%) with one antenna on Bob's array.

19.3.3 Simulation results from dual sense LTE signals

The application of the practical SKG scheme to LTE is straightforward as the scheme only needs access to the channel estimates in the frequency domain, which are readily available in the physical layer. In order to assess its performance in a LTE system, Monte-Carlo simulations have been performed using MATLAB®.

19.3.3.1 Simulators

For performance assessment, we use the MATLAB-based LTE link-level simulators [25] developed by Technical University of Vienna. The simulators implement standard-compliant LTE downlink (DL) and LTE uplink (UL) transceivers with their main features, i.e., basic channel models, modulation and coding, multiple-antenna transmission and reception, channel estimation, multiple-user scenarios and scheduling. The LTE link-level simulators include, among other basic channel models, the QuaDRiGa channel model [26], which can model realistic distance-dependent correlation of radio propagation between Alice–Bob, Alice–Eve and Bob–Eve channels. In the simulations indeed, only the large-scale channels parameters are spatially correlated.

Table 19.2 Entropy estimates

Min-entropy estimates

	Antennas (GHz)	Bob$_1$(%)	Bob$_2$(%)	Eve$_1$(%)	Eve$_2$(%)	Eve$_3$(%)	Eve$_4$(%)
Min-entropy	LTE LOS 2.6	19.5	50	32.4	22.6	28.9	32.4
	WiFi NLOS 2.4	63.1	65.2	74	69.7	76.2	74

Max mutual information

	Antennas (GHz)	Bob$_1$ – Bob$_2$(%)	Bob$_1$ – Eve$_1$(%)	Bob$_2$ – Eve$_1$(%)	Bob$_1$ – Eve$_4$(%)	Bob$_1$ – Eve$_2$(%)	Eve$_1$ – Eve$_2$(%)	Eve$_3$ – Eve$_4$(%)
Max mutual information	LTE LOS 2.6	19.7	16.5	38.6	24.9	84	73.9	
	WiFi NLOS 2.4	19.8	18	20.6	19.4	79.7	85	

19.3.3.2   Channel coefficient estimates

Estimates of the channel coefficients are computed for each subcarrier carrying known sequences and each transmit and receive antenna pair, then averaged over the subframe to provide only one coefficient per subframe and per subcarrier per antenna pair. In the DL, Bob (and Eve) can use the DL reference signals (RS) over the whole BW and the channel estimates are obtained by dividing the received signal at the pilot tones or reference sequence location by the known transmitted signal. In the UL the situation can be different as the sounding RS are not mandatory and one cannot rely on them. The demodulation RS are used to estimate the channel at Alice. Alice thus has knowledge of the channel limited to the resource allocated for the UL transmission for Bob.

The subcarriers in the frequency domain are selected such that they hold estimates in both DL and UL directions. However, in the simulations herein, all the resource blocks are allocated to Bob and the BW limitation is not taken into account.

Considering now the TDD configuration of LTE RAT, we need to ensure that the reciprocity assumption is still valid. For this, the channel estimates obtained at adjacent DL and UL subframes should be used, which happens when the system switched from UL to DL. The subframe indexes at which the channel coefficients are extracted at Alice and Bob/Eve depend on the TDD configuration. We assume here a TDD configuration that allows us to extract two sets of channel coefficients per frame.

19.3.3.3   Simulations scenarios and parameters

The simulations process is such that the QuaDRiGa channel coefficients are created and then used first in the DL LTE simulator and second in the UL LTE simulator. At the end of this run, the secret keys generated by Alice, Bob and Eve are compared. A simulation run is set to last 100 frames.

Alice is a fixed base station. Bob and Eve are mobile and follow the same track at the same speed, which in the simulations is a straight line. The speed depends on the radio environment. Alice uses a 4-antennas spatially-uniform linear array and both Bob and Eve use a 2-antennas array. The signal BW is set to 10 MHz and the carrier frequency is 2.6 GHz. The channel model is block-fading and the channel estimation uses the least-square methods as provided in the LTE simulators. The SNR is defined as the average SNR at Bob for the duration of the simulation.

Several standard radio propagation environments have been tested: A1 indoor office, B1 urban microcell, and C2 urban macrocell [27]. The minimum distance between Bob and Alice has been set to 1, 10 and 50 m, respectively. The mobiles' speed has been set to 1, 2 and 14 m/s in A1, B1 and C2, respectively. Eve can be placed at various distance from Bob. The radio propagation can either be LOS or NLOS.

The SKG algorithm outputs a fixed key length of 127 bits. Time and frequency decorrelations are always used. After several tests, the decorrelation thresholds T_f and T_f have been set to value 0.5, that achieves a suitable trade-off between the number of extracted keys and their randomness quality.

Results with and without spatial decorrelation are presented. When spatial decorrelation is used in a LOS environment, the LOS component is removed.

The quantization of the real and imaginary part of the preprocessed channel coefficients produces 1 bit each, i.e., two regions are used. The coding rate of the reconciliation Bose–Chaudhuri–Hocquenghem (BCH) code varies in order to correct the errors between Bob's and Alice's keys. This coding rate needs to be tailored to the SNR on the channel estimates in order to correct errors between Bob's and Alice's keys while preventing Eve from correcting the errors in her keys. In other words, the coding rate is set such that Bob is able to correct the maximum number of errors at each simulation.

For each separation distance between Bob and Eve and each SNR values, 100 channel realizations, corresponding to 100 simulation runs, are processed and statistical distributions of various figure-of-merit over those channel realizations can be extracted.

19.3.3.4   Simulation results

In order for the SKG to work well, Eve should not be able to estimate the key that Alice and Bob have extracted from the channel estimates, Alice and Bob should agree on the same keys, and these keys should have good entropy quality.

When considering the security of the keys, the BER between the keys extracted at Alice and Eve is the main figure-of-merit to be measured. When considering the key agreement, the mismatch between the keys estimated at Alice and Bob will also be measured in order to assess the effect of the channel estimation error, after each step of SKG algorithm: quantization, reconciliation, amplification.

When considering the keys' quality, the intrinsic randomness of the key will also be assessed by using the NIST randomness tests (frequency mono-bit and run tests). These tests are performed on all keys obtained by Bob for a specific SNR value over all channel realizations.

An example of simulation results is shown in Figure 19.4 which is relevant to the urban microcell environment B1 for Bob and Eve moving on a straight line at 2 m/s. Figure 19.4 represents the cumulative distribution functions of the BER of Bob's keys (column 1) and Eve's keys (columns 2–4) compared to Alice's keys.

The first column, which shows the results at Bob's side, evaluates the key agreement efficiency. The next columns, which shows the results at Eve's side (for increasing distances between Bob and Eve) evaluate the key security.

The impact of the value of the SNR is represented in each case (4 curves per figure).

The first line shows the results for the LOS scenario with no spatial decorrelation and the second line shows the results for the LOS scenario with spatial decorrelation.

The third and fourth lines replicate the results of the first and second line for the NLOS scenario.

The results of Figure 19.4 show the following:

- The mismatch between Bob and Alice reduces as the SNR increases.
- In the NLOS case, a separation distance of one wavelength (λ) between Eve and Bob is enough to ensure the key's security against Eve. However in the LOS case, the use of spatial decorrelation might be required to achieve the same security.
- When spatial decorrelation is performed, a higher SNR is needed for Bob to estimate the right key.

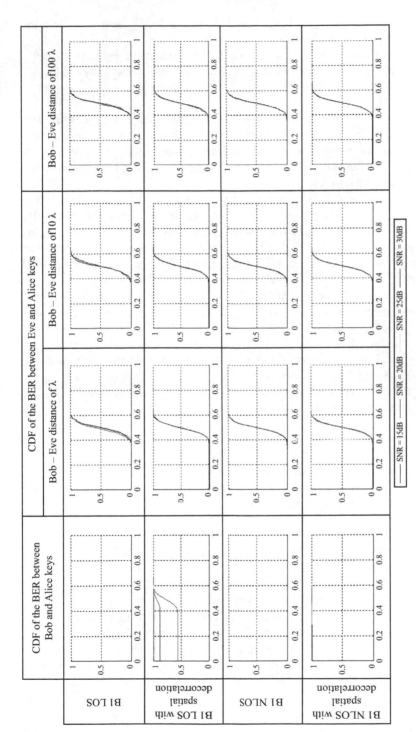

Figure 19.4 BER of Bob and Eve's keys compared to Alice's keys for various environments and SNR values. Movement of Bob and Eve is 2 m/s straight line within the urban microcell environment B1

- Especially in LOS configuration, the use of channel decorrelation increased the number of keys extracted with correct quality.
- In terms of key quality, more than 99% of the keys passed randomness tests after amplification, with or without spatial decorrelation, and for both LOS and NLOS.

19.3.3.5 Discussion

The SKG algorithms implemented in the LTE simulator has proven to work well in most of the simulated radio propagation environments. The minimum distance between Bob and Eve in order to protect the keys extracted at Alice and Bob depends on the radio environments (A1, B1, C2), whether there is LOS or NLOS, and on the use of spatial decorrelation.

Especially in A1, and in B1 too, the distance of one wavelength (λ) is enough to prevent Eve from recovering the secret key in both NLOS and LOS cases, when SKG includes spatial decorrelation in LOS cases. This impacts the needed working SNR, as spatial decorrelation needs a higher SNR to lead to the same match between Alice and Bob compared to the situation when it is not used. NLOS leads to more keys and does not require the use of spatial decorrelation.

In LOS cases of C2, the algorithm still does not protect well Alice's and Bob's keys (against Eve's attempts to recover the keys) at a distance between Bob and Eve of 10λ, even when using spatial decorrelation. However, in NLOS, the keys are protected already at a distance of 10λ.

In C2, spatial decorrelation and channel decorrelation preprocessing improve the secrecy Bob's and Alice's keys in both LOS and NLOS cases. In all simulations, the quality of the key was high after amplification, leading to more than 99% of key satisfying the used randomness tests.

19.3.4 Experimental results from dual sense WiFi signals

In this section, we generate keys from dual sense real signals emitted and received by WiFi chipsets designed by Celeno Communication Ltd. We then evaluate the randomness and secrecy of generated keys.

19.3.4.1 WiFi Test bed and measurement environment

We use here the WiFi-dedicated part of the PHYLAWS's test bed shown in Figure 19.5(a) and described in [13]. Each chipset is based on a Software Defined Radio architecture, using a Digital Signal Processing core that enables to implement algorithms in the physical layer on top a real WiFi system. The test bed supports operation in both 5 and 2.4 GHz bands by using two different chips developed by the fabless semiconductor company Celeno Communications Ltd: the CL2440 is a 4×4 AP chip supporting 5 GHz operation (for up to 80 MHz BW), while the CL2442 is a 4×4 AP chip supporting 2.4 GHz operation (for up to 40 MHz BW). The test bed is also hooked to the local network via Ethernet for control and for data extraction. The antenna spacing on the test bed is always more than half of a wave length (2.7 cm in 5.5 GHz and 6.25 cm in 2.4 GHz) to provide adequate diversity. Experiments are carried out in an testing apartment. The apartment provides a clean testing environment that is relatively interference free. Various indoor NLOS and LOS scenarios can be emulated (Figure 19.5).

19.3.4.2 Processing applied for bidirectional sounding exchange

For channel measurements, the testbed is used as both a transmitter and receiver. The transmitting device transmits a channel sounding frame, as defined in the 802.11 standard. This frame, referred to as Non Data Packet in WiFi standards, is used in 802.11ac/n for explicit sounding exchange. The channel estimates are therefore a good representation of channels as seen by real WiFi devices, including all implementation and RF impairments.

Alice first sends a sounding frame (at 2,462 MHz with a BW of 20 MHz, or at 5,180 MHz with a BW of 80 MHz) which is captured by Bob and Eve. Bob sends back to Alice a sounding frame. Alice, Bob and Eve extract 4×4 channel estimates.

CFR estimates are then processed offline: Alice, Bob and Eve first compensate their channel estimation for timing errors and normalize each channel coefficient.

They extract secret key from estimated CFR with the processing described above: channel decorrelation, quantization with CQA algorithm, information reconciliation with BCH codes, privacy amplification with two-universal family of hash functions (and key length reduction, when necessary).

The randomness of generated keys is evaluated using the Intel Health Check [17] applied on keys after quantization step and after privacy amplification steps. The reciprocity is evaluated by computing the mismatch between Alice and Bob's keys. Finally, the secrecy is evaluated by computing the BER between Bob and Eve generated keys.

19.3.4.3 Practical implantation of SKG from bidirectional channel estimates by WiFi chipsets

Figure 19.5(a) describes the testbed and the experimental testing conditions of the SKG scheme. After channel estimation over real field radio link, an offline MATLAB script runs the SKG scheme on three consecutive channel sounding exchanges between Alice and Bob. For her own attempts to recover Alice's and Bob's keys, Eve also captures the signal sent by Alice. The parametrization of the SKG protocol of Figure 19.5 is the following.

At Alice's side:

- Preprocessing: if applied, selection of low-decorrelated CFR frames with parameters $T_t = 1$ and $T_f = 0.4$
- Quantization of amplitude and phase of CFR into four regions, that provides secret keys of 127 bits length
- Computation of secure sketches used by Bob for information reconciliation using the FEC code BCH (127,15,27)
- Privacy amplification of the secret keys without key length reduction
- Key concatenation to final length of 256 bits
- Test of the key randomness after quantization and amplification with the Intel Heath Check [17]. All keys should pass the test after privacy amplification since a hash function is used during this step.

Geometry of the testing apartment—Components and geometry of the legitimate an attacker devices

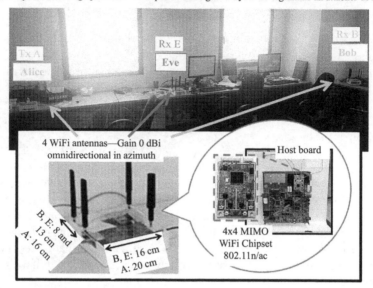

(a)

SKG results in stationary WiFi radio environment without channel decorrelation preprocessing

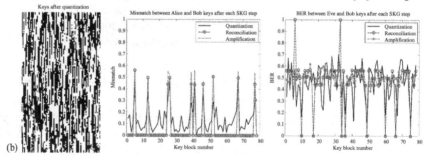

(b)

SKG results in stationary WiFi radio environment with channel decorrelation preprocessing

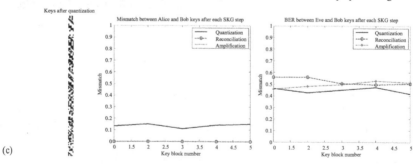

(c)

Figure 19.5   SKG experimental testbed and results from dual sense CFR

- Selection of the 256-bit keys that pass the Intel Health Check after quantization and amplification.
Final secret keys are the output of the privacy amplification step.

Alice also sends over the public channel a message containing indexes of the selected CFR frames and quantization map, secure sketches, hashing parameters and indexes of successful 256-bit secret keys. Although this message helps Bob's to compute same secret keys than Alice, secure sketches sent for reconciliation might leak some information to Eve as it allows her to correct errors she made on Alice's keys. This leaked information can be mitigated by reducing the length of extracted keys during the privacy amplification step.

Bob's and Eve's side (In our experiments, Eve performs exactly the same SKG steps as Bob):

- Preprocessing: selection of CFR frames according to the indexes sent by Alice.
- Quantization of channel measurements taking into account quantization map indexes sent by Alice.
- Information reconciliation step using secure sketches sent by Alice and using BCH (127,15, 27).
- Privacy amplification of the keys using the hashing parameters sent by Alice. Key concatenation to 256 bits.
- Selection of successful 256-bit secret keys according to the indexes sent by Alice.

19.3.4.4   Results without channel decorrelation processing

Figure 19.5(b) shows the keys extracted after quantization by Alice from channel measurements when no preprocessing step is performed. A total of 78 keys of length 127 bits were generated but none of them passed the NIST runs test. A total of 38 keys of length 256 bits were obtained by concatenating previous keys and none of them passed the Intel Health Check.

Figure 19.5(b) also shows the mismatch between Alice and Bob, and the BER between Bob and Eve's keys after each step of the SKG processing: quantization, information reconciliation and privacy amplification.

According to the results, Bob often computes different keys than Alice while Eve manages to recover some of the secret keys: SKG performance is weak in this case.

19.3.4.5   Results with channel decorrelation processing

Figure 19.5(c) shows the keys extracted after quantization by Alice from channel measurements when the channel decorrelation preprocessing is performed with thresholds values $T_t = 1$ (no selection in time domain in this particular test case, because only three time instances were available in the records) and $T_f = 0.4$. Here, 5 keys of length 127 bits were generated and 4 of them passed the NIST runs test. Two keys of length 256 bits were obtained by concatenating previous keys and both of them passed the Intel Health Check.

After privacy amplification, all keys passed both NIST runs test and Intel Health Check.

Using the same representation for Figure 19.5(b) and (c) also shows the BER between Alice and Bob (expected to be close to 0 when the processing is successful), and the BER between Bob and Eve's keys (expected to be close to 0.5 when the processing is successful). Here, Bob successfully computes the same keys than Alice while Eve's BER is always very close to 0.5: the SKG perfectly works. This illustrates the importance of the channel decorrelation preprocessing that selects only frames with low crosscorrelation, increases the available entropy into the selected frames while decreasing the mutual information between Alice and Eve's channel measurements relevant to these frames, leading at the end to a greater number of secure keys.

19.4   Conclusion: security upgrades opportunities for radio access technologies

19.4.1   Existing vulnerabilities

In the current architectures, no protection is applied on the transmission of several crucial parameters that are exchanged during the first access stages with the network (and during roaming procedures). These parameters are used for performing the authentication with privacy, and setting up the integrity and the confidentiality protections of the user and control planes.

Concerning the authentication procedure, the following crucial messages are exchanged in clear text:

- in 2G: RAND, SRES and TMSI;
- in 3/4G: RAND, RES, AUTN, KSI_{ASME} and TMSI.

Radio access of public networks is managed with identification procedures, involving subscriber and network identification numbers, authentication procedures, involving dual sense exchanges of random input parameters, parallel computation and output check at mobile and at core network. In radio-cell networks and in WLAN networks, this processing is performed very early (before the establishment of ciphering keys).

Several crucial parameters (such as IMSI, IP or MAC address) transmitted over the physical layer are not encrypted during the attach or roaming procedures (especially international roaming).

Furthermore, subscriber's and equipment's parameters are used for performing the authentication, setting up the integrity and the confidentiality protections (of both user and control protocol layers). Unfortunately, they are transmitted in clear text with significant temporization times in their transmission procedures. Finally they are very vulnerable to many kind of attacks such as passive monitoring, active hacking (denial of service, replay attack), man in the middle and spoofing.

Concerning the identification procedure of radio-cells, the following crucial subscriber or equipment parameters are exchanged in clear text.

- In 2G, 3G and 4G radio cellular networks: usually TMSI, and when requested (because of international roaming or failure of conventional TMSI identity check), IMSI, IMEI, IMEISV or GUTI
- In WLAN standards: IP Address, SSID and even the MAC address in first attach procedures and in many other dedicated procedures.

Concerning the authentication procedure of radio-cells, the following crucial messages are exchanged in clear text.

- In 2G radio cellular networks: random parameter RAND, SRES at the input and output of the single terminal authentication check by the network.
- In 3/4G radio cellular networks: random parameter RAND, RES, AUTN and KSI$_{ASME}$ at the input and output of the dual sense authentication check.

In general, the interception of identifiers (such as mentioned above) reveals sensitive information such as subscriber identity and location. It thus allows Eve to focus on the monitoring of target messages of given subscribers, to build replay attacks, to spoof and impersonate terminals and nodes, etc. See [13,28] for more details.

Moreover, the hacking of long-term secret keys K/Ki by cyber attackers have been recently reported (for more details see [13]). Therefore, it then becomes easy for a passive eavesdropper to retrieve the other necessary parameters by monitoring the signaling and access messages. First, Eve can recover authentication and cipher keys, then Eve can break all protections (such as the integrity control and the confidentiality of an on-going communication).

Finally, a major security enhancement of existing and future radio-networks would be achieved by preventing the decoding capability by third parties of sensitive message exchanged at the radio air interface between nodes infrastructure and terminals. In particular, the protection of the identification procedures, authentication protocols and cipher establishment should be reinforced, by removing any capability for Eve to intercept and decode the associated parameters that are today given for free at the radio layer. This would strongly enhance privacy and confidentiality, and this would significantly mitigate the consequences of a leakage of K/Ki keys.

19.4.2   *Proposed solutions for securing radio access protocols with secret key generation*

As explained above, the principle of SKG is to reuse radio-channel sounding outputs as common random sources of legitimate radio-devices under an assumption of reciprocity, without any shared secret.

For any radio access using a TDD protocol (such as defined in WLAN, 4G radiocells and expected 5G networks), the SKG technology thus appears very suitable at early stages. As soon as radio channel measurements are enabled from prior frame and slot synchronization, reception of signaling, transmission and reception of access messages, initialization of equalization and Quality of Services procedures, etc., their outputs could be reused of SKG purposes.

SKG can also apply to Frequency Division Duplex (FDD) if the user equipment and the node have the ability to operate on the same carrier frequencies for the access stages.

Moreover, when considering any kind of public radio access protocols, even FDD protocols such as defined in most of 2G and 3G radio cellular, the study carried out in [13] pointed out a key-free secure pairing technology.

This secure pairing procedure is based on dual sense low power self-interfering signals (referred to as tag signals) that allow accurate measurements of CIR and support Interrogation and Acknowledgement Sequences (IAS) between terminals and nodes. Very early in the radio access, IAS provide dual sense paired CIR (which are output by the synchronization and equalization procedures of the dual sense tag signals) at Alice and Bob while they ensure the secure pairing of Alice's and Bob's devices by checking the CIR.

SKG can be input by these paired CIR. Moreover, the exchanges of paired tag signal would offer a native authenticated public channel for exchanging information during the SKG processing: frame index in channel decorrelation preprocessing, plane index in quantization algorithm, secure sketch into reconciliation procedure, etc.

19.4.3 Practical usage of secret key generation into radio access technologies

We finally consider some possible practical usage of SKG.

As seen before, the output keys of 128 or 256 bits, could protect early messages exchanged between Alice and Bob

- as a direct protection of signaling and access messages,
- as a private key (shared only by Alice and Bob) to be used in a traditional cipher scheme applied to sensitive contents of the signaling and access messages.

Keys can be stored in terminal memory and network data base and changed over time when necessary (during on-going communications by using the output or equalization procedures).

Many other potential usages appear for WLAN and radio cellulars are listed below.

- To facilitate new attach procedures and new roaming procedures in idle mode.
- To input and facilitate secure schemes of upper protocols layers during on-going communication. Some examples relevant to in WLAN and radio cellular network are:
  - protection of the headers of IP packets,
  - protection of control frames,
  - protection of return information messages in explicit artificial Noise and Beam forming schemes defined in some WLAN (802.11n/ac),
  - input of the integrity control and cipher schemes of data stream with non-mathematical random,
  - usage of generated keys as temporal identifier,
  - usage as integrity control check to prevent intrusion of messages, rogue and man in the middle attacks of on-going communications,
  - detection of intrusion attempts (including false authentication requests): if the node and the terminal receive in parallel similar messages with uncorrelated keys generated from different channel instances,
  - usage as a pre-shared key or header input in existing ciphering scheme,

o   the use of the secret key to protect ultra-low latency transmission expected in the future, where current stream ciphers are too slow,

o   to protect the unciphered near field communication.

- To cope with the problem of distribution and management of secret keys with the deployment of massive Internet-of-Things.

References

[1]   M. Bloch and J. Barros, *Physical-layer security: From information theory to security engineering*. UK: Cambridge University Press, 2011.

[2]   U. Maurer, "Secret key agreement by public discussion from common information," *IEEE Trans. Inform. Theory*, vol. 39, no. 3, pp. 733–742, May 1993.

[3]   R. Ahlswede and I. Csiszar, "Common randomness in information theory and cryptography. I. Secret sharing," *IEEE Trans. Inform. Theory*, vol. 39, no. 4, pp. 1121–1132, Jul. 1993.

[4]   A. Hassan, W. Stark, J. Hershey, and S. Chennakeshu, "Cryptographic key agreement for mobile radio," *Digital Signal Proc.*, vol. 6, pp. 207–212, Oct. 1996.

[5]   S. Mathur, W. Trappe, N. Mandayam, C. Ye, and A. Reznik, "Radio-telepathy: extracting a secret key from an unauthenticated wireless channel," in *14th ACM international conference on Mobile computing and networking*, September 2008, pp. 128–139.

[6]   S. Jana, S. Premnath, M. Clark, S. Kasera, N. Patwari, and S. Krishnamurthy, "On the effectiveness of secret key extraction from wireless signal strength in real environments," in *15th annual international conference on Mobile computing and networking*, September 2009, pp. 321–332.

[7]   L. Lai, Y. Liang, and W. Du, "Cooperative key generation in wireless networks," *IEEE JSAC*, vol. 30, no. 8, pp. 1578–1588, 2012.

[8]   J. Zhang, R. Woods, T. Duong, A. Marshall, and Y. Ding, "Experimental study on channel reciprocity in wireless key generation," in *IEEE 17th International Workshop on Signal Processing Advances in Wireless Communications (SPAWC)*, July 2016, pp. 1–5.

[9]   T. Mazloum, F. Mani, and A. Sibille, "Analysis of secret key robustness in indoor radio channel measurements," in *Proc. 2015 IEEE 81st Vehicular Technology Conference (VTC-Spring)*, May 2015.

[10]   T. Mazloum and A. Sibille, "Analysis of secret key randomness exploiting the radio channel variability," *Int. J. Antennas Propagation (IJAP)*, vol. 2015, Article ID 106360, 13 pp., 2015.

[11]   J. Wallace and R. Sharma, "Automatic secret keys from reciprocal MIMO wireless channels: Measurement and analysis," *IEEE Trans. Inform. Forensics Security*, vol. 5, no. 3, pp. 381–392, Sep. 2010.

[12]   T. Mazloum, "Analysis and modeling of the radio channel for secret key generation," Ph.D. dissertation, Telecom ParisTech, 2016.

[13] "Phylaws," 2014. http://www.Phylaws-ict.org.

[14] R. Wilson, D. Tse, and R. Scholtz, "Channel identification: Secret sharing using reciprocity in ultrawideband channels," *IEEE Trans. Inform. Forensics Security*, vol. 2, no. 3, pp. 364–375, Sep. 2007.

[15] T. Cover and J. Thomas, *Elements of Information Theory*. New York: Wiley, 1991.

[16] A. Rukhin, J. Soto, J. Nechvatal, *et al.*, *A Statistical Test Suite for Random and Pseudorandom Number Generators for Cryptographic Applications*. Information Technology Laboratory, NIST, Gaithersburg, Maryland, Tech. Rep., 2010.

[17] M. Hamburg, P. Kocher, and M. Marson, *Analysis of Intel's Ivy Bridge Digital Random Number Generator*, 2012.

[18] "Volcano lab," 2012, http://www.siradel.com.

[19] V. Degli-Esposti, F. Fuschini, E. Vitucci, and G. Falciasecca, "Measurement and modelling of scattering from buildings," *IEEE Trans. Antennas Propagation*, vol. 55, no. 1, pp. 143–153, Jan. 2007.

[20] C. Chen and M. Jensen, "Secret key establishment using temporally and spatially correlated wireless channel coefficients," *IEEE Trans. Mobile Comput.*, vol. 10, no. 12, pp. 205–215, 2011.

[21] U. Maurer and S. Wolf, "Secret-key agreement over unauthenticated public channels .ii. privacy amplification," *IEEE Trans. Inf. Theory*, vol. 49, no. 4, pp. 839–851, Apr. 2003.

[22] Y. Dodis, R. Ostrovsky, L. Reyzin, and A. Smith, "Fuzzy extractors: How to generate strong keys from biometrics and other noisy data," *SIAM J. Comput.*, vol. 38, no. 1, pp. 97–139, 2008.

[23] C. H. Bennett, G. Brassard, C. Crepeau, and U. M. Maurer, "Generalized privacy amplification," *IEEE Trans. Inf. Theory*, vol. 41, no. 6, pp. 1915–1923, Nov. 1995.

[24] M. Turan, E. Barker, J. Kelsey, K. McKay, M. Baish, and M. Boyle, *Recommendation for the Entropy Sources Used for Random Bit Generation*. National Institute of Standards and Technology Special Publication 800-90B, NIST, Gaithersburg, Maryland, Tech. Rep., 2016.

[25] C. Mehlfuehrer, J. C. Ikuno, M. Simko, S. S, M. Wrulich, and M. Rupp, "The vienna lte simulators – enabling reproducibility in wireless communications research," *EURASIP Journal on Advances in Signal Processing*, vol. 21, 2011.

[26] S. Jaeckel, L. Raschkowski, K. Börner, and L. Thiele, "Quadriga: A 3-d multicell channel model with time evolution for enabling virtual field trials," *IEEE Trans. Antennas Propag*, vol. 62, pp. 3242–3256, 2014.

[27] "Winner II channel models," 2007. https://www.ist-winner.org/WINNER2-Deliverables/D1.1.2v1.1.pdf.

[28] Y. Zou, J. Zhu, X. Wang, and L. Hanzo, "A survey on wireless security: Technical challenges, recent advances, and future trends," *Proc. IEEE*, vol. 104, no. 9, pp. 1727–1765, 2016.

Chapter 20

Application cases of secrecy coding in communication nodes and terminals

Christiane Kameni Ngassa[1], Cong Ling[2], François Delaveau[1], Sandrine Boumard[3], Nir Shapira[4], Ling Liu[2], Renaud Molière[1], Adrian Kotelba[3], and Jani Suomalainen[3]

20.1 Introduction

The objective of this chapter is to study practical coding techniques to provide security to wireless systems. First, the chapter will briefly introduce theoretical results relevant to low density parity check (LDPC) codes, polar codes and lattice coding for the wiretap channel. Then, it will propose practical secrecy-coding schemes able to provide a reliable and confidential wireless communication link between Alice and Bob. Finally, these practical wiretap codes are implemented in WiFi and long-term evolution (LTE) testbeds, and their confidentiality performance is evaluated using the bit-error rate (BER) as it is a simple and practical the metric for secrecy. The reader is referred to [1] for a throughout survey on recent advances related to the design of wiretap codes for information-theoretic metrics such as strong secrecy and semantic secrecy.

In the first part of this chapter, we use a nested lattice structure to design secrecy codes for Gaussian wiretap channels (GWCs). We design two lattice codes with rates almost equal to the capacities of the legitimate channel (between Alice and Bob) and the wiretapper's channel (between Alice and Eve), respectively. The lattice code for the wiretapper's channel encodes random bits, and the code for the legitimate channel only makes use of the coset leader to send the message.

Furthermore, we will show that

- a similar structure can also be extended to the MIMO channel and fading channel scenarios;
- a simplified wiretap code can be practically implemented in real radio communication standards (such as WiFi, LTE), under state-of-the-art radio devices and processing units, that achieve significant secrecy rate in most of radio-environments.

[1]Thales Communications and Security, USA
[2]Imperial College London, UK
[3]VTT Technical Research Centre of Finland, Finland
[4]Celeno Communications

In Section 20.2, we introduce the wiretap coding scheme for discrete channels, based on the famous LDPC codes [2] and the recently proposed polar codes [3]. Then, we extend the coset wiretap coding scheme for discrete channels to continuous Gaussian channels, by constructing polar lattices. As a result, we show that the proposed scheme is able to achieve strong secrecy and the secrecy capacity. In the meantime, an explicit lattice shaping scheme based on discrete lattice Gaussian distribution is also presented. This shaping scheme is compatible with the wiretap coding structure thanks to the versatility of polar codes, meaning that a convenient implementation is possible, as shown in following sections. Finally, we study in detail the wiretap coding design for the MIMO channels and fading channels, based on a similar structure of nesting lattice codes.

In Section 20.3, we introduce a practical simplified implementation of such wiretap codes, under a radio configuration that provides a slight radio advantage to Bob (a few decibels only), thanks to artificial noise (AN) and beamforming (BF) under to MIMO transmission. Our simplified secrecy scheme consists of

- AN and BF from channel state information measured on the legitimate link,
- error correction with an inner codes identical to forward-error correction (FEC) codes used in wireless standards,
- secrecy with an outer code involving nested polar or Reed–Muller (RM) codes.

An explicit design of the scheme is given for WiFi-like signals and LTE-like signals. Performance is first analyzed with simulations under an additive Gaussian noise channel, then illustrated for real field WiFi links (with real field propagation and real radio devices) and for simulated LTE link scenarios (involving real field propagation records but idealized transmitter and receiver models). Practical considerations about radio network engineering are given, and relevant radio parameters are highlighted.

Section 20.4 concludes the chapter and points out the potential of these techniques for radio standards regarding privacy of subscribers and confidentiality of data stream and suggests tracks for their practical implementation.

20.2   Theoretical aspects of secrecy coding

20.2.1   Wiretap coding for discrete wiretap channels

In this section, we discuss the use of some practical binary codes such as LDPC codes and polar codes to construct efficient secrecy codes for discrete wiretap channels.

The vast majority of work on physical player security is based on nonconstructive random-coding arguments to establish the theoretical results. Such results demonstrate the existence of codes that achieve the secrecy capacity but do not provide any practical method to design these codes. Moreover, the design of wiretap codes is further impaired by the absence of a simple metric, such as the BER, which could numerically evaluated to assess their secrecy performance [4].

In recent years, progress has been made on the construction of practical codes for physical player security, to some extent. The design methodology can be traced back to Wyner's original work on coset coding [5] which suggests that several codewords should represent the same message and that the choice of which codeword to transmit should be random to confuse the eavesdropper. In what follows, we will show how the original concept of the coset coding could be implemented with binary codes to achieve strong secrecy.

20.2.1.1 LDPC codes for discrete wiretap channels

LDPC codes are famous for their capacity-approaching performance on many communication channels. However, wiretap codes built from LDPC codes only had limited success.

When the legitimate channel is noiseless and the wiretapper's channel is the binary erasure channel (BEC), LDPC codes for the BEC were presented in [6,7]. Especially in [7], the authors generalized the link between capacity-approaching codes and weak secrecy capacity. The use of capacity-achieving codes for the wiretapper's channel is a sufficient condition for weak secrecy [8]. This view point provided a clear construction method for coding schemes for secure communication across arbitrary wiretap channels. Then, they used this idea to construct the first secrecy capacity-achieving LDPC codes for a wiretap channel with a noiseless legitimate channel and a BEC under message-passing (MP) decoding, however, in terms of weak secrecy. Later, [4] proved that the same construction can be used to guarantee strong secrecy at lower rates. A similar construction based on two-edge-type LDPC codes was proposed in [9] for the BEC wiretap channel.

The coset coding scheme using LDPC codes for the BEC wiretap channel model can be interpreted as follows: Prior to transmission, Alice and Bob publicly agree on a $(n, n(1 - R))$ binary LDPC code $\mathscr{C}$, where n is the block length of $\mathscr{C}$ and R is the secrecy rate. For each possible value m of the nR bits secret message M, a coset of $\mathscr{C}$ given by $\mathscr{C}(m) = \{x^n \in \{0, 1\}^n : x^n H^T = m\}$ is associated, where H is the parity check matrix of $\mathscr{C}$. To convey the message M to Bob, Alice picks a codeword in $\mathscr{C}(m)$ randomly and transmits it. Bob can obtain the secret message by calculating $X^n H^T$. Suppose that code $\mathscr{C}(m)$ has a generator matrix $G = [a_1, a_2, \ldots, a_n]$, where a_i denotes the ith column of G. Consider that Eve observes u unerased symbols from X^n, with the unerased positions given by $\{i : z_i \neq ?\} = \{i_1, i_2, \ldots, i_u\}$. Message m is secured by $\mathscr{C}$ in the sense that the probability $\Pr\{M = m|Z = z\} = 1/2^{nR}$ if and only if the matrix $G_u = [a_{i_1}, a_{i_2}, \ldots, a_{i_u}]$ has rank u. This is due to the fact that if G_u has rank u, the code $\mathscr{C}$ has all 2^u possible u-tuples in the u unerased positions. Therefore, by linearity, each coset of $\mathscr{C}$ would have all 2^u possible u-tuples in the same positions as well, which means that Eve has no idea which coset Alice has chosen. To guarantee that such G_u is with high probability to be full rank, we need to use the threshold property of LDPC codes, namely for a LDPC code $\mathscr{C}^\perp$ with parity check matrix $H(\mathscr{C}^\perp)$ and belief propagation (BP) decoding threshold ε^{BP} on BEC, a submatrix formed by selecting columns of $H(\mathscr{C}^\perp)$ independently with probability ε would have full column rank with high probability for sufficiently large n, if $\varepsilon < \varepsilon^{BP}$. If $\mathscr{C}^\perp$ is the dual code of $\mathscr{C}$, $H(\mathscr{C}^\perp)$ is equal to the generator matrix G of $\mathscr{C}$, and selecting the columns of $H(\mathscr{C}^\perp)$ with probability ε can be viewed as obtaining G_u from G through a BEC with erasure probability $1 - \varepsilon$. Consequently, the problem of designing a secrecy achieving LDPC code over a binary erasure wiretap channel with erasure probability ε can be converted to the problem of constructing a dual LDPC code over a BEC with erasure probability $1 - \varepsilon$.

Let $P_e^{(n)}(\varepsilon)$ denote the probability of block error probability of a LDPC code $\mathscr{C}$ with block length n over BEC(ε). For a parity check matrix H of $\mathscr{C}$, $1 - P_e^{(n)}(\varepsilon)$ is a lower bound on the probability that the erased columns of H form a full rank submatrix, which means that generator matrix $G_u(\mathscr{C}^\perp)$ resulted by a BEC($1 - \varepsilon$) has full rank with probability larger than $1 - P_e^{(n)}(\varepsilon)$. If $\mathscr{C}^\perp$ is used in the coset coding scheme, the conditional entropy $H(M|Z^n)$ can be bounded as

$$H(M|Z^n) \geq H(M|Z^n, \text{rank}(G_u(\mathscr{C}^\perp))) \tag{20.1}$$

$$\geq H(M|Z^n, G_u(\mathscr{C}^\perp)) \text{ is full rank}) \cdot \Pr\{G_u(\mathscr{C}^\perp)) \text{ is full rank}\} \tag{20.2}$$

$$\geq H(M)(1 - P_e^{(n)}(\varepsilon)). \tag{20.3}$$

Then, we have

$$I(M; Z^n) = H(M) - H(M|Z^n) \leq H(M)P_e^{(n)}(\varepsilon) \leq nRP_e^{(n)}(\varepsilon). \tag{20.4}$$

Therefore, if code $\mathscr{C}$ has a BP threshold ε^{BP} such that $P_e^{(n)}(\varepsilon) = O(\frac{1}{n^\alpha})$ ($\alpha > 1$), for $\varepsilon < \varepsilon^{\text{BP}}$, then the dual code of $\mathscr{C}$ used in the coset coding scheme provides strong secrecy over a binary erasure wiretap channel with erasure probability $\varepsilon > 1 - P_e^{(n)}(\varepsilon)$. Note that weak secrecy is provided, if $\alpha > 0$. The design rate R of $\mathscr{C}$ should satisfy $R < 1 - \varepsilon^{\text{BP}}$, while the secrecy capacity ε of the wiretap channel is larger than $1 - \varepsilon^{\text{BP}}$. To achieve the secrecy capacity, $1 - \varepsilon^{\text{BP}}$ is required to be very close to the code rate R. It is known that LDPC codes achieve capacity over a BEC under the BP decoding [10]. However, the parameter α can only be proved to be positive for the rate arbitrarily close to $1 - \varepsilon^{\text{BP}}$ [6]. To satisfy the strong secrecy requirement $\alpha > 1$, the design rate should be slightly away from $1 - \varepsilon^{\text{BP}}$.

Unfortunately, for other binary memoryless symmetric channels (BMSCs) except BECs, general LDPC codes do not have the capacity achieving property, which means that the coset coding scheme using general LDPC codes cannot achieve the secrecy capacity when the wiretapper's channel is not a BEC. In this case, spatially coupled LDPC (SC-LDPC) codes [11], which have been proved to be able to achieve the capacity of general BMSCs, provide us a promising approach. In [12], a coset coding scheme based on regular two-edge-type SC-LDPC codes is proposed over a BEC wiretap channel, where the main channel is also a BEC. It is shown that the whole rate equivocation region of such BEC wiretap channel can be achieved by using this scheme under weak secrecy condition. Since SC-LDPC codes are universally capacity achieving, it is also conjectured that this construction is optimal for the class of wiretap channel where the main channel and wiretapper's channel are BMSCs, and the wiretapper's channel is physically degraded with respect to the main channel.

In addition, LDPC codes has also been proposed for GWC in [13] but with a different criterion. The proposed coding scheme is asymptotically effective in the sense that it yields a BER very close to 0.5 for an eavesdropper with signal-to-noise ratio (SNR) lower than a certain threshold, even if the eavesdropper has the ability to use a maximum a-posteriori (MAP) decoder. However, this approach does not provide information theoretic strong secrecy.

20.2.1.2 Polar codes for discrete wiretap channels

Polar codes [3] are the first explicit codes which can be proved to achieve the channel capacity of BMSCs with low encoding and decoding complexity. As the block length increases, the capacities of bit-channels polarize to either to 0 or 1. It is shown that as the block length goes to infinity, the proportion of the bit-channels with 1 capacity is equal to the channel capacity. Therefore, polar codes can achieve the channel capacity by just transmitting information bits through these perfect bit-channels.

In addition, polar coding also seems to offer a more powerful approach to design wiretap codes. Recently, there has been a lot of interest in the design of wiretap codes based on polar codes. For example, [14]–[16] use polar codes to build encryption schemes for the wiretap setting with BMSCs. However, these schemes only provide weak security. In [17], it is shown that, with a minor modification of the original design, polar codes achieve strong secrecy (and also semantic security). However, they could not guarantee reliability of the main channel when it is noisy. In [18], a multiblock polar coding scheme was proposed to solve this reliability problem under the condition that the number of block is sufficiently large. In addition, a similar multiblock coding scheme was discussed in [19]. Their polar coding scheme also achieves the secrecy capacity under strong secrecy condition and guarantees reliably for the legitimate receiver. However, the work in [19] only proved the existence of this coding scheme, and thus it might be computationally hard to find the explicit structure. Finally, recent works proposing the use of polar codes for the general nondegraded wiretap channel have being performed in [20,21].

However, in the remaining of this subsection, we concentrate on how polar codes can achieve the secrecy capacity of the binary wiretap channel [17,18]. Define the sets of very reliable and very unreliable indices for a binary channel Q and for $0 < \beta < 0.5$ as

$$\mathcal{G}(Q) = \{i \in [N] : Z(Q_N^{(i)}) \leq 2^{-N^\beta}\}, \text{(Reliability-good indices)} \quad (20.5)$$

$$\mathcal{N}(Q) = \{i \in [N] : I(Q_N^{(i)}) \leq 2^{-N^\beta}\}, \text{(Information-bad indices)}, \quad (20.6)$$

where $Z(Q_N^{(i)})$ and $I(Q_N^{(i)})$ represent the Bhattacharyya parameter and mutual information of the polarized bit-channel $Q_N^{(i)}$ [3]. Let V and W denote the main channel and the wiretapper's channel, respectively. The indices in set $\mathcal{G}(V)$ and set $\mathcal{N}(W)$ are the reliable and the secure indices, respectively. Then, the whole index set $[N]$ can be partitioned into the following four sets:

$$\mathcal{A} = \mathcal{G}(V) \cap \mathcal{N}(W) \quad (20.7)$$

$$\mathcal{B} = \mathcal{G}(V) \cap \mathcal{N}(W)^c \quad (20.8)$$

$$\mathcal{C} = \mathcal{G}(V)^c \cap \mathcal{N}(W) \quad (20.9)$$

$$\mathcal{D} = \mathcal{G}(V)^c \cap \mathcal{N}(W)^c. \quad (20.10)$$

Unlike the standard polar coding, the bit-channels are now partitioned into three parts: the set $\mathcal{A}$ that carries the confidential message bits M, the set $\mathcal{B} \cup \mathcal{D}$ that carries random bits R and the set $\mathcal{C}$ that carries frozen bits F which are known to both Bob and Eve prior to transmission.

According to [17], this assignment introduces a new channel which is also symmetric. The mutual information of this channel can be upper bounded by the sum of the mutual information of bit-channels in $\mathcal{N}(W)$. The threshold of the mutual information on each bit-channel within $\mathcal{N}(W)$ is 2^{-N^β}. Then, the mutual information between the message and the signal Eve received $I(M; Z^N)$ can be upper bounded by $N2^{-N^\beta}$. Therefore, the strong secrecy is achieved since $\lim_{N \to \infty} I(M; Z^N) = 0$. Furthermore, in case of degraded wiretap channels, the achievable secrecy rate is equal to the secrecy capacity. This is due to the facts of polarization theory [3].

This polar coding scheme is also capable of satisfying the reliability condition. According to the construction of polar codes [3], the block error probability at Bob's end is upper bounded by the sum of the Bhattacharyya parameters of those bit-channels that are not frozen. Let $V_N^{(i)}$ denote the ith bit-channel of the main channel V, the decoding error probability P_e^{SC} under the successive cancelation (SC) decoding is upper bounded as

$$P_e^{\text{SC}} \leq \sum_{i \in \mathcal{G}(V) \cup \mathcal{D}} Z(V_N^{(i)}) = \sum_{i \in \mathcal{G}(V)} Z(V_N^{(i)}) + \sum_{i \in \mathcal{D}} Z(V_N^{(i)}). \tag{20.11}$$

The last equation is due to the fact that set $\mathcal{G}(V)$ and set $\mathcal{D}$ are disjoint. By the definition of $\mathcal{G}(V)$, the term $\sum_{i \in \mathcal{G}(V)} Z(V_N^{(i)})$ is bounded by $N2^{-N^\beta}$, which vanishes when N is sufficiently large. However, the bound on the term $\sum_{i \in \mathcal{D}} Z(V_N^{(i)})$ is difficult to derive. To overcome this problem, a modified scheme dividing the message M into several blocks was proposed [18]. For a specific block, $\mathcal{D}$ is still assigned with random bits but transmitted in advance in the set $\mathcal{A}$ of the previous block. By embedding $\mathcal{D}$ in $\mathcal{A}$, we induce some rate loss but obtain an arbitrarily small error probability. Since the size of $\mathcal{D}$ is very small compared with the block length N, the rate loss is negligible and finally the new scheme realizes reliability and strong security simultaneously.

We note that the information-bad set $\mathcal{N}(Q)$ can also be defined by

$$\mathcal{N}(Q) = \{i \in [N] : Z(Q_N^{(i)}) \geq 1 - 2^{-N^\beta}\}, \tag{20.12}$$

which is according to the Bhattacharyya parameter.

20.2.2   Wiretap coding for Gaussian wiretap channels

In this section, we address how to construct polar lattices to achieve the secrecy capacity of GWCs. We will mainly focus on how to obtain the AWGN-good lattice Λ_b and secrecy-good lattice Λ_e for the mod-Λ_s GWCs. The result also holds for the polar lattices when the input distribution is uniform for the genuine GWCs. The setting without power constraint is similar to the Poltyrev setting in the Gaussian point-to-point channel. Then, we introduce an explicit lattice shaping scheme for Λ_b and Λ_e simultaneously and remove the mod-Λ_s front-end. As a result, we develop an explicit wiretap coding scheme based on polar lattices which can be proved to achieve the secrecy capacity of the GWC with no requirement on SNR.

20.2.2.1   Construction of polar lattices codes

From the previous section, we see that polar codes have great potential in solving the wiretap coding problem. The polar coding scheme proposed in [17], combined with the block Markov coding technique [18], was proved to achieve the strong secrecy capacity when W and V are both BMSCs, and W is degraded with respect to V. For continuous channels such as the GWC, there also has been notable progress in lattice wiretap coding. On the theoretical aspect, the existence of lattice codes achieving the secrecy capacity to within 1/2-nat under the strong secrecy as well as semantic security criterion was demonstrated in [22]. On the practical aspect, wiretap lattice codes were proposed in [23,24] to maximize the eavesdropper's decoding error probability.

A lattice is a discrete subgroup of $\mathbb{R}^n$ which can be described by $\Lambda = \{\lambda = Bx : x \in \mathbb{Z}^n\}$, where B is the n-by-n lattice generator matrix, and we always assume that it has full rank in this section.

For a vector $x \in \mathbb{R}^n$, the nearest-neighbor quantizer associated with Λ is $Q_\Lambda(x) = \arg\min_{\lambda \in \Lambda} \|\lambda - x\|$. We define the modulo lattice operation by $x \bmod \Lambda \triangleq x - Q_\Lambda(x)$. The Voronoi region of Λ, defined by $\mathscr{V}(\Lambda) = \{x : Q_\Lambda(x) = 0\}$, specifies the nearest-neighbor decoding region. The Voronoi cell is one example of fundamental region of the lattice. A measurable set $\mathscr{R}(\Lambda) \subset \mathbb{R}^n$ is a fundamental region of the lattice Λ if $\cup_{\lambda \in \Lambda}(\mathscr{R}(\Lambda) + \lambda) = \mathbb{R}^n$ and if $(\mathscr{R}(\Lambda) + \lambda) \cap (\mathscr{R}(\Lambda) + \lambda')$ has measure 0 for any $\lambda \neq \lambda'$ in Λ. The volume of a fundamental region is equal to that of the Voronoi region $\mathscr{V}(\Lambda)$, which is given by $\text{vol}(\Lambda) = |\det(B)|$.

A sublattice $\Lambda' \subset \Lambda$ induces a partition (denoted by Λ/Λ') of Λ into equivalence classes modulo Λ'. The order of the partition is denoted by $|\Lambda/\Lambda'|$, which is equal to the number of cosets. If $|\Lambda/\Lambda'| = 2$, we call this a binary partition. Let $\Lambda/\Lambda_1/\cdots/\Lambda_{r-1}/\Lambda'$ for $r \geq 1$ be an n-dimensional lattice partition chain. For each partition $\Lambda_{\ell-1}/\Lambda_\ell$ ($1 \leq \ell \leq r$ with convention $\Lambda_0 = \Lambda$ and $\Lambda_r = \Lambda'$), a code C_ℓ over $\Lambda_{\ell-1}/\Lambda_\ell$ selects a sequence of representatives a_ℓ for the cosets of Λ_ℓ. Consequently, if each partition is binary, the code C_ℓ is a binary code.

20.2.2.2   Polar lattices for Gaussian wiretap channels

The idea of wiretap lattice coding over the mod-Λ_s GWC [22] can be explained as follows: Let Λ_b and Λ_e be the AWGN-good lattice and secrecy-good lattice designed for Bob and Eve accordingly. Let $\Lambda_s \subset \Lambda_e \subset \Lambda_b$ be a nested chain of N-dimensional lattices in $\mathbb{R}^N$, where Λ_s is the shaping lattice. Note that the shaping lattice Λ_s here is employed primarily for the convenience of designing the secrecy-good lattice and secondarily for satisfying the power constraint. Consider a one-to-one mapping: $\mathscr{M} \to \Lambda_b/\Lambda_e$ which associates each message $m \in \mathscr{M}$ to a coset $\tilde{\lambda}_m \in \Lambda_b/\Lambda_e$. Alice selects a lattice point $\lambda \in \Lambda_e \cap \mathscr{V}(\Lambda_s)$ uniformly at random and transmits $X^{[N]} = \lambda + \lambda_m$, where λ_m is the coset representative of $\tilde{\lambda}_m$ in $\mathscr{V}(\Lambda_e)$. This scheme has been proved to achieve both reliability and semantic security in [22] by random lattice codes. We will make it explicit by constructing polar lattice codes.

Polar lattices are built by "Construction D" [25, p. 232] using a set of nested polar codes $C_1 \subseteq C_2 \subseteq \cdots \subseteq C_r$ [26]. Suppose C_ℓ has block length N and the number of information bits k_ℓ for $1 \leq \ell \leq r$. Choose a basis $\mathbf{g}_1, \mathbf{g}_2, \ldots, \mathbf{g}_N$ from the polar

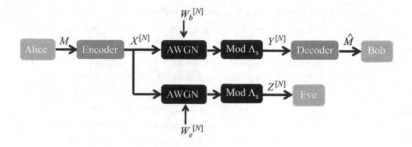

Figure 20.1   The mod-Λ_s Gaussian wiretap channel

generator matrix G_N such that $\mathbf{g}_1, \ldots \mathbf{g}_{k_\ell}$ span C_ℓ. When the dimension $n = 1$, the lattice L admits the form [26]

$$
L = \left\{ \sum_{\ell=1}^{r} 2^{\ell-1} \sum_{i=1}^{k_\ell} u_\ell^i \mathbf{g}_i + 2^r \mathbb{Z}^N \mid u_\ell^i \in \{0, 1\} \right\},
\tag{20.13}
$$

where the addition is carried out in $\mathbb{R}^N$. The fundamental volume of a lattice obtained from this construction is given by

$$
\mathrm{vol}(L) = 2^{-NR_C} \cdot \mathrm{vol}(\Lambda_r)^N,
$$

where $R_C = \sum_{\ell=1}^{r} R_\ell = \frac{1}{N} \sum_{\ell=1}^{r} k_\ell$ denotes the sum rate of component codes. In this section, we limit ourselves to the binary lattice partition chain and binary polar codes for simplicity.

Now, we consider the construction of secrecy-good polar lattices over the mod-Λ_s GWC shown in Figure 20.1. The difference between the mod-Λ_s GWC and the genuine GWC is the mod-Λ_s operation on the received signal of Bob and Eve. With some abuse of notation, the outputs $Y^{[N]}$ and $Z^{[N]}$ at Bob and Eve's ends, respectively, become

$$
\begin{cases}
Y^{[N]} = [X^{[N]} + W_b^{[N]}] \bmod \Lambda_s, \\
Z^{[N]} = [X^{[N]} + W_e^{[N]}] \bmod \Lambda_s.
\end{cases}
$$

Let Λ_b and Λ_e be constructed from a binary partition chain $\Lambda/\Lambda_1/\cdots/\Lambda_{r-1}/\Lambda_r$ and assume $\Lambda_s \subset \Lambda_r^N$ such that $\Lambda_s \subset \Lambda_r^N \subset \Lambda_e \subset \Lambda_b$.[1] Also, denote by $X_{1:r}^{[N]}$, the bits encoding Λ^N/Λ_r^N, which include all information bits for message M as a subset. We have that $[X^{[N]} + W_e^{[N]}] \bmod \Lambda_r^N$ is a sufficient statistic for $X_{1:r}^{[N]}$ [26].

---

[1]This is always possible with sufficient power, since the power constraint is not our primary concern. We will deal with it using lattice Gaussian distribution later.

In our context, we identify Λ with Λ_r^N and Λ' with Λ_s, respectively. Since the bits encoding Λ_r^N / Λ_s are uniformly distributed, the mod-Λ_r^N operation is information-lossless in the sense that

$$I(X_{1:r}^{[N]}; Z^{[N]}) = I(X_{1:r}^{[N]}; [X^{[N]} + W_e^{[N]}] \bmod \Lambda_r^N).$$

As far as mutual information $I(X_{1:r}^{[N]}; Z^{[N]})$ is concerned, we can use the mod-Λ_r^N operator instead of the mod-Λ_s operator here. Under this condition, similarly to the multilevel lattice structure introduced in [26], the mod-Λ_s channel can be decomposed into a series of BMSCs according to the partition chain $\Lambda / \Lambda_1 / \cdots / \Lambda_{r-1} / \Lambda_r$. Therefore, the already mentioned polar coding technique for BMSC channels can be employed. Moreover, the channel resulted from the lattice partition chain can be proved to be equivalent to that based on the chain rule of mutual information. Following this channel equivalence, we can construct an AWGN-good lattice Λ_b and a secrecy-good lattice Λ_e, using the wiretap coding technique (20.5) at each partition level. Finally, the polar wiretap coding for each partition level guarantees reliable and secure communication and the nested polar codes form an AWGN-good polar lattice for Bob and a secrecy-good polar lattice for Eve, respectively.

We then apply Gaussian shaping on the AWGN-good and secrecy-good polar lattices. The idea of lattice Gaussian shaping was proposed in [27] and then implemented in [28] to construct capacity-achieving polar lattices. For wiretap coding, the discrete Gaussian distribution can also be utilized to satisfy the power constraint. In simple terms, after obtaining the AWGN-good lattice Λ_b and the secrecy-good lattice Λ_e, Alice still maps each message m to a coset $\tilde{\lambda}_m \in \Lambda_b / \Lambda_e$ as mentioned previously. However, instead of the mod-Λ_s operation, Alice samples the encoded signal X^N from a lattice Gaussian distribution $D_{\Lambda_e + \lambda_m, \sigma_s}$, where λ_m is the coset representative of $\tilde{\lambda}_m$ and σ_s^2 is arbitrarily close to the signal power. Please see [22,28] for more details of the implementation of lattice Gaussian shaping.

20.2.3 *Wiretap coding for MIMO and fading channels*

In this section, we consider MIMO (multiple input–multiple output) wiretap channels, where Alice is communicating with Bob in the presence of Eve, and communication is done via MIMO channels. We suppose that Alice's strategy is to use a code book which has a lattice structure, which then allows her to perform coset encoding. We analyze Eve's probability of correctly decoding the message Alice meant to Bob, and from minimizing this probability, we derive a code design criterion for MIMO lattice wiretap codes. The case of block fading channels is treated similarly, and fast-fading channels are derived as a particular case.

Similarly to the Gaussian wiretap coding case mentioned in the preceding section, we consider the case where Alice transmits lattice codes using coset encoding, which requires two nested lattices $\Lambda_e \subset \Lambda_b$. Alice encodes her data in the coset representatives of Λ_b / Λ_e. Both Bob and Eve try to decode using coset decoding. For Gaussian channels, it was shown in [29] that a wiretap coding strategy is to design Λ_b for Bob (since Alice knows Bob's channel, she can ensure he will decode with high probability), while Λ_e is chosen to maximize Eve's confusion, characterized by a lattice invariant called secrecy gain, under the assumption that Eve's noise is worse than the one experienced by Bob. We can generalize this approach to MIMO channels (and in fact block and fast-fading channels as particular cases). We compute Eve's probability of making a correct decoding decision, and deduce how the lattice Λ_e should be designed to minimize this probability. A MIMO wiretap channel will then consist of two nested lattices $\Lambda_e \subset \Lambda_b$ where Λ_b is designed to ensure Bob's reliability, while Λ_e is a subset of Λ_b chosen to increase Eve's confusion.

When the channel between Alice and Bob, resp. Eve, is a quasistatic MIMO channel with n_t transmitting antennas at Alice's end, n_b resp. n_e receiving antennas at Bob's, resp. Eve's end and a coherence time T, that is:

$$Y = H_b X + V_b, \tag{20.14}$$

$$Z = H_e X + V_e, \tag{20.15}$$

where the transmitted signal X is a $n \times T$ matrix, the two channel matrices are of dimension $n_b \times n_t$ for H_b and $n_e \times n_t$ for H_e, and V_b, V_e are $n_b \times T$, resp. $n_e \times T$ matrices denoting the Gaussian noise at Bob, respectively, Eve's side, both with coefficients zero mean and respective variance σ_b^2 and σ_e^2. The fading coefficients are complex Gaussian i.i.d. random variables, and in particular, H_e has covariance matrix $\sum_e = \sigma_{H_e}^2 I_{n_e}$. As for the Gaussian case, we assume that Alice transmits a lattice code, via coset encoding, and that the two receivers are performing coset decoding of the lattice, thus $n_b, n_e \geq n_t$. Indeed, if the number of antennas at the receiver is smaller than that of the transmitter, the lattice structure is lost at the receiver. This case will not be treated. Finally, we denote by $\gamma_e = \sigma_{H_e}^2 / \sigma_e^2$ Eve's SNR. We do not make assumption on knowing Eve's channel or on Eve's SNR, since we will compute bounds which are general, though their tightness will depend on Eve's SNR.

In order to focus on the lattice structure of the transmitted signal, we vectorize the received signal and obtain

$$\mathrm{vec}(Y) = \mathrm{vec}(H_b X) + \mathrm{vec}(V_b) = \begin{bmatrix} H_b & & \\ & \ddots & \\ & & H_b \end{bmatrix} (X) + (V_b) \tag{20.16}$$

$$\mathrm{vec}(Z) = \mathrm{vec}(H_e X) + \mathrm{vec}(V_e) = \begin{bmatrix} H_e & & \\ & \ddots & \\ & & H_e \end{bmatrix} (X) + (V_e). \tag{20.17}$$

We now interpret the $n \times T$ codeword X as coming from a lattice. This is typically the case if X is a space-time code coming from a division algebra [30], or more generally if X is a linear dispersion code as introduced in [31] where Tn_t symbols QAM are linearly encoded via a family of Tn_t dispersion matrices. We write $\mathrm{vec}(H_e X) = M_b u$, where $u \in \mathbb{Z}[i]^{Tn_t}$, and M_b denotes the $Tn_t \times Tn_t$ generator matrix of the $\mathbb{Z}[i]$-lattice Λ_b intended to Bob. Thus, in what follows, by a lattice point $x \in \Lambda_b$, we mean that $x = \mathrm{vec}(X) = M_b u$, and similarly for a lattice point $x \in \Lambda_e$, we have $x = \mathrm{vec}(X) = M_e u$.

We now focus on Eve's channel, since we know from [32] how to design a good linear dispersion space-time code, and the lattice Λ_b is chosen so as to correspond to this space-time code. We also know that Eve's probability of correctly decoding is upper bounded by

$$P_{c,e,H_e} \leq \frac{\text{vol}(\Lambda_{b,H_e})}{(2\pi\sigma_e^2)^{n_t T}} \sum_{\mathbf{r} \in \Lambda_{e,H_e}} e^{-\|\mathbf{r}\|^2/2\sigma_e^2} \tag{20.18}$$

$$= \frac{\text{vol}(\Lambda_b)}{(2\pi\sigma_e^2)^{n_t T}} \det(H_e H_e^*)^T \sum_{X \in \Lambda_e} e^{-\|H_e X\|_F^2/2\sigma^2}, \tag{20.19}$$

where $\|H_e X\|_F^2 = \text{Tr}(H_e X X^* H_e^*)$ is the Frobenius norm.

Using the equation of error probability, we derive Eve's average probability of correct decision:

$$\overline{P_{c,e}} = \mathbb{E}_{H_e}[P_{c,e,H_e}] \tag{20.20}$$

$$\leq \frac{\text{vol}(\Lambda_b)}{(2\pi\sigma_e^2)^{n_t T}(2\pi\sigma_{H_e}^2)^{n_e n_t}} \tag{20.21}$$

$$\cdot \sum_{X \in \Lambda_e} \int_{\mathbb{C}^{n_e \times n_t}} \det(H_e H_e^*)^T e^{-\text{Tr}\left(H_e^* H_e\left[\frac{1}{2\sigma_{H_e}^2}\mathbf{I}n_t + \frac{1}{2\sigma_e^2}XX^*\right]\right)} dH_e. \tag{20.22}$$

According to the analysis in [33], we finally obtain an upper bound on the average probability of correct decoding for Eve as

$$\overline{P_{c,e}} \leq C_{\text{MIMO}}\gamma_e^{Tn_t} \sum_{X \in \Lambda_e} \det(\mathbf{I}_{n_t} + \gamma_e XX^*)^{-n_e-T}, \tag{20.23}$$

where we set $C_{\text{MIMO}} = \frac{\text{vol}(\Lambda_b)\Gamma_{n_t}(n_e+T)}{\pi^{n_t T}\Gamma_{n_t}(n_e)}$.

In order to design a good lattice code for the MIMO wiretap channel, we use the so-called rank-criterion of [32]. This means that, if $X \neq 0$ and $T \geq n_t$, we have rank$(X) = n_t$. If we assume now γ_e is high compared to the minimum distance of Λ_e, we get

$$\overline{P_{c,e}} \leq C_{\text{MIMO}}\left[\gamma_e^{Tn_t} + \frac{1}{\gamma_e^{n_e n_t}} \sum_{X \in \Lambda_e \setminus \{0\}} \det(XX^*)^{-n_e-T}\right]. \tag{20.24}$$

We thus conclude that to minimize Eve's average probability of correct decoding, the design criterion is to minimize $\sum_{X \in \Lambda_e \setminus \{0\}} \det(XX^*)^{-n_e-T}$.

For block and fast-fading channels, we cannot use the final result for MIMO channels immediately, since the integral over all positive definite Hermitian matrices does not hold anymore. However, we can start from the generic equation and use a polar coordinates change. Then, following a similar fashion, we obtain an upper bound of the average probability of correct decision for Eve

$$\overline{P_{c,e}} \leq C_{\text{BF}}\gamma_e^{nT} \sum_{X \in \Lambda_e} \Pi_{i=1}^n [1 + \gamma_e\|x_i\|^2]^{-1-T}, \tag{20.25}$$

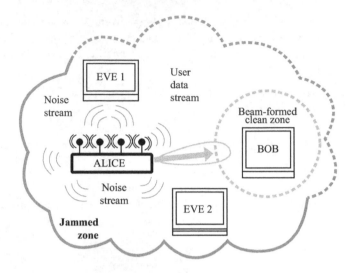

Figure 20.2　Scheme of transmission with AN–BF protections

where $C_{BF} = \frac{(T!)^n \text{vol}(\Lambda_b)}{\pi^{nT}}$ and similarly to the MIMO case, $\gamma_e = \frac{\sigma_{H_e}^2}{\sigma_e^2}$. Then, the design criterion for block and fast-fading channels is to minimize $\sum_{x \in \Lambda_e \setminus \{0\}} \prod_{i=1}^n [\gamma_e \|x_i\|^2]^{-1-T}$.

20.3 Integration of secrecy-coding techniques into existing radio access technologies

20.3.1 Radio advantage establishment—case of MIMO transmission

In MIMO radio access technologies (RATs), the radio advantage is achieved by combining BF of data toward the legitimate receiver with the emission of interfering signals (AN) elsewhere (Figure 20.2). The AN power is controlled, and the user signal (US) is steered to optimize the decoding capability for legitimate receivers while decreasing it for eavesdroppers.

20.3.1.1 Artificial noise and beamforming processing

Most promising AN and BF schemes studied in the literature proceed as follows [34]:

- Estimation of the legitimate channel frequency response (CFR) or channel impulse response (CIR), from Alice to Bob, and extraction of orthogonal directions of the legitimate CFR or CIR.
- Transmission of noise streams on orthogonal directions to the legitimate channel. As Eve cannot estimate the legitimate channel matrix (CM), she is thus forced into low signal-to-interference + noise ratio (SINR) regime and is unable to decode.
- BF of the Alice–Bob data stream for Bob to maximize the legitimate link budget. Bob extracts Alice's channel and suppresses orthogonal noisy channel directions

thanks to BF. In ideal cases, the interference at Bob's side completely vanishes, and the SINR at Bob's side becomes equal to a signal-to-noise ratio ($\text{SINR}_{\text{Bob}} = \text{SNR}_{\text{Bob}}$).

When AN and BF techniques are established, a better SINR is provided to Bob than to Eve in any case and the maximal SINR_{Eve} at Eve's side is controlled by Alice. The relevant radio advantage ($\text{RA} = \text{SINR}_{\text{Bob}} - \text{SINR}_{\text{Eve}}$) is thus guaranteed, and it can be further exploited by the legitimate link to compute secrecy codes described in Section 20.3.2.

20.3.1.2 AN and BF for initiating secrecy-coding schemes

The goal of secrecy codes is to ensure reliable communication for the legitimate link and to avoid any information leakage elsewhere. Secrecy codes conceal the information sent by Alice up to a secrecy capacity. In general case, the secrecy capacity is always larger than or equal the difference of channel capacities at Bob and Eve's side; in simplest cases such as AWGN channel, it is directly driven by the radio advantage. Without a positive radio advantage, secrecy capacity is null. Moreover, in real environment with time and space varying channel, shadowing and fading, the radio advantage should be controlled to guarantee a minimum secrecy capacity, so that Alice and Bob can properly choose and control the secrecy-coding parameters.

20.3.1.3 Power of the jamming signal—case of colocated transmit antennas

Several access methods have been developed across the years in the different standards: frequency division multiple access (FDMA), time division multiple access (TDMA), code division multiple access (CDMA) or orthogonal frequency division multiple access (OFDMA). Regardless of the access method, one has to consider the limit of the decoding sensitivity which is required in the standard in term of received signal-to-interference + noise ratio (noted SINR in decibel). In linear value, this SINR is noted $\rho_{\text{SINR}} = S_{\text{Rx}}/(J_{\text{Rx}} + N_{\text{Rx}})$, where S_{Rx} is the received US power, N_{Rx} the receiving noise power and J_{Rx} the received interference power.

- Without interference, SINR reduces to a Signal-to-Noise Ratio (SNR), meaning $\rho_{\text{SINR}} = \rho_{\text{SNR}} = S_{\text{Rx}}/N_{\text{Rx}}$.
- When receiving noise is negligible, SINR reduces to a Signal-to-Interference Ratio (SIR), meaning $\rho_{\text{SINR}} = \rho_{\text{SIR}} = S_{\text{Rx}}/J_{\text{Rx}}$.
- In any case, $\text{SINR} \leq \text{SIR}$ and $\rho_{\text{SINR}} \leq \rho_{\text{SIR}}$, $\text{SINR} \leq \text{SNR}$ and $\rho_{\text{SINR}} \leq \rho_{\text{SNR}}$.

When the sources of AN and user data stream are strictly colocated (i.e., same transmitting antenna elements), Alice has to adjust her jamming signal $J_{\text{Tx,Alice}}$, to the user transmit signal $S_{\text{Tx,Alice}}$ to a minimum value of $\rho_{\text{SIR,Rx,Eve}}$, called ρ_{min} that makes decoding impossible for Eve even in optimal reception condition (e.g., whatever her receiving noise is). For any RAT, a relevant sufficient condition is $\rho_{\text{SIR,Tx,Alice}} = S_{\text{Tx,Alice}}/J_{\text{Tx,Alice}} \leq \rho_{\text{min}}$ leading to $J_{\text{Tx,Alice}} \geq S_{\text{Tx,Alice}}/\rho_{\text{min}}$.

In the case of FDMA TDMA or OFDMA RATs, such a colocated jamming signal at each orthogonal direction with power roughly equal to the US ($\rho_{\text{SIR,Tx,Alice}} \approx 1$) is enough to limit most eavesdropping risks. More generally, a jamming signal 6 dB above the data stream should avoid any eavesdropping risk of FDMA TDMA or OFDMA link. Extremal ρ_{min} should thus verify $\rho_{min} \geq 0.25$ (-6 dB).

In the case of pseudo noise (PN) or CDMA schemes such as UMTS, the situation is more complex because the lowest data rate signals have spreading factors of 256 (24 dB), Thus, practical value of ρ_{min} is around -18 dB. Nevertheless, the power control and the global received noise at Eve's side highly depend on the network engineering practices regarding signaling and data communication streams under the carrier of the serving and neighbor cells. As a consequence, a practical $\rho_{SIR,Tx,Alice}$ value of -12 dB for a colocated jamming signal achieves a significant advantage in most of 3G network engineering and traffic scenarios.

20.3.1.4 Impact of the locations of antennas that transmit user and jamming signals

When the sources of AN and user data stream are not strictly colocated (i.e., different antennas or different antenna elements, such as in many scenarios of cooperative jamming), the efficiency of AN–BF is highly dependent on the spatial correlation at Bob's side, and on the source separation capabilities at Eve's side [35].

Indeed, the channel estimation performed by Bob over dedicated frames sent by Alice may not match perfectly for AN issued from different locations of antenna.

Moreover, several questions relevant to Eve's capabilities regarding source separation remain open. Even with a very small distance between transmitting antennas (lower than a quarter of wavelength), several laboratory experiments mentioned in [36] and performed in [35] showed that an accurate location of Eve's multiple antennas combined with analog mitigation techniques and digital power inversion may achieve practical discrimination of the user data stream and mitigation of the AN with significant performance.

Nevertheless, even if discrimination of Alice's streams can be performed at Eve's side, real field signal recorded and processed in [36] show that recovering of the legitimate channel by Eve is almost impossible when propagation is dispersive.

As a conclusion, we should consider that the most resilient AN–BF schemes would use colocated and even same antenna elements for AN and user data stream.

20.3.2 Description of the practical secrecy-coding scheme

Our goal is to design a low-complexity and practical secrecy-coding scheme for current and next-generation RATs.

Since polar codes provide strong security for discrete channels [17], a first idea is to concatenate them to a capacity approaching code. The capacity approaching code should be the inner code so that the channel between the polar encoder and the polar decoder can be viewed as a binary symmetric channel (BSC). Thus, we propose a scheme which is initially composed of a LDPC code as inner code and of a polar code as outer code 20.3a. The inner code can also be any FEC codes employed currently for practical wireless communications such as turbo codes or binary convolutional codes (BCC). The design of the inner code is therefore straightforward as we only follow the requirements defined in standards. In this work, we consider particularly LDPC codes defined in the 802.11 standard (WiFi).

20.3.2.1 Construction of the outer code using polar codes

We first consider two nested polar codes of length $N = 2^n$ as the outer code. The rate of the first polar code is the target rate for Eve, denoted R_E, and the rate of the second polar code is the target rate for Bob, denoted, R_B. Since we suppose that legitimate users have a radio advantage over Eve, $R_E < R_B$ (meaning that the wiretap channel is degraded). Therefore, Eve can perfectly decode $N.R_E$ bits and Bob $N.R_B$ bits. In order to confuse Eve and to ensure 0.5 error probability at her side, we send random bits over $N.R_E$ perfect bit-channels.

The design strategy of the outer code is then the following:

- Bhattacharyya parameters are computed for Bob target's error probability at the output of the inner decoder.
- Bit-channels are sorted in ascending order of their Bhattacharyya parameters.
- Random bits are sent over the first $N.R_E$ bit-channels.
- Information bits are sent over the following $N(R_B - R_E)$ bit-channels.
- Frozen bits (i.e., zeros) are sent over the remaining bit-channels.

20.3.2.2 Construction of the outer code using Reed–Muller codes

We propose to use RM codes as an alternative to polar codes in the design of the outer code [37]. The constructions of RM codes and polar codes are similar. The main difference is the selection criteria of bit-channels. Indeed, for polar codes, the selection criteria is the Bhattacharyya parameter, whereas this selection criteria is the Hamming weight of rows of the generator matrix for the RM codes. Consequently, at same length, the RM code usually has a larger minimum distance and better performance than the corresponding polar code.

The design strategy for the outer RM code is then modified as follows:

- Hamming weights of generator matrix's rows are computed.
- Bit-channels are sorted in ascending order of their Hamming weight.
- Random bits are sent over the $N.R_E$ first bit-channels.
- Information bits are sent over the $N(R_B - R_E)$ following bit-channels.
- Frozen bits (i.e., zeros) are sent over the remaining bit-channels.

20.3.2.3 Decoding algorithm for polar and Reed–Muller codes

When Arikan introduced polar codes, he also proposed a low-complexity decoding algorithm named the SC decoding algorithm [3]. However, the SC decoder has limited performance at moderate block length. In [38], Tal and Vardy proposed an improved version of the SC decoder referred to as SC list decoder. We use the LLR-based SC list decoding algorithm presented in [39] (with list size of 8).

20.3.2.4 Practical metrics for secrecy

The security provided by each secrecy code is evaluated by computing the BER. Secrecy is considered to be achieved when BER = 0.5.

Table 20.1   Designed secrecy codes

Secrecy code	SC 1	SC 2	SC 3	SC 4
Inner code	*LDPC code of length 1,296 and rate 5/6 defined in the 802.11 standard*			
Outer code	Polar code	Polar code	Reed–Muller code	Reed–Muller code
Eve's target rate	0.05	0.13	0.05	0.05
Bob's target rate	0.55	0.52	0.5	0.4
(R, F, I)	(51, 512, 461)	(133, 399, 492)	(56, 430, 538)	(56, 330, 638)
Secrecy-coding rate	0.4	0.3	0.33	0.25

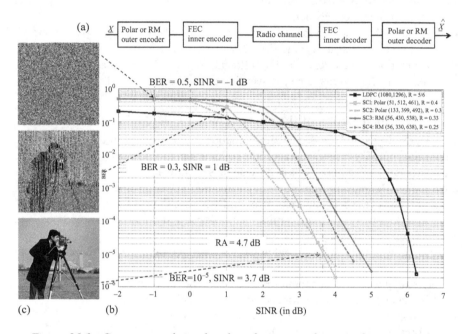

Figure 20.3   Structure and simulated performance of practical secrecy codes

20.3.2.5   Practical designed secrecy codes

We use the LDPC code of length 1,296 and rate 5/6 defined in the 802.11n/ac standard as the inner code. The outer code is either a polar code of length $2^{10} = 1{,}024$ or a RM code of the same length. For simulation purpose, four outer codes were designed using polar and RM codes of different rates. The parameters of these four secrecy codes are presented in Table 20.1. Note that R, I and F denote respectively the number of random bits, information bits and frozen bits.

20.3.3   Performance analysis of designed secrecy codes

Figure 20.3(a) presents the architecture of the proposed secrecy codes.

Simulations were carried out with MATLAB, and messages were sent over an AWGN channel using a QPSK modulation.

Taking into account characteristics of WiFi encoders, Figure 20.3(b) shows the performance of the designed secrecy codes.

- The black curve with square markers represents the BER at the output of the LDPC decoder.
- The dark gray curves represent the BER at the output of secrecy polar decoders.
- The light gray curves represent the BER at the output of secrecy RM decoders.

The BP algorithm is used for LDPC decoders, and the SC list decoding algorithm is used for polar and RM decoders.

The results show that

- Polar-based secrecy codes have better reliability performance than RM-based secrecy codes of similar rates.
- When SINR ≤ -1 dB, the BER at the output of the four secrecy codes is equal to 0.5. Meaning that, all secrecy codes guarantee no information leakage if Eve's SINR is less than -1 dB.
- When SINR ≤ 0 dB, the BER at the output of secrecy codes SC3 and SC4 is equal to 0.5, while the BER at the output of secrecy codes SC1 and SC3 is above 0.45. Meaning that if Eve's SINR is less than 0 dB, SC3 and SC4 guarantee no information leakage, while only a limited amount of information (less than 5%) is leaked for SC1 and SC2.
- For a target error probability of 10^{-5} for Bob, the required radio advantage to ensure no information leakage is limited to 4.4–4.7 dB.

These simulation results demonstrate that Eve cannot retrieve any transmitted information when a slight radio advantage (<5 dB) is provided to legitimate users. The secrecy is achieved with a limited increase in coding and decoding complexity.

Figure 20.3 illustrates the performance of the secrecy-coding scheme by simulating the transmission of the cameraman image over an AWGN channel using the polar-based secrecy codes of rate 0.4 (SC1), for different values of the SINR.

Figure 20.3(c) shows that

- when the SINR ≤ -1 dB, no clue on the transmitted image can be deduced from the received image. The BER at the output of the secrecy code is equal to 0.5.
- When SINR $= 1$ dB, BER $= 0.3$ and Eve manages to successfully decode enough information on the transmitted image. Although 0.3 is a high value for a BER, too much information is leaked. Consequently, Eve's BER should be as close as possible of 0.5 to guarantee no information leakage.
- When SINR ≥ 3.7 dB, BER $= 10^{-5}$ and Bob can perfectly decode the transmitted information.

20.3.4   Simulation results on LTE signals

20.3.4.1   Configuration of simulations

The simulations described below are relevant to LTE-cellular-based links at frequency 2.6 GHz in the downlink transmission direction mode referred as Transmission Mode 7 (TM7) which support BF.

For performance assessment of the proposed secrecy-coding scheme, we use MATLAB-based LTE link-level simulators [40] developed by Technical University of Vienna. The simulators implement standard-compliant LTE downlink and LTE uplink transceivers with their main features, i.e., basic channel models, modulation and coding, multiple-antenna transmission and reception, channel estimation and scheduling. For reliable performance assessment, the channels seen by Bob and Eve need to show a distance-dependent correlation, which WINNER II model cannot model. For that reason, the QuaDRiGa channel model [41], which can produce correlation between Alice–Bob, Alice–Eve and Bob–Eve channels, is used. The configuration of the simulation and its main parameters are synthesized Figure 20.4(a).

- The LTE carrier frequency is 2.6 GHz and the channel bandwidth is 10 MHz. Alice transmits data to Bob using QPSK modulation with coding rate 602/1,024, which corresponds to channel quality indicator (CQI) value of 6. Bob's SNR is assumed to be 10 dB.
- We consider an outdoor urban microcell radio environment with line-of-sight (LOS) component, so called B1 [42], with LOS component (delay spread: 36 ns, shadow fading: 3 dB) and NLOS component (delay spread: 76 ns, shadow fading: 4 dB). Alice uses a 4-element circular antenna array circular antennas array, Bob and Eve are single antenna each, and they use the same processing for CM estimation (least-squares method). Similarly, in the uplink direction, Alice and Eve use least squares method to estimate, respectively, Bob–Alice and Bob–Eve channels.
- Distance between Alice and Bob is 15 m, and the distance between Bob and Eve is 11.5 m, which corresponds to 100 wavelengths at carrier frequency of 2.6 GHz, Eve being located at one of four possible locations denoted by P1, P2, P3 and P4 lying on the circle of radius 11.5 m (100 wavelength) centered at Bob's position.

20.3.4.2   Simulation of transmitting and processing of the secret encoded LTE signals

Assuming Time-Division-Duplexing (TDD) transmission mode, the BF coefficients are determined from the CM estimated by the eNodeB Alice from the uplink transmission of reference signals by intended user equipment (UE) Bob. A single BF coefficient is used per resource block. The AN signal is generated such that it lies in the null space of the Alice-to-Bob CM, and it is added to all symbols. Following the discussion in Section 20.3.1.3, the AN signal lies 6 dB above information-bearing signal to reduce any eavesdropping risk.

Besides, in LTE systems, turbo codes are used for FEC. Thus, following the architecture of Figure 20.3(a), the secrecy-coding scheme is implemented by concatenating an outer RM code with the inner standard-compliant turbo code. We use $(56, 330, 638)$ RM-based secrecy code defined in Table 20.1 as the secrecy code.

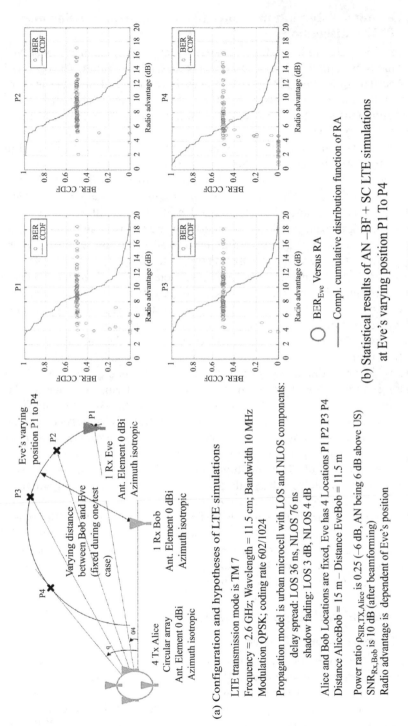

(a) Configuration and hypotheses of LTE simulations

LTE transmission mode is TM 7
Frequency = 2.6 GHz; Wavelength = 11.5 cm; Bandwidth 10 MHz
Modulation QPSK; coding rate 602/1024

Propagation model is urban microcell with LOS and NLOS components:
    delay spread: LOS 36 ns, NLOS 76 ns
    shadow fading: LOS 3 dB, NLOS 4 dB

Alice and Bob Locations are fixed, Eve has 4 Locations P1 P2 P3 P4
Distance AliceBob = 15 m – Distance EveBob = 11.5 m

Power ratio $\rho_{SIR,TX,Alice}$ is 0.25 (–6 dB, AN being 6 dB above US)
$SNR_{Rx,Bob}$ is 10 dB (after beamforming)
Radio advantage is dependent of Eve's position

○ BER_{Eve} Versus RA

—— Compl. cumulative distribution function of RA

**(b) Statistical results of AN –BF + SC LTE simulations
at Eve's varying position P1 To P4**

Figure 20.4   Configuration, parameters and results of LTE simulations

20.3.4.3 Results of simulations under LTE carrier transmission mode TM7—discussion

At each Eve's location, we take 100 independent snapshots of channel model, and for each of them, we simulate the transmission of 20 LTE subframes. The observed figures-of-merit include Eve's and Bob's BER as well as established radio advantage of Bob over Eve. In Figure 20.4(b), we plot of the empirical complementary cumulative distribution function of Bob's radio advantage over Eve as well as Eve's BER as a function of the radio advantage.

These results first demonstrate that radio advantage of 5–6 dB is sufficient to preclude Eve from reliably decoding the transmitted signal.

Nevertheless, the location of Eve with respect to Alice and Bob significantly affects the radio advantage. For example, if Eve is closer to Alice than Bob (P4), her signal is obviously stronger than Bob's signal, and the probability of achieving a sufficient RA is significantly reduced. For example, when Eve is in position P2 or P3, the probability of achieving at least 5 dB of radio advantage is above 90%, when Eve is in position P4 the respective probability drops to 60% only.

Furthermore, establishing and maintaining sufficient radio advantage is a challenging engineering task in fading channels, because fading can affect the CM measurement and the BF establishment: while channel state is changing, any channel estimation errors reduce the effectiveness of the AN–BF processing.

Thus, the AN–BF and secrecy-coding scheme should be designed for the worst case scenario and applied for high mean SNR regimes where channel estimation errors are smaller, whatever is the fading into the transmission. It can thus be expected that non-LOS long-range radio propagation should be more difficult to handle in LTE networks than short-range propagation because the AN and the BF values are fixed for the whole resource block (performance suffers when channel changes occur during the block). Nevertheless, in any case, the power control can contribute to the AN–BF + SC scheme by ensuring that the $SINR_{Rx,Bob}$ at Bob's side is sufficiently large over the block to allow successful CM estimation, efficient BF establishment at Alice's side and reliable decoding at Bob's side, while AN still prevents Eve's decoding attempts.

Finally, the simulation results above demonstrate that the secrecy schemes of Figure 20.3 should well apply to real world radio-cellular networks (significant performances with limited RA value). Besides, to achieve significant performances in most difficult NLOS propagation conditions, these results also show that the network engineering (SINR threshold of the legitimate link, power control, etc.) has to be adapted in the same time of the tuning of the AN–BF + SC scheme.

20.3.5 Experimental results on WiFi signals

20.3.5.1 Configuration of experiments

The experiments described below are relevant to 802.11ac WiFi links at frequency 5.2 GHz, with standard modulation coding schemes at transmitter Alice and at receivers Bob and Eve. The geometry is indoor and LOS.

The access point Alice is implemented on a 4-antenna dedicated chipset (CL 2400), developed by the Company Celeno Communications. Through the IPERF test application (commonly used to generate TCP and USP traffic), Alice transmits a predefined bit pattern as US to facilitated BER evaluation. In addition, Alice adds AN to the data part of the US bit pattern and beam forms it toward Bob.

Bob is implemented by using a single-antenna smartphone device (XIAOMI's MI5). Eve is implemented by using a 3-antenna MacBook Pro, working in sniffer mode with the Wireshark application. The Wireshark application outputs packet error rates and stores Rx signals frames. The BER at Eve side is then computed offline (using a MATLAB script) by comparing the stored received packets to the known transmitted pattern.

The overall geometry and locations of Alice Bob and Eve are represented into Figure 20.5(a).

The overall hardware and software components hosting the AN–BF application are represented Figure 20.5(b) (CL 2400 WiFi chipsets and host board). The AN–BF processing is based on a spatial multiplexing (SM) transmit matrix which is computed from a single value decomposition (SVD) of the CM issued from channel sounding exchanges. Note that when antennas are calibrated at Alice and Bob's side, AN–BF can be based on channel reciprocity assumption, without any added information exchanged over the air.

During computations, Alice has to restrict Rx or Tx operations and match numerous technological constraints. Thus, several compressions, acceleration and parametrization capabilities are added to support AN–BF:

- QR decomposition and size reduction of the matrix involved in the computations,
- adjustment of the number of noise spatial streams (NAN = 3 among 4) and user spatial stream (NSS = 1 among 4),
- adjustment of the power ratios $\rho_{SIR,Tx,Alice}$ between the data and the noise streams,
- uniform distribution of independent noise samples over all transmitting antennas,
- gain scaling of the entire signal to ensure that the total Tx power matches the required digital back-off level and avoids saturation of the digital-to-analog converter, etc.

The WiFi transmitting and receiving radio parameters are recalled Figure 20.5(c).

Figure 20.5(d) provides the values of the power ratios $\rho_{SIR,Tx,Alice}$ and the relevant values of packet error rates (PER at Bob's side) that lead to the experimental results shown in the following sections.

20.3.5.2 Transmission and processing of the secret encoded WiFi signals

To experiment the decoding of secret codes by Eve and Bob, a fixed frame is still sent by Alice over repeated transmissions (by using the same IPERF application as above). These frame is now offline precomputed from the initial bit pattern and one of the secret encoder described in Section 20.3.3. The parameters of the secrecy code used in the experiments results below is the polar-based secrecy code $(R,I,F) = (102, 409, 513)$.

Note that the code-word length of 1,024 bits perfectly matches the WiFi frame length. While the new secret encoded bit pattern now replaces the initial one, the decoding at Bob's and Eve's side is done offline from signal frames records by using a MATLAB script.

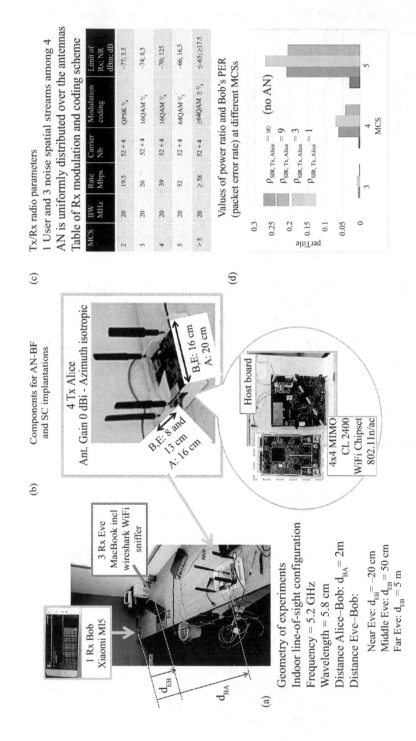

Figure 20.5 Configuration of the AN–BF and SC experiments in indoor line-of-sight (LOS) environment

The whole procedure allows to estimate the efficiency of secrecy coding through BER estimates at Eve's side. Moreover, comparison of Eve's BER when using first the native the WiFi FEC scheme (either LDPC or BCC) and then the concatenated secret coding scheme described in Section 20.3.3 allow to analyze first the basic protection of the radio advantage provided by AN–BF alone and the security enhancement due to the secrecy-coding scheme itself.

Recall that Eve has 3 Rx antennas and is supposed to have the complete information about the secret code. Moreover, she can test any modulation and coding scheme in her attempts to recover parts of the legitimate user information, MCS2 being the best for Eve regarding the resilience of her decoding when facing AN.

20.3.5.3 Experimental results in line-of-sight geometry (LOS)—Discussion

Figure 20.6 shows the results of the AN–BF scheme and of the combined AN–BF + SC scheme on recorded WiFi frames. Two (low and middle) values of the power ratio $\rho_{\text{SIR,Tx,Alice}}$ are taken into account ($\rho_{\text{SIR,Tx,Alice}} = 3$ in Figure 20.6(a), while the AN power is 25% of the total power and $\rho_{\text{SIR,Tx,Alice}} = 1$ in Figure 20.6(b), while the AN power is 50% of the total power).

In any cases, Bob uses the MCS4 decoder with $\text{PER}_{\text{Bob}} \approx 0$, $\text{BER}_{\text{Bob}} \leq 0.1$, $\text{SINR}_{\text{Rx,Bob}} \geq 12.5$ dB, while Eve attempts to decode the signal frames by using the MCS2 decoder (that advantages her by decreasing the radio advantage of Bob–Eve gets about 4 dB more compared to the MCS4). The radio advantage indications in Figure 20.6 are given with respect of one received antenna at Eve's side.

When comparing the result of Figure 20.6 to analyses of Section 20.3.3, remembering the particular propagation properties of indoor LOS configurations and considering the low and medium values of power ratio $\rho_{\text{SIR,Tx,Alice}}$, we can note the following trends:

- At far Eve's locations, even in LOS geometry when the power ratio $\rho_{\text{SIR,Tx,Alice}}$ remains low, the radio advantage is very significant. This very favorable situation for security occurs mainly thanks to the BF that achieves significant BF rejection performances (12 dB and more in the experiments reported in Figure 20.6).
- We can be confident that similar trends would occur in any NLOS environments, whatever is Eve's location, because the BF rejection should be enhanced thanks to the positive effects of propagation reflectors in the neighborhood of Alice and Bob.
- When coming back to LOS configuration and considering now Eve locations closer to Bob. One has to interpret the decreasing performances of the AN–BF + SC scheme in the following sense. First, a main lobe is most often the result of LOS propagation impact to BF processing. Second, this main lobe can be intercepted by Eve. In addition, the 3 Rx of Eve in our experimental configuration can provide some array discrimination and processing gain on the data US. Finally, the effect of BF at Alice side can be partially mitigated by Eve close to Bob. To counter this, the power ratio $\rho_{\text{SIR,Tx,Alice}}$ should be decreased down to value $\rho_{\text{SIR,Tx,Alice}} = 1/4$ (that correspond to an AN power that is 6 dB over the US power as mentioned in Section 20.3.1.3), and the antenna aperture at Alice's side should be enlarged to decrease the main lobe size and improve the rejection performances of BF.

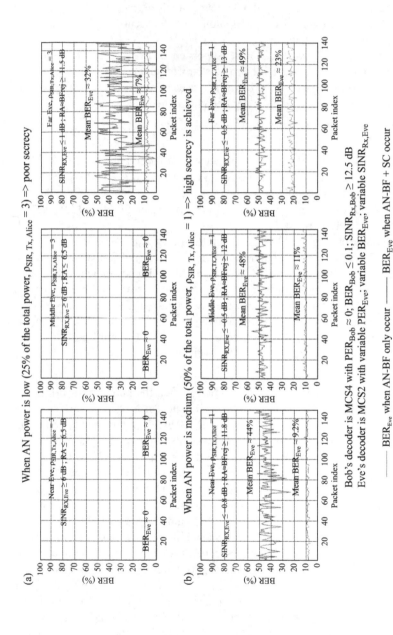

Figure 20.6 *Experimental results of secrecy-code schemes in LOS environment—comparison between AN–BF alone and AN–BF + SC for several values of $\rho_{SIR,Tx,Alice}$*

Finally, the experimental results above are evidence that the proposed secrecy schemes well applies to real world WLAN chipsets and propagation with limited AN power in most of practical NLOS and LOS configurations. It is also evident that even when very adverse conditions occur (LOS configurations, Eve very close to Bob or very close to Alice), secrecy efficiency should be achieved through a suitable tuning of the radio parameters (increasing of the AN noise, enlarging of the Alice antenna array).

20.3.6 Tuning of the radio advantage for OFDM/QPSK wave forms such as WiFi and LTE signals—considerations on radio engineering

Analysis and experimental results show that the BER at the output of the polar decoder is 0.5 up to a given *attacker threshold* of the SINR ($\text{SINR}_{\text{Rx,Eve}}$), depending on the modulation and concatenated coding scheme, that ensures no information leakage.

When the $\text{SINR}_{\text{Rx,Bob}}$ increases, the BER at the output of the polar decoder vanishes. When $\text{SINR}_{\text{Rx,Bob}}$ is high enough (greater than a user threshold $\text{SINR}_{\text{user,min}}$), the BER at the output of the polar decoder approaches zero.

In all the presented simulation and tests, only a few dB of radio advantage (typically 3–5 dB) is required to provide both reliability and secrecy to legitimate users. These reasonable values ensure the compatibility of secrecy-code schemes with exiting AN–BF schemes and other means for providing the radio advantage (such as directive antennas for transmission, full duplex communications technologies).

For these secrecy-coding schemes, the typical value of $\text{SINR} = -1$ dB (0.8) in linear should be considered the maximum $\text{SINR}_{\text{Rx,Eve}}$ tolerated at Eve's receiver as demonstrated in Section 20.3.3.

Therefore, the corresponding ρ_{min} should be equal to 0.8 to be used at Alice's transmitters in order to tune the AN power from the value of the user data stream power with a ratio $(J_{\text{Tx}}/S_{\text{Tx}}) \geq 1/\rho_{\text{min}}$.

Then, the tuning of the power of the user (signaling or data) stream and of the BF performance is set in order to achieve reliable communication for Bob. When considering the mean channel propagation losses (noted l_{AB} in linear values, L_{AB} in dB), the BF rejection (noted bf_{rej} in linear value, BF_{rej} in dB), the receiving noise at Bob's side (noted N_{Rx} in linear values) and the SNR threshold of Bob's receiver (noted $\rho_{\text{Thres,Rx}}$ in linear value),

- the global signal-to-noise + interference ratio at Bob side's is given by $\rho_{\text{SINR,Rx,Bob}} = [S_{\text{Tx}}/l_{\text{AB}}]/[(J_{\text{Tx}}/l_{\text{AB}}/\text{bf}_{\text{rej}}) + N_{\text{Rx}}]$ and
- the global signal-to-noise ratio by $\rho_{\text{SNR,Rx,Bob}} = [S_{\text{Tx}}/l_{\text{AB}}]/[N_{\text{Rx}}]$.

In practice, one has to

- define two margin values η_1 and η_2 such that $1 < \eta_2 < \eta_1$ to tune of the radio engineering,
- tune the user stream power S_{Tx} to achieve enough receiving power such that $\rho_{\text{SNR,Rx,Bob}} \geq \rho_{\text{Thres,Rx}} \cdot \eta_1$,
- turn the BF rejection bf_{rej} such that $\rho_{\text{SINR,Rx,Bob}} \geq \rho_{\text{Thres,Rx}} \cdot \eta_2$.

As a summary, it appears that two main radio parameters are necessary to implant the secrecy-coding schemes:

- A "minimum $SINR_{user,min}$" for the legitimate link, which is relevant to the performance of the modulation and coding schemes for Bob (when taking into account some margin, typical values are a few dBs, 3–5 dB in the secret codes considered above). Achieving SINR values greater than $SINR_{user,min}$ for on-going legitimate radio communications involves some network engineering activities: management of the network topology (real field path losses, transmit power, energy budget link and control of the BF performance BF_{rej} in the established AN scheme (input by channel state information). Note that all these parameters are involved in the equalization processing and in the Quality of Service Management.
- A "SINR Security gap" $SINR_{SG}$ that represents the lower bound of the radio advantage to be provided to the legitimate link by increasing the interference J_{Tx} emitted by Alice. BF_{rej} being controlled by Alice and Bob, $SINR_{SG}$ drives the tuning of the AN power to ensure the radio advantage independently of Eve's location.

In general case, Alice and Bob have thus to manage parameters S_{Tx}, J_{Tx} and bf_{rej} so that $SINR_{user,min}$ and input $SINR_{SG}$ values are between 3 and 5 dB depending on the applied secrecy-coding scheme. Nevertheless, the exact radio advantage remains dependent on Eve's receiver capabilities.

In simplified optimal case where jamming signal transmission is colocated with US transmission (Eve has no spatial rejection capabilities whatever is her receiver performance), and receiving noise at Bob is negligible (thus $SINR_{Rx,Bob} \approx SIR_{Rx,Bob} = SIR_{Tx,Alice} + BF_{rej}$ while we have in any case $SINR_{Rx,Eve} \leq SIR_{Rx,Eve} = SIR_{Tx,Alice}$), Alice and Bob get facilitated radio management of the link with parameters $SIR_{Tx,Alice}$ and BF_{rej} such that $SIR_{Tx,Alice} + BF_{rej} \geq SINR_{user,min}$ and $SIR_{Tx,Alice} \leq SNR_{user,min} - SINR_{SG}$.

Hence, the radio advantage verifies $RA \geq BF_{rej}$ (nonequality occurs when Eve's receiver noise is significant, what increasing the advantage of Bob), and a simplified requirement for BF_{rej} is achieved by considering $BF_{rej} \geq \max\{SINR_{SG}, SINR_{user,min} - SIR_{Tx,Alice}\}$.

20.4 Conclusion: security upgrades provided to future radio access technologies

As described above, any secrecy-coding schemes applies under the assumption that a prior radio advantage is provided. For achieving this radio advantage, several tacks are followed into the *PHYLAWS* project [29]:

- Use of AN–BF schemes into MIMO architectures, as described above,
- Use of directive antennas, and use of directive array of antennas (with BF technologies),
- Use of full duplex radio technologies as described in [43],

- Secure pairing and interrogation technologies such as in system for identification friend or foe. A particular key-free application of such a technology is developed and studied into [34] for public RATs. It is based on low power self-interfered signals (named Tag Signals—TS) and on interrogation and acknowledgment sequences (IAS) supported by these Tag Signals. Very early in the radio access, IAS achieves the security pairing of Alice's and Bob's devices, then provides dual sense TS with a controlled radio advantage. Then, secret codes can be applied to these TSs in order to achieve subscriber identity authentication and further nego-tiation of the communications services without any disclosure risk of subscriber private data on the physical layer.

As soon a slight radio advantage is achieved, the results provided above prove that the proposed secrecy-coding schemes are efficient.

- The provided secrecy rate is significant (as shown Table 20.1 and Figure 20.3), even if they remain suboptimal when compared to theoretical results in ideal case without any constraint on code length and complexity.
- Realistic constraints apply to code length and computation operations that make the technique fully compatible with existing wireless standards.
- The simulation of LTE links (Section 20.3.4) and the experimental WiFi results in real field (Section 20.3.5) give the feasibility proof of the technique and show that it can be readily implemented in existing wireless MIMO or MISO communication systems that propose BF services (such as in WLAN 802.11ac, LTE and for emerging 5G standards that would involve massive MIMO technologies):
  - the AN–BF scheme being activated, only minor modifications of the software architecture of the nodes and terminals are required for the implementation of the secrecy-coding scheme
  - all modifications are only located at the coding stage and remain transparent for upper protocol layers.

Besides, AN and BF Schemes, we can be confident that same kind of simplified implantation architecture of secret codes schemes should apply to most of "radio advantage technologies":

- MIMO and massive MIMO architectures (evolution of existing standard toward usages for Internet of Things, for public safety applications; new WLAN standards and new radio-cellular standards)
- Directive antenna patterns, and especially
  - Microwave links and satellite links in C band (4–8 GHz) and in upper bands
  - Many of automatic radio-command of planes and of unmanned aircraft trans-port vehicles in C band that should be deployed in the future airborne traffic control (ATC) standards and systems [44].
- Full duplex technologies, when they will be mature and deployed in radio networks [43].

Finally, secrecy coding appears to be an accessible technology to implant in radio-communications systems once a radio advantage is established. SC can apply at numerous stages of the radio protocol:

- First, to enhance the weakly secure transmission of signaling and access messages (use of clear text) that exist in public radio cellular networks and in wireless access networks of nowadays.
- Then, to limit the disclosure risk of subscriber and network parameters that are relevant to identification, authentication and ciphering procedures.
- Finally, to complete the protection of on-going communication by adding a security protection at the physical layer in addition to the traditional cipher schemes of the user data stream.

References

[1]   M. Bloch, M. Hayashi, and A. Thangaraj, "Error control coding for physical layer secrecy," *Proceedings of the IEEE*, vol. 103, no. 10, pp. 1725–1746, Oct. 2015.

[2]   R. Gallager, "Low density parity check codes," Ph.D. dissertation, MIT Press, Cambridge, 1963.

[3]   E. Arikan, "Channel polarization: A method for constructing capacity-achieving codes for symmetric binary-input memoryless channels," *IEEE Transactions on Information Theory*, vol. 55, no. 7, pp. 3051–3073, Jul. 2009.

[4]   M. Bloch and J. Barros, *Physical-Layer Security: From Information Theory to Security Engineering*, 1st ed. New York, NY: Cambridge University Press, 2011.

[5]   A. D. Wyner, "The wire-tap channel," *The Bell System Technical Journal*, vol. 54, no. 8, pp. 1355–1387, Oct. 1975.

[6]   A. T. Suresh, A. Subramanian, A. Thangaraj, M. Bloch, and S. W. McLaughlin, "Strong secrecy for erasure wiretap channels," in *Information Theory Workshop (ITW), 2010 IEEE*, Dublin, Ireland, Aug. 2010, pp. 1–5.

[7]   A. Thangaraj, S. Dihidar, A. R. Calderbank, S. W. McLaughlin, and J. M. Merolla, "Applications of LDPC codes to the wiretap channel," *IEEE Transactions on Information Theory*, vol. 53, no. 8, pp. 2933–2945, Aug. 2007.

[8]   A. Subramanian, A. Thangaraj, M. Bloch, and S. W. McLaughlin, "Strong secrecy on the binary erasure wiretap channel using large-girth ldpc codes," *IEEE Transactions on Information Forensics and Security*, vol. 6, no. 3, pp. 585–594, Sep. 2011.

[9]   V. Rathi, M. Andersson, R. Thobaben, J. Kliewer, and M. Skoglund, "Performance analysis and design of two edge-type LDPC codes for the BEC wiretap channel," *IEEE Transactions on Information Theory*, vol. 59, no. 2, pp. 1048–1064, Feb. 2013.

[10] T. Richardson and R. Urbanke, *Modern Coding Theory*. New York, NY: Cambridge University Press, 2008.

[11]   S. Kudekar, T. Richardson, and R. L. Urbanke, "Spatially coupled ensembles universally achieve capacity under belief propagation," *IEEE Transactions on Information Theory*, vol. 59, no. 12, pp. 7761–7813, Dec. 2013.

[12]   V. Rathi, R. Urbanke, M. Andersson, and M. Skoglund, "Rate-equivocation optimal spatially coupled LDPC codes for the BEC wiretap channel," in *Information Theory Proceedings (ISIT), 2011 IEEE International Symposium on*, St. Petersburg, Russia, July 2011, pp. 2393–2397.

[13]   D. Klinc, J. Ha, S. W. McLaughlin, J. Barros, and B. J. Kwak, "LDPC codes for the Gaussian wiretap channel," *IEEE Transactions on Information Forensics and Security*, vol. 6, no. 3, pp. 532–540, Sep. 2011.

[14]   M. Andersson, V. Rathi, R. Thobaben, J. Kliewer, and M. Skoglund, "Nested polar codes for wiretap and relay channels," *IEEE Communications Letters*, vol. 14, no. 8, pp. 752–754, Aug. 2010.

[15]   E. Hof and S. Shamai, "Secrecy-achieving polar-coding," in *Information Theory Workshop (ITW), 2010 IEEE*, Dublin, Ireland, Aug. 2010, pp. 1–5.

[16]   O. O. Koyluoglu and H. E. Gamal, "Polar coding for secure transmission and key agreement," *IEEE Transactions on Information Forensics and Security*, vol. 7, no. 5, pp. 1472–1483, Oct. 2012.

[17]   H. Mahdavifar and A. Vardy, "Achieving the secrecy capacity of wiretap channels using polar codes," *IEEE Transactions on Information Theory*, vol. 57, no. 10, pp. 6428–6443, Oct. 2011.

[18]   E. Şaşoğlu and A. Vardy, "A new polar coding scheme for strong security on wiretap channels," in *Information Theory Proceedings (ISIT), 2013 IEEE International Symposium on*, Istanbul, Turkey, July 2013, pp. 1117–1121.

[19]   D. Sutter, J. M. Renes, and R. Renner, "Efficient one-way secret-key agreement and private channel coding via polarization," April 2013. [Online]. Available: https://arxiv.org/abs/1304.3658.

[20]   T. Gulchu and A. Barg, "Achieving secrecy capacity of the general wiretap channel and broadcast channel with a confidential component," Nov. 2016. [Online]. Available: https://arxiv.org/abs/1410.3422.

[21]   Y.-P. Wei and S. Ulukus, "Polar coding for the general wiretap channel with extensions to multiuser scenarios," *IEEE Journal on Selected Areas in Communications*, vol. 34, no. 2, pp. 278–291, Feb. 2016.

[22]   C. Ling, L. Luzzi, J. C. Belfiore, and D. Stehl, "Semantically secure lattice codes for the Gaussian wiretap channel," *IEEE Transactions on Information Theory*, vol. 60, no. 10, pp. 6399–6416, Oct. 2014.

[23]   F. Oggier, P. Sol, and J.-C. Belfiore, "Lattice codes for the wiretap Gaussian channel: Construction and analysis," Mar. 2011. [Online]. Available: http://arxiv.org/abs/1103.4086.

[24]   A. M. Ernvall-Hytönen and C. Hollanti, "On the eavesdropper's correct decision in Gaussian and fading wiretap channels using lattice codes," in *Information Theory Workshop (ITW), 2011 IEEE*, Paraty, Brazil, Oct. 2011, pp. 210–214.

[25]   J. H. Conway and N. J. A. Sloane, *Sphere Packings, Lattices, and Groups*. New York: Springer, 1993.

[26] G. D. Forney Jr., M. Trott, and S.-Y. Chung, "Sphere-bound-achieving coset codes and multilevel coset codes," *IEEE Transactions on Information Theory*, vol. 46, no. 3, pp. 820–850, May 2000.

[27] C. Ling and J. Belfiore, "Achieving AWGN channel capacity with lattice Gaussian coding," *IEEE Transactions on Information Theory*, vol. 60, no. 10, pp. 5918–5929, Oct. 2014.

[28] Y. Yan, L. Liu, C. Ling, and X. Wu, "Construction of capacity-achieving lattice codes: Polar lattices," Nov. 2014. [Online]. Available: http://arxiv.org/abs/1411.0187.

[29] J. C. Belfiore and F. Oggier, "Secrecy gain: A wiretap lattice code design," in *Information Theory and its Applications (ISITA), 2010 International Symposium on*, Oct. 2010, pp. 174–178.

[30] B. A. Sethuraman, B. S. Rajan, and V. Shashidhar, "Full-diversity, high-rate space-time block codes from division algebras," *IEEE Transactions on Information Theory*, vol. 49, no. 10, pp. 2596–2616, Oct. 2003.

[31] B. Hassibi and B. M. Hochwald, "High-rate codes that are linear in space and time," *IEEE Transactions on Information Theory*, vol. 48, no. 7, pp. 1804–1824, July 2002.

[32] V. Tarokh, N. Seshadri, and A. R. Calderbank, "Space-time codes for high data rate wireless communication: Performance criterion and code construction," *IEEE Transactions on Information Theory*, vol. 44, no. 2, pp. 744–765, Mar. 1998.

[33] N. R. Goodman, "Statistical analysis based on a certain multivariate complex Gaussian distribution (an introduction)," *Annals of Mathematical Statistics*, vol. 34, no. 1, pp. 152–177, Mar. 1963. [Online]. Available: http://dx.doi.org/10.1214/aoms/1177704250.

[34] N. Romero-Zurita, M. Ghogho, and D. McLernon, "Physical layer security of MIMO-OFDM systems by beamforming and artificial noise generation," *PHYCOM: Physical Communication*, vol. 4, no. 4, pp. 313–321, 2011.

[35] N. O. Tippenhauer, L. Malisa, A. Ranganathan, and S. Capkun, "On limitations of friendly jamming for confidentiality," in *Security and Privacy (SP), 2013 IEEE Symposium on*, May 2013, pp. 160–173.

[36] "PHYLAWS," 2014. http://www.Phylaws-ict.org.

[37] E. Arikan, "A performance comparison of polar codes and Reed–Muller codes," *IEEE Communications Letters*, vol. 12, no. 6, pp. 447–449, Jun. 2008.

[38] I. Tal and A. Vardy, "List decoding of polar codes," *IEEE Transactions on Information Theory*, vol. 61, no. 5, pp. 2213–2226, May 2015.

[39] A. Balatsoukas-Stimming, M. B. Parizi, and A. Burg, "Llr-based successive cancellation list decoding of polar codes," *IEEE Transactions on Signal Processing*, vol. 63, no. 19, pp. 5165–5179, Oct. 2015.

[40] C. Mehlfuehrer, J. C. Ikuno, M. Simko, S. S, M. Wrulich, and M. Rupp, "The vienna LTE simulators – enabling reproducibility in wireless communications research," *EURASIP Journal on Advances in Signal Processing*, vol. 21(1), 29 pp., 2011.

[41] S. Jaeckel, L. Raschkowski, K. Brner, and L. Thiele, "Quadriga: A 3-d multi-cell channel model with time evolution for enabling virtual field trials," *IEEE Transactions on Antennas and Propagation*, vol. 62, pp. 3242–3256, 2014.

[42] "Winner II channel models," 2007. https://www.ist-winner.org/WINNER2-Deliverables/D1.1.2v1.1.pdf.

[43] A. V. V. Z. Zhang, K. Long and L. Hanzo, "Full-Duplex wireless communications: Challenges, solutions and future research directions," Proceedings of the IEEE, 2015.

[44] "SESAR," 2007. http://www.sesarju.eu/.

Index

Printed in the USA
CPSIA information can be obtained
at www.ICGtesting.com
JSHW011507221024
72173JS00005B/1230

9 781785 612350